Applied Control Systems Design

Magdi S. Mahmoud · Yuanqing Xia

Applied Control Systems Design

Springer

Magdi S. Mahmoud
Department of Systems Engineering
King Fahad Univ. of Petroleum & Minerals
Dhahran, Saudi Arabia

Prof. Yuanqing Xia
Dept. Automatic Control
Beijing Institute of Technology
Beijing, China, People's Republic

ISBN 978-1-4471-2878-6 e-ISBN 978-1-4471-2879-3
DOI 10.1007/978-1-4471-2879-3
Springer London Dordrecht Heidelberg New York

British Library Cataloguing in Publication Data
A catalogue record for this book is available from the British Library

Library of Congress Control Number: 2012936365

Printed on acid-free paper

Springer is part of Springer Science+Business Media (www.springer.com)

Preface

An integral part of control systems engineering is the development of controller design methods to achieve some prescribed performance criteria. It has been one of the most active research areas in the past decades. A common denominator of these methods is the availability of a mathematical model that is derived from physical laws, practical consideration or identification of real data. Several important issues arises including the quality and nature of the model, the performance criteria and control design approach.

This book provides a guided tour of applied control system design. The starting point is the construction of system models based on real experimental data. These models will be evaluated and tested using standard signals. A wide spectrum of control design methods will be applied to these models. Closed-loop system responses will be obtained and compared. The end result is to provide an experience-based recipe that can serve as check-list for researchers or control designers. In this regard, the book unifies the methods for developing feedback controllers and filters for a wide class of dynamical systems and reports on the recent advances in design methodologies. Throughout the book, the use of MATLAB is the vehicle for all methods of analysis and design.

After an introductory chapter, the book is divided into eight self-contained chapters with each chapter being equipped with illustrative examples, problems and questions. The book will be supplemented by some design problems, appropriate appendices and index.

It is planned while organizing the material that this book would be appropriate for use either as graduate-level textbook in applied mathematics as well as different engineering disciplines (electrical, mechanical, civil, chemical, systems), a good volume for independent study or a reference for practicing engineers, interested readers, researchers and students.

KFUPM, Saudi Arabia — Magdi S. Mahmoud
BIT, China — Yuanqing Xia

Acknowledgements

Although the material contained in this volume is an outgrowth of our academic teaching and research activities over the past several years, the idea of writing the book arose and developed during Summer 2010 when Magdi Mahmoud was visiting BIT based on an invitation from Yuanqing Xia.

In writing this book, we took the approach of referring within the text to papers and/or books which we believed taught us some ideas and methods. We then complement this by adding some notes and questions at the end of each chapter to shed some light on other related results. We apologize in advance in case we committed injustice and assure our colleagues that any mistake was not made in purpose.

We are immensely pleased for many stimulating discussions with colleagues, students and friends throughout our technical careers which have definitely enriched our knowledge and experience. In particular, we owe a measure of gratitude to *Professor Michael A. Johnson*, University of Strathclyde, for his unfailing guidance, critical review and constructive criticism on earlier draft of the manuscript. We gratefully acknowledge helpful suggestions and assistance by *Oliver Jackson* and *Charlotte Cross* at Springer-London.

It is a great pleasure to acknowledge the financial funding afforded by the deanship of scientific research (DSR) through project No. IN101024 and for providing superb competitive environment and overall support of research activities at KFUPM. Magdi Mahmoud owe a measure of gratitude to KFUPM management for the continuous encouragements and facilitating all sources of help.

Magdi Mahmoud had the privilege of teaching various graduate courses at KFUPM (Saudi Arabia). The course notes, updated and organized, were instrumental in generating different chapters of this book and valuable comments and/or suggestions by graduate students were greatly helpful, particularly those attended the courses SE 507, SE 513 and SE 514 offered at the Systems Engineering Department over the period 2007–2011.

Most of all however, we would wholeheartedly like to thank all the members of our families. Without their constant love, incredible amount of patience and (mostly) enthusiastic support this volume would not have been finished.

Magdi S. Mahmoud
Yuanqing Xia

Contents

List of Notations[1]

List of Symbols

I^+	The set of positive integers
$\Re$	The set of real numbers
$\Re_+$	The set of non-negative real numbers
$\Re^n$	The set of all n-dimensional real vectors
$\Re^{n \times m}$	The set of $n \times m$-dimensional real matrices
$\mathcal{C}^-$	The open right-half complex plane
$\mathcal{C}^+$	The closed right-half complex plane
$\in$	Belong to or element of
$\subset$	Subset of
$\bigcup$	Union
$\bigcap$	Intersection
$\gg$	Much greater than
$\ll$	Much less than
A^t	The transpose of matrix A
A^{-1}	The inverse of matrix A
I	An identity matrix of arbitrary order
I_s	The identity matrix of dimension $s \times s$
e_j	The jth column of matrix I
x^t or A^t	The transpose of vector x or matrix A
$\lambda(A)$	An eigenvalue of matrix A
$\varrho(A)$	The spectral radius of matrix A
$\lambda_j(A)$	The jth eigenvalue of matrix A
$\lambda_m(A)$	The minimum eigenvalue of matrix A, where $\lambda(A)$ are real
$\lambda_M(A)$	The maximum eigenvalue of matrix A, where $\lambda(A)$ are real

[1]Throughout this book, the following terminologies, conventions and notations have been adopted. All of them are quite standard in the scientific media and only vary in form or character. Matrices, if their dimensions are not explicitly stated, are assumed to be compatible for algebraic operations. In symmetric block matrices or complex matrix expressions, we use the symbol $\bullet$ to represent a term that is induced by symmetry.

A^{-1}	The inverse of matrix A
$A^{\dagger}$	The Moore–Penrose-inverse of matrix A
$P > 0$	Matrix P is real symmetric and positive-definite
$P \geq 0$	Matrix P is real symmetric and positive semi-definite
$P < 0$	Matrix P is real symmetric and negative-definite
$P \leq 0$	Matrix P is real symmetric and negative semi-definite
$A(i,j), A_{ij}$	The ijth element of matrix A
$\mathbf{det}(A)$	The determinant of matrix A
$\mathbf{trace}(A)$	The trace of matrix A
$\mathbf{rank}(A)$	The rank of matrix A
$\mathcal{L}_2(-\infty,\infty)$	Space of time domain square integrable functions
$\mathcal{L}_2[0,\infty)$	Subspace of $\mathcal{L}_2(-\infty,\infty)$ with functions zero for $t > 0$
$\mathcal{L}_2(-\infty,0]$	Subspace of $\mathcal{L}_2(-\infty,\infty)$ with functions zero for $t < 0$
$\mathcal{L}_2(j\Re)$	Square integrable functions on $\mathcal{C}_0$ including at ∞
$\mathcal{H}_2$	Subspace of $\mathcal{L}_2(j\Re)$ with functions analytic in $Re(s) > 0$
$\mathcal{L}_\infty(j\Re)$	Subspace of functions bounded on $Re(s) = 0$ including at ∞
$\mathcal{H}_\infty$	The set of $\mathcal{L}_\infty(j\Re)$ functions analytic in $Re(s) > 0$
$\lvert a \rvert$	The absolute value of scalar a
$\lVert x \rVert$	The Euclidean norm of vector x
$\lVert A \rVert$	The induced Euclidean norm of matrix A
$\lVert x \rVert_p$	The ℓ_p norm of vector x
$\lVert A \rVert_p$	The induced ℓ_p norm of matrix A
$Im(A)$	The image of operator/matrix A
$Ker(A)$	The kernel of operator/matrix A
$\max \mathbf{D}$	The maximum element of set $\mathbf{D}$
$\min \mathbf{D}$	The minimum element of set $\mathbf{D}$
$\sup \mathbf{D}$	The smallest number that is larger than or equal to each element of set $\mathbf{D}$
$\inf \mathbf{D}$	The largest number that is smaller than or equal to each element of set $\mathbf{D}$
$\arg\max \mathbf{D}$	The index of maximum element of ordered set $\mathbf{S}$
$\arg\min \mathbf{D}$	The index of minimum element of ordered set $\mathbf{S}$
$\mathbf{B}_r$	The ball centered at the origin with radius r
$\mathbf{R}_r$	The sphere centered at the origin with radius r
$\mathcal{N}$	The fixed index set $\{1, 2, \ldots, N\}$
$[a, b)$	The real number set $\{t \in \Re : a \leq t < b\}$
$[a, b]$	The real number set $\{t \in \Re : a \leq t \leq b\}$
$\mathbf{S}$	The set of modes $\{1, 2, \ldots, s\}$
iff	If and only if
$\otimes$	The Kronecker product
$\mathbf{O}(.)$	Order of (.)
$\mathrm{diag}(\ldots)A$	Diagonal matrix with given diagonal elements
$\mathbf{spec}(A)$	The set of eigenvalues of matrix A (spectrum)
$\mathbf{min\text{-}poly}(A)(s)$	The minimal polynomial of matrix A

List of Abbreviations

ARE	Algebraic Riccati equation
LMI	Linear matrix inequality
SISO	Single-input single-output
MIMO	Multi-input multi-output
BIBS	Bounded-input bounded-state
iISS	Integral-input-to-state stable
UGAS	Uniformly globally asymptotically stable
OLD	Overlapping decomposition
SVD	Singular value decomposition
LBD	Lyapunov-based design
DTS	Discrete-time systems
LQC	Linear quadratic control
LMCR	Liquid-metal cooled reactor
DSMP	Decentralized servomechanism problem
DIP	Distributed information processing
CIP	Centralized information processing
N4SID	Numerical algorithms for subspace state-space system identification

Chapter 1
Introduction

1.1 Overview

In this introductory chapter, we briefly review the basic concepts behind identifying linear time-invariant (LTI) systems, or systems identification (SI). We then proceed to shed some lights about control design (CD) as applied to multivariable dynamic systems. In general, dynamic models for prediction and control include transfer functions, state space models, time-series models, which are parametrized in terms of finite number of parameters. Hence, these dynamic models are referred to as *parametric models*. There are also *non-parametric models* such as impulse responses, and frequency responses, spectral density functions, etc. In this book, we focus on the parametric models with the main thrust evolve around integrating system identification and control design in one pool toward developing effective tools for researchers and designers. In what follows, some brief accounts of common terms are provided.

System: A system is a collection of objects arranged in an ordered form to serve some purpose. Everything not belonging to the system is part of the environment. One may characterize the system by input–output (cause and effect) relations. What constitutes a system depends on the point of view of the observer. The system may be, for example, an amplifier consisting of electronic components, or a control loop including that amplifier as one of its parts, or a chemical processing unit having many such loops, or a plant consisting of a number of units or a number of plants operating together as a system in the environment of a global economy.

Process: A process is a processing plant that serves to manufacture homogeneous material or energy products. Industries that use such processing plants are called process industries. The common process industries are oil, chemicals, electrical power, paper, glass, mining, metals, cement, drugs, food and beverages. A common characteristic of process industries is that their products can be made to flow.

From a control point of view, different kinds of variables in a process interact and produce observable variables. The observable variables of interest to us are usually called outputs. The process is also affected by external variables. External variables that can be manipulated by us are inputs of the process. Other external variables are

M.S. Mahmoud, Y. Xia, *Applied Control Systems Design*,
DOI 10.1007/978-1-4471-2879-3_1,

called disturbances. Disturbances can be divided into two kinds: the measured disturbances which can be directly measured, and the unmeasured disturbances which are only observed through their influence on the outputs. A process is said to be dynamic when the current output value depends not only on the current external stimuli but also on their earlier values.

Model: A model is a representation of the essential aspects of a system (process) which presents knowledge of that system (process) in a usable form.

For the application of modern systems and control theory, it is necessary to use *mathematical models* that describe the relationships among the system variables in terms of difference or differential equations.

Within the major topics of this book, identification is about how to obtain mathematical models of systems (processes) from observations and measurements and use these models in designing controllers to achieve prescribed criteria.

The input–output data are usually collected from an identification test or experiment that is designed to make the measured data maximally informative about the system properties that are of interest to the user.

A set of candidate models is obtained by specifying their common properties; a suitable model is searched for within this set. This is the most theoretical part of the system identification procedure. It is here that a priori knowledge and engineering intuition and insight have to be combined with the formal (mathematical) properties of models. In this book, we will use linear/nonlinear, time-invariant and finite dimensional models of multi-input multi-output (MIMO) systems that are suitable for modeling a large class of industrial processes.

When the data are available and the model set is determined, the next step is to find the best model in this set. For model parameter estimation, an error criterion (loss function) is specified. Often the sum of the squares of some error signals (residuals) is used as the criterion. The values of the parameters are determined by minimizing the loss function.

When a model is identified, the next step is model validation. This step is to test whether the estimated model is sufficiently good for the intended use of the model. First of all, a check to see if the model is in agreement with the a priori knowledge of the system. Then a check if the model can fit the test or experimental data well, preferably using a data sequence that has not been used in model estimation. The final validation of the model is the application of the model.

1.2 Modern Automation Structure

The process industries include the following major sectors: food, textiles, paper, chemicals, petroleum, rubber and plastics, glass, metal and electricity. Due to world-wide competition, shortage of natural resources and environmental pollution, the present process industries face very dynamic and unpredictable market conditions and have to produce under very strict national and international regulations. Computer-based automation systems have been developed in process industries in order to increase the production safety, quality and flexibility, and to reduce energy and material consumption as well as environmental pollution. Some process industries, such as the refinery and petrochemical industry, recognize plant automation more and more as a cost effective way of responding to changes of market conditions and production regulations. A modern automation system in process industries can be depicted as a pyramid and consists of the following layers.

- Instrumentation and primary control: This layer is usually a distributed control system (DCS) which gathers process measurements and performs simple monitoring of the measurements. The measurements include basic process variables such as temperature, flow, pressure, level and valve position. A DCS also performs PID based controls on some of the process variables. Usually one controller only takes care of a single primary variable such as flow or temperature.
- Advanced process control (APC): This part of the system performs multivariable model based control that will ensure stable unit operation and push the process to its operational limits for maximum economic benefit. Here one APC controller can control a whole process such as a distillation column, a reactor. In general, identified dynamic models (most often linear) are used for APC controllers. This layer is usually present in a computer.
- Diagnosis and supervision: This part of the system is to improve the safety and reliability of the unit operation. A diagnosis system performs fault detection and classification and gives suggestions for maintenance and remedies. Early methods are mainly based on limit value checking of easily measurable signals and their derivatives; a recent trend is to use process models for more accurate and quicker diagnosis. The system can also evaluate the performance of the controllers at different levels. This layer is usually present in a minicomputer.
- Optimization: An optimization system manipulates the process degrees of freedom that are left after meeting the requirements of a safe and stable operation, to meet the unit economic objectives such as saving energy and expensive material and/or increasing throughput. The optimizer determines the best set points for APC controllers. Usually the optimization is carried out based on rigorous (first principle) nonlinear static models. Sometimes identified models are also used for optimization, because the cost of using and maintaining a rigorous model can be too high. Usually the optimizer is executed at a slow rate such that the APC controls are at steady state with respect to the previous set point change. The optimization can be performed for a single process as well as a combination of processes. An optimizer is usually located in a minicomputer.
- Planning and scheduling: This part may cover many units of processes and it provides decision support in production planning, allocation of raw materials and scheduling of plant operation for realizing the company's program and for maximizing profits. It is used to respond to the market changes as well as production regulation changes. This part can be located in a minicomputer or a main frame computer.

Each layer plays an unique and complementary role in the total automation system and that allows the company to react rapidly to changes. At present, most process industries have instrumentation and primary control. Only some capital intensive sectors use higher level layers such as APC, optimization and scheduling. To our knowledge, refinery and petrochemical industries take the lead in applying computer automation systems. Due to the availability of affordable and reliable computers and to development of computing and control sciences, the time is coming that more process industries can benefit from this multi-disciplinary technology.

1.3 Systems Identification

Fundamental to most physical sciences is the concept of a mathematical model. For example, models are essential for prediction and control purposes. The type and accuracy of the model depends upon the application in mind, including models for aerospace applications which usually need to be very precise, whereas models for industrial processes, such as blast furnaces, can often be very crude. Models can be obtained from physical reasoning or by analyzing experimental data from the system. In the latter case, our ability to obtain an accurate model is limited by the presence of random fluctuations such as unmeasured disturbances and measurement errors. The problem of obtaining mathematical models of physical systems from real experimental data constitutes a major part of this book. In particular, we study the problem of estimation of the parameters within models of dynamic systems. We also investigate the effects of various experimental conditions upon model accuracy, see [1–48].

In the majority of practical situations, it is necessary to implement a methodology for direct identification of these dynamic (control) models from experimental data. We note that there are two types of dynamic models:

1. *Non-parametric models* (example: frequency response, step response).
2. *Parametric models* (example: transfer function, differential or difference equation).

Henceforward, we will be concerned with the identification of parametric dynamic models. In this regard, system identification is an experimental approach for determining the dynamic model of a system. It includes four steps:

1. Input/output data acquisition under an experimentation protocol.
2. Selection or estimation of the "model" structure (complexity).
3. Estimation of the model parameters.
4. Validation of the identified model (structure and values of the parameters).

A complete identification operation must necessarily comprise the four stages indicated above. The specific methods used at each stage depend on the type of model desired (parametric or non-parametric, continuous-time or discrete-time) and on the experimental conditions (for example, hypothesis on the noise, open loop or closed loop identification). The *validation* is the mandatory step to decide if the identified model is acceptable or not.

In what follows, we adopt the approach that filtering and system identification are powerful techniques for building models of complex systems in communications, signal processing, control, and other engineering disciplines.

1.4 Control Design

System theory, in particular, automatic control and system identification have experienced a fast evolution in the past decades. New methods have been developed, performance requirements in traditional engineering areas have significantly increased and in addition, numerous demanding applications in other areas of engineering and science have appeared. On one hand, the use of the so-called "practical" examples has been a dominant feature of many textbooks. On the other hand, other books focus directly on the practical issues involved, leaving the theory out. However, there is still an important gap between practical model building and the control design tools.

The extraordinary development of digital computing devices including microprocessors and micro controllers and their extensive use in control systems in all fields of applications has brought about important changes in the design of control systems. Their performance and low cost make them quite suitable for use in control systems of various types which, in turn, places a demand for better capabilities and performances than those provided by analog controllers. From a practical standpoint, to take advantage of the capabilities of microprocessors, it is not enough to reproduce the behavior of analog controllers. One really needs to implement specific and high-performance model based-control techniques developed for computer-controlled systems. In this context, *identification of a plant dynamic model from data is a fundamental step in the design of the control system.*

It is increasingly apparent that the association of books with software and on-line material is radically changing the teaching methods of the control engineering field. Computer-aided control design software requires the understanding of a number of concepts in order to be used efficiently. The use of software for illustrating the various concepts and algorithms helps understanding and rapidly gives a feeling of the various phenomena. Details concerning effective implementation and on-site optimization of the control systems designed have been provided.

1.5 Outline of the Book

The chief objective of this book is to provide a complete description of the application of linear system identification and linear control design to practical systems. Thus, this book brings together advanced methods of modern, robust and resilient control on one hand and applied system identification methods on the other hand. The starting point is a set of real experimental data of some industrial processes. For generality in exposition, the main focus is on merits and demerits of different control and identification methods. Through extensive simulation studies, several conclusions will be drawn and useful design toolkit will be delineated.

1.5.1 Methodology

Throughout the textbook, our methodology is composed of five-steps:

- *Mathematical Modeling* in which we focus on the use of system identification techniques to generate transfer-function and/or state-space models based on real-data and subsequently discuss the main ingredients of the derived models under consideration.
- *Definitions and/or Assumptions* here we introduce the definitions and/or constraints on the model variables then proceed to methods of system analysis.
- *Examples and Illustrations* this represent the backbone of the book around which the material of the different sections and subsections evolve. This material includes some solved examples based on MATLAB environment to demonstrate the effectiveness of the various algorithms and techniques.
- *Remarks* which are given to shed some light of the relevance of the developed results vis-a-vis published work. These also help in identifying pertinent features and properties.
- *Methods* which are provided most of the time in the form of algorithms and/or MATLAB procedures.

In the sequel, definitions (assumptions, remarks) are keyed to chapters and stated in roman font with **bold titles**, for example, **Definition 2.3** means Definition 3 in Chap. 2 and so on. For convenience, we have added references and problems at the end of each chapter. Relevant notes and issues are offered as well at the end of each chapter for the purpose of stimulating the reader.

1.5.2 Chapter Organization

The material covered is divided into nine chapters whereby continuous-time results go in parallel with discrete-time results. Every chapter includes the corresponding MATLAB hints, file names along with some pertinent statements which illustrate how the algorithms can be used in simulation, computation and implementation. A problem section for practice of the design is included as well. All the developed results are conveniently expressed in MATLAB-based procedures.

In Chap. 2, we start our guided tour through the book by presenting several processes that are commonly used in industrial applications. These processes include:

1. Steam Generation Unit.
2. Small-Power Wind Turbine.
3. Unmanned Marine Vehicles.
4. Industrial Evaporation Unit.
5. Distillation Column.
6. Falling Film Evaporator.

7. Vapor Compression Cycle Systems.
8. Flutter of an Aircraft F-18.
9. A Hydraulic Pumping System.

The purpose is to provide a wide pool of information about practical systems and to acquaint the reader with the properties of the processes.

The objective of Chap. 3 is to establish a solid foundation of system identification methods that will be used and experimented in the subsequent chapters. The material covered is subdivided as follows:

1. Parameter estimation approach: with focus on estimation algorithms, gradient algorithm, least squares algorithm, choice of the adaptation gain.
2. Transfer-function methods: these include autoregressive (AR) method, ARX method, autoregressive moving average (ARMAX) method, Box–Jenkins method, prediction error method (PEM).
3. Subspace identification method: attention is paid to state space models, block Hankel matrices and state sequences, model matrices, geometric tools, orthogonal and oblique projections. Deterministic, stochastic and combined deterministic-stochastic subspace identification schemes are presented.

Chapter 4 contains detailed computer experiments and MATLAB simulation results of applying the identification methods of Chap. 3 to the industrial processes of Chap. 2. Focus is placed on data analysis, validation and model fitness. State-space order determinations and comparison of MIMO models are prime factors. In addition, conclusions drawn from comparisons among these methods are delineated.

The core techniques in the design of linear control methods are described in Chap. 5. In preparation, some basic definitions and features of feedback control systems (open-loop versus closed-loop control systems, standard representations)are introduced. Next, the main structural properties of stability, controllability and observability are established. Feedback control methods (state-feedback, output-feedback, observer-based feedback) are developed and applied to several practical systems. We then provides classifications of industrial controllers and emphasize on the control actions (P-control, I-control, PI-control, PD control and PID control). Discussions are given to closed-loop system subjected to a disturbance. Methods of linear optimal control (continuous-time and discrete-time), model predictive control, the Kalman filter and linear quadratic Gaussian regulator design are analysis in detailed and demonstrated on typical systems examples.

Applications of the linear control design methods are detailed in Chap. 6 and implemented on:

1. Shaping Process of Automobile Belt.
2. An Unmanned Helicopter.
3. Reverse Osmosis Desalination Plant.
4. Turbocharged Diesel Engine.
5. A Rotational Hydraulic Drive.
6. The Falling Film Evaporator.
7. Vapor Compression Cycle Systems.

8. Control of Quadrotor Unmanned Air Vehicle.
9. Stabilization of F-8 Fly-by-Wire Aircraft.
10. Air Conditioning System.
11. Three-Degree-of-Freedom Helicopter Model.
12. Design of Aircraft Controller.

Chapter 7 introduces an introductory material to advanced control design methods including $\mathcal{H}_2$, $\mathcal{H}_\infty$ and mixed $\mathcal{H}_2/\mathcal{H}_\infty$ performance criteria. It paves the way by defining norm measures of signals and systems to construct the problems under consideration. In addition, it discusses integral control for robust tracking. Both optimal and stabilizing solutions are given along with some examples.

Adaptive methods in the form of control design algorithms are illustrated in Chap. 8.

Throughout the book, MATLAB implementation and simulation results are emphasized. Each chapter includes some selected solved examples and/or case studies and is supplemented by relevant questions and problems. An Appendix containing some relevant mathematical tools and basic results is provided as Chap. 9.

References

1. Abdelazim, T., Malik, O.: Identification of nonlinear systems by Takagi-Sugeno fuzzy logic grey box modeling for real-time control. Control Eng. Pract. **13**(12), 1489–1498 (2005)
2. Aguirre, L.A.: A nonlinear correlation function for selecting the delay time in dynamical reconstructions. Phys. Lett. **203A**(2–3, 88–94 (1995)
3. Aguirre, L.A., Donoso-Garcia, P.F., Santos-Filho, R.: Use of a priori information in the identification of global nonlinear models—A case study using a buck converter. IEEE Trans. Circuits Syst. I, Regul. Pap. **47**(7), 1081–1085 (2000)
4. Aguirre, L.A., Barroso, M.F.S., Saldanha, R.R., Mendes, E.M.A.M.: Imposing steady-state performance on identified nonlinear polynomial models by means of constrained parameter estimation. IEE Proc. Part D. Control Theory Appl. **151**(2), 174–179 (2004)
5. Aguirre, L.A., Coelho, M.C.S., Corrêa, M.V.: On the interpretation and practice of dynamical differences between Hammerstein and Wiener models. IEE Proc. Part D. Control Theory Appl. **152**(4), 349–356 (2005)
6. Astrom, K.J., Eykhoff, P.: System identification—A survey. Automatica **7**(2), 123–162 (1971)
7. Baker, J.E.: Reducing bias and inefficiency in the selection algorithm. In: Proc. 2nd Int. Conf. Genetic Algorithms Genetic Algorithms Their Appl., Mahwah, N.J., pp. 14–21 (1987)
8. Bakker, H.H.C., Marsh, C., Paramalingam, S., Chen, H.: Cascade controller design for concentration in a falling film evaporators. Food Control **17**(5), 325–330 (2006)
9. Barbosa, B.H.: Instrumentation, modelling, control and supervision of a hydraulic pumping system and turbine–generator module (in Portuguese). Master's thesis, Sch. Elect. Eng., Federal Univ. Minas Gerais, Belo Horizonte, Brazil (2006)
10. Barroso, M.S.F., Takahashi, R.H.C., Aguirre, L.A.: Multi-objective parameter estimation via minimal correlation criterion. J. Process Control **17**(4), 321–332 (2007)
11. Billings, S.A., Voon, W.S.F.: Least squares parameter estimation algorithms for nonlinear systems. Int. J. Syst. Sci. **15**(6), 601–615 (1984)
12. Billings, S.A., Chen, S., Korenberg, M.J.: Identification of MIMO nonlinear systems using a forward-regression orthogonal estimator. Int. J. Control **49**(6), 2157–2189 (1989)
13. Bingulac, S., Sinha, N.K.: On the identification of continuous-time systems from the samples of input–output data. In: Proc. Seventh Int. Conf. on Mathematical and Computer Modeling, Chicago, IL, pp. 231–239 (1989)

14. Bucharles, A., Cassan, H., Roubertier, J.: Advanced parameter identification techniques for near real = time flight flutter test analysis. AIAA, Paper 90-1275, May 1990
15. Burl, J.B.: Linear Optimal Control, 3rd edn. Prentice Hall, New York (1998)
16. Chankong, V., Haimes, Y.Y.: Multiobjective Decision Making: Theory and Methodology. North-Holland (Elsevier), New York (1983)
17. Chen, S., Billings, S.A., Luo, W.: Orthogonal least squares methods and their application to nonlinear system identification. Int. J. Control **50**(5), 1873–1896 (1989)
18. Connally, P., Li, K., Irwing, G.W.: Prediction and simulation error based perceptron training: Solution space analysis and a novel combined training scheme. Neurocomputing **70**, 819–827 (2007)
19. Cooper, J.: Parameter estimation methods for the flight flutter testing. In: Proc. the 80th AGARD Structures and Materials Panel, CP-566, AGARD, Rotterdam, The Netherlands, 1995
20. Correa, M.V., Aguirre, L.A., Saldanha, R.R.: Using steady-state prior knowledge to constrain parameter estimates in nonlinear system identification. IEEE Trans. Circuits Syst. I, Regul. Pap. **49**(9), 1376–1381 (2002)
21. Cunningham, P., Canty, N., O'Mahony, T., O'Connor, B., O'Callagham, D.: System identification of a falling film evaporator in the dairy industry. In: Proc. of SYSID'94, Copenhagen, Denmark, vol. 1, 234–239 (1994)
22. Ghiaus, C., Chicinas, A., Inard, C.: Grey-box identification of air-handling unit elements. Control Eng. Pract. **15**(4), 421–433 (2007)
23. Goldberg, D.E.: Genetic Algorithms in Search, Optimization and Machine Learning. Addison-Wesley, New York (1989)
24. Hsia, T.C.: On sampled-data approach to parameter identification of continuous-time linear systems. IEEE Trans. Autom. Control **AC-17**, 247–249 (1972)
25. Hsia, T.: System Identification: Least-Squares Methods. Lexington Books, Lexington (1977)
26. Jakubek, S., Hametner, C., Keuth, N.: Total least squares in fuzzy system identification: An application to an industrial engine. Eng. Appl. Artif. Intell. **21**, 1277–1288 (2008)
27. Karimi, M., Jahanmiri, A.: Nonlinear modeling and cascade control design for multieffect falling film evaporator. Iran. J. Chem. Eng. **3**(2) (2006)
28. Kehoe, M.W.: A historical overview of flight flutter testing, NASA TR 4720, Oct. 1995
29. Leontaritis, I.J., Billings, S.A.: Input–output parametric models for nonlinear systems. Part II: Deterministic nonlinear system. Int. J. Control **41**(2), 329–344 (1985)
30. Miranda, V., Simpson, R.: Modelling and simulation of an industrial multiple-effect evaporator: Tomato concentrate. J. Food Eng. **66**, 203–210 (2005)
31. Neilsen, K.M., Pedersen, T.S., Nielsen, J.F.D.: Simulation and control of multieffect evaporator
32. Nepomuceno, E.G., Takahashi, R.H.C., Aguirre, L.A.: Multiobjective parameter estimation: Affine information and least-squares formulation. Int. J. Control **80**(6), 863–871 (2007)
33. Norgaard, M.: Neural network based system identification—TOOLBOX, Tech. Univ. Denmark, Lyngby, Tech. Rep. 97-E-851 (1997)
34. Ogata, K.: MATLAB for Control Engineers. Prentice-Hall, New York (2008)
35. Pan, Y., Lee, J.H.: Modified subspace identification for long-range prediction model for inferential control. Control Eng. Pract. **16**(12), 1487–1500 (2008)
36. Piroddi, L.: Simulation error minimization methods for NARX model identification. Int. J. Model. Identif. Control **3**(4), 392–403 (2008)
37. Piroddi, L., Spinelli, W.: An identification algorithm for polynomial NARX-models based on simulation error minimization. Int. J. Control **76**(17), 1767–1781 (2003)
38. Quaak, P., van Wijck, M.P.C.M., van Haren, J.J.: Comparison of process identification and physical modeling for falling film evaporators. Food Control **5**(2), 73–82 (1994)
39. Rangaiah, G., Saha, P., Tade, M.: Nonlinear model predictive control of an industrial four-stage evaporator system via simulation. Chem. Eng. J. **87**, 285–299 (2002)
40. Roffel, B., Betlem, B.: Process Dynamics and Control. Wiley, London (2006)
41. Sinha, N.K.: Estimation of transfer function of continuous-time systems from samples of input–output data. Proc. Inst. Electr. Eng. **119**, 612–614 (1972)

42. Sinha, N.K., Kuszta, B.: Modelling and Identification of Dynamic Systems. Von-Nostrand Reinhold, New York (1983)
43. Sinha, N.K., Rao, G.P. (eds.): Identification of Continuous-Time Systems. Kluwer Academic, Dordrecht (1991)
44. Sjoberg, J., Zhang, Q., Ljung, L., Beneviste, A., Delyon, B., Glorennec, P., Hjalmarsson, H., Juditsky, A.: Non-linear black-box modeling in system identification: A unified overview. Automatica **31**, 31–1961 (1995)
45. Soderstrom, T., Stoica, P.: System Identification. Prentice-Hall, New York (1989)
46. Stefanov, Z., Hoo, K.A.: Control of a multiple-effect falling-film evaporator plant. Ind. Eng. Chem. Res. **44**, 3146–3158 (2005)
47. Van Wijck, M.P., Quaak, P., van Haren, J.J.: Multivariable supervisory control of a four-effect falling film evaporator. Food Control **5**(2), 234–243 (1994)
48. Zwillinger, D.: Standard Mathematical Tables and Formulae, 31st edn. Chapman & Hall/CRC, Boca Raton (2002)

Chapter 2
Some Industrial Systems

2.1 Introduction

Identification of process parameters for control purposes must often be done using a digital computer, from samples of input–output observations. On the other hand, the process is usually of continuous-time nature, and its dynamical model is most aptly described in terms of differential equations. Thus, our problem may be stated as determining a continuous-time model from samples of input–output data.

During the past few decades, several approaches have been developed [30, 46–48]. For the sake of simplicity, these can be classified as

- direct methods,
- indirect methods.

Methods belonging to the first type attempt to estimate the parameters of a continuous-time model directly from the samples of the observations, mostly using some type of numerical integration. In methods of the latter group, the problem is conveniently divided into two subproblems:

The first subproblem consists of estimating the parameters of a discrete-time model from the samples of the input–output observations.

The second subproblem, on the other hand, consists of determining a suitable continuous-time model that is equivalent to the discrete-time model obtained for a given sampling interval.

Generally speaking, the problem of system identification may now be stated as the estimation of the elements of the matrices A, B, C, D associated with the linear time-invariant system

$$\begin{aligned} \dot{x}(t) &= Ax(t) + Bu(t), \\ y(t) &= Cx(t) + Du(t) \end{aligned} \tag{2.1}$$

from a record of the samples of the input output data

$$\{u(kT), y(kT)\}, \quad \text{for } k = 0; 1; 2; \ldots; N$$

where N is a suitable large number.

M.S. Mahmoud, Y. Xia, *Applied Control Systems Design*,
DOI 10.1007/978-1-4471-2879-3_2, © Springer-Verlag London Limited 2012

It may be noted that the matrix D represents direct coupling between the input and the output, and will be zero for strictly proper transfer functions. Without any loss of generality and unless otherwise stated, this will be assumed to be the case throughout this book. It should be noted that none of the matrices A, B, C in (2.1) are unique for a system with a given input–output description. Given a special canonical form for the system state equations in either the continuous-time or the equivalent discrete time models overcomes this problem and also minimizes the number of parameters to be estimated. It should also be noted that it is tacitly assumed that the order of the linear state space model is known, and that the sampling interval has been suitably selected. In practice, both of these are important, and have been subjects of considerable research [24, 25, 48].

In fact, the problem is further complicated by the fact that the available data are usually contaminated with random noise that are produced either by disturbances or introduced in data acquisition and measurement. The literature on system identification abounds in papers devoted to methods for estimating the parameters in the presence of noise, see [47] for a detailed list of references.

2.2 Steam Generation Unit

There are two types of configurations in the electricity generation using drum boilers and steam turbines:

1. A single boiler is used to generate steam that is directly fed to a single turbine. This configuration is usually referred to as a boiler–turbine unit.
2. A header is used to accommodate all the steam produced from several boilers, and the steam is then distributed to several turbines through the header. The steam can be used to generate electricity as well as other purposes. This configuration is commonly used in industrial utility plants.

Boiler–turbine units are nowadays preferred over header systems, because they can achieve quick response to electricity demands from a power grid or network. It is generally accepted that a boiler–turbine unit is a highly nonlinear and strongly coupled complex system. However, there is no definite quantification of the complexity of a unit. Specifically, how nonlinear is it? Can a linear controller be used to cover the whole operating range? These are fundamental issues in the control system design for a boiler–turbine unit. Without a thorough understanding, modeling and identification of the system, the operating range and performance of a linear controller cannot be guaranteed. Figure 2.1 shows the schematic diagram of the steam generator model.

2.2.1 System Dynamics

For the system considered here, the input/output experimental data has been obtained from [20] in which the modeling of a steam generator at Abbot power plant

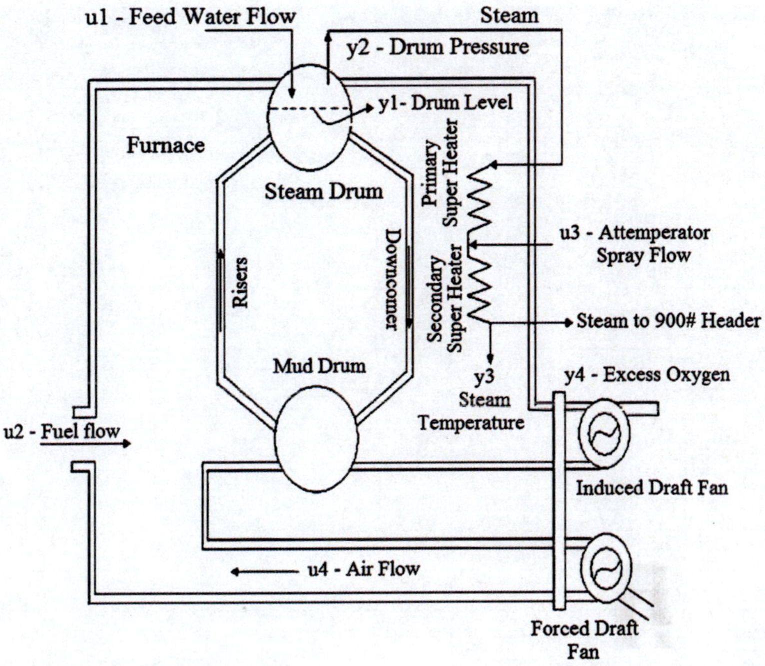

Fig. 2.1 Steam generating unit

in Champaign IL is considered. The data comes from a model of this steam generator. The inputs are listed as follows:

- U1: Fuel scaled 0–1,
- U2: Air scaled 0–1,
- U3: Reference level,
- U4: Disturbance defined by the load level.

The outputs are

- Y1: Drum pressure,
- Y2: Excess oxygen in exhaust gases,
- Y3: Level of oxygen in the drum,
- Y4: Steam flow.

The data values are presented in Fig. 2.2.

The simulation data constitutes 9600 samples at a sampling rate of 3 s, which characterizes a MIMO process. In implementation, a set of 4000 samples (5000 : 9000) are used for testing, another set of 4000 samples (2500 : 6500) for validation purpose. The important statistical parameters of all inputs and outputs are listed in Table 2.1.

time-steps	input fuel	input air	input level ref.	input disturbance	drum pressure	excess oxygen	water level	steam flow
0	0.723567501	0.69162867	-2.283056412	0.017982753	320.0823889	2.506773853	0.0327013	9.302970053
3	0.527147387	0.383020194	-3.716481388	0.01871789	321.7109902	2.545907948	0.284799436	9.662620584
6	0.589603123	0.705814627	-1.530992972	0.020756328	320.9133077	2.360561543	0.203651646	10.99095548
9	0.364854222	0.713277726	-3.173065004	0.022927443	325.0025177	0.027054182	0.326186879	12.43010713
12	0.576540101	0.361978097	-1.796211886	0.020370994	326.6527646	0.285649429	0.753776251	13.68166563
15	0.643106594	0.559504307	-2.80426768	0.023209106	326.1869164	2.631328071	1.851854064	14.60888638
18	0.495364003	0.690460919	-3.207444006	0.016374061	326.5893955	0.997855279	2.302339595	15.71319958
21	0.585075453	0.470516735	-1.041846551	0.015338046	328.734886	0.004474148	2.447356587	16.5363068
24	0.664653688	0.360282868	-1.931675122	0.017923064	329.609686	0.755694209	3.036620233	17.02536888
27	0.643043139	0.574080989	-1.478481089	0.011513461	330.1852705	0.150455487	3.567548546	17.61613049
30	0.464561148	0.387073874	-3.913301903	0.017580064	331.7391065	0.01263697	3.462808629	18.10196376
33	0.404921448	0.660802771	-1.59869963	0.015723326	333.1073413	0.012963788	3.423381121	18.6178379
36	0.724379922	0.446190083	-3.989163444	0.014141722	332.2526813	0.00143049	3.542126862	18.6357234
39	0.731302053	0.697717252	-1.150864528	0.02067146	331.9641902	1.261865971	3.542697727	18.75132224
42	0.716108207	0.470846609	-2.509671086	0.023348293	332.2584412	-0.004533128	3.033212727	19.50889423
45	0.519158864	0.510866342	-0.97339551	0.017045047	332.698191	-0.036224068	2.155174268	20.45663212
48	0.595279473	0.642852826	-3.337995093	0.018983813	333.7106987	-0.028905017	2.130121016	20.89645496
51	0.48403137	0.377171236	-2.516360051	0.010279153	334.9233445	0.032445181	2.267387406	21.02755549
54	0.406252116	0.686653761	-1.666630649	0.020761998	332.9221974	0.066819178	2.219327997	20.83617396
57	0.515172894	0.730148283	-1.756067442	0.01314353	330.9151141	-0.017981385	1.573688265	20.70063463
60	0.353699529	0.487394405	-2.117513651	0.02007896	328.3267396	1.317601729	1.296664701	20.39427621
63	0.51216968	0.64642243	-3.74016714	0.011627354	326.4036469	3.151273589	0.551315927	20.22160655
66	0.453003993	0.718432285	-1.327949705	0.021740968	323.6588589	3.726752428	-0.115948509	19.74100857
69	0.648760825	0.343399712	-2.045945519	0.013691541	322.0601255	3.548648872	-0.543213656	19.84917364
72	0.533849639	0.709241376	-1.831108838	0.012666236	320.4334403	4.010865711	-1.184954324	19.75339354
75	0.399869705	0.348706015	-1.082696127	0.020909792	321.2297867	0.600347579	-1.232502023	19.77856981
78	0.578235907	0.709798053	-1.374318147	0.010350625	319.5896845	0.359864199	-1.678045037	19.79985922
81	0.443618032	0.665604084	-0.861456561	0.022472615	318.0445196	0.5383361	-1.522747247	19.43914125
84	0.63785371	0.535170707	-3.6838049	0.016782261	316.7543457	0.592062835	-1.699705982	19.58416198
87	0.485757786	0.454677147	-1.137821862	0.018268976	314.4978796	2.209165353	-1.99815298	19.49900196
90	0.360064787	0.39484178	-2.856183924	0.012938862	313.3338226	1.636450336	-1.993150363	19.64935081
93	0.66066444	0.704633989	-1.060758232	0.019753636	311.8162066	-0.007715276	-2.187443944	19.32839398
96	0.515891469	0.645825437	-2.579230429	0.013735595	309.1821627	0.087425935	-1.859827468	19.1695092
99	0.612877778	0.356006904	-0.957565452	0.017618584	307.8668511	0.299624857	-2.316410273	19.23562254
102	0.608584721	0.493893118	-2.141022066	0.015179435	309.1451192	0.925039703	-2.456011943	19.27840083
105	0.619788222	0.60421281	-3.952446283	0.014320419	308.4521475	0.020171863	-2.032363803	19.34002846
108	0.369935157	0.695078504	-1.396240297	0.009636202	307.1421935	-0.010024978	-2.204217845	19.34207649
111	0.536819511	0.56283553	-2.278566849	0.020844465	306.7214071	-0.016982704	-2.212139324	19.11273381
114	0.579805316	0.436789359	-3.081064258	0.018543932	304.9138787	1.727690747	-2.153424917	18.86746672
117	0.698441545	0.433841554	-1.304343744	0.017743059	304.230753	2.123490842	-2.020366963	19.07370268
120	0.695132002	0.540081641	-1.354954605	0.010611542	303.2929753	-0.004211119	-2.67449837	19.23870373
123	0.633049299	0.579749551	-3.794079204	0.023729927	303.0388137	-0.024038614	-2.607590229	19.33070714
126	0.666181679	0.713919546	-1.539948697	0.01281085	303.5975859	-0.018974644	-2.552692993	19.74181993
129	0.692100112	0.429825546	-0.853122616	0.010553078	304.0344998	-0.003489681	-2.394882293	19.82902142
132	0.730122251	0.531272784	-1.597177347	0.012054672	305.1232434	0.019294486	-2.115191494	19.73419025

Fig. 2.2 Statistical data pattern

Table 2.1 Statistical data

Input/output	Type	Mean	Standard deviation	Min	Max
$I1$	Fuel scaled 0–1	0.504	0.229	0.000	1.07
$I2$	Air scaled 0–1	0.528	0.295	0.000	1.07
$I3$	Reference level	0.554	2.460	−4.00	4.53
$I4$	Disturbance	0.004	0.010	−0.015	0.023
$O1$	Drum pressure	329.4	85.94	154	534
$O2$	Excess oxygen in air	4.544	6.157	−0.069	21
$O3$	Drum oxygen level	0.552	2.849	−9.55	12.3
$O4$	Steam flow	14.85	7.571	1.99	34.6

2.3 Small-Power Wind Turbine

Wind energy is a fast-growing interdisciplinary field that encompasses many different branches of engineering and science. Despite the amazing growth in the installed capacity of wind turbines in recent years, engineering and science challenges still exist. Because larger wind turbines have power capture and economical advantages, the typical size of utility-scale wind turbines has grown dramatically over the last three decades. Modern wind turbines are large, flexible structures operating in uncertain environments and lend themselves nicely to advanced control solutions. Advanced controllers can help achieve the overall goal of decreasing the cost of wind energy by increasing the efficiency, and thus the energy capture, or by reducing structural loading and increasing the lifetimes of the components and turbine structures. In what follows, our goal is to introduce control engineers to the technical challenges that exist in the wind energy industry and to encourage new control systems research in this area.

2.3.1 Wind Turbine Basics

The main components of a horizontal-axis wind turbine that are visible from the ground are its tower, nacelle, and rotor. The nacelle houses the generator, which is driven by the high-speed shaft. The high speed shaft is in turn usually driven by a gear box, which steps up the rotational speed from the low-speed shaft. The low-speed shaft is connected to the rotor, which includes the airfoil-shaped blades. These blades capture the kinetic energy in the wind and transforms it into the rotational kinetic energy of the wind turbine. The description of the wind turbine system depends on the designs of the wind turbine either horizontal-axis or vertical axis, see Fig. 2.3.

Vertical-axis wind turbines (VAWTs) are pretty rare and the only one currently in commercial production is the Darrieus turbine, which looks kind of like an egg figure. In a VAWT, the shaft is mounted on a vertical axis, perpendicular to the ground. VAWTs are always aligned with the wind, unlike their horizontal-axis counterparts, so there's no adjustment necessary when the wind direction changes. On the other hand, a VAWT is not normally self starting, it needs energy from its electrical system to get started. Instead of a tower, it typically uses wires for support, so the rotor elevation is lower. Lower elevation means slower wind due to ground interference, so VAWTs are generally less efficient than horizontal-axis wind turbines (HAWTs). On the upside, all equipment is at ground level for easy installation and servicing, but that means a larger footprint for the turbine, which is a big negative in farming areas. VAWTs may be used for small-scale turbines and for pumping water in rural areas, but all commercially produced, utility-scale wind turbines are (HAWTs), see Figs. 2.4–2.5.

From its name, the HAWT shaft is mounted horizontally, parallel to the ground. HAWT needs to continuously align itself with the wind speed by using a yaw-adjustment mechanism. The yaw system typically consists of electric motors and

Fig. 2.3 VAWT and HAWT

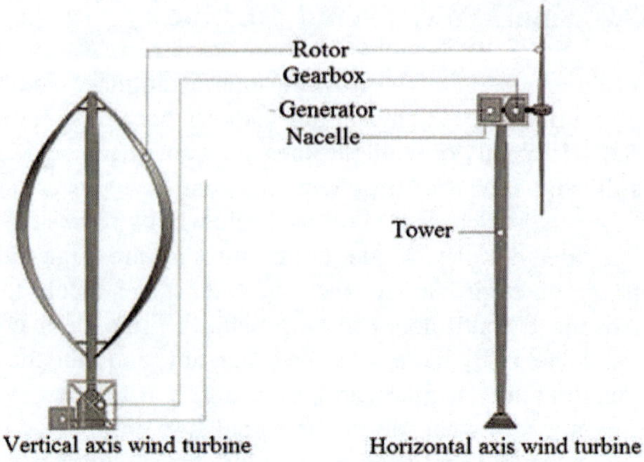

Fig. 2.4 The main components of HAWT

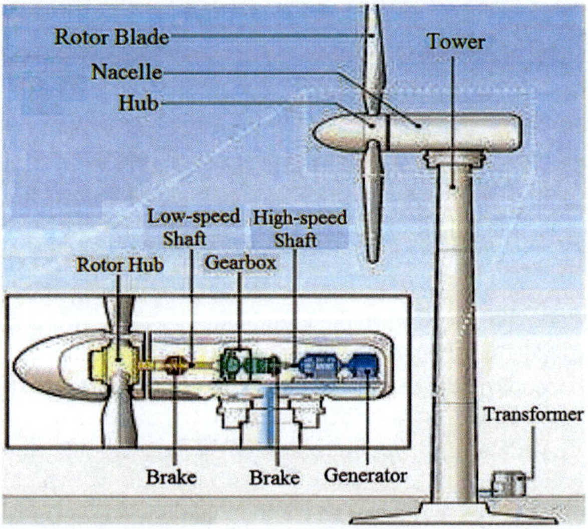

gearboxes which move the whole rotor left or right in small increments to hold the higher speed. The turbine's electronic controller reads the position of a wind vane device either mechanical or electronic and adjusts the position of the rotor to capture the most wind energy available [26]. HAWTs use a tower to lift the turbine components to an optimum elevation for wind speed and so the blades can take up very little ground space since wind velocities increase at higher altitudes due to surface aerodynamic drag and the viscosity of the air. Horizontal-axis wind turbines have the main rotor shaft and electrical generator at the top of a tower and must be pointed into the wind. Small turbines are pointed by a simple wind vane, while large turbines generally use a wind sensor coupled with a servo motor. Most of HAWTs have a gearbox which turns the slow rotation of the blades into a quicker rotation that is more appropriate to drive an electrical generator. The main components of

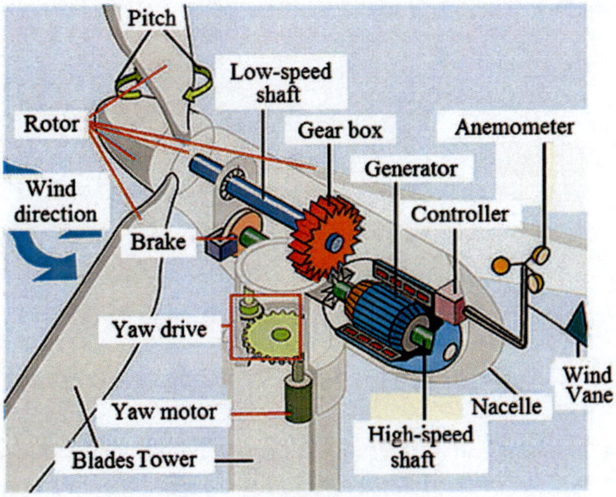

Fig. 2.5 Parts inside the wind turbine

HAWTs are Rotor blades which capture wind's energy and convert it to rotational energy of low speed shaft and Shaft that transfers rotational energy into generator. Also, Nacelle casing that holds Gearbox which increases speed of shaft between rotor hub and generator, Generator that uses rotational energy of shaft to generate electricity using electromagnetism and usually an induction generator that produces AC electricity is used. Moreover, Electronic control unit that monitors system and starts up the machine at wind speeds of about 3–8 m/s and shuts down the machine at about 20 m/s which turbines do not operate at wind speeds above about 20 m/s because they might be damaged by the high winds, Yaw controller is used to keep the rotor facing into the wind as the wind direction changes, and Brakes that stop rotation of shaft in case of power overload or system failure.

In addition to these components, the tower that used to support rotor and nacelle and lifts entire setup to higher elevation where blades can safely clear the ground and towers are made from tubular steel, concrete, or steel lattice. Wind speed increases with height and this mean, taller tower enable turbines to capture more energy and generate more electricity. The electrical equipment that is used to transmit electricity from generator down through tower and controls many safety elements of turbine, and anemometer that measures the wind speed and transmits these readings to the controller. The most commonly activated safety system in a turbine is the braking system, which is triggered by above-threshold wind speeds. These setups use a power-control system that essentially hits the brakes when wind speeds get too high and then release the brakes when the wind is coming back.

2.4 Unmanned Surface Marine Vehicle

The Atlantis is assumed to be traveling upon a straight line, conveniently assumed to be coincident with the x-axis, through water at a constant velocity, V_x. The dis-

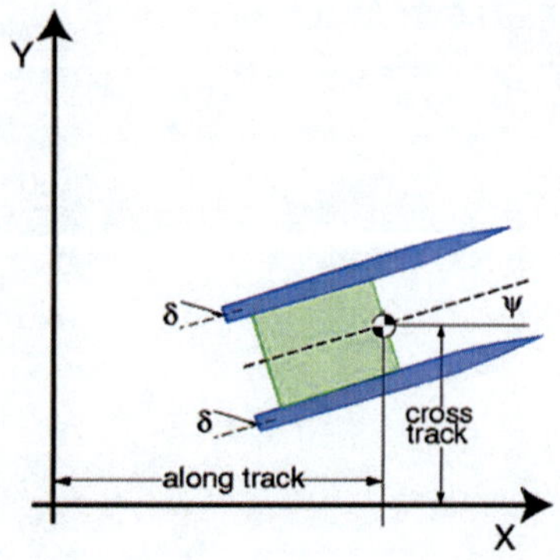

Fig. 2.6 A schematic model of the assumed path of the Atlantis

tance along that line is X (meters), the perpendicular distance to the line is Y (meters), the cross-track error, and the angle that the center-line of the Atlantis makes with the x-axis is Ψ, the angular error (radians). Figure 2.6 illustrates a schematic model of the assumed path of the Atlantis. The coordinate frame can always be rotated to have the x-axis aligned to the desired path of the Atlantis, and so the assumption that the Atlantis travels down the x-axis is a good one. The assumption of constant velocity, however, is not appropriate since velocity is a function of the wind speed. Wind speed, of course, cannot be controlled and is highly variable.

2.4.1 Dynamic Model

The continuous-time state-space equations for the kinematic model can be represented as

$$\begin{bmatrix} \dot{Y} \\ \dot{\Psi} \\ \dot{\delta} \end{bmatrix} = \begin{bmatrix} 0 & V_x & 0 \\ 0 & 0 & \frac{V_x}{L} \\ 0 & 0 & 0 \end{bmatrix} \begin{bmatrix} Y \\ \Psi \\ \delta \end{bmatrix} + \begin{bmatrix} 0 \\ 0 \\ 1 \end{bmatrix} u \tag{2.2}$$

where δ is the angle of the rudders with respect to the hull center-line (radians). The distance L is from the boat center of mass to the center of pressure of the rudders (in meters), and the input, u, is the slew rate of the rudders (in radians/second). This kinematic model assumes that the boat is running on constant V_x. This assumption is known to be poor, since unless the wind can be controlled, the velocity will always be dependent on the speed of the wind. Azimuth and cross-track error in fact do not integrate with time, but rather with distance traveled upon the line. This has great implications, since this is exactly the cause of instability with increasing velocity present in the simple kinematic model. By introducing two new variable,

$$\tilde{Y} \equiv \frac{Y}{V_x}, \qquad \tilde{\Psi} \equiv \frac{\Psi}{V_x}. \tag{2.3}$$

Substituting (2.3) back into (2.2), the kinematic model can be rewritten in the following velocity-invariant form:

$$\begin{bmatrix} \dot{\tilde{Y}} \\ \dot{\tilde{\Psi}} \\ \dot{\delta} \end{bmatrix} = \begin{bmatrix} 0 & 1 & 0 \\ 0 & 0 & \frac{1}{L} \\ 0 & 0 & 0 \end{bmatrix} \begin{bmatrix} \tilde{Y} \\ \tilde{\Psi} \\ \delta \end{bmatrix} + \begin{bmatrix} 0 \\ 0 \\ 1 \end{bmatrix} u. \tag{2.4}$$

2.5 Industrial Evaporation Unit

An identification experiment is performed by exciting the system with appropriate signals and observing its input and output over a time interval. These signals are normally recorded in a computer mass storage for subsequent information processing. Then one proceeds to fit a parametric model of the process from the recorded input and output sequences. The first step is to determine an appropriate form of the model (typically a linear difference equation of a certain order). As a second step some statistically based method is used to estimate the unknown parameters of the model (such as the coefficients in the difference equation). In practice, the estimations of structure and parameters are often done iteratively. This means that a tentative structure is chosen and the corresponding parameters are estimated. The model obtained is then tested to see whether it is an appropriate representation of the system. If this is not the case, some more complex model structure must be considered, its parameters estimated, the new model validated, etc. Note that the 'restart' after the model validation gives an iterative scheme.

2.5.1 Mathematical Models

Models and/or systems can be roughly divided into classes such as linear and nonlinear time invariant or time varying discrete time or continuous time with lumped or with distributed parameters etc. While at first sight the class of linear time invariant models with lumped parameters seems to be rather restricted it turns out in practice that many real life input output behaviors of practical industrial processes can be approximated very well by such a model.

Mathematical models of dynamical systems are used for analysis simulation prediction optimization monitoring fault detection training and control. There are several approaches to generate a model of a system. One could for instance start from first principles such as writing down the basic physical or chemical laws that generate the behavior of the system. This so called white box approach works for simple examples but its complexity increases rapidly for real world systems. In some cases the systems equations are known up to within some unknown parameters, which are estimated using some parameter estimation method gray-box modeling.

Another approach is provided by system identification in which first measurements or observations are collected from the system which are then modeled using a so-called black-box identification approach. Such an approach basically consists of first defining a parameterization of the model, and then determining the model parameters in such a way that the measurements are explained as accurately as possible by the model. Typically, this is done by formulating the identification problem as an optimization problem in which the variables are the unknown parameters of the model the constraints are the model equations and the objective function a measure of the deviation between the observations and the predictions or simulations obtained from the model.

The field of linear system identification is certainly not new although we can safely say that it only started to blossom in the 1970s. Yet, 20-years of research have generated a lot of results and practical hands on experience. Among the key references of identification are [6, 20, 31, 49].

In what follows, we use data for industrial evaporator from [27].

2.5.2 Multistage Evaporator System

The selected evaporator system is the first step in the liquor burning process associated with the Bayer process for alumina production at the Wager up alumina refinery in western Australia. It consists of one falling film, three forced-circulation and a super-concentration evaporators in series [44].

The main components of each stage are a flash tank (FT), a flash pot and a heater (HT). A simplified schematic of the evaporator system is depicted in Fig. 2.7. Flash pots are not shown in this figure for simplicity of the schematic. Spent liquor, which is recovered after precipitation of the alumina from its solution, is fed to the

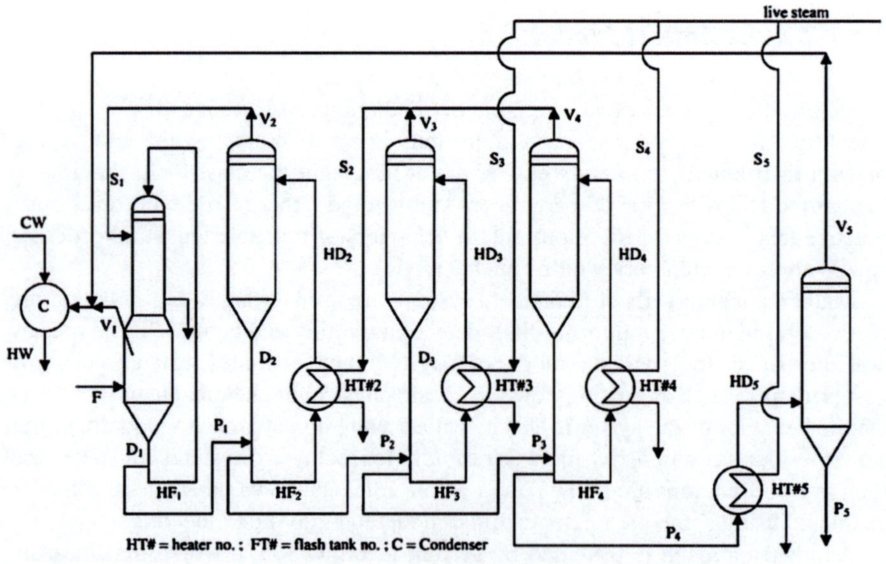

Fig. 2.7 A simplified schematic of the evaporator system

falling film stage (FT #1). The volatile component, water in this case, is removed under high recycle rate and the product is further concentrated through the three forced-circulation stages (FT #2–4). The super-concentration stage (FT #5) is used to remove the residual 'flashing' of the concentrated liquor without recycle. In each of the forced-circulation and super-concentration stages, the spent liquor is heated through a shell and tube heat exchanger (heater) and water is removed as vapor at lower pressure in the FT. The vapor given off is used as the heating medium in the heaters upstream. The flashed vapor from FT #3 and 4 are combined and used in HT #2 while the vapor from FT #2 is used in HT #1. The flashed vapor from FT #5 is sent directly to the condenser, C in Fig. 2.6. The steam condensates from the heaters are collected in the flash pots. Live steam is used as the heating medium for HT #3, 4 and 5. Live steam to HT #3 is set in ratio to the amount of live steam entering HT #4, while the amount of live steam to HT #5 is set depending on the amount of residual 'flashing' to be removed. The cooling water flow to the contact condenser, C is set such that all remaining flashed vapor is condensed. The evaporator system is crucial in the aluminum refinery operation and is difficult to control due to recycle streams, strong process interactions and nonlinearities.

2.6 Distillation Tower

Distillation towers are widely used in the chemical process industries where large quantities of liquids have to be distilled. Industrial distillation towers are usually operated at a continuous steady state. From a practical viewpoint, the most important manipulated variables are the bottom supply energy, the top energy removal, the reflux ratio, which influence the tower operating pressure, the tray load and degree of separation. Concerning the system outputs, a distinction must be made between the controlled and the uncontrolled variables. If the underlying task is to produce a required product quality, then the top and bottom qualities are the most important controlled variables. At a tray only the temperature can be continuously measured and this yields a good indication of the condition of the tower.

There are several assumptions that are commonly made in order not to complicate matters unnecessarily. These assumptions include that the vapor mass at a tray is negligible compared to the liquid mass and the energy content of the vapor mass at a tray is neglected.

2.6.1 A Particular Tower

In this section, we focus our study on a class of distillation towers commonly used in natural gas plants, an example of which is in *Aramco-Saudi Arabia*. It must be noted for this class that unless disturbed by changes in feed, heat, ambient temperature, or condensing, the amount of feed being added normally equals the amount of product being removed. A typical physical layout of distillation tower (DT) is portrayed in Fig. 2.8.

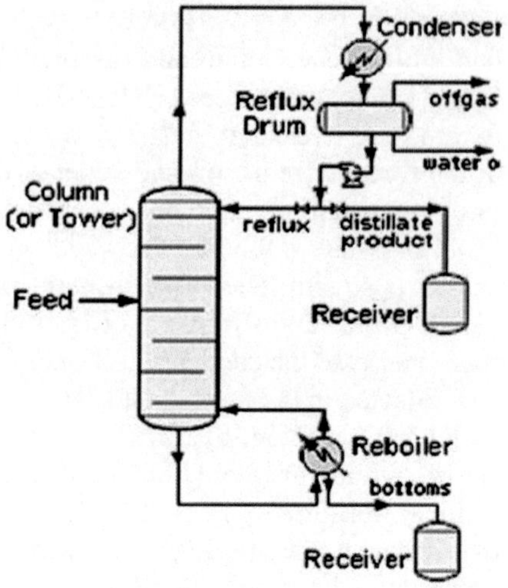

Fig. 2.8 Distillation unit

For simplicity in exposition, the identification studies carried out in the subsequent chapter are based on one input and one output data set each of 10080 samples with a sampling period of 60 s:

- Input: Feed inlet temperature in F°.
- Output: Tower outlet compound of C_2 in mol %.

2.7 Falling Film Evaporator

The most common used evaporator in the dairy industry is the falling film evaporator, for the concentration of products like milk, skimmed milk and whey. A four stage evaporator is used to reduce the water content of the product, that is, milk. The data was taken from [21]. The identification scheme used for the data is the N4SID subspace based identification. The data consists of 6305 samples with three inputs, feed flow, vapor flow to the first evaporator stage and cooling water flow and three outputs, dry matter content, the flow and the temperature of the out coming product.

The solution containing the desired product is fed to the evaporator and passes a heat source. The applied heat converts the water in the solution to vapor. The vapor is removed from the rest of the solution and is condensed while the now concentrated solution is either fed into the second evaporator is removed. The evaporator generally as a machine consists of four sections. The heating section consists of the heating medium. Steam is fed into this section. The concentrating and separating section removes the vapor being produced from the solution. The condenser condensates the separated vapor, then the vacuum or pump provides pressure to increase the circulation.

Evaporation is used basically in the dairy industry for the concentration of products like milk, skimmed milk etc. Concentration involves the removal of water from the product. To minimize the cost, evaporation is usually performed in multiple effect evaporators where two or more effects operate at progressively lower boiling points. In this type of arrangement, the vapor produced in the previous effect can be used as the heating medium in the next. The evaporator considered here is a four falling film effects and has a water evaporation capacity of 800 kg/h. The evaporators most commonly are used in the split effect mode, where only the third effect and the finishing effect are used.

2.7.1 A Single Effect Evaporator

In what follows, for simplicity, we will consider a single effect falling film evaporator to outline the operating principles.

A single effect evaporator consists of a balance tank, a condenser, a preheater, an evaporator calandria, a separator and a vacuum pump, see Fig. 2.9. The process can

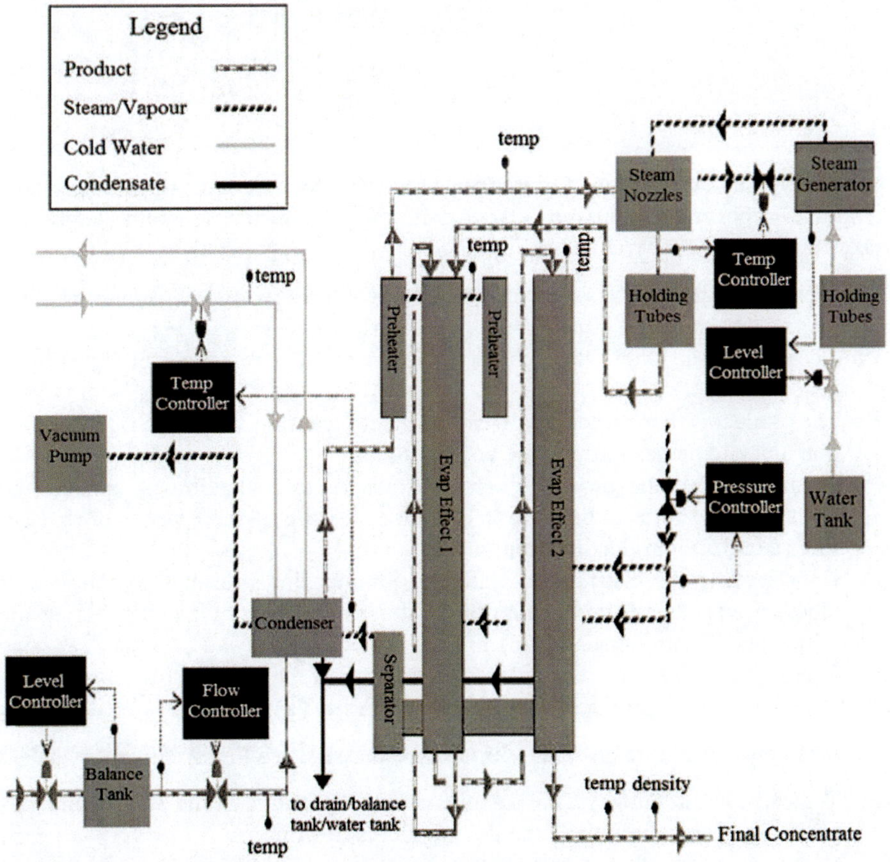

Fig. 2.9 Schematic diagram of evaporator in split effect

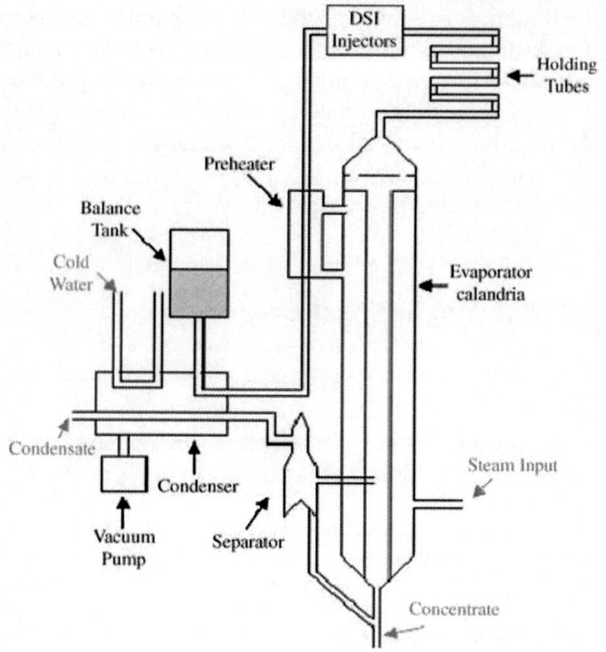

Fig. 2.10 Block diagram of single effect falling film evaporator

be decomposed into a product route (steps **Pa–Pf**), a steam route (steps **Sa–Sc**) and a product vapor route (steps **Va–Vd**). Firstly, we will consider the path the product takes through the evaporator, see Fig. 2.10.

Pa From the balance tank, the concentrate flows through the condenser where it gets its first injection of heat—see (**Vc**) overleaf.

Pb The product then flows through the preheater where it gets a second injection of heat (see **Sc**).

Pc The product is then pasteurized via the Direct Steam Injection (DSI) pasteurization unit and passes through the holding tubes.

Pd From the DSI, the product enters the evaporator calandria. A nozzle and spreader plate form a distribution system at the top of the evaporator that ensures a uniform product distribution.

Pe Upon leaving the distribution plate, the product flows through stainless steel tubes. The product forms a thin film on the inside of the tube while the outside of the tube is surrounded by steam.

Pf The product from the tubes reaches the bottom of the calandria where it is collected along with product from the separator (see **Va**).

Next, consider the steam's path through the process, see Fig. 2.11.

Sa Typically, but not always, the steam enters the calandria at the bottom and surrounds the tubes through which the product is flowing.

Sb Heat is then transferred from the steam to the product. This transfer of heat causes the water in the product to boil and produce vapor inside the tubes.

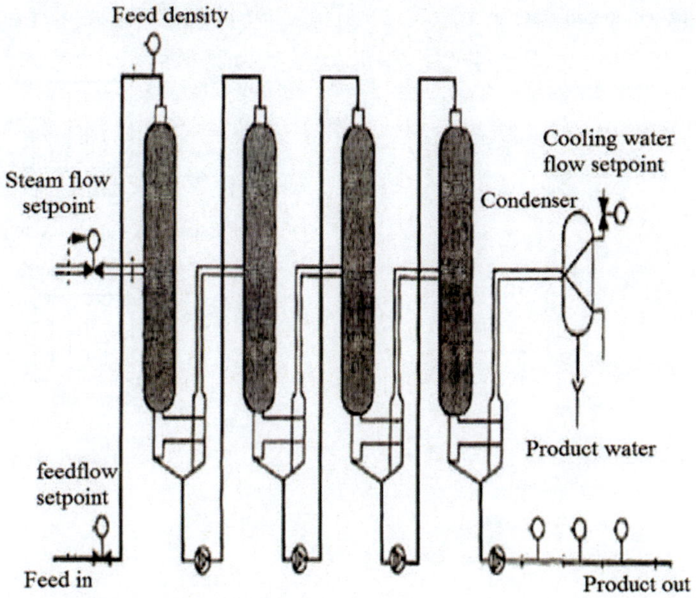

Fig. 2.11 Four-effect falling film evaporator

Sc Some steam from the calandria shell enters the preheater and is used as the heating medium in the preheater (see **Pb**).

Finally, consider the route of the product vapor through the process.

Va The product vapor exits the bottom of the calandria and enters the separator where product is removed from the vapor and returned to the product stream.
Vb The vapor then enters the condenser.
Vc In the condenser, the vapor acts as a heating medium for the product (see **Pa**).
Vd The vapor then passes the cold water pipes and condenses.

2.8 Vapor Compression Cycle Systems

In vapor compression cycle systems, it is desirable to effectively control the thermodynamic cycle by controlling the thermodynamic states of the refrigerant. By controlling the thermodynamic states with an inner loop, supervisory algorithms can manage critical functions and objectives such as maintaining superheat and maximizing the coefficient of performance.

The primary goal of any air-conditioning or refrigeration system is to move energy from one location to another. An idealized vapor compression cycle (VCC) system, as shown in Fig. 2.12, is a thermodynamic system driven by the phase characteristics of the refrigerant that is flowing through it. Therefore, it is useful to describe the system in terms of the state of its refrigerant, as shown on a pressure-enthalpy (P–H) diagram, see Fig. 2.13.

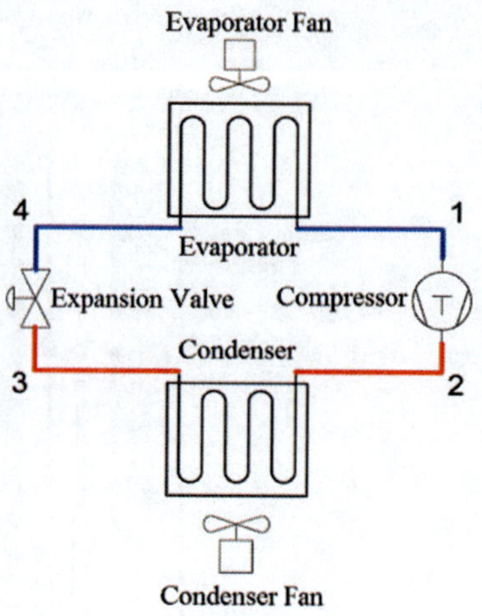

Fig. 2.12 Schematic diagram of VCC system

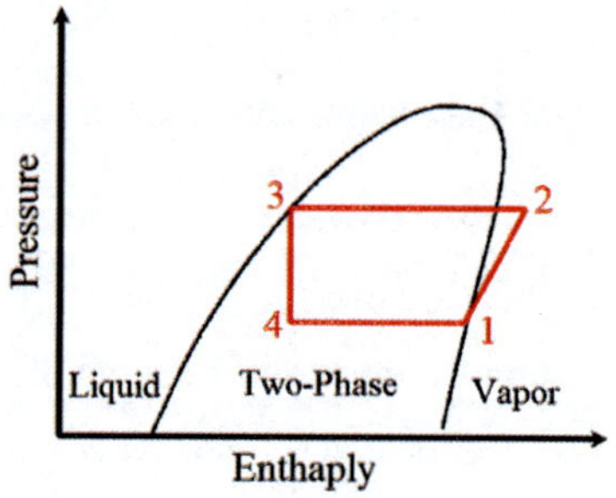

Fig. 2.13 *P–H* cycle diagram

2.8.1 A Typical System

An ideal VCC system assumes isentropic compression, isenthalpic expansion, and isobaric condensation and evaporation. The basic control objectives of a VCC system can be conceptualized visually via Fig. 2.13. For example, the difference between and represents the increase in enthalpy across the evaporator, that is, the amount of energy removed from the cooled space. This is a measure of evaporator capacity. The difference between and represents the increase in enthalpy across the compressor, that is, the amount of work done by the compressor to increase the pressure of the refrigerant vapor. The system coefficient of performance (COP), a measure of system efficiency, is defined as the ratio between these two changes in enthalpy.

The focus of this study is to present a comprehensive controller design approach, that is, one that covers displacement and velocity control, addresses the nonlinearities present in the vapor compression system and considers practical issues such as transient response and real-time implementation.

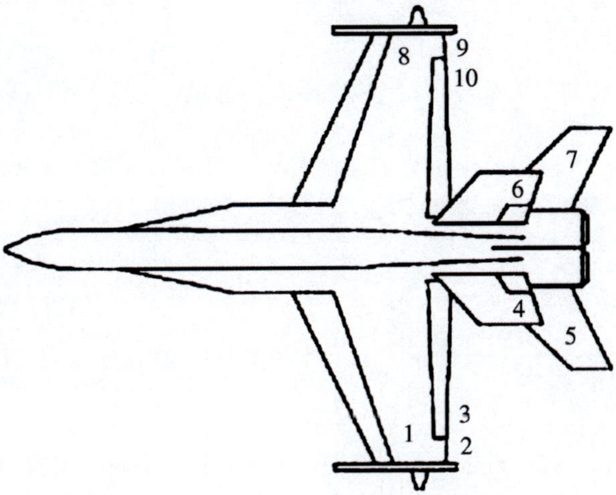

Fig. 2.14 F-18 sensor configuration

2.9 Flutter of an Aircraft F-18

The Flutter is a self-feeding and potentially destructive vibration where aerodynamic forces on an object couple with a structure's natural mode of vibration to produce rapid periodic motion [14]. Flutter can occur in any object within a strong fluid flow, under the conditions that a positive feedback occurs between the structure's natural vibration and the aerodynamic forces, see Fig. 2.14. That is, that the vibration movement of the object increases an aerodynamic loads which in turn drives the object to move further [17, 34]. If the energy during the period of aerodynamic excitation is larger than the natural damping of the system, the level of vibration will increase, resulting in self-exciting oscillation. The vibration levels can thus build up and are only limited when the aerodynamic or mechanical damping of the object match the energy input, this often results in large amplitudes and can lead to rapid failure. Because of this, structures exposed to aerodynamic forces—including wings, aerofoil, but also chimneys and bridges—are designed carefully within known parameters to avoid flutter. It is however not always a destructive force; recent progress has been made in small-scale wind generators for under served communities in developing countries, designed specifically to take advantage of this effect.

2.9.1 Flutter Input and Output Data

The data comprises of one input and one output which has a sampling time of 1 s, the number of samples in the data are 1024, see Fig. 2.15. In this section, the date in subdivided into the *estimation* and *validation* data parts, each part is comprised

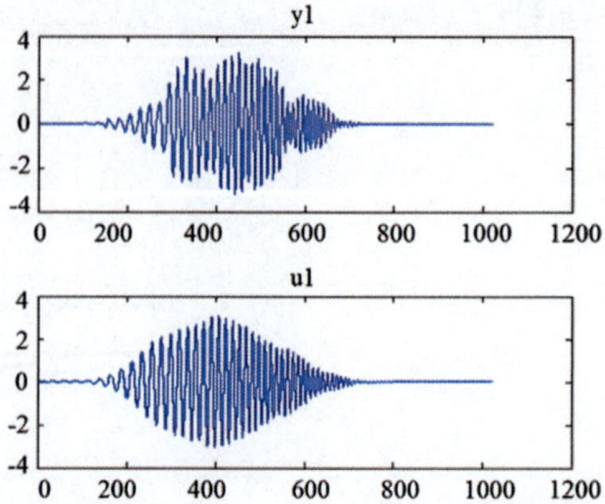

Fig. 2.15 Flutter input and output data

of 512 samples. As we shall see in later chapters, applications of the identification techniques are employed on the estimation data and then the estimated models are evaluated on the validation data.

2.10 A Hydraulic Pumping System

It is often desirable to find parsimonious models with good static and dynamical responses [32]. The estimation of nonlinear models with such features is quite hard mainly because static and dynamic information are not equally weighed in a single set of data. In this respect, static and dynamic information can be thought of as being conflicting. Flexible black-box structures are able to accurately fit a single piece of data. However, there are two main drawbacks with most of such structures. First, once such models are estimated, the static information (e.g., static nonlinearity) is not readily available analytically. Second, not all such model structures and algorithms have been adapted to permit the effective use of static information during training (parameter estimation). It should be noticed that black-box identification does not necessarily guarantee correct steady-state performance when the model is nonlinear [3].

When the data sets are conflicting in some way, it is advisable to use multi-objective approaches which yield a set of optimal solutions called the Pareto set. Bi-objective algorithms have proved to be quite useful in combining both static and dynamic data during model identification [10].

In what follows, we aim to identify models of a 15 kW hydraulic pumping system. There has been a clear increase of variable frequency drives as the final control element for such systems. This has enabled the implementation of fast and automatic control systems. Models of such systems are highly desirable for characterization and control. Such models should, ideally, represent the system accurately both in transient and steady-state regimes over a wide range of operating conditions. This requires, more often than not, the use of nonlinear models.

We focus in this book to obtain models that perform well both in transient and steady-state regimes, different identification approaches were implemented to "guarantee" a good balance between such features. In order to improve the model steady-state performance, the measured static curve of the pumping system was used as auxiliary information. Such information was used in different intensities, depending on the model representation used. An improved bi-objective identification approach is presented and a new decision-maker is defined. In this brief, we used and compared polynomial and neural nonlinear autoregressive with moving average and exogenous variables (NARMAX) models.

2.10.1 Hydraulic Process and the Data

In a full-scale hydroelectric power plant (over 80% of Brazilian electrical energy is produced in such plants), the water head can be considered constant over reasonably long periods of time. At testing plants, however, the turbines are fed by powerful hydraulic systems and not by a water head. Because of the characteristics of the centrifugal pumps used in such plants, the pressure on the turbine decreases as the water flow increases. Therefore, in realistic testing plants, pressure must be controlled over a wide range of operating conditions. Mathematical models are desired to simulate and to design the closed-loop control of the real pumping system, where the models output is the system pressure and the models input is the pumps reference speed.

The hydraulic plant described in this section is composed by two centrifugal pumps that feed a hydraulic turbine. The hydraulic plant should be seen by the turbine as a water head. The static and dynamic data used in this brief were measured from this plant, composed by two centrifugal pumps coupled to induction motors of 7.5 kW and variable speed drive systems (see Fig. 2.16). The pumps can be operated alone, in parallel or in a series configuration, always at the same speed. In this work, the pumps were set in a parallel configuration working at the same instantaneous speed with a Francis turbine as load [9].

Fig. 2.16 Hydraulic pumping system

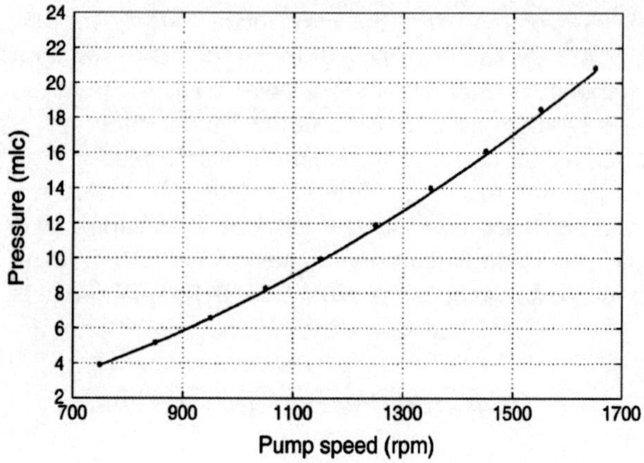

Fig. 2.17 Static curve of the hydraulic pumping system and its approximation

The modeling data presented in this work were collected from a data acquisition system. The piezo-resistive pressure transmitter error is ± 0.175 mlc (meter of liquid column).

2.10.2 Static Behavior

The static curve of the system was measured by:

1. Setting the turbine distributor blade to 50% and
2. Maintaining the pumps speed fixed at the chosen values—the speed references of both pumps were maintained the same during this procedure. After transients died out, the output pressure was recorded for each reference speed.

During this test, the pumps speed was varied from 750 to 1650 r/min. The static curve is shown in Fig. 2.17 as well as the second-order polynomial approximation

$$H(\bar{u}) = \beta \bar{u}^2 + \alpha \bar{u} + \kappa \tag{2.5}$$

with $\beta = 7.2652 \times 10^{-6}$, $\alpha = 1.4933 \times 10^{-3}$, $\kappa = -1.3312$, and where is the pressure in the output pipe and is the steady-state pump speed. This static curve will be useful during the gray-box modeling and will also be used to evaluate the identified models.

In Chap. 4, we will perform identification methods to generate appropriate models.

2.11 Notes and References

In this introductory chapter, some representative system applications were presented to help in motivating the readers to the upcoming topics. It must be emphasized that the target goal is to launch an information-based approach to control system design.

Being an applied design approach, we start by examining some industrial systems and shed light into their input/output variables. Indeed, there are many similar systems in practice and hence we encourage the readers to look at these systems and apply the methods of this book. We will make every effort to produce the subsequent chapters as a self-contained examination of the background and methods of industrial dynamical systems. For a good introduction to the subject matter, the reader is referred to [1, 2, 4, 5, 7, 8, 11–13, 15–19, 22, 23, 28, 29, 33–41]. For a MATLAB tool box, it is advisable to consult [40, 42, 43].

References

1. Abdelazim, T., Malik, O.: Identification of nonlinear systems by Takagi–Sugeno fuzzy logic grey box modeling for real-time control. Control Eng. Pract. **13**(12), 1489–1498 (2005)
2. Aguirre, L.A.: A nonlinear correlation function for selecting the delay time in dynamical reconstructions. Phys. Lett. **203A**(2–3), 88–94 (1995)
3. Aguirre, L.A., Donoso-Garcia, P.F., Santos-Filho, R.: Use of a priori information in the identification of global nonlinear models—A case study using a buck converter. IEEE Trans. Circuits Syst. I, Regul. Pap. **47**(7), 1081–1085 (2000)
4. Aguirre, L.A., Barroso, M.F.S., Saldanha, R.R., Mendes, E.M.A.M.: Imposing steady-state performance on identified nonlinear polynomial models by means of constrained parameter estimation. IEE Proc. Part D. Control Theory Appl. **151**(2), 174–179 (2004)
5. Aguirre, L.A., Coelho, M.C.S., Corrêa, M.V.: On the interpretation and practice of dynamical differences between Hammerstein and Wiener models. IEE Proc. Part D. Control Theory Appl. **152**(4), 349–356 (2005)
6. Astrom, K.J., Eykhoff, P.: System identification—A survey. Automatica **7**(2), 123–162 (1971)
7. Baker, J.E.: Reducing bias and inefficiency in the selection algorithm. In: Proc. 2nd Int. Conf. Genetic Algorithms Genetic Algorithms Their Appl., Mahwah, N.J., pp. 14–21 (1987)
8. Bakker, H.H.C., Marsh, C., Paramalingam, S., Chen, H.: Cascade controller design for concentration in a falling film evaporators. Food Control **17**(5), 325–330 (2006)
9. Barbosa, B.H.: Instrumentation, modelling, control and supervision of a hydraulic pumping system and turbine–generator module (in Portuguese). Master's thesis, Sch. Elect. Eng., Federal Univ. Minas Gerais, Belo Horizonte, Brazil (2006)
10. Barroso, M.S.F., Takahashi, R.H.C., Aguirre, L.A.: Multi-objective parameter estimation via minimal correlation criterion. J. Process Control **17**(4), 321–332 (2007)
11. Billings, S.A., Voon, W.S.F.: Least squares parameter estimation algorithms for nonlinear systems. Int. J. Syst. Sci. **15**(6), 601–615 (1984)
12. Billings, S.A., Chen, S., Korenberg, M.J.: Identification of MIMO nonlinear systems using a forward-regression orthogonal estimator. Int. J. Control **49**(6), 2157–2189 (1989)
13. Bingulac, S., Sinha, N.K.: On the identification of continuous-time systems from the samples of input–output data. In: Proc. Seventh Int. Conf. on Mathematical and Computer Modeling, Chicago, IL, pp. 231–239 (1989)
14. Bucharles, A., Cassan, H., Roubertier, J.: Advanced parameter identification techniques for near real-time flight flutter test analysis. AIAA, Paper 90-1275, May 1990
15. Chen, S., Billings, S.A., Luo, W.: Orthogonal least squares methods and their application to nonlinear system identification. Int. J. Control **50**(5), 1873–1896 (1989)
16. Connally, P., Li, K., Irwing, G.W.: Prediction and simulation error based perceptron training: Solution space analysis and a novel combined training scheme. Neurocomputing **70**, 819–827 (2007)
17. Cooper, J.: Parameter estimation methods for the flight flutter testing. In: Proc. the 80th AGARD Structures and Materials Panel, CP-566, AGARD, Rotterdam, The Netherlands (1995)

18. Correa, M.V., Aguirre, L.A., Saldanha, R.R.: Using steady-state prior knowledge to constrain parameter estimates in nonlinear system identification. IEEE Trans. Circuits Syst. I, Regul. Pap. **49**(9), 1376–1381 (2002)
19. Cunningham, P., Canty, N., O'Mahony, T., O'Connor, B., O'Callagham, D.: System identification of a falling film evaporator in the dairy industry. In: Proc. of SYSID'94, Copenhagen, Denmark, vol. 1, pp. 234–239 (1994)
20. De Moor, B.L.R. (ed.): DaISy: Database for the Identification of Systems. Department of Electrical Engineering, ESAT/SISTA, K.U.Leuven, Belgium. http://www.esat.kuleuven.ac.be/sista/daisy
21. De Moor, B.L.R., Ljung, L., Zhu, Y., Van Overschee, P.: Comparison of three classes of identification methods. In: Proc. of SYSID'94, Copenhagen, Denmark, vol. 1, 175–180 (1994)
22. Draper, N.R., Smith, H.: Applied Regression Analysis, 3rd edn. Wiley, New York (1998)
23. Ekawati, E., Bahri, P.A.: Controllability analysis of a five effects evaporator system. In: Proc. Foundations of Computer-Aided Process Operations, FOCAPO2003, pp. 417–420 (2003)
24. El-Sherief, H., Sinha, N.K.: Identification and modelling for linear multivariable discrete-time systems: A survey. J. Cybern. **9**, 43–71 (1979)
25. El-Sherief, H., Sinha, N.K.: Determination of the structure of a canonical model for the identification of linear multivariable systems. IEEE Trans. Syst. Man Cybern. **SMC-12**, 668–673 (1982)
26. Energy Efficiency and Renewable Energy, U.S. Department of Energy. www.energy.gov
27. Favoreel, W., De Moor, B.L.R., Van Overschee, P.: Subspace state-space system identification for industrial processes. J. Process Control **10**(2–3), 149–155 (2000)
28. Ghiaus, C., Chicinas, A., Inard, C.: Grey-box identification of air-handling unit elements. Control Eng. Pract. **15**(4), 421–433 (2007)
29. Goldberg, D.E.: Genetic Algorithms in Search, Optimization and Machine Learning. Addison-Wesley, New York (1989)
30. Hsia, T.C.: On sampled-data approach to parameter identification of continuous-time linear systems. IEEE Trans. Autom. Control **AC-17**, 247–249 (1972)
31. Hsia, T.: System Identification: Least-Squares Methods. Lexington Books, Lexington (1977)
32. Jakubek, S., Hametner, C., Keuth, N.: Total least squares in fuzzy system identification: An application to an industrial engine. Eng. Appl. Artif. Intell. **21**, 1277–1288 (2008)
33. Karimi, M., Jahanmiri, A.: Nonlinear modeling and cascade control design for multieffect falling film evaporator. Iran. J. Chem. Eng. **3**(2) (2006)
34. Kehoe, M.W.: A historical overview of flight flutter testing, NASA TR 4720, Oct. 1995
35. Leontaritis, I.J., Billings, S.A.: Input–output parametric models for nonlinear systems. Part II: Deterministic nonlinear system. Int. J. Control **41**(2), 329–344 (1985)
36. Miranda, V., Simpson, R.: Modelling and simulation of an industrial multiple-effect evaporator: Tomato concentrate. J. Food Eng. **66**, 203–210 (2005)
37. Neilsen, K.M., Pedersen, T.S., Nielsen, J.F.D.: Simulation and control of multieffect evaporator
38. Nepomuceno, E.G., Takahashi, R.H.C., Aguirre, L.A.: Multiobjective parameter estimation: Affine information and least-squares formulation. Int. J. Control **80**(6), 863–871 (2007)
39. Norgaard, M.: Neural network based system identification—TOOLBOX, Tech. Univ. Denmark, Lyngby, Tech. Rep. 97-E-851 (1997)
40. Ogata, K.: MATLAB for Control Engineers. Prentice-Hall, New York (2008)
41. Pan, Y., Lee, J.H.: Modified subspace identification for long-range prediction model for inferential control. Control Eng. Pract. **16**(12), 1487–1500 (2008)
42. Piroddi, L.: Simulation error minimization methods for NARX model identification. Int. J. Model. Identif. Control **3**(4), 392–403 (2008)
43. Piroddi, L., Spinelli, W.: An identification algorithm for polynomial NARX-models based on simulation error minimization. Int. J. Control **76**(17), 1767–1781 (2003)
44. Rangaiah, G., Saha, P., Tade, M.: Nonlinear model predictive control of an industrial four-stage evaporator system via simulation. Chem. Eng. J. **87**, 285–299 (2002)
45. Roffel, B., Betlem, B.: Process Dynamics and Control. Wiley, London (2006)

46. Sinha, N.K.: Estimation of transfer function of continuous-time systems from samples of input–output data. Proc. Inst. Electr. Eng. **119**, 612–614 (1972)
47. Sinha, N.K., Kuszta, B.: Modelling and Identification of Dynamic Systems. Von-Nostrand Reinhold, New York (1983)
48. Sinha, N.K., Rao, G.P. (eds.): Identification of Continuous-Time Systems. Kluwer Academic, Dordrecht (1991)
49. Soderstrom, T., Stoica, P.: System Identification. Prentice-Hall, New York (1989)

Chapter 3
System Identification Methods

3.1 Introduction

System identification is concerned with the estimation of a system on the basis of observed data. This involves specification of the model structure, estimation of the unknown model parameters, and validation of the resulting model. Least squares and maximum likelihood methods are discussed, for stationary processes (without inputs) and for input–output systems.

In most practical applications, the system is not known and has to be estimated from the available information. This is called *the identification problem*. The identification method will depend on the intended model use, as this determines what aspects of the system are of relevance. The three main choices in system identification are the following.

1. *Data*: In some situations, it is possible to generate a large amount of reliable data by carefully designed experiments. In other situations, the possibilities to obtain data are much more limited and it is not possible to control for external factors that influence the outcomes. That is, the magnitude of outside disturbances ('noise') may differ widely from one application to another.
2. *Model Class*: A model describes relations between the observed variables. For practical purposes, the less important aspects are neglected to obtain sufficiently simple models. The identified model should be validated to test whether the imposed simplifications are acceptable.
3. *Criterion*: The criterion reflects the objectives of the modeler. It expresses the usefulness of models in representing the observed data.

Generally speaking, system identification should be then considered as an iterative procedure as illustrated in Fig. 3.1. The "classic" identification methodology used to obtain parametric models based on non-parametric models of the type "step response" is illustrated in Fig. 3.2. This methodology, initially used to obtain continuous time parametric models, has been extended to the identification of discrete-time models.

From the shape of the plant step response, one selects a type of model and the parameters of this model are graphically determined. As the sampling frequency

M.S. Mahmoud, Y. Xia, *Applied Control Systems Design*,
DOI 10.1007/978-1-4471-2879-3_3,

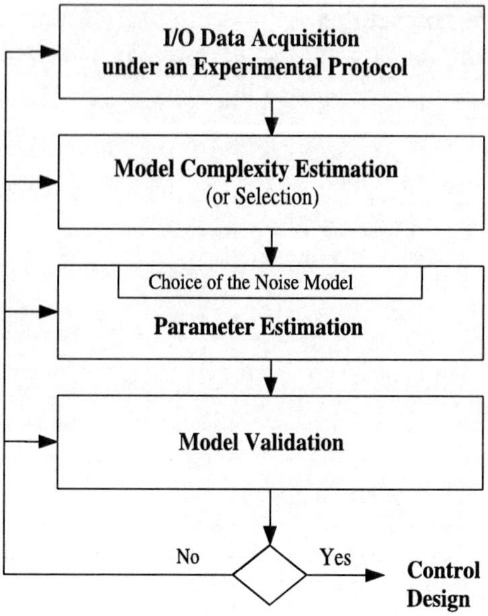

Fig. 3.1 System identification methodology

is known, one can obtain the corresponding discrete time model from conversion tables.

This methodology has several disadvantages:

- Test signals with large magnitude (seldom acceptable in the industrial systems).
- Reduced accuracy.
- Bad influence of disturbances.
- Models for disturbances are not available.
- Lengthy procedure.
- Absence of model validation.

3.2 Parameter Estimation Approach

The availability of a digital computer permits the implementation of algorithms that automatically estimate the parameters of the discrete time models. It should be emphasized that the identification of the parametric discrete time models allows to obtain (by simulation) non-parametric models of the step-response or *frequency-response* type, *with a far higher degree of accuracy with respect to a direct approach, and using extremely weak excitation signals.* The identification of parametric sampled data models leads to models of a very general use and offers several advantages over the other approaches.

High performance identification algorithms, which have a recursive formulation tailored to real-time identification problems and to their implementation on microcomputer, have been developed. The fact that these identification methods can op-

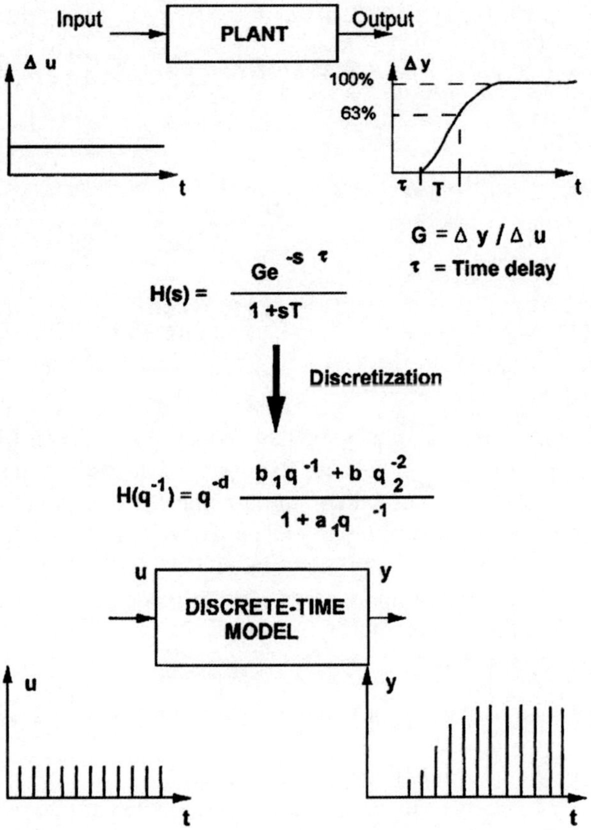

Fig. 3.2 "Classic" identification methodology

erate with extremely weak excitation signals is a very much appreciated quality in practical situations.

The parameter estimation principle for discrete time models is illustrated in Fig. 3.3. A sampled input sequence $u(t)$ (where t is the discrete time) is applied to the physical system (the cascade actuator–plant–transducer) by means of a digital-to-analog converter (DAC) followed by a zero-order hold block (ZOH). The measured sampled plant output $y(t)$ is obtained by means of an analog-to-digital converter (ADC).

A discrete-time model with adjustable parameters is implemented on the computer. The error between the system output $y(t)$ at instant t, and the output $y(t)$ predicted by the model (known as the prediction error) is used by a parameter adaptation algorithm that, at each sampling instant, will modify the model parameters in order to minimize this error on the basis of a chosen criterion.

The input is, in general, a very low level pseudo-random binary sequence generated by the computer (sequence of rectangular pulses with randomly variable duration). Once the model is obtained, an objective validation can be made by carrying out statistical tests on the prediction error $\varepsilon(t)$ and the predicted output $\hat{y}(t)$. The

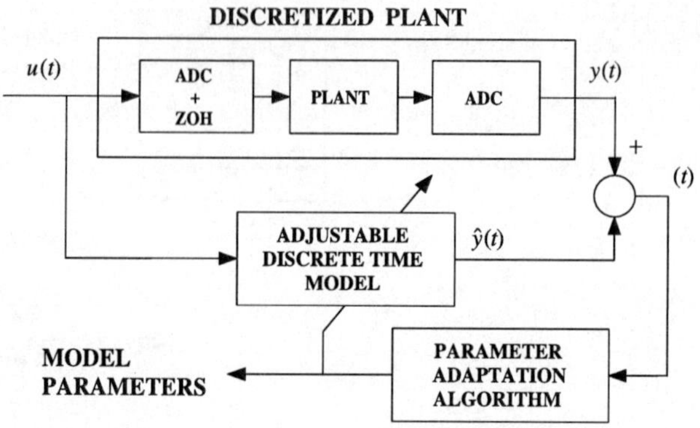

Fig. 3.3 Principle of model parameter estimation

validation test enables the best model to be chosen (for a given plant), *that is* the best structure and the best algorithm for the estimation of the parameters.

Finally, by computing and graphically representing the step responses and the frequency response of the identified model, the characteristics of the continuous-time model (step response or frequency response) can be extracted.

This modern approach to system model identification avoids all the problems related to the previously mentioned "classical" methods and also offers other possibilities such as:

- Tracking of the variations of the system parameters in real time allowing returning of controllers during operation.
- Identification of disturbances models.
- Modeling of the transducer noises in view of their elimination.
- Detection and measurement of vibration frequencies.
- Spectral analysis of the signals.

One of the key elements for implementing this system model identification approach is the *parameter adaptation algorithm* (PAA) that drives the parameters of the adjustable prediction model from the data collected on the system at each sampling instant. This algorithm has a "recursive" structure, that is, the new value of the estimated parameters is equal to the previous value plus a correction term that will depend on the most recent measurements.

A "parameter vector" is defined, in general, as the vector of the different parameters that must be identified. All the parameter adaptation algorithms have the following structure:

$$\begin{bmatrix}\text{New parameters}\\ \text{estimation}\\ \text{(vector)}\end{bmatrix} = \begin{bmatrix}\text{Old parameters}\\ \text{estimation}\\ \text{(vector)}\end{bmatrix} + \begin{bmatrix}\text{Adaptation}\\ \text{gain}\\ \text{(matrix)}\end{bmatrix}$$
$$\times \begin{bmatrix}\text{Measurement}\\ \text{function}\\ \text{(vector)}\end{bmatrix} \times \begin{bmatrix}\text{Error prediction}\\ \text{function}\\ \text{(scalar)}\end{bmatrix}.$$

The measurement function vector is also known as the "observation vector".

Note that nonrecursive parametric identification algorithms also exist (which process as a one block the input/output data files obtained over a certain time horizon). Recursive identification offers the following advantages with respect to these nonrecursive techniques:

- Obtaining an estimated model as the system evolves.
- Considerable data compression, since the recursive algorithms process at each instant only one input/output pair instead of the whole input/output data set.
- Much lower requirements in terms of memory and central-processing unit (CPU) power.
- Easy implementation on microcomputers.
- Possibility to implement real-time identification systems.
- Possibility to track the parameters of time variable systems.

Section 3.2 introduces the main types of parameter estimation (identification) algorithms in their recursive form. The effect of the noise on the parameter estimation algorithms will be discussed in Sect. 3.4.

Model Validation Different points of view can be considered for the choice of a model validation procedure. The goal is to verify that the output model excited by the same input applied to the plant reproduce the variations of the output caused by the variations of the input regardless the effect of the noise.

3.2.1 Estimation Algorithms

We will illustrate the principles of parametric identification presented in Fig. 3.3 by an example. Consider the discrete-time model of a plant described by

$$\begin{aligned} y(t+l) &= -a_1 y(t) + b_1(t)u(t) \\ &:= \theta^t \phi(t) \end{aligned} \tag{3.1}$$

where a_1 and b_1 are the unknown parameters.

The model output can be also written under the form of a scalar product between the unknown *parameter vector*

$$\theta^t = [a_1, b_1] \tag{3.2}$$

and the vector of measures termed measurement vector or plant model regressor vector

$$\phi^t(t) = \big[-y(t), u(t)\big]. \tag{3.3}$$

This vector representation is extremely useful since it allows easy consideration of models of any order.

Following the diagram given in Fig. 3.3, one should construct an adjustable prediction model, which will have the same structure as the discrete-time model of the plant given in (3.1):

$$\hat{y}^o(t+1) = \hat{y}\big(t+1|\hat{\theta}(t)\big) = -\hat{a}_1(t)y(t) + \hat{b}_1(t)u(t) = \hat{\theta}(t)^t\phi(t) \tag{3.4}$$

where $\hat{y}^o(t+1)$ is the predicted output at the instant t based on the knowledge of the parameters estimated at time $t\,(\hat{a}_1(t), \hat{b}_1(t))$. $\hat{y}^o(t+1)$ is called the a priori prediction. Note in (3.4) that

$$\theta(t)^t = \left[\hat{a}_1(t), \hat{b}_1(t)\right] \tag{3.5}$$

is the *vector of estimated parameters* at time t.

One can define now the prediction error (a priori) as in Fig. 3.4:

$$\varepsilon^o(t+1) = y(t+1) - \hat{y}^o(t+l) = \varepsilon^o\big(t+1, \hat{\theta}(t)\big). \tag{3.6}$$

The term $\hat{y}^o(t+1)$ is effectively computed between the sampling instants t and $t+1$ once $\hat{\theta}(t)$ is available, $\varepsilon^o(t+1)$ is computed at the instant $t+1$ after the acquisition of $y(t+1)$ (between $t+1$ and $t+2$). Note that $\varepsilon^o(t+1)$ depends on $\hat{\theta}(t)$.

Now it will be necessary to define a criterion in terms of the prediction error, which will be minimized by an appropriate evolution of the parameters of the adjustable prediction model, driven by the parameter adaptation algorithm. Since the objective is to minimize the magnitude of the prediction error independently of its sign, the choice of a quadratic criterion is natural. A first approach can be the synthesis of a parameter adaptation algorithm which at each instant minimizes the square of the a priori prediction error. This can be expressed as finding an expression for $\hat{\theta}(t)$ such that at each sampling one minimizes

$$J(t+1) = \left[\varepsilon^o(t+1)\right]^2 = \left[\varepsilon^o\big(t+1, \hat{\theta}(t)\big)\right]^2. \tag{3.7}$$

The structure of the parameter adaptation algorithm will be of the form

$$\begin{aligned}\hat{\theta}(t+1) &= \hat{\theta} + \Delta\hat{\theta}(t+1)\\ &= \hat{\theta}(t) + f\big(\hat{\theta}(t), \phi(t), \varepsilon^o(i+1)\big).\end{aligned} \tag{3.8}$$

The correction term $f(\hat{\theta}(t), \phi(t), \varepsilon^o(i+1))$ should only depend upon the information available at the instant $t+1$ (last measurement $y(t+1)$, parameter vector $\hat{\theta}(t)$ and a finite number of measurements or information at $t, t-1, \dots, t-n$). The solution to this problem will be given in Sect. 3.2.2. A recursive adaptation algorithm will be derived enabling both on-line and off-line implementation.

The criterion in (3.7) is not the only one step ahead criterion which can be considered and this aspect will also be discussed in Sect. 3.2.2.

When a set of input/output measurements over a time horizon t $(i = l, 2, \dots, t)$ is available, and we are looking for an off line identification, one may ask how to use this set of data optimally. The objective will be to search for a vector of parameters $\hat{\theta}(t)$ using the available data up to instant t and that minimizes a criterion of the form

$$J(t+1) = \sum_{i=1}^{t}\left[\varepsilon^o\big(i, \hat{\theta}(t)\big)\right] \tag{3.9}$$

that means the minimization of the sum of the squares of the prediction errors over the time horizon t. This point of view will lead to the least squares algorithm which will be presented in Sect. 3.2.3 (under the non-recursive and recursive form).

3.2.2 Gradient Algorithm

The aim of the gradient parameter adaptation algorithm is to minimize a one step quadratic criterion in terms of the prediction error (one-step ahead).

Consider the same example as in Sect. 3.2.1. The discrete time model of the plant is expressed by

$$\begin{aligned} y(t+1) &= -a_1 y(t) + b_1 u(t) \\ &= \theta^t \phi(t) \end{aligned} \tag{3.10}$$

where

$$\theta^t = [a_1, b_1] \tag{3.11}$$

is the parameter vector and

$$\phi(t)^t = \left[-y(t), u(t)\right] \tag{3.12}$$

is the vector of measures (pant model regressor vector).

The adjustable prediction model (a priori) is described by

$$\begin{aligned} \hat{y}^o(t+1) &= \hat{y}\left(t+1 \,|\, \hat{\theta}(t)\right) \\ &= -\hat{a}_1 y(t) + \hat{b}_1(t) u(t) \\ &= \hat{\theta}(t)^t \phi(t) \end{aligned} \tag{3.13}$$

where $\hat{y}^o(t+1)$ represents the a priori prediction depending on the values of the parameters estimated at instant t and

$$\hat{\theta}(t)^t = \left[\hat{a}_1(t), \hat{b}_1(t)\right] \tag{3.14}$$

is the estimated parameter vector.[1]

The a priori prediction error is given by

$$\varepsilon^o(t+1) = y(t+1) - \hat{y}^o(t+1). \tag{3.15}$$

To evaluate the quality of the new estimated parameter vector $\hat{\theta}(t+1)$, which will be provided by the parameter adaptation algorithm, it is useful to define the a posteriori output of the adjustable predictor, which corresponds to re-computing (3.15) with the new values of the parameters estimated at $t+1$.

The a posteriori predictor output is defined by

$$\begin{aligned} \hat{y}(t+1) &= \hat{y}\left(t+1 | \hat{\theta}(t+1)\right) \\ &= -a_1(t+1) y(t) + \hat{b}_1(t+1) u(t) \\ &= \hat{\theta}(t+1)^t \phi(t). \end{aligned} \tag{3.16}$$

One also defines an a posteriori prediction error:

$$\varepsilon(t+1) = y(t+1) - \hat{y}(t+1). \tag{3.17}$$

[1] In this case, the predictor regressor vector is identical to the measurement vector.

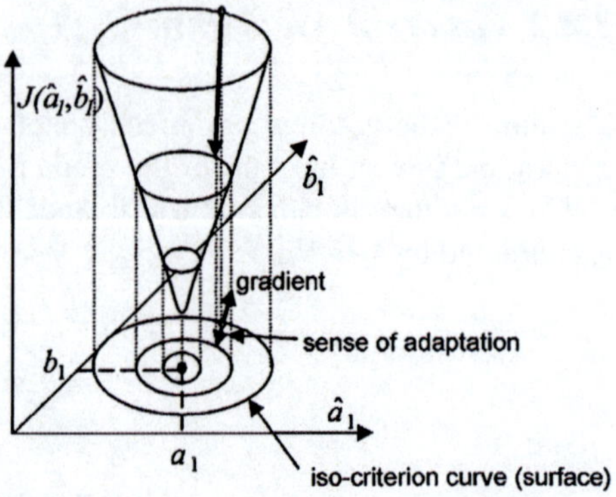

Fig. 3.4 Principle of gradient method

A recursive parametric adaptation algorithm with memory is desired. The structure of such an algorithm is[2]

$$\begin{aligned}\hat{\theta}(t+1) &= \hat{\theta}(t) + \Delta\hat{\theta}(t+1) \\ &= \hat{\theta}(t) + f\big(\hat{\theta}(t), \phi(t), \varepsilon^o(t+1)\big).\end{aligned} \tag{3.18}$$

The correction term $f(\hat{\theta}(t), \phi(t), \varepsilon^o(t+1))$ must only depend upon the information available at instant $t+1$ (last measure $y(t+1)$, parameters of $\hat{\theta}(t)$ and eventually a finite number of information at instants $t, t-1, t-2, \dots, n$). The correction term should allow one to minimize at each step the a priori prediction error with respect to the criterion

$$\min_{\hat{\theta}(t)} J(t+1) = \big[\varepsilon^o(t+l)\big]^2. \tag{3.19}$$

If one represents the criterion J and the parameters $\hat{a}_1$ and $\hat{b}_1$ in three-dimensional space, one gets the form represented in Fig. 3.5 (a reversed conic surface). The optimum of the criterion will correspond to the bottom of the cone and the projection of this point on the plane $\hat{a}_1$, $\hat{b}_1$ will give us the optimal values of the plant parameters: a_1 and b_1. It is obvious that, in order to reach as quickly as possible this point (the optimum of the criterion), it will be advantageous to go down along the steepest descent. This solution is analytically given by the *gradient technique*.

The horizontal sections of the surface correspond to curves along which the criterion has a constant value (isocriterion curves). If one represents the projection of the isocriterion curves ($J = \text{const}$) in the plane of the parameters $\hat{a}_1, \hat{b}_1$, one obtains concentric closed curves around the point a_1, b_1 (the parameters of the plant model) which minimizes the criterion. As the value of the criterion $J(= \text{const})$ increases, the isocriterion curves move further and further away from the minimum. This is illustrated in Fig. 3.5.

In order to minimize the value of the criterion, one moves in the direction of the steepest descent that, see Fig. 3.4, corresponds to move in the opposite direction

[2]Effectively, if the correction term is null, one holds the previous value of the estimated parameters.

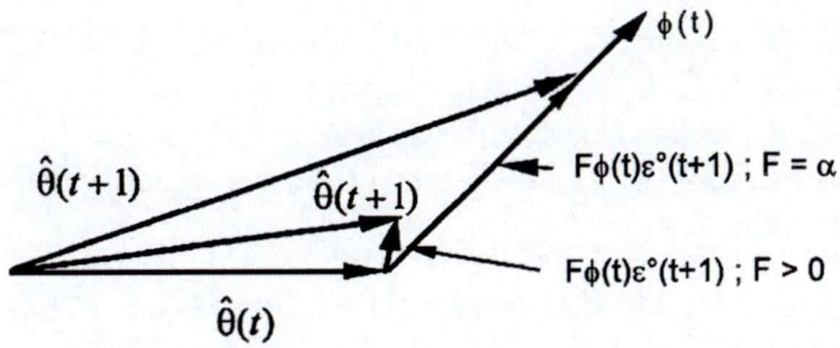

Fig. 3.5 Geometric interpretation of the gradient adaptation algorithm

of the gradient associated to the isocriterion curve. This will lead us to a curve corresponding to $J = \text{const}$ of a smaller value, as shown in Fig. 3.4.

The corresponding parametric adaptation algorithm will have the form

$$\hat{\theta}(t+1) = \hat{\theta}(t) - F\frac{\partial J(t+1)}{\partial \hat{\theta}(t)} \tag{3.20}$$

where $F = \alpha I$ ($\alpha > 0$) is the adaptation matrix gain (I—identity matrix) and $\partial J(t+1)/\partial\hat{\theta}(t)$ is the gradient of the criterion of (3.20) with respect to $\hat{\theta}(t)$.

From (3.20), one gets

$$\frac{1}{2}\frac{\partial J(t+1)}{\partial \hat{\theta}(t)} = \frac{\partial \varepsilon^o(t+1)}{\partial \hat{\theta}(t)}\varepsilon^o(t+1). \tag{3.21}$$

Since

$$\begin{aligned}\varepsilon^o(t+1) &= y(t+l) - \hat{y}^o(t+1)\\ &= y(t+1) - \hat{\theta}(t)^t\phi(t)\end{aligned} \tag{3.22}$$

and then

$$\frac{\partial \varepsilon^o(t+1)}{\partial \hat{\theta}(t)} = -\phi(t). \tag{3.23}$$

Substituting (3.23) into (3.20), the parametric adaptation algorithm of (3.20) becomes

$$\hat{\theta}(t+1) = \hat{\theta}(t) + F\phi(t)\varepsilon^o(t+l) \tag{3.24}$$

where F is the adaptation matrix gain.[3] Two choices are possible:

1. $F = \alpha I;\ \alpha > 0$.
2. $F > 0$ (positive definite matrix).[4]

The geometric interpretation of the parametric adaptation algorithm expressed by (3.24) is given in Fig. 3.5.

---

[3]In equations of the form of (3.24) the vector ϕ is generally called the *observation vector*. In this particular case, it corresponds to the measurement vector.

[4]A positive definite matrix is characterized by: (i) each diagonal term is positive; (ii) the matrix is symmetric; (iii) the determinants of all principal matrix minors are positive. See the Appendix.

The parametric adaptation algorithm given by (3.24) presents some instability possibilities if the adaptation gain (respectively, α) is large (this can be well understood with the support of Fig. 3.4).

Let consider (3.17) of the a posteriori error. By using (3.13) and (3.13), it can be re-written as

$$\begin{aligned}\varepsilon(t+1) &= y(t+1) - \hat{y}(t+1) \\ &= y(t+1) - \hat{\theta}(t)T\phi(t) + \left[\hat{\theta}(t) - \hat{\theta}(t+1)\right]^t \phi(t).\end{aligned} \tag{3.25}$$

From (3.24) it yields

$$\hat{\theta}(t) - \hat{\theta}(t+1) = -F\phi(t)\varepsilon^o(t+1) \tag{3.26}$$

and by also taking into account (3.15), (3.25) becomes

$$\varepsilon(t+1) = \varepsilon^o(t+1) - \phi(t)^t F\phi(t)\varepsilon^o(t+1). \tag{3.27}$$

In case that $F = \alpha I$, it becomes:

$$\varepsilon(t+1) = \left(1 - \alpha\phi(t)^t\phi(t)\right)\varepsilon^o(t+1). \tag{3.28}$$

If $\hat{\theta}(t+1)$ is a better estimation than $\hat{\theta}(t)$ (which means that the estimation of the parameters goes in the good sense) one should get $\varepsilon(t+l)^2 < \varepsilon^o(t+1)^2$. Therefore, it results from (3.28) that the adaptation gain α should satisfy the (necessary) condition

$$\alpha < 2/\phi(t)^t\phi(t). \tag{3.29}$$

In this algorithm, in other words, the adaptation gain must be chosen as a function of the magnitude of the signals.[5]

In order to avoid the possible instabilities, and the dependence of the adaptation gain with respect to the magnitude of the measured signals, one uses the same gradient approach but with a different criterion, which has as objective the minimization of the a posteriori prediction error at each step according to

$$\min_{\hat{\theta}(t+1)} J(t+1) = \left[\varepsilon(t+1)\right]^2. \tag{3.30}$$

Thus, one gets:

$$\frac{1}{2}\frac{\partial J(t+1)}{\partial\hat{\theta}(t+1)} = \frac{\partial\varepsilon(t+1)}{\partial\hat{\theta}(t+1)}\varepsilon(t+1). \tag{3.31}$$

From (3.16) and (3.17), it follows that

$$\varepsilon(t+1) = y(t+1) - \hat{y}(t+1) = y(t+1) - \hat{\theta}(t+1)^t\phi(t) \tag{3.32}$$

and respectively, that

$$\frac{\partial\varepsilon(t+1)}{\partial\hat{\theta}(t+1)} = -\phi(t). \tag{3.33}$$

[5] One can derives from (3.28) that an optimal value for α is $\alpha \approx 1/\phi(t)^t\phi(t)$.

Substituting (3.33) into (3.31), the parameter adaptation algorithm of (3.20) becomes

$$\hat{\theta}(t+1) = \hat{\theta}(t) + F\phi(t)\varepsilon(t+1). \tag{3.34}$$

This algorithm depends on $\varepsilon(t+1)$, which is a function $\hat{\theta}(t+1)$. In order to implement this algorithm, it is necessary to express $\varepsilon(t+1)$ as a function of $\varepsilon^o(t+1) : (\varepsilon(t+1) = f(\hat{\theta}(t), \phi(t), \varepsilon^o(t+1)))$.

Observe that (3.32) can be rewritten as

$$\varepsilon(t+1) = y(t+1) - \hat{\theta}(t)^t\phi(t) - \left[\hat{\theta}(t+1) - \hat{\theta}(t)\right]^t\phi(t). \tag{3.35}$$

The first two terms of the right side correspond to $\varepsilon^o(t+1)$ and, from (3.34), one gets

$$\hat{\theta}(t+1) - \hat{\theta}(t) = F\phi(t)\varepsilon(t+1) \tag{3.36}$$

which allows one to write (3.35) in the form

$$\varepsilon(t+1) = \varepsilon^o(t+1) - \phi(t)^t F\phi(t)\varepsilon(t+1) \tag{3.37}$$

from which one derives the desired relation between $\varepsilon(t+1)$ and $\varepsilon^o(t+1)$

$$\varepsilon(t+1) = \frac{\varepsilon^o(t+1)}{1 + \phi(t)^t F\phi(t)} \tag{3.38}$$

and the algorithm of (3.34) becomes

$$\hat{\theta}(t+1) = \hat{\theta} + \frac{F\phi(t)\varepsilon^o(t+1)}{1 + \phi(t)^t F\phi(t)} \tag{3.39}$$

that is a stable algorithm regardless of the gain F (positive definite matrix). The division by $1 + \phi(t)^t F\phi(t)$ introduces a normalization that reduces the sensitivity of the algorithm with respect to F and $\phi(t)$.

The sequence of operation corresponding to the recursive estimation algorithms can be summarized as follows:

1. Before $t+1$: $u(t), u(t-1), \ldots, y(t), y(t-1), \ldots, \phi(t), \hat{\theta}(t), F$ are available.
2. Before $t+1$ one computes: $\frac{F\phi(t)}{1+\phi(t)^t F\phi(t)}$ and $y^o(t+1)$ (given by (3.13)).
3. At instant $t+1$ $y(t+1)$ is acquired and $u(t+1)$ is applied.
4. The parametric adaptation algorithm is implemented.
   a. One computes $\varepsilon^o(t+1)$ by using (3.15).
   b. One computes $\hat{\theta}(t+1)$ from (3.39).
   c. (Optionally) one computes $\varepsilon(t+1)$.
5. Return to step 1.

3.2.3 Least Squares Algorithm

By using the gradient algorithm, at each step $\varepsilon^2(t+1)$ is minimized or, more precisely, one moves in the steepest decreasing direction of the criterion, with a step

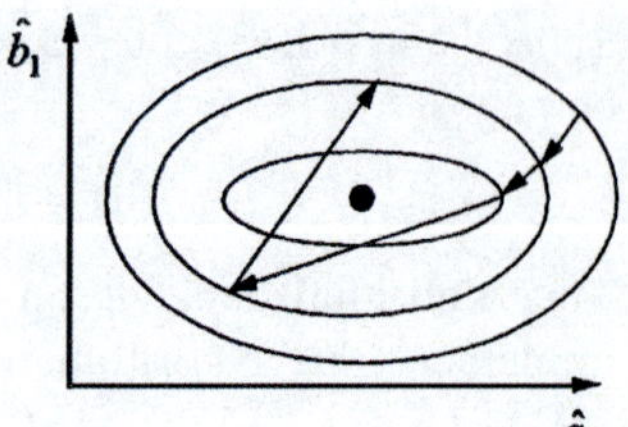

Fig. 3.6 Evolution of an adaptation algorithm of the gradient type

update depending on F. The minimization of $\varepsilon^a 2(t+1)$ at each step does not necessarily lead to the minimization of

$$\sum_{i=1}^{t} \varepsilon^2(i)$$

on a t-steps time horizon, as illustrated in Fig. 3.6. In fact, in the proximity of the optimum, if the gain is not small enough, oscillations may occur around the minimum. On the other hand, in order to obtain a satisfactory convergence speed at the beginning, when the current estimation is theoretically far from the optimum, a high adaptation gain is preferable. The least squares algorithm offers, in fact, such a variation profile for the adaptation gain.

The same equations, as in the gradient algorithm, are considered for the plant, the prediction model and the prediction errors, namely (3.15) to (3.22).

The aim is to find a recursive algorithm of the form of (3.18) that minimizes the "least squares" criterion

$$\begin{aligned}\min_{\hat{\theta}(t)} J(t) &= \frac{1}{t}\sum_{i=1}^{t}\left[y(i) - \hat{\theta}(t)^t \phi(i-1)\right]^2 \\ &= \frac{1}{t}\sum_{i=1}^{t} \varepsilon^2\left(i, \hat{\theta}(t)\right).\end{aligned} \tag{3.40}$$

The term $\hat{\theta}(t)^t \phi(i-1)$ corresponds to

$$\begin{aligned}\hat{\theta}(t)\phi(i-1) &= -\hat{a}_1(t)y(i-1) + \hat{b}_1(t)u(i-l) \\ &= \hat{y}|\hat{\theta}(t).\end{aligned} \tag{3.41}$$

This is the prediction of the output at instant i $(i \leq t)$ based on the parameter estimate at instant t obtained using t measurements. The objective is therefore the minimization of the sum of the squares of the prediction errors.

First, a parameter $\hat{\theta}$ must be estimated at instant t, so that it minimizes the sum of the squares of the differences between the output of the plant and the output of the prediction model over a horizon of t measurements. The value of $\hat{\theta}(t)$ that minimizes the criterion of (3.40) is obtained by looking for the value that cancels $\partial J(t)/\partial\theta(t)$:[6]

$$\frac{\partial J(t)}{\partial\hat{\theta}(t)} = -2\sum_{i-1}^{t}\left[y(i) - \hat{\theta}(t)^t\phi(i-1)\right]\phi(i-1)$$
$$= 0. \tag{3.42}$$

From (3.42), taking into account that

$$\left[\hat{\theta}(t)^t\phi(i-1)\right]\phi(i-1) = \phi(i-1)\phi(i-1)^t\hat{\theta}(t)$$

one readily obtains

$$\left[\sum_{i=1}^{t}\phi(i-1)\phi(i-1)^t\right]\hat{\theta}(t) = \sum_{i=1}^{t}y(i)\phi(i-1).$$

Multiplying the left both terms of this equation by

$$\left[\sum_{i=1}^{t}\phi(i-1)\phi(i-1)^t\right]^{-1}$$

it results in

$$\hat{\theta}(t) = \left[\sum_{i=1}^{t}\phi(i-1)\phi(i-1)^t\right]^{-1}\sum_{i=1}^{t}y(i)\phi(i-1)$$
$$= F(t)\sum_{i=1}^{t}y(i)\phi(i-1) \tag{3.43}$$

where

$$F(t)^{-1} = \sum_{i=1}^{t}\phi(i-1)\phi(i-1)^t. \tag{3.44}$$

Observed that this estimation algorithm is not recursive. In order to obtain a recursive algorithm, the estimation of $\hat{\theta}(t+1)$ is first considered:

$$\hat{\theta}(t+1) = F(t+1)\sum_{i=1}^{t+1}y(i)\phi(i-1), \tag{3.45}$$

$$F(t+1)^{-1} = \sum_{i=1}^{t+1}\phi(i-1)\phi(i-1)^t$$
$$= F(t)^{-1} + \phi(t)\phi(t)^t. \tag{3.46}$$

---

[6] This is the real minimum with the condition that the second derivative of the criterion, with respect to $\hat{\theta}(t)$ is positive, that is $\frac{\partial^2 J(t)}{\partial\hat{\theta}(t)^2} = 2\sum_{i=1}^{t}\phi(i-1)\phi(i-1)^t > 0$, as it is in general the case for $t \geq \dim\theta$ (see also Sect. 3.3).

Next, one should express it as a function of $\hat{\theta}(t)$:

$$\hat{\theta}(t+1) = \hat{\theta}(t) + \Delta\hat{\theta}(t+1). \tag{3.47}$$

From (3.45), adding and subtracting $\phi(t)\phi(t)^t\hat{\theta}(t)$), one gets

$$\begin{aligned}\sum_{i=1}^{t+1} y(i)\phi(i-1) &= \sum_{i=1}^{t} y(i)\phi(i-1) + y(t+1)\phi(t) \\ &= \phi(t)\phi(t)^t\hat{\theta}(t).\end{aligned} \tag{3.48}$$

Taking into account (3.43), (3.45) and (3.46), then (3.48) can be rewritten as

$$\begin{aligned}&\sum_{i=1}^{t+1} y(i)\phi(i-1) \\ &\quad = F(t+1)^{-1}\hat{\theta}(t+1) \\ &\quad = F(t)^{-1}\hat{\theta}(t) + \phi(t)\phi(t)^t\hat{\theta}(t) + \phi(t)\big[y(t+1) - \hat{\theta}(t)^t\phi(t)\big].\end{aligned} \tag{3.49}$$

On the other hand, on the basis of (3.15) and (3.46), one obtains

$$F(t+1)^{-1}\hat{\theta}(t+1) = F(t+1)^{-1}\hat{\theta}(t) + \phi(t)\varepsilon^o(t+1). \tag{3.50}$$

Multiplying on the left by $F(t+1)$, one gets

$$\hat{\theta}(t+1) = \hat{\theta}(t) + F(t+1)\phi(t)\varepsilon^o(t+1). \tag{3.51}$$

The adaptation algorithm of (3.51) has a recursive form similar to the gradient algorithm given in (3.24), with the difference that the gain matrix $F(t+1)$ is now time varying since it depends on the measurements (it automatically corrects the gradient direction and the step length).

A recursive formula for $F(t+1)$ remains to be provided starting from the recursive formula for $F^{-1}(t+1)$ given in (3.46). This is obtained by using the *matrix inversion lemma* (given below in a simplified form, see the Appendix for a general form).

Lemma 3.1 *Let F be a regular matrix of dimension $(n \times n)$ and ϕ a vector of dimension n; then*

$$\left(F^{-1} + \phi\phi^t\right)^{-1} = F - \frac{F\phi\phi^t F}{1 + \phi^t F\phi}. \tag{3.52}$$

Observe that to verify the inversion formula, one can simply multiply both terms by $F^{-1} + \phi\phi^t$.

From (3.46) and (3.52), one gets

$$F(t+1) = F(t) - \frac{F(t)\phi(t)\phi(t)^t F(t)}{1 + \phi(t)^t F(t)\phi(t)} \tag{3.53}$$

and, regrouping the different equations, a first formulation of the recursive least squares (RLS) parameter adaptation algorithm (PAA) is given by

$$\hat{\theta}(t+1) = \hat{\theta}(t) + F(t+1)\phi(t)\varepsilon^o(t+1), \tag{3.54}$$

$$F(t+1) = F(t) - \frac{F(t)\phi(t)^t F(t)}{1+\phi(t)^t F(t)\phi(t)}, \tag{3.55}$$

$$\varepsilon^o(t+1) = y(t+1) - \hat{\theta}^t\phi(t). \tag{3.56}$$

An equivalent form of this algorithm is obtained by substituting the expression of $F(t+1)$ given by (3.55) into (3.54). It yields

$$\left[\hat{\theta}(t+1)\hat{\theta}(t)\right] = F(t+1)\phi(t)\varepsilon^o(t+1) = F(t)\phi(t)\frac{\varepsilon^o(t+1)}{1+\phi(t)^t F(t)\phi(t)}. \tag{3.57}$$

On the other hand from (3.15)–(3.17) and (3.57), one obtains:

$$\begin{aligned}\varepsilon(t+1) &= y(t+1) - \hat{\theta}(t+1)\phi(t)\\ &= y(t+1) - \hat{\theta}(t)\phi(t) - \left[\hat{\theta}(t+1) - \hat{\theta}(t)\right]^t\phi(t)\\ &= \varepsilon^o(t+1) - \phi(t)^t F(t)\phi(t)\frac{\varepsilon^o(t+1)}{1+\phi(t)^t F(t)\phi(t)}\\ &= \frac{\varepsilon^o(t+1)}{1+\phi(t)^t F(t)\phi(t)}.\end{aligned} \tag{3.58}$$

This expresses the relation between the a posteriori prediction error and the a priori prediction error. Using this relation in (3.57), an equivalent form of the parameter adaptation algorithm for the recursive least squares is obtained[7]

$$\hat{\theta}(t+1) = \hat{\theta}(t) + F(t)\phi(t)\varepsilon(t+1), \tag{3.59}$$

$$F(t+1)^{-1} = F(t)^{-1} + \phi(t)\phi(t)^t, \tag{3.60}$$

$$F(t+1) = F(t) - \frac{F(t)\phi(t)\phi(t)^t F(t)}{1+\phi(t)^t\phi(t)F(t)\phi(t)}, \tag{3.61}$$

$$\varepsilon(t+1) = \frac{y(t+1) - \hat{\theta}(t)^t\phi(t)}{1+\phi(t)^t F(t)\phi(t)}. \tag{3.62}$$

For the recursive least squares algorithm to be exactly equivalent to the nonrecursive least squares algorithm, it must be started at instant $t_0 = \dim\phi(t)$, since normally $F(t)^{-1}$ given by (3.44) becomes non-singular for $t = t_0$.

In practice, the algorithm is initialized at $t = 0$ by choosing

$$F(0) = \frac{1}{\delta}I = (GI)I; \quad 0 < \delta < 1 \tag{3.63}$$

a typical value being $\delta = 0.001$ ($GI = 1000$). It can be observed, from the expression of $F(t+1)^{-1}$ given by (3.60) that the influence of this initial error decreases with time. A rigorous analysis (based on the stability theory—see [49]) shows nevertheless that for any positive definite matrix $F(0)$ ($F(0) > 0$),

$$\lim_{t\to 0} \varepsilon(t+1) = 0.$$

[7] This equivalent form is particularly useful in analyzing and understanding the algorithm.

The recursive least squares algorithm is an algorithm with a decreasing adaptation gain. This is clearly seen if the estimation of a single parameter is considered. In this case $F(t)$ and $\phi(t)$ are scalars, then (3.61) becomes

$$F(t+1) = \frac{F(t)}{1+\phi(t)^2 F(t)} \le F(t).$$

The recursive least squares algorithm gives, in fact, less and less weight to the new prediction errors, and thus to the new measurements.

It can be readily seen that this type of variation of the adaptation gain is not suitable for the estimation of time varying parameters, and other variation profiles must therefore be considered for the adaptation gain.

One must emphasize that he least squares algorithm, presented up to now for $\theta(t)$ and $\phi(t)$ of dimension 2, may be generalized to the n-dimensional case on the basis of the description of discrete-time systems of the form

$$y(t) = \frac{q^{-d}B(q^{-1})}{A(q^{-1})}u(t) \tag{3.64}$$

where

$$A(q^{-1}) = 1 + a_1 q^{-1} + \cdots + a_{n_A} q^{-n_A}, \tag{3.65}$$

$$B(q^{-1}) = b_1 q^{-1} + \cdots + b_{n_B} q^{-n_B} \tag{3.66}$$

which can further be rewritten as

$$y(t+1) = -\sum_{i=1}^{n_A} a_i(t+1-i) + \sum_{i=1}^{n_B} b_i u(t-d-i+1) = \theta^t \phi(t) \tag{3.67}$$

where

$$\theta^t = [a_1, \ldots, a_{n_A}, b_1, \ldots, b_{n_B}], \tag{3.68}$$

$$\begin{aligned} \phi(t)^t = \big[&-y(t), \ldots, -y(t-n_A+1), \\ &u(t-d), \ldots, u(t-d-n_B+1)\big]. \end{aligned} \tag{3.69}$$

The a priori adjustable predictor is given in the general case by

$$\hat{y}^o(t+1) = -\sum_{i=1}^{n_A} \hat{a}_i y(t+1-i) + \sum_{i=1}^{n_B} \hat{b}_i u(t-d-i+1) = \hat{\theta}^t \phi(t) \tag{3.70}$$

where

$$\hat{\theta}(t)^t = \big[\hat{a}_1(t), \ldots, \hat{a}_{n_A}(t), \hat{b}_1(t), \ldots, \hat{b}_{n_B}(t)\big] \tag{3.71}$$

and, for the estimation of $\hat{\theta}(t)$, the algorithm given in (3.54)–(3.56) is used with the appropriate dimension for $\hat{\theta}(t)$, $\phi(t)$ and $F(t)$.

3.2.4 Choice of the Adaptation Gain

The recursive formula for the inverse of the adaptation gain $F(t+1)^{-1}$ given by (3.46) or (3.60) is generalized by introducing two weighting sequences $\lambda_1(t)$and $\lambda_2(t)$, as indicated below:

$$F(t+1)^{-1} = \lambda_1(t)F(t)^{-1} + \lambda_2(t)\phi(t)\phi(t)^t,$$
$$0 < \lambda_1(t) \leq 1; \quad 0 \leq \lambda_2(t) < 2; \quad F(0) > 0. \tag{3.72}$$

Note that $\lambda_1(t)$ and $\lambda_2(t)$ in (3.72) have the opposite effect: $\lambda_1(t) < 1$ tends to increase the adaptation gain (the gain inverse decreases), $\lambda_2(t) > 0$ tends to decrease the adaptation gain (the gain inverse increases). For each choice of sequences $\lambda_1(t)$ and $\lambda_2(t)$, a different variation profile of the adaptation gain is found and, consequently, an interpretation in terms of the error criterion that is minimized by the PAA. Using the matrix inversion lemma given by (3.52), one obtains from (3.72):

$$F(t+1) = \frac{1}{\lambda_1(t)}\left[F(t) - \frac{F(t)\phi(t)\phi(t)^t F(t)}{\frac{\lambda_1(t)}{\lambda_2(t)} + \phi(t)^t F(t)\phi(t)}\right]. \tag{3.73}$$

Next a selection of choices for $\lambda_1(t)$ and $\lambda_2(t)$ and their interpretations will be given.

1. *Decreasing Gain (RLS)*
   In this case,

$$\lambda_1(t) = \lambda_1 = 1, \qquad \lambda_2(t) = 1 \tag{3.74}$$

   and $F(t+1)^{-1}$ is given by (3.60) which leads to a decreasing adaptation gain. The minimized criterion is expressed by (3.40).

   This type of profile is suited for the identification of stationary systems (with constant parameters).
2. *Constant Forgetting Factor*
   In this case,

$$\lambda_1(t) = \lambda_1 = 1, \qquad 0 < \lambda_1 < 1, \qquad \lambda_2(t) = \lambda_2 = 1. \tag{3.75}$$

   Typical values for λ_1 are: $\lambda_1 = 0.95, \ldots, 0.99$.

   The criterion to be minimized will be

$$J(t) = \sum_{i=1}^{t} \lambda_1^{t-i}\left[y(i) - \hat{\theta}(t)^t\phi(i-1)\right]^2. \tag{3.76}$$

   The effect of $\lambda_1 < 1$ is to introduce a decreasing weighting on the past data ($i < t$). This is why λ_1 is known as the forgetting factor. The maximum weight is given to the most recent error. This type of profile is suited for the identification of slowly time varying systems.

Remark 3.2 Note that $F(t+1)^{-1}$ given by (3.72) can be interpreted as the output of a filter characterized b the pulse transfer operator $H(q^{-1}) = \lambda_2(t)/(1 -$

$\lambda_1(t)q^{-1})$ whose input is $\phi\phi^t$. In addition, when an excitation is not provided ($\phi(t)\phi(t)^t = 0$), $F(t+1)^{-1}$ goes towards zero (because in this case $F(t+1)^{-1} = \lambda_1 F(t)^{-1}$, $\lambda_1 < 1$, leading to very high adaptation gains, a situation that should be avoided.

3. *Variable Forgetting Factor*
   In this case

$$\lambda_2(t) = \lambda_2 = 1 \tag{3.77}$$

and the forgetting factor λ_1 is given by

$$\lambda_1 t = \lambda_0 \lambda_1(t-1) + 1 - \lambda_0; \quad 0 < \lambda_0 < 1 \tag{3.78}$$

typical values being: $\lambda_1(0) = 0.95, \ldots, 0.99$; $\lambda_0 = 0.95, \ldots, 0.99$.

Observe that (3.78) leads to a forgetting factor that asymptotically tends towards 1. The criterion minimized will be

$$J(t) = \sum_{i=1}^{t} \left[\sum_{j=1}^{t-1} \lambda_1(j-i) \right] \left[y(i) - \hat{\theta}(t)^t \phi(i-1) \right]^2. \tag{3.79}$$

As $\lambda_1(t)$ tends towards 1 for large i, only the initial data are forgotten (the adaptation gain tends towards a decreasing gain). This type of profile is highly *recommended for the identification of stationary systems*, since it avoids a too rapid decrease of the adaptation gain, thus generally resulting in an acceleration of the convergence (by maintaining a high gain at the beginning when the estimates are far from the optimum).

4. *Constant Trace*
   In this case, $\lambda_1(t)$ and $\lambda_2(t)$ are automatically chosen at each step in order to ensure a constant trace of the gain matrix (constant sum of the diagonal terms)

$$\operatorname{Tr} F(t+1) = \operatorname{Tr} F(t) = \operatorname{Tr} F(0) = nG_i \tag{3.80}$$

in which n is the number of parameters and G_i the initial gain (typical values: $GI = 0, 1 \ldots, 4$), the matrix $F(0)$ having the form

$$F(0) = \begin{bmatrix} G_i & & 0 \\ & \ddots & \\ 0 & & G_i \end{bmatrix}. \tag{3.81}$$

The minimized criterion is of the form

$$J(t) = \sum_{i=1}^{t} f(t,i) \left[y(i) - \hat{\theta}(t)^t \phi(i-1) \right]^2 \tag{3.82}$$

in which $f(t,i)$ represents the forgetting profile.

Using this technique, at each step there is a movement in the optimal direction of the RLS but the gain is maintained approximately constant (*reinflation* of the RLS gain).

The values of $\lambda_1(t)$ and $\lambda_2(t)$ are determined from

$$\begin{aligned}\operatorname{Tr} F(t+1) &= \frac{1}{\lambda_1(t)} \operatorname{tr}\left[F(t) - \frac{F(t)\phi(t)\phi(t)^t F(t)}{\alpha(t) + \phi(t)^t F(t)\phi(t)}\right] \\ &= \operatorname{Tr} F(t).\end{aligned} \tag{3.83}$$

It is easy to see that by imposing the ratio $\alpha(t) = \lambda_1(t)/\lambda_2(t)$, (3.83) is obtained from (3.73). This type of profile is suited for the identification of systems with time varying parameters.

5. *Decreasing Gain + Constant Trace*
   In this case, there is a switch from $A1$ to $A4$ when

$$\operatorname{Tr} F(t) \leq nG, \qquad G = 0.1 \to 4 \tag{3.84}$$

   where G is fixed at the beginning. This profile is suited for the identification of time variable systems in the absence of initial information on the parameters.
6. *Variable Forgetting Factor + Constant Trace*
   In this case, there is a switch from $A3$ to $A4$ when

$$\operatorname{Tr} F(t) \leq nG. \tag{3.85}$$

   The domain of application is the same as for item 5.
7. *Constant Gain (Improved Gradient Algorithm)*
   In this case,

$$\lambda_1(t) = \lambda_1 = 1; \qquad \lambda_2(t) = \lambda_2 = 0 \tag{3.86}$$

   and thus from (3.83), it results that

$$F(t+1) = F(t) = F(0). \tag{3.87}$$

   The improved gradient adaptation algorithm given by (3.34) or (3.39) is thus obtained.

   This algorithm can be used to identify stationary or time varying systems with few parameters (≤ 3), and in the presence of a reduced noise level.

This type of adaptation gain results in performances which are inferior to those provided by profiles 1 through 4, but it is simpler to implement.

Choice of the Initial Adaptation Gain $F(0)$ The initial adaptation gain $F(0)$ is of the form given by (3.63), respectively (3.81).

In the absence of initial information, upon the parameters to be estimated (a typical choice is to set the initial estimation to zero), a high initial gain (G_i) is chosen for reasons that have been explained by (3.63) in Sect. 3.2.3. A typical value is $G_i = 1000$.

On the other hand, if an initial parameter estimation is available (resulting for example from a previous identification), a low initial gain is chosen. In general, in this case $G_i \leq 1$.

Since the adaptation gain decreases as the correct model parameter estimations are approached (a significant index is its trace), the adaptation gain may be interpreted as an index of the accuracy of the estimation (or prediction). This explains the choices of $F(0)$ proposed above. Note that, under certain hypotheses, $F(0)$ is effectively an index of the quality of the estimation because it represents the covariance of the parameter error vector $\tilde{\theta}(t) = \hat{\theta}(t) - \theta$. This property can give some information on the evolution of an estimation procedure. If the trace of $F(t)$ is not significantly decreasing, the parameter estimation, in general, is bad. This phenomenon occurs, for example, when the amplitude and the type of the input used are not suited for the identification. The importance of the nature of the identification signal will be discussed in the following section.

3.3 Transfer-Function Methods

Fundamental to most physical sciences is the concept of a mathematical model. Models are essential for prediction and control purposes. The type and accuracy of the model depends upon the application in mind. For example, models for aerospace applications usually need to be very precise, whereas models for industrial processes, such as blast furnaces, can often be very crude. Models can be obtained from physical reasoning or by analyzing experimental data from the system. In the latter case, our ability to obtain an accurate model is limited by the presence of random fluctuations such as unmeasured disturbances and measurement errors. The problem of obtaining mathematical models of physical systems from noisy observations is the subject of this book. In particular, we study the problem of estimation of the parameters within models of dynamic systems. We also investigate the effects of various experimental conditions upon model accuracy.

In the terminology of system identification, parametric models include transfer function, differential or difference equation. Henceforward, we will be concerned with the identification of parametric dynamic models, which are the most suitable for the design and tuning of applied industrial control systems. Basic methods include AutoRegressive (AR) method, AutoRegressive with eXogenous input (ARX) method, AutoRegressive Moving Average (ARMA) method, AutoRegressive Moving Average with eXogenous input (ARMAX) method, the Box–Jenkins method and prediction error method (PEM). We initially focus on the latter method leaving the remaining methods to later sections.

3.3.1 Prediction Error Method (PEM)

The prediction error method (PEM) is sometimes called the generalized least squares (GLS) method, although GLS originally was associated with a certain numerical minimization procedure [95]. This method was proposed in [17], where he extended the equation error model and assumed that the true process is given by

$$A^o(q)y(t) = B^o(q)u(t) + \frac{1}{D^o(q)}e(t) \tag{3.88}$$

or

$$y(t)=\frac{B^o(q)}{A^o(q)}u(t)+\frac{1}{A^o(q)D^o(q)}e(t) \tag{3.89}$$

where

$$\begin{aligned} A^o(q) &= 1+a_1^o q^{-1}+a_2^o q^{-2}+\cdots+a_{n_a}^o q^{-n_a}, \\ B^o(q) &= b_1^o q^{-1}+b_2^o q^{-2}+\cdots+b_{n_b}^o q^{-n_b}, \\ D^o(q) &= 1+d_1^o q^{-1}+d_2^o q^{-2}+\cdots+d_{n_d}^o q^{-n_d} \end{aligned}$$

and $e(t)$ is white noise with zero mean and variance λ.

So the equation disturbance is assumed to be an AR (autoregressive) process. Then, (3.88) can be written as

$$D^o(q)A^o(q)y(t)=D^o(q)B^o(q)u(t)+e(t). \tag{3.90}$$

This enlarged equation has a white noise disturbance $e(t)$. From the study of the least-squares method, we know that consistent and efficient estimates of a_i, b_i, d_i can be obtained by minimizing the loss function

$$\begin{aligned} V_{\mathrm{PEM}} &= \frac{1}{N}\sum_{t=1}^{N}\varepsilon^2(t) \\ &= \frac{1}{N}\sum_{t=1}^{N}\big(D(q)\big[A(q)y(t)-B(q)u(t)\big]\big)^2. \end{aligned} \tag{3.91}$$

This implies that, in the identification a model should be used which has the same structure as the true process

$$D(q)A(q)y(t)=D(q)B(q)u(t)+\varepsilon(t) \tag{3.92}$$

where $\varepsilon(t)$ is the residual; see Sect. 3.4.2. When $D(t)=I$, then (3.92) can be written using $\varphi(t)$ and θ,

$$y(t)=\varphi^*(t)\theta+\varepsilon(t) \tag{3.93}$$

where

$$\begin{aligned} \varphi^*(t) &= \big[-y(t-1)\cdots-y(t-n_a)\,u(t-1)\cdots u(t-n_b)\big] \\ \theta &= (a_1\cdots a_{n_a}\,b_1\cdots b_{n_a})^* \end{aligned}$$

and for computing $\hat{\theta}$

$$\hat{\theta}=\left[\frac{1}{N}\sum_{t=1}^{N}\varphi(t)\varphi^*(t)\right]^{-1}\left[\frac{1}{N}\sum_{t=1}^{N}\varphi(t)y(t)\right]. \tag{3.94}$$

Note that all the discussions about algorithms for computing $\hat{\theta}$ will remain valid. The results derived there depend only on the 'algebraic structure' of the estimate (3.94). For the statistical properties, though, it is of crucial importance whether

$\varphi(t)$ is an a priori given quantity, or whether it is a realization of a stochastic process. The reason why this difference is important is that for the dynamic models, when taking expectations of various quantities, it is no longer possible to treat Φ as a constant matrix.

3.4 Subspace Identification Method

This section contains a description of the central ideas pertaining to subspace identification method. First, we describe state space models, which is the type of models that is delivered by subspace identification algorithms. Then we explain how subspace identification algorithms work.

3.4.1 State Space Models

Models in the sequel are lumped, discrete time, linear, time-invariant, state space models. It is interesting to observe that many industrial processes can be described very accurately by this type of models, especially locally in the neighborhood of a working point. Moreover, there is a large number of control system design tools available to build controllers for such systems and models. These models are described mathematically by the following set of difference equations:

$$\begin{cases} x_{k+1} = Ax_k + Bu_k + w_i, \\ y_k = Cx_k + Du_k + v_k, \end{cases} \tag{3.95}$$

with

$$\mathbb{E}\left[\begin{bmatrix} w_p \\ v_p \end{bmatrix} \begin{bmatrix} w_q^t & v_q^t \end{bmatrix}\right] = \begin{bmatrix} Q & S \\ S^t & R \end{bmatrix} \delta_{pq} \geq 0 \tag{3.96}$$

where $\mathbb{E}$ denotes the expected value operator and δ_{pq} the Kronecker delta. In this model, we denote by the vectors $u_k \in \Re^m$ and $y_k \in \Re^\ell$ the observations at time instant k of respectively, the m inputs and ℓ outputs of the process. The vector $x_k \in \Re^n$ is the state vector of the process at discrete time instant k and contains the numerical values of n states. $v_k \in \Re^\ell$ and $w_k \in \Re^n$ are unobserved vector signals, usually called the measurement, respectively, process noise. It is assumed that they are zero mean, stationary, white noise vector sequences. (The Kronecker delta in (3.96) means $\delta_{pq} = 0$ if $p \neq q$, and $\delta_{pq} = 1$ if $p = q$, The effect of the process w_k is different from that of v_k : w_k as an input will have a dynamic effect on the state x_k and output y_k, while v_k only affects the output y_k directly and therefore is called a measurement noise.)

In addition, $A \in \Re^{n\times n}$ is called the system matrix. It describes the dynamics of the system (as characterized by its eigenvalues), $B \in \Re^{n\times m}$ is the input matrix, which represents the linear transformation by which the deterministic inputs influence the next state and $C \in \Re^{\ell\times n}$ is the output matrix, which describes how the internal state is transferred to the outside world in the observations y_k. The term with

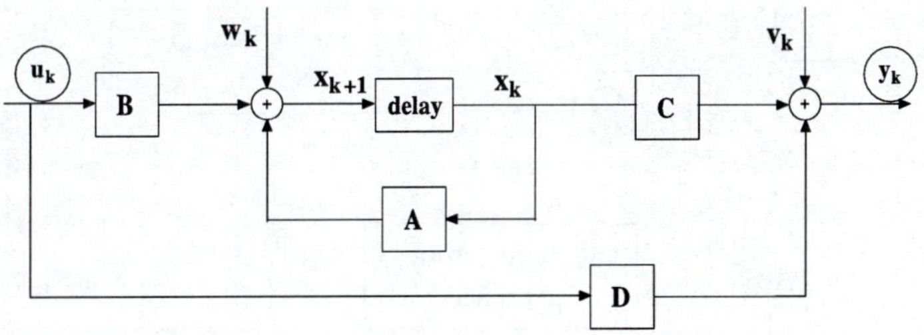

Fig. 3.7 Discrete system under consideration

the matrix $D \in \Re^{l \times m}$ is called the direct feed through term. The matrices $Q \in \Re^{n \times n}$, $S \in \Re^{n \times \ell}$ and $R \in \Re^{\ell \times \ell}$ are the covariance matrices of the noise sequences w_k and v_k. The block matrix in (3.96) is assumed to be positive definite, as is indicated by the inequality sign. The matrix pair $\{A, C\}$ is assumed to be observable, which implies that all modes in the system can be observed in the output y_k and can thus be identified. The matrix pair $\{A, [BQ^{1/2}]\}$ is assumed to be controllable, which in its turn implies that all modes of the system can be excited by either the deterministic input u_k and/or the stochastic input w_k.

A graphical representation of the system can be found in Fig. 3.7.

The main mathematical problem here is phrased as follows: *Given s consecutive input and output observations $u_o, \ldots, u_{s-1}$, and $y_o, \ldots, y_{s-1}$. Find an appropriate order n and the system matrices A, B, C, D, Q, R, S.*

Subspace identification algorithms are based on concepts from system theory, numerical linear algebra and statistics. The main concepts in subspace identification algorithms are:

1. The *state sequence of the dynamical system* is determined first, directly from input/output observations, without knowing the model. That this is possible for the model class (3.95) is one of the main contributions of subspace algorithms, as compared to "classical" approaches that are based on an input–output framework. The difference is illustrated in Fig. 3.8. So an important achievement of the research in subspace identification was to demonstrate how the Kalman filter states can be obtained directly from input–output data using linear algebra tools (QR and singular value decomposition) without knowing the mathematical model. An important consequence is that, once these states are known, the identification problem becomes a linear least squares problem in the unknown system matrices, and the process and measurement noise covariance matrices follow from the least squares residuals, as is easy to see from (3.95):

$$\underbrace{\begin{bmatrix} x_{i+1} & x_{i+2} & \cdots & x_{i+j} \\ y_i & y_{i+1} & \cdots & y_{i+j-1} \end{bmatrix}}_{\text{known}} = \begin{bmatrix} A & B \\ C & D \end{bmatrix} \underbrace{\begin{bmatrix} x_i & x_{i+1} & \cdots & x_{i+j-1} \\ u_i & u_{i+1} & \cdots & u_{i+j-1} \end{bmatrix}}_{\text{known}} + \begin{bmatrix} w_i & w_{i+1} & \cdots & w_{i+j-1} \\ v_i & v_{i+1} & \cdots & v_{i+j-1} \end{bmatrix}. \tag{3.97}$$

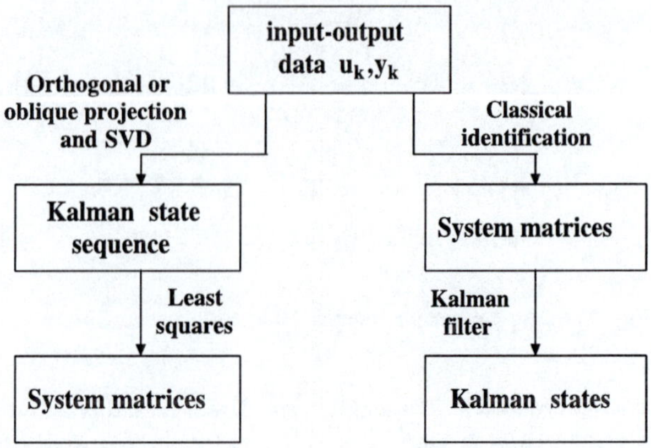

Fig. 3.8 Subspace and identification classical approaches

The meaning of the parameters i and j will become clear henceforth. Even though the state sequence can be determined explicitly, in most variants and implementations, this is not done explicitly but rather implicitly. Putting it differently, the set of linear equations above can be solved 'implicitly' as will become clear below, without an explicit calculation of the state sequence itself. Of course, when needed, the state sequence can be computed explicitly.

The two main steps that are taken in subspace algorithms are the following:

a. Determine the model order n and a state sequence $\hat{x}_i, \hat{x}_{i+1}, \ldots, \hat{x}_{i+j}$ (estimates are denoted by a $\hat{\cdot}$). They are typically found by first projecting row spaces of data block Hankel matrices, and then applying a singular value decomposition (see Sects. 3.4.6, 3.4.7, 3.4.8).
b. Solve a least squares problem to obtain the state space matrices:

$$\begin{bmatrix} \hat{A} & \hat{B} \\ \hat{C} & \hat{D} \end{bmatrix} = \min_{A,B,C,D} \left\| \begin{bmatrix} \hat{x}_{i+1} & \hat{x}_{i+2} & \cdots & \hat{x}_{i+j} \\ \hat{y}_i & \hat{y}_{i+1} & \cdots & \hat{y}_{i+j-1} \end{bmatrix} - \begin{bmatrix} A & B \\ C & D \end{bmatrix} \begin{bmatrix} \hat{x}_i & \hat{x}_{i+1} & \cdots & \hat{x}_{i+j-1} \\ \hat{u}_i & \hat{u}_{i+1} & \cdots & \hat{u}_{i+j-1} \end{bmatrix} \right\|_F^2, \tag{3.98}$$

where $\|\cdot\|_F$ denotes the Frobenius-norm of a matrix. The estimates of the noise covariance matrices follow from

$$\begin{bmatrix} \hat{Q} & \hat{S} \\ \hat{S}^t & \hat{R} \end{bmatrix} = \frac{1}{j} \begin{bmatrix} \rho_{w_i} & \rho_{w_{i+1}} & \cdots & \rho_{w_{i+j-1}} \\ \rho_{v_i} & \rho_{v_{i+1}} & \cdots & \rho_{v_{i+j-1}} \end{bmatrix} \begin{bmatrix} \rho_{w_i} & \rho_{w_{i+1}} & \cdots & \rho_{w_{i+j-1}} \\ \rho_{v_i} & \rho_{v_{i+1}} & \cdots & \rho_{v_{i+j-1}} \end{bmatrix}^t, \tag{3.99}$$

where

$$\rho_{w_k} = \hat{x}_{k+1} - \hat{A}\hat{x}_k - \hat{B}u_k, \quad \rho v_k = y_k - \hat{C}\hat{x}_k - \hat{D}u_k \quad (k = i, \ldots, i+j-1)$$

are the least squares residuals.

2. Subspace system identification algorithms make full use of the well developed body of concepts and algorithms from numerical linear algebra. Numerical robustness is guaranteed because of the well-understood algorithms, such as the

QR-decomposition, the singular value decomposition and its generalizations. Therefore, they are very well suited for large data sets ($s \to \infty$) and large scale systems (m, ℓ, n large). Moreover, subspace algorithms are not iterative. Hence, there are no convergence problems. When carefully implemented, they are computationally very efficient, especially for large datasets.

3. The conceptual layout of subspace identification algorithms translates into user friendly software implementations. Recall that there is no explicit need for parameterizations in the geometric framework of subspace identification. Thus, the user is not confronted with highly technical and theoretical issues such as canonical parameterizations. The number of user choices is greatly reduced when using subspace algorithms because we use full state space models and the only parameter to be specified by the user, is the order of the system, which can be determined by inspection of certain singular values.

3.4.2 Block Hankel Matrices and State Sequences

Block Hankel matrices with output and/or input data play an important role in subspace identification algorithms. These matrices can be easily constructed from the given input–output data. Input block Hankel matrices are defined as

$$U_{0|2i-1} =: \left[\begin{array}{ccccc} u_0 & u_1 & u_2 & \cdots & u_{j-1} \\ u_1 & u_2 & u_3 & \cdots & u_j \\ \vdots & \vdots & \vdots & \ddots & \vdots \\ u_{i-1} & u_i & u_{i+1} & \cdots & u_{i+j-2} \\ \hline u_i & u_{i+1} & u_{i+2} & \cdots & u_{i+j-1} \\ u_{i+1} & u_{i+2} & u_{i+3} & \cdots & u_{i+j} \\ \vdots & \vdots & \vdots & \ddots & \vdots \\ u_{2i-1} & u_{2i} & u_{2i+1} & \cdots & u_{2i+j-2} \end{array}\right] = \left[\begin{array}{c} U_{0|i-1} \\ \hline U_{i|2i-1} \end{array}\right] = \left[\begin{array}{c} U_p \\ \hline U_f \end{array}\right] \tag{3.100}$$

$$= \left[\begin{array}{ccccc} u_0 & u_1 & u_2 & \cdots & u_{j-1} \\ u_1 & u_2 & u_3 & \cdots & u_j \\ \vdots & \vdots & \vdots & \ddots & \vdots \\ u_{i-1} & u_i & u_{i+1} & \cdots & u_{i+j-2} \\ u_i & u_{i+1} & u_{i+2} & \cdots & u_{i+j-1} \\ \hline u_{i+1} & u_{i+2} & u_{i+3} & \cdots & u_{i+j} \\ \vdots & \vdots & \vdots & \ddots & \vdots \\ u_{2i-1} & u_{2i} & u_{2i+1} & \cdots & u_{2i+j-2} \end{array}\right] = \left[\begin{array}{c} U_{0|i} \\ \hline U_{i+1|2i-1} \end{array}\right] = \left[\begin{array}{c} U_p^+ \\ \hline U_f^- \end{array}\right] \tag{3.101}$$

where:

- The number of block rows (i) is a user-defined index which is large enough, that is, it should at least be larger than the maximum order of the system one wants to identify. Note that, since each block row contains in (number of inputs) rows, the matrix $U_{0|2i-1}$ consists of $2mi$ rows.

- The number of columns (j) is typically equal to $s-2i+1$, which implies that all s available data samples are used. In any case, j should be larger than $2i-1$. Throughout the paper, for statistical reasons, we will often assume that $j, s \to \infty$. For deterministic (noiseless) models, that is, where $v_k \equiv 0$ and $w_k \equiv 0$, this will however not be needed.
- The subscripts of $U_{0|2i-1}$, $U_{0|i-1}$, $U_{0|i}$, $U_{i|2i-1}$, etc., denote the subscript of the first and last element of the first column in the block Hankel matrix. The subscript "p" stands for "past" and the subscript "f" for "future". The matrices U_p (the past inputs) and U_f (the future inputs) are defined by splitting $U_{0|2i-1}$ in two equal parts of i block rows. The matrices U_p^+ and U_f^- on the other hand are defined by shifting the border between past and future one block row down. They are defined as $U_p^+ = U_{0|i}$ and $U_f^- = U_{i+1|2i-1}$, where the superscript "+" stands for "add one block row" while the superscript "−" stands for "delete one block row". The output block Hankel matrices $Y_{0|2i-1}$, Y_p, Y_f, Y_p^+, Y_f^- are defined in a similar way. State sequences play an important role in the derivation and interpretation of subspace identification algorithms. The state sequence X_i is defined as:

$$X_i := (x_i \quad x_{i+1} \quad \cdots \quad x_{i+j-2} \quad x_{i+j-1}) \in \Re^{n \times j}, \tag{3.102}$$

where the subscript i denotes the subscript of the first element of the state sequence.

3.4.3 Model Matrices

Subspace identification algorithms make extensive use of the observability and of its structure. The extended ($i > n$) observability matrix Γ_i (where the subscript i denotes the number of block rows) is defined as:

$$\Gamma_i := \begin{bmatrix} C \\ CA \\ CA^2 \\ \vdots \\ CA^{i-1} \end{bmatrix} \in \Re^{\ell i \times n}. \tag{3.103}$$

We assume henceforth that the pair $\{A, C\}$ to be observable, which implies that the rank of Γ_i is equal to n.

3.4.4 Orthogonal Projections

In the following sections, we introduce the main geometric tools used to reveal some system characteristics. They are described from a linear algebra point of view, independently of the subspace identification framework. For simplicity in exposition, we assume that the matrices $A \in \Re^{p\times j}$, $B \in \Re^{q\times j}$ and $C \in \Re^{r\times J}$ are given and are of local use in this section. We also assume that $j \geq \max(p, q, r)$, which will always be the case in the identification algorithms.

Recall that the orthogonal projection of the row space of A into the row space of B is denoted by A/B and its matrix representation is

$$A/B := AB^t\left(BB^t\right)^{\dagger}B, \tag{3.104}$$

where $\dagger$ denotes the Moore–Penrose pseudo-inverse of the matrix and $A/B^{\perp}$ is the projection of the row space of A into $B^{\perp}$, the orthogonal complement of the row space of B, for which we have $A/B^{\perp} = A - A/B = A(I_j - B(BB^t)^{\dagger}B)$. The projections Π_B and $\Pi_{B^{\dagger}}$ decompose a matrix A into two matrices, the row spaces of which are orthogonal:

$$A = A\Pi_B + A\Pi_{B^{\perp}}. \tag{3.105}$$

The matrix representations of these projections can be easily computed via the *LQ* decomposition of

$$\begin{bmatrix} B \\ A \end{bmatrix}$$

which is the numerical matrix version of the Gram–Schmidt orthogonalization procedure.

Let A and B be matrices of full row rank and let the *LQ* decomposition of $\left[\begin{smallmatrix} B \\ A \end{smallmatrix}\right]$ be denoted by

$$\begin{bmatrix} B \\ A \end{bmatrix} = LQ^t = \begin{bmatrix} L_{11} & 0 \\ L_{21} & L_{22} \end{bmatrix} \begin{bmatrix} Q_1^t \\ Q_2^t \end{bmatrix}, \tag{3.106}$$

where $L \in \Re^{(p+q)\times(p+q)}$ is lower triangular, with $L_{11} \in \Re^{q\times q}$, $L_{21} \in \Re^{p\times q}$, $L_{22} \in \Re^{p\times p}$ and $Q \in \Re^{j\times(p+q)}$ is orthogonal, that is,

$$Q^t Q = \begin{bmatrix} Q_1^t \\ Q_2^t \end{bmatrix} \begin{bmatrix} Q_1 & Q_2 \end{bmatrix} = \begin{bmatrix} I_q & 0 \\ 0 & I_p \end{bmatrix}.$$

Then, the matrix representations of the orthogonal projections can be written as

$$A/B = L_{21}Q_1^t, \tag{3.107}$$

$$A/B^{\perp} = L_{22}Q_2^t. \tag{3.108}$$

3.4.5 Oblique Projections

Instead of decomposing the rows of A as in (3.105) as a linear combination of the rows of two orthogonal matrices (Π_B and $\Pi_{B^{\dagger}}$), they can also be decomposed as a linear combination of the rows of two non-orthogonal matrices B and C and of the orthogonal complement of B and C. This can be written as $A = L_B B + L_c C +$

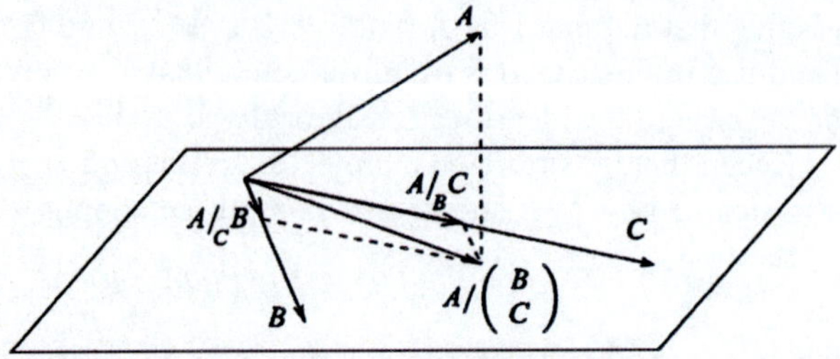

Fig. 3.9 Interpretation of oblique projection in the j-dimensional space ($j = 3$)

$L_{B^\perp, C^\perp}\left[\begin{smallmatrix} B \\ C \end{smallmatrix}\right]^\perp$. The matrix $L_C C$ is defined as the oblique projection of the row space of A along the row space of B into the row space of C:

$$A/_B C := L_C C. \tag{3.109}$$

Note that L_B and L_c are only unique when B and C are of full row rank and when the intersection of the row spaces of B and C is $\{0\}$, said in other words, $\operatorname{rank}\left[\begin{smallmatrix} B \\ C \end{smallmatrix}\right] = \operatorname{rank}(B) + \operatorname{rank}(C) = q + r$. The oblique projection can also be interpreted through the following recipe: project the row space of A orthogonally into the joint row space of B and C and decompose the result along the row space of B and C. This is illustrated in Fig. 3.9 for $j = 3$ and $p = q = r = 1$, where $A/\left[\begin{smallmatrix} B \\ C \end{smallmatrix}\right]$ denotes the orthogonal projection of the row space of A into the joint row space of B and C, $A/_B C$ is the oblique projection of A along B into C and $A/_C B$ is the oblique projection of A along C into B.

Let the LQ decomposition of $\left[\begin{smallmatrix} B \\ C \\ A \end{smallmatrix}\right]$ be given by

$$\begin{bmatrix} B \\ C \\ A \end{bmatrix} = \begin{bmatrix} L_{11} & 0 & 0 \\ L_{21} & L22 & 0 \\ L_{31} & L_{32} & L_{33} \end{bmatrix} \begin{bmatrix} Q_1^t \\ Q_2^t \\ Q_3^t \end{bmatrix}.$$

Then, the matrix representation of the orthogonal projection of the row space of A into the joint row space of B and C is equal to:

$$A/\begin{bmatrix} B \\ C \end{bmatrix} = \begin{bmatrix} L_{31} & L_{32} \end{bmatrix} \begin{bmatrix} Q_1^t \\ Q_2^t \end{bmatrix}. \tag{3.110}$$

It is obvious that the orthogonal projection of A into $\left[\begin{smallmatrix} B \\ C \end{smallmatrix}\right]$ can also be written as a linear combination of the rows of B and C:

$$A/\begin{bmatrix} B \\ C \end{bmatrix} = L_B B + L_C C = \begin{bmatrix} L_B & L_C \end{bmatrix} \begin{bmatrix} L_{11} & 0 \\ L_{21} & L22 \end{bmatrix} \begin{bmatrix} Q_1^t \\ Q_2^t \end{bmatrix}. \tag{3.111}$$

Equating (3.110) and (3.111) leads to

$$\begin{bmatrix} L_B & L_C \end{bmatrix} \begin{bmatrix} L_{11} & 0 \\ L_{21} & L22 \end{bmatrix} = \begin{bmatrix} L_{31} & L_{32} \end{bmatrix}. \tag{3.112}$$

The oblique projection of the row space of A along the row space of B into the row space of C can thus be computed as

$$A/_B C = L_C C = L_{32} L_{22}^{-1} \begin{bmatrix} L_{21} & L_{22} \end{bmatrix} \begin{bmatrix} Q_1^t \\ Q_2^t \end{bmatrix}. \tag{3.113}$$

Note that when $B = 0$ or when the row space of B is orthogonal to the row space of $C(BC^t = 0)$ the oblique projection reduces to an orthogonal projection, in which case $A/_B C = A/C$.

3.4.6 Deterministic Subspace Identification

In what follows, we treat subspace identification of purely time- invariant deterministic systems, with no measurement nor process noise ($v_k \equiv 0$ and $w_k \equiv 0$ in Fig. 3.7).

Calculation of a State Sequence

The state sequence of a deterministic system can be found by computing the intersection of the past input and output and the future input and output spaces. This can be seen as follows. Consider w_k and v_k in (3.95) to be identically 0, and derive the following matrix input–output equations:

$$Y_{0|i-1} = \Gamma_i X_i + H_i U_{0|i-1}, \tag{3.114}$$

$$Y_{i|2i-1} = \Gamma_i X_{2i} + H_i U_{i|2i-1}, \tag{3.115}$$

in which H_i is an $li \times mi$ lower block Triangular Toeplitz matrix with the so-called Markov parameters of the system:

$$H_i = \begin{bmatrix} D & 0 & 0 & \cdots & 0 \\ CB & D & 0 & \cdots & 0 \\ CAB & CB & D & \cdots & 0 \\ \vdots & \vdots & \vdots & \ddots & \vdots \\ CA^{i-2}B & CA^{i-3}B & \cdots & \cdots & D \end{bmatrix}.$$

From this, we find that

$$\begin{bmatrix} Y_{0|i-1} \\ U_{0|i-1} \end{bmatrix} = \begin{bmatrix} \Gamma_i & H_i \\ 0 & I_{mi} \end{bmatrix} \begin{bmatrix} X_i \\ U_{0|i-1} \end{bmatrix}, \tag{3.116}$$

from which we get

$$\operatorname{rank} \begin{bmatrix} Y_{0|i-1} \\ U_{0|i-1} \end{bmatrix} = \operatorname{rank} \begin{bmatrix} X_i \\ U_{0|i-1} \end{bmatrix}.$$

Hence,

$$\operatorname{rank}\begin{bmatrix} Y_{0|i-1} \\ U_{0|i-1} \end{bmatrix} = mi + n$$

provided that $U_{0|i-1}$ is of full row rank. In the sequel, we assume throughout that $j \gg mi$, that there is no intersection between the row spaces of X_i and that of $U_{0|i-1}$ and that the state sequence is of full row rank as well *'full state space excited'*. These are experimental conditions that are generically satisfied and that can be considered as 'persistency-of-excitation' requirements for subspace algorithms to work.

A similar derivation under similar conditions can be done for

$$\operatorname{rank}\begin{bmatrix} Y_{i|2i-1} \\ U_{i|2i-1} \end{bmatrix} = mi + n, \qquad \operatorname{rank}\begin{bmatrix} Y_{0|2i-1} \\ U_{0|2i-1} \end{bmatrix} = 2mi + n.$$

We can also relate X_{2i} to X_i as

$$X_{2i} = A^i X_i + \Delta_i^r U_{0|i-1}, \tag{3.117}$$

in which $\Delta_i^r = (A^{i-1}BA^{i-2}B \cdots ABB)$ is a reversed extended controllability matrix. Assuming that the model is observable and that $i \geq n$, we find from (3.115) that

$$X_{2i} = \begin{bmatrix} -\Gamma_i^\dagger H_i & \Gamma_i^\dagger \end{bmatrix} \begin{bmatrix} U_{i|2i-1} \\ Y_{i|2i-1} \end{bmatrix},$$

which implies that the row space of X_{2i} is contained within the row space of

$$\begin{bmatrix} U_f \\ Y_f \end{bmatrix}.$$

Similarly, from (3.117) and (3.114) we find that

$$\begin{aligned} X_{2i} &= A^i \begin{bmatrix} \Gamma_i^\dagger Y_{0|i-1} & -\Gamma_i^\dagger H_i U_{0|i-1} \end{bmatrix} + \Delta_i^r U_{0|i-1} \\ &= \begin{bmatrix} \Delta_i^r - A^i \Gamma_i^\dagger H_i & A^i \Gamma_i^\dagger \end{bmatrix} \begin{bmatrix} U_{0|i-1} \\ Y_{0|i-1} \end{bmatrix}, \end{aligned}$$

which implies that the row space of X_{2i} is equally contained within the row space of

$$\begin{bmatrix} U_p \\ Y_p \end{bmatrix}.$$

Let's now apply Grassmann's dimension theorem (under the generic assumptions on persistency of excitation)

$$\dim\left[\text{row space}\begin{bmatrix} U_p \\ Y_p \end{bmatrix} \cap \text{row space}\begin{bmatrix} U_f \\ Y_f \end{bmatrix}\right] \operatorname{rank}\begin{bmatrix} U_p \\ Y_p \end{bmatrix} \tag{3.118}$$

$$+ \operatorname{rank}\begin{bmatrix} U_f \\ Y_f \end{bmatrix} - \operatorname{rank}\begin{bmatrix} U_p \\ Y_p \\ U_f \\ Y_f \end{bmatrix} \tag{3.119}$$

$$= (mi + n) + (mi + n) - (2mi + n) = n. \tag{3.120}$$

Indeed, above we have shown that any basis for the intersection between 'past' and 'future' represents a valid state sequence X_i. The state sequence X_{i+1} can be obtained analogously. Different ways to compute the intersection have been proposed. A first way, is by making use of a singular value decomposition of a concatenated Hankel matrix

$$\begin{bmatrix} U_{0|2i-1} \\ Y_{0|2i-1} \end{bmatrix}.$$

This allows to estimate the model order n and to calculate the linear combination of the rows of

$$\begin{bmatrix} U_p \\ Y_p \end{bmatrix}$$

or equivalently of

$$\begin{bmatrix} U_f \\ Y_f \end{bmatrix}$$

that generate the intersection. A second way is by taking as a basis for the intersection the principal directions between the row space of the past inputs and outputs and the row space of the future inputs and outputs. A nonempty intersection between two subspaces is characterized by a number of principal angles equal to zero, and the principal directions corresponding to these zero angles form a basis for the row space of the intersection.

Computing the System Matrices

As soon as the order of the model and the state sequences X, and X_{i+1} are known, the state space matrices A, B, C, D can be solved from

$$\underbrace{\begin{bmatrix} x_{i+1} \\ y_{i|i} \end{bmatrix}}_{\text{known}} = \begin{bmatrix} A & B \\ C & D \end{bmatrix} \underbrace{\begin{bmatrix} x_i \\ U_{i|i} \end{bmatrix}}_{\text{known}}, \tag{3.121}$$

where $U_{i|i}$, $Y_{i|i}$ are block Hankel matrices with only one block row of inputs respectively outputs, namely $U_{i|i} = (u_i \;\; u_{i+1} \;\; \cdots \;\; u_{i+j-1})$ and similarly for $Y_{i|i}$. This set of equations can be solved. As there is no noise, it is consistent.

3.4.7 Stochastic Subspace Identification

In this section, we treat subspace identification of linear time-invariant stochastic systems with no external input ($u_k \equiv 0$). The stochastic identification problem thus consists of estimating the stochastic system matrices A, C, Q, S, R from given output data only. We show how this can be done using geometric operations. In the next part, we show how a state sequence can be found and in the following part the system matrices are computed.

Calculation of a State Sequence

The state sequence of a stochastic model can be obtained in two steps: first, the future output space is projected orthogonally into the past output space and next, a singular value decomposition is carried out.

1. *Orthogonal Projection*: As explained in Sect. 3.4.4, we will use the *LQ* decomposition to compute the orthogonal projection. Let $Y_{0|2i-1}$ be the $2\ell i \times j$ output block Hankel matrix. Then, we partition the *LQ* decomposition of $Y_{0|2i-1}$ as follows

$$\begin{bmatrix} Y_{0|i-1} \\ Y_{i|i} \\ Y_{i+1|2i-1} \end{bmatrix} = \begin{bmatrix} L_{11} & 0 & 0 \\ L_{21} & L_{22} & 0 \\ L_{31} & L_{32} & L_{33} \end{bmatrix} \begin{bmatrix} Q_1^t \\ Q_2^t \\ Q_3^t \end{bmatrix} \tag{3.122}$$

where $Y_{0|i-1} \in \Re^{\ell i}$, $Y_{i|i} \in \Re^{\ell}$, $Y_{i+1|2i-1} \in \Re^{\ell(i-1)}$, $L_{11} \in \Re^{\ell i \times \ell i}$, $L_{21} \in \Re^{\ell \times \ell i}$, $L_{22} \in \Re^{\ell \times \ell}$, $L_{31} \in \Re^{\ell(i-1)\times \ell i}$, $L_{32} \in \Re^{\ell(i-1)\times \ell}$, $L_{33} \in \Re^{\ell(i-1)\times \ell(i-1)}$.

At this stage, we need two projections. The orthogonal projection Y_f/Y_p of the future output space into the past output space, which is denoted by O_i, and the orthogonal projection Y_f^-/Y_p^+ of Y_f^- into Y_p^+, denoted by O_{i-1}, see Sect. 3.4.2 for the definitions of Y_p, Y_f, Y_p^+ and Y_f^-. Now, applying (3.107) leads to

$$\begin{aligned} O_i &= Y_f/Y_p = \begin{bmatrix} L_{21} \\ L_{31} \end{bmatrix} Q_1^t, \\ O_{i-1} &= Y_f^-/Y_p^+ = \begin{bmatrix} L_{31} & L_{32} \end{bmatrix} \begin{bmatrix} Q_1^t \\ Q_2^t \end{bmatrix}. \end{aligned} \tag{3.123}$$

It can be shown that the matrix O_i, is equal to the product of the extended observability matrix and a matrix $\hat{X}_i$, which contains certain Kalman filter states. Thus,

$$O_i = \Gamma_i \hat{X}_i, \tag{3.124}$$

where Γ_i is the $li \times n$ observability matrix, see (3.103) and

$$\hat{X}_i = \begin{bmatrix} \hat{x}_i^{[0]} & \hat{x}_i^{[1]} & \cdots & \hat{x}_i^{[j-1]} \end{bmatrix}.$$

Similarly, O_{i-1} is equal to

$$O_{i-1} = \Gamma_{i-1} \hat{X}_{i+1}, \tag{3.125}$$

where $\hat{X}_{i+1} = [\hat{x}_{i+1}^{[0]} \; \hat{x}_{i+1}^{[1]} \; \cdots \; \hat{x}_{i+1}^{[j-1]}]$.

2. *Singular Value Decomposition*: The singular value decomposition of 0_i, allows us to find the order of the model (the rank of O_i), and the matrices Γ_i and $\hat{X}_i$.

Let the singular value decomposition of $\left[\begin{smallmatrix} L_{21} \\ L_{31} \end{smallmatrix}\right]$ be equal to

$$\begin{bmatrix} L_{21} \\ L_{31} \end{bmatrix} = \begin{bmatrix} U_1 & U_2 \end{bmatrix} \begin{bmatrix} S_1 & 0 \\ 0 & 0 \end{bmatrix} \begin{bmatrix} V_1^t \\ V_2^t \end{bmatrix} = U_1 S_1 V_1^t, \tag{3.126}$$

$$
\begin{array}{llllllll}
\hat{X}_0 = [& 0 & \cdots & 0 & \cdots & 0 &] & \\
\hat{P}_0 = 0 & & & & & & & \text{Kalman Filter} \\
 & y_0 & \downarrow & y_q & \downarrow & y_{j-1} & & \downarrow \\
Y_p & \vdots & \downarrow & \vdots & \downarrow & \vdots & & \downarrow \\
 & y_{i-1} & \downarrow & y_{i+q-1} & \downarrow & y_{i+j-2} & & \downarrow \\
\hat{X}_i = [& \hat{x}_i^{[0]} & \cdots & \hat{x}_i^{[q]} & \cdots & \hat{x}_i^{[j-1]} &] &
\end{array}
$$

Fig. 3.10 Interpretation of the sequence $\hat{X}_i$

where $U_1 \in \Re^{li \times n}$, $S_1 \in \Re^{n \times n}$, and $V_1 \in \Re^{li \times n}$. Then, we can choose $\Gamma_i = U_1 S_1^{1/2}$ and $\hat{X}_i = S_1^{1/2} V_1^t Q_1^t$. This state sequence is generated by a bank of non-steady state Kalman filters working in parallel on each of the columns of the block Hankel matrix of past outputs Y_p. The j Kalman filters run in a vertical direction (over the columns). It should be noted that each of these j Kalman filters only uses partial output information. The qth Kalman filter ($q = 0, \ldots, j-1$)

$$\hat{x}_{k+1}^{[q]} = (A - K_k C)\hat{x}_k^{[q]} + K_k y_{k+q}, \tag{3.127}$$

runs over the data in the qth column of Y_p, for $k = 0, 1, \ldots, i-1$.

The "shifted" state sequence $\hat{X}_{i+1}$, on the other hand, can be obtained as

$$\hat{X}_{i+1} = (\hat{\Gamma}_i)^{\dagger} O_{i-1}, \tag{3.128}$$

where $\hat{\Gamma}_i = \Gamma_{i-1}$ denotes the matrix Γ_i without the last ℓ rows, which is also equal to $\hat{U}_1 S_1^{1/2}$. In Fig. 3.10, an interpretation of the sequence $\hat{X}_i$ as a sequence of nonsteady state Kalman filter estimates based upon i observations of y_k is given. When the system matrices A, C, Q, R, S were known, the state $\hat{x}_i^{[q]}$ could be determined from a non-steady state Kalman filter as follows: Start the filter at time q, with an initial state estimate 0. Next, iterating the nonsteady state Kalman filter over i time steps (as indicated by the vertical arrow down). The Kalman filter will then return a state estimate $\hat{x}_i^{[q]}$. This procedure could be repeated for each of the j columns, and thus we speak about a bank of non-steady state Kalman filters. The major observation in subspace algorithms is that the system matrices A, C, Q, R, S do not have to be known to determine the state sequence $\hat{X}_i$. It can be determined directly from output data through geometric manipulations.

Computing the System Matrices

At this moment, we have calculated $\hat{X}_i$ and $\hat{X}_{i+1}$, using geometrical and numerical operations on output data only. We can now form the following set of equations:

$$\underbrace{\begin{bmatrix} \hat{X}_{i+1} \\ Y_{i|i} \end{bmatrix}}_{\text{known}} = \begin{bmatrix} A \\ C \end{bmatrix} \underbrace{[\hat{X}_i]}_{\text{known}} + \underbrace{\begin{bmatrix} \rho_w \\ \rho_v \end{bmatrix}}_{\text{residuals}}, \tag{3.129}$$

where $Y_{i|i}$ is a block Hankel matrix with only one block row of outputs. This set of equations can be easily solved for A, C. Since the Kalman filter residuals ρ_w, ρ_v, (the innovations) are uncorrelated with $\hat{X}_i$, solving this set of equations in a least squares sense (since the least squares residuals are orthogonal and thus uncorrelated with the regressors $\hat{X}_i$) results in an asymptotically (as $j \to \infty$) unbiased estimate $\hat{A}$, $\hat{C}$ of A, C as

$$\begin{bmatrix} \hat{A} \\ \hat{C} \end{bmatrix} = \begin{bmatrix} \hat{X}_{i+1} \\ Y_{i|i} \end{bmatrix} \hat{X}_i^{\dagger}.$$

An estimate $\hat{Q}_i$, $\hat{S}_i$, $\hat{R}_i$ of the noise covariance matrices Q, S and R can be obtained from the residuals:

$$\begin{bmatrix} \hat{Q}_i & \hat{S}_i \\ \hat{S}_i^t & \hat{R}_i \end{bmatrix} = \frac{1}{j} \begin{bmatrix} \rho_w \\ \rho_v \end{bmatrix} [\rho_w^t \ \rho_v^t]$$

where the subscript i indicates that the estimated covariances are biased, with however an exponentially decreasing bias as $i \to \infty$.

By making the following substitutions:

$$\hat{X}_i = \Gamma_i^{\dagger} O_i = S_1^{1/2} V_1^t Q_1^t, \tag{3.130}$$

$$\hat{X}_{i+1} = \Gamma_{i-1}^{\dagger} O_{i-1} = (\hat{\Gamma}_i)^{\dagger} O_{i-1} = [\hat{U}_1 S_1^{1/2}]^{\dagger} (L_{31} \quad L_{32}) \begin{bmatrix} Q_1^t \\ Q_2^t \end{bmatrix}, \tag{3.131}$$

$$Y_{i|i} = (L_{21} \quad L_{22}) \begin{bmatrix} Q_1^t \\ Q_2^t \end{bmatrix} \tag{3.132}$$

the least squares solution reduces to

$$\begin{bmatrix} \hat{A} \\ \hat{C} \end{bmatrix} = \begin{bmatrix} [\hat{U}_1 S_1^{1/2}]^{\dagger} L_{31} \\ L_{21} \end{bmatrix} V_1 S_1^{-1/2}, \tag{3.133}$$

and the noise covariances are equal to

$$\begin{aligned} \begin{bmatrix} \hat{Q}_i & \hat{S}_i \\ \hat{S}_i^t & \hat{R}_i \end{bmatrix} &= \frac{1}{j} \begin{bmatrix} [\hat{U}_1 S_1^{1/2}]^{\dagger} L_{31} & [\hat{U}_1 S_1^{1/2}]^{\dagger} L_{32} \\ L_{21} & L_{22} \end{bmatrix} \begin{bmatrix} I - V_1 V_1^t & 0 \\ 0 & I \end{bmatrix} \\ &\times \begin{bmatrix} L_{31}^t [S_1^{1/2} (\hat{U}_1)^t]^{\dagger} & L_{21}^t \\ L_{32}^t [S_1^{1/2} (\hat{U}_1)^t]^{\dagger} & L_{22}^t \end{bmatrix}. \end{aligned} \tag{3.134}$$

Note that the Q-matrices of the LQ factorization cancel out of the least-squares solution and the noise covariances. This implies that in the first step, the Q-matrix should never be calculated explicitly. Since typically $j \gg 2mi$, this reduces the computational complexity and memory requirements significantly.

3.4.8 Combined Deterministic-Stochastic Algorithm

Hereafter, we give one variant of subspace algorithms, for the identification of

$$A, \ B, \ C, \ D, \ Q, \ R, \ S.$$

Other variants can be found in the literature. The algorithm works in two main steps. First, the row space of a Kalman filler state sequence is obtained directly from the input–output data, without any knowledge of the system matrices. In the second step, the system matrices are extracted from the state sequence via a least squares problem.

Calculation of a State Sequence

The state sequence of a combined deterministic-stochastic model can again be obtained from input output data in two steps. First, the future output row space is projected along the future input row space into the joint row space of past input and past output. A singular value decomposition is carried out to obtain the model order, the observability matrix and a state sequence, which has a very precise and specific interpretation.

1. *Oblique projection*: We will use the LQ decomposition to compute the oblique projection $Y_f/_{U_f}\left[\begin{smallmatrix}U_p\\Y_p\end{smallmatrix}\right]$. Let $U_{0|2i-1}$ be the $2mi \times j$ and $Y_{0|2i-1}$ the $2li \times j$ block Hankel matrices of the input and output observations. Then, we partition the LQ decomposition of $\left[\begin{smallmatrix}U\\Y\end{smallmatrix}\right]$ as follows

$$\begin{bmatrix} U_{0|i-1} \\ U_{i|i} \\ U_{i+1|2i-1} \\ Y_{0|i-1} \\ Y_{i|i} \\ Y_{i+1|2i-1} \end{bmatrix} = \begin{bmatrix} L_{11} & 0 & 0 & 0 & 0 & 0 \\ L_{21} & L_{22} & 0 & 0 & 0 & 0 \\ L_{31} & L_{32} & L_{33} & 0 & 0 & 0 \\ L_{41} & L_{42} & L_{43} & L_{44} & 0 & 0 \\ L_{51} & L_{52} & L_{53} & L_{54} & L_{55} & 0 \\ L_{61} & L_{62} & L_{63} & L_{64} & L_{65} & L_{66} \end{bmatrix} \begin{bmatrix} Q_1^t \\ Q_2^t \\ Q_3^t \\ Q_4^t \\ Q_5^t \\ Q_6^t \end{bmatrix}. \tag{3.135}$$

The matrix representation of the oblique projection $Y_f/_{U_f}\left[\begin{smallmatrix}U_p\\Y_p\end{smallmatrix}\right]$ of the future output row space along the future input row space into the joint space of past input and past output is denoted by O_i. Analogously to the derivation in Sect. 3.4.5, the oblique projection can be obtained as

$$O_i = Y_f/_{U_f}\begin{bmatrix} U_p \\ Y_p \end{bmatrix} = L_{U_p} L_{11} Q_1^t + L_{Y_p} \begin{bmatrix} L_{41} & L_{42} & L_{43} & L_{44} \end{bmatrix} \begin{bmatrix} Q_1^t \\ Q_2^t \\ Q_3^t \\ Q_4^t \end{bmatrix}, \tag{3.136}$$

where

$$\begin{bmatrix} L_{U_p} & L_{U_f} & L_{Y_p} \end{bmatrix} \left[\begin{array}{c|cc|c} L_{11} & 0 & 0 & 0 \\ \hline L_{21} & L_{22} & 0 & 0 \\ L_{31} & L_{32} & L_{33} & 0 \\ \hline L_{41} & L_{42} & L_{43} & L_{44} \end{array}\right] = \left[\begin{array}{c|cc|c} L_{51} & L_{52} & L_{53} & L_{54} \\ L_{61} & L_{62} & L_{63} & L_{64} \end{array}\right] \tag{3.137}$$

from which L_{U_p}, L_{U_f} and L_{Y_p} can be calculated. On the other hand, the oblique projection

$$Y_f^- /_{U_f^-} \begin{bmatrix} U_p^+ \\ Y_p^+ \end{bmatrix}$$

denoted by O_{i-1}, is equal to

$$O_i = L_{U_p^+} \begin{bmatrix} L_{11} & 0 \\ L_{21} & L_{22} \end{bmatrix} \begin{bmatrix} Q_1^t \\ Q_2^t \end{bmatrix} + L_{Y_p^+} \begin{bmatrix} L_{41} & L_{42} & L_{43} & L_{44} & 0 \\ L_{51} & L_{52} & L_{53} & L_{54} & L_{55} \end{bmatrix} \begin{bmatrix} Q_1^t \\ Q_2^t \\ Q_3^t \\ Q_4^t \\ Q_5^t \end{bmatrix}, \tag{3.138}$$

where

$$\begin{bmatrix} L_{U_p^+} & L_{U_f^-} & L_{Y_p^+} \end{bmatrix} \left[\begin{array}{cc|c|c|c} L_{11} & 0 & 0 & 0 & 0 \\ L_{21} & L_{22} & 0 & 0 & 0 \\ \hline L_{31} & L_{32} & L_{33} & 0 & 0 \\ \hline L_{41} & L_{42} & L_{43} & L_{44} & 0 \\ L_{51} & L_{52} & L_{53} & L_{54} & L_{55} \end{array}\right] = \left[L_{61} \;\; L_{62} \,\middle|\, L_{63} \,\middle|\, L_{64} \;\; L_{65} \right]. \tag{3.139}$$

Under the assumptions that
a. the process noise W_k and measurement noise v_k are uncorrelated with the input u_k,
b. the input u_k is persistently exciting of order $2i$, that is, the input block Hankel matrix $U_{0|2i-1}$ is of full row rank,
c. the sample size goes to infinity: $j \to \infty$,
d. the process noise w_k and the measurement noise v_k are not identically zero,

one can show that the oblique projection O_i is equal to the product of the extended observability matrix Γ_i and a sequence of Kalman filter states, obtained from a bank of nonsteady state Kalman filters, in essence the same as in Fig. 3.10:

$$O_i = \Gamma_i \tilde{X}_i. \tag{3.140}$$

Similarly, the oblique projection O_{i-1} is equal to

$$O_{i-1} = \Gamma_{i-1} \tilde{X}_{i+1}. \tag{3.141}$$

2. *Singular value decomposition*: Let the singular value decomposition of

$$L_{U_p} \begin{bmatrix} L_{11} & 0 & 0 & 0 \end{bmatrix} + L_{Y_p} \begin{bmatrix} L_{41} & L_{42} & L_{43} & L_{44} \end{bmatrix}$$

be equal to

$$\begin{aligned} & L_{U_p} \begin{bmatrix} L_{11} & 0 & 0 & 0 \end{bmatrix} + L_{Y_p} \begin{bmatrix} L_{41} & L_{42} & L_{43} & L_{44} \end{bmatrix} \\ & = \begin{bmatrix} U_1 & U_2 \end{bmatrix} \begin{bmatrix} S_1 & 0 \\ 0 & 0 \end{bmatrix} \begin{bmatrix} V_1^t \\ V_2^t \end{bmatrix} \qquad (3.142) \\ & = U_1 S_1 V_1^t. \qquad (3.143) \end{aligned}$$

Then, the order of the system (3.95) is equal to the number of singular values in (3.142) different from zero. The extended observability matrix Γ_i can be taken to be

$$\Gamma_i = U_1 S_1^{1/2}, \tag{3.144}$$

and the state sequence $\tilde{X}_i$ is equal to

$$\tilde{X}_i = \Gamma_i^{\dagger} O_i = S_1^{1/2} V_1^t \begin{bmatrix} Q_1^t \\ Q_2^t \\ Q_3^t \\ Q_4^t \end{bmatrix}. \tag{3.145}$$

The "shifted" state sequence $\tilde{X}_{i+1}$, on the other hand, can be obtained as

$$\tilde{X}_{i+1} = (\tilde{\Gamma}_i)^{\dagger} O_{i-1}, \tag{3.146}$$

where $\tilde{\Gamma}_i = \Gamma_{i-1}$ denotes the matrix Γ_i without the last l rows.

There is an important observation to be made. Corresponding columns of $\tilde{X}_i$ and of $\tilde{X}_{i+1}$ are state estimates of X_i and X_{i+1} respectively, obtained from the same Kalman filters at two consecutive time instants, but with different initial conditions. This is in contrast to the stochastic identification algorithm, where the initial states are equal to 0, see Fig. 3.10.

Computing the System Matrices

From Sect. 3.4.8, we find that:

- The order of the system from inspection of the singular values of (3.142).
- The extended observability matrix Γ_i; from (3.144) and the matrix Γ_{i-1} as $\tilde{\Gamma}_i$, where $\tilde{\Gamma}_i$ denotes the matrix Γ_i without the last ℓ rows.
- The state sequences $\tilde{X}_i$ and $\tilde{X}_{i+1}$.

The state space matrices A, B, C and D can now be found by solving a set of over-determined equations in a least squares sense:

$$\begin{bmatrix} \tilde{X}_{i+1} \\ Y_{i|i} \end{bmatrix} = \begin{bmatrix} \hat{A} & \hat{B} \\ \hat{C} & \hat{D} \end{bmatrix} \begin{bmatrix} \tilde{X}_i \\ U_{i|i} \end{bmatrix} + \begin{bmatrix} \rho_w \\ \rho_v \end{bmatrix}, \tag{3.147}$$

where ρ_w and ρ_v are residual matrices. The estimates of the covariances of the process and measurement noise are obtained from the residuals ρ_w and ρ_v of (3.147) as:

$$\begin{bmatrix} \hat{Q}_i & \hat{S}_i \\ \hat{S}_i^t & \hat{R}_i \end{bmatrix} = \frac{1}{j} \begin{bmatrix} \rho_w \\ \rho_v \end{bmatrix} \begin{bmatrix} \rho_w^t & \rho_v^t \end{bmatrix}, \tag{3.148}$$

where i again indicates that the estimated covariances are biased, with an exponentially decreasing bias as $i \to \infty$. As in the stochastic identification algorithm, the

Q-matrices of the LQ factorization cancel out in the least-squares solution and the computation of the noise covariances. This implies that the Q-matrix of the LQ factorization should never be calculated explicitly. Note however, that corresponding columns of $\tilde{X}_i$ and of $\tilde{X}_{i+1}$ are state estimates of X_i and X_{i+1} respectively, obtained with different initial conditions. As a consequence, the set of relations (3.147) is not theoretically consistent, which means that the estimates of the system matrices are slightly biased. It can however be proven that the estimates of A, B, C and D are unbiased if at least one of the following conditions is satisfied:

- $i \to \infty$,
- the system is purely deterministic, that is, $v_k = w_k = 0, \forall k$,
- the deterministic input u_k is white noise.

If none of the above conditions is satisfied, one obtains biased estimates. However, there exist more involved algorithms that provide consistent estimates of A, B, C and D, even if none of the above conditions is satisfied, for which we refer to the literature.

3.4.9 Variations

Several variants on the algorithm that was explained above, exist. First, we note that the oblique projection O_i can be weighted left and right by user defined weighting matrices $W_1 \in \Re^{li \times li}$ and $W_2 \in \Re^{j \times j}$ respectively, which should satisfy the following conditions: W_1 should be of full rank and the rank of

$$\begin{bmatrix} U_p \\ Y_p \end{bmatrix} W_2$$

should be equal to the rank of

$$\begin{bmatrix} U_p \\ Y_p \end{bmatrix}.$$

Furthermore, one can distinguish between two classes of subspace identification algorithms. The first class uses the state estimates $\tilde{X}_i$ (the right singular vectors of $W_1 O_1 W_2$) to find the system matrices. The algorithm in Sect. 3.146 belongs to this class. The second class of algorithms uses the extended observability matrix Γ_i (the left singular vectors of $W_1 O_i W_2$) to first determine estimates of A and C and subsequently of B, D and Q, S, R.

Remark 3.3 It can be shown that three subspace algorithms that have been described in the literature (N4SID, MOESP and CVA) all start from $W_1 O_i W_2$ with for each of the algorithms a specific choice of weighting matrices W_l and W_2. The results are summarized in Table 3.1. From this table, it is clear that the algorithm described above is the N4SID algorithm ($W_1 = I_{li}$ and $W_2 = I_j$). The acronym N4SID stands for "Numerical algorithms for Subspace State Space System IDentification", MOESP for "Multivariable Output-Error State sPace" and CVA is the acronym of "Canonical Variate Analysis".

Table 3.1 Interpretations of different existing subspace identification algorithms

Acronym	W_1	W_2
N4SID	I_{li}	I_j
CVA	$(\lim_{j\to\infty}[(Y_f/U_f^{\perp})(Y_f/U_f^{\perp})^t])^{-1/2}$	$\Pi_{U_f^{\perp}}$
MOESP	I_{li}	$\Pi_{U_f^{\perp}}$

Remark 3.4 In Table 3.1, we give interpretations of different existing subspace identification algorithms in a unifying framework. All these algorithms first calculate an oblique projection O_i followed by an SVD of the weighted matrix $W_1 O_i W_2$. The first two algorithms, N4SID and CVA, use the state estimates $\tilde{X}_i$ (the right singular vectors) to find the system matrices, while MOESP is based on the extended observability matrix Γ_i (the left singular vectors). The matrix $U_f^{\perp}$ in the weights of CVA and MOESP represents the orthogonal complement of the row space of U_f.

3.5 Output-Error Parametric Model Identification

After studying this chapter, you will be able to

- describe the output-error model-estimation problem;
- parameterize the system matrices of a MIMO LTI state-space model of fixed and known order such that all stable models of that order are presented;
- formulate the estimation of the parameters of a given system parameterization as a nonlinear optimization problem;
- numerically solve a nonlinear optimization problem using gradient-type algorithms;
- evaluate the accuracy of the obtained parameter estimates via their asymptotic variance under the assumption that the signal-generating system belongs to the class of parameterized state space models; and
- describe two ways for dealing with a nonwhite noise acting on the output of an LTI system when estimating its parameters.

3.5.1 Introduction

Hereafter, we move another step forward in our exploration of how to retrieve information about linear time-invariant (LTI) systems from input and output measurements. The step forward is taken by analyzing how we can estimate (part of) the system matrices of the signal-generating model from acquired input and output data. We first tackle this problem as a complicated estimation problem by attempting to estimate both the state vector and the system matrices.

This section presents an introduction to estimating the parameters in a user-defined LTI model. In this chapter, we start with the determination of a model to approximate the deterministic relation between measurable input and output sequences. The uncertainties due to noises acting on the system are assumed to be lumped together as an additive perturbation at the output. Therefore, the estimation methods presented in this chapter are referred to as the *output-error methods*. In Chap. 8, we deal with the approximation of both the deterministic and the stochastic parts of the system's response, using an innovation model.

The reason for dealing with output-error methods for the analysis of estimating the parameters of a parametric model of an LTI system is twofold. First, in a number of applications, only the deterministic transfer from the measurable input to the output is of interest. An example is identification-based fault diagnosis, in which the estimated parameters of the deterministic part of the model are compared with their nominal "fault-free" values. Second, the restriction to the deterministic part simplifies the discussion and allows us to highlight how the estimation of parameters in an LTI model can be approached systematically. This systematic approach. which lies at the heart of many identification methods, is introduced in Sect. 3.5.2 and consists of the following four steps. The *first step* is parameterizing the model; that is, the selection of which parameters to estimate in the model. For MIMO LTI state-space models, some parameterizations and their properties are discussed in Sect. 3.5.3. *Step two* consists of formulating the estimation of the model parameters as an optimization problem. Section 3.5.4 presents such an optimization problem with the widely used least-squares cost function. *Step three* is the selection of a numerical procedure to solve the optimization problem iteratively. Methods for minimizing a least-squares cost function are presented in Sect. 3.5.5. The final step is evaluation of the accuracy of the obtained estimates via the covariance matrix of the estimates. This is discussed in Sect. 3.5.6. In these four steps, it is assumed that the additive error to the output is a zero-mean white noise. Section 3.5.7 discusses the treatment of colored additive noise.

3.5.2 Problems in Estimating Parameters

Consider the signal-generating LTI system to be identified, given by

$$y(k) = G(q)u(k) + v(k), \tag{3.149}$$

where $v(k)$ represents measurement noise that is statistically independent from the input $u(k)$. Then a general formulation of the output-error (OE) model-estimation problem is as follows.

Given a finite number of samples of the input signal $u(k)$ and the output signal $y(k)$ and the order of the following predictor,

$$\hat{x}(k+1) = A\hat{x}(k) + Bu(k), \tag{3.150}$$

$$\hat{y}(k) = C\hat{x}(k) + Du(k) \tag{3.151}$$

the goal is to estimate a set of system matrices A, B, C, and D in this predictor such that the output $y\hat{(k)}$ approximates the output of the system (3.149).

First, we consider the case in which $v(k)$ is a white-noise sequence. In Sect. 3.5.7, we then consider the more general case in which $v(k)$ is colored noise.

A common way to approach this problem is to assume that the entries of the system matrices depend on a parameter vector θ and to estimate this parameter vector. The parameterized predictor model based on the system (3.150)–(3.151) becomes

$$\hat{x}(k+1,\theta) = A(\theta)\hat{x}(k,\theta) + B(\theta)u(k), \tag{3.152}$$

$$\hat{y}(k,\theta) = C(\theta)\hat{x}(k,\theta) + D(\theta)u(k). \tag{3.153}$$

Note that the output data $\hat{y}(k,\theta)$ depends not only on the input and the parameters θ used to parameterize the system matrices $A(\theta)$, $B(\theta)$, $C(\theta)$, and $D(\theta)$, but also on the initial state $\hat{x}(0)$ of the model (3.152)–(3.153). Therefore, the initial state is often also regarded as a parameter and added to the parameter vector θ. The notation $\hat{x}(0,\theta)$ is used to denote the treatment of the initial state as a part of the parameter vector θ.

The problem of estimating the parameter vector θ can be divided into four parts.

1. This concerns the determination of a parameterization. A parameterization of the system (3.152)–(3.153) is the specification of the dependence of the system matrices on the parameter vector θ. One widely used approach to parameterize systems is to use unknown physical constants in a mathematical model derived from the laws of physics, such as Newton's or Kirchoff's laws. An example of such a parameterization is given below in identification example 3.1.
2. This concerns the selection of a criterion to judge the quality of a particular value of θ. In the foregoing sections, we consider a quadratic error criterion of the form

$$\frac{1}{N}\sum_{k=0}^{N-1}\left\|y(k) - \hat{y}(k,\theta)\right\|_2^2, \tag{3.154}$$

   with $\hat{y}(k,\theta)$ given by (7.4) and (7.5). For each particular value of the parameter vector θ, this criterion has a positive value. The optimality may therefore be expressed by selecting that parameter value that yields the minimal value of (3.154). Though such a strategy is a good starting point, a more detailed consideration is generally necessary in order to find the most appropriate model for a particular application.
3. This concerns the numerical minimization of the criterion (3.154). Let the "optimal" parameter vector $\hat{\theta}_N$ be the argument θ of the cost function (3.154) that minimizes this cost function; this is denoted by

$$\hat{\theta}_N = \arg\min \frac{1}{N}\sum_{k=0}^{N-1}\left\|y(k) - \hat{y}(k,\theta)\right\|_2^2. \tag{3.155}$$

   As indicated by (3.152)–(3.153), the prediction $\hat{y}(k,\theta)$ of the output is a filtered version of the input $u(k)$ only. A method that minimizes a criterion of the form (3.154), where $\hat{y}(k,\theta)$ is based on the input only, belongs to the class of output-error methods [54]. The Kalman filter discussed in Chap. 5 determines a prediction of the output by filtering both the input $u(k)$ and the output $y(k)$. A specific interpretation of the criterion (3.155) will be given in due course.

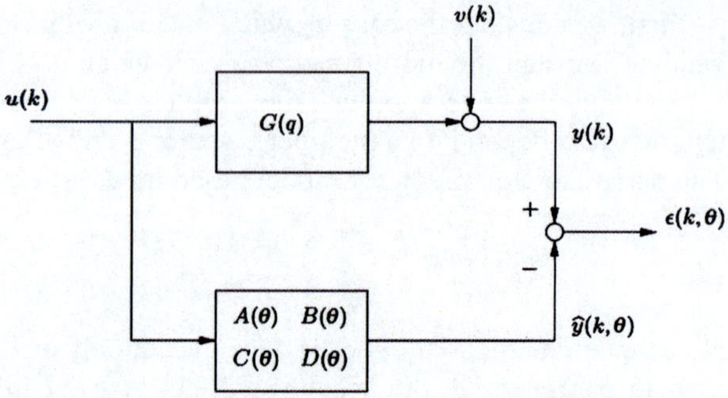

Fig. 3.11 The output-error model-estimation method

4. This concerns the analysis of the accuracy of the estimate θ_N. Since the measurements $y(k)$ are assumed to be stochastic processes, the derived parameter estimate $\hat{\theta}_N$ obtained via optimizing (3.154) will be a random variable. Therefore, a measure of its accuracy could be its bias and covariance.

The above four problems, which are analyzed in the listed order in Sects. 3.5.3–3.5.6, aim, loosely speaking, at determining the "best" predictor such that the difference between the measured and predicted output is made "as small as possible." The output-error approach is illustrated in Fig. 3.11.

3.5.3 Identification Example 3.1

The electrical-mechanical equations describing a permanent-magnet synchronous motor (PMSM) were derived in [72]. These equations are used to obtain a model of a PRISM and summarized below, Fig. 3.12 shows a schematic drawing of the PMSM, The magnet, marked with its north and south poles, is turning and along with it is the rotor reference frame indicated by the d-axis and q-axis. In the model, the following physical quantities are used:

- (i_d, i_q) are the currents and (v_d, v_q) are the voltages with respect to the rotor reference frame;
- α is the rotor position and ω its velocity;
- T_L represents the external load;
- N is the number of magnetic pole pairs in the motor;
- R is the phase resistance;
- L_d and L_q are the direct- and quadrature-axis inductances, respectively;
- ϕ_a is the permanent magnetic constant; and
- J is the rotor inertia.

On the basis of these definitions the physical equations describing a PMSM are [72]

$$\dot{i}_d = -\frac{R}{L_d} i_d + \frac{N\omega L_q}{L_d} i_q + \frac{1}{L_d} v_d, \tag{3.156}$$

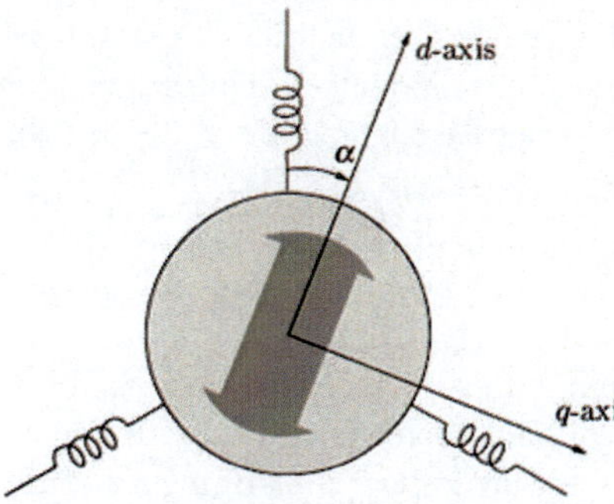

Fig. 3.12 A schematic representation of the permanent-magnet synchronous motor

$$\dot{i}_q = -\frac{R}{L_q} i_q - \frac{N\omega L_d}{L_q} i_d - \frac{N\phi_a}{L_q}\omega + \frac{1}{L_q} v_q, \tag{3.157}$$

$$\dot{\omega} = \frac{N\phi_a}{J} i_q - \frac{1}{J} T_L, \tag{3.158}$$

$$\dot{\alpha} = N\omega. \tag{3.159}$$

The state of this system equals $[i_d \;\; i_q \;\; \omega \;\; \alpha]^t$. The parameters that would allow us to simulate this state, given the (input) sequences T_L, v_d, and v_q, are

$$\{N, R, L_d, \phi_a, J\}.$$

Hence, a parameterization of the PMSM model (3.156)–(3.159) corresponds to the mapping from the parameter set $\{N, R, L_d, \phi_a, J\}$ to the model description (3.156)–(3.159). Note that a discrete-time model of the PMSM can be obtained by approximating the derivatives in (3.156)–(3.159) by finite differences.

In this chapter, we assume that the order of the LTI system, that is, the dimension of the state vector, is known. In practice, this is often not the case. Estimating the order from measurements is discussed in Chap. 8, together with some relevant issues that arise in the practical application of system identification.

3.5.4 Parameterizing a MIMO Model

Finding a model to relate input and output data sequences in the presence of measurement errors and with lack of knowledge about the physical phenomena that relate these data is a highly nonunique, nontrivial problem. To address this problem one specializes to specific *models*, *model sets*, and *parameterizations*. These notions are defined below for MIMO state-space models of finite order given by (3.152)–(3.153).

Let p be the dimension of the parameter vector θ. The set $\Omega \subset \mathfrak{R}^p$ that constrains the parameter vector, in order to guarantee that the parameterized models comply with prior knowledge about the system, such as the system's stability or the positiveness of its DC gain, is called the parameter set. By taking different values of θ from the set Ω, we get state-space models of the form (3.152)–(3.153) with different system matrices. A state-space *model set* is a collection or enumeration of state-space models of the form given by (3.152)–(3.153).

The transfer function of the nth-order system (3.152)–(3.153) is of the form

$$G(q,\theta) = D(\theta) + C(\theta)\big(qI_n - A(\theta)\big)^{-1} B(\theta). \tag{3.160}$$

Thus, for each particular value of θ we get a certain transfer function. From Sect. 3.4.4, we know that this transfer function is an $l \times m$ proper rational function with a degree of at most n. We use $\mathcal{R}_n^{l\times m}$ to denote the set of all $l \times m$ proper rational transfer functions with real coefficients and a degree of at most n.

A *parameterization* of the nth-order state-space model (3.152)–(3.153) is a mapping from the parameter set $\Omega \in \Re^p$ to the space of rational transfer functions $\mathcal{R}_n^{l\times m}$. This mapping is called the state-space *model structure* and is denoted by $\mathcal{M} : \Omega \to \mathcal{R}_n^{l\times m}$, thus $G(q,\theta) = \mathcal{M}(\theta)$. Since the structure of the transfer function is fixed and given by (7.12), the parameterization defined in this way is nothing but a prescription of how the elements of the system matrices A, B, C, and D are formed from the parameter vector θ.

Before we continue, we recall some properties of a mapping The map $f : X \to Y$ maps the set X onto the set Y. The set X is called the *domain* of f and Y is called the *range* of f. The map f is called *surjective* if for every $y \in Y$ there exists an $x \in X$ such that $f(x) = y$. In other words, to every point in its range there corresponds at least one point in its domain. It is important to realize that the surjective property of a map depends on the definitions of its domain X and its range Y. The map f is called *injective* if $f(x_1) = f(x_2)$ implies $x_1 = x_2$, that is, to every point in its range there corresponds at most one point in its domain. Finally, if the map f is both surjective and injective, it is called *bijective*.

Since a similarity transformation of the state vector does not alter the transfer function, not all parameterizations need to be injective. A parameterization that is not injective gives rise to a nonunique correspondence between the parameter vector and the transfer function. This is illustrated in the following example.

3.5.5 Identification Example 3.2

Consider the LTI system

$$x(k+1) = \begin{bmatrix} 1.5 & 1 \\ -0.7 & 0 \end{bmatrix} x(k) + \begin{bmatrix} 1 \\ 0.5 \end{bmatrix} u(k),$$
$$y(k) = \begin{bmatrix} 1 & 0 \end{bmatrix} x(k).$$

We parameterize this system using all the entries of the system matrices; this results in the following *parametric* model with $\theta \in \Re^8$:

$$\tilde{x}(k+1) = \begin{bmatrix} \theta(1) & \theta(2) \\ \theta(3) & \theta(4) \end{bmatrix} \tilde{x}(k) + \begin{bmatrix} \theta(5) \\ \theta(6) \end{bmatrix} u(k),$$
$$y(k) = \begin{bmatrix} \theta(7) & \theta(8) \end{bmatrix} \tilde{x}(k).$$

However, this parameterization is not injective. since we can find more than one parameter vector θ that results in the same transfer function between the input $u(k)$ and the output $y(k)$. For example, the following two values of the parameter vector θ give rise to the same transfer function:

$$\theta_1^t = \begin{bmatrix} 0 & -0.7 & 1 & 1.5 & 0.5 & 1 & 0 & 1 \end{bmatrix},$$
$$\theta_2^t = \begin{bmatrix} 2.9 & 6.8 & -0.7 & -1.4 & 0 & 0.5 & 1 & 2 \end{bmatrix}.$$

The reason for this nonuniqueness is that the transfer function from input to output remains unchanged when a similarity transformation is applied to the state vector $x(k)$. To obtain the parameter values θ_1, the following similarity transformation of the state vector was used:

$$x(k) = \begin{bmatrix} 0 & 1 \\ 1 & 0 \end{bmatrix} \tilde{x}(k);$$

and for θ_2 we made use of

$$x(k) = \begin{bmatrix} 1 & -2 \\ 0 & 1 \end{bmatrix} \tilde{x}(k).$$

To be able to identify uniquely a model from input and output data requires an injective parameterization. However, often the main objective is to find a state-space model that describes the input and output data, and uniqueness is not needed. In a system identification context, it is much more important that each transfer function with an order of at most n given by (3.160) can be represented by at least one point in the parameter space Ω. In other words, we need to have a parameterization with domain $\Omega \subset \Re^p$ and range $\mathcal{R}_n^{l\times m}$ that is surjective. An example of a surjective parameterization results on taking all entries of the system matrices A, B, C, and D as elements of the parameter vector θ as in identification example 3.1. This vector then has dimension p equal to

$$p = n^2 + n(l+m) + ml.$$

Since this number quickly grows with the state dimension n, alternative parameterization have been developed. For example, for multiple-input, single-output systems, the observable canonical form can be used; it is given by [54].

$$\hat{x}(k+1) = \begin{bmatrix} 0 & 0 & \cdots & 0 & -a_0 \\ 1 & 0 & \cdots & 0 & -a_1 \\ 0 & 1 & \cdots & 0 & -a_2 \\ \vdots & \vdots & \ddots & \vdots & \vdots \\ 0 & 0 & \cdots & 1 & -a_{n-1} \end{bmatrix} \hat{x}(k) + \begin{bmatrix} b_{11} & \cdots & b_{1m} \\ b_{21} & \cdots & b_{2m} \\ \vdots & \ddots & \vdots \\ b_{n1} & \cdots & b_{nm} \end{bmatrix} u(k), \quad (3.161)$$

$$\hat{y}(k) = \begin{bmatrix} 0 & 0 & 0 & \cdots & 1 \end{bmatrix} \hat{x}(k) + \begin{bmatrix} d_{11} & \cdots & d_{1m} \end{bmatrix} u(k). \quad (3.162)$$

The parameter vector (without incorporating the initial state) is given by

$$\theta^t = \begin{bmatrix} a_0 & \cdots & a_{n-1} & \cdots & b_{11} & \cdots & b_{nm} & d_{11} & \cdots & d_{1m} \end{bmatrix}.$$

The size of θ is

$$p = n + nm + m.$$

This parameterization $\mathcal{M} : \Omega \rightarrow \mathcal{R}^{1\times m}$ is surjective, the reason for this being that, although the observer canonical form is always observable, it can be not reachable. When it is not reachable, it is not minimal and the state dimension can be reduced, the order of the system becomes less than n. For a SISO transfer function it means that roots of the numerator polynomial (the zeros of the system) cancel out those of the denominator (the poles of the system). Different pole zero cancellations correspond to different parameter values that represent the same transfer function, hence the conclusion that the parameterization is surjective.

Apart from the size of the parameter vector θ and the surjective and/or injective property of the mapping $\mathcal{M}(\theta)$, the consequences of selecting a parameterization on the numerical calculations performed with the model need to be considered as well. Some examples of the numerical implications of a parameterization are the following.

1. In estimating the parameter vector θ by solving the optimization problem indicated in (3.155), it may be required that the mapping is differentiable, such that the Jacobian

$$\frac{\partial y(k,\theta)}{\partial\theta}$$

exists on a subset in $\Re^p$.

2. In case the mapping is surjective, the parameter optimization (3.155) may suffer from numerical problems due to the redundancy in the entries of the parameter vector. A way to avoid such numerical problems is *regularization* [57], which is discussed in Sect. 3.5.4.
3. Restrictions on the set of transfer functions $\mathcal{M}(\theta)$ need to be translated into constraints on the parameter set in $\Re^p$. For example, requiring asymptotic stability of the model leads to restrictions on the parameter set. In this respect it may be more difficult to impose such restrictions on one chosen parameterization than on another. Let Ω denote this constraint region in the parameter space, that is, $\Omega \subset \Re^p$; then we can formally denote the model set M as

$$M = \{\mathcal{M}(\theta) | \theta \in \Omega\}. \tag{3.163}$$

An example of constraining the parameter space is given below.

4. The numerical sensitivity of the model structure $\mathcal{M}(\theta)$ with respect to the parameter vector may vary dramatically between parameterizations. An example of numerical sensitivity is given later on in identification example 3.4.

3.5.6 *Identification Example 3.3*

Consider the transfer function

$$G(p) = \frac{q+2}{q^2 + a_1 q + a_0} \tag{3.164}$$

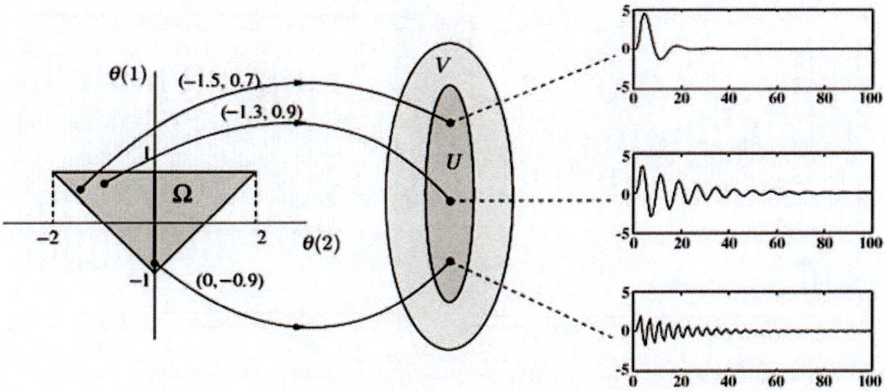

Fig. 3.13 Imposing stability on the second-order transfer function

parameterized by $\theta = [a_0, a_1]^t$. To impose stability on the transfer function $G(q)$, we need to find a set Ω such that $\theta \in \Omega$ results in a stable transfer function of the form (3.164). In other words, we need to determine a suitable domain for the mapping $\mathcal{M} : \Omega \to U$, with U the set of all stable transfer functions of the form (3.164). For this particular second-order example, the determination of the set Ω is not that difficult and is requested later on in the problems. The set Ω is mapped onto the set U of all stable second-order transfer functions of the form (3.164). The set V is the set of all stable second-order transfer functions. On the right are the impulse responses for the three indicated points in the parameter space Ω. Figure 3.13 shows the set Ω. Every point in the set Ω corresponds uniquely to a point in the set U, and thus the parameterization is injective. The parameterization is bijective with respect to the set U (with the particular choice of zeros in (3.164), no pole-zero cancellation can occur for stable poles), but not with respect to the set V that consists of all stable second-order transfer functions.

Note that Fig. 3.13 shows impulse responses of three systems that correspond to three different choices of the parameter θ from the set Ω. These impulse responses are quite different. which illustrates the richness of the set of systems described by Ω.

3.5.7 Identification Example 3.4

The system matrix A in the observer canonical form (3.161)–(3.162) is called a companion matrix [34]. A companion matrix is a numerically sensitive representation of the system dynamics; its eigenvalues are very sensitive to small changes in the coefficients $a_0, a_2, \ldots, a_{n-1}$.

We use the observer canonical form (3.161)–(3.162) to represent a system with transfer function

$$G(q) = \frac{1}{q^4 + a_3 q^3 + a_2 q^2 + a_1 q + a_0}.$$

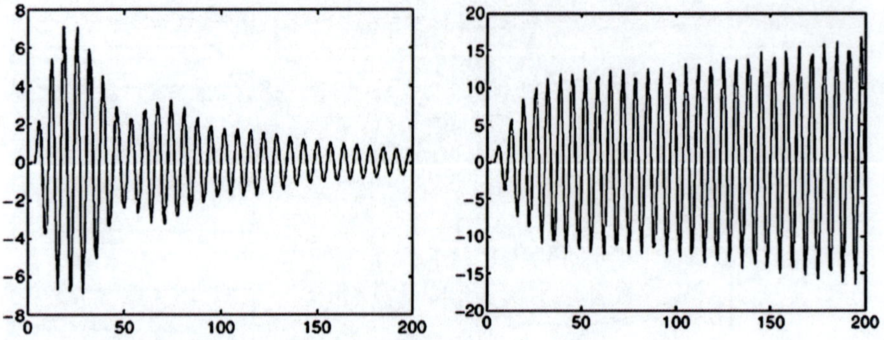

Fig. 3.14 Impulse responses of the stable (left) and the unstable system (right)

In this case, the parameter vector is equal to

$$\theta^t = \begin{bmatrix} a_0 & a_1 & a_2 & a_3 \end{bmatrix}.$$

If we take the parameter vector equal to

$$\theta^t = \begin{bmatrix} 0.915 & -2.1 & 3.11 & -2.2 \end{bmatrix},$$

the matrix A has two eigenvalues with a magnitude equal to 0.9889 up to four digits and two eigenvalues with a magnitude equal to 0.9673 up to four digits. Figure 3.14 shows the impulse response of the system $G(q)$ for this choice of θ.

If we change the parameter $\theta(3) = a_2$ into 3.12, the properties of the system become very different. For this slightly different choice of parameters, the matrix A has two eigenvalues with a magnitude equal to 1.0026 up to four digits and two eigenvalues with a magnitude equal to 0.9541 up to four digits. Hence, even only a small change in the parameter a_2 makes the system unstable. The impulse response of the system with $a_2 = 3.12$ is also shown in Fig. 3.14. We clearly see that the impulse response has changed dramatically. It should be remarked that, for systems of larger order, results similar to those illustrated in the example can be obtained with perturbations of magnitude the order of the machine precision of the computer.

In the following subsections, we present two particular parameterizations that are useful for system identification, namely the output normal form and the tridiagonal form.

3.5.8 The Output Normal Form

The output-normal form parameterization was first introduced for continuous-time state-space models by Hanzon and Ober [35, 36], and later extended for MIMO discrete-time state-space models [37, 38]. A big advantage of the output normal form is that the parameterized model is guaranteed to be asymptotically stable without the need for additional constraints on the parameter space. A definition of the output normal parameterization of the pair (A, C) in the case of a state-space model determined by the system matrices A, B, C, and D is as follows.

Definition 3.5 The output-normal-form parameterization of the pair (A, C) with $A \in \Re^{l \times n}$ is given as

$$\begin{bmatrix} C(\theta) \\ A(\theta) \end{bmatrix} = T_1\big(\theta(1)\big)T_2\big(\theta(1)\big)\cdots T_{nl}\big(\theta(nl)\big)\begin{bmatrix} 0 \\ I_n \end{bmatrix} \tag{3.165}$$

where $\theta \in \Re^{nl}$ is the parameter vector with entries in the interval $[-1, 1]$, and where the matrices $T_I(\theta(i))$ are based on the 2×2 matrix

$$U(\alpha) = \begin{bmatrix} -\alpha & \sqrt{1-\alpha^2} \\ \sqrt{1-\alpha^2} & \alpha \end{bmatrix}$$

with $\alpha \in \Re$ in the interval $[-1, 1]$; the matrices $T_i(\theta(i)) \in \Re^{(n+\ell)\times(n+\ell)}$ are given by

$$T_i\big(\theta(1)\big) = \begin{bmatrix} I_{n-1} & 0 & 0 \\ 0 & U(\theta(1)) & 0 \\ 0 & 0 & I_{l-1} \end{bmatrix},$$

$$\vdots$$

$$T_l\big(\theta(l)\big) = \begin{bmatrix} I_{n+l-2} & 0 \\ 0 & U(\theta(l)) \end{bmatrix},$$

$$T_{l+1}\big(\theta(l+1)\big) = \begin{bmatrix} I_{n-2} & 0 & 0 \\ 0 & U(\theta(l+1)) & 0 \\ 0 & 0 & 1 \end{bmatrix},$$

$$\vdots$$

$$T_{2l}\big(\theta(2\ell)\big) = \begin{bmatrix} I_{n+l-3} & 0 & 0 \\ 0 & U(\theta(2l)) & 0 \\ 0 & 0 & 1 \end{bmatrix},$$

$$\vdots$$

$$T_{(n-1)l+1}\big(\theta((n-1)\ell+1)\big) = \begin{bmatrix} U(\theta((n-1)l+1)) & 0 \\ 0 & I_{n+l-2} \end{bmatrix},$$

$$\vdots$$

$$T_{nl}\big(\theta(n\ell)\big) = \begin{bmatrix} I_{l-1} & 0 & 0 \\ 0 & U(\theta(nl)) & 0 \\ 0 & 0 & I_{n-1} \end{bmatrix}.$$

The next lemma shows that the parameterized pair of matrices in Definition 3.5 has the identity matrix as observability Grammian.

Lemma 3.6 *Let an asymptotically stable state-space model be given by*

$$x(k+1) = Ax(k) + Bu(k),$$
$$y(k) = Cx(k) + Du(k),$$

with the pair (A, C) *given by the output-normal-form parameterization* (3.165) *of Definition* 3.5, *then the observability Grammian* Q, *defined as the solution of*

$$A^t Q A + C^t C = Q,$$

is the identity matrix.

The proof follows from the fact that the matrices $U(\alpha)$ satisfy $U(\alpha)^t U(\alpha) = I_2$.
The output-normal-form parameterization of the pair (A, C) can be used to parameterize any stable state-space model, as shown in the following lemma.

Lemma 3.7 *Let an asymptotically stable and observable state-space model be given as*

$$\hat{x}(k+1) = A\hat{x}(k) + Bu(k), \tag{3.166}$$

$$\hat{y}(k) = C\hat{x}(k) + Du(k) \tag{3.167}$$

then a surjective parameterization is obtained by parameterizing the pair (A, C) *in the output normal form given in Definition* 3.5 *with the parameter vector* $\theta_{AC} \in \Re^{nl}$ *and parameterizing the pair of matrices* (B, D) *with the parameter vector* $\theta_{BD} \in \Re^{m(n+l)}$ *that contains all the entries of the matrices* B *and* D.

Proof The proof is constructive and consists of showing that any stable, observable state-space system of the form (3.166)–(3.167) can be transformed via a similarity transformation to the proposed parameterization.

Since A is asymptotically stable and since the pair (A, C) is observable, the solution Q to the Lyapunov equation

$$A^t Q A + C^t C = Q,$$

is positive-definite. Therefore, a Cholesky factorization can be carried out:

$$Q = T_q T_q^t.$$

The matrix $T_t = T_q^{-t}$ is the required similarity transformation. Note that T_t exists, because Q is positive-definite. The equivalent matrix pair $(T_t^{-1} A T_t, C T_t)$ then satisfies

$$A_t^t A_t + C_t^t C_t = I_n.$$

In other words, the columns of the matrix

$$\begin{bmatrix} C_t \\ A_t \end{bmatrix}$$

are orthogonal. To preserve this relationship under a second similarity transformation on the matrices A_t and C_t, this transformation needs to be orthogonal. As revealed by solving identification example 3.1, for any pair (A_t, C_t) there always exists an orthogonal similarity transformation T_h, such the pair $(T_h^{-1} A_t T_h, C_t T_h)$ is in the so-called observer Hessenberg form [82]. The observer Hessenberg form has a particular pattern of nonzero entries, which is illustrated below for the ease $n = 5$, $l = 2$

$$\left[\begin{array}{c} C_t T_h \\ \hline T_h^{-1} A_t T_h \end{array}\right] = \left[\begin{array}{c} C_h \\ \hline A_h \end{array}\right] = \left[\begin{array}{ccccc} \star & 0 & 0 & 0 & 0 \\ \star & \star & 0 & 0 & 0 \\ \hline \star & \star & \star & 0 & 0 \\ \star & \star & \star & \star & 0 \\ \star & \star & \star & \star & \star \\ \star & \star & \star & \star & \star \\ \star & \star & \star & \star & \star \end{array}\right],$$

with $\star$ denoting a possibly nonzero matrix entry.

The pair (A_h, C_h) in observer Hessenberg form can always be represented by a series of real numbers $\theta(i) \in G[-1, 1]$ for $i = 1, 2, \dots, nl$ that define an output-normal-form parameterization as in Definition 3.5. This is illustrated for the case $n = 2$ and $l = 2$. From (3.165) it follows that we need to show that the pair (A_h, C_h) satisfies

$$T_{nl}^t\big(\theta(nl)\big) \cdots T_2^t\big(\theta(2)\big) T_1^t\big(\theta(1)\big) \begin{bmatrix} C_h \\ A_h \end{bmatrix} = \begin{bmatrix} 0 \\ I_n \end{bmatrix}.$$

The first transformation, $T_1^t(\theta(1))$, is applied as

$$\begin{bmatrix} 1 & 0 & 0 \\ 0 & U^t(\theta(1)) & 0 \\ 0 & 0 & 1 \end{bmatrix} \left[\begin{array}{c} C_h \\ \hline A_h \end{array}\right] = \begin{bmatrix} 1 & 0 & 0 \\ 0 & U^t(\theta(1)) & 0 \\ 0 & 0 & 1 \end{bmatrix} \left[\begin{array}{cc} x_{11} & 0 \\ x_{21} & x_{22} \\ \hline x_{31} & x_{32} \\ x_{41} & x_{42} \end{array}\right]$$

$$= \left[\begin{array}{cc} x_{11} & 0 \\ x_{21}' & 0 \\ \hline x_{31}' & x_{32}' \\ x_{41} & x_{42} \end{array}\right],$$

with $U(\theta(1))$ such that

$$U^t\big(\theta(1)\big) \begin{bmatrix} x_{22} \\ x_{32} \end{bmatrix} = \begin{bmatrix} 0 \\ x_{32}' \end{bmatrix}$$

and primes denoting modified entries, The second transformation, $T_2^t(\theta(2))$, yields

$$\begin{bmatrix} I_2 & 0 \\ 0 & U^t(\theta(2)) \end{bmatrix} \left[\begin{array}{cc} x_{11} & 0 \\ x_{21}' & 0 \\ \hline x_{31}' & x_{32}' \\ x_{41} & x_{42} \end{array}\right] = \left[\begin{array}{cc} x_{11} & 0 \\ x_{21}' & 0 \\ \hline x_{31}'' & 0 \\ x_{41}'' & x_{42}'' \end{array}\right],$$

with double primes denoting modified entries. Since the matrices $U(\theta(1))$ and $U(\theta(2))$ are orthogonal, and the pair (A_h, C_h) satisfies $A_h^t A_h + C_h^t C_h = I_n$, we have

$$\begin{bmatrix} x_{11} & x'_{21} & x''_{31} & x''_{41} \\ 0 & 0 & 0 & x''_{42} \end{bmatrix} \begin{bmatrix} x_{11} & 0 \\ x'_{21} & 0 \\ \hline x''_{31} & 0 \\ x''_{41} & x''_{42} \end{bmatrix} = I_2.$$

This implies $x''_{41} = 0$ and $(x''_{42})^2 = 1$. The value of x''_{42} can thus be taken as -1 or 1; in the sequel, the positive value is used. We see that the rightmost column and bottom row of the transformed matrix are already in the correct form. Subsequently, the first column is transformed into the correct form by annihilating the entries x_{11} and x'_{21}. This is done using the orthogonal Givens rotations $U(\theta(3))$ and $U(\theta(4))$. We obtain

$$\begin{bmatrix} 1 & 0 & 0 \\ 0 & U^t(\theta(4)) & 0 \\ 0 & 0 & 1 \end{bmatrix} \begin{bmatrix} U^t(\theta(3)) & 0 \\ 0 & I_2 \end{bmatrix} \begin{bmatrix} I_2 & 0 \\ 0 & U^t(\theta(2)) \end{bmatrix}$$

$$\times \begin{bmatrix} 1 & 0 & 0 \\ 0 & U^t(\theta(1)) & 0 \\ 0 & 0 & 1 \end{bmatrix} \begin{bmatrix} x_{11} & 0 \\ x_{21} & x_{22} \\ \hline x_{31} & x_{32} \\ x_{41} & x_{42} \end{bmatrix} = \begin{bmatrix} 0 & 0 \\ 0 & 0 \\ 1 & 0 \\ 0 & 1 \end{bmatrix}.$$

To complete the parameterization of the state-space system (3.166)–(3.167), the matrices $(B_h, D) = (T_h^{-1} T_t^{-1} B, D)$ of the transformed state-space system are parameterized by all their entries. This completes the proof. □

The total number of parameters for the output normal parameterization of the state-space model (3.166)–(3.167) is

$$p = n\ell + nm + m\ell.$$

3.5.9 *Identification Example 3.5*

Consider a second-order state-space model with system matrices

$$A = \begin{bmatrix} 1.5 & -0.7 \\ 1 & 0 \end{bmatrix}, \qquad B = \begin{bmatrix} 1 \\ 0 \end{bmatrix}, \qquad C = \begin{bmatrix} 1 & 0.5 \end{bmatrix}, \qquad D = 0.$$

Since A is asymptotically stable and the pair (A, C) is observable, we can apply Lemma 7.2. We start by finding a similarity transformation T_t such that the transformed pair $(T_t^{-1} A T_t, C T_t) = (A_t, C_t)$ satisfies $A_t^t A_t + C_t^t C_t = I$. Since the pair (A, C) is observable and the system matrix A is asymptotically stable, the solution Q of the Lyapunov equation

$$A^t Q A + C^t C = Q$$

is positive-definite. Therefore, the matrix Q has a Cholesky factorization $T_q T_q^t$ that defines the necessary similarity transformation $T_t = T_q^{-t}$

$$T_q = \begin{bmatrix} 4.3451 & 0 \\ -2.6161 & 1.6302 \end{bmatrix}, \qquad T_t = \begin{bmatrix} 0.2301 & 0.3693 \\ 0 & 0.6134 \end{bmatrix}.$$

By applying the transformation T_t to the quartet of system matrices we obtain a similarly equivalent quartet. The pair (A_t, C_t) of this quartet reads

$$\begin{bmatrix} C_t \\ \hline A_t \end{bmatrix} = \begin{bmatrix} 0.2301 & 0.6760 \\ \hline 0.8979 & -0.4248 \\ 0.3752 & 0.6021 \end{bmatrix}.$$

This pair (A_t, C_t) already satisfies $A_t^t A_t + C_t^t C_t = I_2$. However, to obtain the factorization in (3.165), we have to perform a number of additional transformations. First, we perform an orthogonal similarity transformation T_h, such that

$$\begin{bmatrix} C_t \\ \hline C_t A_t \end{bmatrix} T_h$$

is lower triangular. This transformation can be derived from the Q factor of the RQ factorization of the matrix

$$\begin{bmatrix} C_t \\ \hline C_t A_t \end{bmatrix}.$$

It follows that

$$T_h = \begin{bmatrix} 0.3233 & 0.9466 \\ 0.9466 & -0.3233 \end{bmatrix}.$$

Applying the similarity transformation T_h yields the following transformed pair:

$$\begin{bmatrix} C_h \\ \hline A_h \end{bmatrix} = \begin{bmatrix} 0.7141 & 0 \\ 0.6176 & 0.4706 \\ -0.3294 & 0.8824 \end{bmatrix}.$$

To yield the factorization (3.165), we search for a transformation T_1 such that

$$T_1^t \begin{bmatrix} C_h \\ \hline A_h \end{bmatrix} = \begin{bmatrix} \star & 0 \\ \hline \star & 0 \\ 0 & 1 \end{bmatrix},$$

where the $\star$ indicate a number not of interest in this particular step. The required transformation T_1 is based on the rotation $U(\alpha)$ in Definition 3.5 that transforms the lower-right elements $[0.4706, 0.8823]^t$ into $[0, 1]^t$, and is given by

$$T_1 = \begin{bmatrix} 1 & 0 & 0 \\ 0 & -0.8824 & 0.4706 \\ 0 & 0.4706 & 0.8824 \end{bmatrix}.$$

Finally, the matrix T_2^t transforms the upper-left elements $[0.7141, -0.7]^t$ into $[0, 1]^t$ and again is based on a Givens rotation. The transformation T_2 is given by

$$\begin{bmatrix} 0.7 & 0.7141 & 0 \\ 0.7141 & -0.7 & 0 \\ 0 & 0 & 1 \end{bmatrix},$$

defining $\theta(2)$ equal to -0.7.

The parameter vector $\theta_{AC} = 0$ to parameterize the transformed pair (A, C) then equals

$$\begin{bmatrix} 0.8824 \\ -0.7 \end{bmatrix}.$$

To complete the parameterization in output normal form, the vector θ_{BD} is defined equal to

$$\theta_B D = \begin{bmatrix} T_h^{-1} T_t^{-1} \\ 0 \end{bmatrix} = \begin{bmatrix} 1.4003 \\ 4.1133 \\ 0 \end{bmatrix}.$$

3.5.10 The Tridiagonal Form

The tridiagonal parameterization exploits the numerical property that for every square matrix A there exists a (nonsingular) similarity transformation T, such that $T^{-1}AT$ is a tridiagonal matrix [34]. A tridiagonal matrix has nonzero entries only on the diagonal and one layer above and below the diagonal. An illustration of the form is given for $n = 4$:

$$A(\theta) = \begin{bmatrix} \theta(1) & \theta(2) & 0 & 0 \\ \theta(3) & \theta(4) & \theta(5) & 0 \\ 0 & \theta_6 & \sigma & \theta \\ 0 & 0 & \theta 9 & \theta(10) \end{bmatrix}.$$

To complete the parameterization of the LTI system (3.152)–(3.153), we add the entries of the matrices B, C, and D. The total number of parameters equals in this case

$$p = 32n - 2 + n(m + \ell) + m\ell,$$

which is an excess of $3n - 2$ parameters compared with the number of parameters required before. The surjective property of this parameterization requires that special care is taken during the numerical search for the parameter vector θ [57]. This special care is called *regularization* and will be discussed later on.

3.5.11 The Output-Error Cost Function

As stated earlier, in order to estimate a state-space model of the form (3.152)–(3.153) from input and output data we consider the quadratic cost function

$$J_N(\theta) = \frac{1}{N}\sum_{k=0}^{N-1} \| y(k) - \hat{y}(k,\theta)\|_2^2 \tag{3.168}$$

where $y(k)$ is the measured output signal, and $\hat{y}(k,\theta)$ is the output signal of the model (3.152)–(3.153). The cost function $J_N(\theta)$ is scalar-valued and depends on the parameter vector θ. In mathematical terms it is a *functional* [65]. Taking the constraints on the parameter vector θ into account, we denote the optimization problem as

$$\min_\theta J_N(\theta) \text{ subject to } \theta \in \Omega \subset \Re^p \text{ and (3.152)–(3.153)}. \tag{3.169}$$

Properties such as convexity of the functional $J_N(\theta)$ have a great influence on the numerical way of finding the optimum of (3.168). In general, we are able to find only a local minimum and finding the global minimum, when it exists, requires either special properties of $J_N(\theta)$ or an immense computational burden.

For state-space models, a more specific form of $J_N(\theta)$, including the effect of the initial state, is given in the following theorem.

Theorem 3.8 *For the state-space model* (3.152)–(3.153), *the functional* $J_N(\theta)$ *can be written as*

$$J_N(\theta_{AC},\theta_{BD}) = \frac{1}{N}\sum_{k=0}^{N-1} \| y(k) - \phi(k,\theta_{AC})\theta_{BD}\|_2^2 \tag{3.170}$$

with θ_{AC} *the parameters necessary to parameterize the pair* (A, C) *and*

$$\theta_{BD} = \begin{bmatrix} \hat{x}(0) \\ \text{vec}(B) \\ \text{vec}(D) \end{bmatrix}.$$

The matrix $\phi(k,\theta_{AC}) \in \Re^{\ell\times(n+m(\ell+n))}$ *is explicitly given as*

$$\phi(k,\theta_{AC}) = \left[C(\theta_{AC})^k, \sum_{\tau=0}^{k-1} u^\tau(\tau)\otimes C(\theta_{AC})A(\theta_{AC})^{k-1-\tau}, u^t(k)\otimes I_l \right].$$

Proof The parameterized state-space model (3.152)–(3.153) is given by

$$\begin{aligned} \hat{x}(k+1,\theta_{AC},\theta_{BD}) &= A(\theta_{AC})\hat{x}(k,\theta_{AC},\theta_{BD}) + B(\theta_{BD})u(k), \\ \hat{y}(k,\theta_{AC},\theta_{BD}) &= C(\theta_{AC})\hat{x}(k,\theta_{AC},\theta_{BD}) + D(\theta_{BD})u(k). \end{aligned}$$

The output of this state-space model can explicitly be written in terms of the input and the initial state $\hat{x}(0,\theta_{BD})$ as (see Sect. 3.4.2)

$$\hat{y}(k,\theta_{AC},\theta_{BD}) = C(\theta_{AC})A(\theta_{AC})^k\hat{x}(0,\theta_{BD}) + \sum_{\tau=0}^{k-1} C(\theta_{AC})A(\theta_{AC})^{k-1-\tau}B(\theta_{BD})u(\tau)D(\theta_{BD})u(k).$$

Application of the property that $\text{vec}(XYZ) = (Z^t \otimes X)\,\text{vec}(Y)$ and writing down the resulting equation for $k = 1, 2, \ldots, N$ completes the proof.

The parameter vector θ in the original state-space model (3.152)–(3.153) could be constructed by simply stacking the vectors θ_{AC} and θ_{BD} of Theorem 3.8 as

$$\theta = \begin{bmatrix} \theta_{AC} \\ \theta_{BD} \end{bmatrix}.$$

The output normal form presented in Lemma 3.7 will give rise to the formulation of the functional as expressed in Theorem 3.8. If the parameters θ_{AC} are fixed, the cost function (7.22) is linear in the parameters θ_{BD}. This fact can be exploited by applying the principle of separable least squares [33] in the search for the minimum of the cost function. Separable least squares first eliminates the parameters θ_{BD} from the cost function and searches for a minimum with respect to the parameters θ_{AC} only. Once the optimal value of the parameter vector θ_{AC} has been found, the parameter values θ_{BD}G are derived by simply solving a linear least-squares problem. The critical requirement is that there are no parameters in common between those contained in θ_{AC} and θ_{BD}. This is the case for the output normal form, defined in Sect. 3.5.2 but not for the tridiagonal form of Sect. 3.5.3. The application of separable least squares for the identification of LTI state-space models is dismissed by Bruls et al. [11] and Haverkamp [39]. □

The influence of the choice of the parameterization on the shape of the cost function $J_N(\theta)$. and therefore on the numerical optimization process (3.170), is illustrated in the example below.

3.5.12 Identification Example 3.6

Consider the state-space system from identification example 3.4. We demonstrate that the shape of the cost function $J_N(\theta)$ depends on the parameterization of the state-space system. We consider three cases.

- The system is converted into observer canonical form. For this particular system, we just have to switch the two states to arrive at

  $$A = \begin{bmatrix} 0 & -a_0 \\ 0 & -a_1 \end{bmatrix}, \qquad B = \begin{bmatrix} 0.5 \\ 1 \end{bmatrix}, \qquad C = \begin{bmatrix} 0 & 1 \end{bmatrix}$$

  where $a_0 = 0.7$ and $a_1 = -1.5$. We parameterize the system with the parameter vector $\theta = [a_0, a_1]^t$.

  Figure 3.15 shows how the cost function varies with the parameter vector θ. The minimum value of the cost function occurs for $\theta = [0.7, -1.5]$. This function is clearly nonlinear, it has several local minima.

Fig. 3.15 The cost $J_N(\theta)$ as a function of the parameters $\theta(1), \theta(2)$: case 1

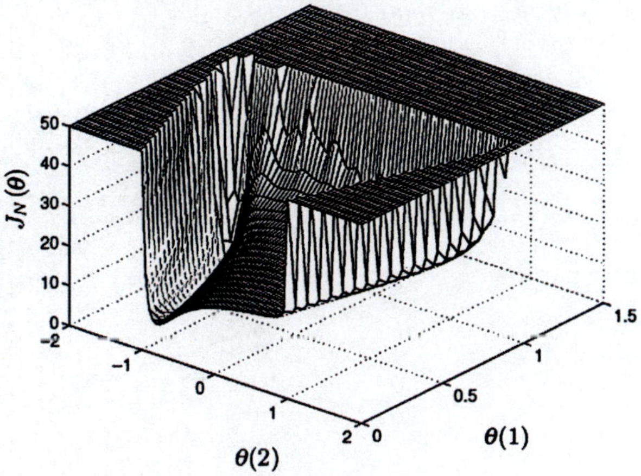

Fig. 3.16 The cost $J_N(\theta)$ as a function of the parameters $\theta(1), \theta(2)$: case 2

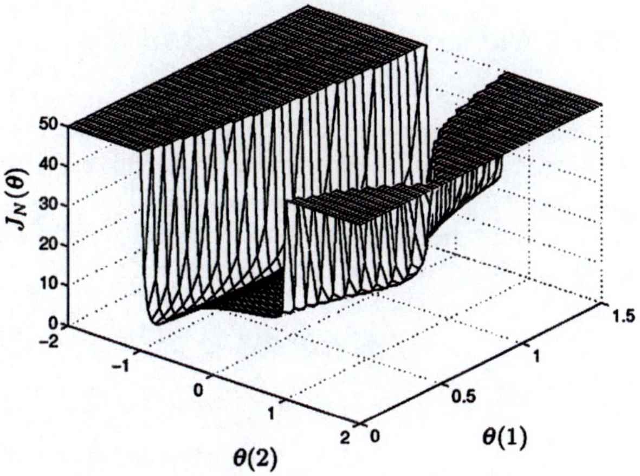

- We take again the observer canonical form, but now take the parameter vector θ equal to $[a_0/a_1, a_1]^t$. This means that we parameterize the A matrix as follows:

$$A = \begin{bmatrix} 0 & \theta(1)\theta(2) \\ 1 & -\theta(2) \end{bmatrix}.$$

Figure 3.16 shows how the cost function varies with the parameter vector θ. The minimum value of the cost function occurs for $\theta \approx [0.47, -1.5]$.
- The system is converted to the output normal form, as explained in identification example 3.5. We vary the two parameters that parameterize the matrices A and C. The minimum value of the cost function occurs for $\theta \approx [0.8824, -0.7]$. The cost function is displayed in Fig. 3.17. Again, we see that the cost function is nonlinear. Unlike in the previous cases, it always remains bounded, since with the output-normal parameterization the system can never become unstable. However, we still observe that the cost function is nonconvex.

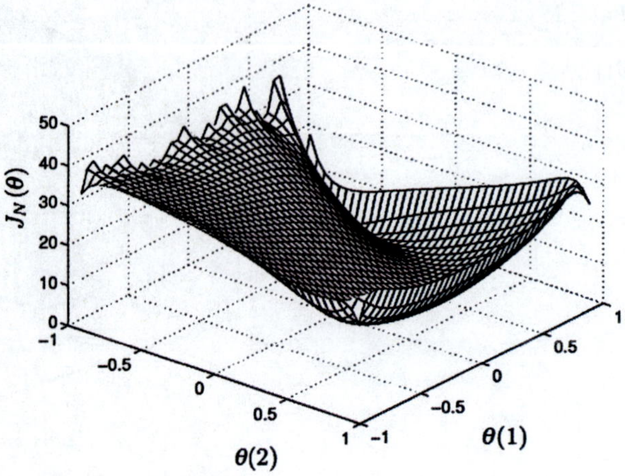

Fig. 3.17 The cost $J_N(\theta)$ as a function of the parameters $\theta(1)$, $\theta(2)$: case 3

3.5.13 Numerical Parameter Estimation

To determine a numerical solution to the parameter-optimization problem (3.170) of the previous section, the cost function $J_N(\theta)$ is expanded in a Taylor series around a given point $\theta^{(i)}$ in the parameter space Ω. This point $\theta^{(i)}$ may be the starting point of the optimization process or an intermediate estimate obtained during the search for the minimum of $J_N(\theta)$. The Taylor-series expansion is given by

$$\begin{aligned} J_N(\theta) = {} & J_N\big(\theta^{(i)}\big) + \big(J'_N\big(\theta^{(i)}\big)\big)^t\big(\theta - \theta^{(i)}\big) \\ & + \frac{1}{2}\big(\theta - \theta^{(i)}\big)^t J''_N\big(\theta^{(i)}\big)\big(\theta - \theta^{(i)}\big) \\ & + \text{higher-order terms,} \end{aligned}$$

where $J'_N(\theta^{(i)})$ is Jacobian and $J''_N(\theta^{(i)})$ the Hessian of the functional $J_N(\theta)$ at $\theta^{(i)}$, given by

$$J'_N(\theta) = \frac{\partial J_N(\theta)}{\partial \theta} = \begin{bmatrix} \frac{\partial J_N(\theta)}{\partial \theta(1)} \\ \frac{\partial J_N(\theta)}{\partial \theta(2)} \\ \vdots \\ \frac{\partial J_N(\theta)}{\partial \theta(p)} \end{bmatrix},$$

$$J''_N(\theta) = \frac{\partial^2 J_N(\theta)}{\partial \theta \partial \theta^t} = \begin{bmatrix} \frac{\partial J_N(\theta)}{\partial \theta(1)\partial \theta(1)} & \frac{\partial J_N(\theta)}{\partial \theta(1)\partial \theta(2)} & \cdots & \frac{\partial J_N(\theta)}{\partial \theta(1)\partial \theta(p)} \\ \frac{\partial J_N(\theta)}{\partial \theta(2)\partial \theta(1)} & \frac{\partial J_N(\theta)}{\partial \theta(2)\partial \theta(2)} & \cdots & \frac{\partial J_N(\theta)}{\partial \theta(2)\partial \theta(p)} \\ \vdots & \vdots & \ddots & \vdots \\ \frac{\partial J_N(\theta)}{\partial \theta(p)\partial \theta(1)} & \frac{\partial J_N(\theta)}{\partial \theta(p)\partial \theta(2)} & \cdots & \frac{\partial J_N(\theta)}{\partial \theta(p)\partial \theta(p)} \end{bmatrix}.$$

We approximate $J_N(\theta)$ as

$$J_N(\theta) \approx J_N(\theta^{(i)}) + \big(J'_N(\theta^{(i)})\big)^t(\theta - \theta^{(i)}) \\ + \frac{1}{2}(\theta - \theta^{(i)})^t J''_N(\theta^{(i)})(\theta - \theta^{(i)}). \tag{3.171}$$

The necessary condition for minimizing this approximation of $J_N(\theta)$ becomes

$$J'_N(\theta^{(i)}) + J''_N(\theta^{(i)})(\theta - \theta^{(i)}) = 0.$$

Therefore, provided that the Hessian at $\theta^{(i)}$ is invertible, we can update the parameter vector $\theta^{(i)}$ to θ by the update equation

$$\theta = \theta^{(i)} - J''_N(\theta^{(i)})^{-1} J'_N(\theta^{(i)}). \tag{3.172}$$

This type of parameter update is called the Newton method. To arrive at explicit expressions for $J'_N(\theta)$ and $J''_N(\theta)$, we introduce the error vector

$$E_N(\theta) = \begin{bmatrix} \epsilon(0,\theta) \\ \epsilon(1,\theta) \\ \vdots \\ \epsilon(N-1,\theta) \end{bmatrix},$$

with $\epsilon(k,\theta) = y(k) - \bar{y}(k,\theta)$. We can denote the cost function $J_N(\theta)$ as

$$J_N(\theta) = \frac{1}{N}\sum_{k=0}^{N-1} \|y(k) - \hat{y}(k,\theta)\|_2^2 \\ = \frac{1}{N} E_N^t(\theta) E_N(\theta). \tag{3.173}$$

Using the calculus of differentiating functionals outlined in [10], and using the notation

$$\Psi_N(\theta) = \frac{\partial E_N(\theta)}{\partial \theta^t} \tag{3.174}$$

the Jacobian and Hessian of $J_N(\theta)$ can be expressed as

$$J'_N(\theta) = \frac{1}{N}\frac{\partial E_N^t(\theta)}{\partial \theta} E_N(\theta) + \frac{1}{N}\big(I_p \otimes E_N^t(\theta)\big)\frac{\partial E_N(\theta)}{\partial \theta} \\ = \frac{2}{N}\frac{\partial E_N^t(\theta)}{\partial \theta} E_N(\theta) \\ = \frac{2}{N}\left(\frac{\partial E_N(\theta)}{\partial \theta^t}\right)^t E_N(\theta) \\ = \frac{2}{N}\Psi_N^t(\theta) E_N(\theta), \tag{3.175}$$

$$J''_N(\theta) = \frac{2}{N}\frac{\partial^2 E_N^t(\theta)}{\partial \theta^t \partial \theta}\big(I_p \otimes E_N(\theta)\big) + \frac{2}{N}\frac{\partial E_N^t(\theta)}{\partial \theta}\frac{\partial E_N(\theta)}{\partial \theta^t} \\ = \frac{2}{N}\frac{\partial^2 E_N^t(\theta)}{\partial \theta^t \partial \theta}\big(I_p \otimes E_N(\theta)\big) + \frac{2}{N}\left(\frac{\partial E_N(\theta)}{\partial \theta^t}\right)^t \frac{\partial E_N(\theta)}{\partial \theta^t} \\ = \frac{2}{N}\frac{\partial^2 E_N^t(\theta)}{\partial \theta^t \partial \theta}\big(I_p \otimes E_N(\theta)\big) + \frac{2}{N}\Psi_N^t(\theta)\Psi_N(\theta). \tag{3.176}$$

3.5.14 The Gauss–Newton Method

The Gauss–Newton method consists of approximating the Hessian $J''_N(\theta^{(i)})$ by the matrix $H_N(\theta^{(i)})$:

$$H_N(\theta^{(i)} = \frac{2}{N}\Psi^t_N(\theta)\Psi_N(\theta).$$

Such an approximation of the Hessian holds in the neighborhood of the optimum where the second derivative of the error and the error itself are weakly correlated. In that case, the first term of (3.176) can be neglected. This results in considerable computational savings. When the matrix $H_N(\theta^{(i)})$ is invertible, we can write the parameter update equation for the Gauss–Newton method as

$$\theta^{(i+1)} = \theta^{(i)} - H_N\big(\theta^{(i)}\big)^{-1} J'_N\big(\theta^{(i)}\big). \tag{3.177}$$

A different way to derive this update equation is by using a Taylor-series expansion on $E_N(\theta)$ in the neighborhood of $\theta^{(i)}$ as follows:

$$\begin{aligned} J_N\big(\theta^{(i)} + \delta\theta^{(i)}\big) &= \frac{1}{N}\big\| E_N\big(\theta^{(i)} + \delta\theta\big)\big\|_2^2 \\ &\approx \frac{1}{N}\big\| E_N\big(\theta^{(i)}\big) + \Psi_N\big(\theta^{(i)}\big)\delta\theta^{(i)}\big\|_2^2 \end{aligned} \tag{3.178}$$

where $\Psi^t_N(\theta)$ is given by (3.174). The parameter update $\delta\theta^{(i)} = \theta^{(i+1)} - \theta^{(i)}$ follows on solving the ‘following linear least-squares problem:

$$\min_{\delta\theta^{(i)}} \big\| E_N\big(\theta^{(i)}\big) + \Psi_N\big(\theta^{(i)}\big)\delta\theta^{(i)}\big\|_2^2,$$

and we get

$$\begin{aligned} \theta^{(i+1)} &= \theta^{(i)} - \big(\Psi_N\big(\theta^{(i)}\big)^t \Psi_N\big(\theta^{(i)}\big)\big)^{-1} \Psi_N\big(\theta^{(i)}\big)^t E_N\big(\theta^{(i)}\big) \\ &= \theta^{(i)} - H_N\big(\theta^{(i)}\big)^{-1} J'_N\big(\theta^{(i)}\big) \end{aligned} \tag{3.179}$$

which equals (3.177).

According to (3.177), at every iteration we need to calculate the approximate Hessian $H_N(\theta^{(i)})$ and the Jacobian $J'_N(\theta^{(i)})$. To ease the computational burden, it is important to have an efficient way of calculating these quantities. Note that (3.175) and (3.176) show that in fact we need calculate only $E_N(\theta)$ and $\Psi_N(\theta)$. To compute $E_N(\theta)$, we need to compute $\hat{y}(k,\theta)$ for $k = 1, 2, \dots, N$. This can be done efficiently by simulating the following system:

$$\hat{x}(k+1,\theta) = A(\theta)\hat{x}(k,\theta) + B(\theta)u(k), \tag{3.180}$$

$$\hat{y}(k,\theta) = C(\theta)\hat{x}(k,\theta) + D(\theta)u(k). \tag{3.181}$$

This will also yield the signal $\hat{x}(k,\theta)$ which we need to compute $\Psi_N(\theta)$, as explained below. Note that $\Psi_N(\theta)$ is given by

$$\Psi_N(\theta) = \begin{bmatrix} \frac{\partial \epsilon(0,\theta)}{\partial^t} \\ \frac{\partial \epsilon(1,\theta)}{\partial^t} \\ \vdots \\ \frac{\partial \epsilon(N-1,\theta)}{\partial^t} \end{bmatrix} = \begin{bmatrix} & \frac{\partial \hat{y}(0,\theta)}{\partial \theta^t} \\ & \frac{\partial \hat{y}(1,\theta)}{\partial \theta^t} \\ - & \vdots \\ & \frac{\partial \hat{y}(N-1,\theta)}{\partial \theta^t} \end{bmatrix},$$

and that

$$\frac{\partial \hat{y}(k,\theta)}{\partial \theta^t} = \begin{bmatrix} \frac{\partial \hat{y}(k,\theta)}{\partial \theta(1)} & \frac{\partial \hat{y}(k,\theta)}{\partial \theta(2)} & \cdots & \frac{\partial \hat{y}(k,\theta)}{\partial \theta(p)} \end{bmatrix},$$

where $\theta(i)$ denotes the ith entry of the vector θ. It is easy to see that for every parameter $\theta(i)$ we have

$$\begin{aligned} \frac{\partial \hat{x}(k+1,\theta)}{\partial \theta(i)} &= A(\theta)\frac{\partial \hat{x}(k,\theta)}{\theta(i)} + \frac{\partial A(\theta)}{\partial \theta(i)}\hat{x}(k,\theta) + \frac{\partial B(\theta)}{\partial \theta(i)}u(k), \\ \frac{\partial \hat{y}(k,\theta)}{\partial \theta(i)} &= C(\theta)\frac{\partial \hat{x}(k,\theta)}{\theta(i)} + \frac{\partial C(\theta)}{\partial \theta(i)}\hat{x}(k,\theta) + \frac{\partial D(\theta)}{\partial \theta(i)}u(k). \end{aligned}$$

On taking $X_i(k,\theta) = \partial \hat{x}(k,\theta)/\partial \theta(i)$, this becomes

$$X_i(k+1,\theta) = A(\theta)X_i(k,\theta) + \frac{\partial A(\theta)}{\partial \theta(i)}\hat{x}(k,\theta) + \frac{\partial B(\theta)}{\partial \theta(i)}u(k), \tag{3.182}$$

$$\frac{\partial \hat{y}(k,\theta)}{\partial \theta(i)} = C(\theta)X_i(k,\theta) + \frac{\partial C(\theta)}{\partial \theta(i)}\hat{x}(k,\theta) + \frac{\partial D(\theta)}{\partial \theta(i)}u(k). \tag{3.183}$$

The previous two equations show that the derivative of $\hat{y}(k,\theta)$ with respect to $\theta(i)$ can be obtained by simulating a linear system with state $X_i(k,\theta)$ and inputs $\hat{x}(k,\theta)$ and $u(k)$. Note that the matrices

$$\frac{\partial A(\theta)}{\partial \theta(i)}, \quad \frac{\partial B(\theta)}{\partial \theta(i)}, \quad \frac{\partial C(\theta)}{\partial \theta(i)}, \quad \frac{\partial D(\theta)}{\partial \theta(i)}$$

are fixed and depend only on the particular parameterization that is used to describe the system. We conclude that the calculation of $\Psi_N(\theta)$ boils down to simulating a linear system for every element of the parameter vector θ. Therefore, if θ contains p parameters, we need to simulate $p+1$ linear systems in order to compute both $E_N(\theta)$ and $\Psi_N(\theta)$.

3.5.15 Identification Example 3.7

Let the model output be given by $\hat{y}(k,\theta) = \phi(k)^t\theta$, with $y(k) \in \Re$ and $\phi(k) \in \Re^p$; then the cost function $J_N(\theta)$ is

$$J_N(\theta) = \frac{1}{N}\sum_{k=0}^{N-1}\left(y(k) - \phi(k)^t\theta\right)^2 \tag{3.184}$$

and the vector $E_N(\theta)$ is

$$E_N(\theta) = \begin{bmatrix} y(0) - \phi(0)^t\theta \\ y(1) - \phi(1)^t\theta \\ \vdots \\ y(N-1) - \phi(N-1)^t\theta \end{bmatrix}.$$

Let $\phi_i(j)$ denote the ith entry of the vector $\phi(j)$, then

$$\frac{\partial E_N^t(\theta)}{\partial\theta(i)} = -\begin{bmatrix} -\phi_i(0) & \phi_i(1) & \cdots & \phi_i(N-1) \end{bmatrix}.$$

Hence,

$$\Psi_N(\theta)^t = -\begin{bmatrix} \phi(0) & \phi(1) & \cdots & \phi(N-1) \end{bmatrix},$$

$$\begin{aligned} \left[\frac{\partial E_N(\theta)}{\partial\theta^t}\right]^t E_N(\theta) &= -\begin{bmatrix} \phi(0) & \phi(1) & \cdots & \phi(N-1) \end{bmatrix} \\ &\quad \times \begin{bmatrix} y(0) - \phi(0)^t\theta \\ y(1) - \phi(1)^t\theta \\ \vdots \\ y(N-1) - \phi(N-1)^t\theta \end{bmatrix} \\ &= -\Phi_N^t(Y_N - \Phi_N\theta), \end{aligned}$$

with

$$\begin{aligned} Y_N &= \begin{bmatrix} y(0) & y(1) & \cdots & y(N-1) \end{bmatrix}^t, \\ \Phi_N &= \begin{bmatrix} \phi(0) & \phi(1) & \cdots & \phi(N-1) \end{bmatrix}^t. \end{aligned}$$

Assuming that the matrix $\Phi_N^t\Phi_N/N$ is invertible, we can write the parameter update (3.179) as

$$\begin{aligned} \theta^{(i+1)} &= \theta^{(i)} + \left(\frac{1}{N}\Phi_N^t\Phi_N\right)^{-1}\frac{1}{N}\Phi_N^t\left(Y_N - \Phi_N\theta^{(i)}\right) \\ &= \left(\frac{1}{N}\Phi_N^t\Phi_N\right)^{-1}\frac{1}{N}\Phi_N^t Y_N. \end{aligned} \tag{3.185}$$

The assumed invertibility condition depends on the vector time sequence $\phi(k)$. A systematic framework has been developed to relate this invertibility condition to the notion of persistency of excitation of the time sequence [54].

The updated parameter vector $\theta^{(i+1)}$ becomes independent from the initial one $\theta^{(i)}$. Therefore, the iterative parameter-update rule (3.185) can be stopped after one iteration (one cycle) and the estimate becomes

$$\hat{\theta}_N = \left(\frac{1}{N}\Phi_N^t\Phi_N\right)^{-1}\frac{1}{N}\Phi_N^t Y_N. \tag{3.186}$$

The underlying reason for this is that the functional (3.185) is quadratic in θ. The latter is a consequence of the model output $\phi(k)^t\theta$ being *linear* in the unknown parameter vector θ.

Note that the derived solution of the quadratic cost function (3.186) equals the one obtained by solving the normal equations for a linear least-squares problem, see Sect. 2.6.

3.5.16 Regularization in the Gauss–Newton Method

The matrix $H_N(\theta^{(i)}$ used in the Gauss–Newton update equation (3.177) to approximate the Hessian may be singular. This will, for example, be the ease when the parameterization selected is non-injective; different sets of parameters yield the same value of the cost function $J_N(\theta)$ and thus the θ that minimizes $J_N(\theta)$ no longer need be unique. One possible means of rescue to cope with this singularity is via regularization, which leads to a numerically more attractive variant of the Gauss–Newton method. In regularization, a penalty term is added to the cost function to overcome the nonuniqueness of the minimizing θ. Instead of just minimizing $J_N(\theta)$, the minimization problem becomes

$$\min_{\theta} J_N(\theta) + \lambda\|\theta\|_2^2.$$

The real number λ is positive and has to be selected by the user. Using the same approximation of the cost function $J_N(\theta)$ as in (3.178), the regularized Gauss–Newton update can be derived as

$$\theta^{(i+1)} = \theta^{(i)} - \left(H_N\left(\theta^{(i)}\right) + \lambda I_p\right)^{-1} J_N'\left(\theta^{(i)}\right).$$

By adding λI_p to $H_N(\theta^{(i)})$, the matrix $H_N(\theta^{(i)}) + \lambda I_p$ is made nonsingular for $\lambda > 0$. However, the selection of the regularization parameter λ is far from trivial. A systematic approach that is widely used is known as the *Levenberg–Marquardt* method [61].

3.5.17 The Steepest Descent Method

The steepest-descent method does not compute or approximate the Hessian, it just changes the parameters into the direction of the largest decrease of the cost function. This direction is, of course, given by the Jacobian. Hence, the steepest-descent algorithm updates the parameters as follows:

$$\theta^{(i+1)}(\mu) = \theta^{(i)} - \mu J_N'\left(\theta^{(i)}\right) \tag{3.187}$$

where an additional step size $\mu \in [0, 1]$ is introduced. This step size is usually determined via the additional scalar optimization problem,

$$\hat{\theta}^{(i+1)} = \arg\min_{\mu\in[0,1]} J_N\left(\theta^{(i+1)}(\mu)\right).$$

In general, the iteration process of the steepest-descent algorithm has a lower convergence speed than that of the iteration in the Gauss–Newton method. However, the steepest-descent algorithm results in considerable computational savings in each individual iteration step. This is due to the fact that, to compute $J'_N(\theta)$, we compute the product $\Psi_N^t(\theta)E_N(\theta)$ directly, without computing $\Psi_N(\theta)$ separately. This requires only two simulations of an nth-order system, as explained below. Recall that

$$\Psi_N^t(\theta)E_N(\theta) = \sum_{k=0}^{N-1}\left(-\frac{\partial\hat{y}(k,\theta)}{\partial\theta}\right)^t \epsilon(k,\theta).$$

Using (3.183), we can write the right-hand side as

$$\begin{aligned}\sum_{k=0}^{N-1}\left(\frac{\partial\hat{y}(k,\theta)}{\partial\theta(i)}\right)^t \epsilon(k,\theta) &= \sum_{k=0}^{N-1} X_i(k,\theta)^C(\theta)^t\epsilon(k,\theta)\\ &\quad+\sum_{k=0}^{N-1}\hat{x}(k,\theta)^t\left(\frac{\partial C(\theta)}{\theta(i)}\right)^t\epsilon(k,\theta)\\ &\quad+\sum_{k=0}^{N-1}u(k)^t\left(\frac{\partial D(\theta)}{\partial\theta(i)}\right)^t\epsilon(k,\theta).\end{aligned}$$

To obtain $\hat{x}(k,\theta)$, one simulation of the state equation (3.180) is required. From the discussion in Sect. 3.5.1, it follows that, to compute $X_i(k,\theta)$, the p systems defined by (3.182) and (3.183) need to be simulated. However, for the steepest-descent method $X_i(k,\theta)$ is not needed; only the sum

$$\sum_{k=0}^{N-1} X_i(k,\theta)^t C(\theta)^t\epsilon(k,\theta)$$

is needed. This sum can be computed by just one (backward) simulation of the system

$$\bar{X}(k-1,\theta) = A(\theta)^t\bar{X}(k,\theta) + C(\theta)^t\epsilon(k,\theta) \tag{3.188}$$

involving the *adjoint* state $\bar{X}(k,\theta)$, because

$$\sum_{k=0}^{N-1} X_i(k,\theta)^t C(\theta)^t\epsilon(k,\theta) = \sum_{k=0}^{N-1} W_i(k,\theta)^t\bar{X}(k,\theta) \tag{3.189}$$

where

$$W_i(k,\theta) = \frac{\partial A(\theta)}{\partial\theta(i)}\hat{x}(k,\theta) + \frac{\partial B(\theta)}{\partial\theta(i)}u(k).$$

The equality (3.189) can be derived by writing (3.182) as

$$X_i(k+1,\theta) = A(\theta)X_i(k,\theta) + W_i(k,\theta).$$

Taking $X_i(0,\theta) = 0$, we can write

$$\begin{bmatrix} X_i(0,\theta) \\ X_i(1,\theta) \\ X_i(2,\theta) \\ \vdots \\ X_i(N-1,\theta) \end{bmatrix} = \begin{bmatrix} 0 & 0 & \cdots & \cdots & 0 \\ I_n & 0 & \cdots & \cdots & 0 \\ A & I_n & 0 & \cdots & 0 \\ \vdots & \vdots & \vdots & \ddots & \vdots \\ A^{N-2} & \cdots & A & I_n & I_n \end{bmatrix} \begin{bmatrix} W_i(0,\theta) \\ W_i(1,\theta) \\ W_i(2,\theta) \\ \vdots \\ W_i(N-1,\theta) \end{bmatrix}. \quad (3.190)$$

For the adjoint state $\bar{X}(N-1,\theta) = 0$, we have

$$\begin{bmatrix} \bar{X}_i(0,\theta) \\ \bar{X}_i(1,\theta) \\ \bar{X}_i(2,\theta) \\ \vdots \\ \bar{X}(N-1,\theta) \end{bmatrix} = \begin{bmatrix} 0 & I_n & A^t & \cdots & (A^t)^{N-2} \\ 0 & 0 & I_n & \cdots & \vdots \\ \vdots & \vdots & 0 & \ddots & \vdots \\ \vdots & \vdots & \vdots & \ddots & I_n \\ 0 & 0 & 0 & \cdots & 0 \end{bmatrix} \begin{bmatrix} C(\theta)^t \epsilon(0,\theta) \\ C(\theta)^t \epsilon(1,\theta) \\ C(\theta)^t \epsilon(2,\theta) \\ \vdots \\ C(\theta)^t \epsilon(N-1,\theta) \end{bmatrix}. \quad (3.191)$$

On combining (3.190) and (3.191), it is easy to see that (3.189) holds. We can conclude that only *two simulations* of an nth-order system are required for the steepest-descent method, instead of $p+1$ simulations.

3.5.18 Gradient Projection

When a chosen parameterization is non-injective, the Hessian needs to be regularized as discussed in Sect. 3.5.2. For the special case when the surjective parameterization consists of taking all entries of the system matrices A, B, C, and D, the singularity of the Hessian due to similarity transformations of the state-space system can be dealt with in another way. This parameterization that has all the entries of the system matrices in the parameter vector θ is called the *full parameterization*.

Consider the system given by the matrices $\bar{A}$, $\bar{B}$, $\bar{C}$, and $\bar{D}$ obtained by applying a similarity transformation $T \in \Re^{n\times n}$ to the matrices A, B, C, and D as

$$\begin{bmatrix} \bar{A} & \bar{B} \\ \bar{C} & \bar{D} \end{bmatrix} = \begin{bmatrix} T^{-1}AT & T^{-1}B \\ CT & D \end{bmatrix}. \quad (3.192)$$

The system given by $\bar{A}$, $\bar{B}$, $\bar{C}$, and $\bar{D}$ has the same transfer function, and thus the same input–output behavior, as the system defined by A, B, C, and D.

By taking all possible nonsingular similarity transformations T, we obtain a set of systems that have the same input–output behavior, and can thus not be distinguished on the basis of input and output data. This set of similar systems forms a manifold M in the parameter space θ, as pictured schematically in Fig. 3.18. By changing the parameters along the manifold M, we do not change the input–output behavior of the system and we therefore do not change the value of the cost function $J_N(\theta)$.

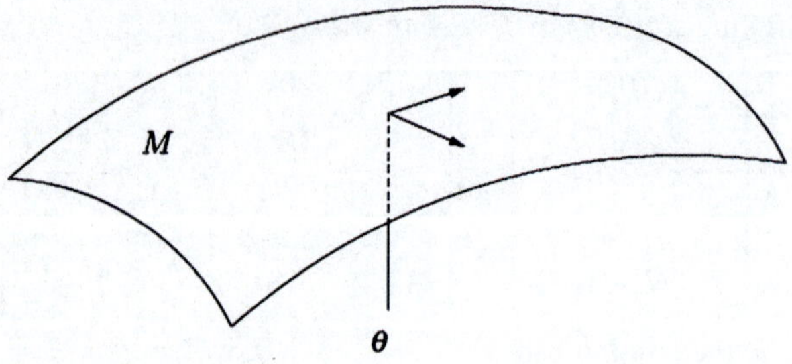

Fig. 3.18 A schematic representation of the manifold M of similar systems

To avoid problems with the numerical parameter update in minimizing $J_N(\theta)$, we should avoid modifying the parameters such that they stay on this manifold. This idea has been put forward by McKelvey and Helmersson [59] and by Lee and Poolla [51]. At a certain point θ on the manifold M we can determine the tangent plane (see Fig. 3.18). The tangent plane contains the directions in the parameter space along which an update of the parameters does not change the cost function $J_N(\theta)$. The tangent plane of the manifold is determined by considering similar systems for small perturbations of the similarity transformation around the identity matrix, that is $T = I_n + \Delta T$. A first-order approximation of similarly equivalent systems is then given by

$$\begin{bmatrix} \bar{A} & \bar{B} \\ \bar{C} & \bar{D} \end{bmatrix} = \begin{bmatrix} T^{-1}AT & T^{-1}B \\ CT & D \end{bmatrix} \approx \begin{bmatrix} A & B \\ C & D \end{bmatrix} + \begin{bmatrix} A\Delta T - \Delta T A & -\Delta T B \\ C\Delta T & 0 \end{bmatrix}. \tag{3.193}$$

If the entries of the system matrices are stacked in column vectors as

$$\theta = \begin{bmatrix} \mathrm{vec}(A) \\ \mathrm{vec}(B) \\ \mathrm{vec}(C) \\ \mathrm{vec}(D) \end{bmatrix}, \qquad \hat{\theta} = \begin{bmatrix} \mathrm{vec}(\bar{A}) \\ \mathrm{vec}(\bar{B}) \\ \mathrm{vec}(\bar{C}) \\ \mathrm{vec}(\bar{D}) \end{bmatrix}$$

applying the vec operator to (3.193) and using the relation $\mathrm{vec}(XYZ) = (Z^t \otimes X)\,\mathrm{vec}(Y)$ (see the Appendix) shows that the parameters of the similar systems are related as

$$\bar{\theta} = \theta + Q(\theta)\,\mathrm{vec}(\Delta T) \tag{3.194}$$

with the matrix $Q(\theta)$ defined by

$$Q(\theta) = \begin{bmatrix} I_n \otimes A - A^t \otimes I_n \\ -B^t \otimes I_n \\ I_n \otimes \\ 0 \end{bmatrix}.$$

The matrix Q depends on θ, since θ contains the entries of the system matrices A, B, C, and D. Note that (3.194) shows that the columns of the matrix $Q(\theta)$ span the tangent plane at the point θ on the manifold of similar systems. If we update the parameters θ along the directions of the orthogonal complement of the matrix

$Q(\theta)$, we will avoid the criterion that we do not change the cost function $J_N(\theta)$. The orthogonal complement of $Q(\theta)$ follows from an SVD of the matrix $Q(\theta)$:

$$Q(\theta) = \begin{bmatrix} U(\theta) & U_\perp(\theta) \end{bmatrix} \begin{bmatrix} \Sigma(\theta) & 0 \\ 0 & 0 \end{bmatrix} \begin{bmatrix} V(\theta)^t \\ V(\theta)^t \end{bmatrix},$$

with $\Sigma(\theta) > 0$ and $U_\perp(\theta) \in \Re^{p \times p-r}$, with $p = n^2 + n(l+m) + lm$ and $r = \text{rank}(Q(\theta))$. The columns of the matrix $U(\theta)$ form a basis for the column space of $Q(\theta)$; the columns of the matrix $U_\perp(\theta)$ form a basis for the orthogonal complement of the column space of $Q(\theta)$. The matrices $U(\theta)$ and $U_\perp(\theta)$ can be used to decompose the parameter vector θ into two components:

$$\theta = U(\theta)U(\theta)^t\theta + U_\perp(\theta)U_\perp(\theta)^t\theta \tag{3.195}$$

where the first component corresponds to directions that do not influence the cost function (the column space of Q) and the second component to the directions that change the value of the cost function (the orthogonal complement of the column space of Q).

In solving the optimization problem (3.169), the parameters θ are updated according to the rule

$$\theta^{(i+1)} = \theta^{(i)} + \delta\theta^{(i)},$$

where $\delta\theta^{(i)}$ is the update. For the steepest-descent method (3.187), this update equals $\delta\theta^{(i)} = -\mu J_N'(\theta^{(i)})$. Preventing an update of the parameters in directions that do not change the cost function is achieved by decomposing $\delta\theta^{(i)}$ similarly to in (3.195) and discarding the first component. On the basis of this observation, the parameter update of the steepest-descent method (3.187) becomes

$$\theta^{(i+1)} = \theta^{(i)} - \mu U_\perp(\theta^{(i)})U_\perp(\theta^{(i)})^t J_M'(\theta^{(i)}),$$

and the update of the Gauss–Newton method (3.177), which is implemented by imposing an update in the direction of the range space of $U_\perp(\theta^{(i)})$ only, is given by

$$\begin{aligned} \theta^{(i+1)} = \theta^{(i)} &- \mu U_\perp(\theta^{(i)})\big(U_\perp(\theta^{(i)})^t H_N(\theta^{(i)})U_\perp(\theta^{(i)})\big)^{-1} \\ &\times U_\perp(\theta^{(i)})^t J_N'(\theta^{(i)}). \end{aligned}$$

This insight can be obtained by solving problem 3.10.

3.5.19 Analyzing the Accuracy of the Estimates

The result of the numerical optimization procedure described in the previous section is

$$\hat{\theta}_N = \arg\min \frac{1}{N}\sum_{k=0}^{N-1} \|y(k) - \hat{y}(k,\theta)\|_2^2.$$

A possible way to characterize the accuracy of the estimate $\hat{\theta}_N$ is via an expression for its mean and covariance matrix. In this section, we derive this covariance matrix for the case that the system to be identified belongs to the model class. This means that $G(q)$ of the system

$$y(k) = G(q)u(k) + v(k)$$

belongs to the parameterized model set $\mathcal{M}(\theta)$.

The Gauss–Newton optimization method approximates the cost function as in (3.171). This approximation holds exactly in the special case of a model output that is linear in the parameters as treated in identification example 3.7. Therefore, we study the asymptotic variance first for the special case when $J_N(\theta)$ is given by

$$J_N(\theta) = \frac{1}{N}\sum_{k=0}^{N-1}\bigl(y(k) - \theta(k)^t\theta\bigr)^2. \tag{3.196}$$

We assume that the system is in the model class, thus the measured output $y(k)$ is assumed to be generated by the system

$$y(k) = \theta(k)^t\theta_0 + e(k) \tag{3.197}$$

where θ_0 are the true parameter values, and $e(k)$ is a zero-mean white-noise sequence with variance σ_e^2 that is statistically independent from $\phi(k)$.

Expanding the cost function (3.196) and using the expression for $y(k)$ yields

$$\begin{aligned}
J_N(\theta) = {} & \frac{1}{N}\sum_{k=0}^{N-1} e(k)^2 + \frac{1}{N}\sum_{k=0}^{N-1} e(k)\phi(k)(\theta_0 - \theta) \\
& + \frac{1}{N}\sum_{k=0}^{N-1}(\theta_0 - \theta)^t\phi(k)\phi(k)^t(\theta_0 - \theta),
\end{aligned}$$

which is exactly the right-hand side of (3.171). The parameter vector $\hat{\theta}_N$ that minimizes this criterion for $\mu = 1$ was derived in identification example 3.7 and equals

$$\begin{aligned}
\hat{\theta}_N = {} & \left(\frac{1}{N}\Phi_N^t\Phi_N\right)^{-1}\frac{1}{N}\Phi_N Y_N \\
& \times \left[\frac{1}{N}\sum_{k=0}^{N-1}\phi(k)\phi(k)^t\right]^{-1}\left[\frac{1}{N}\sum_{k=0}^{N-1}\phi(k)y(k)\right].
\end{aligned}$$

Again using the expression for $y(k)$, we get

$$\hat{\theta}_N - \theta_0 = \left[\frac{1}{N}\sum_{k=0}^{N-1}\phi(k)\phi(k)^t\right]^{-1}\left[\frac{1}{N}\sum_{k=0}^{N-1}\phi(k)e(k)\right].$$

Since $e(k)$ and $\phi(k)$ are independent, $E[\hat{\theta}_N - \theta_0] = 0$ and thus the estimated parameters $\hat{\theta}_N$ are unbiased. The covariance matrix of $\hat{\theta}_N - \theta_0$ equals

$$\mathbb{E}\bigl[[\hat{\theta}_N - \theta_0][\hat{\theta}_N - \theta_0]^t\bigr]$$

$$
\begin{aligned}
&= \left[\frac{1}{N}\sum_{k=0}^{N-1}\phi(k)\phi(k)^t\right]^{-1} E\left[\frac{1}{N^2}\sum_{k=0}^{N-1}\phi(k)e(k)\sum_{j=0}^{N-1}\phi(j)^t e(j)\right] \\
&\quad \times \left[\frac{1}{N}\sum_{k=0}^{N-1}\phi(k)\phi(k)^t\right]^{-1} \\
&= \left[\frac{1}{N}\sum_{k=0}^{N-1}\phi(k)\phi(k)^t\right]^{-1} \frac{1}{N^2}\sum_{k=0}^{N-1}\phi(k)\phi(k)^t\sigma_e^2 \times \left[\frac{1}{N}\sum_{k=0}^{N-1}\phi(k)\phi(k)^t\right]^{-1} \\
&= \frac{\sigma_e^2}{N}\left[\frac{1}{N}\sum_{k=0}^{N-1}\phi(k)\phi(k)^t\right]^{-1},
\end{aligned}
$$

when the matrix

$$
\left[\frac{1}{N}\sum_{k=0}^{N-1}\phi(k)\phi(k)^t\right]^{-1}
$$

converges to a constant bounded matrix Σ_ϕ, the last equation shows that the covariance matrix of $\hat{\theta}_N$ goes to zero asymptotically (as $N \to \infty$). In this case the estimate is called *consistent*. The fact that $y(k)$ is given by (3.197) indicates that the system used in optimizing (3.196) is in the model set. In this case the output-error method is able to find the unbiased and consistent estimates of the parameter vector θ.

Now, we take a look at the more general case in which the cost function is given by

$$
J_N(\theta_{AC}, \theta_{BD}) = \frac{1}{N}\sum_{k=0}^{N-1}\left\| y(k) - \phi(k, \theta_{AC})\theta_{BD}\right\|_2^2,
$$

as in Theorem 3.8. We assume again that the system to be identified is in the model class; that is, the system to be identified can be described by the parameters $\theta_{AC,0}$, and $\theta_{BD,0}$ such that the measured output satisfies

$$
y(k) = \phi(k, \theta_{AC,0})\theta_{BD,0} + e(k),
$$

where $e(k)$ is a zero-mean white-noise sequence with variance σ_e^2 that is statistically independent from $\phi(k, \theta_{AC})$. Denoting the true parameters by

$$
\theta_0 = \begin{bmatrix} \theta_{AC,0} \\ \theta_{BD,0} \end{bmatrix},
$$

and the estimated parameters obtained from the output-error method by $\hat{\theta}_N$, it can again be shown that $E[\hat{\theta}_N - \theta_0]^t = 0$ [54] and thus the estimated parameters $\hat{\theta}_N$ are unbiased. The covariance matrix of this unbiased estimate is [54]

$$
\mathbb{E}[\hat{\theta}_N - \theta_0][\hat{\theta}_N - \theta_0]^t = \frac{\sigma_e^2}{N}\left[J''(\theta_0)\right]^{-1}
$$

and it can be approximated as

$$
\mathbb{E}[\hat{\theta}_N]. \tag{3.198}
$$

The approximation of the covariance matrix of the estimated parameters holds only asymptotically in N. This needs to be taken into account when using the approximation to describe the model error.

3.5.20 Dealing with Colored Measurement Noise

At the beginning of this chapter, we considered the signal model

$$y(k) = G(q)u(k) + v(k) \tag{3.199}$$

where $v(k)$ is a white-noise sequence. In this section, we investigate the more general case in which $v(k)$ is nonwhite noise. Consider the cost function

$$\begin{aligned} J_N\theta &= \frac{1}{N}\sum_{k=0}^{N-1}\|y(k) - \hat{y}(k,\theta)\|_2^2 \\ &= \frac{1}{N}\sum_{k=0}^{N-1}\|\epsilon(k,\theta)\|_2^2 \\ &= \frac{1}{N}E_N^t E_N. \end{aligned} \tag{3.200}$$

If v_k is a white-noise sequence, the residual vector $\epsilon(k,\theta)$ will also be a white-noise sequence if the following two conditions are satisfied:

1. The transfer function $G(q)$ of (3.199) belongs to the parameterized model set $\mathcal{M}(\theta)$; and
2. The estimate $\hat{\theta}$ is the global minimizing argument of (3.200) in the limit $N \to \infty$.

In this case, all temporal information has been modeled; there is no correlation between different samples of error $\epsilon(k,\theta)$. If the output measurements are perturbed by colored noise, the error $\epsilon(k,\theta)$ can never become a white-noise sequence. The consequence is that, although the estimated parameter θ can still be unbiased, it no longer has minimum variance. This is illustrated in the following example.

3.5.21 Identification Example 3.8

Consider the quadratic cost function of Example 3.7 given by

$$J_N(\theta) = \frac{1}{N}\sum_{k=0}^{N-1}\left(y(k) - \phi(k)^{\theta}\right)^2. \tag{3.201}$$

We assume that the system is in the model class, thus the measured output $y(k)$ is assumed to be generated by the system

$$y(k) = \phi(k)^t\theta_0 + v(k) \tag{3.202}$$

where θ_0 are the true parameter values, and $v(k)$ is a zero-mean random sequence that is statistically independent from $\phi(k)$.

Adopting the notation of identification example 3.7, we can write the minimization of $J_N(\theta)$ as the least-squares problem

$$\min_{\theta} V_N^t V N \quad \text{subject to } Y_N = \Phi_N \theta + V_N \tag{3.203}$$

where

$$V_N = \begin{bmatrix} v(0) & v(1) & \cdots & v(N-1) \end{bmatrix}^t.$$

We learn before that, to obtain a minimum-variance estimate of θ, we have to solve the weighted least-squares problem

$$\min_{\theta} E_N^t E_N \quad \text{subject to } Y_N = \Phi_N \theta + L E_N,$$

where $E(E_N E_N^t) = I_N$. On comparing this with (3.203), we see that, to obtain a minimum-variance estimate, we need to have $LE_N = V_N$ with $L = \Sigma_v^{1/2}$ such that $E(V_N V_N^t) = \Sigma_v$. If no information about $v(k)$ is available, this is not possible. It follows that simply setting $L = I$ will lead to a minimum-variance estimate only if $v(k)$ is white noise; for colored noise $v(k)$ the minimum variance is obtained for $L = \Sigma_v^{1/2}$.

3.5.22 Weighted Least Squares

One way to obtain a minimum-variance parameter estimate when the additive noise $v(k)$ at the output in (3.150) is nonwhite requires that we know its covariance matrix. Let the required covariance matrix be denoted by Σ_v and equal to

$$\Sigma_v = E \begin{bmatrix} v(0) \\ v(1) \\ \vdots \\ v(N-1) \end{bmatrix} \begin{bmatrix} v(0) & v(1) & \cdots & v(N-1) \end{bmatrix}.$$

Then, if we assume that $\Sigma_v > 0$, we adapt the cost function (3.173) to the following weighted least-squares sum:

$$J_N(\theta, \Sigma_v) = \frac{1}{N} E_N^t \Sigma_v^{-1} E_N = \frac{1}{N} \bigl(\Sigma_v^{-T/2} E_N\bigr)^t \bigl(\Sigma_v^{-T/2} E_N\bigr). \tag{3.204}$$

The numerical methods outlined in Sect. 3.5 can be adapted in a straightforward manner by replacing E_N by $\Sigma_v^{-T/2} E_N$ and Ψ_N by $\Sigma_v^{-T/2} \Psi_N$.

In general, the covariance matrix is a full $Nl \times Nl$ matrix, and, therefore, for large N its formation and inversion requires a prohibitive amount of memory. However, recent work by David [20] provides a way to circumvent this problem, by employing an analytic and sparse expression for the inverse covariance matrix based on the

Gohberg-Heinig inversion theorem. This sparsity can be taken into account to derive computationally efficient methods [8].

A practical procedure for applying the weighting discussed above is the following.

1. Minimize the output-error cost function (3.200) and compute the corresponding residual vector E_N for the optimum.
2. Use the residual vector from the previous step to estimate a multivariable AR model of the noise, and use that model to compute the Cholesky factor of the inverse covariance matrix as described by David [20].
3. Minimize the weighted cost function (3.204).

After step 3, again the residual vector E_N can be computed, and steps 2 and 3 can be repeated. This can be done several times, but in our experience two iterations are usually sufficient, which corresponds to the observations made by David and Bastin [21].

3.5.23 Prediction-Error Methods

Another way to improve the accuracy of the estimates of a parametric model of $G(q)$ in (3.199) when the perturbation $v(k)$ is nonwhite noise consists of incorporating a model of this noise into the estimation procedure. We assume that $v(k)$ can be described by a filtered white-noise sequence $e(k)$, such that

$$y(k) = G(q)u(k) + H(q)e(k).$$

Prediction-error methods (PEM) aim at finding parameters of a model that models both of the transfer functions $G(q)$ and $H(q)$. Making use of the Kalman-filter theory of Sect. 3.5, the above transfer-function model can be described together with the following innovation state-space model:

$$\begin{aligned} \hat{x}(k+1) &= A\hat{x}(k) + Bu(k) + Ke(k), \\ y(k) &= C\hat{x}(k) + Du(k) + e(k), \end{aligned}$$

where $e(k)$ is a white-noise sequence. Note that, in general, the dimension of the state vector can be larger than the order n of the transfer function $G(q)$, to incorporate the dynamics of $H(q)$; the dimension equals n only in the special case in which $G(q)$ and $H(q)$ have the same system poles.

We recall the one-step-ahead predictor of the innovation representation,

$$\begin{aligned} \hat{x}(k+1|k) &= (A - KC)x(k|k-1) + (B - KD)u(k) + Ky(k), \\ \hat{y}(k|k-1) &= Cx(k|k-1) + Du(k). \end{aligned}$$

If we can parameterize this predictor by the parameter vector, we are able to use a number of the instruments outlined in this chapter to estimate these parameters by means of minimizing a cost function based on the one-step-ahead prediction error

$$J_N(\theta) = \frac{1}{N} \sum_{k=0}^{N-1} \| y(k) - \hat{y}(k|k-1, \theta) \|_2^2.$$

The resulting prediction-error methods are widely used and so important that we will devote the next chapter to them.

3.6 Prediction-Error Parametric Model Estimation

In this section, we are going to

- describe the prediction-error model-estimation problem;
- parameterize the system matrices of a Kalman filter of fixed and known order such that all stable MIMO Kalman filters of that order are presented;
- formulate the estimation of the parameters of a given Kalman filter parameterization via the solution of a nonlinear optimization problem;
- evaluate qualitatively the bias in parameter estimation for specific SISO parametric models, such as ARX, ARMAX, output-error, and Box–Jenkins models, under the assumption that the signal-generating system does not belong to the class of parameterized Kalman filters; and
- describe the problems that may occur in parameter estimation, when using data generated in closed-loop operation of the signal-generating system.

3.6.1 Introduction

This section continues the discussion started in Sect. 3.5, on estimating the parameters in an LTI state-space model. It addresses the determination of a model of both the deterministic and the stochastic part of an LTI model.

The objective is to determine, from a finite number of measurements of the input and output sequences, a one-step-ahead predictor given by the stationary Kalman filter without using knowledge of the system and covariance matrices of the stochastic disturbances. In fact, these system and covariance matrices (or alternatively the Kalman gain) need to be estimated from the input and output measurements. The restriction imposed on the derivation of a Kalman filter from the data is the assumption of a stationary one-step-ahead predictor of a known order. The estimation of a Kalman filter from input and output data is of interest in problems where predictions of the output or the state of the system into the future are needed. Such predictions are necessary in model-based control methodologies such as predictive control [18, 31, 70]. Predictions can be made from state-space models or from transfer function models. The estimation problems related to both model classes are treated in this chapter.

We start in Sect. 3.6.2 with the estimation of the parameters in a state-space model of the one-step-ahead predictor given by a stationary Kalman filter. As in Sect. 3.5, we address the four steps of the systematic approach to estimating the parameters in a state-space model, but now for the case in which this state-space model is a Kalman filter. Although the output-error model can be considered as a special case of the Kalman filter, it will be shown that a lot of insight about parameterizations, numerical optimization, and analysis of the accuracy of the estimates acquired in Sect. 3.5 can be reused here.

In Sect. 3.6.3 specific and widely used SISO transfer-function models, such as ARMAX, ARX, output-error, and Box–Jenkins, are introduced as special parameterizations of the innovation state-space model. This relationship with the Kalman-filter theory is used to derive the one-step-ahead predictors for each of these specific classical transfer-function models.

When the signal-generating system does not belong to the class of parameterized models, the predicted output has a *systematic error* or bias even when the number of observations goes to infinity. Section 3.6.4 presents, for several specific SISO parameterizations of the Kalman filter given in Sect. 3.6.3, a qualitative analysis of this bias. A typical example of a case in which the signal-generating system does not belong to the model class is when the signal-generating system is of higher order than the parameterized model. The bias analysis presented here is based on the work of Ljung [53] and Wahlberg and Ljung [92].

We conclude this chapter in Sect. 3.6.5 by illustrating points of caution when using output-error or prediction-error methods with input and output measurements recorded in a feedback experiment. Such closed-loop data experiments in general require additional algorithmic operations to get consistent estimates, compared with the case in which the data are recorded in open-loop mode. The characteristics of a number of situations advocate the need to conduct parameter estimation with data acquired in a feedback experiment. An example is the identification of an F-16 fighter aircraft that is unstable without a feedback control system. In addition to this imposed need for closed-loop system identification, it has been shown that models identified with closed-loop data may result in improved feedback controller designs [22, 32, 73]. The dominant plant dynamics in closed-loop mode are more relevant to designing an improved controller than the open-loop dynamics are.

3.6.2 Prediction-Error Methods

In Sect. 3.5.7, we briefly introduced prediction-error methods. When the output of an LTI system is disturbed by additive colored measurement noise, the estimates of the parameters describing the system obtained by an output-error method do not have minimum variance. The second alternative presented in that section as a means by which to obtain minimum-variance estimates was the use of prediction-error methods.

The signal-generating system that is considered in this chapter represents the colored-noise perturbation as a filtered white-noise sequence. Thus, the input–output data to be used for identification are assumed to be generated in the following way:

$$y(k) = G(q)u(k) + H(q)e(k) \tag{3.205}$$

where $e(k)$ is a zero-mean white-noise sequence that is statistically independent from $u(k)$, and $G(q)$ represents the deterministic part and $H(q)$ the stochastic part of the system. If we assume a set of input–output data sequences on a finite time interval, then a general formulation of the prediction-error model-estimation problem is as follows.

Given a finite number of samples of the input signal $u(k)$ and the output signal $y(k)$, and the order of the predictor

$$\hat{x}(k+1) = A\hat{x}(k) + Bu(k) + K\big(y(k) - C\hat{x}(k) - Du(k)\big), \tag{3.206}$$

$$\hat{y}(k) = C\hat{x}(k) + Du(k) \tag{3.207}$$

the goal is to estimate the system matrices A, B, C, D, and K in this predictor such that the output $\hat{y}(k)$ approximates the output of (3.205).

Recall that the postulated model (3.206)–(3.207) represents a stationary Kalman filter. If we assume that the entries of the system matrices of this filter depend on the parameter vector θ, then we can define the underlying innovation model as

$$\hat{x}(k+1|k,\theta) = A(\theta)\hat{x}(k|k-1,\theta) + B(\theta)u(k) + K(\theta)\epsilon(k), \tag{3.208}$$

$$y(k) = C(\theta)\hat{x}(k|k-1,\theta) + D(\theta)u(k) + \epsilon(k). \tag{3.209}$$

If we denote this innovation model by means of transfer functions, then, in analogy with the signal-generating system (3.205), we get the following parameterizations of the deterministic and stochastic part:

$$G(q,\theta) = D(\theta) + C(\theta)\big(qI - A(\theta)\big)^{-1}B(\theta),$$

$$H(q,\theta) = I + C(\theta)\big(qI - A(\theta)\big)^{-1}K(\theta).$$

Note that the matrix A appears both in $G(q)$ and in $H(q)$, therefore it characterizes the dynamics both of the deterministic and of the stochastic part of (3.205).

The four problems involved in estimating the parameters of a model defined in Sect. 3.5.2 will be addressed in the following subsections for the prediction-error problem. The prediction-error approach is illustrated in Fig. 3.19. In this figure, $\hat{y}(k,\theta)$ is derived from (3.209) as $C(\theta)\hat{x}(k|k-1,\theta) + D(\theta)u(k)$.

3.6.3 Parameterizing an Innovation State-Space Model

Corresponding to the innovation state-space model (3.208)–(3.209), we could represent conceptually the following parameterization of the one-step-ahead predictor:

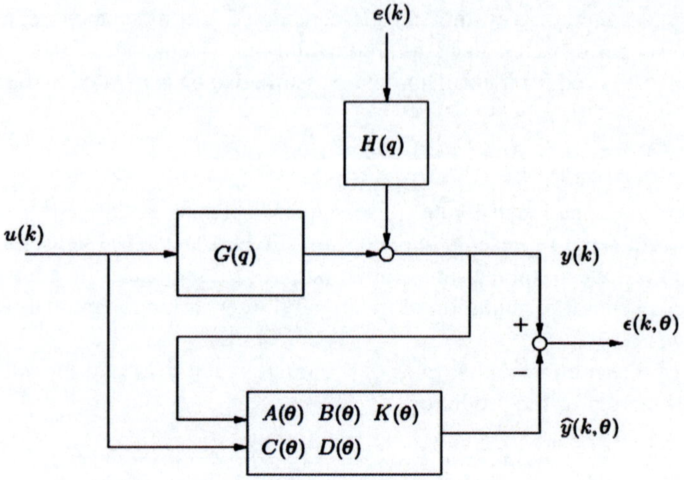

Fig. 3.19 The prediction-error model-estimation method

$$\hat{x}(k+1|k,\theta) = \big(A(\theta) - K(\theta)C(\theta)\big)\hat{x}(k|k-1,\theta) + \big(B(\theta) - K(\theta)D(\theta)\big)u(k) + K(\theta)y(k), \qquad (3.210)$$

$$\hat{y}(k|k-1,\theta) = C(\theta)\hat{x}(k|k-1,\theta) + D(\theta)u(k). \qquad (3.211)$$

Various choices of parameterization for this predictor exist. The parameterization introduced in Sect. 3.5.3 for the output-error case can be used for the prediction-error case if the "A" matrix is taken as $A - KC$ and the "B" matrix as $[B - KD, K]$ and we use $[u(k), y(k)]^t$ as the input to the system.

On making the evident assumption that the model derived from input–output data is reachable and observable, Theorem 3.4 may be used to impose on the system matrix $A - KC$ the additional constraint of asymptotic stability. This constraint then leads to the definition of the set Ω in the model structure $\mathcal{M}(\theta)$ in (3.163). Depending on the parameterization selected, the additional constraints in the parameter space on the one hand may be cumbersome to determine and on the other may complicate the numerical parameter search. In identification example 3.3, it was illustrated how challenging it is to construct the constraints on the parameter set while restricting the parameterization to yield a stable model. Furthermore, extending the example to third- or fourth-order systems indicates that the analysis needs to be performed individually for each dedicated model parameterization. For such models of higher than second order, the parameter set Ω becomes *nonconvex*. This increases the complexity of the optimization problem involved in estimating the parameters. The advantage of the output normal form is that it inherently guarantees the asymptotic stability of the system matrix $A - KC$ of the one-step-ahead predictor as detailed in the following lemma.

Lemma 3.9 *Let a predictor of the innovation model be given by*

$$\hat{x}(k+1) = (A - KC)\hat{x}(k) + (B - KD)u(k) + Ky(k), \tag{3.212}$$

$$\hat{y}(k) = C\hat{x}(k) + Du(k) \tag{3.213}$$

with the matrix $\bar{A} = A - KC$ *asymptotically stable and the pair* (A, C) *observable, then a surjective parameterization is obtained by parameterizing the pair* (A, C) *in the output normal form given in Definition 3.5 with the parameter vector* $\theta_{\bar{A}C} \in \Re^{nl}$ *and parameterizing the triple of matrices* $(\bar{B}, D, K)$ *with the parameter vector* $\theta_{\bar{B}DK} \in \Re^{(m+l)+ml}$ *that contains all the entries of the matrices* $\bar{B}$, D, *and* K, *with* $B = B - KD$.

The proof goes along the same lines as the proof of Lemma 3.7.

To complete the parameter vector parameterizing (3.212)–(3.213) including the initial state conditions $\hat{x}(0)$, we simply extend $\theta_{\bar{A}C}$ and $\theta_{\bar{B}DK}$ in the above lemma with these initial conditions to yield the parameter vector θ as

$$\theta = \begin{bmatrix} \hat{x}(0) \\ \theta_{\bar{A}C} \\ \theta_{\bar{B}DK} \end{bmatrix}.$$

The total number of parameters in this case is $p = n(2l + m) + ml + n$.

3.6.4 The Prediction-Error Cost Function

The primary use of the innovation model structure (3.212)–(3.213) is to predict the output (or state) by making use of a particular value of the parameter vector θ and of the available input–output data sequences. To allow for on-line use of the predictor, the predictor needs to be causal. In off-line applications, we may also operate with mixed causal, anticausal predictors, such as the Wiener optimal filter [40] and the Kalman-filter/smoothing combination. In what follows, we restrict the discussion to the causal multi-step-ahead prediction.

Definition 3.10 For the innovation state-space model structure (3.210)–(3.211), the N_p multi-step-ahead prediction of the output is a prediction of the output at a time instant $k + N_p$ making use of the input measurements $u(l)$, $l \le k + N_p$ and the output measurements $y(l)$, $l \le k$. This estimate is denoted by

$$\hat{y}(k + N_p | k, \theta).$$

The definition does not give a procedure for computing a multi-step-ahead prediction. The following lemma gives such a procedure based on the Kalman filter.

Lemma 3.11 *Given the model structure* (3.210)–(3.211) *and the quantities* $\hat{x}(k|k-1, \theta)$, $u(k)$, *and* $y(k)$ *at time instant* k, *then the one-step-ahead prediction at time instant* k *is given as*

$$\hat{x}(k+1|k,\theta) = \big(A(\theta) - K(\theta)C(\theta)\big)\hat{x}(k|k-1,\theta) + \big(B(\theta) - K(\theta)D(\theta)\big)u(k) + K(\theta)y(k), \tag{3.214}$$

$$\hat{y}(k+1|k,\theta) = C(\theta)\hat{x}(k+1,k,\theta) + D(\theta)u(k) \tag{3.215}$$

and, on the basis of this one-step-ahead prediction, the multi-step-ahead prediction for $N_p > 1$ is given as

$$\hat{x}(k+N_p|k,\theta) = A(\theta)^{N_p-1}\hat{x}(k+1|k,\theta) + \sum_{i=0}^{Np-2} A(\theta)^{N_p-i-2}B(\theta)u(k+i+1), \tag{3.216}$$

$$\hat{y}(k+N_p|k,\theta) = C(\theta)\hat{x}(k+N_p|k,\theta) + D(\theta)u(k+N_p). \tag{3.217}$$

The one-step-ahead prediction model (3.216)–(3.217) in this lemma directly follows from the parameterized innovation model (3.210)–(3.211). On the basis of this estimate, the multi-step-ahead prediction can be found by computing the response to the system,

$$z(k+j,\theta) = A(\theta)z(k+j-1,\theta) + B(\theta)u(k+j-1),$$

for $j > 1$ with initial condition $z(k+1,\theta) = \hat{x}(k+1|k,\theta)$. The multi-step-ahead prediction is then obtained by setting $\hat{x}(k+N_p|k,\theta) = z(k+N_p,\theta)$. Thus, the multi-step-ahead prediction is obtained by iterating the system using the one-step-ahead predicted state as initial condition. It can be proven that the multi-step-ahead predictor in the lemma is the optimal predictor, in the sense that it solves the so-called Wiener problem. The interested reader is referred to the book of Hayes [40, Chapter 7].

Given a finite number of measurements N of the input and output sequences of the data-generating system, we can estimate the parameters θ of the multi-step-ahead predictor (3.216)–(3.217) by minimizing a least-squares cost function

$$\min_{\theta} J_N(\theta, N_p) = \min_{\theta} \frac{1}{N}\sum_{k=0}^{N-1} \big\| y(k) - \hat{y}(k|k-N_p,\theta)\big\|_2^2. \tag{3.218}$$

This least-squares criterion is inspired by the minimum-variance state-reconstruction property of the Kalman filter. To reveal this link, consider the data-generating system in innovation form for the case $N_p = 1$,

$$\hat{x}(k+1,\theta) = A(\theta)\theta(k,\theta) + B(\theta)u(k) + K(\theta)e(k),$$
$$y(k) = C(\theta)\hat{x}(k,\theta) + e(k),$$

with $\hat{x}(0,\theta)$ given and with $K(\theta)$ derived from the solution of the discrete algebraic Riccati equation (DARE) about which we will learn more in Chap. 5. From this innovation representation, we can directly derive the Kalman filter as

$$\hat{x}(k+1,\theta_0) = A(\theta_0)\hat{x}(k,\theta) + B(\theta_0)u(k) + K(\theta) + \big(y(k) - C(\theta_0)\hat{x}(k,\theta_0)\big),$$
$$\hat{y}(k,\theta) = C(\theta)\hat{x}(k,\theta).$$

The minimum-variance property of the estimates obtained by use of the Kalman filter means that the variance of the prediction error $y(k) - \hat{y}(k, \theta_0)$ is minimized. Therefore, if we denote $\hat{y}(k, \theta)$ as the output of a Kalman filter as above but determined by the parameter vector θ instead of by θ_0, then the latter satisfies

$$\theta_0 = \arg\min \operatorname{Tr} \mathbb{E}\big[\big(y(k) - \hat{y}(k, \theta)\big)\big(y(k) - \hat{y}(k, \theta)\big)^t\big].$$

Generally, it was shown that the Kalman filter is time-varying and, therefore, that the variance of the prediction error will change over time. However, if we make the assumption that the variance is constant and the prediction error is an ergodic sequence, an estimate of θ_0 may be obtained by means of the following optimization problem:

$$\hat{\theta}_0 = \arg\min \lim_{N\to\infty} \frac{1}{N} \sum_{k=0}^{N-1} \big\| y(k) - \hat{y}(k|k-1, \theta) \big\|_2^2.$$

The foregoing parameter-optimization problem will be referred to as the *prediction-error estimation problem*. It forms a small part of the complete procedure of system identification, since it implicitly assumes the order of the state-space model (n) and the parameterization to be given.

Henceforth, we will concentrate on the one-step-ahead prediction error, and thus consider the optimization problem

$$\min_\theta J_N(\theta) = \min_\theta \sum_{k=0}^{N-1} \big\| y(k) - \hat{y}(k|k-1, \theta) \big\|_2^2. \tag{3.219}$$

For innovation models and recalling Lemma 3.9, for the innovation model (3.212)–(3.213), a more specific form of $J_N(\theta)$ is given in the following theorem:

Theorem 3.12 *The functional $J_N(\theta)$ can be written as*

$$J_N(\theta_{\bar{A}C}, \theta_{\bar{B}DK}) = \frac{1}{N} \sum_{k=0}^{N-1} \big\| y(k) - \phi(k, \theta_{\hat{A}C})\theta_{\hat{B}DK} \big\|_2^2 \tag{3.220}$$

with $\theta_{\hat{A}C}$ the parameters necessary to parameterize the pair $(\bar{A}, C)$ with $\bar{A} = A - KC$ and

$$\begin{bmatrix} \hat{x}(0) \\ \operatorname{vec}(\bar{B}) \\ \operatorname{vec}(K) \\ \operatorname{vec}(D) \end{bmatrix},$$

with $\bar{B} = B - KD$. The matrix $\phi(k, \theta_{\bar{A}C}) \in \Re^{l\times(n+m(l+n)+nl)}$ is explicitly given as

$$\phi(k, \theta_{\bar{A}C}) = \Bigg[C(\theta_{\bar{A}C})\bar{A}(\theta_{\bar{A}C})^k \quad \sum_{\tau=0}^{k-1} u^t(\tau) \otimes C(\theta_{\bar{A}C})\bar{A}(\theta_{\bar{A}C})^{k=1-\tau}$$
$$\sum_{\tau=0}^{k-1} y^t(\tau) \otimes C(\theta_{\bar{A}C})\bar{A}(\theta_{\bar{A}C})^{k-1-\tau} \quad u^t(k) \otimes I_l \Bigg].$$

Proof The one-step-ahead predictor related to the parameterized innovation model (3.212)–(3.213) is

$$\begin{aligned}\hat{x}(k+1,\theta_{\bar{A}C},\theta_{\bar{B}D}) &= \bar{A}(\theta_{\bar{A}C})\hat{x}(k,\theta_{\bar{B}DK}) + \bar{B}(\theta_{\bar{B}DK})u(k) + K(\theta_{\bar{B}DK})y(k),\\ \hat{y}(k,\theta_{\bar{A}C},\theta_{\bar{B}DK}) &= C(\theta_{\bar{A}C})\hat{x}(k,\theta_{\bar{A}C,\theta_{B}DK}) + D(\theta_{\bar{B}DK})u(k),\end{aligned}$$

with an initial state $\hat{x}(0,\theta_{\bar{B}DK})$. The output of this state-space model can explicitly be written in terms of the input, output, and initial state $\hat{x}(0,\theta_{\bar{B}DK})$ as:

$$\begin{aligned}\hat{y}(k,\theta_{\bar{A}C},\theta_{\bar{B}DK}) = {} & C(\theta_{\bar{A}C})\bar{A}(\theta_{\bar{A}C})^{k}\hat{x}(0.\theta_{\bar{B}DK})\\ & + \sum_{\tau=0}^{k-1} C(\theta_{\bar{A}C})\bar{A}(\theta_{\bar{A}C})^{k-1-\tau}B(\theta_{\bar{B}DK})u(\tau)\\ & + D(\theta_{\bar{B}DK})u(k)\\ & + \sum_{\tau=0}^{k-1} C(\theta_{\bar{A}C})\bar{A}(\theta_{\bar{A}C})^{k-1-\tau}K(\theta_{\bar{B}DK})y(\tau).\end{aligned}$$

Application of the property that $\text{vec}(XYZ) = (Z^t \otimes X)\,\text{vec}(Y)$ completes the proof. □

The parameter vector θ in the original innovation model (3.212)–(3.213) could be constructed by simply stacking the vectors $\theta_{\bar{A}C}$ and $\theta_{\bar{B}DK}$ of Lemma 3.9 as

$$\theta = \begin{bmatrix}\theta_{\bar{A}C}\\ \theta_{\bar{B}DK}\end{bmatrix}.$$

The output normal form presented in Lemma 3.11 can be used to parameterize the formulation of the functional as expressed in Lemma 3.9.

3.6.5 Numerical Parameter Estimation

To solve the prediction-error problem, the iterative methods can be generally used. Of course, some minor adjustments are necessary. For example, if the one-step-ahead prediction is used, the cost function is computed by simulating the predictor given by the system (3.212)–(3.213), and the dynamic system (3.182)–(3.82) that needs to be simulated to obtain the Jacobian in the Gauss–Newton method becomes

$$\begin{aligned}X_i(k+1,\theta) &= \bar{A}(\theta)X_i(k,\theta) + \frac{\partial \bar{A}(\theta)}{\partial \theta(i)}\hat{x}(k,\theta) + \frac{\partial \bar{B}(\theta)}{\partial \theta(i)}u(k)\\ &\quad + \frac{\partial K(\theta)}{\partial \theta(i)}y(k),\\ \frac{\partial \hat{y}(k,\theta)}{\partial \theta(i)} &= C(\theta)X_i(k,\theta) + \frac{\partial C(\theta)}{\partial \theta i}\hat{x}(k,\theta) + \frac{\partial D(\theta)}{\partial \theta(i)}u(k),\end{aligned}$$

with

$$\bar{A}(\theta) = A(\theta) - K(\theta)C(\theta),$$
$$\bar{B}(\theta) = B(\theta) - K(\theta)D(\theta).$$

Similar straightforward adjustments are needed in the other numerical methods discussed in Sect. 3.5.5.

3.6.6 Analyzing the Accuracy of the Estimates

To analyze the accuracy of the estimates obtained, the covariance matrix of the solution $\hat{\theta}_N$ to the prediction-error optimization problem can be used. The theory presented in Sect. 3.5.6 for the output-error methods applies also to the prediction-error methods. Using the covariance matrix to analyze the accuracy of the estimated model is done under the assumption that the system to be identified belongs to the assumed model set $\mathcal{M}(\theta)$ (3.163). Generally, in practice this assumption does not hold and the model parameters will be biased.

Using an output-error or prediction-error method, the estimates of the model parameters are obtained from a finite number of input and output measurements as

$$\hat{\theta}_N = \arg\min J_N(\theta).$$

The best possible model $\theta_\star$ within a given model structure is given by the minimizing parameter vector of the cost function

$$\theta_\star = \arg\min \lim_{N\to\infty} J_N(\theta) = \arg\min \bar{J}(\theta).$$

The quality of an estimated model $\hat{\theta_N}$ can now be measured using [54, 68]

$$\mathbb{E}\bar{J}(\hat{\theta}_N) \tag{3.221}$$

where the expectation E is with respect to the model $\bar{\theta}_N$. The measure (3.221) describes the expected fit of the model to the true system, when the model is applied to a new set of input and output measurements that have the same properties (distributions) as the measurements used to determine $\hat{\theta}_N$. This measure can be decomposed as follows [54, 68]:

$$\mathbb{E}\bar{J}(\bar{\theta}_N) \approx \underbrace{E\|y(k) - y_0(k,\theta_0)\|_2^2}_{\text{noise}} + \underbrace{E\|y_0(k,\theta_0) - \hat{y}(k,\theta_\star)\|_2^2}_{\text{bias}}$$
$$+ \underbrace{E\|\hat{y}(k,\theta_\star) - k, \hat{\theta_N}\|_2^2}_{\text{variance}},$$

where $y_0(k,\theta_0)$ is the output of the predictor based on the true model, that is, $y(k) = y_0(k,\theta) + e(k)$, with $e(k)$ white-noise residuals. The three parts in this decomposition will now be discussed.

- *Noise part*: The variance of the error between the measured output and a predictor based on the true model 0. This error is a white-noise sequence.

- *Bias part*: The model structures of the true predictor $y_0(k, \theta_0)$ and of the model class adopted can be different. The bias error expresses the difference between the true predictor and the best possible approximation of the true predictor within the model class adopted.
- *Variance part*: The use of a finite number of samples N to determine the model $\hat{\theta}_N$ results in a difference from the best possible model (within the model class adopted) $\theta_\star$ based on an infinite number of samples.

3.6.7 Some Model Parameterizations for SISO Systems

For identification of SISO systems, various parameterizations of the innovation representation (3.212)–(3.213) are in use [9, 44, 54, 69]. It is shown in this section that these more-classical model parameterizations can be treated as *special cases* of the MIMO innovation model parameterization discussed in Sect. 3.6.2. We adopt the common practice of presenting these special SISO parameterizations in a transfer-function setting.

3.6.8 The ARMAX and ARX Model Structures

The ARMAX, standing for *Auto-Regressive Moving Average with eXogenous input*, model structure considers the following specific case of the general input–output description (3.205):

$$y(k) = \frac{b_1 q^{-1} + \cdots + b_n q^{-n}}{1 + a_1 q^{-1} + \cdots + a_n q^{-n}} u(k) + \frac{1 + c_1 q^{-1} + \cdots + c_n q^{-n}}{1 + a_1 q^{-1} + \cdots + a_n q^{-n}} e(k) \quad (3.222)$$

where $e(k) \in \Re$ is again a zero-mean white-noise sequence that is independent from $u(k) \in \Re$ and a_i, b_i, and c_i $(i = 1, 2, \ldots, n)$ are real-valued scalars. It is common practice to use negative powers of q in the description of the ARMAX model.

A more general ARMAX representation exists, in which the order of the numerators and denominators may be different, and the transfer from $u(k)$ to $y(k)$ may contain an additional dead-time. To keep the notation simple, these fine-tunings are not addressed in this book. When the order n is known, we can define an estimation problem to estimate the parameters a_i, b_i, and c_i $(i = 1, 2, \ldots, n)$ from a finite number of input–output measurements. The formulation and the solution of such an estimation problem is discussed next and is addressed by establishing a one-to-one correspondence between the ARMAX transfer-function description (3.222) and a particular minimal parameterization of the state-space system (3.212)–(3.213), as summarized in the following lemma.

Lemma 3.13 *There is a one-to-one correspondence between the ARMAX model given by* (3.222) *and the following parameterization of a SISO state-space system in innovation form:*

$$x(k+1) = \begin{bmatrix} -a_1 & 1 & 0 & \cdots & 0 \\ -a_2 & 0 & 1 & \cdots & 0 \\ \vdots & \vdots & \vdots & \ddots & \vdots \\ -a_{n-1} & 0 & \cdots & \cdots & 1 \\ -a_n & 0 & \cdots & \cdots & 0 \end{bmatrix} x(k) + \begin{bmatrix} b_1 \\ b_2 \\ \vdots \\ b_{n-1} \\ b_n \end{bmatrix} u(k)$$

$$+ \begin{bmatrix} c_1 - a_1 \\ c_2 - a_2 \\ \vdots \\ c_{n-1} - a_{n-1} \\ c_n - a_n \end{bmatrix} e(k), \tag{3.223}$$

$$y(k) = \begin{bmatrix} 1 & 0 & 0 & \cdots & 0 \end{bmatrix} x(k) + e(k). \tag{3.224}$$

Proof The proof follows on showing that from the parameterization (3.223)–(3.224) we can obtain in a unique manner the difference equation (3.222). Let $x_i(k)$ denote the ith component of the vector $x(k)$, then (3.223) is equivalent to the following set of equations:

$$\begin{aligned} x_1(k+1) &= -a_1 x_1(k) + x_2(k) + b_1 u(k) + (c_1 - a_1)e(k), \\ x_2(k+1) &= -a_2 x_1(k) + x_3(k) + b_2 u(k) + (c_2 - a_2)e(k), \\ &\vdots \\ x_n(k+1) &= -a_n x_1(k) + b_n u(k) + (c_n - a_n)e(k). \end{aligned}$$

Making the substitution $y(k) = x_1(k) + e(k)$ yields

$$\begin{aligned} x_1(k+1) &= -a_1 y(k) + x_2(k) + b_1 u(k) + c_1 e(k), \quad \star \\ x_2(k+1) &= -a_2 y(k) + x_3(k) + b_2 u(k) + c_2 e(k), \quad \star \\ &\vdots \\ x_{n-1}(k+1) &= -a_{n-1} y(k) + x_n(k) + b_{n-1} u(k) + c_{n-1} e(k), \quad \star \\ x_n(k+1) &= -a_n y(k) + b_n u(k) + c_n e(k). \end{aligned}$$

Increasing the time index of all the equations indicated by a star ($\star$) and subsequently replacing $x_n(k+1)$ by the right-hand side of the last equation yields the following expressions for the indicated equations:

$$\begin{aligned} x_1(k+2) &= -a_1 y(k+1) + x_2(k+1) + b_1 u(k+1) + c_1 e(k+1), \\ x_2(k+2) &= -a_2 y(k+1) + x_3(k+1) + b_2 u(k+1) + c_2 e(k+1), \\ &\vdots \\ x_{n-2}(k+2) &= -a_{n-2} y(k+1) + x_{n-1}(k+1) + b_{n-2} u(k+1) \\ &\quad + c_{n-2} e(k+1), \\ x_{n-1}(k+2) &= -a_{n-1} y(k+1) - a_n y(k) + b_n u(k) + c_n e(k) + b_{n-1} u(k+1) \\ &\quad + c_{n-1} e(k+1). \end{aligned}$$

Implementing the above recipe $n-2$ times yields the single equation

$$\begin{aligned}x_1(k+n) = &-a_1 y(k+n-1) - a_2 y(k+n-2) - \cdots - a_n y(k)\\ &+ b_1 u(k+n-1) + b_2 u(k+n-2) + \cdots + b_n u(k)\\ &+ c_1 e(k+n-1) + c_2 e(k+n-2) + \cdots + a_n e(k).\end{aligned}$$

By making use of the output equation (3.114), we finally obtain

$$\begin{aligned}y(k+n) = &-a_1 y(k+n-1) - a_2 y(k+n-2) - \cdots - a_n y(k)\\ &+ b_1 u(k+n-1) + b_2 u(k+n-2) + \cdots + b_n u(k)\\ &+ e(k+n) + c_1 e(k+n-1) + c_2 e(k+n-2) + \cdots + a_n e(k).\end{aligned}$$

This is the difference equation of (3.222). □

The ARMAX model is closely related to the observer canonical form in linear system theory. The ARMAX model can be converted into the observer canonical form and vice versa by turning the state-vector upside down. The one-step-ahead predictor for the ARMAX model is summarized in the next lemma.

Lemma 3.14 *Let the differences $c_i - a_i$ be denoted by k_i for $i = 1, 2, \dots, n$, then the one-step ahead predictor for the ARMAX model* (3.222) *is given by*

$$\begin{aligned}\hat{y}(k|l-1) = &\frac{b_1 q^{-1} + \cdots + b_n q^{-n}}{1 + c_1 q^{-1} + \cdots + c_n q^{-n}} u(k)\\ &+ \frac{k_1 q^{-1} + \cdots + k_n q^{-n}}{1 + c_1 q^{-1} + \cdots + c_n q^{-n}} y(k).\end{aligned} \tag{3.225}$$

Proof Making use of the state-space parameterization of the ARMAX model given by (3.223) and (3.224), the one-step-ahead prediction based on (3.216) and (3.217) equals

$$\begin{aligned}\hat{x}(k+1|k) = &\left(\begin{bmatrix} -a_1 & 1 & 0 & \cdots & 0 \\ -a_2 & 0 & 1 & \cdots & 0 \\ \vdots & \vdots & \vdots & \ddots & \vdots \\ -a_{n-1} & 0 & \cdots & \cdots & 1 \\ -a_n & 0 & \cdots & \cdots & 0 \end{bmatrix} - \begin{bmatrix} k_1 \\ k_2 \\ \vdots \\ k_{n-1} \\ k_n \end{bmatrix} \begin{bmatrix} 1 & 0 & \cdots & 0 & 0 \end{bmatrix} \right)\\ &\times \hat{x}(k|k-1)\\ &+ \begin{bmatrix} b_1 \\ b_2 \\ \vdots \\ b_{n-1} \\ b_n \end{bmatrix} u(k) + \begin{bmatrix} k_1 \\ k_2 \\ \vdots \\ k_{n-1} \\ k_n \end{bmatrix} y(k),\\ \hat{y}(k|k-1) = &\begin{bmatrix} 1 & 0 & 0 & \cdots & 0 \end{bmatrix} \hat{x}(k|k-1);\end{aligned}$$

with $c_i = k_i + a_i$, this equals

$$\hat{x}(k+1|k) = \begin{bmatrix} -c_1 & 1 & 0 & \cdots & 0 \\ -c_2 & 0 & 1 & \cdots & 0 \\ \vdots & \vdots & \vdots & \ddots & \vdots \\ -c_{n-1} & 0 & \cdots & \cdots & 1 \\ -c_n & 0 & \cdots & \cdots & 0 \end{bmatrix} \hat{x}(k|k-1) + \begin{bmatrix} b_1 \\ b_2 \\ \vdots \\ b_{n-1} \\ b_n \end{bmatrix} u(k) + \begin{bmatrix} k_1 \\ k_2 \\ \vdots \\ k_{n-1} \\ k_n \end{bmatrix} y(k),$$

$$\hat{y}(k|k-1) = \begin{bmatrix} 1 & 0 & 0 & \cdots & 0 \end{bmatrix} \hat{x}(k|k-1).$$

Following the proof of Lemma 3.11, the transfer-function representation of this state-space model equals (3.225). □

Now on introducing the following polynomials in the shift operator q,

$$A(q) = 1 + a_1 q^{-1} + \cdots + a_n q^{-n},$$
$$B(q) = b_1 q^{-1} + \cdots + b_n q^{-n},$$
$$C(q) = 1 + c_1 q^{-1} + \cdots + c_n q^{-n},$$

the ARMAX model can be denoted by

$$y(k) = \frac{B(q)}{A(q)} u(k) + \frac{C(q)}{A(q)} e(k). \tag{3.226}$$

The one-step-ahead predictor is denoted by

$$\hat{y}(k|k-1) = \frac{B(q)}{C(q)} u(k) + \frac{C(q - A(q))}{C(q)} y(k). \tag{3.227}$$

This is a stable predictor, provided that the polynomial $C(q)$ has all its roots within the unit circle.

The Auto-Regressive with eXogenous input (ARX) model is a special case of the ARMAX model structure constraining the parameters $c_i = 0$ for $i = 1, 2, \ldots, n$, and thus $C(q) = 1$. Therefore, the ARX model is given by

$$y(k) = \frac{B(q)}{A(q)} u(k) + \frac{1}{A(q)} e(k),$$

and the associated predictor equals

$$\hat{y}(k|k-1) = B(q)u(k) + \left[1 - A(q)\right] y(k). \tag{3.228}$$

To identify a model in the ARMAX or ARX structure, we minimize the prediction-error cost function $J_N(\theta)$ described in Sect. 3.6.2. The methods for minimizing this cost function were described in Sects. 3.5.5 and 3.6.2. They require the evaluation of the cost function and its Jacobian. This evaluation depends on the particular parameterization of the state-space innovation model. As pointed out in Sect. 3.6.2, the choice of a specific parameterization changes only the following matrices in the evaluation of the Jacobian:

$$\frac{\partial \bar{A}}{\partial \theta(1)}, \quad \frac{\partial B}{\partial \theta(1)}, \quad \frac{\partial C}{\partial \theta(1)}, \quad \frac{\partial D}{\partial \theta(1)}, \quad \frac{\partial K}{\partial \theta(1)}$$

for $i = 1, 2, \ldots, p$. The following example shows that these quantities are easy to compute.

3.6.9 Identification Example 3.9

Given an ARMAX model, with matrices

$$A = \begin{bmatrix} -\theta_1 & 1 \\ -\theta_2 & 0 \end{bmatrix}, \qquad B = \begin{bmatrix} \theta_3 \\ \theta_4 \end{bmatrix},$$
$$C = \begin{bmatrix} 1 & 0 \end{bmatrix}, \qquad K = \begin{bmatrix} \theta_5 \\ \theta_6 \end{bmatrix}$$

it is easy to see that

$$\bar{A} = A - KC = \begin{bmatrix} -\theta_1 - \theta_5 & 1 \\ -\theta_2 - \theta_6 & 0 \end{bmatrix},$$

and therefore

$$\frac{\partial \bar{A}}{\partial \theta_i} = \begin{bmatrix} -1 & 0 \\ 0 & 0 \end{bmatrix}, \qquad i = 1, 2,$$
$$\frac{\partial \bar{A}}{\partial \theta_i} = \begin{bmatrix} 0 & 0 \\ -1 & 0 \end{bmatrix}, \qquad i = 1, 5,$$
$$\frac{\partial \bar{A}}{\partial \theta_i} = \begin{bmatrix} 0 & 0 \\ 0 & 0 \end{bmatrix}, \qquad i = 3, 4.$$

The following example illustrates that, for an ARX model, minimization of the prediction-error cost function $J_N(\theta)$ described in Sect. 3.6.2 leads to a linear least-squares problem.

3.6.10 Identification Example 3.10

The ARX predictor is given by (3.228). Taking

$$A(q) = 1 + a_1 q^{-1} + \cdots + a_n q^{-n},$$
$$B(q) = b_1 q^{-1} + \cdots + b_n q^{-n},$$

we can write

$$\hat{y}(k|k-1) = \phi(k)^t \theta,$$

with

$$\begin{aligned}\theta &= \begin{bmatrix} -a_1 & -a_2 & \cdots & -a_n | b_1 & b_2 & \cdots & b_n \end{bmatrix}^t, \\ \phi(k) &= \begin{bmatrix} y(k-1) & \cdots & y(k-1) | u(k-1) & \cdots & u(k-n) \end{bmatrix}.\end{aligned}$$

Thus, the prediction-error cost function is given by

$$J_N(\theta) = \frac{1}{N} \sum_{k=0}^{N-1} \left(y(k) - \phi(k)^t \theta\right)^2.$$

Identification example 3.10 shows that this form of the cost function leads to a linear least-squares problem.

3.6.11 The Box–Jenkins and Output-Error Model Structures

The Box–Jenkins (BJ) [9] model structure parameterizes the input–output relationship (3.205) as

$$y(k) = \frac{b_1 q^{-1} + \cdots + b_n q^{-n}}{1 + a_1 q^{-1} + \cdots + a_n q^{-n}} u(k) + \frac{1 + c_1 q^{-1} + \cdots + c_n q^{-n}}{1 + d_1 q^{-1} + \cdots + d_n q^{-n}} e(k). \quad (3.229)$$

On introducing the polynomials

$$A(q) = 1 + a_1 q^{-1} + \cdots + a_n q^{-n}, \quad (3.230)$$

$$B(q) = b_1 q^{-1} + \cdots + b_n q^{-n}, \quad (3.231)$$

$$C(q) = 1 + c_1 q^{-1} + \cdots + c_n q^{-n}, \quad (3.232)$$

$$D(q) = 1 + d_1 q^{-1} + \cdots + d_n q^{-n} \quad (3.233)$$

the BJ model can be denoted by

$$y(k) = \frac{B(q)}{A(q)} u(k) + \frac{C(q)}{D(q)} e(k). \quad (3.234)$$

A similar result to that in Lemma 3.11, but now for the BJ model, is given next.

Lemma 3.15 *There is a one-to-one correspondence between the BJ model given by* (3.229) *and the following parameterization of a SISO state-space system in innovation form:*

$$x(k+1)=\left[\begin{array}{ccccc|ccccc} -a_1 & 1 & 0 & \cdots & 0 & 0 & 0 & 0 & \cdots & 0 \\ -a_2 & 0 & 1 & \cdots & 0 & 0 & 0 & 0 & \cdots & 0 \\ \vdots & \vdots & \vdots & \ddots & \vdots & \vdots & \vdots & \vdots & \ddots & \vdots \\ -a_{n-1} & 0 & 0 & \cdots & 1 & 0 & 0 & 0 & \cdots & 0 \\ -a_n & 0 & 0 & \cdots & 0 & 0 & 0 & 0 & \cdots & 0 \\ \hline 0 & 0 & 0 & \cdots & 0 & -d_1 & 1 & 0 & \cdots & 0 \\ 0 & 0 & 0 & \cdots & 0 & -d_2 & 0 & 1 & \cdots & 0 \\ \vdots & \vdots & \vdots & \ddots & \vdots & \vdots & \vdots & \vdots & \ddots & \vdots \\ 0 & 0 & 0 & \cdots & 0 & -d_{n-1} & 0 & 0 & \cdots & 1 \\ 0 & 0 & 0 & \cdots & 0 & -d_n & 0 & 0 & \cdots & 0 \end{array}\right] x(k)$$

$$+\left[\begin{array}{c} b_1 \\ b_2 \\ \vdots \\ b_{n-1} \\ b_n \\ \hline 0 \\ 0 \\ \vdots \\ 0 \\ 0 \end{array}\right] u(k)+\left[\begin{array}{c} 0 \\ 0 \\ \vdots \\ 0 \\ 0 \\ \hline c_1-d_1 \\ c_2-d_2 \\ \vdots \\ c_{n-1}-d_{n-1} \\ c_n-d_n \end{array}\right] e(k), \tag{3.235}$$

$$y(k)=\left[\begin{array}{ccccc|ccccc} 1 & 0 & 0 & \cdots & 0 & 1 & 0 & 0 & \cdots & 0 \end{array}\right] x(k)+e(k). \tag{3.236}$$

The proof is similar to the one given for Lemma 3.11.

On embedding the specific BJ model into the general state-space model considered earlier, we draw the conclusion that the asymptotic stability of the one-step-ahead predictor requires the roots of the deterministic polynomial $A(q)$ to be within the unit circle. This condition is necessary in order to make the pair $(A, Q^{1/2})$ of the BJ model (3.235)–(3.236) corresponding to the state-space model being stabilizable.

The following lemma shows that the one-step-ahead predictor of the BJ model equals

$$\hat{y}(k|k-1)=\frac{D(q)}{C(q)}\frac{D(q)}{A(q)}u(k)+\frac{C(q)-D(q)}{C(q)}y(k).$$

Lemma 3.16 *The one-step-ahead predictor for the BJ model* (3.234) *is given by*

$$\hat{y}(k|k-1)=\frac{D(q)}{C(q)}\frac{B(q)}{A(q)}u(k)+\frac{C(q)-D(q)}{C(q)}y(k) \tag{3.237}$$

where the polynomials $A(q)$, $B(q)$, $C(q)$, *and* $D(q)$ *are given by* (3.229)–(3.233).

Proof Making use of the state-space parameterization of the BJ model given by (3.234)–(3.235) and the definition $k_i=c_i-d_i$, the one-step-ahead prediction based on (3.216)–(3.217) equals

$$\hat{x}(k+1|k) = \left[\begin{array}{ccccc|ccccc} -a_1 & 1 & 0 & \cdots & 0 & 0 & 0 & 0 & \cdots & 0 \\ -a_2 & 0 & 1 & \cdots & 0 & 0 & 0 & 0 & \cdots & 0 \\ \vdots & \vdots & \vdots & \ddots & \vdots & \vdots & \vdots & \vdots & \ddots & \vdots \\ -a_{n-1} & 0 & 0 & \cdots & 1 & 0 & 0 & 0 & \cdots & 0 \\ -a_n & 0 & 0 & \cdots & 0 & 0 & 0 & 0 & \cdots & 0 \\ \hline -k_1 & 0 & 0 & \cdots & 0 & -d_1-k_1 & 1 & 0 & \cdots & 0 \\ -k_2 & 0 & 0 & \cdots & 0 & -d_2-k_2 & 0 & 1 & \cdots & 0 \\ \vdots & \vdots & \vdots & \ddots & \vdots & \vdots & \vdots & \vdots & \ddots & \vdots \\ -k_{n-1} & 0 & 0 & \cdots & 0 & -d_{n-1}-k_{n-1} & 0 & 0 & \cdots & 1 \\ -k_n & 0 & 0 & \cdots & 0 & -d_n-k_n & 0 & 0 & \cdots & 0 \end{array}\right]$$

$$\times \hat{x}(k|k-1) + \left[\begin{array}{c} b_1 \\ b_2 \\ \vdots \\ b_{n-1} \\ b_n \\ \hline 0 \\ 0 \\ \vdots \\ 0 \\ 0 \end{array}\right] u(k) + \left[\begin{array}{c} 0 \\ 0 \\ \vdots \\ 0 \\ 0 \\ \hline k_1 \\ k_2 \\ \vdots \\ k_{n-1} \\ k_n \end{array}\right] y(k),$$

$$\hat{y}(k|k-1) = \left[\begin{array}{ccccc|ccccc} 1 & 0 & 0 & \cdots & 0 & 1 & 0 & 0 & \cdots & 0 \end{array}\right] \hat{x}(k|k-1).$$

This system is denoted briefly by

$$\hat{x}(k+1|k) = \left[\begin{array}{c|c} A_{11} & 0 \\ A_{21} & A_{22} \end{array}\right] \hat{x}(k|k-1) + \left[\begin{array}{c} B \\ \hline 0 \end{array}\right] u(k) + \left[\begin{array}{c} 0 \\ \hline K \end{array}\right] y(k),$$

$$\hat{y}(k|k-1) = [C_1 | C_2] \hat{x}(k|k-1).$$

Since $A_{21} = -KC_1$, we can write the one-step-ahead prediction of the output as

$$\begin{aligned}
\hat{y}(k|k-1) &= [C_1|C_2]\left(qI - \left[\begin{array}{c|c} A_{11} & 0 \\ -KC_1 & A_{22} \end{array}\right]\right)^{-1} \left(\left[\begin{array}{c} B \\ \hline 0 \end{array}\right] u(k) + \left[\begin{array}{c} 0 \\ \hline K \end{array}\right] y(k)\right) \\
&= [C_1|C_2]\left[\begin{array}{c|c} qI - A_{11} & 0 \\ KC_1 & qI - A_{22} \end{array}\right]^{-1} \left(\left[\begin{array}{c} B \\ \hline 0 \end{array}\right] u(k) + \left[\begin{array}{c} 0 \\ \hline K \end{array}\right] y(k)\right) \\
&= \big(C_1(qI - A_{11})^{-1}B - C_2(qI - A_{22})^{-1}KC_1(qI - A_{11})^{-1}B\big)u(k) \\
&\quad + C_2(qI - A_{A22})^{-1}Ky(k) \\
&= \big(I - C_2(qI - A_{22})^{-1}K\big)\big(C_1(qI - A_{11})^{-1}B\big)u(k) \\
&\quad + C_2(qI - A_{22})^{-1}Ky(k). \qquad (3.238)
\end{aligned}$$

Since

$$A_{22} + KC_2 = \begin{bmatrix} -d_1 & 1 & 0 & \cdots & 0 \\ -d_2 & 0 & 1 & \cdots & 0 \\ \vdots & \vdots & \vdots & \ddots & \vdots \\ -d_{n-1} & 0 & 0 & \cdots & 1 \\ -d_n & 0 & 0 & \cdots & 0 \end{bmatrix},$$

and $k_i = c_i - d_i$, it follows from Lemma 3.11 that

$$I - C_2(qI - A_{22})^{-1}K = \left[\frac{C(q)}{D(q)}\right]^{-1}.$$

Therefore, (3.238) can be written in terms of the defined polynomials as (3.237).

On putting the parameters c_i and d_i for $i = 1, 2, \dots, n$ into the BJ model structure, we obtain a model and predictor that fit within the output-error model. The resulting specific transfer-function parameterization has classically been referred to as the output-error (OE) model. In polynomial form, it reads as

$$y(k) = \frac{B(q)}{A(q)}u(k) + e(k),$$

and the associated predictor is given by

$$\hat{y}(k|k-1) = \frac{B(q)}{A(q)}u(k). \tag{3.239}$$

Thus, if the OE model is stable, then also its predictor is stable. □

3.6.12 Qualitative Analysis of the Model Bias

The asymptotic variance analyzed in Sects. 3.5.6 and 3.6.2 can be used as an indication of the accuracy of the estimated parameters if the system that generated the input–output data set belongs to the model set $\mathcal{M}(\theta)$. The latter hypothesis generally does not hold. Examples are when the underlying system has a very large state dimension, whereas for designing a controller one is interested in a low-dimensionality model.

Therefore, in addition to the variance, also the bias in the estimated parameters needs to be considered. In this section, we will analyze the bias for some specific SISO systems. We first introduce some notation. Let $\theta_\star$ be the minimizing parameter vector of the cost function $J_N(\theta)$ for $N \to \infty$

$$\theta_\star = \arg\min \lim_{N\to\infty} J_N(\theta) = \arg\min \bar{J}(\theta),$$

and let the system by which the input–output data were generated be described as

$$\begin{aligned}y(k) &= \frac{B_0(q)}{A_0(q)}u(k) + v(k) \\ &= \frac{b_1^0 q^{-1} + b_2^0 q^{-2} + \cdots + b_n^0 q^{-n}}{1 + a_1^0 q^{-1} + a_2^0 q^{-2} + \cdots + a_n^0 q^{-n}} u(k) + v(k) \qquad (3.240)\end{aligned}$$

with n the order of the system and with $v(k)$ a stochastic perturbation that is independent from $u(k)$. Under these notions the *bias* is the difference between comparable quantities derived from the estimated model and from the true system that persists on taking the limit for $N \to \infty$. One such comparable quantity is the transfer function, which can, for example, be presented as a Bode plot.

To quantify the variance in the estimate $\hat{\theta}_N$ given by

$$\hat{\theta}_N = \arg\min J_N(\theta),$$

we should then analyze

$$\mathbb{E}\big[[\hat{\theta}_N - \theta_\star][\hat{\theta}_N - \theta_\star]^t\big],$$

instead of $\mathbb{E}[[\hat{\theta}_N - \theta_0][\hat{\theta}_N - \theta_0]]$ as was done in Sect. 3.5.6.

The bias of the estimated model is analyzed under the assumption that the time sequences are ergodic. In that case, the following limit holds:

$$\lim_{N\to\infty} \frac{1}{N} \sum_{k=0}^{N-1} \big(y(k) - \hat{y}(k|\,k-1)\big)^2 = E\big[\big(y(k) - \hat{y}(k|\,k-1)\big)^2\big].$$

When the prediction of the output depends on the parameter vector θ, the above equation can be written

$$\lim_{N\to\infty} \frac{1}{N} \sum_{k=0}^{N-1} \big(y(k) - \hat{y}(k|\,k-1,\theta)\big)^2 = \bar{J}(\theta) \qquad (3.241)$$

establishing the link with the cost function $\bar{J}(\theta)$. This cost function is now analyzed for the ARMAX and BJ model structures that were introduced in the previous section.

Lemma 3.17 [54] *Let the LTI system that generates the output $y(k)$ for a given input sequence $u(k)$, $k = 0, 1, 2, \ldots, N-1$, with spectrum $\Phi^u(\omega)$ be denoted by*

$$y(k) = G_0(q)u(k) + v(k),$$

where $v(k)$ is a stochastic perturbation independent from $u(k)$ with spectrum $\Phi^v(\omega)$, and let the time sequences $v(k)$, $u(k)$, and $y(k)$ be ergodic and let the parameters a_i, b_i, and c_i of an ARMAX model be stored in the parameter vector θ, then the parameter vector $\theta_\star$ minimizing the cost function

$$\bar{J}(\theta) = \lim_{N\to\infty} \frac{1}{N} \sum_{k=0}^{N-1} \big(y(k) - \hat{y}(k|\,k-1,\theta)\big)^2$$

satisfies

$$\theta_\star = \arg\min \frac{1}{2\pi} \int_{-\pi}^{\pi} \left| G_0\big(e^{j\omega}\big) - \frac{B(e^{j\omega},\theta)}{A(e^{j\omega},\theta)} \right|^2 \left| \frac{A(e^{j\omega},\theta)}{C(e^{j\omega},\theta)} \right|^2 \Phi^u(\omega) + \left| \frac{A(e^{j\omega},\theta)}{C(e^{j\omega},\theta)} \right|^2 \Phi^v(\omega)\, d\omega. \tag{3.242}$$

Proof The one-step-ahead predictor related to the ARMAX model structure is given by (3.227). Hence, the one-step-ahead prediction error $\epsilon(k|k-1) = y(k) - \hat{y}(k|k-1)$ is given by

$$\epsilon(k|k-1) = \frac{A(q,\theta)}{C(q,\theta)} y(k) - \frac{B(q,\theta)}{C(q,\theta)} u(k).$$

To express $\epsilon(k|k-1)$ as the sum of two statistically independent time sequences, simplifying the calculation of the spectrum of $\epsilon(k|k-1)$, we substitute into the above expression the model of the system that generated the sequence $y(k)$. This yields

$$\epsilon(k|k-1) = \frac{A(q,\theta)}{C(q,\theta)} \left(G_0(q) - \frac{B(q,\theta)}{A(q,\theta)} \right) u(k) + \frac{A(q,\theta)}{C(q,\theta)} v(k).$$

By virtue of the ergodic assumption,

$$\bar{J}(\theta) = E\big[\epsilon(k|k-1)^2\big].$$

Using Parseval's identity, see the Appendix, (assuming a sample time $T = 1$), this can be written as

$$E\big[\epsilon(k|k-1)^2\big] = \frac{1}{2\pi} \int_{-\pi}^{\pi} \Phi^\epsilon(\omega)\, d\omega. \tag{3.243}$$

An expression for $\Phi^\epsilon(\omega)$ can be derived by exploiting the independence between $u(k)$ and $v(k)$:

$$\Phi^\epsilon(\omega) = \left| G_0\big(e^{j\omega}\big) - \frac{B(e^{j\omega},\theta)}{A(e^{j\omega},\theta)} \right|^2 \left| \frac{A(e^{j\omega},\theta)}{C(e^{j\omega},\theta)} \right|^2 \Phi^u(\omega) + \left| \frac{A(e^{j\omega},\theta)}{C(e^{j\omega},\theta)} \right|^2 \Phi^v(\omega).$$

Substitution into (3.243) results in (3.242) as desired. □

Since the ARX model structure is a special case of the ARMAX model structure, we can, with a redefinition of the parameter vector θ, immediately derive the expression for the parameter vector $\theta_\star$ minimizing $\bar{J}(\theta)$ in (3.241) as

$$\theta_\star = \arg\min \frac{1}{2\pi} \int_{-\pi}^{\pi} \left| G_0\big(e^{j\omega}\big) - \frac{B(e^{j\omega},\theta)}{A(e^{j\omega},\theta)} \right|^2 \big|A\big(e^{j\omega},\theta\big)\big|^2 \Phi^u(\omega) + \big|A\big(e^{j\omega},\theta\big)\big|^2 \Phi^v(\omega)\, d\omega. \tag{3.244}$$

The use of Lemma 3.17 in qualitatively analyzing the bias in the estimate obtained with the ARX model structure is highlighted in the following example.

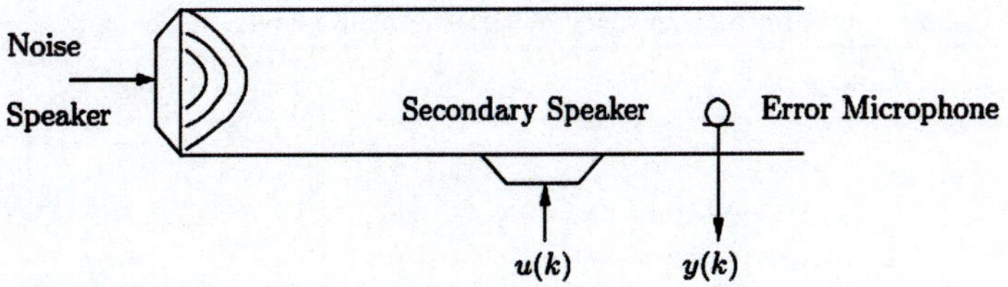

Fig. 3.20 A schematic representation of an acoustical duct

3.6.13 Identification Example 3.11

The system to be modeled is an acoustical duct, depicted in Fig. 3.20, which is used for active-noise-control experiments. At the left-hand end of the duct is mounted a loudspeaker that produces an undesired noise. The goal is to drive the secondary loudspeaker mounted just before the other end of the duct such that at the far-right end of the duct a region of silence is created. Most control algorithms used in active noise control need a model of the transfer from the secondary loudspeaker to the error microphone. A high-order approximation of the acoustical relationship between the speaker activated with the signal u, and the microphone producing the measurements y, is given by the following transfer function:

$$G(q) = \frac{\sum_{j=0}^{19} b_j q^{-j}}{\sum_{j=0}^{19} a_j q^{-j}}$$

with a_j and b_j listed in Table 3.2.

The magnitude of the Bode plot of the transfer function $G(e^{j\omega})$ is depicted by the thick line in the top part of Fig. 3.21. The input sequence $u(k)$ is taken to be a zero-mean unit-variance white-noise sequence of length 10000. With this input sequence, an output sequence $y(k)$ is generated using the high-order transfer function $G(q)$. These input and output sequences are then used to estimate a sixth-order ARX model via the use of a QR factorization to solve the related linear least-squares problem, see identification example 3.7. The estimated transfer function $\hat{G}(e^{j\omega})$ is depicted by the thin line in the top part of Fig. 3.21. We observe that, according to (3.241) with $\Phi^v(\omega) = 0$, the estimated low-order model accurately matches the high-order model for those frequency values for which $|A(e^{j\omega})|$ is large. From the graph of $|A(e^{j\omega})|$ in the lower part of Fig. 3.21, we observe that this holds in the high-frequency region above 100 Hz.

The following lemma gives a result similar to Lemma 3.17, but for the BJ model.

Lemma 3.18 [54] *Let the LTI system that generates the output $y(k)$ for a given input sequence $u(k)$, $k = 0, 1, 2, \dots, N-1$, with spectrum $\Phi^u(\omega)$ be denoted by*

$$y(k) = G_0(q)u(k) + v(k),$$

where $v(k)$ is a stochastic perturbation independent from $u(k)$ with spectrum $\Phi^v(\omega)$, let the time sequences $v(k)$, $u(k)$, and $y(k)$ be ergodic, and let the param-

Table 3.2 Coefficients of the transfer function between u and y in the model of the acoustical duct

θ	Value	θ	Value
a_0	1	b_0	0
a_1	−1.8937219532483E-0	b_1	−5.6534330123106E-6
a_2	9.2020408176247E-1	b_2	5.6870704280702E-6
a_3	8.4317527635808E-13	b_3	7.7870811926239E-3
a_4	−6.9870644340972E-13	b_4	1.3389477125431E-3
a_5	3.2703011891141E-13	b_5	−9.1260667240191E-3
a_6	−2.8053825784320E-14	b_6	1.4435759589218E-8
a_7	−4.8518619047975E-13	b_7	−1.2021568096247E-8
a_8	9.0515016323085E-13	b_8	−2.2746529807395E-9
a_9	−8.9573340462955E-13	b_9	6.3067990166664E-9
a_{10}	6.2104932381850E-13	b_{10}	9.1305924779895E-10
a_{11}	−4.0655443037130E-13	b_{11}	−7.5200613526843E-9
a_{12}	3.8448359402553E-13	b_{12}	1.9549739577695E-9
a_{13}	−4.9321540807220E-13	b_{13}	1.3891832078608E-8
a_{14}	5.3571245452629E-13	b_{14}	−1.6372496840947E-8
a_{15}	−6.7043859898372E-13	b_{15}	9.0003511972213E-3
a_{16}	6.5050860651120E-13	b_{16}	−1.9333235975678E-3
a_{17}	6.6499999999978E-1	b_{17}	−7.0669966879457E-3
a_{18}	−1.2593250989101E-0	b_{18}	−3.7850561971775E-6
a_{19}	6.1193571437226E-1	b_{19}	3.7590122810601E-6

In the above values, E-0 means $\times 10^0$, E-6 means $\times 10^{-6}$, etc.

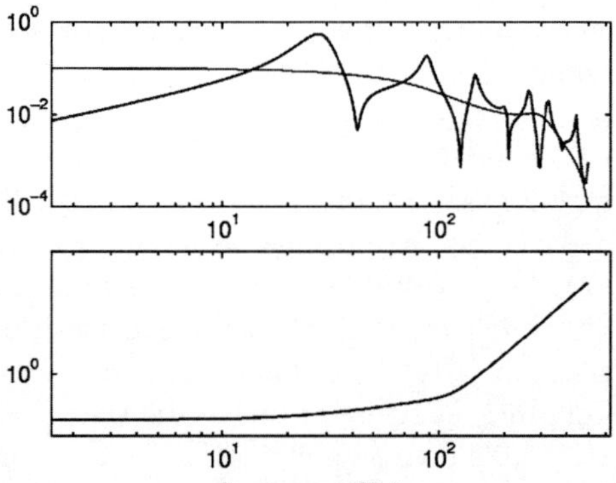

Fig. 3.21 *Top*: A magnitude plot of the transfer function between $u(k)$ and $y(k)$ of the true (*thick line*) and the estimated ARX model (*thin line*). *Bottom*: The weighting function $|A(e^{j\omega})|$

eters a_i, b_i, c_i, and d_i of a BJ model be stored in the parameter vector θ, then the parameter vector $\theta_\star$ minimizing the cost function

$$\bar{J}(\theta) = \lim_{N\to\infty} \frac{1}{N} \sum_{k=0}^{N-1} \left(y(k) - \hat{y}(k|k-1,\theta)\right)^2$$

satisfies

$$\theta_\star = \arg\min \frac{1}{N}\int_{-\pi}^{\pi}\left|G_0(e^{j\omega}) - \frac{B(e^{j\omega},\theta)}{A(e^{j\omega},\theta)}\right|^2\left|\frac{D(e^{j\omega},\theta)}{C(e^{j\omega},\theta)}\right|^2 \Phi^u(\omega) + \left|\frac{D(e^{j\omega},\theta)}{C(e^{j\omega},\theta)}\right|^2 \Phi^v(\omega)\, d\omega. \tag{3.245}$$

The proof is similar to the proof of Lemma 3.17 using the predictor related to the BJ model structure as given by (3.237).

Since the OE model structure is a special case of the BJ model structure, we can with a redefinition of the parameter vector θ immediately derive an expression for the parameter vector $\theta_\star$ of an OE model minimizing the cost function $\bar{J}(\theta)$:

$$\theta_\star = \arg\min \frac{1}{N}\int_{-\pi}^{\pi}\left|G_0(e^{j\omega}) - \frac{B(e^{j\omega},\theta)}{A(e^{j\omega},\theta)}\right|^2 \Phi^u(\omega) + \Phi^v(\omega)\, d\omega. \tag{3.246}$$

The use of Lemma 3.18 in qualitatively analyzing the bias in the estimate obtained with the OE model structure is highlighted with a continuation of identification example 3.11.

3.6.14 Identification Example 3.12

Making use of the same acoustical model of the duct as analyzed in identification example 3.11, we now attempt to estimate a sixth-order output-error model. By generating several realizations of the input and output data sequences with the same statistical properties as outlined in identification example 3.11, a series of sixth-order output-error models was estimated using the tools from the MATLAB System Identification toolbox [56]. Because of the nonquadratic nature of the cost function to be optimized by the output-error method, the numerical search discussed in Sect. 3.6.2 "got stuck" in a local minimum a number of times. The best result obtained out of 30 trials is presented below.

A Bode plot of the transfer function $G(e^{j\omega})$ is depicted by the thick line in Fig. 3.22. The transfer function $\hat{G}(e^{j\omega})$ of one estimated sixth-order OE model is also depicted in Fig. 3.22. Clearly, the most dominant peak around 25 Hz is completely captured. According to the theoretical qualitative analysis summarized by (3.246) for $\Psi^v(\omega) = 0$, it would be expected that the second most dominant peak around 90 Hz would be matched. However, this conclusion assumes that the global minimum of the cost function $\bar{J}(\theta)$ optimized by the output-error method has been found. The fact that the peak around 200 Hz is matched subsequently instead of the one around 90 Hz indicates that the global optimum still is not being found.

The BJ model structure allows us to estimate the parameters a_i and b_i for $i = 1, 2, \dots, n$ *unbiasedly*, irrespective of the values of the parameters c_i, d_i, for $i = 1, 2, \dots, n$, provided that they generate a stable predictor, and provided that n

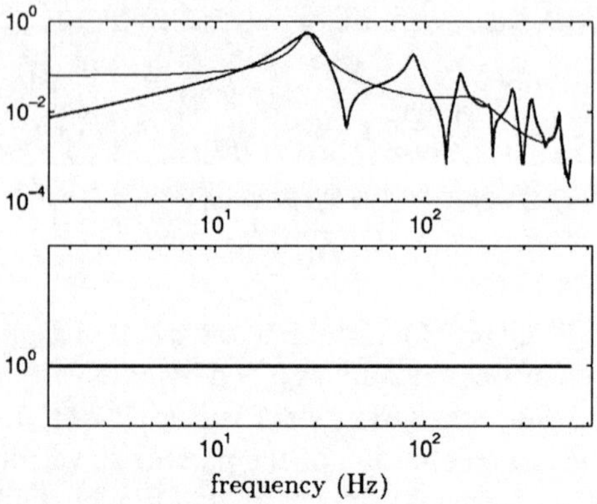

Fig. 3.22 *Top*: A magnitude plot of the transfer function between $u(k)$ and $y(k)$ of the true (*thick line*) and the estimated OE model (*thin line*). *Bottom*: The weighting function of the error on the transfer function estimate

corresponds to the true order of the data-generating system. Let the data-generating system be represented as

$$y(k) = \frac{B_0(q)}{A_0(q)} u(k) + v(k),$$

with $v(k)$ a stochastic zero-mean perturbation that is independent from $u(k)$. The BJ model structure has the ability to estimate the deterministic part,

$$\frac{B(q)}{A(q)} u(k),$$

correctly even if the noise part,

$$\frac{C(q)}{D(q)} e(k),$$

does not correspond to that in the underlying signal-generating system. To see this, let θ_{ab} denote the vector containing the quantities $a_i, b_i, i = 1, 2, \ldots, n$, and let θ_{cd} denote the vector containing the quantities $c_i, d_i, i = 1, 2, \ldots, n$. Consider the noise part of the BJ model to be fixed at some value $\bar{\theta}_{cd}$, then we can denote the criterion $J_N(\theta)$ as

$$\begin{aligned} J_N(\theta_{ab}, \bar{\theta}_{cd}) &= \frac{1}{N} \big(y(k) - \hat{y}(k|k-1) \big)^2 \\ &= \frac{1}{N} \sum_{k=0}^{N-1} \left[\frac{D(q, \bar{\theta}_{cd})}{C(q, \bar{\theta}_{cd})} \left(\frac{B_0(q)}{A_0(q)} u(k) + v(k) - \frac{B(q, \theta_{ab})}{A(q, \theta_{ab})} u(k) \right) \right]^2. \end{aligned}$$

When we take the limit $N \to \infty$ and assume ergodicity of the time sequences, then, by Parseval's identity (9.55) on p. 544, the prediction-error methods will perform the following minimization:

$$\min_{\theta_{ab}} \frac{1}{2\pi} \int_{-\pi}^{\pi} \left| \frac{D(e^{j\omega}, \bar{\theta}_{cd})}{C(e^{j\omega}, \bar{\theta}_{cd})} \right|^2 \underbrace{\left| \frac{B_0(e^{j\omega})}{A_0(e^{j\omega})} - \frac{B(e^{j\omega}, \theta_{ab})}{A(e^{j\omega}, \theta_{ab})} \right|^2}_{} \Phi^u(\omega)$$
$$+ \left| \frac{D(e^{j\omega}, \bar{\theta}_{cd})}{C(e^{j\omega}, \bar{\theta}_{cd})} \right|^2 \Phi^v(\omega) \, d\omega.$$

When n is correctly specified, or, more generally, when the orders of the polynomials $A_0(q)$ and $B_0(q)$ correspond exactly to the orders of the polynomials $A(q)$ and $B(q)$, respectively, the minimum that corresponds to the underbraced term is zero. Therefore, if the global optimum of the above criterion $\bar{J}(\theta_{ab})$ is found, the true values of the polynomials $A_0(q)$ and $B_0(q)$ are estimated.

3.6.15 Estimation Problems in Closed-Loop Systems

This section briefly highlights some of the complications that arise on using the prediction-error method with input and output samples recorded during a closed-loop experiment. We consider the closed-loop configuration of an LTI system P and an LTI controller C as depicted in Fig. 3.23. In general, system identification is much more difficult in closed-loop identification experiments. This will be illustrated by means of a few examples to highlight that, when identifying innovation models, it is necessary to parameterize *both* the deterministic and the stochastic part of the model *exactly equal* to the corresponding parts of the signal-generating system. The first example assumes only a correct parameterization of the deterministic part, whereas in the second example both the stochastic and the deterministic part are correctly parameterized.

3.6.16 Identification Example 3.13

Consider the feedback configuration in Fig. 3.23 driven by the external reference signal $r(k)$, with the system P given as

$$y(k) = b_1^0 u(k-1) + b_2^0 u(k-2) + v(k) \tag{3.247}$$

where $v(k)$ is a zero-mean stochastic sequence that is independent from the external reference $r(k)$. The controller C is a simple proportional controller [27], of the form

$$u(k) = K\big(r(k) - y(k)\big). \tag{3.248}$$

If we were to use an OE model structure with a correctly parameterized deterministic part corresponding to that of the system P, the one-step-ahead prediction error would be

$$\epsilon(k|k-1) = y(k) - \begin{bmatrix} u(k-1) & u(k-2) \end{bmatrix} \begin{bmatrix} b_1 \\ b_2 \end{bmatrix},$$

and with a prediction-error method we would solve the following least-squares problem:

$$\min_{b_1,b_2} \frac{1}{N} \sum_{k=0}^{N-1} \left(y(k) - \begin{bmatrix} u(k-1) & u(k-2) \end{bmatrix} \begin{bmatrix} b_1 \\ b_2 \end{bmatrix} \right)^2.$$

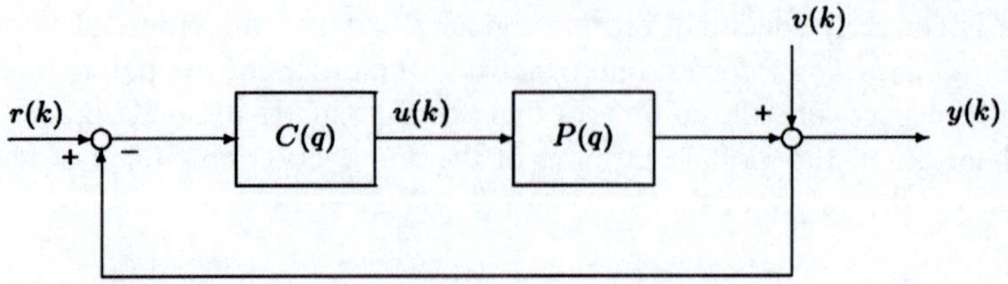

Fig. 3.23 A block scheme of an LTI system P in a closed-loop configuration with a controller C

If we substitute for $y(k)$ the expression given in (3.244), this problem can be written as

$$\min_{\beta} \frac{1}{N} \sum_{k=0}^{N-1} \left(\begin{bmatrix} u(k-1) & u(k-2) \end{bmatrix} \beta + v(k) \right)^2,$$

subject to

$$\beta = \begin{bmatrix} b_1^0 - b_1 \\ b_2^0 - b_2 \end{bmatrix}.$$

We assume the recorded time sequences to be ergodic. If the above least-squares problem has a unique solution in the limit of $N \to \infty$, this solution is zero ($\beta = 0$), *provided that* the following conditions are satisfied:

$$E\big[u(k-1)v(k)\big] = 0, \qquad E\big[u(k-2)v(k)\big] = 0. \tag{3.249}$$

However, substituting (3.247) into (3.248) yields

$$u(k) = \frac{K}{1 + Kb_1^0 q^{-1} + Kb_2^0 q^{-2}} r(k) - \frac{K}{1 + Kb_1^0 q^{-1} + Kb_2^0 q^{-2}} v(k),$$

which clearly shows that, for $K \neq 0$, the input $u(k)$ is not independent from the noise $v(k)$. For $K \neq 0$, the conditions (3.249) are satisfied only if $v(k)$ is a white-noise sequence. This corresponds to the correct parameterization of the stochastic part of the output-error model. If $v(k)$ were colored noise, biased estimates would result. This is in contrast to the open-loop case, for which the assumption that $u(k)$ and $v(k)$ are independent is sufficient to obtain unbiased estimates.

The final example in this section illustrates the necessity that the model set $\mathcal{M}(\theta)$ (3.163) encompasses *both* the deterministic and the stochastic part of the signal-generating system.

3.6.17 Identification Example 3.14

Consider the feedback configuration in Fig. 3.23 driven by the external reference signal $r(k)$, with the system P given as

$$y(k) = a^0 y(k-1) + b^0 u(k-1) + e(k) \tag{3.250}$$

where $e(k)$ is a zero-mean white-noise sequence. The controller C has the following dynamic form:

$$u(k) = fu(k.1) + g\big(r(k) - y(k)\big) \tag{3.251}$$

with $f, g \in \Re$. If we were to use an ARX model structure with correctly parameterized deterministic and stochastic parts for the system P, the one-step-ahead prediction error would be

$$\epsilon(k|k-1) = y(k) - \begin{bmatrix} y(k-1) & u(k-1) \end{bmatrix} \begin{bmatrix} a \\ b \end{bmatrix}.$$

Following identification example 3.12, the conditions for consistency become

$$\mathbb{E}\big[y(k-1)e(k)\big] = 0, \qquad E\big[u(k-1)e(k)\big] = 0. \tag{3.252}$$

These conditions hold since

$$\begin{aligned} u(k) &= \frac{g(1 - a^0 q^{-1})}{1 - (f + a^0 - gb^0)q^{-1} + fa^0 q^{-2}} r(k) \\ &\quad - \frac{g}{1 - (f + a^0 - gb^0)q^{-1} + fa^0 q^{-2}} e(k), \\ y(k) &= \frac{gb^0 q^{-1}}{1 - (f + a^0 - gb^0)q^{-1} + fa^0 q^{-2}} r(k) \\ &\quad + \frac{1 - fq^{-1}}{1 - (f + a^0 - gb^0)q^{-1} + fa^0 q^{-2}} e(k), \end{aligned}$$

and $e(k)$ is a white-noise sequence.

The consistency that is obtained in identification example 3.13 with a correctly parameterized ARX model of a system operating in closed-loop mode can be generalized for the class of MIMO innovation model structures (3.212)–(3.213) when the signal-generating system belongs to the model set.

3.6.18 Software

The described basis algorithms and variants have been implemented utilizing the commercial software standards: for system identification:

- The System Identification Toolbox in MATLAB, developed by L. Ljung, Linkoping, Sweden: http://www.mathworks.com/products/sysid/.
- The system identification package ADAPTx of Adaptics, Inc, developed by W.E. Larimore: http://www.adaptics.com/.
- The ISID-module in Xmath, developed by P. Van Overschee and Prof. B. De Moor and in license sold to ISI Inc. (now Wind River), USA: http://www.windriver.com.
- The software packages RaPID and INCA of IPCOS International: http://www.ipcos.be.
- The package MACEC, developed at the department of Civil Engineering of the K. U. Leuven in Belgium: http://www.kuleuven.ac.be/bwm/macec/.

- Products of LMS International: http://www.lms-international.com.

Additionally, public domain software, as

- SLICOT: http://www.win.tue.nl/niconet/NIC2/slicot.html.
- The SMI toolbox of the Control Laboratory at the T. U. Delft: http://lcewww.et.tudelft.nl/-verdult/smi/.
- The Cambridge University System Identification Toolbox http://www-control.eng.cam.ac.uk/jmm/cuedsid/.
- The website http://www.esat.kuleuven.ac.be/sista-cosic-docarch/ contains subspace identification algorithms.

3.7 Questions

1. For a given vector $y \in \Re^n$, there always exists an orthogonal Householder transformation Q such that

$$Q_y = \begin{bmatrix} \xi \\ 0 \\ \vdots \\ 0 \end{bmatrix},$$

with $\xi = \pm\|y\|_2$. Use this transformation to show that, for any pair of matrices $A \in \Re^{n\times n}$ and $C \in \Re^{l\times n}$, there exists an orthogonal transformation T_h such that the entries above the main diagonal of the matrix

$$\begin{bmatrix} CT_h \\ T_h^{-1}AT_h \end{bmatrix}$$

are zero.

2. Consider a parameterized model with parameters a_0, a_1, b_0, and b_1; and a transfer function given by

$$H(q, a_0, a_1, b_0, b_1) = \frac{b_1 q + b_0}{q^2 + a_1 q + a_0}.$$

For which values of the parameters a_0 and a_1 is this transfer function stable?

3. Consider the following single-input, multiple-output system:

$$y(k) = \begin{bmatrix} \frac{1+aq^{-1}}{(1+aq^{-1})(1+bq^{-1})} \\ \frac{1+bq^{-1}}{(1+aq^{-1})(1+bq^{-1})} \end{bmatrix} u(k).$$

a. Determine a state-space model of this system such that the C matrix of this state-space model equals the identity matrix.

b. Denote the state-space model derived above by

$$\begin{aligned} x(k+1) &= Ax(k) + Bu(k), \\ y(k) &= x(k). \end{aligned}$$

Show that the matrices A and B of this state-space model can be determined from a finite number of input and output measurements by solving a linear least-squares problem.

4. Consider the predictor model

$$\hat{y}(k,\theta) = \frac{b_q^{-1} + b_2 q^{-2}}{1 + a_1 q^{-1} + a_2 q^{-2}} u(k)$$

for $k \geq 2$, with unknown initial conditions $\hat{y}(0)$ and $\hat{y}(1)$. Show that, for

$$\begin{bmatrix} -a_1 & 1 \\ -a_2 & 0 \end{bmatrix}, \quad C = \begin{bmatrix} 1 & 0 \end{bmatrix},$$

the predictor can be written in the following form:

$$\hat{y}(k,\theta) = \phi(k, a_1, a_2) \begin{bmatrix} \hat{y}(0) \\ \hat{y}(1) \\ \hline b_1 \\ b_2 \end{bmatrix},$$

with $\phi(k, a_1, a_2)$ given by

$$\phi(k, a_1, a_2) = \left[CA^k \begin{bmatrix} 1 & 0 \\ a_1 & 1 \end{bmatrix} \sum_{\tau=0}^{k-1} CA^{k-1-\tau} \begin{bmatrix} 1 \\ 0 \end{bmatrix} u(\tau) \sum_{\tau=0}^{k-1-\tau} \begin{bmatrix} 0 \\ 1 \end{bmatrix} u(\tau) \right],$$

for $k \geq 2$.

5. Consider the predictor model

$$\hat{x}(k+1,\theta) = A(\theta)\hat{x}(k,\theta) + B(\theta)u(k),$$
$$\hat{y}(k,\theta) = C(\theta)\hat{x}(k,\theta) + D(\theta)u(k),$$

in observer canonical form with system matrices

$$A = \begin{bmatrix} 0 & -a_0 \\ 1 & -a_1 \end{bmatrix}, \qquad B = \begin{bmatrix} b_0 \\ b_1 \end{bmatrix}, \qquad C = \begin{bmatrix} 0 & 1 \end{bmatrix}, \qquad D = 0,$$

so that the parameter vector equals

$$\theta = \begin{bmatrix} a_0 & a_1 & b_0 & b_1 \end{bmatrix}.$$

a. Determine for this parameterization the system matrices

$$\frac{\partial A(\theta)}{\partial \theta(i)}, \quad \frac{\partial B(\theta)}{\partial \theta(i)}, \quad \frac{\partial C(\theta)}{\partial \theta(i)}, \quad \frac{\partial D(\theta)}{\partial \theta(i)},$$

for $i = 1, 2, 3, 4$, which are needed to compute the Jacobian of the output-error cost function using (3.182) and (3.183).

b. Determine the conditions on the parameter vector θ such that the combination of the above predictor model with the dynamic model (3.182) and (3.183) is asymptotically stable.

6. Consider the predictor model

$$\hat{x}(k+1,\theta) = A(\theta)\hat{x}(k,\theta) + B(\theta)u(k),$$
$$\hat{y}(k,\theta) = C(\theta)\hat{x}(k,\theta) + D(\theta)u(k),$$

with system matrices

$$A = \begin{bmatrix} 0 & 1 & 0 & \cdots & 0 \\ 0 & 0 & 1 & \cdots & 0 \\ \vdots & \vdots & \vdots & \ddots & \vdots \\ 0 & 0 & 0 & \cdots & 0 \end{bmatrix}, \qquad B = \begin{bmatrix} b_1 \\ b_2 \\ \vdots \\ b_n \end{bmatrix},$$
$$C = \begin{bmatrix} 1 & 0 & 0 & \cdots & 0 \end{bmatrix}, \qquad D = 0,$$

and parameter vector $\theta = [b_1, \dots, b_n]$.

a. Show that the predictor model can be written as

$$\hat{y}(k,\theta) = \left(b_1 q^{-1} + b_2 q^{-2} + \cdots b_n q^{-n}\right)u(k).$$

b. Show that the gradients

$$\frac{\partial \hat{y}(k,\theta)}{\partial \theta_i}, \qquad i = 1,2,\dots,n,$$

are equal to their finite-difference approximations given by

$$\frac{y(k,\theta - y(k,\theta + \Delta e_i))}{\Delta}, \qquad i = 1,2,\dots,n,$$

with $\Delta \in \Re$ and $e_i \in \Re^n$ a vector with the ith entry equal to 1 and the other entries equal to zero.

c. Determine the adjoint state-space equation (3.188) and evaluate (3.189).

7. We are given the system described by

$$y(k) = \left(b_0 + b_1 q^{-1}\right)u(k) + e(k),$$

with $u(k)$ and e(k) ergodic, zero-mean, and statistically independent stochastic sequences. The sequence $u(k)$ satisfies

$$E\left[u(k)^2\right] = \sigma_u^2, \qquad E\left[u(k)u(k-1)\right] = \Gamma$$

where $\Gamma \in \Re$ and $e(k)$ is a white-noise sequence with variance σ_e^2. Using input–output measurements of this system, we attempt to estimate the unknown coefficient b of the output predictor given by

$$\hat{y}(k,b) = bu(k-1).$$

a. Determine a closed-form expression for the prediction error criterion for $N \in \infty$, given by

$$\bar{J}(b) = \lim_{N\to\infty} \frac{1}{N} \sum_{k=0}^{N-1} \left(y(k) - \hat{y}(k)\right)^2,$$

in terms of the unknown parameter b.

b. Determine the parameter value of $\hat{b}$ that satisfies

$$\hat{b} = \arg\min \bar{J}(b).$$

c. Use the expression derived for $\hat{b}$ to determine conditions on the input $u(k)$ such that $\hat{b} = b_1$.

8. Show that, for $X \in \Re^{n\times n}$,

$$(I_n + X)^{-1} = I_n - X + X^2 - X^3 + \cdots + (-1)^n X_n (I_n + X)^{-1},$$

and thus that a first-order approximation of $(I_n + X)^{-1}$ equals $I_n - X$.

9. Given the matrices

$$A = \begin{bmatrix} 1.5 & 1 \\ -0.7 & 0 \end{bmatrix}, \qquad \bar{A} = \begin{bmatrix} 1.5 & 1 \\ -\alpha^2 + 1.5\alpha - 0.7 & \alpha \end{bmatrix},$$

with $\alpha \in \Re$,

a. Determine a similarity transformation such that $A = T^{-1} A T$.

b. Approximate the similarity transformation as $I_n + \Delta T$ and determine ΔT as in Sect. 3.5.5.

10. Consider the constrained least-squares problem

$$\min_{\theta \in \text{range}(U)} \quad \|Y - \Phi\theta\|_2^2 \tag{3.253}$$

with the matrices $\Phi \in \Re^{N\times n} (n < N)$, $Y \in \Re^N$, and $\theta \in \Re^n$, and with the matrix $U \in \Re^{n\times p} (p < n)$ of full column rank. Show that, if the product ΦU has full column rank, the solution to (3.253) satisfies

$$\hat{\theta} = U\left(U^t \Phi^t \Phi U\right)^{-1} U^t \Phi^t Y.$$

11. Consider the transfer function

$$\mathcal{M}(z) = \begin{bmatrix} D & 0 \end{bmatrix} + C\left(zI - (A - KC)\right)^{-1} \begin{bmatrix} B & K \end{bmatrix},$$

with arbitrary system matrices $A \in \Re^{n\times n}$, $B \in \Re^{n\times m}$, $C \in \Re^{l\times n}$, $D \in \Re^{l\times m}$, and $K \in \Re^{n\times l}$.

a. Let $a(z)$ be a scalar polynomial of order n given by

$$a(z) = z^n + a_1 z^{n-1} + \cdots + a_n.$$

Let $\mathcal{B}(z)$ and $\mathcal{K}(z)$ be polynomial matrices with polynomial entries of order $n-1$ given as

$$\mathcal{B}(z) = \begin{bmatrix} b_{11}(z) & \cdots & b_{1m}(z) \\ \vdots & \ddots & \vdots \\ b_{l1}(z) & \cdots & b_{lm}(z) \end{bmatrix},$$

$$\mathcal{K}(z) = \begin{bmatrix} k_{11}(z) & \cdots & k_{1l}(z) \\ \vdots & \ddots & \vdots \\ k_{l1}(z) & \cdots & k_{ll}(z) \end{bmatrix}.$$

Show that the transfer function $\mathcal{M}(z)$ can be expressed as

$$\begin{bmatrix} D & 0 \end{bmatrix} + \frac{\begin{bmatrix} \mathcal{B}(z) & \mathcal{K}(z) \end{bmatrix}}{a(z)}.$$

b. For the special case $l = 1$ show that the observable canonical form (3.160) and (3.161) is a surjective parameterization of the transfer function $\mathcal{M}(z)$.

12. Consider the one-step-ahead predictor for a second-order ($n = 2$) ARMAX model as given in Lemma 3.13. Let $c_i = a_i + k_i$ ($i = 1, 2$). The parameters in the one-step-ahead prediction will be estimated using N measurements of the input $u(k)$ and the output $y(k)$ of the system:

$$y(k) = \frac{q^{-1} + 0.5q^{-2}}{1 - 1.5q^{-1} + 0.7q^{-2}} u(k) + v(k),$$

with $u(k)$ and $v(k)$ zero-mean, statistically independent white-noise sequences of unit variance.

a. Determine an expression for the matrix $\Phi(c_1, c_2)$ such that the prediction-error criterion $J_N(c_1, c_2, \theta_{bk})$ can be written as

$$J_N(c_1, c_2, \theta_{bk}) = \frac{1}{N} \left\| Y - \Phi(c_1, c_2)\theta_{bk} \right\|_2^2,$$

with

$$\theta_{bk} = \begin{bmatrix} \hat{x}(0)^t & k_1 & k_2 & b_1 & b_2 \end{bmatrix}^t,$$
$$Y = \begin{bmatrix} y(0) & y(1) & \cdots & y(N-1) \end{bmatrix}^t.$$

b. If the coefficient c_2 is fixed to its true value 0.7, derive the condition on c_1 such that the ARMAX predictor is asymptotically stable.

c. Write a MATLAB program that calculates the matrix $\Phi(c_1, c_2)$, and takes as input arguments the vector $c = [c_1 \ c_2]$, the output sequence Y, and the input sequence stored in the vector $U = [u(1) \ u(2) \ \cdots \ u(N)]^t$.

d. Let δ_S denote the interval on the real axis for which the ARMAX predictor with $c_2 = 0.7$ is asymptotically stable. Plot the prediction-error criterion $J_N(c_1, 0.7, \theta_{bk})$ as a function of $c_1 \in \delta_S$. Does the minimal value of this criterion indicate the correct value of c_1?

13. Consider the ARX predictor given by (3.228). Using the measurements $u(k)$ and $y(k)$ acquired in the closed-loop configuration with an LTI controller with transfer function $C(e^{j\omega})$ as depicted in Fig. 3.23, the task is to estimate an ARX model for the unknown plant P. Show that, in the limit of $N \to \infty$, the prediction-error method attempts to find the following estimate:

$$\begin{aligned} \theta_\star = \arg\min \frac{1}{2\pi} \int_{-\pi}^{\pi} & \left| P(e^{j\omega}) - \frac{B(e^{j\omega}, \theta)}{A(e^{j\omega}, \theta)} \right|^2 \\ & \times \left| \frac{A(e^{j\omega}, \theta) C(e^{j\omega}, \theta)}{1 + P(e^{j\omega}) C(e^{j\omega})} \right|^2 \Phi^r(\omega) \\ & + \left| \frac{1 + \frac{B(e^{j\omega}, \theta)}{A(e^{j\omega}, \theta)} C(e^{j\omega})}{1 + P(e^{j\omega}) C(e^{j\omega})} \right|^2 \left| A(e^{j\omega}, \theta) \right|^2 \Phi^v(\omega)\, d\omega. \end{aligned}$$

14. Let the following state-space model be given:

$$x(k+1) = Ax(k) + Bu(k),$$
$$y(k) = Cx(k) + u(k).$$

a. Show that the transfer function describing the transfer from $u(k)$ to $y(k)$ is given as

$$y(k) = \big(I + C(qI - A)^{-1}B\big)u(k).$$

b. Show that the transfer function describing the transfer from $y(k)$ to $u(k)$ is given as

$$u(k) = \big(I - C\big(qI - (A - BC)\big)^{-1}B\big)y(k).$$

15. Consider the OE predictor given by (3.239). Using the measurements $u(k)$ and $y(k)$ acquired in the closed-loop configuration with the LTI controller with transfer function $C(e^{j\omega})$ as depicted in Fig. 8.5 on p. 471, the task is to estimate an OE model for the unknown plant P:

a. Show that, in the limit of $N \to \infty$, the prediction-error method attempts to find the following estimate:

$$\begin{aligned}\theta_\star = \arg\min \frac{1}{2\pi}\int_{-\pi}^{\pi} &\left|P(e^{j\omega}) - \frac{B(e^{j\omega},\theta)}{A(e^{j\omega},\theta)}\right|^2 \\ &\times \left|\frac{C(e^{j\omega})}{1 + P(e^{j\omega})C(e^{j\omega})}\right|^2 \Phi^2(\omega) \\ &+ \left|\frac{1 + \frac{B(e^{j\omega},\theta)}{A(e^{j\omega},\theta)}C(e^{j\omega})}{1 + P(e^{j\omega})C(e^{j\omega})}\right|^2 \Phi^v(\omega)\,d\omega.\end{aligned}$$

b. Show that, for $v(k) = 0$, the model given by

$$\frac{B(e^{j\omega},\theta)}{A(e^{j\omega},\theta)}$$

approximates the system $P(e^{j\omega})$ accurately in the so-called *cross-over-frequency* region, that is, the frequency region in which the loop gain $P(e^{j\omega})C(e^{j\omega}) \approx -1$.

16. Adapted from [54]. We are given the system described by

$$y(k) = \frac{b_0 q^{-1}}{1 + a_0 q^{-1}}u(k) + \frac{1 + c_0 q^{-1}}{1 + a_0 q^{-1}}e(k),$$

with $u(k)$ and $e(k)$ ergodic, zero-mean and statistically independent white-noise sequences with variances σ_u^2 and σ_e^2, respectively. Using N measurements of the input and the output of this system, we attempt to estimate the two unknown coefficients a and b in a first-order ARX model.

a. Show that, in the limit of $N \to \infty$,

$$\mathbb{E}\big[y^2(k)\big] = \frac{b_0^2\sigma_u^2 + (c_0(c_0 - a_0) - a_0 c_0 + 1)\sigma_e^2}{1 - a_0^2}.$$

b. Show that, in the limit of $N \to \infty$, the prediction-error criterion $\bar{J}(a,b)$ that is minimized by the ARX method is given as

$$\bar{J}(a,b) = \mathbb{E}\big[y^2(k)\big]\big(1 + a^2 - 2aa_0\big) + \big(b^2 - 2bb_0\big)\sigma_u^2 + 2ac_0\sigma_e^2.$$

c. Show that, in the limit of $N \to \infty$, the optimal parameter values for a and b that minimize the above criterion are

$$\hat{a} = a_0 - \frac{c_0}{E[y^2(k)]}\sigma_e^2,$$
$$\hat{b} = b_0.$$

d. Show by explicitly evaluating the criterion values $\bar{J}(\hat{a}, \hat{b})$ and $\bar{J}(a_0, b_0)$ that, in the limit of $N \to \infty$, the following relationship holds:

$$\bar{J}(\hat{a}, \hat{b}) < \bar{J}(a_0, b_0).$$

3.8 Notes and References

In this section, we briefly comment on the relation with other identification methods for linear systems, we elaborate on some important open problems and briefly discuss several extensions. For further elaboration and wide scope, the reader is advised to look up Refs. [1–4, 23–26, 28, 41, 46–48, 60, 62, 67, 81, 94].

As we have shown in Fig. 3.8, the so-called classical identification methods first determine a model (and if needed then proceed via a Kalman filter to estimate a state sequence). A good introduction to these methods including such as least squares methods, instrumental variables, prediction error methods (PEM), and others can be found in this Encyclopedia under *Identification of linear Systems in Time Domain*. Obviously, subspace identification algorithms are just one (important) group of methods for identifying linear systems. But many users of system identification prefer to start from linear input–output models, parametrized by numerator and denominator polynomials and then use maximum likelihood or instrumental variables based techniques. The at first sight apparent advantage of having an input–output parametrization however often turns out to be a disadvantage, as the theory of parameterizations of multivariable systems is certainly not easy nor straightforward and therefore complicates the required optimization algorithms (for example, there is not one single parametrization for a multiple-output system).

In many implementations of PEM-identification, a model obtained by subspace identification typically serves as a good initial guess. Recall that PEMs require a nonlinear nonconvex optimization problem to be solved, for which a good initial guess if required.

Another often mentioned disadvantage of subspace methods is the fact that it does not optimize a certain cost function. The reason for this is that, contrary to input–output models (transfer matrices), we can not (as of this moment) formulate a likelihood function for the identification of the state space model, that also leads to an amenable optimization problem. So, in a certain sense, subspace identification algorithms provide (often surprisingly good) 'approximations' of the linear model, but there is still a lot of ongoing research on how the identified model relates to a maximum likelihood formulation of the problem. In particular, it is also not straightforward at all to derive expressions for the error covariances on the estimates, nor the quantify exactly in what sense the obtained state sequence is an approximation to the 'real' (theoretical) Kalman filter state sequence, if some of the assumptions we made are not satisfied and/or the block dimensions i and/or j are not infinite (which they never are in practice). Yet, it is our experience that subspace algorithms often tend to give very good linear models for industrial data sets. By now, in the literature, many successful implementations and cases have been reported in mechanical engineering (modal and vibrational analysis of mechanical structures such as cars, bridges (civil engineering), airplane wings (flutter analysis), missiles (ESA's Ariane), etc.), process industries (chemical, steel, paper and pulp, ...), data assimilation methods (in which large systems of PDEs are discretized and reconciliated with observations using large scale Kalman filters and subspace identification methods are used in an 'error correction' mode), dynamic texture (reduction of sequences of images that are highly correlated in both space (within one image) and time (over several images).

Since the introduction of subspace identification algorithms, the basic ideas have been extended to other system classes, such as closed-loop systems, linear parameter-varying state-space systems, bilinear systems, continuous-time systems, descriptor systems, periodic systems. We refer the reader to the bibliography for more information. Furthermore, efforts have been made to fine-tune the algorithms as presented in this paper. For example, several algorithms have been proposed to ensure stability of the identified model. For stochastic models, the positive-realness property should hold, which is not guaranteed by the raw subspace algorithms for certain data sets. Also for this problem, extensions have been made.

In this chapter, we discussed the identification of an LTI state-space model based on a finite number of input and output measurements. We assume that the order of the system is given and that the disturbances can be modeled as an additive white-noise signal to the output. The first step in estimating the parameters is the determination of a parameterization of the LTI state-space system. A parameterization is a mapping from the space of parameters to the space of rational transfer functions that describe the LTI system. We discuss injective, surjective, and bijective properties of parameterizations and highlight the numerical sensitivity of certain parameterizations. We describe the output-normal parameterization and the tridiagonal parameterization in detail.

For the estimation of the parameters, we need a criterion to judge the quality of a particular value of the parameters. We introduce the output-error cost function for this purpose and show that the properties of this cost function depend on the particular parameterization that is used. For most parameterizations considered in this chapter, the cost function is non-convex and has multiple local minima.

To obtain the optimal values of the parameters with respect to the output-error cost function, we numerically minimize this cost function. We discuss the Gauss–Newton, regularized Gauss–Newton, and steepest-descent methods. In addition, we present an alternative approach called the gradient-projection method that can be used to deal with full parameterizations. These numerical procedures are guaranteed only to find local minima of the cost function.

To analyze the accuracy of the estimates obtained by minimizing the output-error cost function, we derived an expression for the covariance matrix of the error between the true and the estimated parameters.

If the additive disturbance to the output is a colored, nonwhite noise, then the output-error method does not yield the minimum-variance estimates of the parameters. To deal with this problem, we discussed two approaches. The first approach is to apply a weighting with the inverse of the covariance matrix of the additive disturbance in the output-error cost function. The second approach is to optimize the prediction error instead of the output error. The prediction-error methods will be discussed in greater detail in the next chapter.

In [5], the question of estimating the order in the context of subspace methods is addressed. Three different approaches are presented and the asymptotic properties thereof derived. Two of these methods are based on the information contained in the estimated singular values, while the third method is based on the estimated innovation variance.

Bauer et al. [7] presented states asymptotic normality of subspace estimates. In addition, a consistency result for the system matrix estimates is given. An algorithm to compute the asymptotic variances of the estimates is presented.

The effect of some weighting matrices on the asymptotic variance of the estimates of linear discrete time state space systems estimated using subspace methods was investigated in [6]. The analysis deals with systems with white or without observed inputs and refers to the Larimore type of subspace procedures. The main result expresses the asymptotic variance of the system matrix estimates in canonical form as a function of some of the user choices, clarifying the question on how to choose them optimally. It is shown that the CCA weighting scheme leads to optimal accuracy.

A new structure for subspace identification algorithms is proposed in [12] to help fixing problems when certain experimental conditions cause ill-conditioning.

The major costs in the identification of state-space models still remain because of the need for the singular value (or sometimes QR) decomposition. It turns out that proper exploitation, using results from the theory of displacement structure, of the Toeplitz-like nature of several matrices arising in the procedure reduces the computational effort [13].

In many on-line identification scenarios with slowly time-varying systems, it is desirable to update the model as time goes on with the minimal computational burden. In [14], the results of the batch processing algorithm are extended to allow updating of the identified state space model with few flops.

The problem of identifying multivariable finite dimensional linear time-invariant systems from noisy input/output measurements was considered in [15]. Apart from the fact that both the measured input and output are corrupted by additive white noise, the output may also be contaminated by a term which is caused by a white input process noise; furthermore, all these noise processes are allowed to be correlated with each other.

In [16], algorithms were presented to find stable approximants to a least-squares problem, which are then applied to subspace methods to ensure stability of the identified model.

It is known that certain popular stochastic subspace identification methods may fail for theoretical reasons related to positive realness. In [19], the authors describe how to generate data for which the methods do not find a model.

The paper [24] describes the modification of the family of MOESP subspace algorithms when identifying mixed causal and anti-causal systems.

In [86], two subspace algorithms are presented to realize a finite dimensional, linear time-invariant state-space model from input–output data. Both schemes are versions of the MIMO Output-Error State Space model identification (MOESP) approach.

The elementary MOESP algorithm is analyzed in [85] It is shown that the MOESP implementation yields asymptotically unbiased estimates. Furthermore, the model reduction capabilities of the elementary MOESP schemes are analyzed when the observations are error-free. On the other hand, the ordinary MOESP algorithm is analyzed and extended in [87]. The extension of the ordinary MOESP scheme with instrumental variables increases the applicability of this scheme. Moreover, in [88], the extension of the MOESP family of subspace model identification schemes to the Hammerstein type of nonlinear system is outlined.

In [89], subspace model identification algorithms that allow the identification of a linear, time-varying state space model from an ensemble set of input–output measurements are presented.

An overview of existing subspace-based techniques for system identification was given in [90]. The methods are grouped into the classes of realization-based and direct techniques. Similarities between different algorithms were pointed out, and their applicability is commented upon.

The paper [91] gave a statistical investigation of subspace-based system identification techniques. Explicit expressions for the asymptotic estimation error variances of the corresponding pole estimates were given.

In [93], it is shown that the MOESP class of subspace identification schemes can be extended to identify Wiener systems, a series connection of a linear dynamic system followed by a static nonlinearity.

A simulation study, in which the performances of the subspace and the transfer function approaches are compared [71], shows that the latter can provide more accurate models than the former at a lower computational cost.

The paper [74] shows how one can impose stability to the model that is identified with a subspace algorithm. The method proposed is based on regularization.

In [75], a subspace algorithm is derived to consistently identify stochastic state space models from given output data. Two subspace algorithms for the identification of mixed deterministic-stochastic systems are derived [76]. Similarities between three different subspace algorithms for the identification of combined deterministic-stochastic systems are presented in [77]. It is shown that all three algorithms are special cases of one unifying scheme. In the book [78], the theory of subspace identification algorithms is presented in detail. A subspace identification method is discussed [80] that deals with multivariable linear parameter varying state-space systems with affine parameter dependence. A general overview [30] of subspace system identification methods is given. A comparison between subspace identification and prediction error methods is made on the basis of computational complexity and precision of the methods by applying them on 10 industrial data sets. The class of existing linear subspace identification techniques is generalized to subspace identification algorithms for bilinear systems [29]. In [63], four subspace algorithms which are based on an initial estimate of the state are considered. For the algorithms considered, a consistency result is proved. In a simulation study, the relative (statistical) efficiency of these algorithms in relation to the maximum likelihood algorithm is investigated

The identification of discrete-time bilinear state space systems with multiple inputs and multiple outputs is discussed [79]. The subspace algorithm is modified such that it reduces the dimension of the matrices involved.

In [83], the identification of linear time-invariant (LTI) systems operating in a closed-loop with an LTI compensator is reformulated to an open-loop multi -input-multi-output (MIMO) (state space model) identification problem, followed by a model reduction step. The open-loop identification problem is solved by the MOESP (MIMO output-error state space model) identification technique. Two algorithms to identify a linear, time-invariant, finite dimensional state space model from input–output data are described [84]. The system to be identified is assumed to be excited by a measurable input and an unknown process noise and the measurements are disturbed by unknown measurement noise. Both noise sequences are discrete zero-mean white noise.

The Kullback information is developed as the natural measure of the error in model approximation for general model selection methods including the selection of model state order in large as well as small samples [50]. It also plays a central role in developing statistical decision procedures for the optimal selection of model order as well as structure based on the observed data. The optimality of the canonical variate analysis (CVA) method is demonstrated for both an open and closed-loop multivariable system with stochastic disturbances.

In [52], the authors analyze a class of state space identification algorithms for time-series, based on canonical correlation analysis, in the light of recent results on stochastic systems theory. In this paper, the statistical problem of stochastic modeling from estimated covariances is phrased in the geometric language of stochastic realization theory.

The problem of MIMO recursive identification is analyzed [55] within the framework of subspace model identification and the use of recent signal processing algorithms for the recursive update of the singular value decomposition is proposed.

An identification algorithm which identifies low complexity models of infinite-dimensional systems from equidistant frequency-response data is presented [58]. The new algorithm is a combination of the Fourier transform technique with subspace techniques.

In [64], the stochastic realization of stationary processes with exogenous inputs in the absence of feedback is studied, and its application to identification is briefly discussed.

A method of identification of linear input–output models using canonical variate analysis (CVA) is developed [66] for application to chemical processes identification and compares it with the traditional prediction error methods. The authors present several comparisons between prediction error methods and subspace methods, including comparisons of accuracy and computational effort.

In [42], one shows that state-space subspace system identification (4SID) can be viewed as a linear regression multistep-ahead prediction error method with certain rank constraints.

In [43], the consistency of a large class of methods for estimating the extended observability matrix is analyzed. Persistence of excitation conditions on the input signal are given which guarantee consistent estimates for systems with only measurement noise. For systems with process noise, it is shown that a persistence of excitation condition on the input is not sufficient. More precisely, an example for which the subspace methods fail to give a consistent estimate of the transfer function is given. This failure occurs even if the input is persistently exciting of any order. It is also shown that this problem can be eliminated if stronger conditions on the input signal are imposed.

The Tennessee Eastman challenge process is a realistic simulation of a chemical process that has been widely used in process control studies [45]. In this case study, several identification methods are examined and used to develop MIMO models that contain seven inputs and ten outputs. For a variety of reasons, the only successful models are the state-space models produced by two popular subspace algorithms, N4SID and canonical variate analysis (CVA). The CVA model is the most accurate.

References

1. Akaike, H.: Stochastic theory of minimal realization. IEEE Trans. Autom. Control **19**, 667–674 (1974)
2. Akaike, H.: Markovian representation of stochastic processes by canonical variables. SIAM J. Control **13**(1), 162–173 (1977)
3. Aoki, M.: State Space Modeling of Time Series. Springer-Verlag, Berlin (1987)
4. Astrom, K.J., Eykhoff, P.: System identification—A survey. Automatica **7**(2), 123–162 (1971)
5. Bauer, D.: Order estimation for subspace methods. Automatica **37**, 1561–1573 (2001)
6. Bauer, D., Ljung, L.: Some facts about the choice of the weighting matrices in Larimore type of subspace algorithms. Automatica **38**(5), 763–773 (2002)
7. Bauer, D., Deistler, M., Scherrer, W.: Consistency and asymptotic normality of some subspace algorithms for systems without observed inputs. Automatica **35**, 1243–1254 (1999)
8. Bergboer, N., Verdult, V., Verhaegen, M.: An efficient implementation of maximum likelihood identification of LTI state-space models by local gradient search. In: Proc. 41st IEEE Conference on Decision and Control, Las Vegas, Nevada, pp. 616–621. IEEE Press, Piscataway (2002)
9. Box, G.E.P., Jenkins, G.M.: Time Series Analysis, Forecasting and Control. Holden-Day, San Francisco (1970)
10. Brewer, J.W.: Kronecker products and matrix calculus in system theory. IEEE Trans. Circuits Syst. **25**(9), 772–781 (1978)
11. Bruls, J., Chou, C.T., Haverkamp, B., Verhaegen, M.: Linear and non-linear system identification using separable least-squares. Eur. J. Control **5**(1), 116–128 (1999)
12. Chiuso, A., Picci, G.: Some algorithmic aspects of subspace identification with inputs. Int. J. Appl. Math. Comput. Sci. **11**(1), 55–75 (2001)
13. Cho, Y., Xu, G., Kailath, T.: Fast identification of state space models via exploitation of displacement structure. IEEE Trans. Autom. Control **AC-39**(10), 2004–2017 (1994)
14. Cho, Y., Xu, G., Kailath, T.: Fast recursive identification of state space models via exploitation of displacement structure. Automatica **30**(1), 45–59 (1994)
15. Chou, C.T., Verhaegen, M.: Subspace algorithms for the identification of multivariable dynamic errors-in-variables models. Automatica **33**(10), 1857–1869 (1997)
16. Chui, N.L.C., Maciejowski, J.M.: Realization of stable models with subspace methods. Automatica **32**(11), 1587–1595 (1996)
17. Clarke, D.: Generalized least squares estimation of parameters of a dynamic model. In: Proc. First IFAC Symposium on Identification in Automatic Control Systems, pp. 3–17 (1967)

18. Clarke, D.W., Mohtadi, C., Tuffs, P.S.: Generalized predictive control part I: The basic algorithm. Automatica **23**(2), 137–148 (1987)
19. Dahlen, A., Lindquist, A., Mari, J.: Experimental evidence showing that stochastic subspace identification methods may fail. Syst. Control. Lett. **34**, 303–312 (1998)
20. David, B.: Parameter estimation in nonlinear dynamical systems with correlated noise. Ph D thesis, Universite Catholique de Louvain, Louvain LaNeuve, Belgium (2001)
21. David, B., Bastin, G.: An estimator of the inverse covariance matrix and its application to ML parameter estimation in dynamical systems. Automatica **37**(1), 99–106 (2001)
22. De Bruyne, F., Gevers, M.: Identification for control: Can the optimal restricted complexity model always be indentified. In: Proc. of the 33rd IEEE Conference on Decision and Control, Orlando, Florida, pp. 3912–3917. IEEE Press, Piscataway (1994)
23. De Moor, B., Van Overschee, P.: Graphical User Interface Software for System Identification. Tech. Rep. Report 94-06I, ESAT-SISTA, Department of Electrical Engineering, Katholieke Universiteit Leuven, Belgium (1994)
24. De Moor, B., Van Overschee, P., Favoreel, W.: Algorithms for subspace state-space system identification: An overview. Appl. Comput. Control Signals Circuits. **1**, 247–311 (1999)
25. Desai, U.B., Pal, D.: A transformation approach to stochastic model reduction. IEEE Trans. Autom. Control **AC-29**(12), 1097–1100 (1984)
26. Desai, U.B., Kirkpatrick, R.D., Pal, D.: A realization approach to stochastic model reduction. Int. J. Control **42**(2), 821–838 (1985)
27. Dorf, R., Bishop, R.: Modern Control Systems, 8th edn. Addison-Wesley, New York (1998)
28. Faure, P.: Stochastic realization algorithms. In: Mehra, R., Lainiotis, D. (eds.) System Identification: Advances and Case Studies. Academic Press, San Diego (1976)
29. Favoreel, W., De Moor, B., Van Overschee, P.: Subspace identification of bilinear systems subject to white inputs. IEEE Trans. Autom. Control **44**(6), 1157–1165 (1999)
30. Favoreel, W., De Moor, B., Van Overschee, P.: Subspace state space system identification for industrial processes. J. Process Control **10**, 149–155 (2000)
31. Garcia, C.E., Prett, D.M., Morari, M.: Model predictive control: Theory and practice – a survey. Automatica **25**(3), 335–348 (1989)
32. Gevers, M.: Towards a joint design of identification and control. In: Trentelman, H.L., Willems, J.C. (eds.) Essays on Control: Perspectives in the Theory and Its Applications, pp. 111–151. Birkhauser, Boston (1993)
33. Golub, G.H., Pereyra, V.: The differentiation of pseudo-inverses and nonlinear least squares problems whose variables separate. SIAM J. Numer. Anal. **10**(2), 413–432 (1973)
34. Golub, G.H., Van Loan, C.F.: Matrix Computations, 3rd edn. The Johns Hopkins University Press, Baltimore (1996)
35. Hanzon, B., Ober, R.J.: Overlapping block-balanced canonical forms and parametrizations: The stable SISO case. SIAM J. Control Optim. **35**(1), 228–242 (1997)
36. Hanzon, B., Ober, R.J.: Overlapping block-balanced canonical forms for various classes of linear systems. Linear Algebra Appl. **281**, 171–225 (1998)
37. Hanzon, B., Peeters, R.L.M.: Balanced parameterizations of stable SISO all-pass systems in discrete time. Math. Control Signals Syst. **13**(3), 240–276 (2000)
38. Hanzon, B., Peeters, R., Olivi, M.: Balanced parameterizations of discrete-time stable all-pass systems and the tangential Schur algorithm. In: Proc. of the European Control Conference, Karlsruhe, 1999. Duisburg: Universitat Duisburg (CD Info: http://www.uniduisburg.de/euca/ecc99/proceedi.htm)
39. Haverkamp, B.: Subspace method identification, theory and practice. Ph D Thesis, Delft University of Technology, Delft, The Netherlands (2000)
40. Hayes, M.H.: Statistical Digital Signal Processing and Modeling. John Wiley and Sons, New York (1996)
41. Ho, B.L., Kalman, R.E.: Effective construction of linear state-variable models from input/output functions. Regelungstechnik **14**(12), 545–548 (1966)
42. Jansson, M., Wahlberg, B.: A linear regression approach to state-space subspace system identification. Signal Process. **52**(2), 103–129 (1996)

43. Jansson, M., Wahlberg, B.: On consistency of subspace methods for system identification. Automatica **34**(12), 1507–1519 (1998)
44. Johansson, R.: System Modeling and Identification. Prentice-Hall, Englewood Cliffs (1993)
45. Juricek, B.C., Seborg, D.E., Larimore, W.E.: Identification of the Tennessee Eastman challenge process with subspace methods. Control Engineering Practice **9**(12), 1337–1351 (2001)
46. Kalman, R.E.: A new approach to linear filtering and prediction problems. Trans. Am. Soc. Mech. Eng., J. Basic Eng. **83**(1), 35–45 (1960)
47. Kalman, R.E.: Mathematical description of linear dynamical systems. SIAM J. Control **1**, 152–192 (1963)
48. Kung, S.Y.: A new identification method and model reduction algorithm via singular value decomposition. In: Proc. the 12th Asilomar Conference on Circuits, Systems and Computations, pp. 705–714 (1978)
49. Landau, Y.D.: Adaptive Control: The Model Reference Approach. Marcel Dekker, New York (1979)
50. Larimore, W.E.: Statistical optimality and canonical variate analysis system identification. Signal Process. **52**(2), 131–144 (1996)
51. Lee, L.H., Poolla, K.: Identification of linear parameter-varying systems using nonlinear programming. J. Dyn. Syst. Meas. Control **121**(1), 71–78 (1999)
52. Lindquist, A., Picci, G.: Canonical correlation analysis, approximate covariance extension, and identification of stationary time series. Automatica **32**(4), 709–733 (1996)
53. Ljung, L.: Convergence analysis of parametric identification methods. IEEE Trans. Autom. Control **23**(5), 770–783 (1978)
54. Ljung, J.: System Identification: Theory for the User, 2nd edn. Prentice Hall, New Jersey (1999)
55. Lovera, M., Gustafsson, T., Verhaegen, M.: Recursive subspace identification of linear and non-linear wiener state-space models, Automatica **36**(8), 1639–1650 (2000)
56. Mathworks, System identification toolbox 7.4, available at http://www.mathworks.com/products/sysid.
57. McKelvey, T.: Identification of state-space model from time and frequency data. PhD thesis, Linkoping University, Linkoping, Sweden (1995)
58. McKelvey, T., Akcay, H., Ljung, L.: Subspace-based identification of infinite-dimensional multivariable systems from frequency-response data. Automatica **32**(6), 885–902 (1996)
59. McKelvey, T., Helmersson, A.: System identification using overparametrized model class – improving the optimization algorithm. In: Proc. 36th IEEE Conference on Decision and Control, San Diego, Piscataway, pp. 2984–2989. IEEE Press, New Jersey (1997)
60. Moonen, M., De Moor, B., Vandenberghe, L., Vandewalle, J.: On- and off-line identification of linear state-space models. Int. J. Control **49**(1), 219–232 (1989)
61. More, J.J.: The Levenberg–Marquardt algorithm: Implementation and theory. In: Watson, G.A. (ed.) Numerical Analysis. Lecture Notes in Mathematics, vol. 630, pp. 106–116. Springer-Verlag, Berlin (1978)
62. Overschee, P.V., De Moor, Bart: N4SID: Subspace algorithms for the identification of combined deterministic-stochastic systems. Automatica **30**(1), 75–93 (1994)
63. Peternell, K., Scherrer, W., Deistler, M.: Statistical analysis of novel subspace identification methods. Signal Process. **52**(2), 161–177 (1996)
64. Picci, G., Katayama, T.: Stochastic realization with exogenous inputs and subspace-methods identification. Signal Process. **52**(2), 145–160 (1996)
65. Rudin, W.: Real and Complex Analysis, 3rd edn. McGraw-Hill, New York (1986)
66. Schaper, C.D., Larimore, W.E., Seborg, D.E., Mellichamp, D.A.: Identification of chemical processes using canonical variate analysis. Comput. Chem. Eng. **18**(1), 55–69 (1994)
67. Silverman, L.: Realization of linear dynamical systems. IEEE Trans. Autom. Control **AC-16**, 554–567 (1971)
68. Sjoberg, J., Zhang, Q., Ljung, L., Benveniste, A., Delyon, B., Glorennec, P.Y., Hjalmarsson, H., Juditsky, A.: Nonlinear black-box modeling in system identification: A unified overview. Automatica **31**(12), 1691–1724 (1995)
69. Soderstrom, T., Stoica, P.: System Identification. Prentice-Hall, New York (1989)

70. Soeterboek, R.: Predictive Control: A Unified Approach. Prentice-Hall, New York (1992)
71. Stoica, P., Jansson, M.: MIMO system identification: State-space and subspace approximations versus transfer function and instrumental variables. IEEE Trans. Signal Process. **48**(11), 3087–3099 (2001)
72. Tatematsu, K., Hamada, D., Uchida, K., Wakao, S., Onuki, T.: New approaches with sensorless drives. IEEE Ind. Appl. Mag. **6**(4), 44–50 (2000)
73. Van den Hof, P.M.J., Schrama, R.J.P.: Identification and control – closed loop issues. In: Preprints of the IFAC Symposium on System Identification, Copenhagen, Denmark, pp. 1–13. Elsevier Science Ltd, Oxford (1994)
74. Van Gestel, T., Suykens, J., Van Dooren, P., De Moor, B.: Identification of stable models in subspace identification by using regularization. IEEE Trans. Autom. Control **46**(9), 1416–1420 (2001)
75. Van Overschee, P., De Moor, B.: Subspace algorithms for the stochastic identification problem. Automatica **29**, 649–660 (1993)
76. Van Overschee, P., De Moor, B.: N4SID-subspace algorithms for the identification of combined deterministic-stochastic systems. Automatica **30**(1), 75–94 (1994)
77. Van Overschee, P., De Moor, B.: A unifying theorem for three subspace system identification algorithms. Automatica **31**(12), 1853–1864 (1995)
78. Van Overschee, P., De Moor, B.: Subspace Identification for Linear Systems: Theory, Implementation and Applications. Kluwer Academic Publishers, Dordrecht, The Netherlands (1996)
79. Verdult, V., Verhaege, M.: Identification of multivariable bilinear state space systems based on subspace techniques and separable least squares optimization. Int. J. Control **74**(18), 1824–1836 (2001)
80. Verdult, V., Verhaege, M.: Subspace identification of multivariable linear parameter-varying systems. Automatica **38**(5), 805–814 (2002)
81. Verhaegen, M.: Realization of covariance sequences. In: Proc. the Toeplitz Memorial Conference, Tel Aviv, Israel (1981)
82. Verhaegen, M.: A new class of algorithms in linear system theory. PhD thesis, KU Leuven, Leuven, Belgium (1985)
83. Verhaegen, M.: Application of a subspace model identification technique to identify LTI systems operating in closed-loop. Automatica **29**(4), 1027–1040 (1993)
84. Verhaegen, M.: Identification of the deterministic part of MIMO state space models given in innovations form from input–output data. Automatica **30**(1), 61–74 (1994)
85. Verhaegen, M., Dewilde, P.: Subspace model identification Part 2: Analysis of the elementary output-error state-space model identification algorithm. Int. J. Control **56**(5), 1211–1241 (1992)
86. Verhaegen, M., Dewilde, P.: Subspace model identification Part 1: The output-error state-space model identification class of algorithms. Int. J. Control **56**(5), 1187–1210 (1992)
87. Verhaegen, M., Dewilde, P.: Subspace model identification Part 3: Analysis of the ordinary output-error state-space model identification algorithm. Int. J. Control **56**(3), 555–586 (1993)
88. Verhaegen, M., Westwick, D.: Identifying MIMO Hammerstein systems in the context of subspace model identification methods. Int. J. Control **63**(2), 331–349 (1996)
89. Verhaegen, M., Yu, X.: A class of subspace model identification algorithms to identify periodically and arbitrarily time-varying systems. Automatica **31**(2), 201–216 (1995)
90. Viberg, M.: Subspace-based methods for the identification of linear time-invariant systems. Automatica **31**(12), 1835–1851 (1995)
91. Viberg, M., Wahlberg, B., Ottersten, B.: Analysis of state-space system identification methods based on instrumental variables and subspace fitting. Automatica **33**(9), 1603–1616 (1997)
92. Wahlberg, B., Ljung, L.: Design variables for bias distribution in transfer function estimation. IEEE Trans. Autom. Control **31**(2), 134–144 (1986)
93. Westwick, D., Verhaegen, M.: Identifying MIMO Wiener systems using subspace model identification methods. Signal Process. **52**, 235–258 (1996)
94. Zeiger, H., McEwen, A.: Approximate linear realization of given dimension via Ho's algorithm. IEEE Trans. Autom. Control **19**, 153–154 (1974)
95. Zhu, Y.: Multivariable System Identification for Process Control. Pergamon, Lexington (2001)

Chapter 4
Applications I

4.1 Introduction

The importance of system models in the contemporary paradigm of advanced control design cannot be overestimated. There are numerous volumes and survey papers testify to the pervasive use of system models in different aspects of control engineering and in different application areas. This growth in the use of models to accomplish different objectives in the design of industrial control systems has been accompanied by a similar growth in the science of system identification.

System identification is often classed as a white-box problem or a black-box problem, but when the designer is allowed to introduce a priori system knowledge into the process then more pragmatic grey-box methods emerge. A mainstay of the control system modeling paradigm are continuous-time models because they arise naturally when describing the physical phenomena of systems and processes. These models of physical systems usually involve differential equations that stem from the application of physical and chemical laws. However, the widespread use of digital computing technology and the concomitant sampled data led to an emphasis on the use of discrete system models, discrete control designs and sampled-data-based system identification algorithms. For a wider scope of related technical materials, the reader is advised to consult the reference [3–6, 18, 19].

4.2 Distillation Unit

Distillation columns are widely used in the chemical process industries where large quantities of liquids have to be distilled. Industrial distillation towers are usually operated at a continuous steady state.

4.2.1 Data Analysis

Start by plotting input signal and output signal with respect to time Fig. 4.1. It is clear that the input and output signals are affected by an offset that need to be re-

M.S. Mahmoud, Y. Xia, *Applied Control Systems Design*,
DOI 10.1007/978-1-4471-2879-3_4,

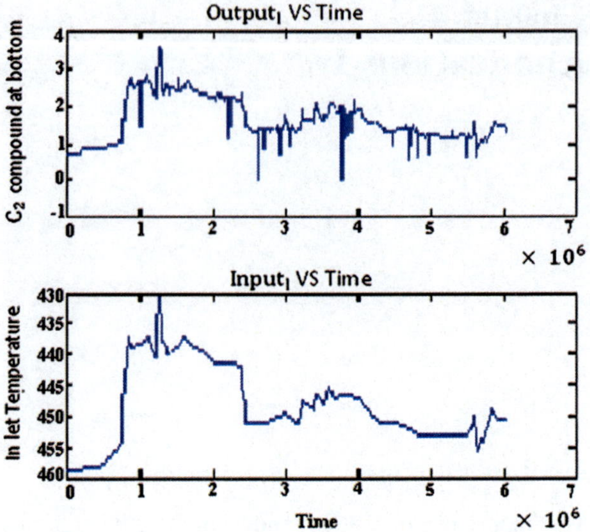

Fig. 4.1 Input and output signals vs. time

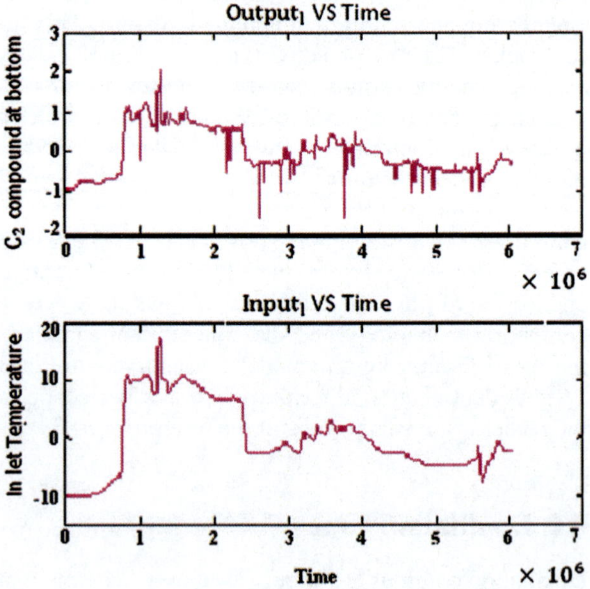

Fig. 4.2 Input and output signals vs. time: zero mean

moved to see how changes in the input give changes in output Fig. 4.2. As mentioned in the foregoing chapter, the available data set consist of 10080 sample the first half (that is, 5040 sample) of which will be used for estimation purpose and the second half is for validation purpose.

The correlation analysis estimate, 4th order ARX model, and state space model are computed and plotted to see the transient response agreement Fig. 4.3.

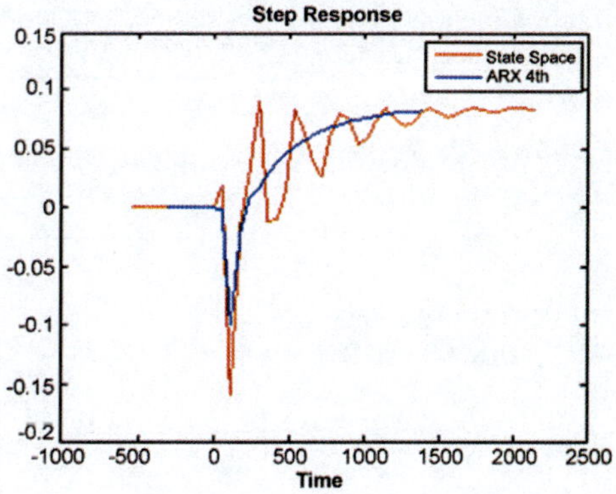

Fig. 4.3 Transient response agreement between ARX and state space

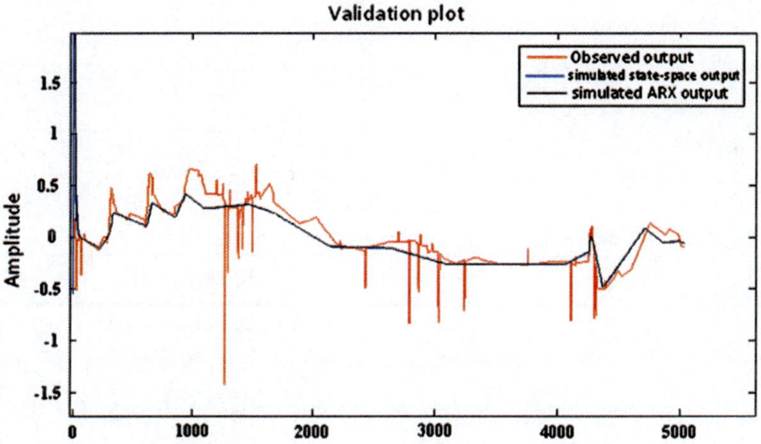

Fig. 4.4 Validation plot for ARX: state-space against measured output

4.2.2 Validation and Model Fitness

The measured validation data output (that is, samples between 5040 to 10080) is compared against the simulated output of the ARX and state space validation model as depicted in Fig. 4.4.

In order to find the closest model structure that represent the on-line real data, a comparison based on the fitness criteria and residual analysis is performed among different model structures. Table 4.1 shows this comparison based on the first data set and an appropriate plot is depicted in Fig. 4.5.

By repeating the above procedure with different data set, we can conclude that BJ model is the best model that represents the highest fit, see Table 4.2 and Fig. 4.6.

Table 4.1 Best fit for the selected model structure

Model	Order	Fitness
ARX	[4 4 1]	−2470
ARMAX	[2 2 2 1]	−2471
BJ	[2 2 2 2 1]	−2743

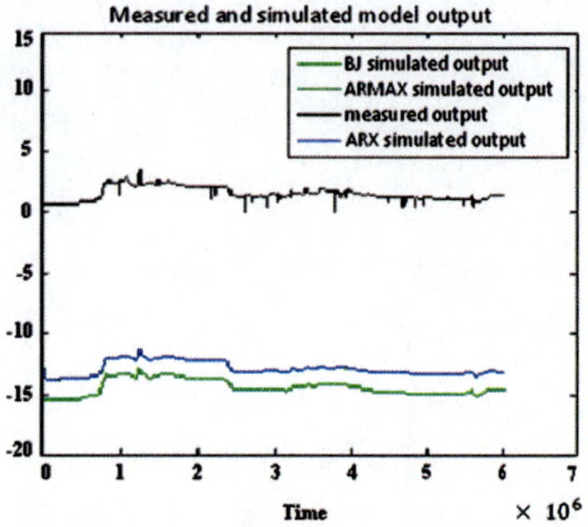

Fig. 4.5 Model outputs of ARX, ARMAX and BJ

Table 4.2 ARX, ARMAX, and BJ model output

Data set	Samples	ARX best fit	ARMAX best fit	BJ best fit
1	5040	−2743	−2744	−2743
2	1000	−5737	−5711	−5449
3	2200	−3189	−3122	−3071
4	2800	−1528	−1644	−1787
5	1810	−2540	−2533	−2627
6	800	−3182	−3165	−3426
7	1470	−2195	−2200	−1795

4.3 Steam Generation Unit

We learned from the previous chapter that the model of steam generation unit is a typical multivariable system. In identification system terminology, multivariable systems are often more challenging to model since the underlying systems with several outputs might be difficult. A basic reason for the difficulties is that the coupling between several inputs and outputs eventually leads to more complex models. The structures involved are richer and more parameters will be required to obtain a good fit. However, models for prediction and control will be able to produce better results if constructed for all outputs simultaneously. In this section, the complete steam generation unit is modeled using MIMO ARX model and MIMO State Space model.

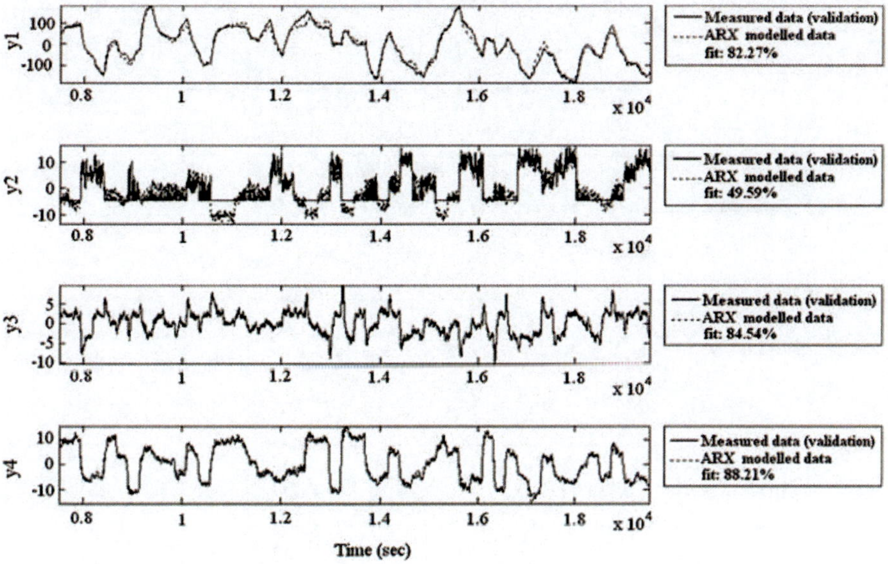

Fig. 4.6 Simulation results: MIMO ARX

4.3.1 MIMO ARX Model

For simulation experiments using the MIMO ARX model, all the four inputs and outputs of the system were considered. The order of the system was specified as

$$n_a = 8 \times ones(4,4), \qquad n_b = 6 \times ones(4,4), \qquad n_c = 3 \times ones(4,4).$$

The coefficients n_a, n_b and n_c were selected on trial and error basis to yield the best fitness levels. The delay coefficients were however not considered as optimum results were available without introducing delay in the system. The model was constructed using samples from 5000–9000. The validation of the model so obtained was carried out on samples from 2500–6500. The results of the simulation have been plotted. The percentage fitness of the various modeled outputs with respect to the measured outputs for the MIMO ARX model is shown in Fig. 4.6.

4.3.2 MIMO State-Space Model

A state space model of order 6 was found to yield optimum results. Further increase in the order yielded only a negligible increase in the fitness of the models. With the model order equal to 9, lower fitness levels were obtained, but it showed better results in the residual analysis. Therefore, to strike a balance between the complexity of the model, the fitness and residuals the order of the state space model was selected as 6. Just as in the previous case, the model was constructed using samples from

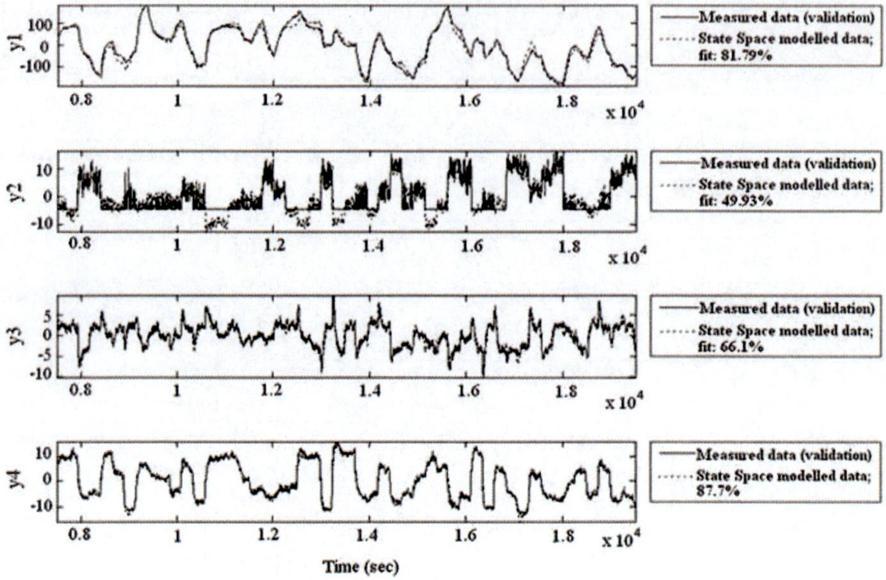

Fig. 4.7 Simulation results: MIMO state-space

5000–9000. The final prediction error (FPE) was found to be 0.00292292 and the loss function was found to be 0.00286093. The results of the simulation have been plotted. The percentage fitness of the various outputs for the State Space Model is shown in Fig. 4.7.

4.3.3 Comparison of MIMO Models

The plot of comparative fitness percentages between the ARX model and the State Space Model can be shown in Table 4.3. Generally speaking, it is found convenient to work with state space models in the multivariable case, since the model structure complexity is easier to deal with. It is essentially a matter of choosing the model order. State-space model also provides a more complete representation of the system than polynomial models. However, when the model order is high, it is better to use an ARX model because the algorithm involved in the estimation is fast and efficient when the number of data points is very large. The state-space model estimation with a large number of data points is slow and requires a large amount of memory and the fitness of the model is also hampered consecutively. Therefore, we observe that the fitness levels for the MIMO ARX model are relatively greater than the state space model. The ARX model therefore is preferable, especially when the model order is high, see Fig. 4.8.

Table 4.3 Comparison of model fitness

Output	MIMO ARX model	State space model
y_1	82.27	81.79
y_2	49.59	49.93
y_3	84.54	66.1
y_4	88.21	87.7

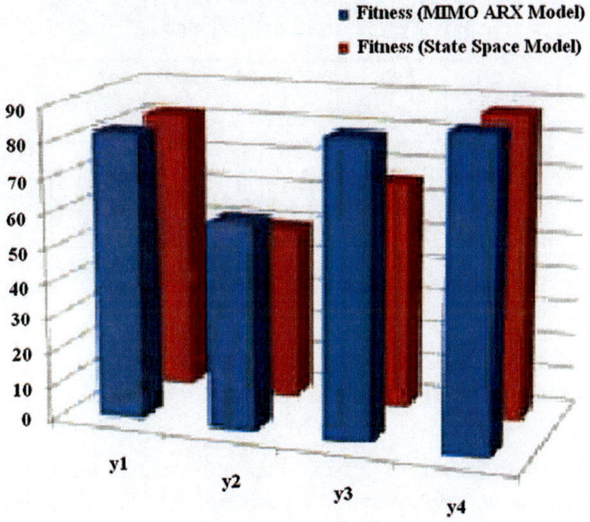

Fig. 4.8 Comparison of MIMO model fitness

4.4 Falling Film Evaporator

The most common used evaporator in the dairy industry is the falling film evaporator, for the concentration of products like milk, skimmed milk and whey. A four stage evaporator is used to reduce the water content of the product, that is, milk. The data was taken from [13]. The identification scheme used for the data is the N4SID subspace based identification. The data consists of 6305 samples with three inputs, feed flow, vapor flow to the first evaporator stage and cooling water flow and three outputs, dry matter content, the flow and the temperature of the out coming product.

The solution containing the desired product is fed to the evaporator and passes a heat source. The applied heat converts the water in the solution to vapor. The vapor is removed from the rest of the solution and is condensed while the now concentrated solution is either fed into the second evaporator is removed. The evaporator generally as a machine consists of four sections. The heating section consists of the heating medium. Steam is fed into this section. The concentrating and separating section removes the vapor being produced from the solution. The condenser condensates the separated vapor, then the vacuum or pump provides pressure to increase the circulation.

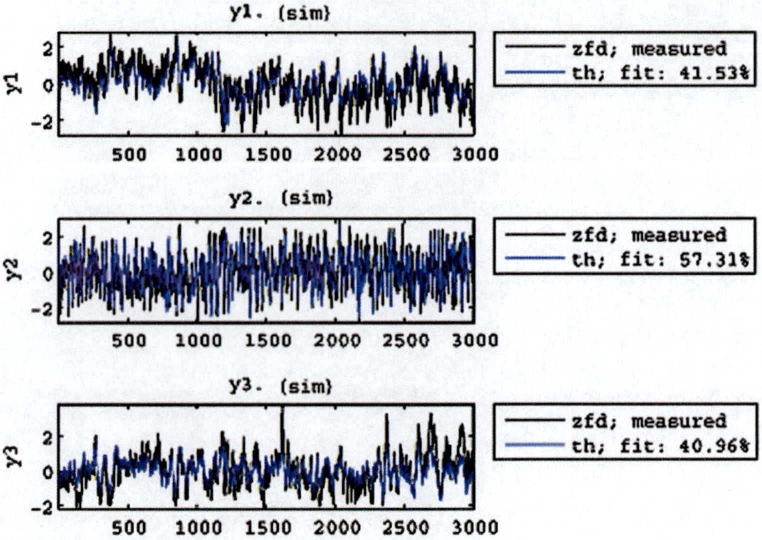

Fig. 4.9 The estimated data

4.4.1 Identification Results

The data for the industrial evaporator is taken into account and the System Identification was done for the same using MATLAB. The N4SID method of identification was performed on the data and the results are shown below for the estimation and validation of the data. The loss function was 0.000414667 and the FPE was 0.000431254 using the N4SID technique and results, shown in Figs. 4.9 and 4.10, are extracted.

4.5 Vapor Compression Cycle Systems

Accurate dynamic models of vapor compression systems play a significant role in the efficient design of systems with optimal component sizes and configurations, and in the development of control strategies to manage these systems. The framework of the dynamic modeling approach is selected through careful consideration of external constraints that limit the usefulness of a particular framework with regard to system design or control development. In the system design phase, the model should accurately predict the performance and behavior of a particular system configuration. The emphasis on accuracy in the design phase has lead to the use of complex models that provide the flexibility to capture the intricate behavior of the fluid flow and heat transfer phenomena common to vapor compression systems. In contrast, the model that is most beneficial for control design is the least complex model that still retains sufficient accuracy to cap-

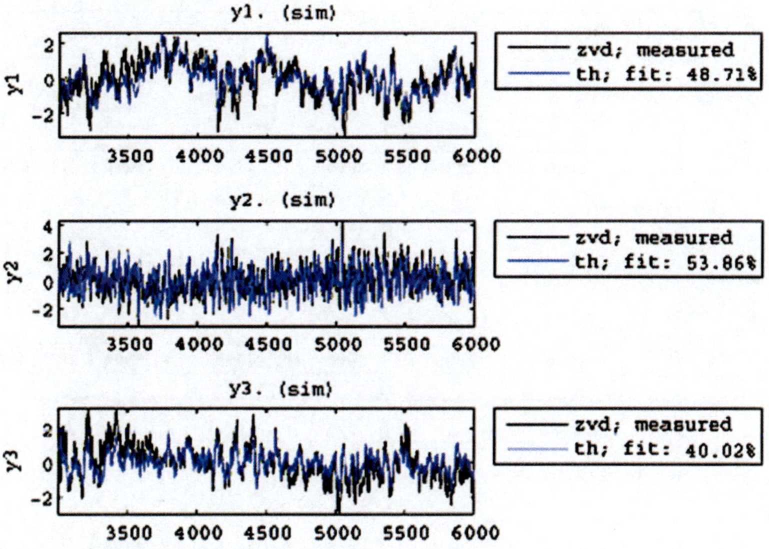

Fig. 4.10 The validated data

ture the gross dynamic behavior of the system. For control design, it is critical to strike a delicate balance between dynamic complexity and accuracy in the model.

With reference to Chap. 2, a representative system model is required to understand which aspects of the thermodynamic cycle are best controlled by which input parameter. In this section, the dynamic response of a VCC system is identified using a time-domain-system identification procedure. Three controllable inputs for a variable-speed VCC are considered: expansion valve opening u_1, compressor speed u_2, and evaporator airflow rate u_3. The condenser airflow rate is considered a disturbance to the system because, in some applications, for example, automotive systems, the condenser airflow rate is a function of vehicle speed and, therefore, is not controlled.

The output measurements consist of six thermodynamic states: two pressures and four temperatures. Recall that, for an idealized cycle, there are two system pressures: $P_2 = P_3$ and $P_1 = P_4$. These correspond to the pressure inside the condenser and the pressure inside the evaporator, respectively. There are four system refrigerant temperatures: T_1, T_2, T_3 and T_4. Again, assuming an idealized cycle with a saturated refrigerant leaving the condenser, these represent evaporator outlet temperature, condenser inlet temperature, condenser saturation temperature, and evaporator saturation temperature, respectively. The output responses to random Gaussian combinations of all three inputs (see Fig. 4.11) around a set of nominal operating conditions, were collected on an air-conditioning and refrigeration experimental test stand. For a more detailed description of the experimental system, [13, 14].

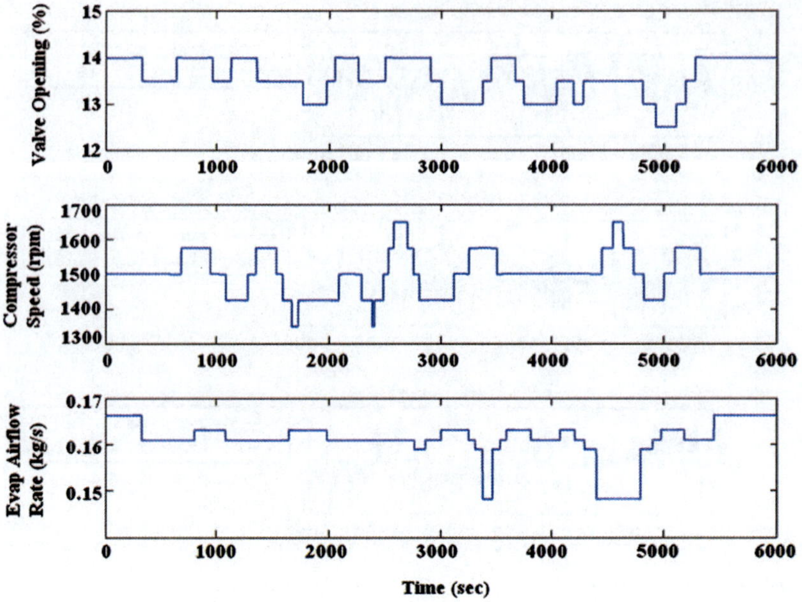

Fig. 4.11 Random Gaussian input signals

4.5.1 Identification Results

A standard prediction error/maximum likelihood system identification algorithm from the MATLAB System Identification Toolbox was used to identify the frequency response between each input and output pair. The identified models were compared with models obtained using polynomial modeling techniques such as MIMO ARX modeling etc. Parameter estimation in polynomial modeling techniques was carried out using recursive least squares approach. With respect to the Complexity of the model, the fitness levels and the residual analysis it was concluded that the Subspace system identification using prediction error method yielded most accurate models. Because system identification is sensitive to scaling, two models were identified, wherein the output parameters within each model shared the same inputs. This means that the first model was identified with all three excited inputs and the two pressure measurements as outputs ($P_2 = P_3$ and $P_1 = P_4$), and a second model was identified with the same inputs and the four temperature measurements as outputs (T_1, T_2, T_3 and T_4). The complete state-space representation of each identified model is included in the Appendix. Note that using instead of provided a better fit with respect to the system identification.

For each identified output, the open-loop system response is compared against the response as predicted by the identified model, Figs. 4.12 and 4.13 show the identified model compared against the data used for the identification. The identified models were then cross-validated using data collected on a different day with a different ambient temperature and humidity level. The fit percentages for each out-

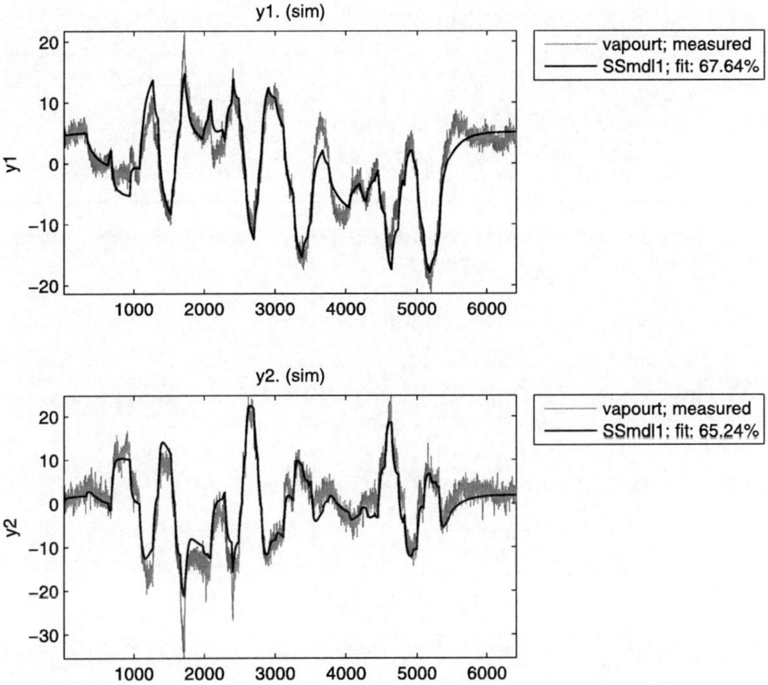

Fig. 4.12 System ID results for P_1; P_{2-1}

put characterizing predictive capability of the models are included in each of the figures.

It was noted that the fitness percentages of the identified model were very low for the model considering the four temperatures T_1, T_2, T_3 and T_4, whereas the fitness percentages for the model considering the pressures P_1 and P_{2-1} were found to be reasonably good that is, approx 65% in each case.

4.6 Unmanned Marine Vehicle

System identification is the art and science of building mathematical models of dynamic systems from observed input–output data. It can be seen as the interface between the real world of applications and the mathematical world of control theory and model abstractions. Identification is a very large topic, with different techniques that depend on the character of the models to be estimated: linear, non linear, hybrid, non parametric etc. Model structure selection is a key step in the identification process. The model structure determines the set in which the model estimation is performed. The complexity of the model structure, of course, affects the accuracy with which the model can approximate the real process. The choice of model structure depends on the noise sequence: *how well is it possible to estimate the noise?*

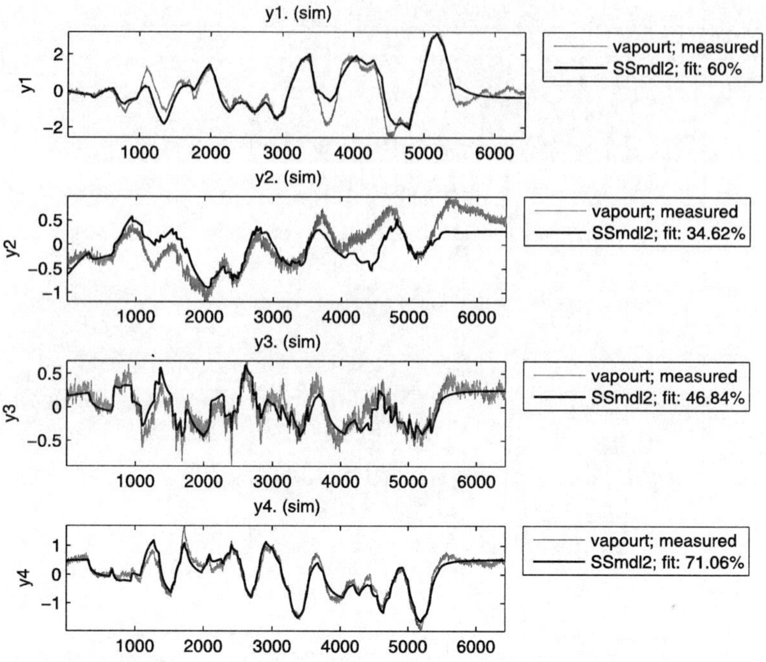

Fig. 4.13 System ID results for T_1; T_2; T_3; T_4

It is not at all necessary that a model with more parameters or more freedom (more polynomials) is better. Finding the best model is a matter of choosing a suitable structure in combination with the number of parameters.

4.6.1 Identification Results

Generally speaking, it is preferable to work with state-space models in the multivariable case, since the model structure complexity is easier to deal with. It is essentially just a matter of choosing the model order. It is observed that the fit gets better when more inputs are included and often gets worse when more outputs are included. To understand the latter fact, realize that a model that has to explain the behavior of several outputs has a tougher job than one that must just account for a single output. Difficulties obtaining good models for a multi-output system might be a sign to model one output at a time, to find out which are the difficult ones to handle.

Models that are just to be used for simulations could very well be built up from single-output models, for one output at a time. However, models for prediction and control will be able to produce better results if constructed for all outputs simultaneously. This follows from the fact that knowing the set of all previous output channels gives a better basis for prediction than just knowing the past outputs in one channel. Also, for systems where the different outputs reflect similar dynamics, using several outputs simultaneously will help estimating the dynamics. Here, some of identification methods used in this paper is reviewed.

ARMAX

There are several elaborations of the basic ARX model, where different noise models are introduced; ARMAX is one of them. The basic disadvantage with the ARX model is the lack of adequate freedom in describing the properties of the disturbance term. ARMAX takes care of this deficiency by describing the equation error as a moving average of white noise. This gives the model:

$$\begin{aligned} y(t) + a_1 y(t-1) + \cdots + a_{n_a} y(t-n_a) \\ = b_1 u(t-1) + \cdots + b_{n_b} u(t-n_b) \\ + e(t) + c_1 e(t-1) + \cdots + c_{n_c} e(t-n_c), \\ C(q) = 1 + c_1 q^{-1} + \cdots + c_{n_c} q^{-n_c} \end{aligned}$$

it can be rewritten as

$$A(q)y(t) = B(q)u(t) + C(q)e(t) \tag{4.1}$$

with

$$G(q,\theta) = \frac{B(q)}{A(q)}, \qquad H(q,\theta) = \frac{C(q)}{A(q)}$$

where now

$$\theta = [a_1 \ \cdots \ a_{n_a} \ \ b_1 \ \cdots \ b_{n_b} \ \ c_1 \ \cdots \ c_{n_c}]^T. \tag{4.2}$$

The ARMAX model has become a standard tool in control and econometrics for both system description and control design. It is a significant tool in controls and simulation purposes but drawing this technique to practical conclusions over the other methods is not relevant.

State Space Models-Based Identifications

In the state space form the relationship between the input, noise, and output signal is written as a system of first order differential equation using an auxiliary vector $x(t)$. For beginning it's easier to construct model in continuous time. Given state space model equations,

$$\dot{x} = A(\theta)x(t) + B(\theta)u(t) \tag{4.3}$$

with A and B are matrices of appropriate dimensions for n states, and m inputs. θ is a vector of parameters, typically correspond to unknown values of physical coefficients. For system identification process, the data available to construct model parameter obviously discrete. Let the measurement result obtained from sensor be the output of state space model corrupted with noise, next called *measurement noise*, and let *process noise* be the noise acting on the state, then next state and measured output can be represented as,

$$x(t+1) = A(\theta)x(t) + B(\theta)u(t) + K(\theta)e(t), \tag{4.4}$$

$$y(t) = C(\theta)x(t) + e(t). \tag{4.5}$$

Next, input output relation can be written in series of polynomial series using shift operator q [15],

$$y(t) = G(q,\theta)u(t) + H(q,\theta)e(t), \tag{4.6}$$

$$G(q,\theta) = C(\theta)\big[qI - A(\theta)\big]^{-1}B(\theta), \tag{4.7}$$

$$H(q,\theta) = C(\theta)\big[qI - A(\theta)\big]^{-1}K(\theta) + I. \tag{4.8}$$

Solving θ for both G and H now can be treated like similar problem in SISO by least square technique. Multiple-output *ARMAX* and *OE* models are covered via state-space representations: *ARMAX* corresponds to estimating the K matrix, while OE corresponds to fixing K to zero. State Space model parameters are computed using iterative *Prediction-Error Minimization.* Once the model structure has been defined, and a data set Z^N has been collected the estimation of the parameter θ is conceptually simple: Minimize the distance between the predicted output (according to parameter θ) and the measured outputs,

$$\hat{\theta}_N = \arg\min_{\theta} V_N(\theta), \tag{4.9}$$

$$V_N(\theta) = \sum_{t=1}^{N} l\big(\bar{y}(t) - y(t)\big). \tag{4.10}$$

Here $\bar{y}$ is the measurement output, and l is suitable distance measure, such as $l(\varepsilon) = \|\varepsilon\|^2$. The connection to the celebrated *maximum likelihood method* is obtained by particular choice of norm. Assume that the data is produced by mechanism

$$\bar{y} = f\big(Z^{t-1},\theta\big) + e(t) \tag{4.11}$$

where $\{e(t)\}$ is a sequence of independent random variables with probability density function $p(x)$, this make $\hat{\theta}_N$ equal to the maximum likelihood estimate. The actual calculation of the minimizing argument is complicated, and possibly a complex search over function with several local minima. The numerical search typically carried out using the *damped Gauss–Newton* method. See [15] for more detail on numerical minimization issue.

Kalman Filter Identifications

Kalman filter identification (KID) is another system identification technique that uses state space modeling to approach the state space problem not from polynomial series of transfer functions, but directly from time domain representations. Kalman filter identification (KID) is an algorithm developed by NASA Langley to model large flexible structures [12]. In this regard, KID is a refined algorithm based on eigen-systems realizations algorithm developed in [12]. Consider a discrete multi-variable linear system,

$$\begin{aligned} x(t+1) &= Ax(t) + Bu(t), \\ y(t) &= Cx(t) + Du(t). \end{aligned} \tag{4.12}$$

Assuming initial conditions, $x(0) = 0$, the set of this equations for a sequence of t can be written as

$$y = \mathbf{YU} \tag{4.13}$$

where

$$\mathbf{Y} = \begin{bmatrix} D & CB & CAB & \cdots & CA^{l-2}B \end{bmatrix}, \tag{4.14}$$

$$\mathbf{U} = \begin{bmatrix} u(0) & u(1) & u(2) & \cdots & u(l-1) \\ & u(0) & u(1) & \cdots & u(l-2) \\ & & u(0) & \cdots & u(l-3) \\ & & & \ddots & \vdots \\ & & & & u(0) \end{bmatrix}. \tag{4.15}$$

Equation (4.13) is a matrix representation of the relationship between input and output histories. The matrix y is a $q \times l$ output data matrix where q is the number of outputs and l is the number f data samples. The $\mathbf{Y}$, of dimension $q \times ml$ with m is the number of inputs, contains all the *Markov* parameters $D, CB, CAB, \ldots, CA^{l-2}B$ to be determined. The $\mathbf{U}$ matrix is an $ml \times l$ upper block triangular input matrix. It is square in the case of a single input system, and otherwise has more rows than columns.

For asymptotically stable system, there is a p such as $A^p \approx 0$, so the $\mathbf{Y}$ and $\mathbf{U}$ can be truncated. Unfortunately, for lightly damped system, p required to make the approximation of (4.14) and (4.15), is impractically large, in the sense that matrix $\mathbf{U}$ is too large to solve for its pseudo inverse $\mathbf{U}^+$ numerically. Dealing with this, a kind of observer feedback loop had been suggested to be added to make the system as stable as desired. Consider

$$\begin{aligned} x(t+1) &= Ax(t) + Bu(t) + My(t) - My(t) \\ &= (A + MC)x(t) + (B + MD)u(t) - My(t) \\ &= \bar{A}x(t) + \bar{B}v(t) \end{aligned} \tag{4.16}$$

where

$$v(t) = \begin{bmatrix} u(t) \\ y(t) \end{bmatrix}. \tag{4.17}$$

When using real data including noise, the eigenvalue of $\bar{A}$ are in fact placed such that $C\bar{A}^i\bar{B} \approx 0$ for $i \leq p$ where p is sufficiently large integer. Using the same approach,

$$y = \bar{\mathbf{Y}}\bar{\mathbf{V}} \tag{4.18}$$

where

$$\bar{\mathbf{Y}} = \begin{bmatrix} \bar{D} & \bar{C}\bar{B} & \bar{C}\bar{A}\bar{B} & \cdots & \bar{C}\bar{A}^{p-1}\bar{B} \end{bmatrix},$$

$$\bar{\mathbf{V}} = \begin{bmatrix} u(0) & u(1) & u(2) & \cdots & u(p) & \cdots & u(l-1) \\ & v(0) & v(1) & \cdots & v(p-1) & \cdots & v(l-1) \\ & & v(0) & \cdots & v(p-2) & \cdots & v(l-2) \\ & & & \ddots & \vdots & \vdots & \vdots \\ & & & & v(0) & \cdots & v(l-p-1) \end{bmatrix}. \tag{4.19}$$

Solving the problem of observer Markov parameters $\bar{\mathbf{Y}}$ using least square algorithm, original Markov parameters $\mathbf{Y}$ can be recovered. The system Markov parameters can then be assembled to form the generalized Hankel matrix. The Hankel matrix can be decomposed into the Observability matrix, a state transition matrix, and the Controllability matrix. The Hankel matrix (which must always be of full rank) can then be truncated using singular value decomposition (SVD) at an order that sufficiently describes the system. The truncated Hankel matrix is then used to reconstruct A, B, and C using a minimum balanced realization algorithm that ensures that the controllability and observability Gramians are equal. This is known as the eigen-system realization algorithm (ERA) and a modified version with data correlation (ERA/DC) can also be used [10].

One of the advantages using KID [12] is that it produces a pseudo Kalman state estimator, which is very useful in control applications. Let (4.12) be extended to include process and measurement noise as

$$\begin{aligned} x(t+1) &= Ax(t) + Bu(t) + w(t), \\ y(t) &= Cx(t) + v(t). \end{aligned} \tag{4.20}$$

It can be shown that any observer satisfying (4.18) can produce the same input output map as a Kalman filter does if the data length is sufficiently long and the order of the observer is sufficiently large so that the truncation error is negligible. Then Kalman steady state gain is given by

$$L = -A^{-1}M. \tag{4.21}$$

The data run available for identification purpose comprises of 14 sets. The longest data run, run 5 will be selected as estimation data, see Fig. 4.14. Validation data selected here, are run 6, run 7, run 9, and run 10. First, best model parameter for each method will be computed from estimation data. Fitness of each model to estimation data then presented. Residual analysis also presented here. At the end, fitness of each model obtained to validation data set will be presented as well.

Fitness are defined as,

$$\text{fitness } (\%) = 100\left(1 - \frac{\|y - \hat{y}\|}{\|y - E(y)\|}\right). \tag{4.22}$$

Residual defined as

$$\begin{aligned} e(t) &= H^{-1}(q)\big(y(t) - G(q)u(t)\big), \\ y(t) &= G(q,\theta)u(t) + H(q,\theta)e(t). \end{aligned} \tag{4.23}$$

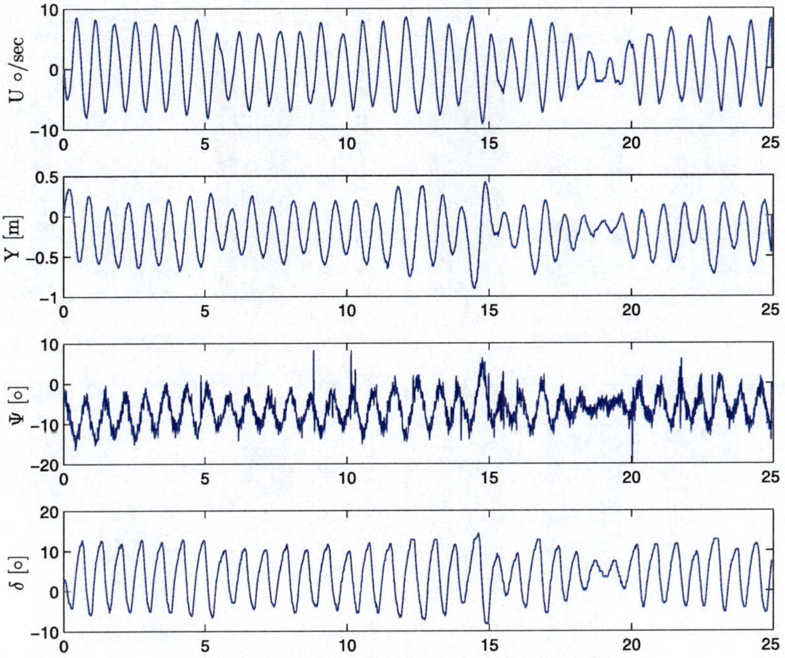

Fig. 4.14 Typical input–output set

Ideally residual should be white and independent of the input signals, this can be examined from autocorrelation plot of residual.

4.6.2 ARMAX Model

The ARMAX identification model describes the equation error as a moving model of white noise. As far as the structuring of the model from the data set is concerned, the model generated for the outputs y_1 and y_3 from the input have a pleasing fit unlike the model for the output y_2. The third order estimation model for y_1, second order estimation model for y_2 and fifth order estimate for y_3 give a satisfactory profile. During validation using run 7, the fourth order model gives the best fit for the output channel u_1–y_1 with a relatively higher amplitude, see Fig. 4.15. Though lower orders give a satisfactory fit as well but do not follow up the profile clearly. For the output channel u_1–y_2, the fit is not high owing to the properties of the output which depend on the speed of the wind. u_1–y_3 channel has a high fitness for the fifth order model upon validation. Upon using run 9, the error in fitness was high showing that the amplitude of the estimated models was higher than the validation data run. Run 10 gives a measure fit for output y_1, for output y_2 the error obtained upon validation is high and for output y_3 the fit was decent enough. The third order model for the first output gives the best results upon validation compared to the

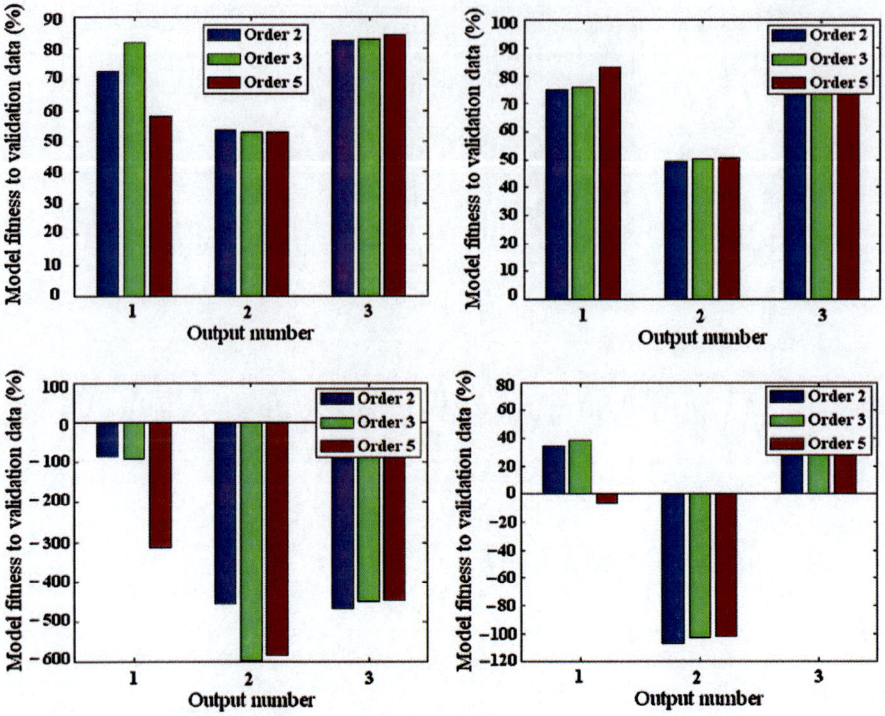

Fig. 4.15 ARMAX model fitness to data runs #6–#10

other orders, see Fig. 4.16. The second order model for the second output is the best choice and the third order model for the third output.

4.6.3 State Space Model

State space model identification will estimate A, B, C, K, and initial state X_0. To determine how many order is sufficient for Atlantis data model, modal singular value will be computed first here, as shown in Fig. 4.17. From this figure, although there was sudden drop between order 4 and order 5, that not big than one in logarithmic scale, but this drop is not adequate to make four order describe behavior of the system. For comparison, here, several model order is selected, 3 as the Atlantis model is built in third order state space model, 4, 10, and 20. For state space model order determination, see Sect. 4.6.6.

Computed state space parameter with state order 20 end up with ill-condition covariance matrix. Trying for order 15 also end up with the same result. This probably come from non linear output error minimization using in PEM technique. However, with the remaining successful identified model, comparison between fitness of state space model to validation data is carried and can be seen in Fig. 4.17.

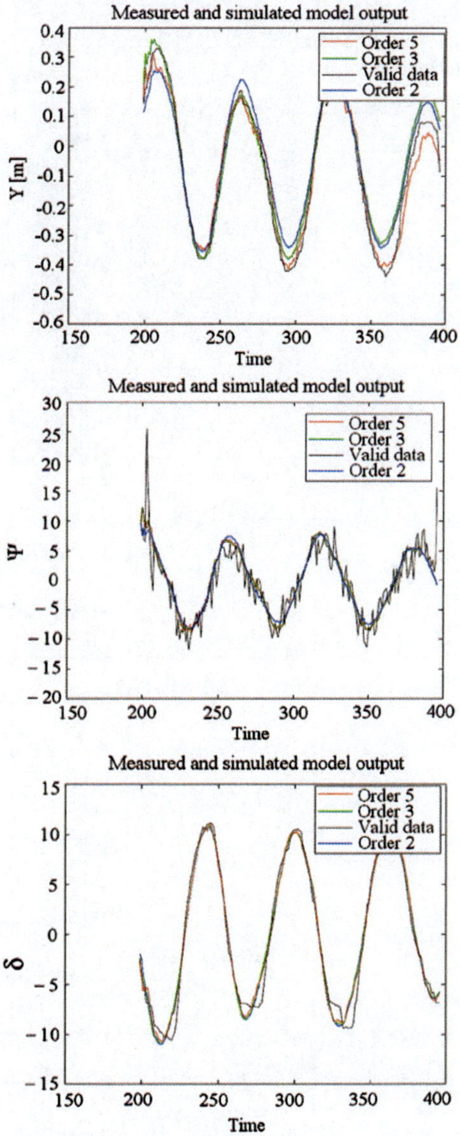

Fig. 4.16 ARMAX model simulated output comparison using data run #7

Model Fitness with Estimation Data PEM estimation give inconsistent fitness improvement as the order number getting higher, this can be seen in output number one, that output estimation using order 3 is better than order 4, also in output number two, estimation using order 4 is better than order 10.

Now, it is clear that state space model built using PEM with order three and four are far from appropriate selection, that because autocorrelation of residual error still have relatively high in sample lag $k \neq 0$. State space model built using PEM order ten, however have residual error near to white noise autocorrelation. It is perhaps

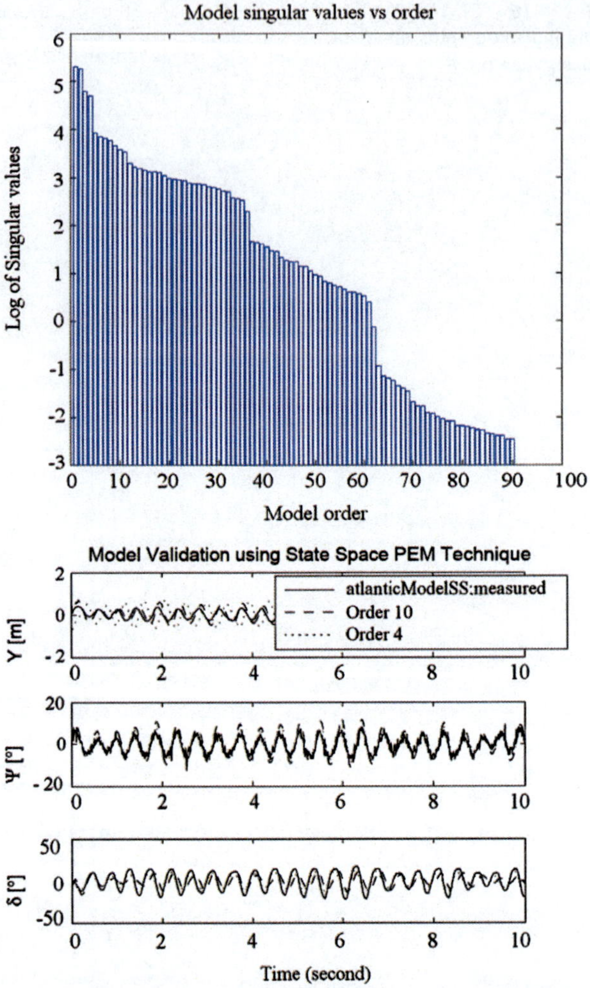

Fig. 4.17 State-space model identification using PEM technique

Table 4.4 Estimation data fitness comparison using PEM

	Order 3	Order 4	Order 10
Output 1	18.2981	−55.0245	77.2154
Output 2	−30.6765	45.9034	44.3733
Output 3	40.1334	69.9786	73.8384

order ten is a good selection. This fact also corresponding to fitness of each model to estimation data, as can be seen in Table 4.4.

Simulation with Validation Data From simulation using four set validation data, state space model obtained using PEM with order three fails to identify all outputs. Fitness to validation data 6, is only 15.826%, −24.589% and 34.253% for output

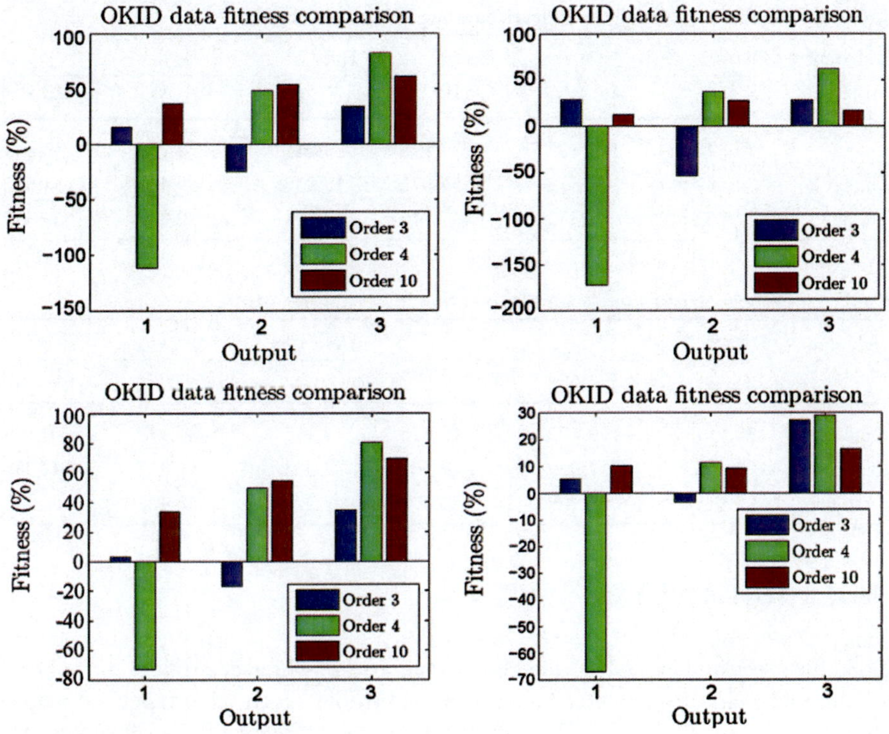

Fig. 4.18 PEM model fitness to validation data

one, two and three. Fitness less than zero is a result of unmatched oscillation period of estimation output, this makes $\|(\hat{y}) - y\|$ greater than $\|y - E(y)\|$, see definition of fitness in (4.22). Nearly the same fitness results are obtained for validation data 7, 9 and 10.

For state space model obtained using PEM with order four, fitness of estimation output to validation data is better than order three for output two, and output three. Fitness of output estimation to validation output one is awful, it reach below -100%, in validation data 6, and also below -50% in validation data 7, 9, and 10.

For state space model obtained using PEM with order ten, fitness of estimation output to validation data is better than the preceding two models. Fitness of this model is always more than zero, that signifies, the ability of the model to track the output in the correct oscillation time, although it may not have correct amplitude. In estimation of validation data 6, this model has the highest fitness result. For validation data 7, 8, and 9, fitness of this model is lower than the fitness of model order 4, at output three and two. Indeed for validation data 9 and 10, fitness to output three is only 16.595% and 16.498%. See Fig. 4.18 and Tables 4.5–4.6 for complete comparison.

Generally, using state space model of order 10 obtained using PEM model, does not give much fitness improvement to validation data.

Table 4.5 Validation data fitness comparison using PEM: runs #6, #7

Val. data	Data 6			Data 7		
Order	3	4	10	3	4	10
Output 1	15.826	−111.800	37.125	33.015	−72.799	33.844
Output 2	−24.589	48.575	54.072	−16.948	49.485	54.652
Output 3	34.253	82.333	61.715	34.691	80.262	69.342

Table 4.6 Validation data fitness comparison using PEM: runs #9, #10

Val. data	Data 9			Data 10		
Order	3	4	10	3	4	10
Output 1	28.951	−172.400	12.316	5.175	−66.977	10.195
Output 2	−53.848	37.059	27.568	−3.338	11.317	9.223
Output 3	28.121	61.787	16.595	27.226	28.922	16.498

4.6.4 KID Model

First thing required in the KID method is to choose p, which determines the number of observer Markov parameters to be identified from a given set of input and output data. In general, p will be sufficiently larger than the effective order of the system, at least four or five times, here p is selected to be 20. Using data run #5, it was difficult to determine effective order of the system, as there was no sudden drop in Hankel singular value matrix. Adding order number also makes a slow improvement in *Model Descriptions* of the data, as defined as *summations of singular value of selected modes divided by summations singular value of all modes in Hankel matrix.* Here, several model orders are selected, 3, 4, 10, and 20. For simulation of KID model, Kalman filter gain given by (4.21) will be incorporated, and each simulation time increment is separated by two step a priori phase and *a posteriori* phase, as below

$$\begin{aligned}
&\text{a priori}\\
&\qquad \bar{x}(t+1) = A\hat{x}(t) + Bu(t),\\
&\qquad \bar{y}(t+1) = C\bar{x}(t+1) + Du(t+1);\\
&\text{a posteriori}\\
&\qquad \bar{e}(t+1) = y(t+1) - \bar{y}(t+1),\\
&\qquad \hat{x}(t+1) = \bar{x}(t+1) + L\bar{e}(t+1),\\
&\qquad \hat{y}(t+1) = C\hat{x}(t+1) + Du(t+1),\\
&\qquad e(t+1) = y(t+1) - \hat{y}(t+1).
\end{aligned} \tag{4.24}$$

Model Fitness with Estimation Data KID estimation, see Fig. 4.19, gives nearly consistent fitness improvement as the order increases, except that for output 2 and

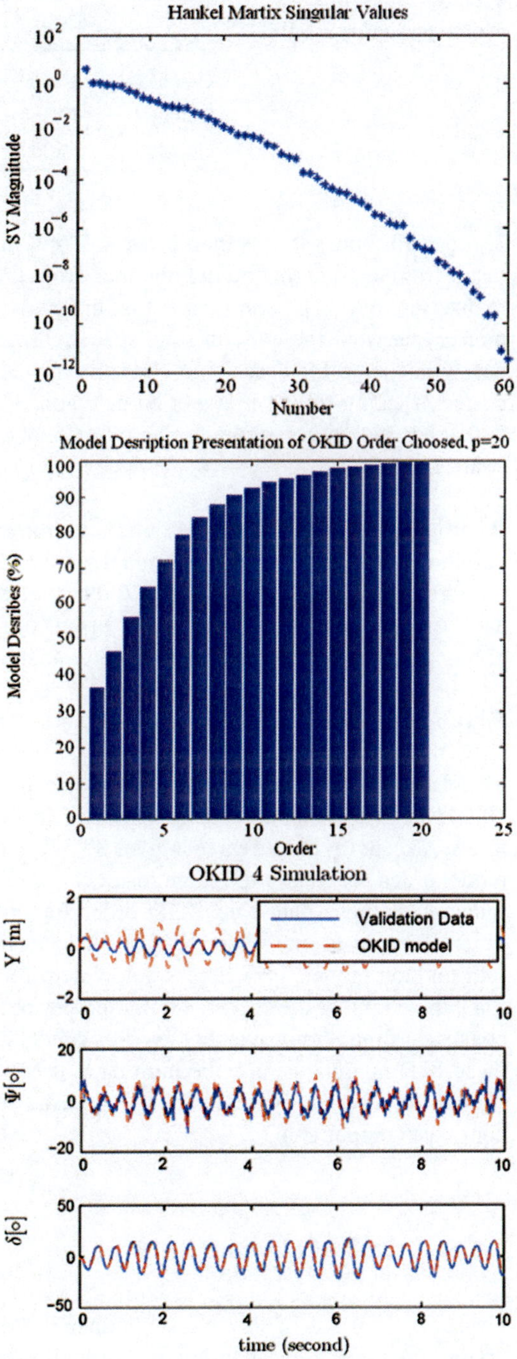

Fig. 4.19 KID system identification

Table 4.7 Estimation data fitness comparison using KID

	Order 3	Order 4	Order 10	Order 20
Output 1	34.209	61.320	68.289	86.773
Output 2	40.388	48.963	33.067	52.154
Output 3	24.060	88.162	85.769	91.016

3, order 10 fitness is less than order 4. Consistency in model fitness seems to be a result from the fact that model obtained from OKID technique is in modal balanced realizations [9, 11]. This means that error truncated in the model is smaller than modes that were included in state space realizations. It can be seen that order 4 is nearly adequate for describing dynamic characteristics of system. It is due to the perfect tracking in fact may not be desirable. Consider for instance output number 2, that has high noise, generating a model that tracks perfectly is not helpful from a control standpoint.

Residual Analysis Using the same equation in Sect. 4.6.4, autocorrelation of each output residual error from each model were carried out. It is found, with regards to the estimation data fitness Table 4.7, that the model that have more fitness, tend to have less autocorrelation magnitude. From four models obtained, none of them are giving autocorrelation plot close to white noise.

Simulation with Validation Data Validation data output estimation using four KID models obtained have shown several interesting facts. For data run 6, all the models are able to track all outputs in correct oscillation time. Model order 3 gives 10.003%, 50.0971%, and 22.29% fitness. Model order four gives nearly the same fitness, except for output three it gives 80.570% fitness. Best fitness is achieved using model order 20. Nearly the same result can also be examined for data runs 7 and 10. Indeed, for these data runs, KID order four model gives 41.227% and 56.269% fitness for output one. One exception occurs in data run 9, that KID order three, four and ten models give fitness below zero for output one, that means they fail to track the output at the correct oscillation period. Generally, KID model tend to have consistent fitness improvement as order model gets higher. Also, from simulation of these KID models using validation data, it can be said that KID order four, is fair enough to describe dynamics of the system. See Fig. 4.20 and Tables 4.8–4.9 for complete comparison.

4.6.5 Result of Comparisons

Fitness comparison of all models obtained before can be seen in Fig. 4.21. ARMAX models have the highest fitness for validation data run 6, KID models also have a good fitness here, except that for output one, low order have small fitness. PEM model's have the smallest fitness level. Nearly the same result are correct for data

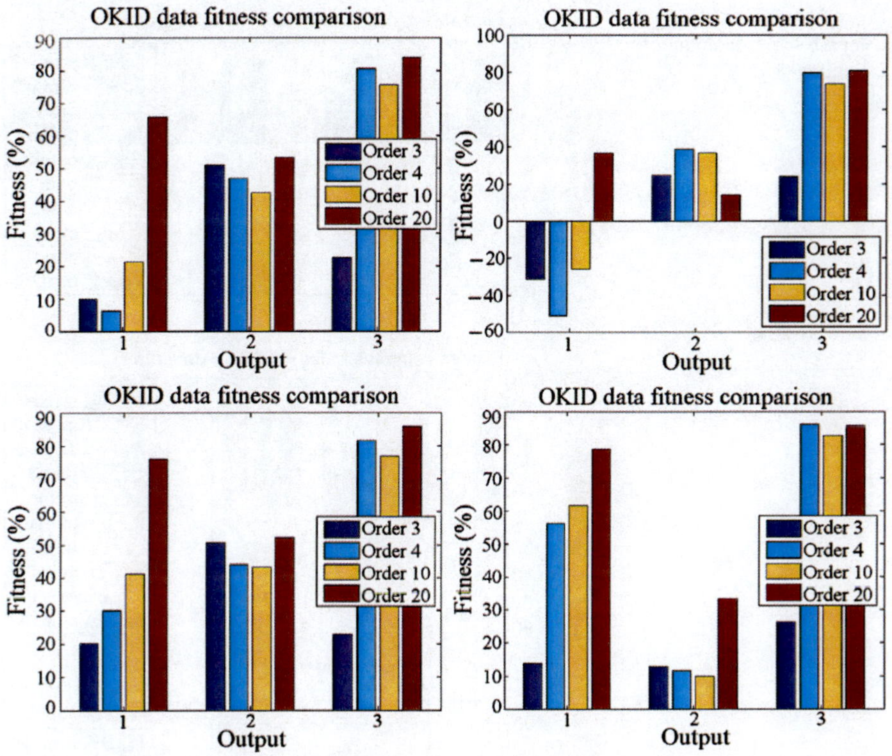

Fig. 4.20 KID model fitness to validation data

Table 4.8 Validation data fitness comparison using OKID: runs #6, #7

Val. data	Data 6				Data 7			
Order	3	4	10	20	3	4	10	20
Output 1	10.003	6.263	21.322	65.703	20.047	30.116	41.227	75.942
Output 2	50.971	46.822	42.512	53.405	50.077	44.210	43.315	52.123
Output 3	22.249	80.570	75.497	84.064	22.804	81.633	76.822	85.977

Table 4.9 Validation data fitness comparison using OKID: runs #9, #10

Val. data	Data 9				Data 10			
Order	3	4	10	20	3	4	10	20
Output 1	−31.229	−51.026	−25.650	36.793	13.984	56.269	61.665	78.635
Output 2	24.776	38.402	36.873	14.031	12.282	11.649	9.856	33.337
Output 3	23.991	79.626	73.734	80.894	26.206	82.706	82.706	85.750

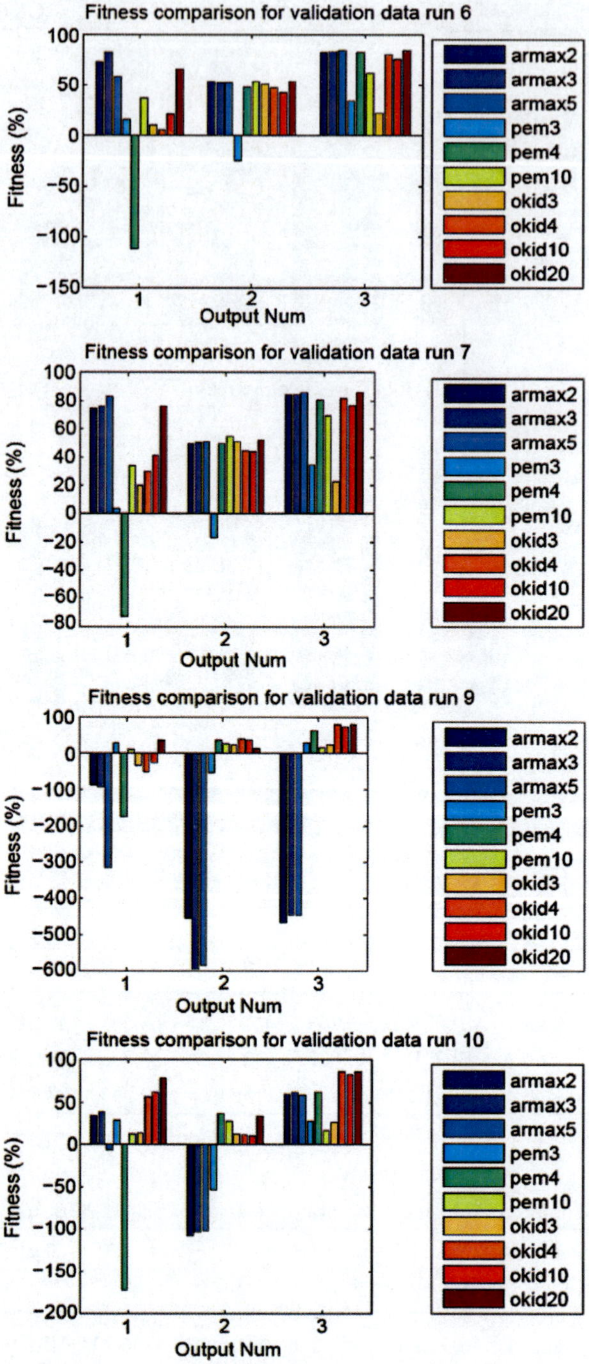

Fig. 4.21 Fitness comparison of techniques to validation data

run 7, except for this data run, KID fitness on output one is better than data run 6. For data run 9, all ARMAX models fail to make estimation, as the fitness runs away very far in all outputs. PEM models have better fitness here, but it still is very small, ten order model of PEM only reaches 12.3% of fitness in output one. KID model is the best here, it has the best fitness value for outputs two and three, except that for output one, model order three, four and ten have fitness below zeros.

For data run 10, KID provides the best fitness, for first output, model order four give 56.3% and order 20 give 78.6%. Fitness for output two is slightly smaller than output two, and fitness of OKID model for output three is high, four order give 86.2%, and 20 order give 85.7%.

As a comparison, Elkaim identification using KID technique on Atlantis boat, with different data set have reconciled that model with order four have enough fitness level [9]. The data run that he used using pseudo random input.

4.6.6 State-Space Order Determinations

This additional section will explain how to determine effective model order of state space model. A *realization* is computation of triplet A, B, C from Markov parameter in (4.14) from which the discrete model in (4.12) is satisfied. It can be shown that any system have an infinite number of realizations which will predict the identical response for any particular output [11]. It's now desired to determine the minimal realizations of model, that means possible smallest state space dimension. System realization begins by forming generalized $\alpha m \times \beta q$ Hankel matrix from Markov parameters from (4.14), where α and β are greater than expected order of minimal realizations,

$$H(k-1) = \begin{bmatrix} Y_k & Y_{k+1} & \cdots & Y_{k+\beta-1} \\ Y_{k+1} & Y_{k+2} & \cdots & Y_{k+\beta} \\ \vdots & \vdots & \ddots & \vdots \\ Y_{k+\alpha-1} & Y_{k+\alpha} & \cdots & Y_{k+\alpha+\beta-2} \end{bmatrix} \tag{4.25}$$

for case $k = 1$,

$$H(0) = \begin{bmatrix} Y_1 & Y_2 & \cdots & Y_\beta \\ Y_2 & Y_3 & \cdots & Y_\beta \\ \vdots & \vdots & \ddots & \vdots \\ Y_\alpha & Y_{1+\alpha} & \cdots & Y_{\alpha+\beta-1} \end{bmatrix}. \tag{4.26}$$

It can be shown that $H(k-1)$ can be decomposed into three matrices, yield,

$$H(k-1) = P_\alpha A^{k-1} Q_\beta \tag{4.27}$$

where,

$$P_\alpha = \begin{bmatrix} C \\ CA \\ \vdots \\ CA^{\alpha-1} \end{bmatrix}, \quad Q_\beta = \begin{bmatrix} B & AB & \cdots & A^{\beta-1}B \end{bmatrix}. \tag{4.28}$$

Here, P_α is observability matrix, whereas Q_β is the controllability matrix. If the realization are minimum, the system will be both controllable and observable, then Q_β, P_α are of rank n, also Hankel matrix will be rank of n, therefore minimum realizations of the system will be n number of states.

In noisy input output data however, Hankel matrix always full rank, however, to determine the true state from noise state, one can determine it from singular value decomposition of Hankel matrix. Sudden drop in one diagonal value of rectangular matrix in singular value decomposition can be sign number of system order. There also another two approach to distinguish between true modes from noise modes, called *Modal Amplitude Coherence* and *Modal Singular Values*, for more detail see [11].

4.7 Industrial Evaporation Unit

We have learned from the foregoing section that system identification is a complex field that can be presented in many deferent ways. In what follows, we provide simulation studies on an industrial evaporation unit, a schematic description of which was given in Chap. 2. The input and output patterns are depicted in Fig. 4.22.

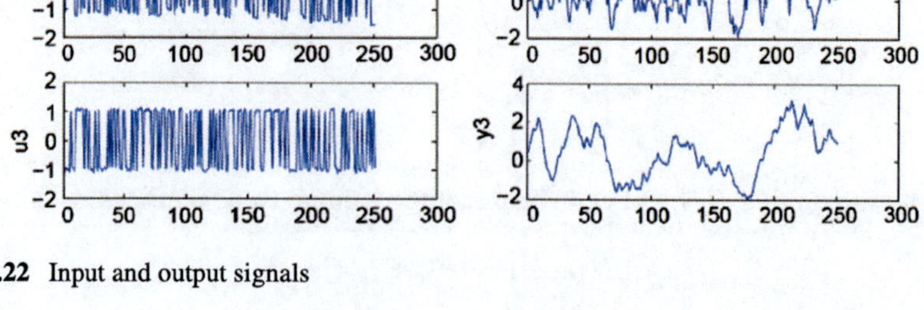

Fig. 4.22 Input and output signals

4.7.1 Continuous-Time Model

For many physical systems, it is natural to work with continuous-time representations, since most basic relationships are expressed in terms of differential equations. It is well known that a linear time invariant, causal system can be described by its impulse response as follows:

$$y(t) = \int_{\tau=0}^{\infty} g(\tau)u(t-\tau)\,d\tau. \tag{4.29}$$

The impulse response $g(\tau)$ gives a complete characterization of the system; Knowing the input signal $u(t)$ at interval $[0; t]$ we can compute the output signal $y(t)$ at interval $[0; t]$.

For continuous systems, we can also use the notation of transfer functions. The result of applying the Laplace transform yields:

$$Y(s) = G(s)U(s). \tag{4.30}$$

4.7.2 Discrete-Time Model

In system identification, we will almost exclusively deal with observations of inputs and outputs in discrete time, since this is the typical data-acquisition mode.

We assume output $y(k)$ and $u(k)$ to be observed at the sampling instants $k = nT$, $n = 1, 2, 3, \ldots$. The interval T will be called the sampling interval.

Equally, we can derive an impulse response notation for the sampled data system. For ease of notation, we assume that T is one time unit;

$$y(k) = \sum_{n=0}^{\infty} g(n)u(k-n). \tag{4.31}$$

4.7.3 Disturbances

According to relation (4.31), we assume that the output can be calculated exactly once the input is known. In most cases, this is unrealistic. There are always (unknown) disturbances affecting the system. In our linear framework, we assume that these disturbances enter the system additively to the output.

$$y(k) = \sum_{n=0}^{\infty} g(n)u(k-n) + v(k). \tag{4.32}$$

Generally, system identification is chosen to represent the noise term $v(k)$ as a filtered white noise. The white noise $e(k)$ emphasizes the unknown (stochastic) nature of the disturbance. By varying the white noise characteristics and choosing different impulse responses $h(k)$, all kinds of disturbances can be represented. Although this description does not give a complete characterization of all possible disturbances, it is good enough for most practical purposes.

$$v(k) = \sum_{n=0}^{\infty} h(n)e(k-n). \tag{4.33}$$

Similar to transfer function description in the continuous time domain, we can use transfer functions in the discrete time domain.

The use of z-transformation offers an elegant method for describing transfer functions in the discrete time domain. z-transformation plays a similar role for discrete processes as Laplace transformation does for continuous processes.

The z-transformation is defined as:

$$G(z) = \sum_{k=0}^{\infty} g(k)z^{-k} \tag{4.34}$$

hence,

$$y(k) = G(z)u(k) \tag{4.35}$$

where $G(z)$ can be called the transfer function of a discrete system.

In some identification studies, the operators z and q are used interchangeably to denote the forward (backward) shift operator, that is, shifting a signal one sampling interval ahead in time. In the same way, z^{-1} and q^{-1} are used interchangeably to denote the backward shift operator, shifting the signal one interval backward in time.

Using the transfer function description, we can define our basic description for a linear system with additive disturbance.

$$y(k) = G(q)u(k) + H(q)e(k). \tag{4.36}$$

4.7.4 The Prediction Error Method (PEM) Method

The prediction error method (PEM) is sometimes called the generalized least squares (GLS) method, although GLS originally was associated with a certain numerical minimization procedure by extending the equation error model and assuming that the true process is given by

$$A^o(q)y(t) = B^o(q)u(t) + \frac{1}{D^o(q)}e(t) \tag{4.37}$$

or

$$y(t) = \frac{B^o(q)}{A^o(q)}u(t) + \frac{1}{A^o(q)D^o(q)}e(t)$$

where

$$A^o(q) = 1 + a_1^o q^{-1} + a_2^o q^{-2} + \cdots + a_{na}^o q^{-n_a},$$
$$B^o(q) = b_1^o q^{-1} + b_2^o q^{-2} + \cdots + b_{nb}^o q^{-n_b},$$
$$D^o(q) = 1 + d_1^o q^{-1} + d_2^o q^{-2} + \cdots + d_{nd}^o q^{-n_d}$$

and $e(t)$ is white noise with zero mean and variance λ.

So the equation disturbance is assumed to be an AR (autoregressive) process. Then, (3.88) can be written as

$$D^o(q)A^o(q)y(t) = D^o(q)B^o(q)u(t) + e(t). \tag{4.38}$$

This enlarged equation has a white noise disturbance $e(t)$. Prom the study of the least-squares method, we know that consistent and efficient estimates of a_i, b_i, d_i can be obtained by minimizing the loss function

$$\begin{aligned} V_{\text{PEM}} &= \frac{1}{N}\sum_{t=1}^{N}\varepsilon^2(t) \\ &= \frac{1}{N}\sum_{t=1}^{N}\big(D(q)\big[A(q)y(t) - B(q)u(t)\big]\big)^2. \end{aligned} \tag{4.39}$$

This implies that, in the identification a model should be used which has the same structure as the true process

$$D(q)A(q)y(t) = D(q)B(q)u(t) + \varepsilon(t) \tag{4.40}$$

where $\varepsilon(t)$ is the residual. When $D(t) = I$, (3.92) can be written using $\varphi(t)$ and θ,

$$y(t) = \varphi^*(t)\theta + \varepsilon(t) \tag{4.41}$$

where

$$\begin{aligned} \varphi^*(t) &= \big[-y(t-1) \quad \cdots \quad -y(t-n_a) \quad u(t-1) \quad \cdots \quad u(t-n_b)\big], \\ \theta &= (a_1 \quad \cdots \quad a_{n_a} \quad b_1 \quad \cdots \quad b_{n_a})^* \end{aligned}$$

and for computing $\hat{\theta}$

$$\hat{\theta} = \left[\frac{1}{N}\sum_{t=1}^{N}\varphi(t)\varphi^*(t)\right]^{-1}\left[\frac{1}{N}\sum_{t=1}^{N}\varphi(t)y(t)\right]. \tag{4.42}$$

Note that all the discussions about algorithms for computing $\hat{\theta}$ will remain valid. The results derived there depend only on the 'algebraic structure' of the estimate (4.42). For the statistical properties, though, it is of crucial importance whether $\varphi(t)$ is an a priori given quantity, or whether it is a realization of a stochastic process. The reason why this difference is important is that for the dynamic models, when taking expectations of various quantities, it is no longer possible to treat Φ as a constant matrix.

4.7.5 Analysis

Consider the least squares estimate (4.42) applied to the model (3.93). Assume that the data obey

$$A^o(q)y(t) = B^o(q)u(t) + v(t) \tag{4.43}$$

or equivalently

$$y(t) = \varphi^*(t)\theta^o + v(t). \tag{4.44}$$

Here, θ^o is called the true parameter vector and $v(t)$ is a stationary stochastic process that is independent of the input signal. If the estimate $\hat{\theta}$ in (4.42) is 'good', it should be close to the true parameter vector θ^o. To examine if this is the case, an expression is derived for the estimation error

$$\begin{aligned}\hat{\theta} - \theta^o &= \left[\frac{1}{N}\sum_{t=1}^{N}\varphi(t)\varphi^*(t)\right]^{-1} \\ &\quad \cdot \left[\frac{1}{N}\sum_{t=1}^{N}\varphi(t)y(t) - \left(\frac{1}{N}\sum_{t=1}^{N}\varphi(t)\varphi^*(t)\right)\theta^o\right] \\ &= \left[\frac{1}{N}\sum_{t=1}^{N}\varphi(t)\varphi^*(t)\right]^{-1}\left[\frac{1}{N}\sum_{t=1}^{N}\varphi(t)v(t)\right].\end{aligned} \tag{4.45}$$

The minimization of the loss function (3.91) has no analytical solution because the error $\varepsilon(t)$ is nonlinear in the parameters. We note, however, that the error $\varepsilon(t)$ of (3.92) has a bilinear feature. For given $D(q)$ it is linear in $A(q)$ and $B(q)$, and vice versa. The bilinear property can be exploited to obtain a simple algorithm for minimizing the loss function (3.91). Specifically, the algorithm consists of repeating the following two steps until convergence.

At iteration $k+1$:

Step Procedure

- For given $\hat{D}^k(q)$ define the residual

$$\varepsilon_1^{k+1}(t) = \hat{D}^k(q)A(q)y(t) - \hat{D}^k(q)B(q)u(t).$$

The error $\varepsilon_1^{k+1}(t)$ is linear in $A(q)$ and $B(q)$, hence we can determine $\hat{A}^{k+1}(q)$ and $\hat{B}^{k+1}(q)$ by solving an LS problem where the loss function

$$\begin{aligned}V_1 &= \frac{1}{N}\sum_{t=1}^{N}\varepsilon_1^{k+1}(t)^2 \\ &= \frac{1}{N}\sum_{t=1}^{N}\left(\hat{D}^k(q)\left[A(q)y(t) - B(q)u(t)\right]\right)^2\end{aligned}$$

is minimized.

- For given $\hat{A}^{k+1}(q)$ and $\hat{B}^{k+1}(q)$ define the residual as

$$\varepsilon_2^{k+1}(t) = D(q)\hat{A}^{k+1}(q)y(t) - D(q)\hat{B}^{k+1}(q)u(t).$$

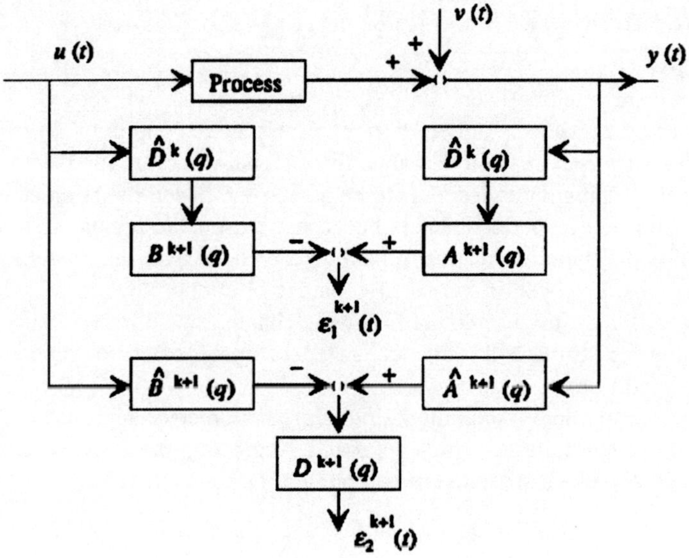

Fig. 4.23 Error generation of the GLS algorithm

Then determine $\hat{D}^{k+1}(q)$ by minimizing

$$\begin{aligned} V_2 &= \frac{1}{N}\sum_{t=1}^{N}\varepsilon_2^{k+1}(t)^2 \\ &= \frac{1}{N}\sum_{t=1}^{N}\big(D(q)[\hat{A}^{k+1}(q)y(t) - \hat{B}^{k+1}(q)u(t)]\big)^2. \end{aligned}$$

This is again an LS problem.

Thus each step of the algorithm solves an LS problem. This is perhaps why the name generalized least-squares (GLS) is given to the algorithm. The iteration can be started with a normal LS estimation. Figure 4.23 shows the block diagram of error generation for the GLS algorithm.

A question to be answered is whether the alternative minimization of V_1 and V_2 will minimize the original loss function V_{GLS} in (3.91). The intuitive answer of the reader may be positive. This is indeed true. The iteration procedure is a special case of the so called separable least-squares problem. Under the persistent excitation condition of the test signal, they can show that, if the iteration converges, it will reach a local minimum of the original loss function V_{GLS} in (3.91). Thus, the iteration is a minimization procedure. Note, however, that the convergence to the global minimum is not guaranteed here.

4.7.6 Modifications

The least squares method is certainly simple to use. As shown above, it gives consistent parameter estimates only under rather restrictive conditions. In some cases, the lack of consistency may be tolerable. If the signal-to-noise ratio is large, the bias will be small. If a regulator design is to be based on the identified model, some bias can in general be acceptable. This is because a reasonable regulator should make the closed loop system insensitive to parameter variations in the open loop part. In other situations, however, it can be of considerable importance to have consistent parameter estimates. In this and the following chapter, two different ways are given of modifying the LS method so that consistent estimates can be obtained under less restrictive conditions.

It is appropriate here to comment on the prediction error approach and why the LS method is a special case of this approach. Neglecting the equation error $\varepsilon(t)$ in the model (3.93), one can predict the output at time t as

$$\begin{aligned}\hat{y}(t) &= -a_1 y(t-1) - \cdots - a_{n_a} y(t-n_a)\\ &\quad + b_1 u(t-1) + \cdots + b_{n_b} u(t-n_b)\\ &= \varphi^*(t)\theta^o. \end{aligned} \tag{4.46}$$

Hence,

$$\varepsilon(t) = y(t) - \hat{y}(t) \tag{4.47}$$

can be interpreted as a prediction error. Therefore, the LS method determines the parameter vector which makes the sum of squared prediction errors as small as possible. Note that the predictor (4.46) is constructed in a rather ad hoe manner. It is not claimed to have any generally valid statistical properties, such as mean square optimality.

There are several ways to modify the GLS algorithm in order to simplify the computation or to speed up the convergence rate. The main idea of these modifications is first to apply the LS method on the model (3.92) with order $n_a + n_d$ in order to obtain consistent estimates of polynomials $D(q)A(q)$ and $D(q)B(q)$, then to perform some kind of model reduction to retrieve $A(q)$, $B(q)$ and $D(q)$. For the problem of model order selection, one can still use the output error criterion. Now, however, there is another possibility. Because the GLS method aims at obtaining white noise residuals, a natural way to order selection is to check the whiteness of residuals for increasing orders. The sample autocorrelation function of the residuals can be used for this test. Note that we have to select both the process order and the order of the disturbance filter. To simplify the procedure, we can let them be equal, i.e., $n = n_d$. More discussions on order selection will be given in a later section.

To see why the GLS method can be called a prediction error method, rewrite the true process (3.88) as

$$y(t) = \frac{B^o(q)}{A^o(q)} u(t) + \frac{1}{A^o(q)D^o(q)} e(t)$$

$$= \frac{B^o(q)}{A^o(q)}u(t) + \left[\frac{1}{A^o(q)D^o(q)} - 1\right]e(t) + e(t). \tag{4.48}$$

Because the coefficients of the highest degree terms of $A^o(q)$ and $D^o(q)$ are 1 (monic polynomials), their product will also have this property:

$$F^o(q) = A^o(q)D^o(q) = 1 + f_1q^{-1} + \cdots + f_{2n}q^{-1}.$$

Thus, the filter

$$\left[\frac{1}{A^o(q)D^o(q)} - 1\right] = \frac{-f_1q^{-1} - \cdots - f_{2n}q^{-1}}{A^o(q)D^o(q)}$$

has one unit delay. This means that the second term in (4.48) is a signal that only depends on the past data up to time $t-1$. When expressing this signal in terms of $u(t)$ and $y(t)$ we have

$$\begin{aligned} y(t) &= \frac{B^o(q)}{A^o(q)}u(t) + \left[\frac{1}{A^o(q)D^o(q)} - 1\right]A^o(q)D^o(q) \\ &\quad \cdot \left[y(t) - \frac{B^o(q)}{A^o(q)}u(t)\right] + e(t) \\ &= B^o(q)D^o(q)u(t) + \left[1 - A^o(q)D^o(q)\right]y(t) + e(t) \\ &= z(t)e(t) \end{aligned} \tag{4.49}$$

where

$$z(t) = B^o(q)D^o(q)u(t) + \left[1 - A^o(q)D^o(q)\right]y(t).$$

Note that $z(t)$ and $e(t)$ are uncorrelated. If $z(t)$ is used as the one step ahead prediction of the output $y(t)$, the prediction error $e(t)$ is white noise. One would expect that this predictor is the best one in some sense, because when the prediction error is white noise, it contains no useful information at all. Indeed, this can be shown more formally. Let $y^*(t)$ be an arbitrary predictor of $y(t)$. Then the variance of the prediction error is

$$\begin{aligned} E\left[y(t) - y^*(t)\right]^2 &= E\left[z(t) + e(t) - y^*(t)\right]^2 \\ &= E\left[z(t) - y^*(t)\right]^2 + Ee^2(t) \\ &\geq Ee^2(t) = E\left[y(t) - z(t)\right]^2. \end{aligned} \tag{4.50}$$

Thus, $z(t)$ is the optimal predictor in the sense of minimum variance. In identification, we will write down the optimal filter in terms of unknown polynomials as

$$y(t|\theta) = B(q)D(q)u(t) + \left[1 - A(q)D(q)\right]y(t) \tag{4.51}$$

and determine the parameters by minimizing the sum of the squares of the prediction errors

$$
\begin{aligned}
V &= \frac{1}{N}\sum_{t=1}^{N}\big[y(t) - y^*(t)\big]^2 \\
&= \frac{1}{N}\sum_{t=1}^{N}\big(D(q)\big[A(q)y(t) - B(q)u(t)\big]\big)^2.
\end{aligned} \tag{4.52}
$$

Again this is the loss function of the GLS method (3.91). The model structure (3.88) is one way to model the equation error noise. Other model structures can also be used. Next, consider optimal prediction for systems given in the state space form

$$
\begin{aligned}
x(t+1) &= A(\theta)x(t) + B(\theta)u(t) + v(t), \\
y(t) &= C(\theta)x(t) + e(t)
\end{aligned} \tag{4.53}
$$

where $v(t)$ and $e(t)$ are mutually uncorrelated white noise sequences with zero means and covariance matrices $\lambda_1(\theta)$ and $\lambda_2(\theta)$, respectively. The optimal one-step predictor of y(t) is given by the Kalman filter,

$$
\begin{aligned}
\hat{x}(t+1|t) &= A(\theta)\hat{x}(t|t-1) \\
&\quad + B(\theta)u(t) + K(\theta)\big[y(t) - C(\theta)\hat{x}(t|t-1)\big], \\
\hat{y}(t|t-1) &= C(\theta)\hat{x}(t|t-1)
\end{aligned} \tag{4.54}
$$

where the gain $K(\theta)$ is given by

$$
K(\theta) = A(\theta)P(\theta)C^*(\theta)\big[C(\theta)P(\theta)C^*(\theta) + \lambda_2(\theta)\big]^{-1}
$$

and where $P(\theta)$ is the solution of the following algebraic Riccati equation:

$$
P(\theta) = A(\theta)P(\theta)A^*(\theta) + \lambda_1(\theta) - K(\theta)C(\theta)P(\theta)A^*(\theta).
$$

This predictor is mean square optimal if the disturbances are Gaussian distributed. For other distributions, it is the optimal linear predictor.

4.7.7 Estimation Using ARX Model

The ARX model is the simplest model incorporating the stimulus signal. The estimation of the ARX model is the most efficient of the polynomial estimation methods because it is the result of solving linear regression equations in analytic form. Moreover, the solution always satisfies the global minimum of the loss function. The ARX model therefore is preferable, especially when the model order is high. The disadvantage of the ARX model is that disturbances are part of the system dynamics.

The parameters of the ARX model structure can be described by a linear difference equation:

$$
\begin{aligned}
&y(t) + a_1y(t-1) + a_2y(t-2) + \cdots + a_{n_a}y(t-n_a) \\
&\quad = b_1u(t-1) + b_2u(t-2) + \cdots + b_{n_b}u(t-n_b) + e(t).
\end{aligned} \tag{4.55}
$$

The adjustable parameters can be lumped in vector θ

$$\theta = \begin{bmatrix} a_1 & a_2 & \cdots & a_{n_a} & b_1 & b_2 & \cdots & b_{n_b} \end{bmatrix}^T.$$

Defining the orders and delay of the ARX model. Specifically, in discrete time q here working as backward shift operator; this means

$$A(q) = 1 + a_1 q^{-1} + a_2 q^{-2} + \cdots + a_{n_a} q^{-n_a},$$
$$B(q) = b_1 q^{-1} + b_2 q^{-2} + \cdots + b_{n_b} q^{-n_b}$$

define

$$G(q) = \frac{B(q)}{A(q)}, \qquad H(q) = \frac{1}{A(q)}$$

then

$$y(t) = \frac{B(q)}{A(q)} u(t) + \frac{1}{A(q)} e(t)$$

or

$$A(q)y(t) = B(q)u(t) + e(t).$$

We call this model the ARX model, where AR refers to the autoregressive part $A(q)y(t)$ and X to the extra input $B(q)u(t)$. The white noise $e(t)$ is assumed to go through the denominator dynamics of the system. From a physical point of view, this is probably not the most natural way of representation, but this makes it possible to define the predictor as a hear regression model.

Let us introduce vector $\varphi(t)$

$$\varphi^*(t) = \begin{bmatrix} -y(t-1) & \cdots & -y(t-n_a) & u(t-1) & \cdots & u(t-n_b) \end{bmatrix}. \tag{4.56}$$

With vectors $\varphi(t)$ and θ, (4.55) can be rewritten as:

$$y(t) = \varphi^*(t)\theta + e(t). \tag{4.57}$$

If the term $e(t)$ is considered to be very small, which may be the case in a lot of practical situations, then according to (4.57) prediction for $y(t)$ depending on the parameter vector θ, can be written as:

$$y(t|\theta) = \varphi^*(t)\theta. \tag{4.58}$$

The predictor is a scalar product of the known (regression) vector $\varphi^{(}t)$ and the parameter vector θ. This is called a hear regression model. With this linear model, simple estimation methods can be applied for the determination of the parameter vector θ.

4.7.8 The Multivariable ARX Case

If we consider the case where input $u(t)$ is an m-dimensional vector and output $y(t)$ is an n-dimensional vector, we obtain for the ARX description:

$$y(t) + A_1 y(t-1) + A_2 y(t-2) + \cdots + A_{n_a} y(t-n_a)$$
$$= B_1 u(t-1) + B_2 u(t-2) + \cdots + B_{n_b} u(t-n_b) + e(t). \quad (4.59)$$

For a system with n_u inputs and n_y outputs, $A(q)$ is an $n_y \times n_y$ matrix. $A(q)$ can be represented as a polynomial in the shift operator q^{-1}:

$$A(q) = I_{n_y} + A_1 q^{-1} + \cdots + A_{n_a} q^{-n_a}$$
$$= \begin{bmatrix} a_{11}(q) & a_{12}(q) & \cdots & a_{1n_y}(q) \\ a_{21}(q) & a_{22}(q) & \cdots & a_{2n_y}(q) \\ \vdots & \vdots & \ddots & \vdots \\ a_{n_y1}(q) & a_{n_y2}(q) & \cdots & a_{n_yn_y}(q) \end{bmatrix}$$

where the entries a_{kj} are polynomials in the delay operator q^{-1},

$$a_{kj}(q) = \delta_{kj} + a_{kj}^1 q^{-1} + a_{kj}^2 q^{-2} + \cdots + a_{kj}^{n_{a_{kj}}} q^{-n_{a_{kj}}}.$$

This polynomial describes how old values of output number jth are affected by the kth output. Here δ_{kj} is the Kronecker-delta; it equals 1 when $k = j$. The kth row of $A(q)$ represents the contribution of the past output values for predict the current value of the kth output.

$B(q)$ is an $n_y \times n_y$ matrix and can be represented as a polynomial in the shift operator q^{-1}:

$$B(q) = B_0 + B_1 q^{-1} + \cdots + B_{n_b} q^{-n_b}$$
$$= \begin{bmatrix} b_{11}(q) & b_{12}(q) & \cdots & b_{1n_u}(q) \\ b_{21}(q) & b_{22}(q) & \cdots & b_{2n_u}(q) \\ \vdots & \vdots & \ddots & \vdots \\ b_{n_y1}(q) & b_{n_y2}(q) & \cdots & b_{n_yn_u}(q) \end{bmatrix}$$

where the matrix element b_{kj} is a polynomial in the shift operator q^{-1}

$$b_{kj}(q) = b_{kj}^1 q^{-n_{b_{kj}}} + \cdots + b_{kj}^{n_{k_{kj}}} q^{-n_{k_{kj}} - n_{b_{kj}} + 1}$$

where $n_{k_{kj}}$ is the delay from the jth input to the kth output. $B(q)$ represents the contributions of inputs to predicting all output values. The simulation results are given by

$$A_0 = \begin{bmatrix} 1 & 0 & 0 \\ 0 & 1 & 0 \\ 0 & 0 & 1 \end{bmatrix},$$

$$A_1 = \begin{bmatrix} -0.516 + 0.012i & 0.104 + 0.006i & -0.008 + 0.037i \\ 0.069 + 0.021i & -0.706 + 0.012i & -0.158 + 0.068i \\ -0.010 + 0.007i & -0.010 + 0.004i & -0.868 + 0.022i \end{bmatrix},$$

$$A_2 = \begin{bmatrix} -0.385 + 0.011i & 0.0192 + 0.006i & 0.015 + 0.036i \\ -0.103 + 0.021i & -0.1595 + 0.012i & 0.126 + 0.067i \\ -0.015 + 0.007i & -0.0060 + 0.004i & -0.101 + 0.022i \end{bmatrix},$$

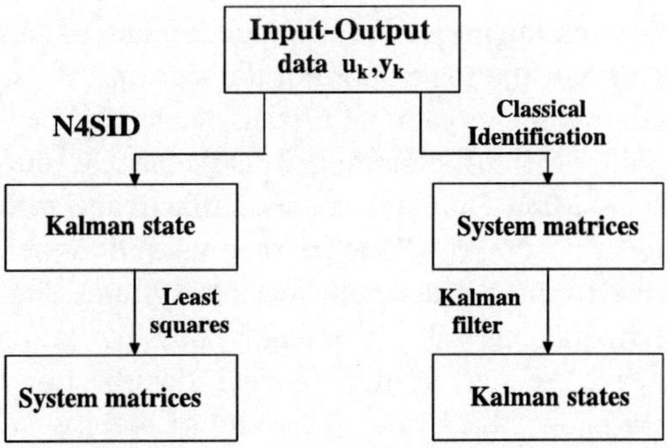

Fig. 4.24 N4S1D approach versus the classical approach

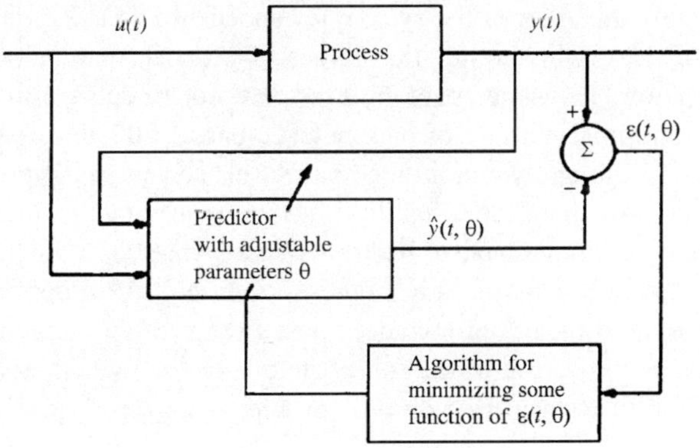

Fig. 4.25 Block diagram of the prediction error method

$$B_0 = \begin{bmatrix} 0 & 0 & 0 \\ 0 & 0 & 0 \\ 0 & 0 & 0 \end{bmatrix},$$

$$B_1 = \begin{bmatrix} -0.006 + 0.003i & 0.006 + 0.003i & 0.104 + 0.006i \\ 0.064 + 0.006i & -0.045 + 0.006i & -0.048 + 0.011i \\ 0.004 + 0.002i & -0.001 + 0.002i & -0.124 + 0.004i \end{bmatrix},$$

$$B_2 = \begin{bmatrix} -0.019 + 0.003i & 0.020 + 0.003i & 0.063 + 0.005i \\ 0.148 + 0.006i & -0.053 + 0.006i & 0.207 + 0.010i \\ 0.006 + 0.002i & -0.004 + 0.002i & -0.020 + 0.003i \end{bmatrix}.$$

4.7.9 Estimated State Space Using N4SID Model

In what follows, we will use N4SID Algorithm 1 and N4SID Algorithm 2 to identify mixed deterministic-stochastic systems. Both algorithms determine state sequences through the projection of input and output data. As we learned from the foregoing sections, the major portion the systems identification literature is concerned with computing polynomial models, which are however known to typically give rise to numerically ill-conditioned mathematical problems, especially for MIMO (Multi-Input Multi-Output) systems. Numerical algorithms for subspace state space system identification (N4SID) are then viewed as the better alternatives. This is especially true for high-order multivariable systems, for which it is not trivial to find a useful parameterizations among all possible parameterizations. This parametrization is needed to start up the classical identification algorithms, which means that a priori knowledge of the order and of the observability (or controllability) indices is required.

With N4SID algorithms, most of this a priori parametrization can be avoided. Only the order of the system is needed and it can be determined through inspection of the dominant singular values of a matrix that is calculated during the identification. The state space matrices are not calculated in their canonical forms (with a minimal number of parameters), but as full state space matrices in a certain, almost optimally conditioned basis (this basis is uniquely determined, so that there is no problem of identifiability). This implies that the observability (or controllability) indices do not have to be known in advance.

Another major advantage is that N4SID algorithms are non-iterative, with no non-linear optimization part involved. For classical identification, an extra parametrization of the initial state is needed when estimating a state space system from data measured on a plant with a nonzero initial condition. A final advantage of the N4SID algorithms, is that there is no difference between zero and nonzero initial states.

In the sequel, we deal with LTI systems subject to input and measurement noises of the type:

$$\begin{aligned} x(t+Ts) &= Ax(t) + Bu(t) + Ke(t), \\ y(t) &= Cx(t) + Du(t) + e, \\ \mathbb{E}\left[\begin{pmatrix} w_k \\ v_k \end{pmatrix} \begin{pmatrix} w_i^t & v_i^t \end{pmatrix}\right] &= \begin{pmatrix} Q^s & S^s \\ (S^s)^t & R^s \end{pmatrix} \delta_{ki} \geq 0^2. \end{aligned} \tag{4.60}$$

System (4.60) can be cast into the standard form

$$\begin{aligned} x_{k+1} &= Ax_k + Bu_k + w_k, \\ y_k &= Cx_k + Du_k + v_k \end{aligned}$$

where a four-stage evaporator system, the three inputs are

- u_1, feed flow,
- u_2, vapor flow to the first evaporator stage,
- u_3, cooling water flow,

and three outputs

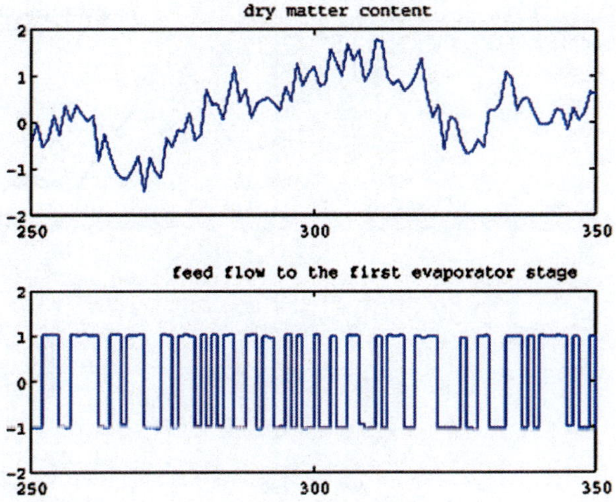

Fig. 4.26 The zero-mean sample 250–350 for u_1–y_1

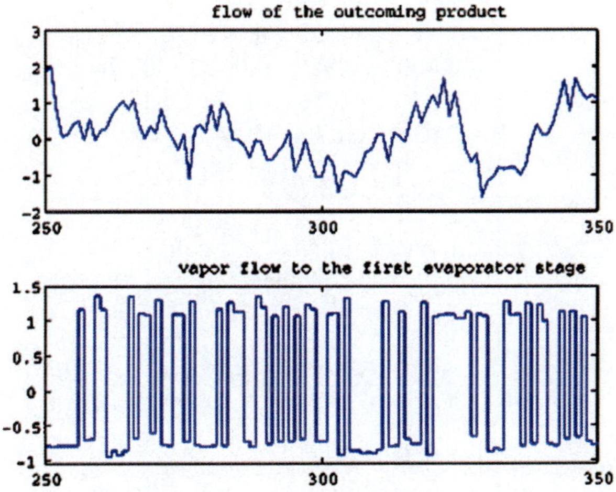

Fig. 4.27 The zero-mean sample 250–350 for u_2–y_2

- y_1, the dry matter content,
- y_2, the flow,
- y_3, the temperature of the out-coming product,

and the number of samples are 6305. From this data, nine different combination can be appeared, namely: (u_1, y_1); (u_1, y_2); (u_1, y_3); (u_2, y_1); (u_2, y_2); (u_2, y_3);(u_3, y_1); (u_3, y_2) and (u_3, y_3). Figure 4.26 shows input (u_1) and output (y_1) data for sample period between 250–350 after removing the constant levels and making zero-mean data. Where input (u_2), output (y_2) and input (u_3), output (y_3) are shown on Figs. 4.27 and 4.28, respectively.

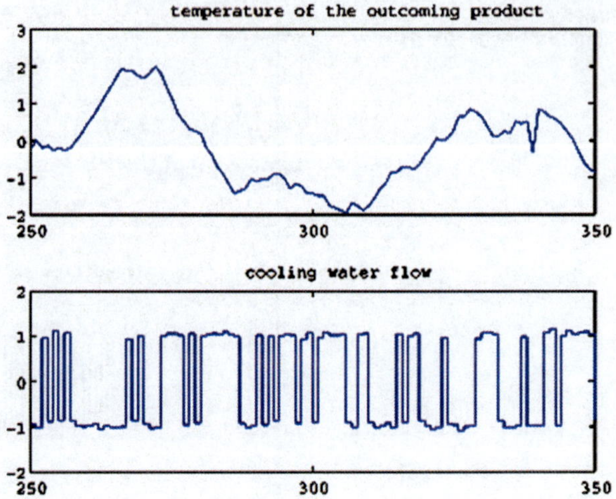

Fig. 4.28 The zero-mean sample 250–350 for u_3–y_3

The model matrices are given below

$$A = \begin{bmatrix} 0.96 & -0.00 & 0.018 & 0.0139 \\ -0.000 & 0.98 & 0.060 & -0.076 \\ -0.006 & -0.10 & 0.93 & -0.18 \\ -0.035 & 0.075 & 0.24 & 0.69 \end{bmatrix},$$

$$B = \begin{bmatrix} -4.2\text{e-}5 & 1.587\text{e-}6 & 0.2 \\ -0.001 & 0.00 & 0.0004 \\ -0.003 & 0.001 & -0.0007 \\ 0.005 & -0.0007 & 0.0004 \end{bmatrix},$$

$$C = \begin{bmatrix} 28.93 & 67.402 & -21.66 & 0.50 \\ -8.32 & -50.44 & -47.416 & -32.04 \\ -77.91 & 1.90 & -2.45 & -0.58 \end{bmatrix}, \tag{4.61}$$

$$D = \begin{bmatrix} 0 & 0 & 0 \\ 0 & 0 & 0 \\ 0 & 0 & 0 \end{bmatrix},$$

$$K = \begin{bmatrix} 0.000 & -0.000 & -0.007 \\ 0.004 & -0.001 & 0.001 \\ -0.000 & -0.001 & -0.001 \\ -0.000 & -0.001 & -0.000 \end{bmatrix}, \quad X(0) = \begin{bmatrix} -8.792\text{e-}5 \\ -0.00576 \\ -0.0004 \\ -0.014 \end{bmatrix}.$$

4.7.10 Numerical Results

PEM method has been used to check performance of this identification. The data samples period from 1 to 3000 have been used to find out parameters of state space

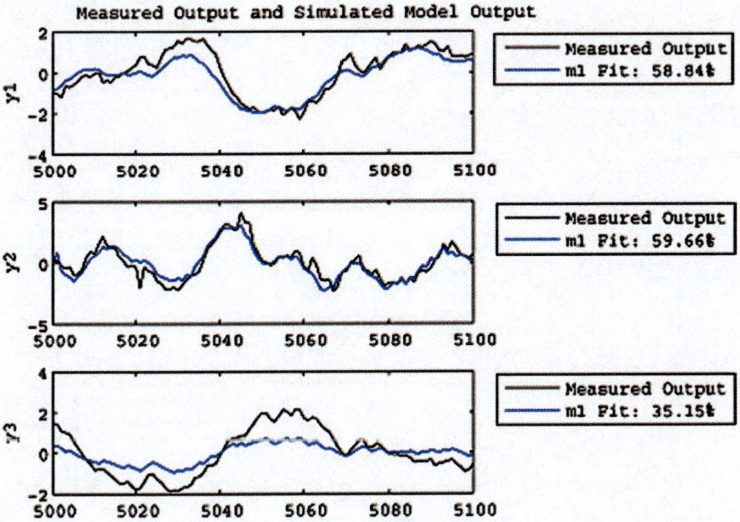

Fig. 4.29 A comparison between original and estimated data using iterative prediction-error minimization method, order 5

equation (4.60), Fig. 4.29. In this case, we use order 5 (4.62) but in Fig. 4.30 order 7 have been used.

$$A = \begin{bmatrix} 0.963 & 0.003 & -0.017 & -0.008 & -0.008 \\ 0.001 & 0.926 & -0.058 & -0.140 & 0.007 \\ 0.013 & 0.138 & 0.960 & 0.099 & 0.057 \\ 0.008 & 0.200 & -0.166 & 0.698 & -0.227 \\ -0.167 & -0.068 & 0.047 & -0.187 & -0.251 \end{bmatrix},$$

$$B = \begin{bmatrix} -0.0001 & 0.0000 & 0.0025 \\ -0.0055 & 0.0015 & 0.0004 \\ 0.0041 & -0.0005 & -0.0001 \\ 0.0085 & -0.0003 & 0.0017 \\ -0.0002 & -0.0025 & 0.0217 \end{bmatrix},$$

$$C = \begin{bmatrix} 18.72 & 27.71 & 36.25 & -1.94 & 1.74 \\ -6.32 & -46.72 & 7.25 & -22.34 & -2.15 \\ -52.37 & 0.31 & 0.17 & -0.49 & 0.8006 \end{bmatrix}, \tag{4.62}$$

$$D = \begin{bmatrix} 0 & 0 & 0 \\ 0 & 0 & 0 \\ 0 & 0 & 0 \end{bmatrix},$$

$$K = \begin{bmatrix} -0.0003 & -0.0003 & -0.0167 \\ 0.0030 & -0.0043 & 0.0079 \\ 0.0049 & 0.0019 & -0.0014 \\ -0.0029 & 0.0003 & -0.0059 \\ -0.0037 & -0.0171 & 0.0046 \end{bmatrix}, \quad X(0) = \begin{bmatrix} 0 & 0 & 0 & 0 & 0 \end{bmatrix}^T.$$

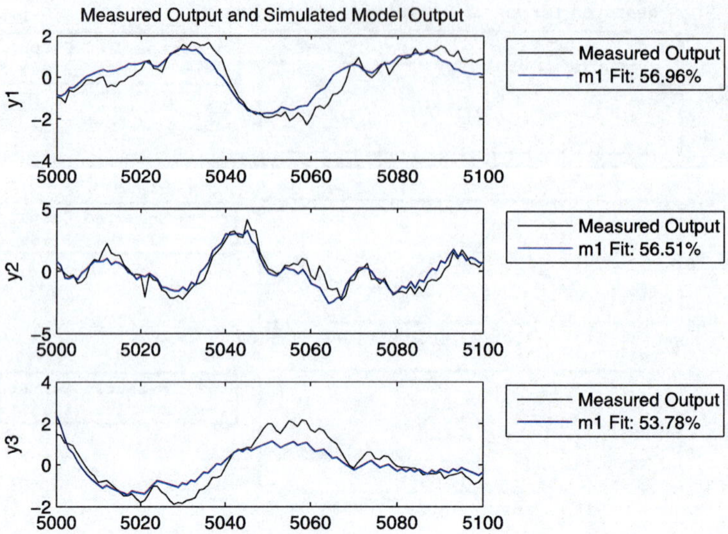

Fig. 4.30 A comparison between original and sampled estimated data using iterative prediction-error minimization method, order 7

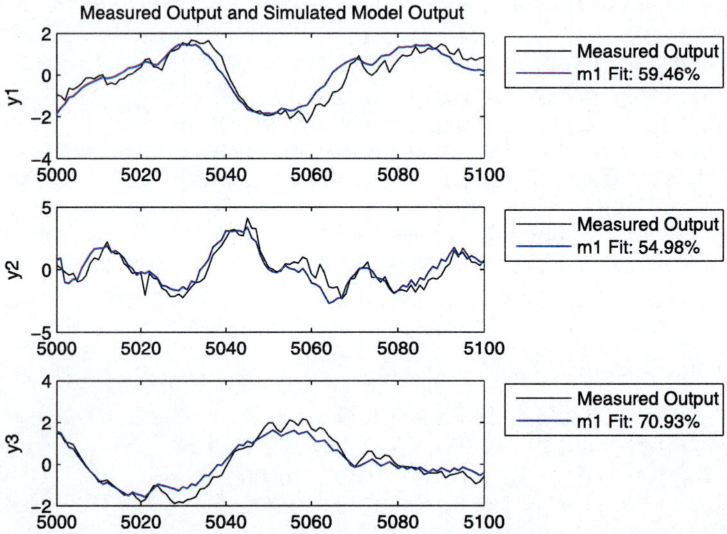

Fig. 4.31 A comparison between original and sampled estimated data using iterative prediction-error minimization method, order 10

From Figs. 4.29–4.46, we observe that for each method with different order we have different fitness the goodness of chosen order depend on singular value. From Fig. 4.34, we observe PEM method is the best for output 1 and 2 where ARX is the best of output 3.

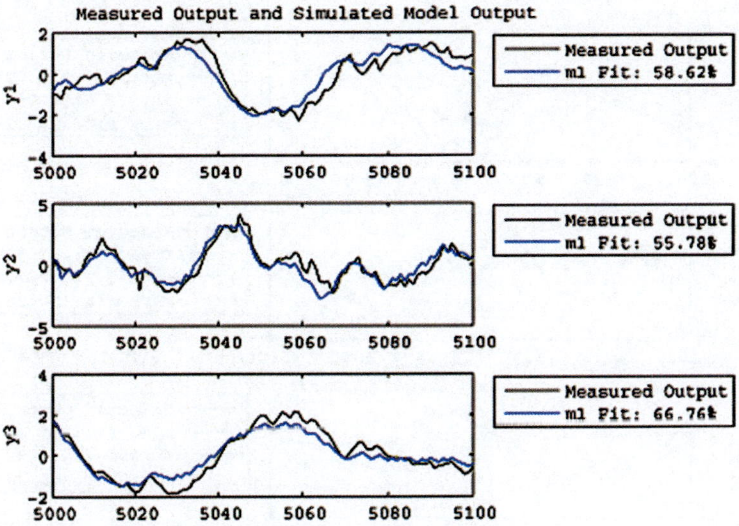

Fig. 4.32 A comparison between original and sampled estimated data using iterative prediction-error minimization method, order 15

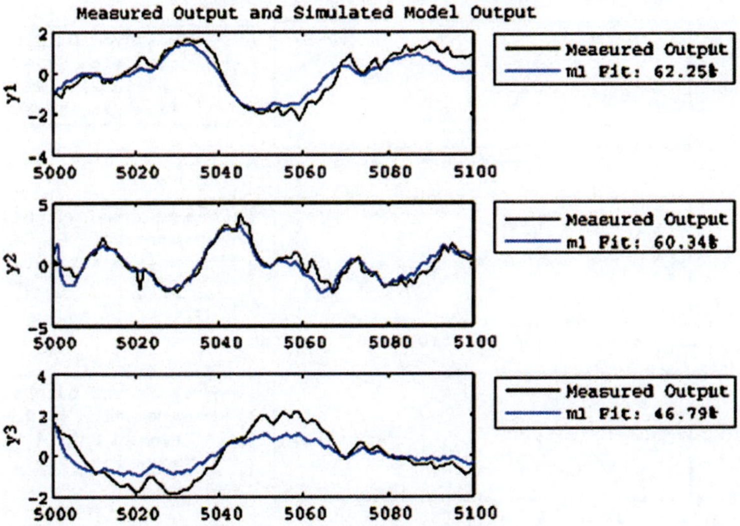

Fig. 4.33 A comparison between original and estimated data using iterative prediction-error minimization method

4.8 A Hydraulic Pumping System

Modeling techniques of industrial processes can be classified into the following three categories [17].

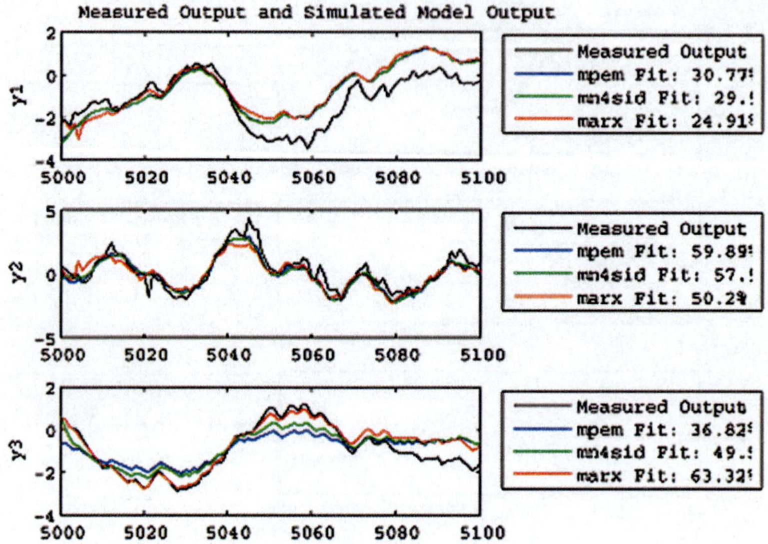

Fig. 4.34 A comparison between three types of system identifications: set 1

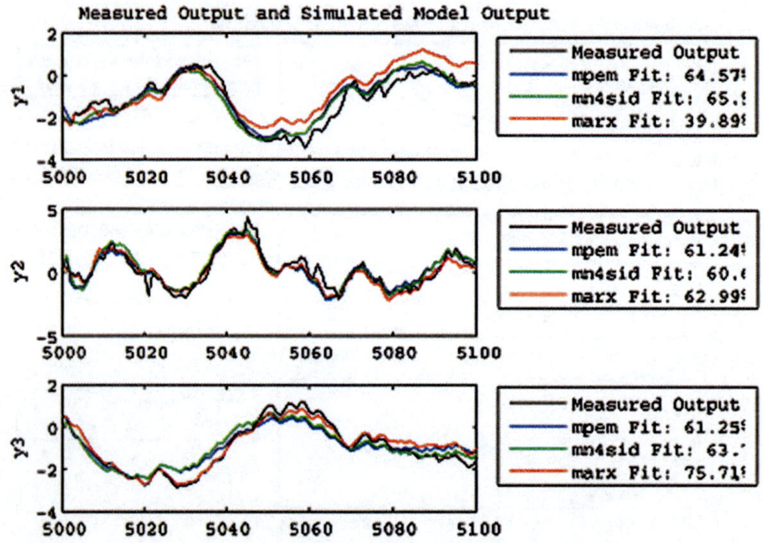

Fig. 4.35 A comparison between three types of system identifications: set 2

1) *White-Box Modeling*: the model is obtained taking into account physical equations that govern the process. In this class, a deep knowledge of the system is necessary.
2) *Gray-Box Modeling*: Prior or auxiliary knowledge of the system is used. Such auxiliary knowledge may be available in the form of steady-state data.

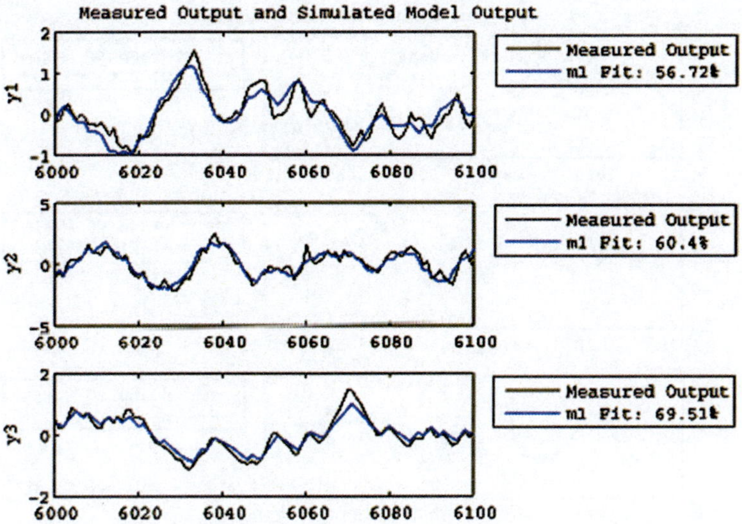

Fig. 4.36 A different comparison between original and estimated data using iterative prediction-error minimization method

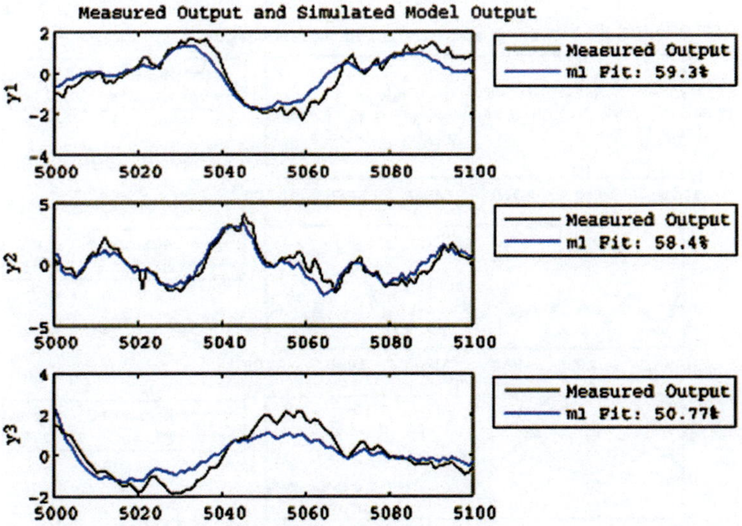

Fig. 4.37 A different comparison between original and sampled estimated data using iterative prediction-error minimization method, order 7

3) *Black-Box Modeling*: The model is identified only using the data set acquired from the process during a dynamical test. In this case, no other source of knowledge is used.

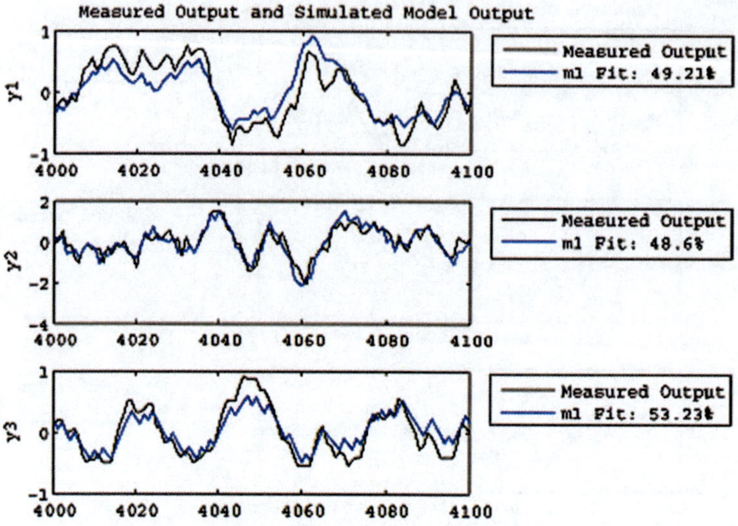

Fig. 4.38 A different comparison between original and sampled estimated data using iterative prediction-error minimization method, order 10

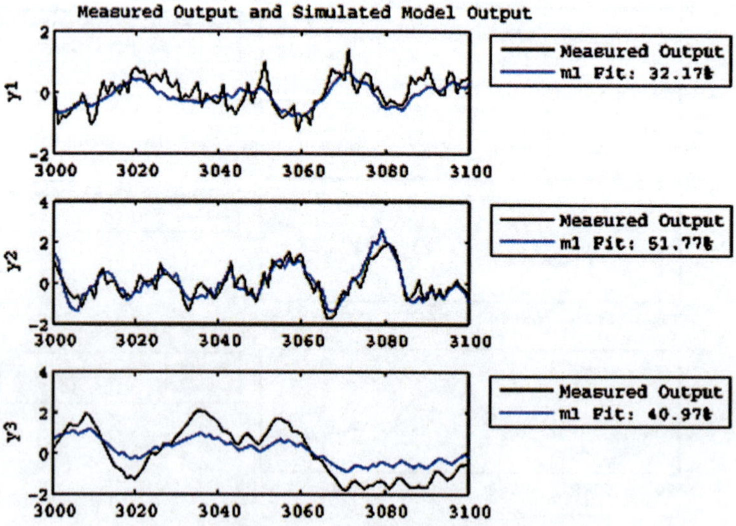

Fig. 4.39 An alternative comparison between original and sampled estimated data using iterative prediction-error minimization method, order 10

In this section, we are concerned with black and gray-box procedures using different model classes.

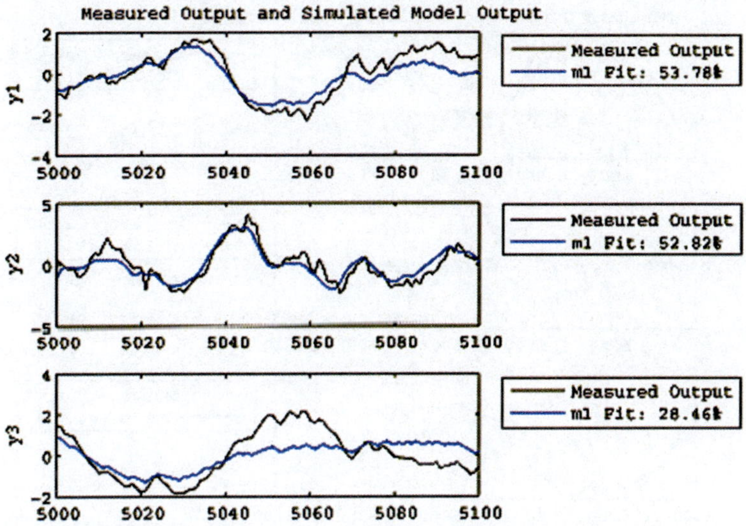

Fig. 4.40 An alternative comparison between original and sampled estimated data using iterative prediction-error minimization method, order 15

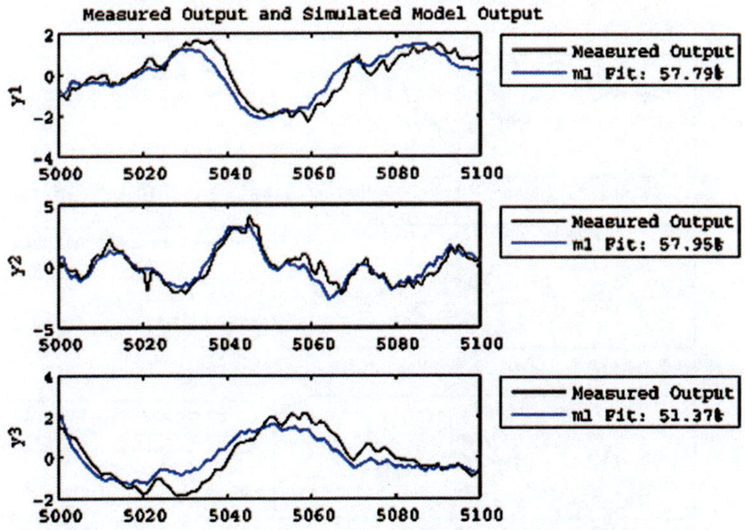

Fig. 4.41 An alternative comparison between original and sampled estimated data using N4SID method, order 15

4.8.1 Dynamical Data

One important task that has to be developed during the identification process is the input signal selection as it can influence not only parameter estimation, but also structure selection in the case of nonlinear systems [16].

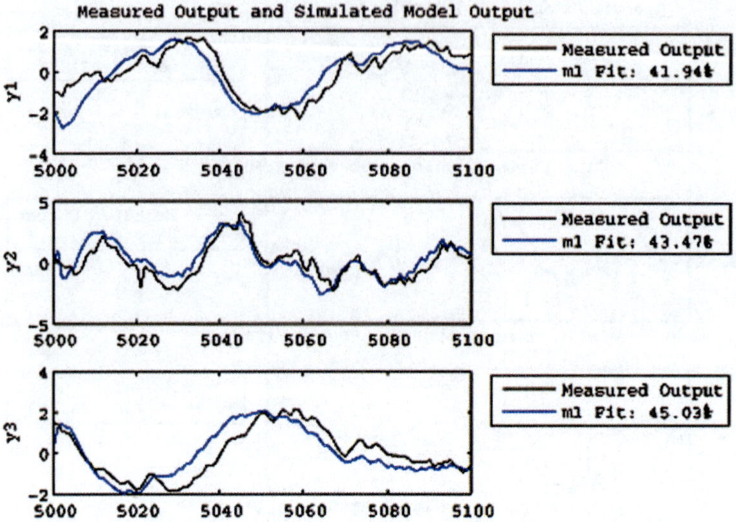

Fig. 4.42 An alternative comparison between original and sampled estimated data using N4SID method, order 10

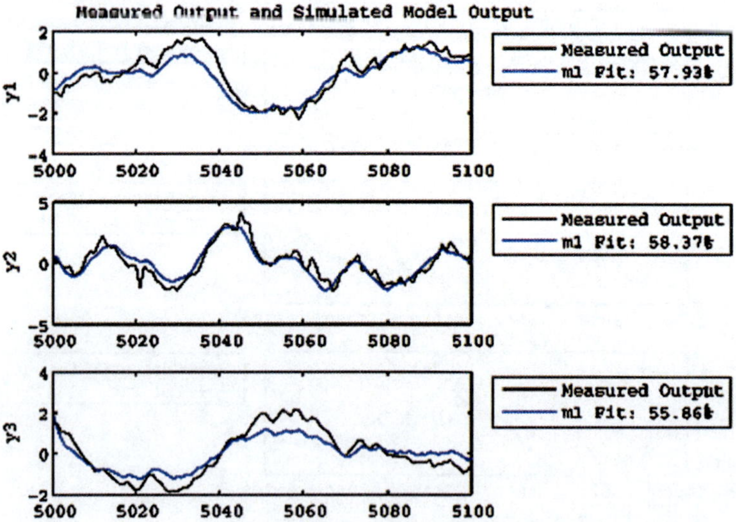

Fig. 4.43 An alternative comparison between original and sampled estimated data using N4SID method, order 5

Since the presence of a "variable time-constant" in the pumping system dynamics was verified in an earlier work [7], the input signal was chosen to excite the system at different operating points using different step sizes. The sampling time $T_s = 50$ ms was selected according to the criterion defined in [1]. Examples of input–output data

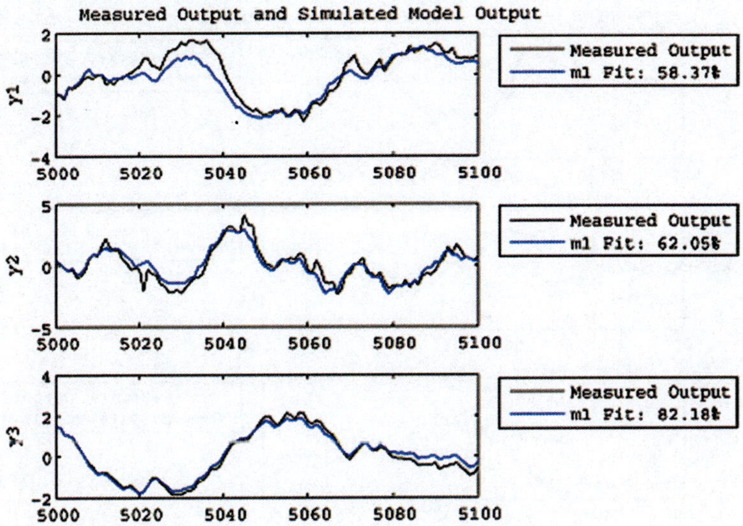

Fig. 4.44 Compare between original data (sample 5000 to 5100) and estimated data from (sample data 1 to 3000) using ARX method, $n_a = [9\ 9\ 9;\ 9\ 9\ 9;\ 9\ 9\ 9]$, $n_b = [9\ 9\ 9;\ 9\ 9\ 9;\ 9\ 9\ 9]$, $n_k = [0\ 0\ 1;\ 1\ 0\ 0;\ 0\ 1\ 0]$

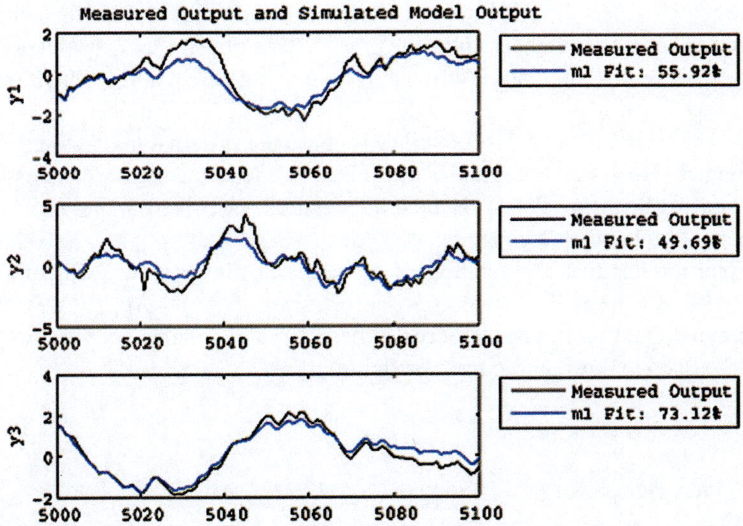

Fig. 4.45 Compare between original data (sample 5000 to 5100) and estimated data from (sample data 1 to 3000) using ARX method, $n_a = [5\ 5\ 5;\ 5\ 5\ 5;\ 5\ 5\ 5]$, $n_b = [3\ 3\ 3;\ 3\ 3\ 3;\ 3\ 3\ 3]$, $n_k = [0\ 0\ 1;\ 1\ 0\ 0;\ 0\ 1\ 0]$

are shown in Fig. 4.47. In this work, $N = 3200$ data points from the dynamical data set were used for model identification and $N = 800$ were used for validation.

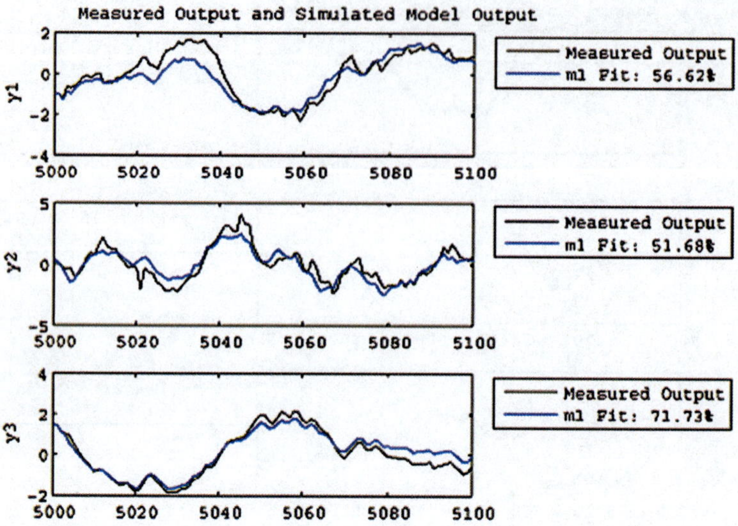

Fig. 4.46 Compare between original data (sample 5000 to 5100) and estimated data from (sample data 1 to 3000) using ARX method, $n_a = [3\ 3\ 3;\ 3\ 3\ 3;\ 3\ 3\ 3]$, $n_b = [5\ 5\ 5;\ 5\ 5\ 5;\ 5\ 5\ 5]$, $n_k = [0\ 0\ 1;\ 1\ 0\ 0;\ 0\ 1\ 0]$

Parametric models describe systems in terms of differential equations and transfer functions. This provides insight into the system physics and a compact model structure. Generally, you can describe a system using an equation, which is known as the general-linear polynomial model or the general-linear model Fig. 4.48. The linear model structure provides flexibility for both the system dynamics and stochastic dynamics. However, a nonlinear optimization method computes the estimation of the general-linear model. This method requires intensive computation with no guarantee of global convergence.

Simpler models that are a subset of the General Linear model structure shown in Fig. 4.48 are possible. By setting one or more of $A(q)$, $B(q)$, $C(q)$ or $D(q)$ polynomials equal to 1, you can create these simpler models such as AR, ARX, ARMAX, Box–Jenkins, and output-error structures.

4.8.2 ARX Modeling

The essential characteristic of the linear regression model is that a residual component e is defined which is a linear function of the unknown model coefficients. In the SISO (single input single output) situation, we can write:

$$\begin{aligned} y(t) + a_1 y(t-1) + \cdots + a_{n_a} y(t-n_a) \\ = b_1 u(t-1) + \cdots + b_{n_b} u(t-n_b) + e(t) \end{aligned} \tag{4.63}$$

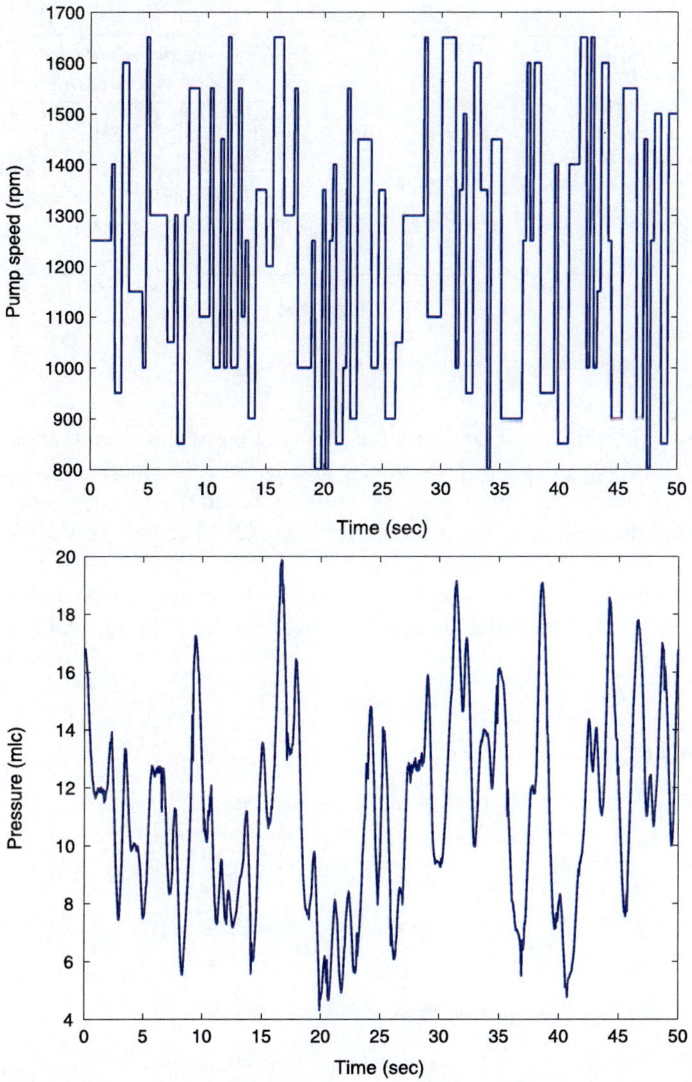

Fig. 4.47 Dynamical data: (*top*) pumps speed reference and (*bottom*) system output pressure

Fig. 4.48 General polynomial model

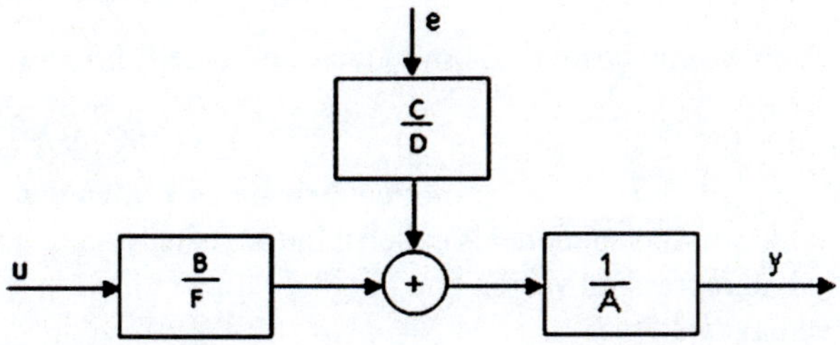

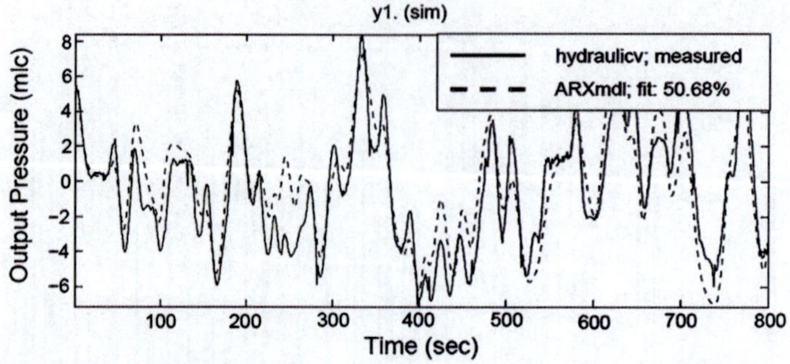

Fig. 4.49 ARX modeled data (- - -) v/s actual data (—)

with $y(t)$ the output signal, and $u(t)$ the input signal of the model, and $a_1, a_2, \ldots, a_{n_a}, b_1, b_2, \ldots, b_{n_b}$ unknown parameters. The use of these kinds of models in estimation and identification problems is essentially based on the argument that a least squares identification criterion is an optimization problem that is analytically solvable.

Since the white noise term $e(t)$ here enters as a direct error in the difference equation, the model is often called an equation error model. The adjustable parameters in this case are:

$$\theta = \begin{bmatrix} a_1 & \cdots & a_{n_a} & b_1 & \cdots & b_{n_b} \end{bmatrix}.$$

If we introduce

$$A(q) = 1 + a_1 q^{-1} + \cdots + a_{n_a} q^{-n_a},$$
$$B(q) = 1 + b_1 q^{-1} + \cdots + b_{n_b} q^{-n_b},$$

we see that the model corresponds to

$$G(q,\theta) = \frac{B(q)}{A(q)}; \qquad H(q,\theta) = \frac{1}{A(q)}.$$

Computing the predictor for the system above, we get

$$\hat{y}(t|\theta) = B(q)u(t) + \big[1 - A(q)\big]y(t). \tag{4.64}$$

Now, we introduce the vector

$$\begin{aligned}\varphi(t) = \big[-y(t-1) \quad \cdots \quad &- y(t-n_a)\\ u(t-1) \quad \cdots \quad &u(t-n_b)\big].\end{aligned}$$

Then we can write the above equation in the following form

$$\hat{y}(t|\theta) = \theta^t.\varphi(t) = \varphi^T(t).\theta. \tag{4.65}$$

The predictor is a scalar product between a known data vector $\varphi(t)$ and a parameter vector θ. Such a model is called a linear regression in statistics and the vector $\varphi(t)$ is called regression vector. See Fig. 4.49 for a comparison of the ARX modeled versus actual data.

4.8.3 ARMAX Modeling

The basic problem with the ARX model is the lack of adequate freedom in describing the properties of the disturbance term. We could add flexibility to that by describing the equation error as a moving average of white noise. This gives the model:

$$\begin{aligned} y(t) + a_1 y(t-1) + \cdots + a_{n_a} y(t-n_a) \\ = b_1 u(t-1) + \cdots + b_{n_b} u(t-n_b) \\ + e(t) + c_1 e(t-1) + \cdots + c_{n_c} e(t-n_c). \end{aligned} \tag{4.66}$$

It can be rewritten as

$$A(q)y(t) = B(q)u(t) + C(q)e(t)$$

where

$$C(q) = 1 + c_1 q^{-1} + \cdots + c_{n_c} q^{-} n_c$$

and

$$G(q,\theta) = \frac{B(q)}{A(q)}; \qquad H(q,\theta) = \frac{C(q)}{A(q)}.$$

The predictor for the ARMAX model can be obtained as

$$\begin{aligned} \hat{y}(t|\theta) = B(q)u(t) + \big[1 - A(q)\big]y(t) \\ + \big[C(q) - 1\big]\varepsilon(t,\theta) \end{aligned} \tag{4.67}$$

where

$$\varepsilon(t,\theta) = y(t) - \hat{y}(t|\theta).$$

In this case, our regression vector would be

$$\begin{aligned} \varphi(t) = \big[-y(t-1) \quad \cdots \quad -y(t-n_a) \\ u(t-1) \quad \cdots \quad u(t-n_b) \\ \varepsilon(t-1,\theta) \quad \cdots \quad \varepsilon(t-n_c,\theta)\big]. \end{aligned}$$

See Fig. 4.50 for a comparison of the ARMAX modeled versus actual data.

4.8.4 Box–Jenkins Model

A natural development of the output error model is to further the properties of the output error. This can be done by assuming that the true process is

$$y(t) = \frac{B(q)}{F(q)}u(t) + \frac{C(q)}{D(q)}e(t) \tag{4.68}$$

where

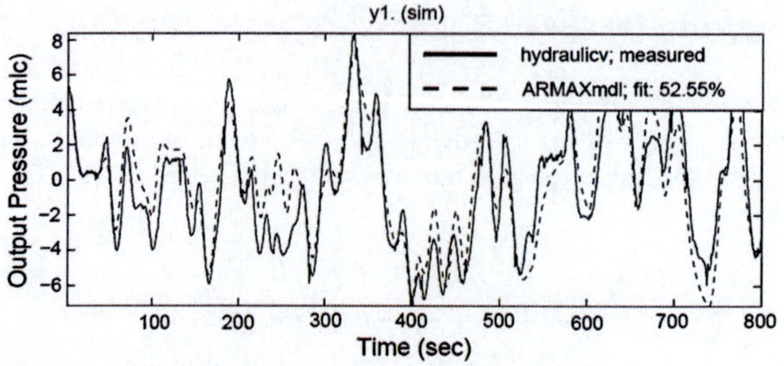

Fig. 4.50 ARMAX modeled data (- - -) v/s Actual data (—)

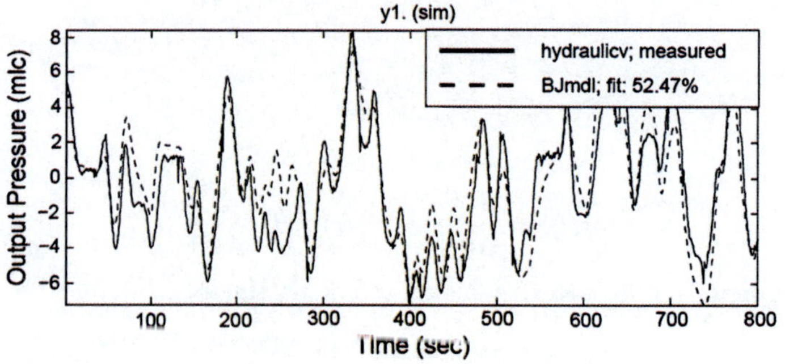

Fig. 4.51 BJ modeled data (- - -) v/s Actual data (—)

$$F(q) = 1 + f_1 q^{-1} + \cdots + f_{n_f} q^{-n_f},$$
$$D(q) = 1 + d_1 q^{-1} + \cdots + d_{n_d} q^{-n_d}.$$

In a sense, this is the most natural finite-dimensional parameterization and the transfer functions G and H are independently parameterized as rational functions. This model was suggested and treated in [8]. In this case, the parameter vector is given by

$$\theta = \begin{bmatrix} b_1 & \cdots & b_{n_b} & f_1 & \cdots & f_{n_f} & c_1 & \cdots & c_{n_c} & d_1 & \cdots & d_{n_d} \end{bmatrix}.$$

See Fig. 4.51 for a comparison of the BJ modeled versus actual data.

4.8.5 State Space Model

In state-space form, the relationship between the input, noise and output signals is written as a system of first order differential or difference equations using an auxiliary state vector $x(t)$. For most physical systems, it is easier to construct models with physical insight in continuous time than in discrete time, simply because most laws of physics are expressed in continuous time. This means that the modeling normally leads to a representation

$$\dot{x}(t) = F(\theta)x(t) + G(\theta)u(t). \tag{4.69}$$

Here F and G are matrices of appropriate dimensions ($n \times n$ and $n \times m$, respectively for an n-dimensional system and an m-dimensional input). Moreover, θ is a vector of parameters that correspond to the unknown values of physical coefficients, material constants, and the like.

Let $\eta(t)$ be the measurements that would be obtained with ideal, noise-free sensors:

$$\eta(t) = Hx(t).$$

Using p as the differential operator, the above state representation can be written as

$$\big[pI - F(\theta)\big]x(t) - G(\theta)u(t).$$

Which means that the transfer function from u to η is

$$\eta(t) = G_c(p, \theta)u(t),$$
$$G_c(p, \theta) = H\big[pI - F(\theta)\big]^{-1}G(\theta).$$

Let the measurements be sampled at the time instants $t = kT$, $k = 1, 2, \ldots$ and the disturbance effects at those time instants be $v_T(kT)$. Hence, the measured output is

$$y(kT) = G_c(p, \theta)u(t) + v_T(kT).$$

There are several ways of transporting $G_c(p, \theta)$ to a representation that is explicitly discrete time. Suppose that the input is constant over the sampling interval T

$$u(t) = u_k = u(kT), \qquad kT \le t < (k+1)T.$$

Then (4.69) can be solved from $t = kT$ to $t = kT + T$, yielding

$$x(kT + T) = A_T(\theta)x(kT) + B_T(\theta)u(kT)$$

where

$$A_T(\theta) = e^{F(\theta)T},$$
$$B_T(\theta) = \int_{\tau=0}^{T} e^{F(\theta)\tau}G(\theta)\,d\tau.$$

The model identified using state space modeling is represented in the form of matrices as follows:

$$A = \begin{bmatrix} 1.0324 & -0.1613 \\ 0.1567 & 0.7600 \end{bmatrix},$$
$$B = \begin{bmatrix} -0.0136 \\ -0.1191 \end{bmatrix}, \qquad C^t = \begin{bmatrix} 199.8214 \\ -7.0361 \end{bmatrix}. \tag{4.70}$$

See Fig. 4.52 for a comparison of the state-space modeled versus actual data.

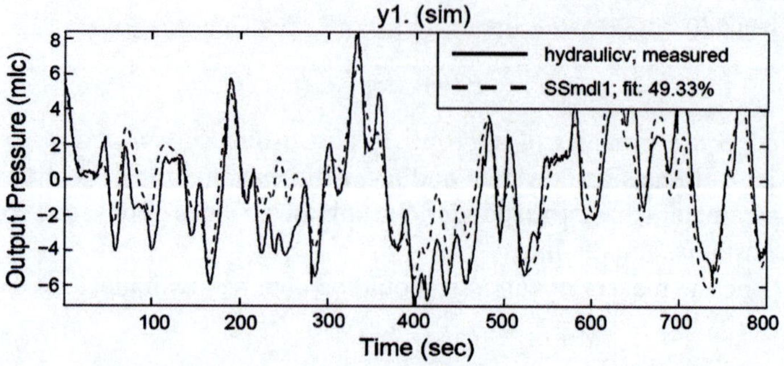

Fig. 4.52 State-space modeled data (- - -) v/s Actual data (—)

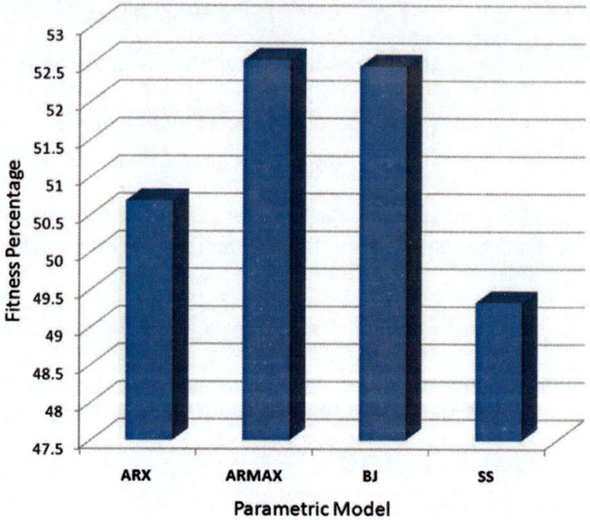

Fig. 4.53 Comparison of fitness percentages using various parametric model structures

4.8.6 Linear Identification Results

The fitness level of the ARMAX modeled data was found to be the best as shown in Fig. 4.53. One possible reason is the influence of disturbance. Unlike the ARX model, the ARMAX model structure includes disturbance dynamics. ARMAX models are useful when you have dominating disturbances that enter early in the process, such as at the input. The ARMAX model has more flexibility in the handling of disturbance modeling than the ARX model. The Box–Jenkins (BJ) structure provides a complete model with disturbance properties modeled separately from system dynamics. The Box–Jenkins model is useful when you have disturbances that enter late in the process. For example, measurement noise on the output is a disturbance late in the process.

As we have discussed, there are a variety of parametric model structures available to assist in modeling a system. The choice of model structure is based upon an understanding of the system identification method and insight and understanding into the system undergoing identification. The characteristics of both system and disturbance dynamics play a role is the proper model selection. These system identification methods can handle a wide range of system dynamics without knowledge of the actual system physics, thereby reducing the engineering effort required to develop models. With respect to the Complexity of the model, the fitness levels and the residual analysis it is concluded that the ARMAX model suits the given system best for the data history provided.

4.9 Flutter for F-18: Estimation and Validation

In what follows, the simulation of different identification techniques used for the flutter of an aircraft F-18 is presented. The parametric methods like ARX, ARMAX, PEM, OE and BJ are used for identification. The non-parametric method used is the N4SID subspace method. All of these methods are estimated and validated for the data for the flutter of an aircraft F-18.

4.9.1 PEM Method

With sampling interval = 1 s, the generated model is given by

$$\begin{aligned}
A(q)y(t) &= [B(q)/F(q)]u(t) + [C(q)/D(q)]e(t),\\
A(q) &= 1 - 2.552q^{-1} + 2.351q^{-2} - 0.7877q^{-3},\\
B(q) &= 0.01881 - 0.04905q^{-1},\\
C(q) &= 1 + 1.988q^{-1} + 0.992q^{-2},\\
D(q) &= 1 - 1.791q^{-1} + 0.9382q^{-2},\\
F(q) &= 1 - 1.05q^{-1} + 1.02q^{-2}
\end{aligned}$$

and the associated simulation results are summarized in Fig. 4.54, for the estimation and validation data.

4.9.2 ARX Method

With sampling interval = 1 s, the generated model is given by

$$\begin{aligned}
A(q)y(t) &= B(q)u(t) + e(t),\\
A(q) &= 1 - 2.498q^{-1} + 2.206q^{-2} - 0.6675q^{-3},\\
B(q) &= -0.3968q^{-4} + 0.7111q^{-5} - 0.3645q^{-6}
\end{aligned}$$

and the associated simulation results are summarized in Fig. 4.55, for the estimation and validation data.

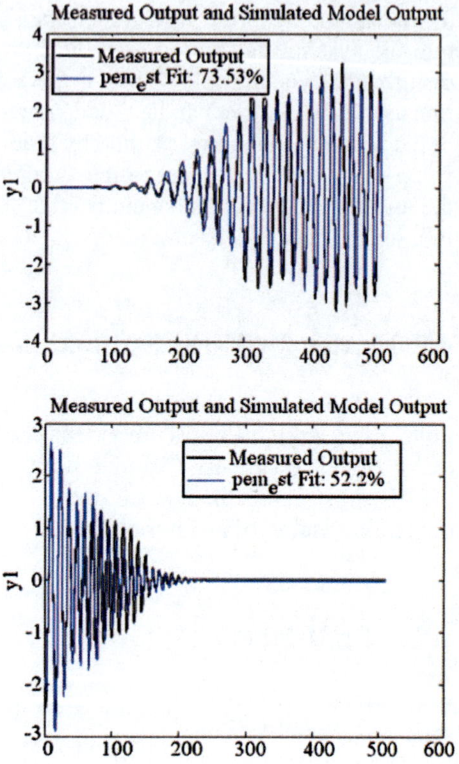

Fig. 4.54 Estimation of PEM method: Estimation (*top*) data and validation data (*bottom*)

4.9.3 ARMAX Method

With sampling interval = 1 s, the generated model is given by

$$
\begin{aligned}
&A(q)y(t) = B(q)u(t) + C(q)e(t),\\
&\quad A(q) = 1 - 1.842q^{-1} + 0.972q^{-2},\\
&\quad B(q) = -0.2457q^{-1} + 1.487q^{-2} - 2.743q^{-3} + 1.106q^{-4} + 1.266q^{-5}\\
&\qquad\qquad - 0.9713q^{-6},\\
&\quad C(q) = 1 + 3.163q^{-1} + 3.818q^{-2} + 2.138q^{-3} + 0.4837q^{-4}
\end{aligned}
$$

and the associated simulation results are summarized in Fig. 4.56, for the estimation and validation.

4.9.4 BJ Method

With sampling interval = 1 s, the generated model is given by

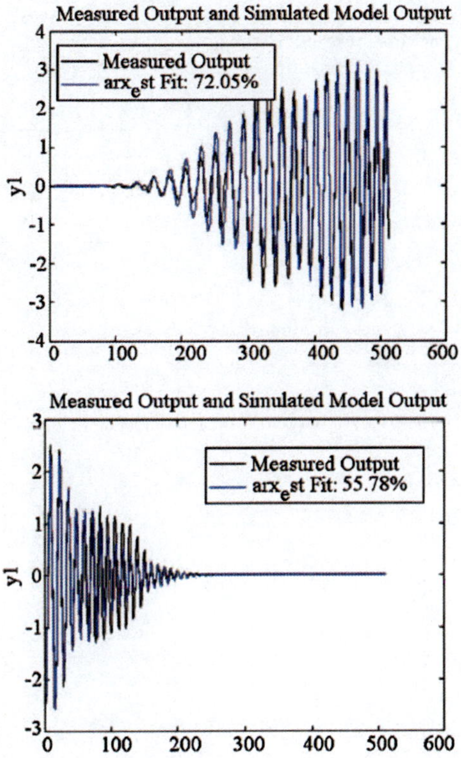

Fig. 4.55 Estimation of ARX method: Estimation (*top*) and validation (*bottom*)

$$
\begin{aligned}
y(t) &= \left[B(q)/F(q)\right]u(t) + \left[C(q)/D(q)\right]e(t),\\
B(q) &= 0.7822 - 1.623q^{-1} + 0.8608q^{-2},\\
C(q) &= 1 + 2.59q^{-1} + 2.578q^{-2} + 0.9919q^{-3},\\
D(q) &= 1 - 2.706q^{-1} + 2.591q^{-2} - 0.8684q^{-3},\\
F(q) &= 1 - 1.762q^{-1} + 0.8845q^{-2} - 0.07037q^{-3}
\end{aligned}
$$

and the associated simulation results are summarized in Fig. 4.57, for the estimation and validation.

4.9.5 Output Equation Method

With sampling interval = 1 s, the generated model is given by

$$
\begin{aligned}
y(t) &= \left[B(q)/F(q)\right]u(t) + e(t),\\
B(q) &= 2.447q^{-1} - 9.462q^{-2} + 14.05q^{-3} - 9.491q^{-4} + 2.464q^{-5},\\
F(q) &= 1 - 2.448q^{-1} + 1.175q^{-2} + 1.737q^{-3} - 2.083q^{-4} + 0.6412q^{-5}
\end{aligned}
$$

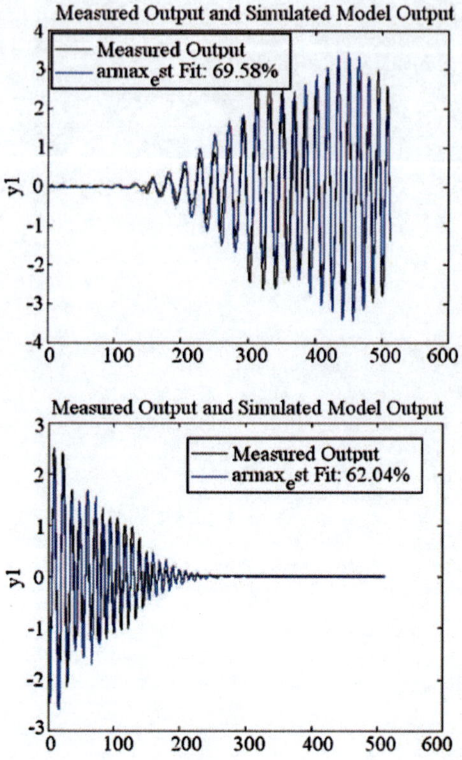

Fig. 4.56 Estimation of ARMAX method: Estimation (*top*) and validation (*bottom*)

and the associated simulation results are summarized in Fig. 4.58, for the estimation and validation.

4.9.6 N4SID Method

With sampling interval = 1 s, the generated model is given by

$$x(t+1) = Ax(t) + Bu(t) + Ke(t),$$

$$y(t) = Cx(t) + Du(t) + e(t),$$

$$A = \begin{bmatrix} 0.78366 & 0.45662 \\ -0.19106 & 0.88406 \end{bmatrix}, \quad B = \begin{bmatrix} 0.007195 \\ 0.023348 \end{bmatrix}, \quad C^t = \begin{bmatrix} -6.8573 \\ -0.75961 \end{bmatrix},$$

$$K = \begin{bmatrix} -0.13446 \\ -0.091591 \end{bmatrix}, \quad x(0) = \begin{bmatrix} 0.00084913 \\ 0.00099388 \end{bmatrix}$$

and the associated simulation results are summarized in Fig. 4.59, for the estimation and validation.

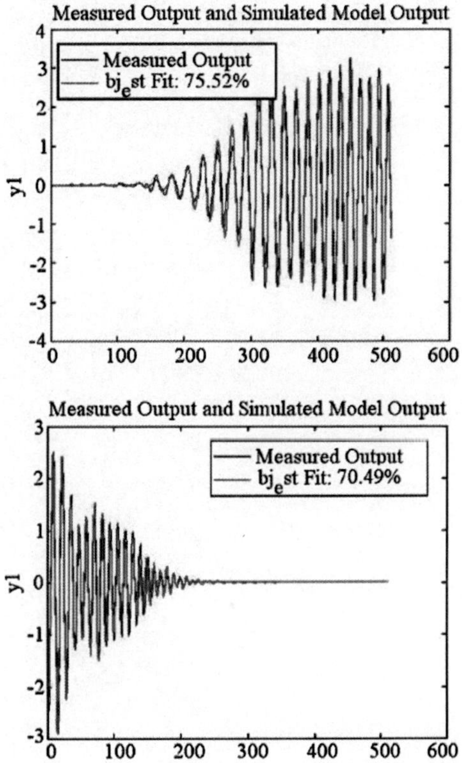

Fig. 4.57 Estimation of ARX method: Estimation (*left*) and validation (*right*)

Table 4.10 Comparisons of flutter models

Model	Loss function	FPE	% Estimation	% Validation
PEM	2.41673×10^{-7}	2.59285×10^{-7}	73.53	52.2
ARX	0.000137359	0.000140617	72.05	55.78
ARMAX	1.97461×10^{-5}	2.07252×10^{-5}	69.58	62.04
BJ	1.90325×10^{-6}	1.99762×10^{-6}	75.52	70.49
OE	0.0282611	0.029417	87.01	50.82
N4SID	0.00880892	0.00908857	59.44	43.69

4.10 Notes and References

The use of a priori information to identify nonlinear systems is usually justified when the system is not well represented in all operating points by the available dynamical data sets, which often occurs in practical situations. For instance, [2] show that information about the static curve of a system can be useful during the dynamic model identification process when this information is not completely available in the dynamic data.

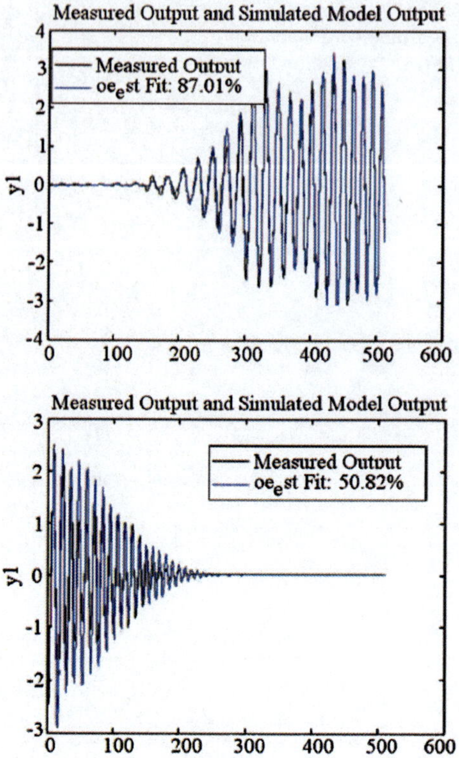

Fig. 4.58 Estimation of OE method: Estimation (*top*) and validation (*bottom*)

Nevertheless, measured static curves and dynamic data were used even though the dynamic data set might supply by itself enough information to arrive at models with good approximation of the static curve of the system. Thus, these data sets could be seen as carrying redundant information.

This brief addressed the problem of identification of nonlinear systems using different methods that use auxiliary information in various degrees. Using data from a 15 kW pumping plant, it was shown that *steady-state information* and *free-run simulation error* criteria can be useful during the identification process.

In this brief a novel multi-objective approach to system identification was proposed: it uses the static curve as the additional source of information and the simulation error criterion instead of the prediction error criterion. Besides, a new decision-maker that takes into account the measurement uncertainty was also introduced. This approach arrived at models with better static curve and dynamic response, being possible to find a model that outperformed the black-box counterpart in the dynamic and static performance criteria.

As far as the simulation error bi-objective approach is concerned, it would be interesting to develop an algorithm to find the Pareto set without having to use the free-run simulation which is very computationally demanding. In spite of its high computational cost, it is also desired in future work to apply the simulation error criterion to detect the model structure of the process studied in this work as in [16].

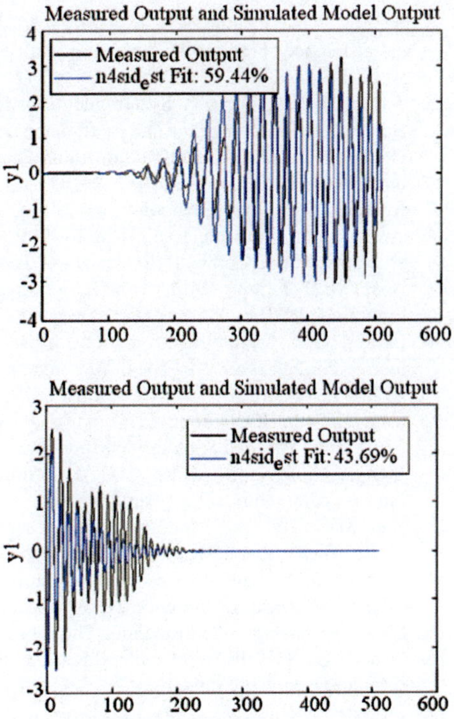

Fig. 4.59 Estimation of N4SID method: Estimation (*top*) and validation (*bottom*)

The greatly different orders of magnitude of the estimated parameters deserves a remark. It must be realized that the parameters multiply regressors variables which, in this case, are usually nonlinear. A large average value of a variable that appears to the cubic power will require a much smaller parameter value to compensate. For all the models in this brief, the contribution of each term multiplied by the respective parameter is of the same order of magnitude. One way of avoiding this situation is to normalize the data. This was not done in this brief in order to maintain the engineering units and therefore to facilitate a physical interpretation of the simulated data.

References

1. Aguirre, L.A.: A nonlinear correlation function for selecting the delay time in dynamical reconstructions. Phys. Lett. **203A**(2, 3), 88–94 (1995)
2. Aguirre, L.A., Donoso-Garcia, P.F., Santos-Filho, R.: Use of a priori information in the identification of global nonlinear models—A case study using a buck converter. IEEE Trans. Circuits Syst. I, Reg. Pap. **47**(7), 1081–1085 (2000)
3. Aguirre, L.A., Barroso, M.F.S., Saldanha, R.R., Mendes, E.M.A.M.: Imposing steady-state performance on identified nonlinear polynomial models by means of constrained parameter estimation. Proc. IEE Part D: Control Theory Appl. **151**(2), 174–179 (2004)

4. Aguirre, L.A., Coelho, M.C.S., Corrêa, M.V.: On the interpretation and practice of dynamical differences between Hammerstein and Wiener models. Proc. IEE Part D: Control Theory Appl. **152**(4), 349–356 (2005)
5. Astrom, K.J., Eykhoff, P.: System identification—A survey. Automatica **7**(2), 123–162 (1971)
6. Baker, J.E.: Reducing bias and inefficiency in the selection algorithm. In: Proc. 2nd Int. Conf. Genetic Algorithms Genetic Algorithms Their Appl., Mahwah, NJ, pp. 14–21. Lawrence Erlbaum Associates, Inc. (1987)
7. Barbosa, B.H.: Instrumentation, modelling, control and supervision of a hydraulic pumping system and turbine–generator module (in Portuguese). Master's thesis, Sch. Elect. Eng., Federal Univ. Minas Gerais, Belo Horizonte, Brazil (2006)
8. Box, G.E.P., Jenkins, G.M.: Time Series Analysis, Forecasting and Control. Holden-Day, San Francisco (1970)
9. Elkaim, G.H.: System identification for precision control of a wing-sailed GPS-guided catamaran. Ph.D. dissertation, Stanford University (December 2001)
10. Evans, J., Elkaim, G., Lo, S., Parkinson, B.: System identification of an autonomous aircraft using GPS. In: ION Global Positioning System Conference, pp. 1065–1074 (1997)
11. Juang, J.N.: Applied System Identification. Prentice Hall, New York (1994)
12. Juang, J.N., Phan, M., Horta, L.G., Longman, R.W.: Identification of Observer/Kalman filter Markov parameters: Theory and experiments. In: *NASA Technical Memorandum*, June 1991
13. Keir, M.C.: Dynamic Modeling, Control and Fault Detection in Vapor Compression Systems, M. Sc. Thesis, Dept. Mech. Eng., Univ. Illinois. Urbana-Champaign, Urbana, IL, 2006
14. Keir, M.C., Alleyne, A.: Feedback structures for vapor compression cycle systems. In: Proc. American Control Conference, New York, pp. 5052–5058 (2007)
15. Ljung, L.: Systems Identification: Theory for the User. Prentice Hall, New York (1999)
16. Piroddi, L., Spinelli, W.: An identification algorithm for polynomial NARX models based on simulation error minimization. Int. J. Control **76**(17), 1767–1781 (2003)
17. Sjoberg, J., Zhang, Q., Ljung, L., Beneviste, A., Delyon, B., Glorennec, P., Hjalmarsson, H., Juditsky, A.: Nonlinear black-box modeling in system identification: A unified overview. Automatica **31**, 31–1961 (1995)
18. Skogestad, S., Postlethwaite, I.: Multivariable Feedback Control. Wiley, New York (1996)
19. Zhu, Y.: Multivariable System Identification for Process Control. Pergamon, Lexington (2001)

Chapter 5
Nominal Control Design

5.1 Introduction

It is increasing apparent that the application of control engineering concepts and techniques has resulted in numerous benefits which manifest our life. This include, but not restricted to, improved product/life quality, minimized waste materials, reduced pollution, increased safety, reduced energy consumption, to name a few. One can observe that the notions of feedback and control play significance roles in most societal and technological aspects. Nowadays, it is becoming widely accepted that control is more *engineering* than *science*, but it certainly requires a concrete theoretical underpinning for it to be successfully applied to ever more challenging projects. This will gradually help in bridging the so-called *theory/practice gap*.

The development of efficient computer software for control has provided many benefits for teaching, research, and the development of control systems design in industry. MATLAB and Simulink are considered the dominant software platforms for control system analysis and design, with numerous off-the-shelf toolboxes dedicated to control systems and related topics. It is clear that MATLAB provides a suitable implement for control engineering.

Feedback and control are almost everywhere. One can virtually link the powerful word *control* to almost anything, such as *diet control*, *financial control*, *motor control*, *pest control*, *and robot control*, to name a few. One can additionally say that *power is nothing without control*, which is believed to be correct in both social and technological contexts. Feedback is an intuitive means for control. For example, when you feel cold (sensing), you add one more layer of cloth (decision and then control action) to keep yourself warm and comfortable (objective). This is biological feedback due to a change in the environment. In technological systems, the loop *sensing-feedback-decision-control* is implemented to change the system behavior into a desirable one. In most cases in this book, we shall focus on the "feedback control" for a given system described by ordinary differential equations (ODEs) with a single input–single output (SISO). More specifically, we will mainly concentrate on analytical and simulation methods for linear feedback control systems and a few aspects of simulation for nonlinear systems. For multiple input–multiple output (MIMO) linear systems, good references are [2, 3, 6, 7, 10, 11, 16, 39].

M.S. Mahmoud, Y. Xia, *Applied Control Systems Design*,
DOI 10.1007/978-1-4471-2879-3_5,

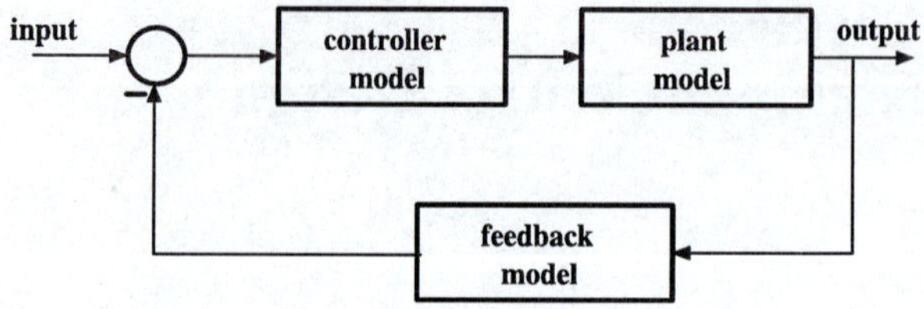

Fig. 5.1 Standard representation

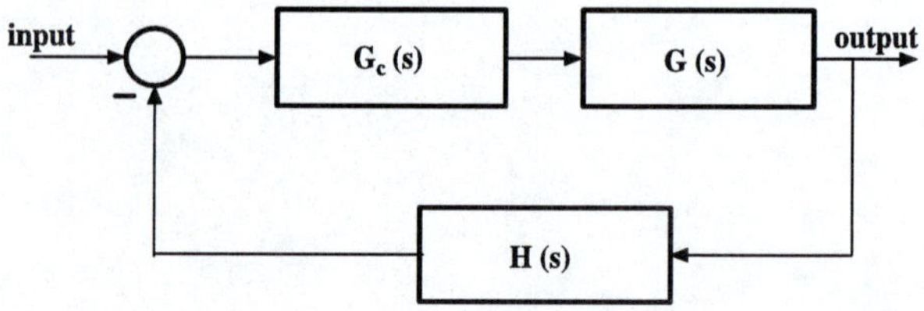

Fig. 5.2 Linear feedback system

Figure 5.1 shows a typical feedback control structure with three blocks, namely, the plant block, the controller block, and the feedback block. In this typical feedback control structure, the plant and the controller blocks form the forward path and the feedback path normally includes the sensor and, possibly, signal conditioning. This system structure is quite commonly seen in process control and other control applications. For simplicity, throughout the book only the paths with negative actions will be labeled in the block diagram, and the ones with positive actions will have the plus sign omitted by default, as in Fig. 5.1. If all three blocks are linear, the feedback control structure can be redrawn, as shown in Fig. 5.2. This model structure will be extensively used in the book.

In control systems, the concept of "feedback" is very important. If we assume that there is no feedback path, the system will be driven solely by the input signal, and after the effect of the control block, the output signal of the system will be generated. This kind of system structure is usually referred to as an open-loop control structure. Under ideal circumstances, an open-loop control strategy will work, but this is based on having an accurate plant model, which never exists in practice due to modeling errors and system disturbances. Thus, for accurate control a closed-loop system structure must be used instead. Closed-loop systems are often referred to as feedback control systems.

5.1.1 Basic Definitions

Among the prevailing trends is that engineering is concerned with understanding and controlling the materials and forces of nature for the benefit of humankind. In our way to address control systems, we provide in the following some essential definitions:

Systems: A system is a combination of components that act together and perform a prescribed objective. It must be noted that a system is not limited to physical ones, rather the concept of system can be equally applied to abstract, dynamic phenomena such as those encountered in economics, biology and the like.

Control Systems: A control system is an interconnection of components forming a system configuration that will provide a desired system response. It must be recorded that the basis for analysis of a system is the foundation provided by linear system theory, which emerges from the representation of individual elements as a cause-effect relationship. This asserts the notion that a system transforms or processes the input signal to provide an output signal.

Systems: A system is a combination of components that act together and perform a prescribed objective. It must be noted that a system is not limited to physical ones, rather the concept of system can be equally applied to abstract, dynamic phenomena such as those encountered in economics, biology and the like.

Control Systems: A control system is an interconnection of components forming a system configuration that will provide a desired system response. It must be recorded that the basis for analysis of a system is the foundation provided by linear system theory, which emerges from the representation of individual elements as a cause-effect relationship. This asserts the notion that a system transforms or processes the input signal to provide an output signal.

Plants: A plant may be a piece of equipment or any physical object, perhaps just a set of machine parts connecting together, the purpose of which is to perform a particular operation. An alternative phrase to the plant is *process*.

For the purpose of this book, a component or process to be controlled can be represented by a single block, as shown in Fig. 5.2.

Controlled Variable: The controlled variable is a quantity or condition that is measured and controlled. Normally, the controlled variable is the output of the system.

Manipulated Variable: The manipulated variable is a quantity or condition that is varied by the controller so as to affect the value of controlled variable.

Disturbances: A disturbance is a signal that tends to adversely affect the value of the output of the system. If a disturbance is generated within the system, it is called *internal*, while an *external* disturbance is generated outside the system and hence, treated as an additional input.

Feedback Control: Feedback control refers to an operation that, in the presence of disturbances, tends to reduce the difference between the output of a system some reference input.

In what follows, we discuss some of the basic configurations usually encountered in control systems.

5.1.2 Feedback Control Systems

Simply stated, a *feedback control system* refers to a configuration that maintains a prescribed relationship between the output and the reference input by comparing them and using the difference in generating a control. Typically in room-temperature system, the actual room temperature is measured and compared with the desired (reference) temperature. Based on the temperature difference, the thermostat turns the heating or cooling equipment on or off so as to keep the room temperature at a comfortable level irrespective of the surrounding conditions.

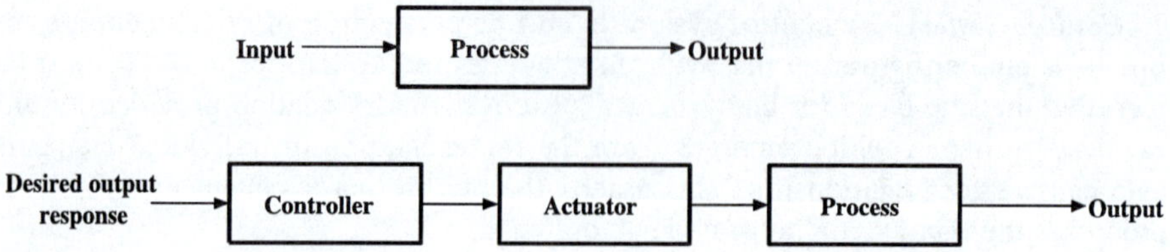

Fig. 5.3 Open-loop system

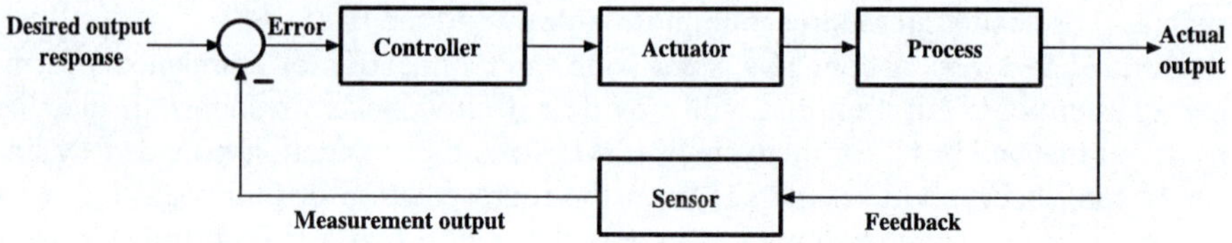

Fig. 5.4 Closed-loop system

5.1.3 Open-Loop Control Systems

An open-loop control system utilizes a controller and an actuator to control the process and obtain the desired response without feedback, see Fig. 5.3. Obviously, the use of open-loop control systems is severely limited in practice.

5.1.4 Closed-Loop Control Systems

In contrast to open-loop control system, a closed-loop control system depicted in Fig. 5.4 uses an additional measure (feedback signal) of the actual output in order to compare the actual output with the desired output response (reference or command). For obvious reasons, a closed-loop control system is frequently labeled as *feedback control system*. Thus, a feedback control system tends to maintain a prescribed relationship of a system variable to another variable by comparing functions of these variables and utilizing the difference as a means of control.

It is fair to admit feedback control is nowadays a fundamental fact of modern industry and society. Several numerous examples in textbooks [2, 6, 9–11, 16, 39] emphasize this fact and illuminate the properties of feedback control systems.

5.1.5 Control Systems Design

Engineering design is the central task of the engineer. It is a complex process in which both creativity and analysis play major roles. Design is the process of conceiving or inventing the forms, parts, and details of a system to achieve a specified purpose. Design activity can be thought of as planning for the emergence of a particular product or system. Design is an innovative act whereby the engineer creatively uses knowledge and materials to specify the shape, function, and material content of a system.

An important factor in realistic design is the limitation of time. Design takes place under imposed schedules, and we eventually settle for a design that may be less than ideal but considered "good enough." In many cases, time-is the only competitive advantage. A major challenge for the designer is writing the specifications for the technical product. Specifications are statements that explicitly state what the device or product is to be and do. The design of technical systems aims to provide appropriate design specifications and rests on four characteristics: complexity, trade-offs, design gaps, and risk. Complexity, trade-off, gaps, and risk are inherent in designing new systems and devices. Although they can be minimized by considering all the effects of a given design, they are always present in the design process.

Design is a process that may proceed in many directions before the desired one is found. It is a deliberate process by which a designer creates something new in response to a recognized need while recognizing realistic constraints. The design process is inherently iterative-we must start somewhere! Successful engineers learn to simplify complex systems appropriately for design and analysis purposes. A gap between the complex physical system and the design model is inevitable. Design gaps are intrinsic in the progression from the initial concept to the final product. We know intuitively that it is easier to improve an initial concept incrementally than to try to create a final design at the start. In this respect, engineering design is not a linear process. It is an iterative, nonlinear, creative process. The design process consists of [10, 11, 30]:

1. establishing the system goals,
2. identifying the variables that we desire to control,
3. writing the specifications in terms of the accuracy we must attain, like good regulation against disturbances, desirable responses to commands, realistic actuator signals, low sensitivities, and robustness.

Briefly stated, the controller design problem is as follows: *Given a model of the system to be controlled (including its sensors and actuators) and a set of design goals, find a suitable controller, or determine that none exists.* As with most of engineering design, the design of a feedback control system is an iterative and nonlinear process. A successful designer must consider the underlying physics of the plant under control, the control design strategy, the controller design architecture. In practice, solving a control problem generally involves

- Choosing sensors to measure the plant output.
- Choosing actuators to drive the plant.
- Developing the plant, actuator, and sensor equations (models).
- Designing the controller based on the models developed and the control criteria.
- Evaluating the design analytically, by simulation, and finally, by testing the physical system.
- If the physical tests are unsatisfactory, iterating the foregoing steps.

5.1.6 Standard Representations

Broadly speaking, for systems control there are three major steps, that is, modeling, analysis and design, also known as the *mad* process. If one is given a system to control, one probably has to go through this "mad" process or loop to achieve a satisfactory control performance. The structure of this book follows a similar *mad* process. For a systematic analysis and design of a control system, mathematical models of the components are usually required. For linear system models (both continuous-time and discrete-time), there are usually four kinds of mathematical models, namely, the transfer function model, the zero-pole-gain model, the block diagram model and more generally the state space model which will be the central theme of this book.

For a class of linear time-invariant (LTI) systems, the state space model is described by

$$\begin{aligned} \dot{x}(t) &= Ax(t) + Bu(t), \\ y(t) &= Cx(t) + Du(t) \end{aligned} \tag{5.1}$$

for continuous-time case with t being the continuous-time instant and

$$\begin{aligned} \dot{x}(k+1) &= Ax(k) + Bu(k), \\ y(k) &= Cx(k) + Du(k) \end{aligned} \tag{5.2}$$

for discrete-time case with k being the discrete-time instant. In (5.1) and (5.2), $x(.) \in \Re^n$, $u(.) \in \Re^m$ and $y(.) \in \Re^p$ are the state, the input and the output vectors, respectively. The corresponding transfer function matrix $T(.)$ from u to y, obtained by Laplace transform of (5.1)–(5.2) at zero-initial condition, is given by

$$T(r) = C(rI - A)^{-1}B + D \tag{5.3}$$

where $r = s$ in the continuous-time case and $r = z$ in the discrete-time case. One possible phrase to both (5.1)–(5.3) is the realization $\{A, B, C, D\}$. An alternative short notation is

$$\begin{bmatrix} A & \cdots & B \\ \vdots & \ddots & \vdots \\ C & \cdots & D \end{bmatrix} = C(sI - A)^{-1}B + D. \tag{5.4}$$

5.2 Basic Properties

Before embarking on the different methods for feedback control design, the goal of this section is to consider the basic structural properties of linear MIMO systems and explore how the feedback action affects them. For more elaborate mathematical treatment, the reader is advised to consult [16, 44, 47, 52, 54].

5.2.1 Stability

Consider the continuous system (5.1) with $u \equiv 0$, it is easy to show that

$$\begin{aligned} x(t) &= e^{A(t-t_o)} x(t_o) \\ &= \Phi(t, t_o) x(t_o) \end{aligned} \tag{5.5}$$

where t_o is the initial time and $\Phi(t, t_o)$ is often called the continuous state-transition matrix. For the discrete system (5.2) with $u \equiv 0$, it is easy to show that

$$\begin{aligned} x(k) &= A^{k-k_o} x(k_o) \\ &= \Phi(k, k_o) x(k_o) \end{aligned} \tag{5.6}$$

where k_o is the initial discrete-instant and $\Phi(k, k_o)$ is often called the discrete state-transition matrix.

By virtue of the Cayley–Hamilton theorem, see the Appendix, that $\Phi(t, t_o)$ or $\Phi(k, k_o)$ can be expressed as polynomial in A. Then by Frobenius' theorem, see the Appendix, the eigenvalue α_i of $\Phi(.,.)$ are related to the eigenvalues λ_i of matrix A by

$$\alpha_i = \begin{cases} e^{\lambda_i (t-t_o)} & \text{Continuous-time,} \\ \lambda_i^k & \text{Discrete-time} \end{cases} \tag{5.7}$$

for the continuous-time and discrete-time cases, respectively. Letting the eigenvalue $\lambda_i = \beta_i \pm j\omega_i$, it is a simple task to express the stability criteria for linear constant systems as follows:

Continuous-time: $\dot{x}(t) = Ax(t)$

- If $\beta_i > 0$ for any simple root or $\beta_i \geq 0$ for any repeated root,
- If $\beta_i \leq 0$ for all simple roots and $\beta_i < 0$ for all repeated root,
- If $\beta_i < 0$ for all roots.

Discrete-time: $x(k+1) = Ax(k)$

- If $|\lambda_i| > 1$ for any simple root or $|\lambda_i| \geq 1$ for any repeated root,
- If $|\lambda_i| \leq 1$ for all simple roots and $|\lambda_i| < 1$ for all repeated root,
- If $|\lambda_i| < 1$ for all roots.

On the other hand, it is readily seen that the origin $\mathbf{0}$ is an *equilibrium point* since $\dot{x} \equiv 0$, or $x(k+1) = x(k)$ yields $x_e = 0$.

In case of MIMO systems, Lyapunov stability theory provides a powerful tool for system analysis and design. The basic theory makes use of a *Lyapunov function* $V(x)$. This scalar function of the state x may be thought of as a generalized energy. A single-valued function $V(x)$ which is continuous and has continuous partial derivatives is said to be *positive definite* is some region Ω about the origin of the state space if

1. $V(\mathbf{0}) = 0$,
2. $V(x) > 0$ for all nonzero $x \in \Omega$.

If condition 2) is relaxed to $V(x) \geq 0$ for all nonzero $x \in \Omega$, then $V(x)$ is said to be *positive semidefinite*. The Lyapunov function $V(x)$ is not unique; rather, many different Lyapunov functions may be found for a given system. Likewise, the inability to find a satisfactory Lyapunov function does not mean that the system is unstable. The basic Lyapunov stability theory is phrased as follows:

Theorem 5.1 *If a positive-definite function $V(x)$ can be found such that*

1. $V(x) > 0, \forall x \neq \mathbf{0}$ *and* $V(\mathbf{0}) = 0$,
2. *Either* $\dot{V}(x) < 0\ \forall x \neq \mathbf{0}$ *for the continuous-time case and*
3. $\Delta V(x) < 0\ \forall x \neq \mathbf{0}$ *for the discrete-time case*

then the origin $\mathbf{0}$ *is asymptotically stable.*

5.2.2 Controllability

Controllability is a property of the coupling between the input and the state, and thus involves the matrices A and B.

Definition 5.2 A linear system is said to be controllable at time t_o if it is possible to find some input function (or sequence in the discrete case) $u(t)$ defined over $t \in \Im$, which will transfer the initial state $x(t_o)$ to the origin at some finite time $t_1 \in \Im$, $T_1 > t_o$. That is there is some input $u_{[t_o,t_1]}$, which gives $x(t_1) = 0$ at a finite $t_1 \in \Im$. If this is true for all initial time t_o and all initial states $x(t_o)$, the system is completely controllable.

As we see later, the full significance of controllability is realized in the course of feedback design. It will be seen there that if a linear system is controllable, it is possible to design a linear state feedback control law that will give arbitrarily specified closed-loop eigenvalues. Thus, an unstable system can be stabilized, a slow system can be speeded up, the natural frequencies can be changed, and so on, if the system is controllable. The existence of solutions to certain optimal control problems can be assured if the system is controllable.

A controllability criterion is stated below.

Definition 5.3 A linear time-invariant (LTI) system with realization A, B, C, D is completely controllable if and only if the $n \times nm$ matrix

$$P_c := \left[B \,\vdots\, AB \,\vdots\, A^2B \,\vdots\, \cdots \quad \cdots \,\vdots\, A^{n-1}B\right]$$

has rank n.

The form of the foregoing condition is exactly the same for both continuous and discrete-time systems.

5.2.3 Control Example 5.1

Consider the continuous system

$$A = \begin{bmatrix} -2 & -2 & 0 \\ 0 & 0 & 1 \\ 0 & -3 & -4 \end{bmatrix}, \qquad B = \begin{bmatrix} 1 & 0 \\ -0 & 1 \\ -1 & 1 \end{bmatrix}.$$

Simple computations yield

$$P_c = \begin{bmatrix} -1 & 0 & \vdots & -2 & -2 & \vdots & 2 & 2 \\ 0 & 1 & \vdots & 1 & 1 & \vdots & -4 & -7 \\ 1 & 1 & \vdots & -4 & -7 & \vdots & 13 & 25 \end{bmatrix}.$$

Since the determinant of the first three columns is nonzero ($= -3$), it means that the rank of P_c is 3. Therefore, this system is completely controllable.

5.2.4 Observability

Observability is a property of the coupling between the state and the output and thus involves the matrices A and C.

Definition 5.4 A linear system is said to be observable at time t_o if $x(t_o)$ can be determined from the output function $y_{[t_o,t_1]}$ (or output sequence) for $t_o \in \Im$, $t_o \leq t_1$, where t_1 is some finite time belonging to $\Im$. If this is true for all initial time t_o and all initial states $x(t_o)$, the system is completely controllable.

Clearly the observability of a system will be a major requirement in filtering and state estimation or reconstruction problems. In many feedback control problems, the controller must use output variables y rather than the state vector x in forming the feedback signals. If the system is observable, then y contains sufficient information about the internal states so that most of the power of state feedback can still be realized.

An observability criterion is stated below.

Definition 5.5 A linear time-invariant (LTI) system with realization A, B, C, D is completely observable if and only if the $n \times np$ matrix

$$P_o := \left[C^t \;\vdots\; A^t C^t \;\vdots\; A^{2^t} C^t \;\vdots\; \cdots \quad \cdots \;\vdots\; A^{n-1^t} C^t\right]$$

has rank n.

The form of the foregoing condition is exactly the same for both continuous and discrete-time systems.

5.2.5 Control Example 5.2

Consider the continuous system

$$A = \begin{bmatrix} 0 & 1 \\ 8 & -2 \end{bmatrix}, \qquad C = \begin{bmatrix} 4 & 1 \end{bmatrix}.$$

Simple computations yield

$$P_o = \begin{bmatrix} 4 & \vdots & 8 \\ 1 & \vdots & 2 \end{bmatrix}.$$

Since the second column is twice the first, rank of P_o is $1 < 2$, it implies that this system is not completely observable.

5.2.6 Control Example 5.3

The roll-angle dynamics of an aircraft is described by the state model

$$\dot{x} = \begin{bmatrix} \dot{\theta} \\ \dot{\omega} \\ \dot{\tau} \end{bmatrix} = \begin{bmatrix} 0 & 1 & 0 \\ 0 & -0.875 & -20 \\ 0 & 0 & -50 \end{bmatrix} \begin{bmatrix} \theta \\ \omega \\ \tau \end{bmatrix} + \begin{bmatrix} 0 \\ 0 \\ 50 \end{bmatrix} u,$$

$$y = \begin{bmatrix} 1 & 0 & 0 \\ 0 & 1 & 0 \end{bmatrix} \begin{bmatrix} \theta \\ \omega \\ \tau \end{bmatrix}.$$

Simple computations yield

$$P_c = \begin{bmatrix} 0 & 0 & -1000 \\ 0 & -1000 & 50875 \\ 50 & -2500 & 125000 \end{bmatrix}, \qquad P_o = \begin{bmatrix} 1 & 0 & 0 \\ 0 & 1 & 0 \\ 0 & 1 & 0 \\ 0 & -0.875 & -20 \\ 0 & -0.875 & -20 \\ 0 & 0.7656 & 1017.5 \end{bmatrix}.$$

Both matrices have rank 3 so the system is both completely controllable and observable. Note that the eigenvalues of the A matrix are $\{0, -0.875, -50\}$ so the system is not asymptotically stable.

5.2.7 Important Notes

One primary reason for feedback control systems design is to stabilize systems that may be unstable. Although our earlier results show that a reachable but unstable system can have its state controlled by appropriate choice of control input, these results were obtained under some critical assumptions:

1. the control must be unrestricted (as our reachability results assumed the control could be chosen freely);
2. the system must be accurately described (that is, we must have an accurate model of it);
3. the initial state must be accurately known.

The trouble with unstable systems is that they are unforgiving when assumptions such as the above do not hold. Even if the first assumption above is assumed to hold, there will undoubtedly be modeling errors, such as improperly modeled dynamics or incompletely modeled disturbances (thus, violating the second assumption). And even if we assume that the dynamics are accurately modeled, the initial state of the system is unlikely to be known precisely (violating the third assumption). It is thus clear that we need ongoing feedback of information about the state of the system, in order to have a hope of stabilizing an unstable system. Feedback can also improve the performance of a stable system. We shall come to understand these issues better over the remaining sections. How, then, can we design feedback controllers that stabilize a given system (or plant—the word used to describe the system that we are interested in controlling)? To answer this, we have to address the issues of what kind of feedback variables are available for our controller. There are, in general, two types of feedback:

- state feedback;
- output feedback.

With state feedback, all of the state variables (for example, x) of a system are available for use by the controller, whereas with output feedback, a set of output are available y. The state feedback problem is easier than the output feedback one, and richer in the sense that we can do more with control. In the following section, we examine eigenvalue placement by state feedback. All our discussion here will be for the case of a known LTI plant. The issue of uncertainty and unmodeled dynamics should be dealt with in previous subsequent chapters; namely. Our development in this section will use the notation of continuous-time (CT) systems—but there is no essential difference for the discrete-time (DT) case.

5.3 State Feedback

In the case of state feedback in Fig. 5.5, we measure all of the state variables. Thus the plant specification is $(A; B; I; 0)$ we omit the direct-feed through matrix, D, for simplicity, because including it would introduce only notational complications, without changing any conclusions. Our plant specification implies that the output equation is simply $y = x$. In many applications, direct measurement of all system state variables is either impossible or impractical. We address the important topic of output feedback a little later in this section.

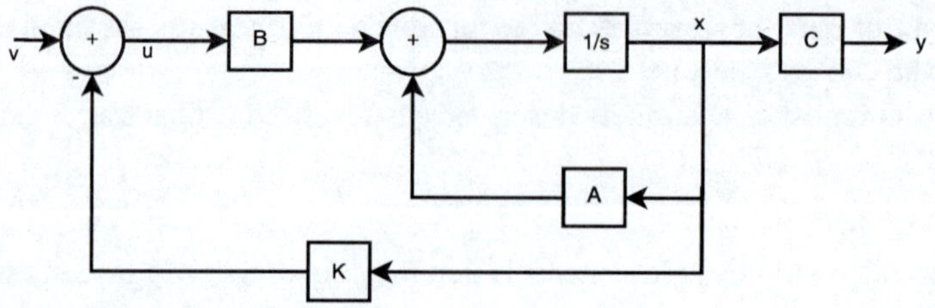

Fig. 5.5 Block diagram of state feedback

5.3.1 Introduction

For now, let us examine state feedback in further detail. Let our control, u, be specified by $u = Kx + v$, where K is a constant matrix, and v is an external input. This corresponds to *LTI state feedback*. Combining this control law with the state-space description for our nth-order plant, namely,

$$\begin{aligned} \delta x &= Ax + Bu, \\ y &= C \end{aligned} \tag{5.8}$$

we find that the closed-loop dynamics are described by

$$\delta x = (A + BK)x + Bv \tag{5.9}$$

where we adopt the notation

$$\delta x = \begin{cases} \dot{x} & \text{for CT systems,} \\ x(k+1) & \text{for DT systems.} \end{cases} \tag{5.10}$$

As is apparent from (5.9), the closed-loop system is stable if and only if the eigenvalues of $A + BK$ are all in the stable region. In other words, K stabilizes this system if and only if

$$\lambda(A + BK) \subset \begin{cases} \text{open left half of the complex plane in continuous-time,} \\ \text{open unit disc in discrete-time,} \end{cases} \tag{5.11}$$

where $\lambda(A+BK)$ is the *spectrum* (set of eigenvalues) of $(A+BK)$. It is interesting to ask: Can K be chosen so that the eigenvalues of $(A + BK)$ are placed at arbitrary desired locations? The answer is provided by the following theorem.

Theorem 5.6 (Eigenvalue placement) *There exists a matrix K such that*

$$\det\bigl(\lambda I - [A + BK]\bigr) = \prod_{i=1}^{n} (\lambda - \mu_i) \tag{5.12}$$

for any arbitrary self-conjugate set of complex numbers $\mu_1, \ldots, \mu_n \in \mathbb{C}$ if and only if $(A; B)$ is reachable.

The proof of this theorem can be found in [30].

5.3.2 Control Example 5.4

In what follows, we consider the case of constant input disturbances and integral feedback. Suppose that in our model we have a constant, but unknown, disturbance vector w,

$$\dot{x}(t) = Ax(t) + Bu(t) + w, \qquad x(0) = x_0,$$
$$y(t) = Cx(t).$$

If we use state feedback, $u(t) = -Kx(t)$, to stabilize the original system, then the presence of w will yield a nonzero steady-state value. This can be reduced by increasing K, but this has limits, because of saturation and noise effects.

A reasonable approach might be to attempt to at least estimate the unknown w in some fashion and use this estimate to cancel out the disturbance. Here we may note that the effects of constant disturbance vectors can often be eliminated by using the so-called integral-error feedback. Thus, introduce an additional state variable

$$\dot{q}(t) = y(t)$$

and use the feedback

$$u(t) = -Kx(t) - K_q q(t).$$

The augmented closed-loop system is

$$\begin{bmatrix} \dot{x}(t) \\ \dot{q}(t) \end{bmatrix} = \begin{bmatrix} A - BK & -BK_q \\ C & 0 \end{bmatrix} \begin{bmatrix} x(t) \\ q(t) \end{bmatrix} + \begin{bmatrix} w \\ 0 \end{bmatrix}$$

and if $\{K, K_q\}$ are chosen to make this system stable, then the steady-state value of $y(\cdot)$ will be zero, since the second equation gives $0 = Cx(\infty) = y(\infty)$.

It is worth noting that the steady-state error (or bias) has been brought to zero without any knowledge of the disturbance w. We should note that by using a command input v_d in addition to integral feedback we can obtain a desired nonzero set point [that is, a desired value of $y(\infty)$].

Some remarks are in order.

Remark 5.7 Suppose $\{A, b\}$ is controllable and we make a change of state variables such that $T^{-1}b = b = [b_1 \; 0 \; \cdots \; 0]'$, say and $T^{-1}AT = \bar{A}$, where

$$\bar{A} = \begin{bmatrix} A_{11} & A_{12} \\ A_{21} & A_{22} \end{bmatrix}$$

and A_{11} is a scalar. It can be verified that $\{A_{22}, A_{21}\}$ is controllable.

Remark 5.8 It can be shown that the relative order of a linear system, which is defined as the difference between the degrees of the denominator and numerator polynomials of its transfer function, is not affected by state-variable feedback.

Remark 5.9 Let $b(s)/a(s)$ be an irreducible transfer function, and write $a(s)\xi(s) = u(s)$, $y(s) = b(s)\xi(s)$.

- The knowledge of $\xi(\cdot)$ and its derivatives completely determines the state variables of any minimal realization of $b(s)/a(s)$. Therefore, $\xi(s)$ is often called the partial state of the system.
- A constant state-feedback corresponds to polynomial feedback of the partial state $\xi(\cdot) : v(s) = u(s) - g(s)\xi(s)$ for some polynomial $g(s)$ of degree less than or equal to $n - 1$. Such feedback the new transfer function is $b(s)/[a(s) + g(s)]$.

Let us now consider an illustrative example.

5.3.3 Control Example 5.5

As a typical example, consider the act of balancing, say, a pointer on your fingertip as in Fig. 5.6. Let us assume that the bottom end of the pointer is moving along the x axis, with your input $u(t)$ being the acceleration of this point: $u(t) = \xi(t)$. The length of the stick is L; assume that its mass m is concentrated at the top end. Suppose φ is small (so that $\sin\varphi \approx \varphi$ and $\cos\varphi \approx 1$). A force from your fingertip can be applied only in the direction of the stick. Therefore, by equating the forces acting, vertically (no acceleration along the vertical axis), we get $mg = F$, and the force component acting in the x direction is

$$mg\varphi(t) = F_x(t) = m\ddot{x}(t)$$

while the center of gravity has the x coordinate given by

$$x(t) = \xi(t) + L\varphi(t).$$

Considering the balancing as a dynamical system with input $u(.)$ (the acceleration of fingertip in the x direction) and output $y(t) = \varphi(t)$ (the angle of the stick to the vertical).

To proceed further, we introduce $z_1 = \varphi(t)$, $z_2 = \dot{\varphi}(t)$ as state variables. It is not difficult to show that $g\varphi(t) = \ddot{\xi}(t) + L\ddot{\varphi}(t)$, which in turn yields the state model

$$\dot{z}(t)A = \begin{bmatrix} 0 & 1 \\ g/L & 0 \end{bmatrix} + \begin{bmatrix} 0 \\ -1 \end{bmatrix} \frac{\ddot{\xi}}{L},$$

$$y(t) = \begin{bmatrix} 1 & 0 \end{bmatrix} z(t).$$

Let

$$u = \frac{\ddot{\xi}}{L}$$

be the normalized input. It is easy to see that the characteristic polynomial is

$$a(s) = s^2 - \frac{g}{L}$$

and the associated transfer function

$$H(s) = \frac{-1}{(s^2 - \frac{g}{L})}.$$

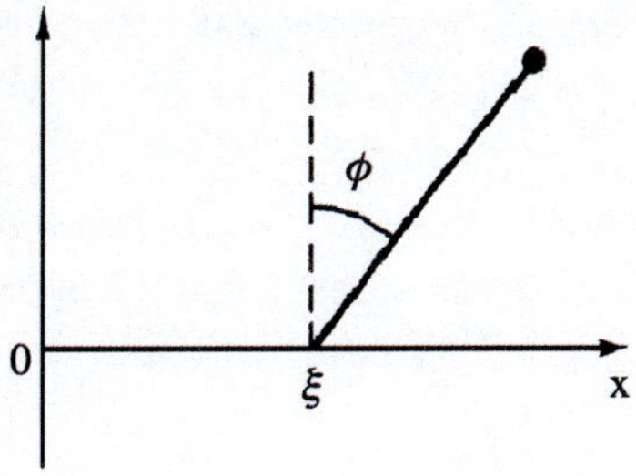

Fig. 5.6 Balancing a pointer

Checking the structural properties, it is found that

$$P_c = \begin{bmatrix} 0 & -1 \\ -1 & 0 \end{bmatrix}, \qquad P_o = \begin{bmatrix} 1 & 0 \\ 0 & 1 \end{bmatrix}$$

so the system is controllable and observable. The system eigenvalues are at $\pm\sqrt{\frac{g}{L}}$ so the system is unstable.

Introducing

$$u(t) = -Ky(t) = \begin{bmatrix} -K & 0 \end{bmatrix} z(t)$$

we get a new system that has eigenvalues at $\pm\sqrt{1 + \frac{g}{L}}$. However, there is a positive eigenvalue and the system remains understandable under constant output feedback.

Alternatively, setting $K = [K_1 \;\; K_2]$ and seek to position the closed-loop eigenvalues at $-1, -1$. A straightforward application of Ackermann yields

$$K_1 = \frac{g}{L} - 1, \qquad K_2 = -2.$$

The following example provides an interesting extension of the foregoing paradigm.

5.3.4 *Control Example 5.6*

For a system driven by a constant unknown disturbance w, design an observer to estimate w, and use this to compensate for the disturbance. We have

$$\begin{aligned} \dot{x} &= Ax + bu + bw, \\ \dot{w} &= 0, \\ y &= cx \end{aligned}$$

where u is the control input and y the observed output. The constant disturbance w is modeled as the output of an undriven integrator. We then have the augmented system shown in Fig. 5.7. If now we had an estimate $\hat{w}$ of w, we could set $u = -\hat{w}$ to attempt to cancel out the disturbance. This motivates us to set up an observer to estimate w.

Fig. 5.7 Augmented system

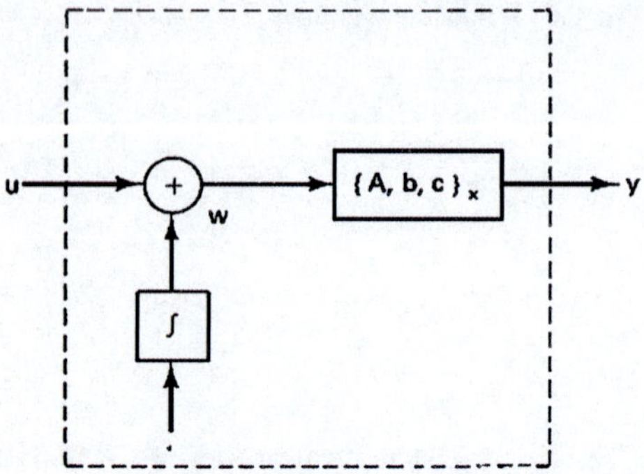

Fig. 5.8 Integral feedback

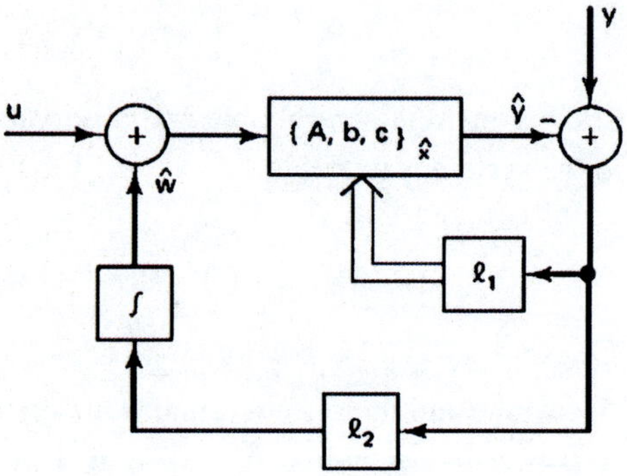

An observer for the augmented system is given by

$$\begin{bmatrix} \dot{\hat{x}} \\ \dot{\hat{w}} \end{bmatrix} = \begin{bmatrix} A & b \\ 0 & 0 \end{bmatrix} \begin{bmatrix} \hat{x} \\ \hat{w} \end{bmatrix} + \begin{bmatrix} b \\ 0 \end{bmatrix} u + l(y - c\hat{x}),$$

$$\hat{x}(0) = 0, \qquad \hat{w}(0) = 0$$

where l is an $(n+1) \times 1$ vector. Partitioning l as $[l_1' \;\; l_2']'$, with l_2 a scalar, we get

$$\begin{bmatrix} \dot{\hat{x}} \\ \dot{\hat{w}} \end{bmatrix} = \begin{bmatrix} A - l_1 c & b \\ -l_2 c & 0 \end{bmatrix} \begin{bmatrix} \hat{x} \\ \hat{w} \end{bmatrix} + \begin{bmatrix} b \\ 0 \end{bmatrix} u + \begin{bmatrix} l_1 \\ l_2 \end{bmatrix} y.$$

The observer structure is then as shown in Fig. 5.8. Now if the augmented system is observable, we can choose l so as to obtain arbitrary error decay modes and thus ensure that $\hat{w}$ approaches w asymptotically. Let us temporarily ignore the question of observability

We also make a particular choice of ℓ as $[0 \rightarrow \ell_2]$, which will simplify the observer design considerably; for the moment we shall not worry about whether this can still ensure that the error-decay modes are stable. Now, on setting $u = -\hat{w}$, our observer equation reduces to

$$\begin{bmatrix} \dot{\hat{x}} \\ \dot{\hat{w}} \end{bmatrix} = \begin{bmatrix} A & 0 \\ -l_2 c & 0 \end{bmatrix} \begin{bmatrix} \hat{x} \\ \hat{w} \end{bmatrix} + \begin{bmatrix} 0 \\ \ell_2 \end{bmatrix} y, \qquad \hat{x}(0) = 0, \qquad \hat{w}(0) = 0.$$

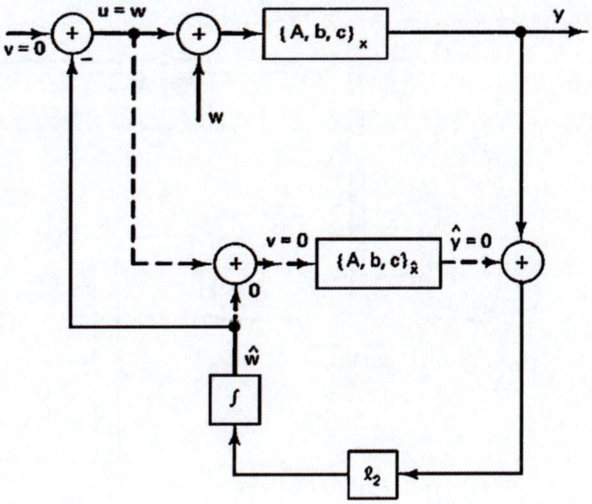

Fig. 5.9 Observer-based feedback controller

Since the equation for $\hat{x}$ is undriven and the initial condition is zero, we have $\hat{x} \equiv 0$, and our observer is simply

$$\dot{\hat{w}} = \ell_2 y, \qquad \hat{w}(0) = 0.$$

The resulting overall compensation scheme is shown in Fig. 5.9 (where the dashed lines indicate parts that drop out of the compensator).

The result of the above procedure is thus precisely the technique that was presented earlier for compensation of constant unknown disturbances, namely *integral feedback*, see Fig. 5.8. It arises here in a more natural and motivated manner. The question we have so far avoided is whether proper choice of ℓ_2 can ensure that $\hat{w}$ approaches w.

Our earlier observer equation shows that the observer error behavior is determined by the roots of

$$\alpha(s) = \det\begin{bmatrix} sI - A & -b \\ l_2 c & s \end{bmatrix} = \det(sI - A) = \det\left[s + l_2 c(sI - A)^{-} b\right]$$
$$= sa(s) + l_2 b(s) = 0.$$

We assume now that the original system $\{A, b, c\}$ was stable (or stabilized) and hence that $a(s)$ is stable, i.e., has roots with strictly negative real parts. It can then be shown that proper choice of ℓ_2 can give stable $\alpha(s)$ if and only if $b(s)$ has no root at origin.

5.3.5 Control Example 5.7

A passive feedback control system does not use any external power source for sensing, error detection, amplification, or actuation. Only the energy available in the

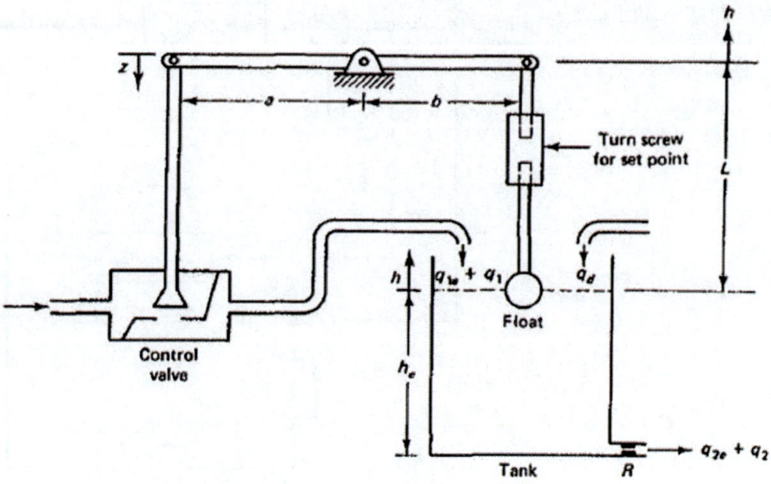

Fig. 5.10 Liquid-level regulator system

input to each and every component of the system is used to produce its output. A conceptual design of a passive mechanical-feedback system for regulating the liquid level in a tank is shown in Fig. 5.10. Here, q_{1e}, q_{2e} and h_e are the constant equilibrium values. Let q_1, q_2, and h denote the deviations from their equilibrium values and q_d be a disturbance flow. In steady-state equilibrium, $q_{1e} = q_{2e}$, $q_d = 0$, and h_e is a constant. The control system is a regulator whose purpose is to maintain the head of the liquid equal to its desired or reference value h_e when there is a disturbance flow q_d Obtain its linear mathematical model. A change h in the liquid level is sensed by a float that is connected by a mechanical lever to a control valve. A turn screw in the float-lever mechanism is used to change the length L when a change h_r in the set point corresponding to the desired level h_e is required. Here, we assume that $h_r = 0$, that is, there is no change in the desired value of h_e. If the head increases by h, the valve moves an amount z and reduces the flow to the tank and vice versa. For small displacements, the valve displacement z is related to the float displacement h by

$$z = \frac{a}{b}h \tag{5.13}$$

where a and b are the lever lengths shown in Fig. 5.10. For a small deviation, the linearized equation for the flow control valve is given by

$$q_1 = -c_1 z \tag{5.14}$$

where $c_1 > 0$. The negative sign in (5.14) indicates that when z increases, the flow q_1 decreases and vice versa. The continuity equation for the tank yields

$$q_1 + q_d - q_2 = A\frac{dh}{dt} \tag{5.15}$$

where A is the tank cross-sectional area and the outflow q_2 is obtained as $q_2 = (\rho g/R)h$. Here, R is the hydraulic resistance of the outlet orifice. Defining a time constant $\tau_1 = AR/\rho g$, then (5.15) becomes

$$q_1 + q_d = \frac{\rho g}{R}(\tau_1 s + 1)h. \tag{5.16}$$

The block diagram is obtained from (5.12), (5.14), and (5.16) and shown in Fig. 5.11(top). Letting $k_1 = ac_1/b$, the block diagram of Fig. 5.11(top) can be expressed as shown in Fig. 5.11(bottom), where h_r which is the reference or desired change in the liquid level, has been set to zero and $-h$ becomes the error. When modeling a regulator, we can represent all variables as deviations from the equilibrium state that is required to be maintained in the presence of disturbances. When the set point is not changed, the reference input is then set to zero.

For this system, which is of first order, we choose one state variable $x = h$. From Fig. 5.11, we obtain

$$(\tau_1 s + 1)x = \frac{R}{\rho g}(q_1 + q_d)$$

i.e.,

$$\dot{x} = -\left(\frac{1}{\tau_1}\right)x + \left(\frac{R}{\rho g \tau_1}\right)q_1 + \left(\frac{R}{\rho g \tau_1}\right)q_d \tag{5.17}$$

and

$$q_1 = -k_1 x. \tag{5.18}$$

We note that (5.17) can also be obtained directly from (5.16) and (5.18) from (5.13) and (5.14) with k_1 as defined in the preceding. On comparing (5.17) and (5.18) to the generic equations (5.1)–(5.3), we note that $x = h$, $u = q_1$, $v = q_d$, $r = h_r = 0$, and $y = h$. Also, A, B, $B_1 K$, and C are scalars and are given by

$$A = -1/\tau_1, \qquad B = R/\rho g \tau_1, \qquad B_1 = R/\rho g \tau_1, \qquad K = k_1, \qquad C = 1.$$

The state equation for the closed-loop system is obtained by substituting for the control law from (5.18) in (5.17) as

$$\dot{x} = -\frac{1}{\tau_1}\left(1 + \frac{Rk_1}{\rho g}\right)x + \left(\frac{R}{\rho g \tau_1}\right)q_d \tag{5.19}$$

where the scalar $A - BK$ of (5.9) becomes

$$A - BK = -\frac{1}{\tau_1}\left(1 + \frac{Rk_1}{\rho g}\right). \tag{5.20}$$

In case the set point is changed, that is, the desired change in the liquid level h_r is not zero, the control law becomes $q_1 = k_1(h_r - h)$ and (5.19) is modified to

$$\dot{x} = -\frac{1}{\tau_1}\left(1 + \frac{Rk}{\rho g}\right)x + \left(\frac{Rk_1}{\rho g \tau_1}\right)h_r + \left(\frac{R}{\rho g r_1}\right)q_d. \tag{5.21}$$

It will be seen later on that this control law, where q_1 is proportional to the error, does not possess a good disturbance-rejection property. The closed-loop transfer

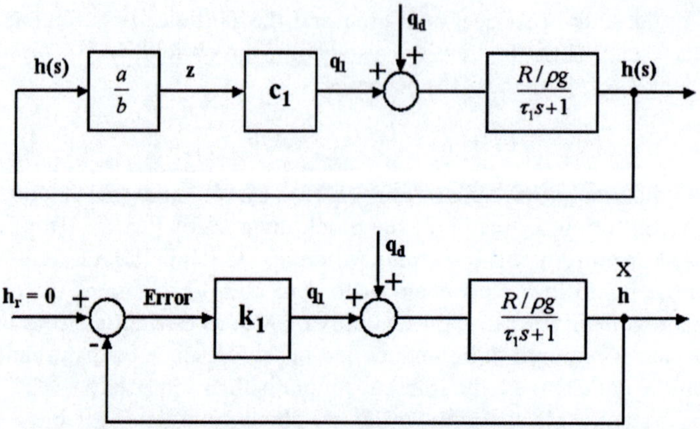

Fig. 5.11 (*Top*) Block diagram, and (*bottom*) standard block diagram

functions relating the output $h(s)$ to the command input $h_r(s)$ and the disturbance q_d can be obtained as:

$$H(s)=\left(\frac{k_1 R/\rho g}{\tau_1 s+1+k_1 R/\rho g}\right)h_r(s)+\left(\frac{R/\rho g}{\tau_1 s+1+k_1 R/\rho g}\right)q_d(s). \tag{5.22}$$

5.3.6 Control Example 5.8

In a class of mechanical, passive, feedback control systems, a *boom* is modeled as a uniform beam of length L and is held in a bearing at its lower end. A passive regulator is to be designed to maintain the boom in its vertical, unstable equilibrium position. The conceptual design uses a spring as a sensor and actuator as shown in Fig. 5.12(bottom). Obtain its linear mathematical model. A freebody diagram of the system is shown in Fig. 5.12(bottom). It is assumed that for small θ, the spring displacement is $L\theta$, the spring constant is k, that there is a viscous friction torque at the bearing with coefficient c, and T_d is the disturbance torque. The mass moment of inertia of the beam about the bearing is $(l/3)mL^2$. Taking moments about the bearing, we obtain

$$T_d+mg\frac{L}{2}sin(\theta)-c\dot{\theta}-kL^2\theta=\frac{1}{3}mL^2\ddot{\theta}. \tag{5.23}$$

For small θ, $\sin(\theta)\approx\theta$ and it follows from (5.23) that

$$\left[\frac{1}{3}mL^2s^2+cs-\frac{1}{2}mgL\right]\theta=-kL^2\theta+T_d. \tag{5.24}$$

Note that the left-hand side of (5.24) represents the system to be controlled and on the right-hand side, $u=-kL^2\theta$, is the control law produced by the spring. Figure 5.13 depicts a block diagram of (5.24) where θ_r is the command or desired change in the angular position. It is set to zero and thus $-\theta$ represents the error.

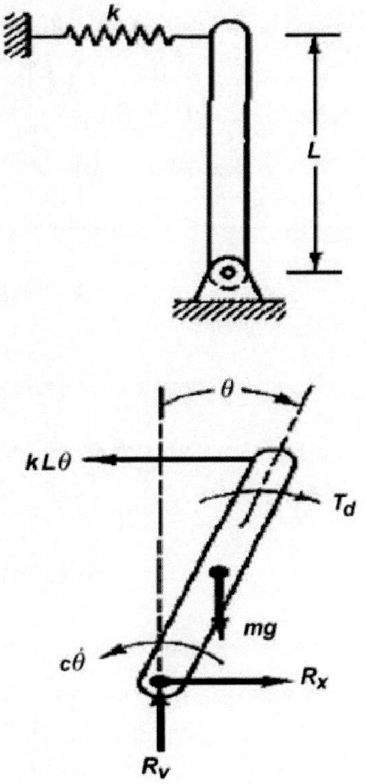

Fig. 5.12 (*Top*) A passive regulator, and (*bottom*) its freebody diagram

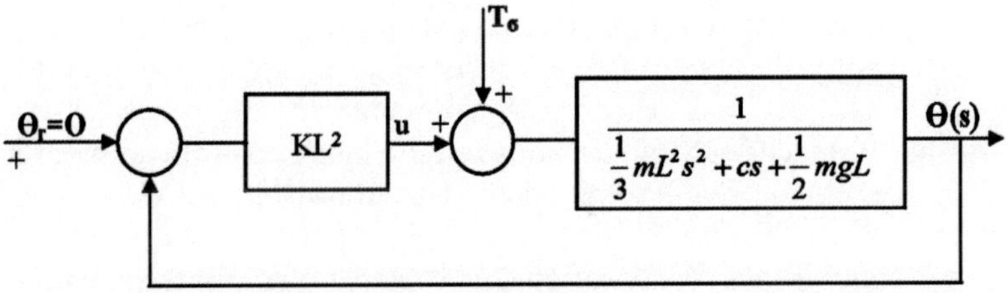

Fig. 5.13 Block diagram of the control system

Choosing $x_1 = \theta$ and $x_2 = \dot{\theta}$ as the state variables, we represent (5.24) in the form

$$\dot{x}_1 = x_2,$$
$$\dot{x}_2 = \frac{3g}{2L}x_1 - \frac{3c}{mL^2}x_2 + \frac{3}{mL^2}u + \frac{3}{mL^2}T_d, \tag{5.25}$$

$$u = -kL^2\theta,$$
$$y = x_1 \tag{5.26}$$

which has the standard state model with

$$A = \begin{bmatrix} 0 & 1 \\ \frac{3g}{2L} & -\frac{3c}{mL^2} \end{bmatrix}, \qquad B = \begin{bmatrix} 0 \\ \frac{3}{mL^2} \end{bmatrix}, \qquad E = \begin{bmatrix} 0 \\ \frac{3}{mL^2} \end{bmatrix}, \tag{5.27}$$
$$K = \begin{bmatrix} KL^2 & 0 \end{bmatrix}, \qquad C = \begin{bmatrix} 1 & 0 \end{bmatrix}.$$

Hence, the closed-loop matrix is given by

$$A - BK = \begin{bmatrix} 0 & 1 \\ \frac{3g}{2L} - \frac{3K}{m} & -\frac{3c}{mL^2} \end{bmatrix},$$
$$\det[sI - A + BK] = s^2 + \left(\frac{3c}{mL^2}\right)s - \frac{3g}{2L} + \frac{3K}{m} \tag{5.28}$$

and the associated transfer function relating the output θ to the disturbance torque T_d is given by

$$G_d(s) = \frac{\frac{3}{mL^2}}{s^2 + (\frac{3c}{mL^2})s - \frac{3g}{2L} + \frac{3K}{m}}. \tag{5.29}$$

Some relevant comments are:

1. In the case that the pair $(A; B)$ is not reachable, then the reachable modes, and only these, can be changed by state feedback.
2. The pair $(A; B)$ is said to be stabilizable if its unreachable modes are all stable, because in this case, and only in this case, K can be selected to change the location of all unstable modes to stable locations.
3. Despite what the theorem says we can do, there are good practical reasons why one might temper the application of the theorem. Attempting to make the closed-loop dynamics very fast generally requires large K, and hence large control effort—but in practice there are limits to how much control can be exercised. Furthermore, unmodeled dynamics could lead to instability if we got too ambitious with our feedback.

   As we shall see later, the linear-quadratic regulator or LQR formulation of the controller problem for linear systems uses an integral-square (that is quadratic) cost criterion to pose a compromise between the desire to bring the state to zero and the desire to limit control effort. In the LTI case, and with the integral extending over an infinite time interval, the optimal control turns out to be precisely an LTI state feedback. The solution of the LQR problem for this case enables computation of the optimal feedback gain matrix K^* (most commonly through the solution of an algebraic Riccati equation).
4. State feedback cannot change reachability, but it can affect observability—either destroying it or creating it.
5. State feedback can change the poles of an LTI system, but does not affect the zeros (unless the feedback happens to induce unobservability, in which case what has occurred is that a pole has been shifted to exactly cancel a zero). Note that, if the open-loop and closed-loop descriptions are minimal, then their transmission

zeros are precisely the values of s where their respective system matrices drop rank. These system matrices are related by a nonsingular transformation

$$\begin{bmatrix} sI-(A+BK) & -B \\ C & 0 \end{bmatrix} = \begin{bmatrix} sI-A & -B \\ C & 0 \end{bmatrix} \begin{bmatrix} I & 0 \\ K & I \end{bmatrix}. \tag{5.30}$$

Hence, the closed-loop and open-loop zeros are identical.

The main purpose of state feedback is to relocate the open loop eigenvalues to pre-determined locations in the s-plane by using some pole placement methods. In control design, placing poles is desirable objective subject to the controllability of the pair (A, B) because the location of the poles (equivalently the eigenvalues of the system) has some effective relations to the characteristics of the response of the system.

Then the poles of the open loop system are the roots of the characteristic equation given by

$$|sI - A| = 0.$$

Full state feedback is utilized by expressing the input vector u in the linear form

$$u = Kx.$$

Substituting into the state space model, we get the closed-loop system

$$\begin{aligned} \dot{x} &= [A - B * K]x, \\ y &= [A - D * K]x. \end{aligned} \tag{5.31}$$

The closed-loop eigenvalues system are the roots of the characteristic equation,

$$\det\big[s\mathbf{I} - (\mathbf{A} - \mathbf{BK})\big] = 0. \tag{5.32}$$

Comparing the terms of (5.32) with those of the desired characteristic equation yields the elements of the feedback matrix K which force the closed-loop eigenvalues to the pole locations specified by the desired characteristic equation.

5.3.7 Control Example 5.9

A cart of mass M slides on a frictionless surface. The cart is pulled by a force $u(t)$. On the cart a pendulum of mass m is attached via a frictionless hinge, as shown in Fig. 5.14. The pendulum's center of mass is located at a distance l from either end. The moment of inertia of the pendulum about its center of mass is denoted by I. The position of the center of mass of the cart is at a distance $s(t)$ from a reference point. The angle $\theta(t)$ is the angle that the pendulum makes with respect to the vertical axis which is assumed to increase clockwise.

First, we write the equations of motion that result from the free-body diagram of the cart. The vertical forces P, R and Mg balance out. For the horizontal forces, we have the following equation

$$M\ddot{s} = u - N. \tag{5.33}$$

Fig. 5.14 Inverted pendulum

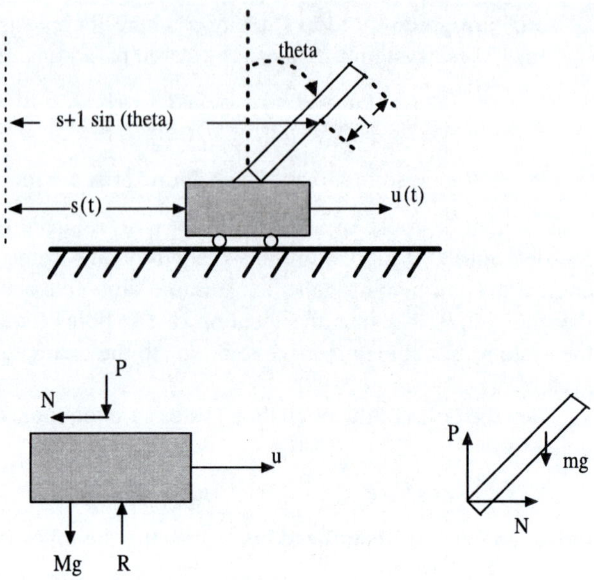

From the free-body diagram of the pendulum, the balance of forces in the horizontal direction gives the equation

$$\begin{aligned} m\frac{d^2}{dt^2}\big(s+\ell\sin(\theta)\big) &= N, \\ m\frac{d}{dt}\big(\dot{s}+\ell\cos(\theta)\dot{\theta}\big) &= N, \\ m\big(\ddot{s}-l\sin(\theta)(\dot{\theta})^2+\ell\cos(\theta)\ddot{\theta}\big) &= N \end{aligned} \tag{5.34}$$

and the balance of forces in the vertical direction gives the equation

$$\begin{aligned} m\frac{d^2}{dt^2}\big(\ell\cos(\theta)\big) &= P-mg, \\ m\frac{d}{dt}\big(-\ell\sin(\theta)\dot{\theta}\big) &= P-mg, \\ m\big(-l\cos(\theta)(\dot{\theta})^2-\ell\sin(\theta)\ddot{\theta}\big) &= P-mg. \end{aligned} \tag{5.35}$$

Finally by balancing the moments around the center of mass, we get

$$I\ddot{\theta} = P\ell\sin(\theta) - N\ell\cos(\theta). \tag{5.36}$$

From (5.33) and (5.34), we can eliminate the force N to obtain

$$(M+m)\ddot{s}+m\big(\ell\cos(\theta)\ddot{\theta}-\ell\sin(\theta)(\dot{\theta})^2\big)=u. \tag{5.37}$$

Substituting (5.34) and (5.35) into (5.36) gives us

$$\begin{aligned} I\ddot{\theta} = {} & \ell\big(mg-m\ell\cos(\theta)(\dot{\theta})^2-m\ell\sin(\theta)\ddot{\theta}\big)\sin(\theta) \\ & -\ell\big(m\ddot{s}-m\ell\sin(\theta)(\dot{\theta})^2+m\ell\cos(\theta)\ddot{\theta}\big)\cos(\theta). \end{aligned}$$

Simplifying the above expression yields

$$(I + m\ell^2)\ddot{\theta} = mg\ell\sin(\theta) - m\ell\ddot{s}\cos(\theta). \tag{5.38}$$

The equations that describe the system are (5.37) and (5.38). We can have a further simplification of the system of equations by removing the term $\ddot{\theta}$ from (5.37), and the term $\ddot{s}$ from (5.38). Define the constants

$$M_t = M + m, \qquad L = \frac{I + m\ell^2}{m\ell}.$$

Substituting $\ddot{\theta}$ from (5.38) into (5.37), we get

$$\left(1 - \frac{m\ell}{M_t L}\cos(\theta)^2\right)\ddot{s} + \frac{m\ell}{M_t L} g\sin(\theta)\cos(\theta) - \frac{m\ell}{M_t L}\sin(\theta)(\dot{\theta})^2 = \frac{1}{M_t}u. \tag{5.39}$$

Similarly we can substitute $\ddot{s}$ from (5.37) into (5.38) to get

$$\left(1 - \frac{m\ell}{M_t L}\cos(\theta)^2\right)\ddot{\theta} - \frac{g}{L}\sin(\theta) + \frac{m\ell}{M_t L}\sin(\theta)\cos(\theta)(\dot{\theta})^2 = -\frac{1}{M_t L}\cos(\theta)u. \tag{5.40}$$

These are nonlinear equations due to the presence of the terms $\sin(\theta)$, $\cos(\theta)$, and $(\dot{\theta})^2$. We can linearize these equations around $\theta = 0$ and $\dot{\theta} = 0$, by assuming that $\theta(t)$ and $\dot{\theta}(t)$ remain small. Recall that for small θ

$$\sin(\theta) \approx \theta - \frac{1}{6}\theta^3, \qquad \cos(\theta) \approx 1 - \frac{1}{2}\theta^2,$$

and using these relations we can linearize (5.39) and (5.40). The linearized system of equations take the form

$$\left(1 - \frac{m\ell}{M_t L}\right)\ddot{s} + \frac{m\ell}{M_t L}\frac{g}{L}\theta = \frac{1}{M_t}u,$$
$$\left(1 - \frac{m\ell}{M_t L}\right)\ddot{\theta} - \frac{g}{L}\theta = -\frac{1}{M_t L}u.$$

Choose the following state variables $x = [s \;\; \dot{s} \;\; \theta \;\; \dot{\theta}]^t$, to write a state space model for the invert ed pendulum. Using these state variables, the following state space model can be easily obtained

$$\frac{d}{dt}\begin{bmatrix} x_1 \\ x_2 \\ x_3 \\ x_4 \end{bmatrix} = \begin{bmatrix} 0 & 1 & 0 & 0 \\ 0 & 0 & -\alpha\frac{m\ell}{M_t L}g & 0 \\ 0 & 0 & 0 & 1 \\ 0 & 0 & \alpha\frac{g}{L} & 0 \end{bmatrix}\begin{bmatrix} x_1 \\ x_2 \\ x_3 \\ x_4 \end{bmatrix} + \begin{bmatrix} 0 \\ \frac{\alpha}{M_t} \\ 0 \\ -\frac{\alpha}{LM_t} \end{bmatrix}u,$$
$$y = \begin{bmatrix} 1 & 0 & 0 \end{bmatrix}x,$$

where the constant α is given by

$$\alpha = \frac{1}{(1 - \frac{m\ell}{M_t L})}.$$

Intuitively it is clear that the equilibrium point $[s = \text{const}, \dot{s} = 0, \theta = 0, \dot{\theta} = 0]$ is an unstable equilibrium point. To verify this, we compute the eigenvalues of the matrix A by solving the equation $\det(\lambda I - A)$. The eigenvalues are

$$\left\{ 0 \quad 0 \quad \sqrt{\frac{\alpha q}{L}} \quad -\sqrt{\frac{\alpha q}{L}} \right\}.$$

Therefore we have two eigenvalues at the $j\omega$ axis and one eigenvalue in the open right half of the complex plane, which indicates instability.

Consider the case where $M = 2$ kg, $m = 0.1$ kg, $l = 0.5$ m, $I = 0.025$ kg m^2, and of course $g = 9.8$ m/s^2. Assume that we can directly measure the state variables, s, $\dot{s}$, θ and $\dot{\theta}$. We want to design a feedback control law $u = F\hat{x} + r$ to stabilize this system. In order to do that, we will choose a feedback matrix F to place the poles of the closed-loop system at $\{-1, -1, -3, -3\}$. Using Ackermann's formula

$$F = -\begin{bmatrix} 0 & 0 & 0 & 1 \end{bmatrix} R_n^{-1} \alpha^d(A)$$

where $\alpha^d(\lambda) = (\lambda + 1)(\lambda + 1)(\lambda + 3)(\lambda + 3)$ which is the polynomial whose roots are the desired new pole locations, and R_n is the reachability matrix. In specific using the parameters of the problem, we have

$$\begin{aligned} F = &-[0 \quad 0 \quad 0 \quad 1] \begin{bmatrix} 0 & 0.4878 & 0 & 0.1166 \\ 0.4878 & 0 & 0.1166 & 0 \\ 0 & -0.4878 & 0 & -4.8971 \\ 0.4878 & 0 & -4.8971 & 0 \end{bmatrix}^{-1} \\ &\times \begin{bmatrix} 9.0 & 24.0 & -7.7 & -1.9 \\ 0 & 9.0 & -24.9 & -7.7 \\ 0 & 0 & 330.6 & 104.3 \\ 0 & 0 & 1047.2 & 330.6 \end{bmatrix} \\ = &\begin{bmatrix} 1.8827 & 5.0204 & 67.5627 & 21.4204 \end{bmatrix}. \end{aligned}$$

The closed-loop system is given by

$$\begin{aligned} \frac{d}{dt} \begin{bmatrix} x_1 \\ x_2 \\ x_3 \\ x_4 \end{bmatrix} = &\begin{bmatrix} 0 & 1.0 & 0 & 0 \\ 0.9184 & 2.449 & 32.7184 & 10.449 \\ 0 & 0 & 0 & 1.0 \\ -0.9184 & -2.4490 & -22.9184 & -10.4490 \end{bmatrix} \begin{bmatrix} x_1 \\ x_2 \\ x_3 \\ x_4 \end{bmatrix} \\ &+ \begin{bmatrix} 0 \\ 0.4878 \\ 0 \\ -0.4878 \end{bmatrix} r. \end{aligned}$$

In Fig. 5.15, we show the time trajectories of the closed-loop linearized system when the reference input $r(t)$ is identically zero and the initial angular displacement of the pendulum is 1.0 radians. In this simulation, the initial conditions on all the other state variables are zero.

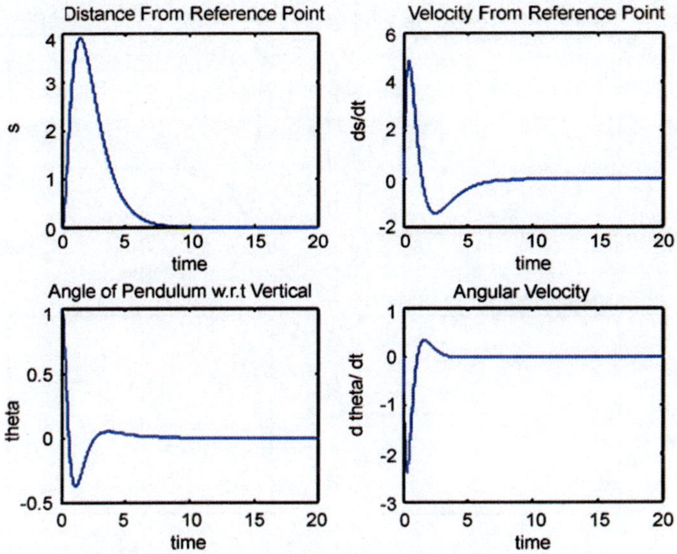

Fig. 5.15 Plot of state variables of the closed-loop linearized system

We can also look at the performance of this controller if it is applied to the nonlinear system. In this case, we should simulate the dynamics of the following nonlinear system of equations

$$\frac{d}{dt}\begin{bmatrix} x_1 \\ x_2 \\ x_3 \\ x_4 \end{bmatrix} = \begin{bmatrix} x_2 \\ -\frac{mlg}{M_t L}\frac{1}{\alpha(x_3)}\sin(x_3)\cos(x_3) + \frac{ml}{M_t}\frac{1}{\alpha(x_3)}\sin(x_3)(x_4)^2 \\ x_4 \\ \frac{g}{L}\frac{1}{\alpha(x_3)}\sin(x_3) - \frac{ml}{M_t L}\frac{1}{\alpha(x_3)}\sin(x_3)\cos(x_3)(x_4)^2 \end{bmatrix}$$

$$+\begin{bmatrix} 0 \\ \frac{1}{M_t}\frac{1}{\alpha(x_3)} \\ 0 \\ \frac{1}{M_t L}\frac{\cos(x_3)}{\alpha(x_3)} \end{bmatrix} u,$$

$$u = \begin{bmatrix} 1.8827 & 5.0204 & 67.5627 & 21.4204 \end{bmatrix}\begin{bmatrix} x_1 \\ x_2 \\ x_3 \\ x_4 \end{bmatrix} + r,$$

where $\alpha(x_3)$ is defined as

$$\alpha(x_3) = \left(1 - \frac{ml}{M_t L}\cos(x_3)^2\right).$$

In Fig. 5.16, we show the time trajectories of the nonlinear closed-loop system when the reference input $r(t)$ is identically zero and the initial angular displacement of

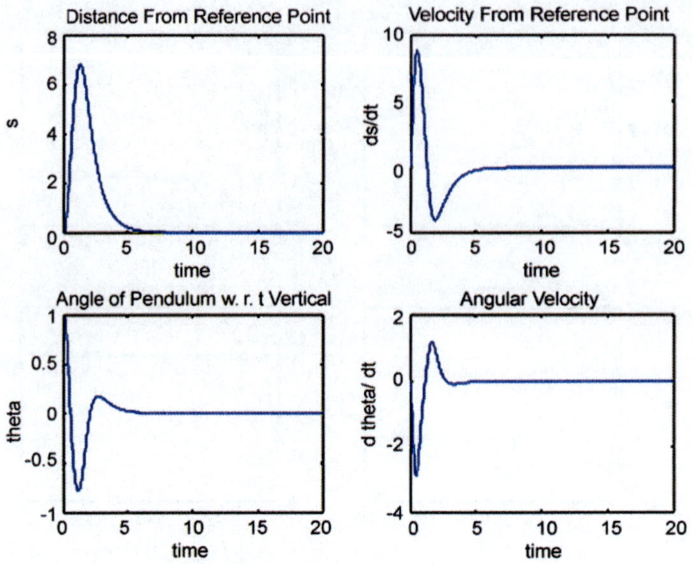

Fig. 5.16 Plot of state variables of the nonlinear closed-loop system

the pendulum is 1.0 radians. In this simulation, the initial conditions on all the other state variables are zero.

5.3.8 State-Feedback in MATLAB

Invoking the MATLAB software, we could apply the command

$$K = \text{place}(A, B, V)$$

to determine the feedback matrix gain K where V is the set of desired eigenvalues.

5.4 Observer-Based Feedback

When some or all of the state-variables are not accessible for measurements, we use an alternative method based on estimating the states. Thus, in observer-based feedback it is required to construct a device or system that generates a good replica of the state, see Fig. 5.17.

5.4.1 Basics

Considering the plant itself, the close-loop dynamics is expressed as

$$\dot{x} = Ax + Bu = Ax + B(K\hat{x} + \upsilon) = Ax + BK\hat{x} + B\upsilon.$$

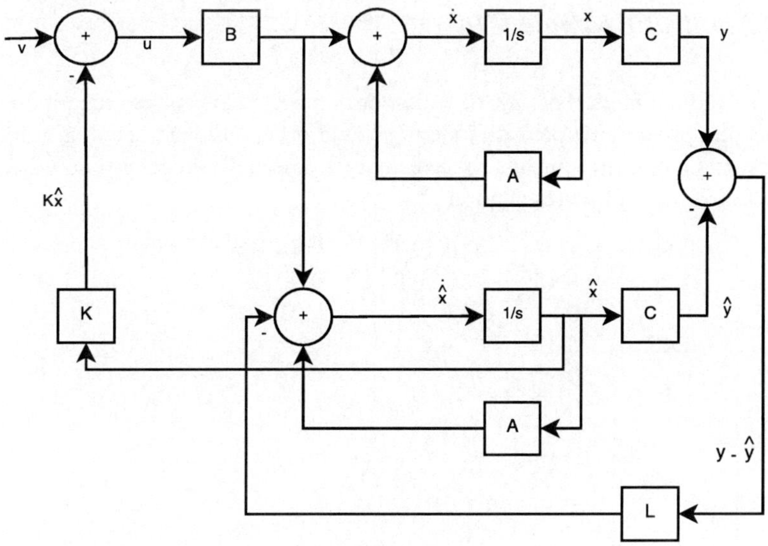

Fig. 5.17 Block diagram of observer-based feedback

In addition, the observer equation is

$$\begin{aligned}\dot{\hat{x}} &= (A - LC)\hat{x} + (B - LD)u + Ly \\ &= (A - LC)\hat{x} + (B - LD)(K\hat{x} + \upsilon) + L(Cx + Du) \\ &= \big[A - LC + (B - LD)K\big]\hat{x} + (B - LD)\upsilon + LCx + LD(K\hat{x} + \upsilon) \\ &= \big[A - LC + (B - LD)K + LDK\big]\hat{x} + LCx + [B - LD + LD]\upsilon \\ &= [A - LC + BK]\hat{x} + LCx + B\upsilon.\end{aligned}$$

The two systems together are written in augmented state space form as:

$$\begin{bmatrix} \dot{x} \\ \dot{\hat{x}} \end{bmatrix} = \begin{bmatrix} A & BK \\ LC & A - LC + BK \end{bmatrix} \begin{bmatrix} x \\ \hat{x} \end{bmatrix} + \begin{bmatrix} B \\ B \end{bmatrix} \upsilon. \tag{5.41}$$

Now to design a controller, we consider a state space system of the form

$$\begin{aligned}\dot{x} &= Ax + Bu, \\ y &= Cx + Du.\end{aligned}$$

Let us choose V_1 as it was in the state feedback case, and V_2 as the desired eigenvalues for the observer. Subject to the structural properties (controllability of the pair (A, B) and observability of the pair (A, C)), the separation theorem holds, and therefore we could apply the foregoing MATLAB command in two separate stages:

$$K = \text{place}(A, B, V_1), \qquad L = \text{place}\big(A^t, C^t, V_2\big).$$

5.4.2 Control Example 5.10

The longitudinal motion of a flexible bombor aircraft [1] is conventionally modeled as a second-order short-period mode, a second-order fuselage bending mode, and two first-order control control-surface actuators. The sixth-order system is described by the following LTI representation:

$$A = \begin{bmatrix} 0.4158 & 1.025 & -0.00267 & -0.0001106 & -0.08021 & 0 \\ -5.5 & -0.8302 & -0.06549 & -0.0039 & -5.115 & 0.809 \\ 0 & 0 & 0 & 1.0 & 0 & 0 \\ -1040 & -78.35 & -34.83 & -0.6214 & -865.6 & -631 \\ 0 & 0 & 0 & 0 & -75 & 0 \\ 0 & 0 & 0 & 0 & 0 & -100 \end{bmatrix},$$

$$B = \begin{bmatrix} 0 & 0 \\ 0 & 0 \\ 0 & 0 \\ 0 & 0 \\ 75 & 0 \\ 0 & 100 \end{bmatrix}, \qquad C^t = \begin{bmatrix} -1491 & 0 \\ -146.43 & 1 \\ 140.2 & 0 \\ -0.9412 & 0 \\ -1285 & 0 \\ -564.66 & 0 \end{bmatrix}.$$

The inputs are the desired elevator deflection (rad), $u_1(t)$, and the desired canard deflection (rad), $u_2(t)$, while the outputs are the sensor location's normal acceleration m/s^2, $y_1(t)$, and the pitch-rate (rad/s), $y_2(t)$.

Testing the system controllability and observability using MATLAB indicates that the system is fully controllable and fully observable. This means that it is easy to place any of it's eigenvalues into new desired position. Using MATLAB code *place*, the desired poles of the system will placed to three different desired position. The parameter γ will be used to multiply the location of the poles. The desired poles defined as:

$$P = \gamma * \begin{bmatrix} -1+1i & -1-1i & -2+2i & -2-2i & -3 & -4 \end{bmatrix}. \tag{5.42}$$

The parameter γ will be selected as 10, 20, and 30. More far the location of the poles to the left side will yields different response of the system. From Fig. 5.18, it can be seen that desired poles with large magnitude yields more stable and nonoscillatory closed-loop system.

An observer-based state feedback controller will now be considered. The state-feedback gain K obtained from the foregoing pole-placement method. Once again, three sets of pole positions will be selected. The parameters that is used for this selection is the multiplier α.

$$P_{\text{obs}} = \alpha * \begin{bmatrix} -1+1i & -1-1i & -2+2i & -2-2i & -3 & -4 \end{bmatrix}. \tag{5.43}$$

The step response of the closed-loop system is shown in Figs. 5.19–5.22, which is the response is very different between the selected poles position. The more far the poles located, the system faster to become stable, but the overshoot is become very large.

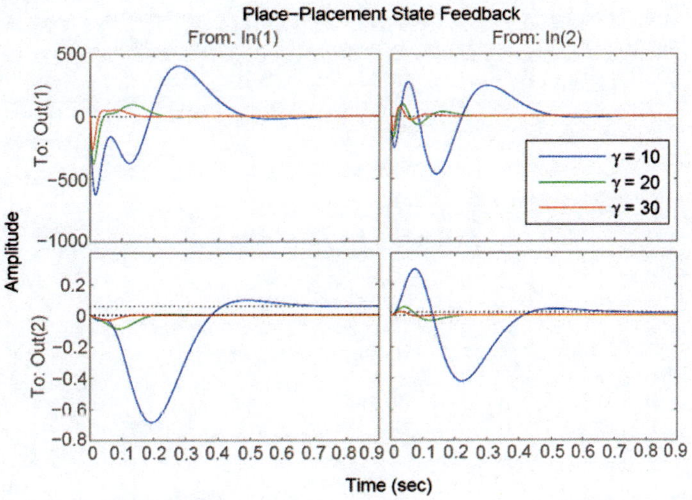

Fig. 5.18 Pole-placement state feedback with different γ

Fig. 5.19 Observer-based controller (Input 1–Output 1) with different α

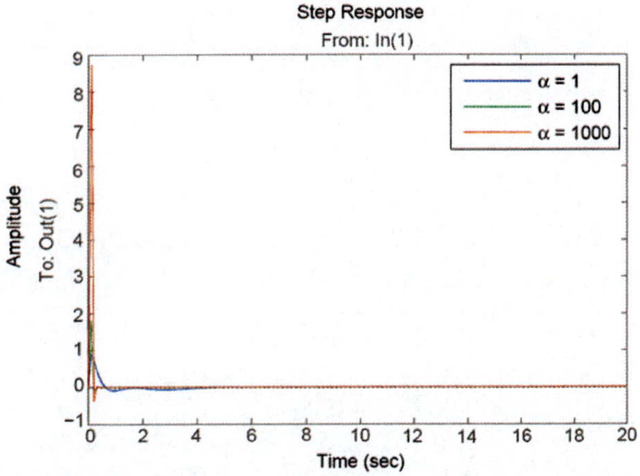

5.4.3 Control Example 5.11

A linearized mathematical model for a direct expansion (DX) A/C system was described in [48], where it is shown that the physical system consists of six states, two inputs and two outputs. The model was developed to be able to capture the transient characteristics of the DX A/C system. A simplified schematic of the model is shown in Fig. 5.23. The system matrices A, B, C are as follows:

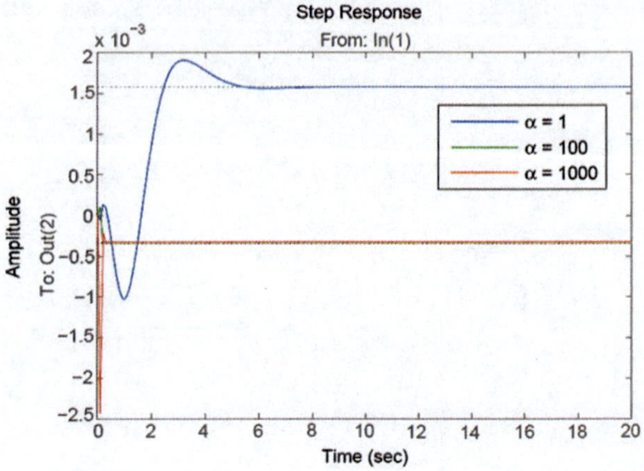

Fig. 5.20 Observer-based controller (Input 1–Output 2) with different α

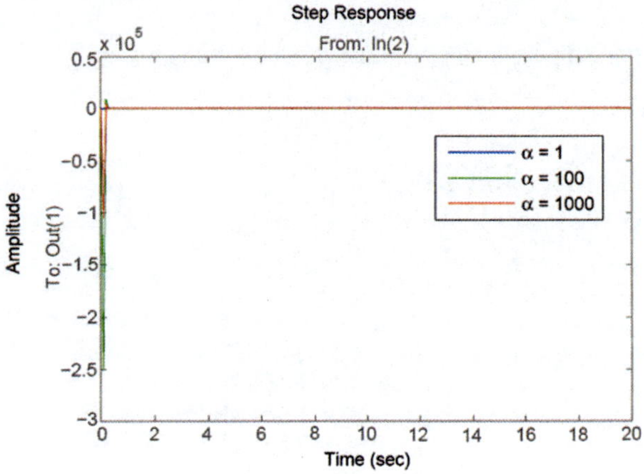

Fig. 5.21 Observer-based controller (Input 2–Output 1) with different α

$$A = \begin{bmatrix} -2.731 & 0 & 0.0756 & 1.1883 & 0.5287 & 5.287 \\ 0.0045 & -0.0045 & 0 & 0 & 0 & 0 \\ 0 & 2.6577 & -1.692 & 2.0346 & 0 & 0 \\ 0.0139 & 0.067 & 0.0206 & -0.0412 & 0 & 0 \\ 0.016 & 0 & 0 & 0 & 0 & 0 \\ 0 & 0 & 0 & 0 & 0.0145 & -0.045 \end{bmatrix},$$

$$B = \begin{bmatrix} 0.035 & 0 \\ -0.098 & 0 \\ 2.5 & 0 \\ 0 & -1.931 \\ 0 & 0 \\ -0.3 & 0 \end{bmatrix}, \qquad C^t = \begin{bmatrix} 0 & 0 \\ 1 & 0 \\ 0 & 0 \\ 0 & 0 \\ 0 & 0 \\ 0 & 1 \end{bmatrix}.$$

We initially observe that the A/C system is controllable and observable. For the purpose of designing observer-based controller, we select one eigenvalue set at

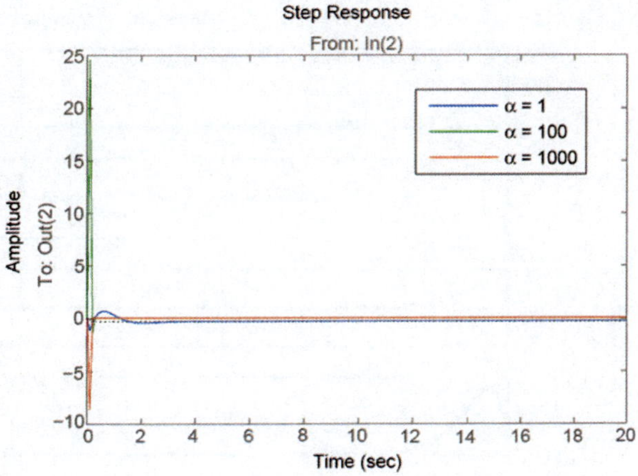

Fig. 5.22 Observer-based controller (Input 2–Output 2) with different α

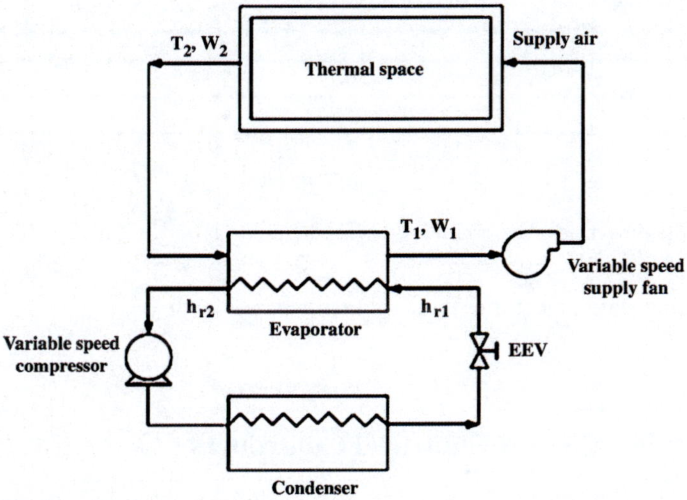

Fig. 5.23 Air-conditioning block-diagram

$V1 = [-0.05 \quad -0.26 \quad -0.28 \quad -0.3 \quad -1.6 \quad -2.8]$ and using the MATLAB command $K = \text{place}(A, B, V1)$ to yield the state-gain matrix

$$K = \begin{bmatrix} -0.0236 & 2.1293 & -0.3942 & 0.5250 & -2.2973 & -0.8044 \\ -0.0280 & 15.3401 & 0.2643 & -0.8959 & -17.7654 & -6.3900 \end{bmatrix}.$$

Then selecting another eigenvalue set at $V2 = [-0.15 \; -0.8 \; -1.3 \; -2.9 \; -3.6 \; -4.8]$ and using the MATLAB command $L = \text{place}(A^t, C^t, V2)$ to yield the observer-gain matrix

$$L = \begin{bmatrix} 49.6 & 4.6 & -379.52 & 497.6 & 86.2 & 8 \\ 307.5 & 7 & 1084.7 & 478.1 & 193.8 & 4.5 \end{bmatrix}.$$

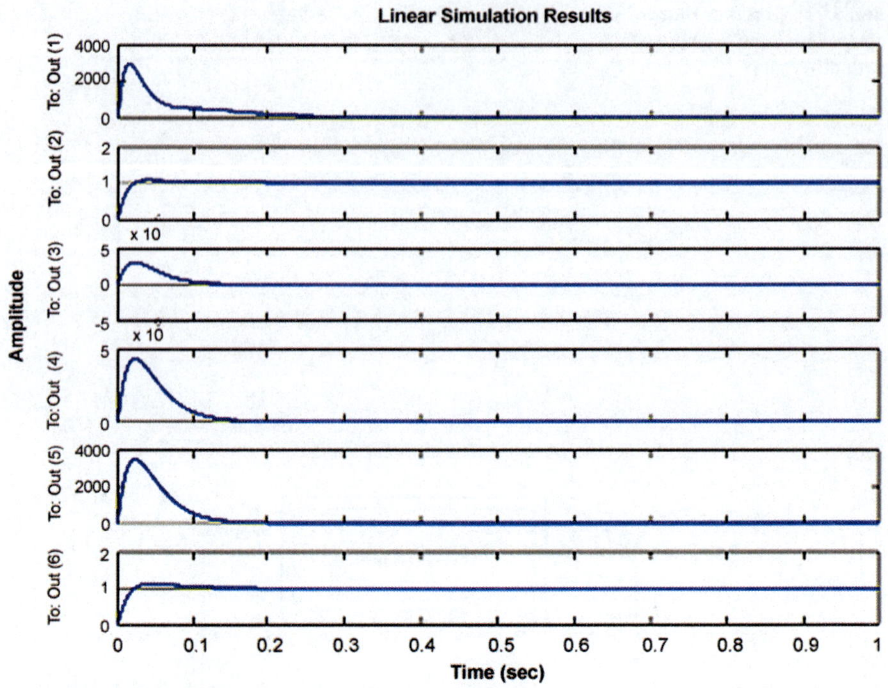

Fig. 5.24 State trajectories under observer-based control

The resulting state trajectories under observer-based feedback control are plotted in Fig. 5.24.

5.5 Classifications of Industrial Controllers

Industrial controllers may be classified according to their control actions as:

- Two-position or on off controllers.
- Proportional controllers.
- Integral controllers.
- Proportional-plus-integral controllers.
- Proportional-plus-derivative controllers.
- Proportional-plus-integral-plus-derivative controllers.

Most industrial controllers [1, 8, 19, 24, 27, 35, 59] use electricity or pressurized fluid such as oil or air as power sources. Consequently, controllers may also be classified according to the kind of power employed in the operation, such as pneumatic controllers, hydraulic controllers, or electronic controllers. What kind of controller to use must be decided based on the nature of the plant and the operating conditions, including such considerations as safety, cost, availability, reliability, accuracy, weight, and size.

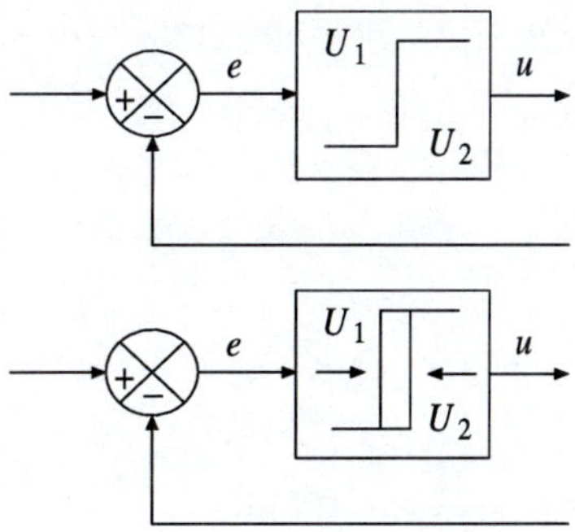

Fig. 5.25 (*Top*) Block diagram of an on–off controller; (*Bottom*) Block diagram of an on–off controller with differential gap

5.5.1 Two-Position Control Action

In a two-position (on–off) control system, the actuating element has only two fixed positions, which are, in many cases, simply on and off. Two-position or on–off control is relatively simple and inexpensive and, for this reason, is very widely used in both industrial and domestic control systems.

Let the output signal from the controller be $u(t)$ and the actuating error signal be $e(t)$. In two-position control, the signal $u(t)$ remains at either a maximum or minimum value, depending on whether the actuating error signal is positive or negative, so that

$$\begin{aligned} u(t) &= U_1, \quad \text{for } e(t) > 0, \\ &= U_2, \quad \text{for } e(t) < 0 \end{aligned}$$

where U_1 and U_2 are constants. The minimum value U_2 is usually either zero or $-U_1$. Two-position controllers are generally electrical devices, and an electric solenoid-operated valve is widely used in such controllers. Pneumatic proportional controllers with very high gains act as two-position controllers and are sometimes called pneumatic two-position controllers.

Figures 5.25(top) and 5.25(bottom) show the block diagrams for two-position or on–off controllers. The range through which the actuating error signal must move before the switching occurs is called the differential gap. A differential gap is indicated in Figs. 5.25(bottom). Such a differential gap causes the controller output $u(t)$ to maintain its present value until the actuating error signal has moved slightly beyond the zero value. In some cases, the differential gap is a result of unintentional friction and lost motion; however, quite often it is intentionally provided in order to prevent too frequent operation of the on–off mechanism.

Consider the liquid-level control system shown in Fig. 5.26(top), where the electromagnetic valve shown in Fig. 5.26(bottom) is used for controlling the inflow rate. This valve is either open or closed. With this two-position control, the water inflow rate is either a positive constant or zero. As shown in Fig. 5.27, the output signal continuously moves between the two limits required to cause the actuating element to move from one fixed position to the other. Notice that the output curve follows one of two exponential curves, one corresponding to the filling curve and the other to

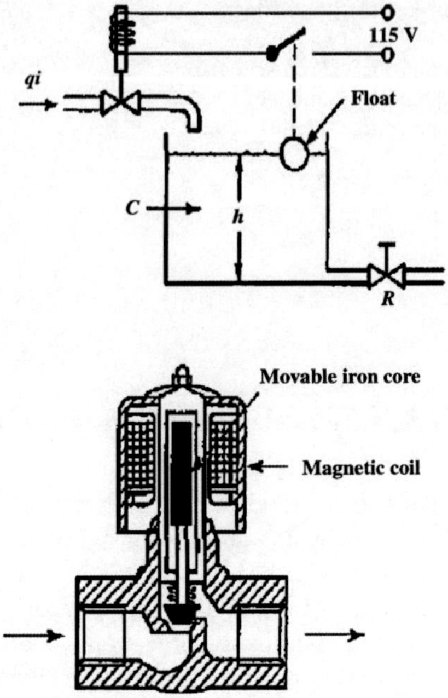

Fig. 5.26 (*Top*) Liquid-level control system; (*Bottom*) Electromagnetic valve

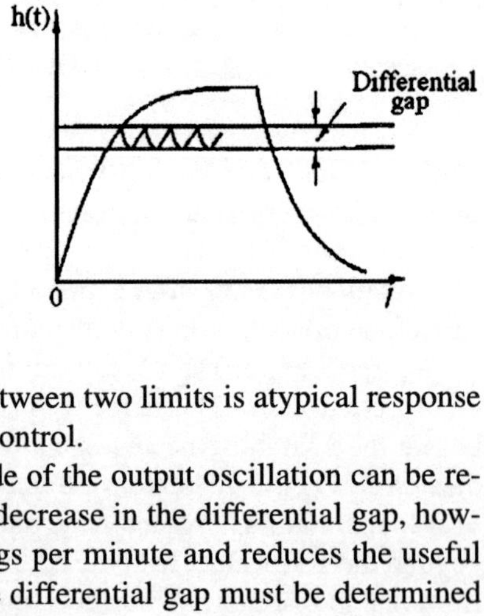

Fig. 5.27 Level $h(t)$ versus ℓ curve

the emptying curve. Such output oscillation between two limits is atypical response characteristic of a system under two position control.

From Fig. 5.27, we notice that the amplitude of the output oscillation can be reduced by decreasing the differential gap. The decrease in the differential gap, however, increases the number of on–off switchings per minute and reduces the useful life of the component 2% be magnitude of the differential gap must be determined from such considerations as the accuracy required and the life of the component.

5.5.2 P-Control Action

For a controller with proportional (P) control action, the relationship between the output of the controller $u(t)$ and the actuating error signal $e(t)$ is

$$u(t) = K_p e(t)$$

or, in Laplace-transformed quantities,

$$\frac{U(s)}{E(s)} = K_p$$

where K_p is termed the proportional gain.

5.5.3 Integral Control Action

In a controller with integral control action, the value of the controller output $u(t)$ is changed at a rate proportional to the actuating error signal $e(t)$. That is,

$$\frac{du(t)}{dt} = K_i e(t)$$

or

$$u(t) = K_i \int_0^t e(t)\,dt$$

where K_i is an adjustable constant. The transfer function of the integral controller is

$$\frac{U(s)}{E(s)} = \frac{K_i}{s}.$$

5.5.4 PI Control Action

The control action of a proportional-plus-integral (PI) controller is defined by

$$u(t) = K_p e(t) + \frac{K_p}{T_1} \int_0^t e(t)\,dt$$

or the transfer function of the controller is

$$\frac{U(s)}{E(s)} = K_p \left(1 + \frac{1}{T_1 s}\right)$$

where T_1 is called the integral time.

5.5.5 PD Control Action

The control action of a proportional-plus-derivative (PD) controller is defined by

$$u(t) = K_p e(t) + K_p T_d \frac{de(t)}{dt}$$

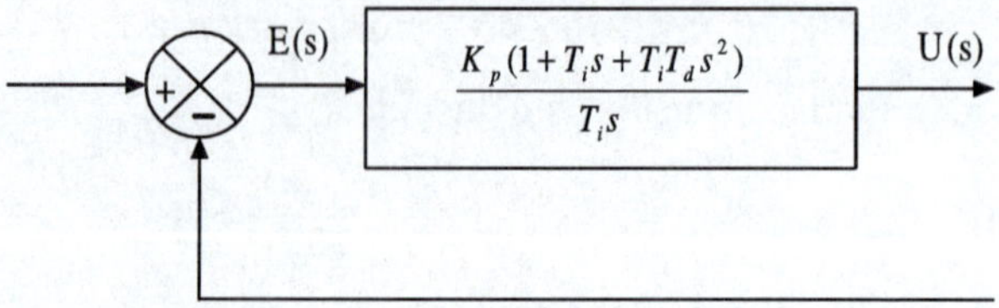

Fig. 5.28 Block diagram of a PID controller

and the transfer function is

$$\frac{U(s)}{E(s)} = K_p(1 + T_d s)$$

where T_d is called the derivative time.

5.5.6 PID Control Action

The combination of proportional (P) control action, integral (I) control action, and derivative (D) control action is termed proportional-plus-integral-plus-derivative (PID) control action. This combined action has the advantages of each of the three individual control actions. The equation of a controller with this combined action is given by

$$u(t) = K_p e(t) + \frac{K_p}{T_1}\int_0^t e(t)\,dt + K_p T_d \frac{de(t)}{dt}$$

or the transfer function is

$$\frac{U(s)}{E(s)} = K_p\left(1 + \frac{1}{T_1 s} + T_d s\right)$$

where K_p is the proportional gain, T_1 is the integral time, and T_d is the derivative time. The block diagram of a proportional-plus-integral-plus-derivative controller is shown in Fig. 5.28.

5.6 Closed-Loop System Subjected to a Disturbance

In what follows, we discuss the performance of control system when subjected to external disturbances. Reference is made to Fig. 5.29 where a closed-loop system subjected to a disturbance is shown.

When two inputs (the reference input and disturbance) are present in a linear system, each input can be treated independently of the other; and the outputs corresponding to each input alone can be added to give the complete output. The way each input is introduced into the system is shown at the summing point by either a plus or minus sign.

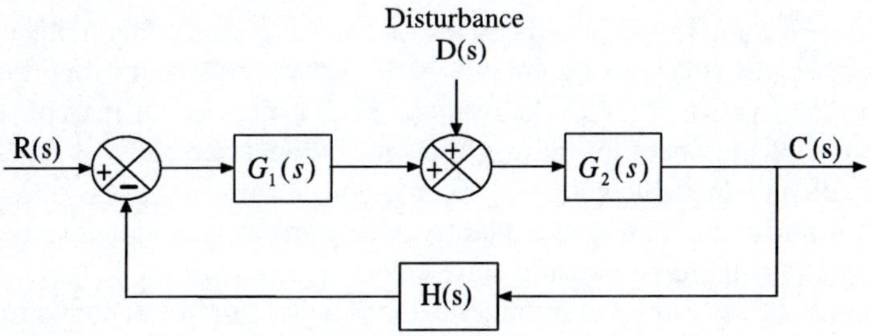

Fig. 5.29 Closed-loop system subjected to a disturbance

5.6.1 Main Issues

Consider the system shown in Fig. 5.29. In examining the effect of the disturbance $D(s)$, we may assume that the reference input is zero; we may then calculate the response $C_D(s)$ to the disturbance only. This response can be found from

$$\frac{C_D(s)}{D(s)} = \frac{G_2(s)}{1 + G_1(s)G_2(s)H(s)}.$$

On the other hand, in considering the response to the reference input $R(s)$, we may assume that the disturbance is zero. Then the response $C_R(s)$ to the reference input $R(s)$ can be obtained from

$$\frac{C_R(s)}{R(s)} = \frac{G_1(s)G_2(s)}{1 + G_1(s)G_2(s)H(s)}.$$

The response to the simultaneous application of the reference input and disturbance can be obtained by adding the two individual responses. In other words, the response $C(s)$ due to the simultaneous application of the reference input $R(s)$ and disturbance $D(s)$ is given by

$$\begin{aligned} C(s) &= C_R(s) + C_D(s) \\ &= \frac{G_2(s)}{1 + G_1(s)G_2(s)H(s)}\big[G_1(s)R(s) + D(s)\big]. \end{aligned}$$

Consider now the case where $|G_1(s)H(s)| \gg 1$ and $|G_1(s)G_2(s)H(s)| \gg 1$. In this case, the closed-loop transfer function $C_D(s)/D(s)$ becomes almost zero, and the effect of the disturbance is suppressed. This is an advantage of the closed-loop system.

On the other hand, the closed-loop transfer function $C_R(s)/R(s)$ approaches $1/H(s)$ as the gain of $G_1(s)G_2(s)H(s)$ increases. This means that if $|G_1(s)G_2(s)H(s)| \gg 1$ then the closed-loop transfer function $C_R(s)/R(s)$ becomes independent of $G_1(s)$ and $G_2(s)$ and becomes inversely proportional to $H(s)$ so that the variations of $G_1(s)$ and $G_2(s)$ do not affect the closed-loop transfer function $C_R(s)/R(s)$. This is another advantage of the closed-loop system. It can easily be seen that any closed-loop system with unity feedback, $H(s) = 1$, tends to equalize the input and output.

Next, we examine the important cases of control action. This includes proportional-plus-integral (PI), proportional-plus-derivative (PD) and proportional-plus-integral-plus-derivative (PID). Whatever the actual mechanism may be and whatever the form of the operating power, the proportional controller is essentially an amplifier with an adjustable gain.

In the proportional control of a plant whose transfer function docs not possess an integrator $1/s$, there is a steady-state error, or offset, in the response to a step input. Such an offset can be eliminated if the integral control action is included in the controller.

In the integral control of a plant, the control signal, the output signal from the controller, at any instant is the area under the actuating error signal curve up to that instant. The control signal $u(t)$ can have a nonzero value when the actuating error signal $e(t)$ is zero, as shown in Fig. 5.30(top). This is impossible in the case of the proportional controller since a nonzero control signal requires a nonzero actuating error signal. (A nonzero actuating error signal at steady state means that there is au offset.) Figure 5.30(bottom) shows the curve $e(t)$ versus t and the corresponding curve $u(t)$ versus t when the controller is of the proportional type.

Note that integral control action, while removing offset or steady-state error, may lead to oscillatory response of slowly decreasing amplitude or even increasing amplitude, both of which arc usually undesirable.

5.6.2 P-Control of Systems

In what follows, we will show that the proportional control of a system without an integrator will result in a steady-state error with a step input. We shall then show that such an error can be eliminated if integral control action is included in the controller.

Consider the system shown in Fig. 5.31. Let us obtain the steady-state error in the unit-step response of the system. Define

$$G(s) = \frac{K}{Ts+1}.$$

Since

$$\frac{E(s)}{R(s)} = \frac{R(s)-C(s)}{R(s)} = 1 - \frac{C(s)}{R(s)} = \frac{1}{1+G(s)}$$

the error $E(s)$ is given by

$$E(s) = \frac{1}{1+G(s)}R(s) = \frac{1}{1+\frac{K}{Ts+1}}R(s).$$

For the unit-step input $R(s) = 1/s$, we have

$$E(s) = \frac{Ts+1}{Ts+1+Ks}\frac{1}{s}.$$

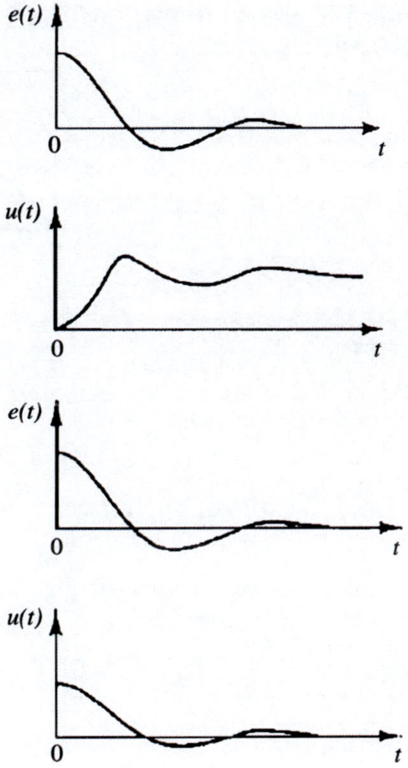

Fig. 5.30 Plots of $e(t)$ and $u(t)$ curves: (*Top*) integral control; (*Bottom*) proportional control

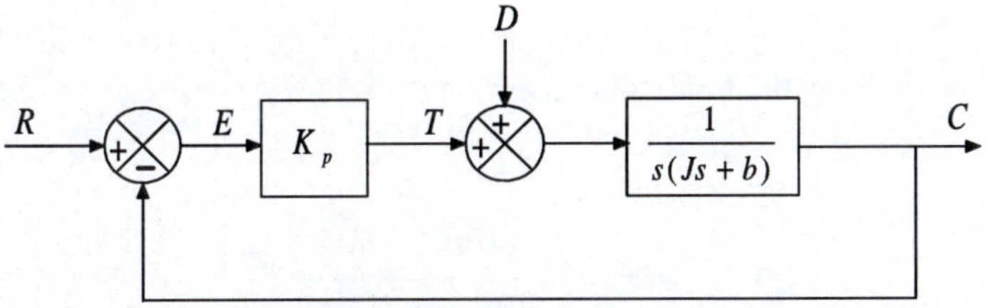

Fig. 5.31 Control system with a torque disturbance

The steady-state error is

$$e_{ss} = \lim_{t\to\infty} e(t) = \lim_{s\to 0} E(s) = \lim_{s\to 0} \frac{Ts+1}{Ts+1+K} = \frac{1}{K+1}.$$

Such a system without an integrator in the feedforward path always has a steady-state error in the step response. Such a steady-state error is called an offset. Figure 5.32 shows the unit-step response and the offset.

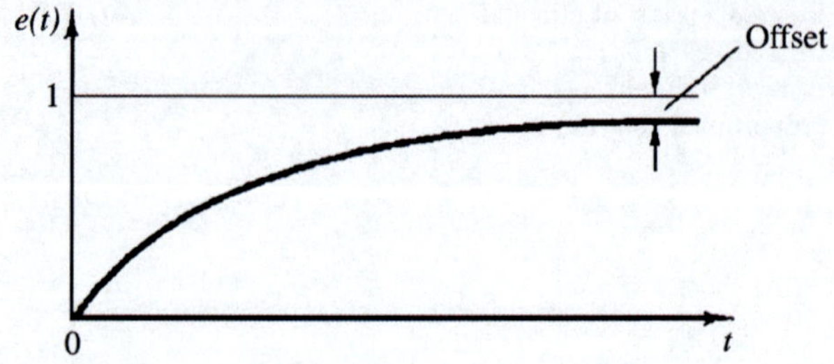

Fig. 5.32 Unit-step response and offset

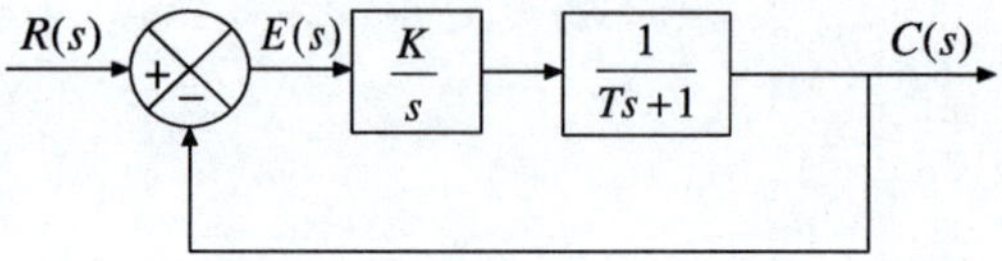

Fig. 5.33 Integral control system

5.6.3 *I-Control of Systems*

Consider the system shown in Fig. 5.33. The controller is an integral controller. The closed-loop transfer function of the system is

$$\frac{C(s)}{R(s)} = \frac{K}{s(Ts+1)+k}.$$

Hence,

$$\frac{E(s)}{R(s)} = \frac{R(s)-C(s)}{R(s)} = \frac{s(Ts+1)}{s(Ts+1)+k}.$$

Since the system is stable, the steady-state error for the unit-step response can be obtained by applying the final-value theorem, as follows:

$$\begin{aligned} e_{ss} &= \lim_{s\to 0} E(s) \\ &= \lim_{s\to 0} \frac{s^2(Ts+1)1}{Ts^2+s+Ks}\frac{1}{s} \\ &= 0. \end{aligned}$$

Integral control of the system thus eliminates the steady-state error in the response to the step input. This is an important improvement over the proportional control alone, which gives offset.

5.7 Response to Torque Disturbances

Let us investigate the effect of a torque disturbance occurring at the load element and for this purpose, consider the system shown in Fig. 5.34.

Fig. 5.34 PI control system with a torque disturbance

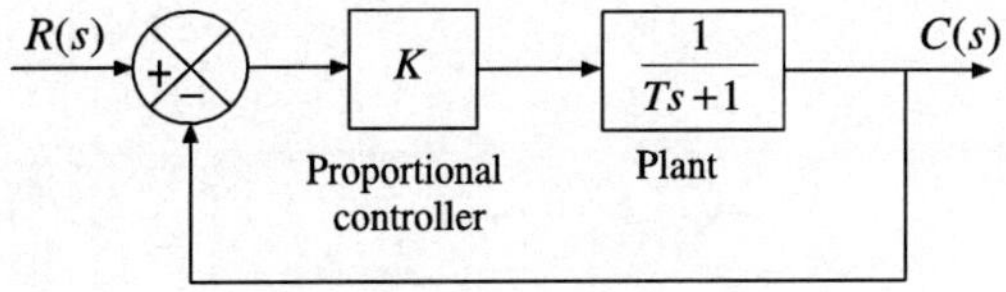

5.7.1 P-Control

The proportional controller delivers torque T to position the load element, which consists of moment of inertia and viscous friction. Torque disturbance is denoted by D. Assuming that the reference input is zero or $R(s) = 0$, the transfer function between $C(s)$ and $D(s)$ is given by

$$\frac{C(s)}{D(s)} = \frac{1}{Js^2 + bs + K_p}.$$

Hence,

$$\frac{E(s)}{D(s)} = -\frac{C(s)}{D(s)} = \frac{1}{Js^2 + bs + K_p}.$$

The steady-state error due to a step disturbance torque of magnitude T_d is given by

$$\begin{aligned} e_{ss} &= \lim_{s\to 0} E(s) \\ &= \lim_{s\to 0} \frac{-s}{Js^2 + bs + K_p} \frac{T_d}{s} \\ &= -\frac{T_d}{K_p}. \end{aligned}$$

At steady state, the proportional controller provides the torque $-T_d$, which is equal in magnitude but opposite in sign to the disturbance torque T_d. The steady-state output due to the step disturbance torque is

$$e_{ss} = -e_{ss} - \frac{T_d}{K_p}.$$

111e steady-state error can be reduced by increasing the value of the gain K_p. Increasing this value, however, will cause the system response to be more oscillatory.

5.7.2 PI-Control

To eliminate offset due to torque disturbance, the proportional controller may be replaced by a proportional-plus-integral (PI) controller. If integral control action is added to the controller, then, as long as there is an error signal, a torque is developed by the controller to reduce this error, provided the control system is a stable one.

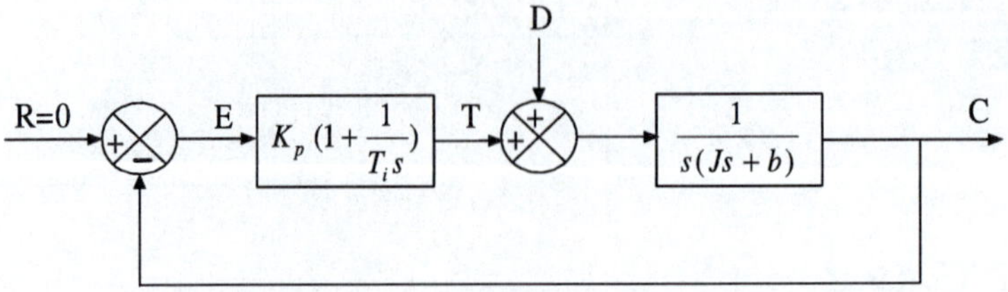

Fig. 5.35 PI control of a load element

Figure 5.35 shows the PI control of the load element, consisting of moment of inertia and viscous friction.

The closed-loop transfer function between $C(s)$ and $D(s)$ is

$$\frac{C(s)}{D(s)} = \frac{s}{Js^3 + bs^2 + K_p s + \frac{K_p}{T_i}}.$$

In the absence of the reference input, or $r(t) = 0$, the error signal is obtained from

$$E(s) = \frac{s}{Js^3 + bs^2 + K_p s + \frac{K_p}{T_i}} D(s).$$

If this control system is stable, that is, if the roots of the characteristic equation

$$Js^3 + bs^2 + K_p s + \frac{K_p}{T_i} = 0$$

have negative real parts, then the steady-state error in the response to a unit-step disturbance torque can be obtained by applying the final-value theorem as follows:

$$\begin{aligned} e_{ss} &= \lim_{s\to 0} E(s) \\ &= \lim_{s\to 0} \frac{-s^2}{Js^3 + bs^2 + K_p s + \frac{K_p}{R_i}} \frac{1}{s} \\ &= 0. \end{aligned}$$

Thus steady-state error to the step disturbance torque can be eliminated if the controller is of the proportional-plus-integral type.

Note that the integral control action added to the proportional controller has converted the originally second-order system to a third-order one. Hence, the control system may become unstable for a large value of K_p since the roots of the characteristic equation may have positive real parts. (The second-order system is always stable if the coefficients in the system differential equation are all positive.)

It is important to point out that if the controller were an integral controller, as in Fig. 5.36, then the system always becomes unstable because the characteristic equation

$$Js^3 + bs^2 + K = 0$$

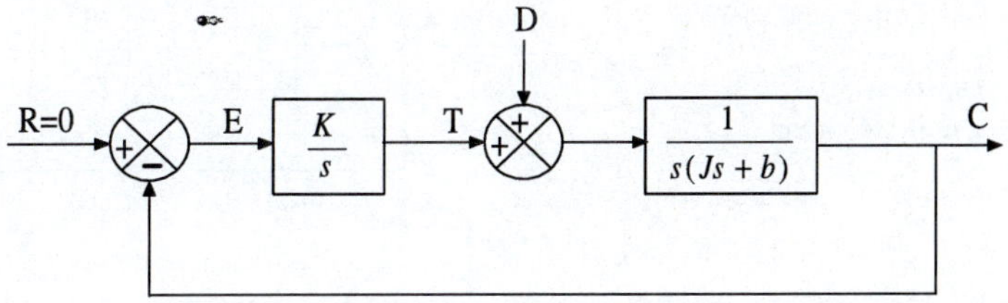

Fig. 5.36 Integral control of a load element

will have roots with positive real parts. Such an unstable system cannot be used in practice.

Note that in the system of Fig. 5.34 the proportional control action tends to stabilize the system, while the integral control action tends to eliminate or reduce steady-state error in response to various inputs.

5.7.3 D-Control Action

Derivative control action, when added to a proportional controller, provides a means of obtaining a controller with high sensitivity. An advantage of using derivative control action is that it responds to the rate of change of the actuating error and can produce a significant correction before the magnitude of the actuating error becomes too large. Derivative control thus anticipates the actuating error, initiates an early corrective action, and tends to increase the stability of the system.

Although derivative control does not affect the steady-state error directly, it adds damping to the system and thus permits the use of a larger value of the gain K, which will result in an improvement in the steady-state accuracy.

Because derivative control operates on the rate of change of the actuating error and not the actuating error itself, this mode is never used alone. It is always used in combination with proportional or proportional-plus-integral control action.

5.7.4 P-Control of Systems with Inertia Load

Before we discuss the effect of derivative control action on system performance, it is convenient to consider the proportional control of an inertia load.

Consider the system shown in Fig. 5.37(top). The closed-loop transfer function is obtained as

$$\frac{C(s)}{R(s)} = \frac{K_p}{Js^2 + K_p}.$$

Since the roots of the characteristic equation

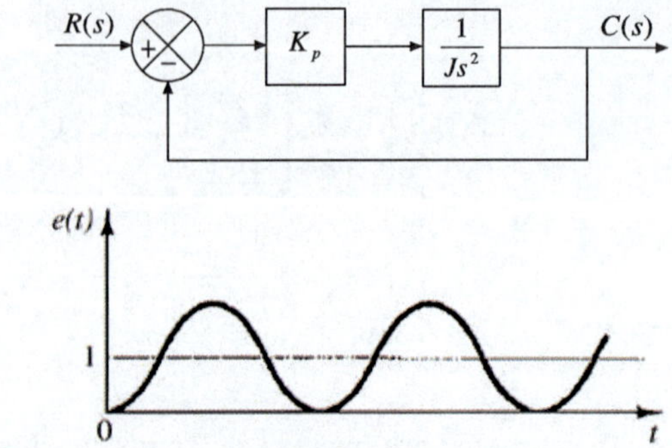

Fig. 5.37 (*Top*) Proportional control of a system with inertia load; (*Bottom*) Response to a unit-step input

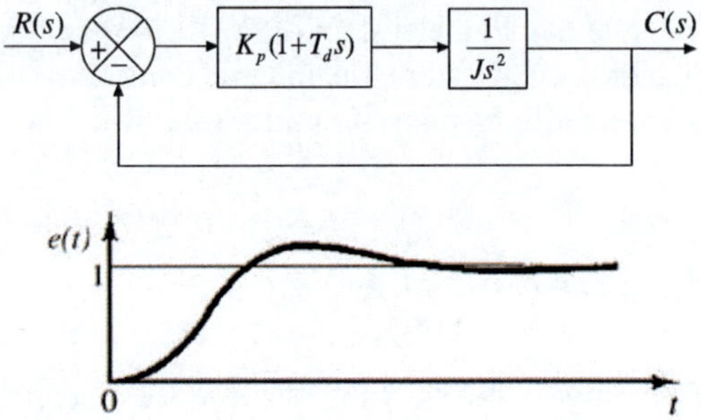

Fig. 5.38 (*Top*) PD control of a system with inertia load; (*Bottom*) Response to a unit-step input

$$Js^2 + K_p = 0$$

are imaginary, the response to a unit-step input continues to oscillate indefinitely, as shown in Fig. 5.38(bottom).

Control systems exhibiting such response characteristics arc not desirable. We will note in the sequel that the addition of derivative control will stabilize the system.

5.7.5 PD-Control of a System with Inertia Load

Let us modify the proportional controller to a proportional-plus-derivative controller whose transfer function is $K_p(1 + T_d s)$. The torque developed by the controller is proportional to $K_p(e + T_d \dot{e})$. Derivative control is essentially anticipatory, measures the instantaneous error velocity, and predicts the large overshoot ahead of time and produces an appropriate counteraction before too large an overshoot occurs.

Consider the system shown in Fig. 5.38(top). The closed-loop transfer function is given by

$$\frac{C(s)}{R(s)} = \frac{K_p(1 + T_f s)}{Js^2 + K_p T_d s + K_p}.$$

The characteristic equation

$$Js^2 + K_p T_d s + K_p = 0$$

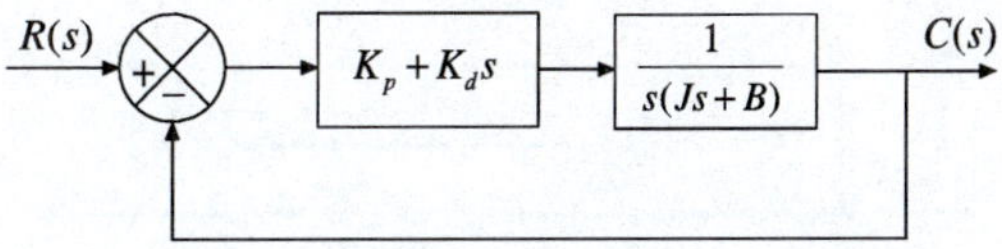

Fig. 5.39 A control system

now has two roots with negative real parts for positive values of J, K_p, and T_d. Thus, derivative control introduces a damping effect. A typical response curve $c(t)$ to a unit-step input is shown in Fig. 5.38(bottom). Clearly, the response curve shows a marked improvement over the original response curve shown in Fig. 5.38(bottom).

5.7.6 PD-Control of Second-Order Systems

A compromise between acceptable transient-response behavior and acceptable steady-state behavior may be achieved by use of proportional-plus-derivative control action.

Consider the system shown in Fig. 5.39. The closed-loop transfer function is

$$\frac{C(s)}{R(s)} = \frac{K_p + K_d s}{J s^2 + (B + K_d)s + K_p}.$$

The steady-state error for a unit-ramp input is

$$e_{ss} = \frac{B}{K_p}.$$

The characteristic equation is

$$J s^2 + (B + K_d)s + K_p = 0.$$

The effective damping coefficient of this system is thus $B + K_d$ rather than B. Since the damping ratio ζ of this system is

$$\zeta = \frac{B + K_d}{2\sqrt{K_p J}}.$$

5.7.7 Control Example 5.12

We consider a mechanical liquid-level control system which incorporates a hydromechanical controller that implements a *PID* control law, for more detailed technical discussions the reader is advised to consult [1, 19, 24, 25, 27, 29, 35, 38, 46, 48, 59]. This system is shown in Fig. 5.40. The displacement z_1 is related to h by

$$z_1 = (a_1/b)h \tag{5.44}$$

and the displacement z_2 of the spool valve by

$$z_2 = (a_2/b)h. \tag{5.45}$$

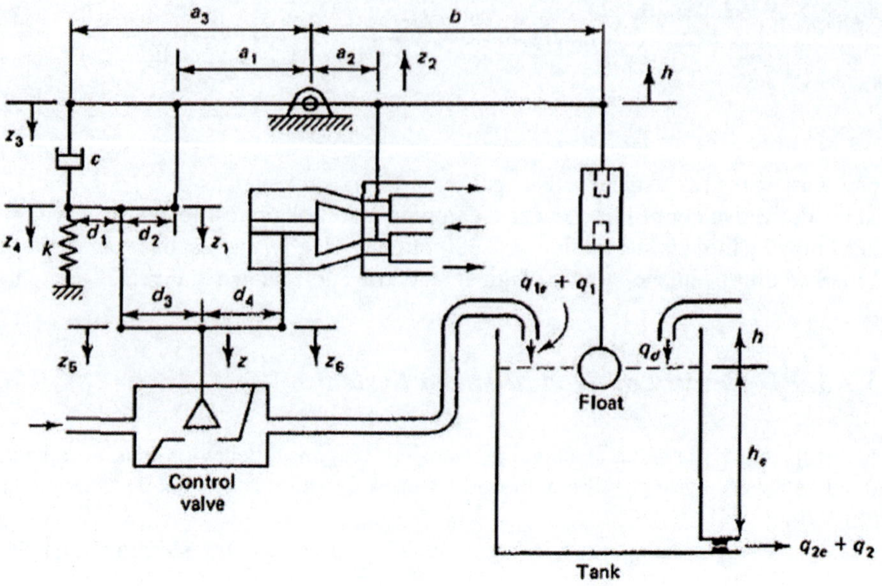

Fig. 5.40 PID controller for a liquid-level control system

The load on the actuator is negligible and hence, we obtain

$$z_6 = (k_1/s)z_2$$

and using (5.45),

$$z_6 = \left(\frac{k_1 a_2}{b}\right)\left(\frac{1}{D}\right)h. \tag{5.46}$$

Equating the damper force to the spring force, we get

$$c(\dot{z}_3 - \dot{z}_4) = kz_4$$

or

$$\begin{aligned} z_4 &= \left(\frac{cs}{cs+k}\right)z_3 \\ &= \left(\frac{\tau s}{\tau s+1}\right)\left(\frac{a_3}{b}\right)h \end{aligned} \tag{5.47}$$

where $\tau = c/k$ and $z_3 = (a_3/b)h$. The valve movement z is obtained as

$$z = \left(\frac{d_4}{d_3+d_4}\right)z_5 + \left(\frac{d_3}{d_3+d_4}\right)z_6 \tag{5.48}$$

where

$$z_5 = \left(\frac{d_1}{d_1+d_2}\right)z_1 + \left(\frac{d_2}{d_1+d_2}\right)z_4. \tag{5.49}$$

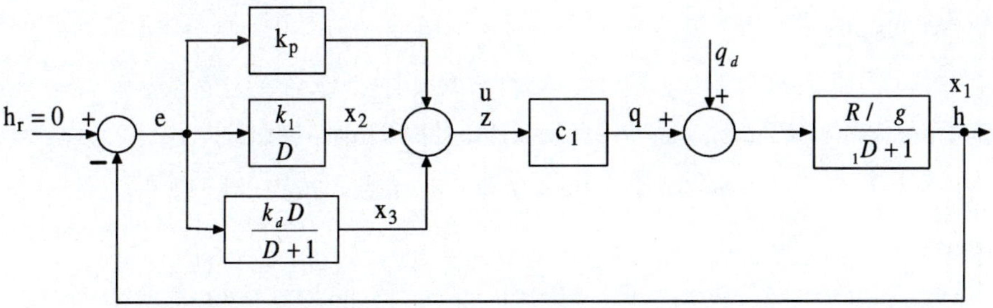

Fig. 5.41 Block diagram of a liquid level control system

Substituting for z_5 in (5.49) from (5.49) and then using (5.44), (5.46), and (5.47), we obtain

$$z = \left(\frac{d_4}{d_3 + d_4}\right)\left(\frac{d_1}{d_1 + d_2}\right)\left(\frac{a_1}{b}\right)h + \left(\frac{d_3}{d_3 + d_4}\right)\left(\frac{k_1 a_2}{b}\right)\frac{1}{D}h$$
$$+ \left(\frac{d_4}{d_3 + d_4}\right)\left(\frac{d_2}{d_1 + d_3}\right)\left(\frac{\tau a_3}{b}\right)\left(\frac{D}{\tau D + 1}\right)h. \tag{5.50}$$

This equation can be expressed as

$$z = k_p h\left(\frac{k_1}{D}\right)h + k_d\left(\frac{D}{\tau D + 1}\right)h \tag{5.51}$$

where the gains k_p, k_1 and k_d are obtained by comparing the corresponding terms in (5.50) and (5.51). The linearized equation for the flow-control valve is

$$q_1 = -cz \tag{5.52}$$

and the mathematical model of the tank has been obtained in Example 3.1. The block diagram may now be completed as shown in Fig. 5.41.

Thus, the hydromechanical controller implements a *PID* control law. The time constant T must be chosen to be small to extend the frequency range of the derivative mode. After summing up the three control actions, we can see that the system is now type 1. The first order of the original system of Example 3.1 has now been raised to the third order. Hence, two additional state variables must be defined as shown in Fig. 5.41 for the state-variables representation.

Noting that the set point has not been changed, that is, $h_r = 0$, we obtain the state equations as follows.

$$\begin{aligned}
\dot{x}_1 &= -\left(\frac{1}{\tau_1}\right)x_1 + \left(\frac{Rc_1}{\tau_1 \rho g}\right)u + \left(\frac{R}{\tau_1 \rho g}\right)q_d,\\
\dot{x}_2 &= -k_1 x_1,\\
\dot{x}_3 &= -\left(\frac{1}{\tau_1}\right)x_3 - \left(\frac{k_d}{\tau}\right)\dot{x}_1\\
&= -\left(\frac{k_d}{\tau\tau_1}\right)x_1 - \left(\frac{1}{\tau}\right)x_3 - \left(\frac{k_d Rc_1}{\tau\tau_1 \rho g}\right)u - \left(\frac{k_d R}{\tau\tau_1 \rho g}\right)q_d
\end{aligned} \tag{5.53}$$

where in the last equation, we have substituted for $\dot{x}_1$ from the first equation.

We also have

$$u = -k_p x_1 + x_2 + x_3.$$

The preceding equations can be expressed in the standard form

$$\dot{x} = Ax + Bu + B_1 v, \qquad u = -Kx$$

where

$$A = \begin{bmatrix} -1/\tau_1 & 0 & 0 \\ -k_1 & 0 & 0 \\ k_d/\tau\tau_1 & 0 & -1/\tau \end{bmatrix}, \qquad B = \begin{bmatrix} Rc_1/\tau_1 \rho g \\ 0 \\ k_d R/\tau\tau_1 \rho g \end{bmatrix}. \tag{5.54}$$

The closed-loop system can now be expressed as

$$\dot{x} = (A - BK)x + B_1 u$$

and its characteristic equation is given by

$$\det|sI - A + BK| = 0.$$

5.8 Linear Optimal Control: Continuous-Time

In Fig. 5.42, the feedback configuration for the linear quadratic regulation (LQR) is shown where we note the negative feedback and the absence of a reference signal. The process is assumed to be a continuous-time LTI system of the form

$$\begin{aligned} \dot{x}(t) &= Ax(t) + Bu(t), \qquad x(0) = x_o, \\ y(t) &= Cx(t), \\ z(t) &= Gx(t) + Hu(t), \end{aligned} \tag{5.55}$$

where $x(t) \in \Re^n$, $u(t) \in \Re^m$, $y(t) \in \Re^p$, $z(t) \in \Re^q$, and it has two distinct outputs.

1. The measured output $y(t)$ corresponds to the signal(s) that can be measured and are therefore available for control.
2. The controlled output $z(t)$ corresponds to the signal(s) that one would like to make as small as possible in the shortest possible time.

Sometimes $z(t) = y(t)$, which means that our control objective is simply to make the measured output very small. At other times one may have

$$z = \begin{bmatrix} y \\ \dot{y} \end{bmatrix}, \tag{5.56}$$

which means that we want to make both the measured output $y(t)$ and its derivative $\dot{y}(t)$ very small. Many other options are possible. The optimal LQR problem consists of finding the control input $u(t)$ that minimizes

$$J_c = \int_0^\infty z^t(t) \mathrm{Q} z(t) + \varrho u^t(t) \mathrm{R} u(t)\, dt, \tag{5.57}$$

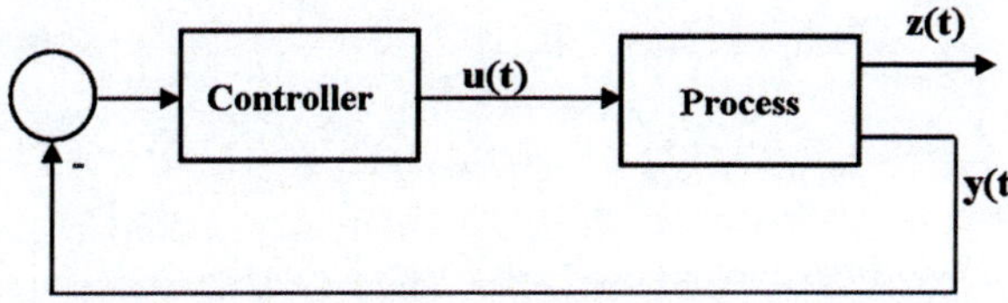

Fig. 5.42 LQR configuration

where $\mathsf{Q} \in \Re^{q\times q}$, $\mathsf{R} \in \Re^{m\times m}$ are symmetric positive-definite matrices and ϱ is a positive constant. The term

$$\int_0^\infty z^t(t)\mathcal{Q}z(t)\,dt$$

corresponds to the energy of the controlled output and the term

$$\int_0^\infty u^t(t)\mathcal{R}u(t)\,dt$$

corresponds to the energy of the control signal. Normally in LQR one seeks a controller that minimizes both energies. However, decreasing the energy of the controlled output will require a large control signal, and a small control signal will lead to large controlled outputs. The role of the constant ϱ is to establish a trade-off between these conflicting goals [2, 3, 6, 7].

1. Choosing ϱ very large, the most effective way to decrease J_c is to employ a small control input, at the expense of a large controlled output.
2. Choosing ϱ very small, the most effective way to decrease J_c is to obtain a very small controlled output, even if this is achieved at the expense of employing a large control input.

The most general form for a quadratic criteria is expressed by

$$J_o = \int_0^\infty \big[x^t(t)\mathcal{Q}x(t) + u^t(t)\mathcal{R}u(t) + 2x^t(t)\mathcal{N}u(t)\big]\,dt. \tag{5.58}$$

It is readily seen on using $z(t) = Gx(t) + Hu(t)$ from (5.55) that (5.57) is a special case of (5.58) with

$$\mathcal{Q} = G^t\mathsf{Q}G, \qquad \mathcal{R} = H^t\mathsf{Q}H + \varrho\mathsf{R}, \qquad \mathcal{N} = G^t\mathsf{Q}H.$$

Associated with system (5.55) a functional

$$H\big(x(.); u(.)\big) := -\int_0^\infty \big[Ax(t) + Bu(t)\big]^t\mathcal{P}x(t) + x^t(t)\mathcal{P}\big[Ax(t) + Bu(t)\big]\,dt,$$

which when computed along a solution of the system, its value depends only on the initial condition x_o as long as

$$\lim_{t\to\infty} x(t) = 0$$

where $\mathcal{P}^t = \mathcal{P}$. This implies that $H(x(.); u(.))$ is *feedback invariant for system* (5.55). To make use of this basic property, we express (5.58) in the form

$$\begin{aligned}J_o &= H\big(x(.); u(.)\big) \\ &\quad + \int_0^\infty x^t(t)\mathcal{Q}x(t) + u^t(t)\mathcal{R}u(t) + 2x^t(t)\mathcal{N}u(t) \\ &\quad + \big[Ax(t) + Bu(t)\big]^t \mathcal{P}x(t) + x^t(t)\mathcal{P}\big[Ax(t) + Bu(t)\big]\,dt \\ &= H\big(x(.); u(.)\big) + \int_0^\infty x^t(t)\big[\mathcal{P}A + A^t\mathcal{P} + \mathcal{Q}\big]x(t) \\ &\quad + u^t(t)\mathcal{R}u(t) + 2u^t(t)\big[B^t\mathcal{P} + \mathcal{N}^t\big]x(t)\,dt. \end{aligned} \tag{5.59}$$

By completing the squares, we have

$$\begin{aligned}&\big[u(t) + Kx(t)\big]^t \mathcal{R}\big[u(t) + Kx(t)\big] \\ &\quad = u^t(t)\mathcal{R}u(t) + [\mathcal{P}B + \mathcal{N}]\mathcal{R}^{-1}\big[B^t\mathcal{P} + \mathcal{N}^t\big]x(t) \\ &\qquad + 2u^t(t)\big[B^t\mathcal{P} + \mathcal{N}^t\big]x(t), \\ &K := \mathcal{R}^{-1}\big[B^t\mathcal{P} + \mathcal{N}^t\big],\end{aligned} \tag{5.60}$$

and therefore we express J_o into the form

$$\begin{aligned}J_o &= H\big(x(.); u(.)\big) \\ &\quad + \int_0^\infty x^t(t)\big(\mathcal{P}A + A^t\mathcal{P} + \mathcal{Q} - [\mathcal{P}B + \mathcal{N}]\mathcal{R}^{-1}\big[B^t\mathcal{P} + \mathcal{N}^t\big]\big)x(t) \\ &\quad + \big[u(t) + Kx(t)\big]^t \mathcal{R}\big[u(t) + Kx(t)\big]\,dt.\end{aligned} \tag{5.61}$$

On selecting the matrix $\mathcal{P}$ such that

$$\mathcal{P}A + A^t\mathcal{P} + \mathcal{Q} - [\mathcal{P}B + \mathcal{N}]\mathcal{R}^{-1}\big[B^t\mathcal{P} + \mathcal{N}^t\big] = 0 \tag{5.62}$$

the minimum of J_o is attained at

$$u^*(t) = -Kx(t), \quad K := \mathcal{R}^{-1}\big[B^t\mathcal{P} + \mathcal{N}^t\big] \tag{5.63}$$

for which the closed-loop system

$$\dot{x}(t) = \big(A - B\mathcal{R}^{-1}\big[B^t\mathcal{P} + \mathcal{N}^t\big]\big)x(t) \tag{5.64}$$

is asymptotically stable and the minimum cost is $J_o^* = x_o^t\mathcal{P}x_o$. It must be noted that (5.62) is called the *algebraic Riccati equation* (*ARE*).

5.8.1 Important Special Case

An important special case of the quadratic criteria (5.58) occurs when $\mathcal{N} \equiv 0$. In this case, the optimal gain and the associated ARE are given by

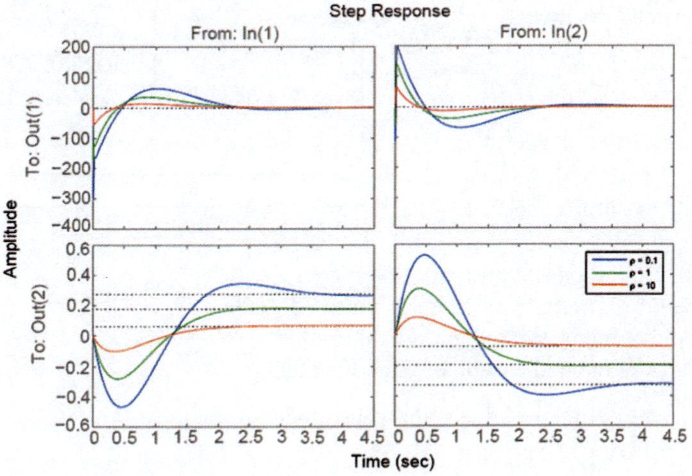

Fig. 5.43 State trajectories under LQR

$$K := \mathcal{R}^{-1} B^t \mathcal{P}, \qquad \mathcal{P}A + A^t\mathcal{P} + \mathcal{Q} - \mathcal{P}B\mathcal{R}^{-1}B^t\mathcal{P} = 0.$$

5.8.2 Control Example 5.13

The model of the longitudinal motion of a flexible bomber aircraft considered earlier is considered hereafter using LQR design. The design is based on MATLAB function $lqr(A, B, Q, R)$. Using this command, we will get feedback gain K and the solution for *Algebraic Riccati Equation*. In this simulation, three different sets of weighting matrices will be selected and the corresponding of close-loop response will be compared. The parameter will be used to multiply the weighting of the states is ρ, that is, $Q = \rho * I_{5\times 5}$ and $R = I_{2\times 2}$.

From Fig. 5.43, it is readily seen that the more weight we put on the states, the states will become more damped and faster to become stable.

5.8.3 Control Example 5.14

The problem of controlling the patient blood gases with the objective of maintaining these blood gases in their physiological ranges during a stable extracorporeal circulation process is of particular interest [42]. An appropriate block diagram is depicted in Fig. 5.44 in which the model set-up is portrayed. In terms of the state and input variables:

- x_1 is the flow rate of oxygen,
- x_2 is the flow rate of carbon dioxide,

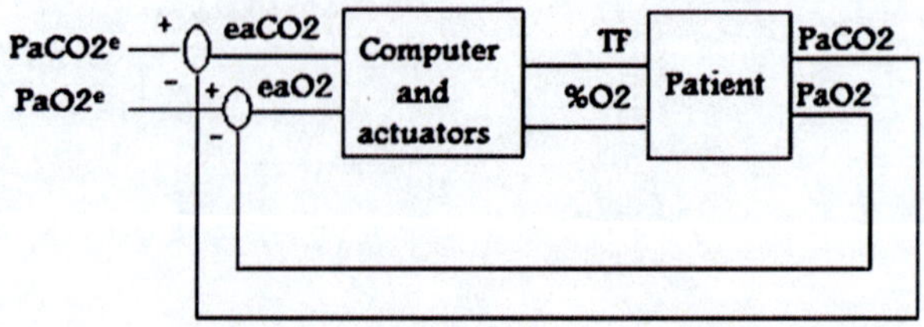

Fig. 5.44 A proposed block diagram

- x_3 is the arterial partial pressure of oxygen,
- x_4 is the arterial partial pressure of carbon dioxide,
- u_1 is the commanded oxygen flow rate and
- u_2 is the commanded carbon dioxide flow rate,

we consider a continuous-time state space model of the form

$$\dot{X}(t) = f\big(X(t), U(t)\big), \qquad Y(t) = g\big(X(t), U(t)\big) \tag{5.65}$$

where $X(t) \in \Re^4$, $U(t) \in \Re^2$, $Y(t) \in \Re^2$ are the state, the control input and the measured output vectors. Let (X_e, U_e) be the reference level of the state and control vectors and introduce

$$x(t) = X(t) - X_e, \qquad u(t) = u(t) - U_e$$

as the corresponding incremental variations. Applying a standard linearization procedure of (5.65) results in a linearized model that can conveniently cast into the format

$$\begin{aligned} \dot{x}(t) &= Ax(t) + Bu(t), \\ y(t) &= Cx(t) \end{aligned} \tag{5.66}$$

where $x(t) \in \Re^4$, $u(t) \in \Re^2$, and $y(t) \in \Re^2$ are the state, the control input and the measured output vectors. The matrices $A \in \Re^{4\times4}$, $B \in \Re^{4\times2}$, $C \in \Re^{2\times4}$ are real constant and describe the dynamics of blood gases during a stable extracorporeal circulation process. In particular, the coefficients of the matrices

$$A = \frac{\partial f(.,.)}{\partial X}\bigg|_{X=X_e, U=U_e}, \qquad B = \frac{\partial f(.,.)}{\partial U}\bigg|_{X=X_e, U=U_e}$$

depend on the conditions of the patient and their nominal values could be evaluated and stored whenever needed. Using reasonable nominal data [42], the model matrices in (5.66) are given by

$$A = \begin{bmatrix} -10.045 & 0.002 & 0.003 & 0.001 \\ 0.001 & -9.989 & 0.001 & 0.001 \\ 6.045 & -3.002 & -4.997 & 0.001 \\ 0.002 & 0.505 & 0.001 & -5.002 \end{bmatrix},$$

$$B = \begin{bmatrix} 10 & 0 \\ 0 & 10 \\ 0 & 0 \\ 0 & 0 \end{bmatrix}, \qquad C = \begin{bmatrix} 0 & 0 & 0 & 1 \\ 0 & 0 & 1 & 0 \end{bmatrix}.$$

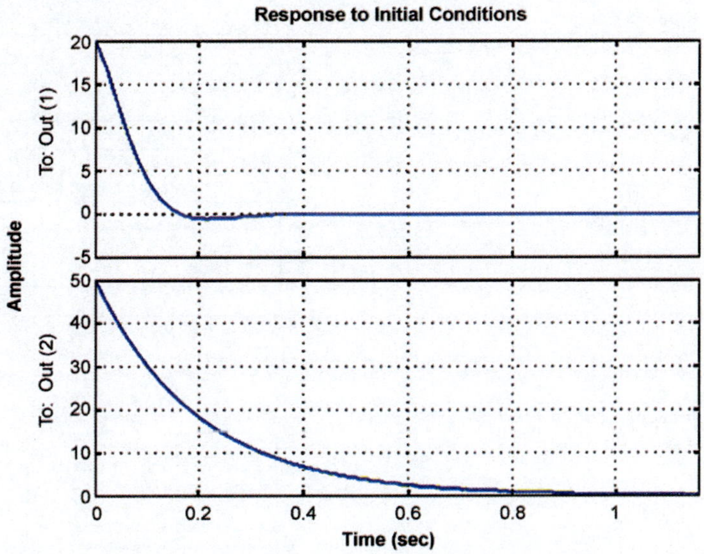

Fig. 5.45 Response of arterial partial pressure of oxygen and carbon dioxide

Fig. 5.46 Input and output response for $p = 2$

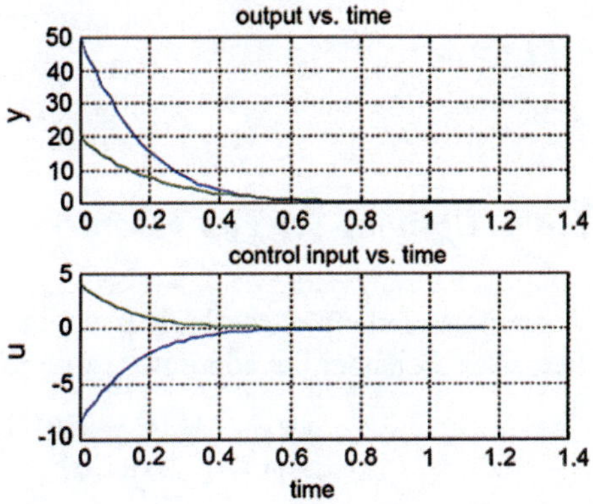

Thus the variables of main concern are x_3 and x_4. At start, we examined the response of the blood gases model to initial impact in the arterial partial pressure of oxygen and carbon dioxide. The result is plotted in Fig. 5.45. In order to illustrate the application of LQR theory, we use the weighting matrices

$$Q = \text{Blockdiag}\begin{bmatrix} 0 & 0 & 1 & 1 \end{bmatrix}, \qquad R = \text{Blockdiag}\begin{bmatrix} p & p \end{bmatrix}$$

for three distinct cases: 1) $p = 2$, 2) $p = 0.02$ and 3) $p = 200$. The ensuing input–output simulation results are depicted in Figs. 5.46–5.48, from which we conclude the input and output variable settles quickly when p is small corresponding to high feedback gain.

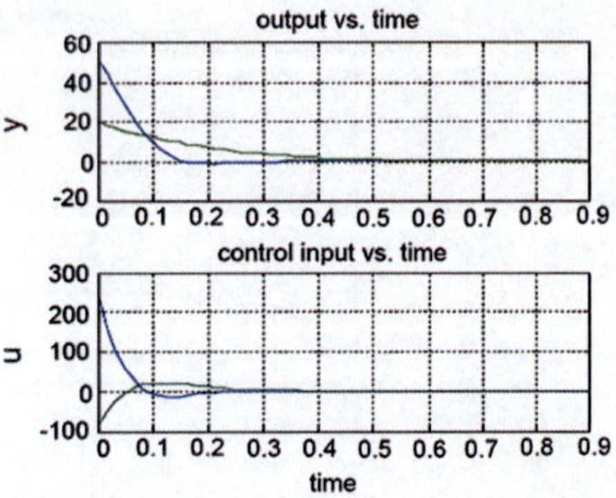

Fig. 5.47 Input and output response for $p = 0.02$

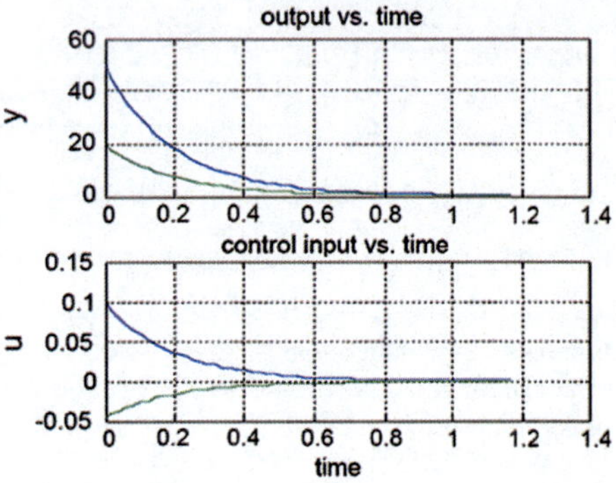

Fig. 5.48 Input and output response for $p = 200$

5.8.4 Optimal Set-Point Control

A practical version of the LQR is the optimal set-point control, which is described hereafter. Consider the continuous-time LTI process

$$\begin{aligned} \dot{x}(t) &= Ax(t) + Bu(t), \qquad x(0) = x_o, \\ z(t) &= Gx(t) + Hu(t) \end{aligned} \tag{5.67}$$

where $x(t) \in \Re^n$, $u(t) \in \Re^m$, $z(t) \in \Re^q$. We wish the controlled output z to converge as fast as possible to a given nonzero constant *set point value* r, corresponding to an equilibrium point (x_e, u_e) of (5.67) for which $z = r$. In light of the foregoing sections, this eventually amounts to an LQR criterion of the form [18]

$$J_s = \int_0^\infty \tilde{z}^t(t)\tilde{\mathsf{Q}}\tilde{z}(t) + \varrho \tilde{u}^t(t)\tilde{\mathsf{R}}\tilde{u}(t)\, dt, \tag{5.68}$$

where $\tilde{\mathsf{Q}} \in \Re^{q\times q}$, $\tilde{\mathsf{R}} \in \Re^{m\times m}$ are symmetric positive-definite matrices and ϱ is a positive constant. In addition, $\tilde{z} := z - r$, $\tilde{u} := u - u_e$. The equilibrium point (x_e, u_e) satisfies

$$\begin{aligned} 0 &= Ax_e + Bu_e, \\ r &= Gx_e + Hu_e \end{aligned}$$

which can be written compactly as

$$\begin{bmatrix} -A & B \\ -G & H \end{bmatrix} \begin{bmatrix} -x_e \\ u_e \end{bmatrix} = \begin{bmatrix} 0 \\ r \end{bmatrix} \tag{5.69}$$

where the block matrix on the left has dimension $(n+q) \times (n+q)$. It must be when the number of inputs m is strictly smaller than the number of controlled outputs q, we have an *underactuated system*. In this case, the system of equations (5.69) generally does not have a solution, because it presents more equations than unknowns. On the other hand, when the number of inputs m is equal to the number of controlled outputs q, (5.69) always has a solution as long as the matrix

$$R(s) := \begin{bmatrix} sI - A & B \\ -G & H \end{bmatrix}$$

is nonsingular for $s = 0$. $R(s)$ is known as Rosenbrock's system matrix [30]. A consequence of this is that $s = 0$ should not be an *invariant zero* of the system (recall that a transmission zero of a transfer matrix is always an invariant zero of its state-space realizations), and therefore it cannot also be a transmission zero of the transfer matrix $T(s) = G(sI - A)^{-1}B + H$. One should expect problems when $s = 0$ is an invariant zero of the system, since as the state $x(t)$ converges to an equilibrium point, the control input $u(t)$ must converge to a constant. By the zero-blocking property, one should then expect the controlled output $z(t)$ to converge to zero and not to r.

It is obvious that when the number of inputs m is strictly larger than the number of controlled outputs q, we have an overactuated system, and (5.69) generally has multiple solutions.

Proceeding further, the optimal set point problem can be reduced to that of optimal regulation by considering an auxiliary system with state $\tilde{x} := x - x_e$. Making use of (5.69) with some manipulations, the dynamics of auxiliary system are expressed by

$$\begin{aligned} \dot{\tilde{x}}(t) &= A\tilde{x}(t) + B\tilde{u}(t), \\ \tilde{z}(t) &= G\tilde{x}(t) + H\tilde{x}(t). \end{aligned} \tag{5.70}$$

At this stage, we can regard (5.68) and (5.70) as an *optimal regulation problem* for which the optimal solution is given by

$$\tilde{u}(t) = -K\tilde{x}(t).$$

Translating this result to the original input and state variables u and x, we conclude that the optimal control for the set-point defined by (5.67) and (5.68) takes the form

$$u(t) = -K\big[x(t) - x_e\big] + u_e, \quad t \ge 0. \tag{5.71}$$

Recall that the solution of (5.69) can be expressed as

$$x_e = Mr, \qquad u_e = Nr$$

for appropriately defined matrices M and N, the control scheme for optimal set-point control is depicted in Fig. 5.49.

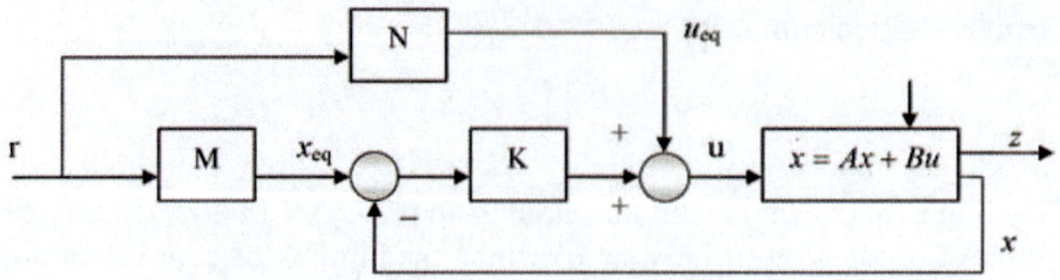

Fig. 5.49 Linear quadratic set point control with state-feedback

5.8.5 An LMI Formulation

With focus on the LQR design, the associated quadratic cost function is

$$J = \int_0^{\infty} \left[\mathbf{y}^t(t) Q \mathbf{y}(t) + \mathbf{u}^t(t) R \mathbf{u}(t)\right] dt \tag{5.72}$$

where $0 < Q$, $0 < R$ are output error and control weighting matrices, which are selected in the course of simulation by observing several sets of criteria of the closed loop-system. In what follows, we present an LMI-based formulation to the LQ control of system (5.66) while minimizing the quadratic cost (5.72). We proceed to determine a linear optimal state-feedback control $\mathbf{u} = L\mathbf{x}$ that achieves this goal. Assume that $V(\mathbf{x})$ has the form $V(\mathbf{x}) = \mathbf{x}^t K_+ \mathbf{x}$, $K_+ > 0$ and satisfies

$$\dot{V}(\mathbf{x}) \le -\left[\mathbf{x}^t C^t Q C \mathbf{x} + \mathbf{u}^t R \mathbf{u}\right]. \tag{5.73}$$

Then, the linear system controlled by $\mathbf{u}$ is asymptotically stable and $J_\infty \le V(x_o)$. With $\mathbf{u} = L\mathbf{x}$, inequality (5.73) is equivalently expressed as

$$\mathbf{x}^t\left[K_+(A + BL) + (A + BL)^t K_+^t\right]\mathbf{x} \le -\mathbf{x}^t\left[C^t Q C + L^t R L\right]\mathbf{x}. \tag{5.74}$$

From (5.74), it is evident that (5.73) is satisfied if there exists L and K_+ such that

$$K_+(A + BL) + (A + BL)^t K_+^t + \left[C^t Q C + L^t R L\right] \le 0. \tag{5.75}$$

Moreover, instead of directly minimizing the cost $x_o^t K_+ x_o$, we proceed to minimize its upper bound. Therefore, we assume that there exists $\gamma_+ > 0$ such that

$$x_o^t K_+ x_o \le \gamma_+. \tag{5.76}$$

In effect, the linear optimal control problem under consideration for given γ_+ can be cast into the format

$$\min_{\gamma_+, K_+, L} \gamma_+ \quad \text{subject to (5.75)–(5.76)}. \tag{5.77}$$

To convexify the above problem, we first express (5.75) as

$$\begin{aligned} \Phi &= K_+(A + BL) + (A + BL)^t K_+^t, \\ \Pi &= \begin{bmatrix} \Phi & C^t & L^t \\ \bullet & -Q^{-1} & 0 \\ \bullet & \bullet & -R^{-1} \end{bmatrix} \le 0. \end{aligned} \tag{5.78}$$

Pre- and post-multiply (5.78) by $\mathrm{diag}\{K_*^{-1}, I, I\}$ and using $Y = K_+^{-1}$, $S = LK_+^{-1}$ it follows that (5.78) is equivalent to

$$\begin{bmatrix} (AY+BS)+(AY+BS)^t & YC^t & YL^t \\ \bullet & -R^{-1} & 0 \\ \bullet & \bullet & -Q^{-1} \end{bmatrix} \le 0. \tag{5.79}$$

Additionally, inequality (5.76) can be expressed as

$$\begin{bmatrix} \gamma_+ & x_o^t \\ \bullet & K_+^{-1} \end{bmatrix} \ge 0 \quad \Longleftrightarrow \quad \begin{bmatrix} \gamma_+ & x_o^t \\ \bullet & Y \end{bmatrix} \ge 0. \tag{5.80}$$

The minimization problem (5.77) is cast into the form

$$\min_{\gamma_+, Y, S} \gamma_+ \quad \text{subject to (5.79)–(5.80).} \tag{5.81}$$

When a feasible solution of problem (5.81) is attained, we get $L = SY^{-1}$, $K_+ = Y^{-1}$.

5.9 Linear Optimal Control: Discrete-Time

In what follows, we direct attention to the class of processes which is assumed to be a discrete-time LTI system of the form

$$x(k+1) = Ax(k) + Bu(k), \qquad x(k_o) = x_o, \tag{5.82}$$

and the performance index

$$L(x_o, u, k) = \sum_{j=k_o+1}^{k} \left[x^t(j)\mathcal{Q}(j)x(j) + u^t(j-1)\mathcal{R}(j)u(j-1)\right], \tag{5.83}$$

where $x(k) \in \Re^n$, $u(k) \in \Re^m$ are the state and control vectors, respectively. The plant (5.82) is initially—that is, at time k_o—in state $x(k_o)$, and the aim is to return the plant state to the origin, or a state close to the origin. To do this, we set up a performance index (5.83), in which $Q(j)$ and $R(j)$ are nonnegative definite symmetric matrices. The performance index has the property that "large" values of the state will tend to make the performance index large. Therefore, by choosing the control sequence $\{u(k_o), u(k_o+1), \ldots\}$, which minimizes the performance index, we can expect to achieve the desired regulator effect.

We shall first solve the optimization problem for the case of finite horizon T. With additional assumptions, we shall then cover the infinite T case, with special reference to time-invariant plants.

The route to a derivation of the optimal control is via the *Principle of Optimality* [2]. Thus, if until time m optimal controls z$\{u(k_o), u(k_o+1), \ldots, u(m-1)\}$ have been applied, leading to a state $x(m)$, then the remaining terms in the optimal control sequence, $\{u(m), u(m+1), \ldots, u(T-1)\}$ must also be optimal in the sense of minimizing $L(x_o, u, k)$.

Now let L^* denote the optimal performance index associated with an initial state $x(t)$ at time t. Then, by the *Principle of Optimality*

$$\begin{aligned} L^*\big(x(k),k\big) &= \min_{u(k)}\big\{\big[Ax(k)+Bu(k)\big]^t \mathcal{Q}(j+1)x(j)\big[Ax(k)+Bu(k)\big] \\ &\quad + u^t(j)\mathcal{R}(j+1)u(j) + L^*\big(Ax(k)+Bu(k),k+1\big)\big\}, \\ &= \min_{u(k)}\big\{u^t(j)\big[B^t\mathcal{Q}(j+1)B+\mathcal{R}(j+1)\big]u(j) \\ &\quad + 2x^t(j)A^t\mathcal{Q}(j+1)Bu(j) + x^t(j)A^t\mathcal{Q}(j+1)Ax(j) \\ &\quad + L^*\big(Ax(k)+Bu(k),k+1\big)\big\}. \end{aligned} \tag{5.84}$$

Bearing in mind the foregoing continuous-time results, it would be reasonable to guess that $L^*(x(k),k)$ would be of the form $x^t(j)P(j)x(j)$. Since it proves convenient to make use of this result almost immediately, we build into the following argument an inductive proof of the result. For this purpose, it is required that

$$\big[B^t\mathcal{Q}(j+1)B+\mathcal{R}(j)\big] > 0 \quad \forall j.$$

Proceeding further, we have

$$L\big(x(T-1),u(.),T-1\big) = x^t(T)\mathcal{Q}(T)x(T) + u^t(T-1)\mathcal{R}(T)u(T-1). \tag{5.85}$$

On using (5.82), we manipulate (5.85) to reach

$$\begin{aligned} L\big(x(T-1),u(.),T-1\big) &= x^t(T-1)A^t\mathcal{Q}(T)Ax(T-1) \\ &\quad + 2x^t(T-1)A^t\mathcal{Q}(T)Bu(T-1) \\ &\quad + u^t(T-1)\big[B^t\mathcal{Q}(T)B+\mathcal{R}(T)\big]u(T-1). \end{aligned} \tag{5.86}$$

It is quite evident that the control $u^*(T-1)$ that minimizes this performance index is a linear function of $x(T-1)$—that is,

$$\begin{aligned} u^*(T-1) &= K(T-1)x(T-1), \\ K(T-1) &= -\big[B^t\mathcal{Q}(T)B+\mathcal{R}(T)\big]^{-1}B^t\mathcal{Q}(T)A. \end{aligned} \tag{5.87}$$

Moreover, the resulting optimal index $L * (x(T-l), T-1)$ becomes quadratic in $x(T-1)$—that is,

$$\begin{aligned} L^*\big(x(T-I),T-1\big) &= x^t(T-1)\mathcal{P}(T-l)x(T-1), \\ \mathcal{P}(T-1) &= A^t\big(\mathcal{Q}(T)-\mathcal{Q}(T)B\big[B^t\mathcal{Q}(T)B \\ &\quad + \mathcal{R}(T)\big]^{-1}B^t\mathcal{Q}(T)\big)A. \end{aligned} \tag{5.88}$$

Our goal is to compute expressions of the matrices $K(j)$, determining the optimal control law, and $P(j)$, determining the optimal performance index, for arbitrary values of j. Building on the foregoing results, we assume that $L^*(x(j+), j+1) =$

$x^t(j+1)P(j+1)x(j+1)$ for a certain matrix $P(j+1)$. Applying the inductive hypothesis to (5.84), we have

$$\begin{aligned} L^*\big(x(k),k\big) = \min_{u(k)} \big\{ & u^t(j)\big[B^t\mathcal{Q}(j+1)B + \mathcal{R}(j+1)\big]u(j) \\ & + 2x^t(j)A^t\mathcal{Q}(j+1)Bu(j) + x^t(j)A^t\mathcal{Q}(j+1)Ax(j) \\ & + x^t(j)A^t\mathcal{P}(j+1)Ax(j) + 2x^t(j)A^t\mathcal{Q}\mathcal{P}(j+1)Bu(j) \\ & + u^t(j)B^t\mathcal{P}(j+1)Bu(j)\big\}. \end{aligned} \tag{5.89}$$

Again, the minimizing $u(j)$, which is the optimal control at time j, is a linear function of $x(j)$,

$$u^*(j) = Kx(j) \tag{5.90}$$

and the optimal performance index $L^*(x(j), j)$, resulting from use of $u*(j)$, is quadratic in $x(j)$—that is,

$$L^*\big(x(j), j\big) = x^t(j)\mathcal{P}(j)x(j).$$

The expression for $K(j)$ is given by

$$\begin{aligned} K(j) &= -\big[B^t\hat{\mathcal{Q}}(j)B + \mathcal{R}(j)\big]^{-1}B^t\hat{\mathcal{Q}}(j)A, \\ \hat{\mathcal{Q}}(j) &= \mathcal{Q}(j) + \mathcal{P}(j). \end{aligned} \tag{5.91}$$

The expression for $\mathcal{P}(j)$ is

$$\mathcal{P}(j) = A^t\big(\hat{\mathcal{Q}}(j) - \hat{\mathcal{Q}}(j)B\big[B^t\hat{\mathcal{Q}}(j)B + \mathcal{R}(j)\big]^{-1}B^t\hat{\mathcal{Q}}(j)\big)A. \tag{5.92}$$

To guarantee that the optimal performance index is finite, we shall require that the pair A, B is controllable. Recall that the forgoing equations have to be solved recursively.

For infinite horizon, the time-varying matrices reach steady state values leading to the following expressions:

$$\begin{aligned} \mathcal{P}^+ &= A^t\big[(\mathcal{Q}^+ + \mathcal{P}^+) \\ &\quad - (\mathcal{Q}^+ + \mathcal{P}^+)B\big[B^t(\mathcal{Q}^+ + \mathcal{P}^+)B + \mathcal{R}^+\big]^{-1}B^t(\mathcal{Q}^+ + \mathcal{P}^+)\big]A, \\ K^+ &= -\big[B^t(\mathcal{Q}^+ + \mathcal{P}^+)B + \mathcal{R}^+\big]^{-1}B^t(\mathcal{Q}^+ + \mathcal{P}^+)A \end{aligned} \tag{5.93}$$

where $\mathcal{Q}^+$, $+\mathcal{R}^+$ are the steady state (constant) values of the weighting matrices $\mathcal{Q}(j)$, $\mathcal{R}(j)$. An assumption guaranteeing asymptotic stability of the closed-loop system

$$x(k+1) = \big[A + BK^+\big]x(k)$$

is that the pair (A, D) is observable where $DD^t = \mathcal{Q}$.

5.9.1 An LMI Formulation

For a discrete-time LQR the linear system under consideration is described by

$$\begin{aligned} x_{k+1} &= Ax_k + Bu_k, \\ y_k &= Cx_k \end{aligned} \tag{5.94}$$

where matrices A, B, C are derived from **A**, **B**, **C** via appropriate discretization scheme [45]. With a performance index given by

$$J = \sum_{k=0}^{\infty} \left(x_k^t Q x_k + u_k^t R u_k \right). \tag{5.95}$$

In what follows, we present an LMI-based formulation to the LQ control of system (5.94) while minimizing the quadratic cost (5.95). Our approach is basically a discrete-version of the foregoing section. We proceed to determine a linear optimal state-feedback control $u_k = Hx_k$ that achieves this goal. Assume that $V(x(k))$ has the form

$$V\big(x(k)\big) = x_k^t K_* x_k, \quad K_* > 0$$

and satisfies

$$V\big(x(k+1)\big) - V\big(x(k)\big) \le -\big[x_k^t Q x_k + u_k^t R u_k\big]. \tag{5.96}$$

Then, the linear system controlled by u_k is asymptotically stable and $J_\infty \le V(x_o)$. With $u_k = Hx_k$, inequality (5.96) is equivalently expressed as

$$x_k^t (A + BH)^t K_* (A + BH) x(k) - x_k^t K_* x_k \le -x_k^t \big[Q + H^t R H\big] x_k. \tag{5.97}$$

From (5.97), it is evident that (5.96) is satisfied if there exists H and K_* such that

$$(A + BH)^t K_* (A + BH) - K_* + \big[Q + H^t R H\big] \le 0. \tag{5.98}$$

Moreover, instead of directly minimizing the cost $x_o^t K_* x_o$, we proceed to minimize its upper bound. Therefore, we assume that there exists $\gamma_* > 0$ such that

$$x_o^t K_* x_o \le \gamma_*. \tag{5.99}$$

In effect, the linear optimal control problem under consideration for given γ_* can be cast into the format

$$\min_{\gamma_*, K_*, H} \gamma_* \quad \text{subject to (5.98)–(5.99)}. \tag{5.100}$$

To convexify the above problem, we first express (5.98) as

$$-K_* + \Theta^t \Pi^{-1} \Theta \le 0, \tag{5.101}$$

$$\begin{aligned} \Theta &= \big[(A + BH)^t \, H^t \, I\big], \\ \Pi &= \begin{bmatrix} K_*^{-1} & 0 & 0 \\ \bullet & R^{-1} & 0 \\ \bullet & \bullet & Q^{-1} \end{bmatrix}. \end{aligned} \tag{5.102}$$

By Schur complements, inequality (5.101) using (5.102) is equivalent to

$$\begin{bmatrix} -K_* & (A+BH)^t & H^t & I \\ \bullet & -K_*^{-1} & 0 & 0 \\ \bullet & \bullet & -R^{-1} & 0 \\ \bullet & \bullet & \bullet & -Q^{-1} \end{bmatrix} \leq 0. \tag{5.103}$$

Pre- and post-multiply (5.103) by $\mathrm{diag}\{K_*^{-1}, I, I, I\}$ and using $X = K_*^{-1}$, $Z = HK_*^{-1}$, it follows that (5.103) is equivalent to

$$\begin{bmatrix} -X & (AX+BZ)^t & Z^t & X \\ \bullet & -X & 0 & 0 \\ \bullet & \bullet & -R^{-1} & 0 \\ \bullet & \bullet & \bullet & -Q^{-1} \end{bmatrix} \leq 0. \tag{5.104}$$

Additionally, inequality (5.99) can be expressed as

$$\begin{bmatrix} \gamma_* & x_o^t \\ \bullet & K_*^{-1} \end{bmatrix} \geq 0 \iff \begin{bmatrix} \gamma_* & x_o^t \\ \bullet & X \end{bmatrix} \geq 0. \tag{5.105}$$

The minimization problem (5.100) is cast into the form

$$\min_{\gamma_*, X, Z} \gamma_* \quad \text{subject to (5.104)–(5.105)}. \tag{5.106}$$

When a feasible solution of the convex minimization problem (5.106) is attained, then we get

$$H = ZX^{-1}, \qquad K_* = X^{-1}.$$

5.9.2 Direct Driven Inverted Pendulum

The inverted pendulum system is a standard problem in the area of control systems and has two equilibria, one of which is stable while the other is unstable. The stable equilibrium corresponds to a state in which the pendulum is pointing downwards, see Fig. 5.50. In the absence of any control force, the system will naturally return to this state. The stable equilibrium requires no control input to be achieved and, thus, is uninteresting from a control perspective. The unstable equilibrium corresponds to a state in which the pendulum points strictly upwards and, thus, requires a control force to maintain this position.

The basic control objective of the inverted pendulum problem is to maintain the unstable equilibrium position when the pendulum initially starts in an upright position. Traditionally, an inverted pendulum was driven by a rotating servo motor which drove the cart via transfer mechanism to keep the balance of the system. The flaw of this configuration was the inclusion of the transmission friction and gap in the system. Transmission by flexible belt would also produce vibration, extension, and delay, and make the control system unpredictable. To overcome the defect of

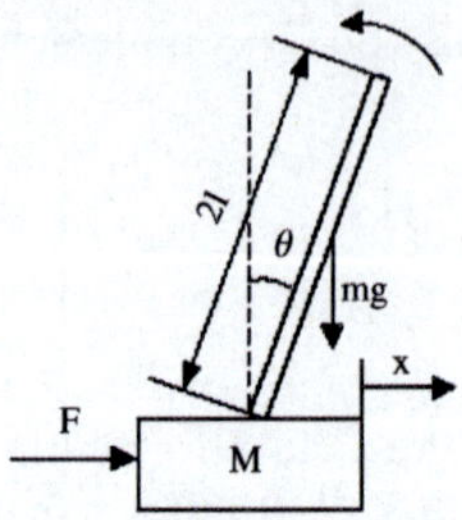

Fig. 5.50 Physical model of single inverted pendulum

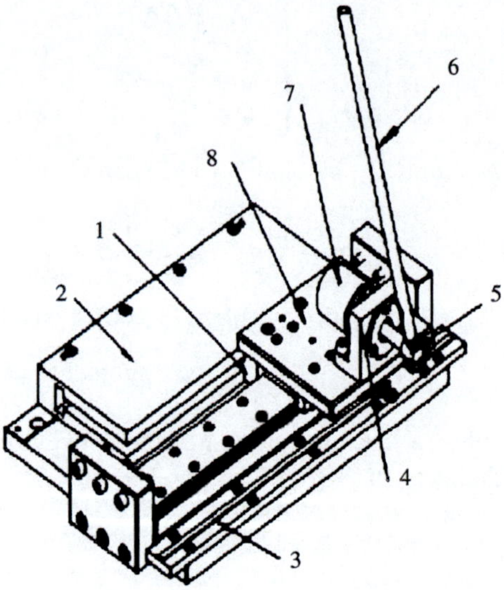

Fig. 5.51 Linear synchronous motor

the inverted pendulum driven by a rotating machine, a new scheme is proposed and named the direct driven inverted pendulum (dDIP).

The dDIP consists of a linear motor, a pendulum, a pedestal and a rotary encoder, as shown in Fig. 5.51. The cart for inverted pendulum is attached to the mover of the linear motor by rigid connection. In this way, the mover can directly drive the cart to achieve linear motion without transfer mechanism.

Linear motor is a new type of driving device which can directly transform electrical energy to mechanical linear motion and is called "direct transmission" or "zero clearance transmission". It has the advantages of high velocity, high acceleration, high accuracy, and no maximal travel length restriction. Linear motor can be used in industry, commercial, military and any other field where linear motion is needed. Linear motor can be classified into linear induction motor, linear synchronous motor etc. The motor used in our system is an ironless permanent magnet linear synchronous motor. Its maximal velocity is 5 m/s; maximal acceleration is 100 m/s^2; rated thrust force is 98 N; the peak thrust force is 280 N and the stage's resolution is 5 μm.

The assumptions for the modeling are as follows: (1) the pendulum and the pedestal are both rigid bodies. (2) air resistance and friction force between pendulum and the bearing are ignored. (3) the direction of the arrowhead is positive direction of the vector.

Analyzing the physical model of the single IP, we can obtain the mathematical expression of IP as follows:

$$(M+m)\ddot{x}+ml\ddot{\theta}\cos\theta+ml\dot{\theta}^2\sin\theta=F,$$
$$\left(I+ml^2\right)\ddot{\theta}+ml\ddot{x}\cos\theta=mgl\sin\theta.$$

While the IP is running, normally θ (radian) hardly changes at the equilibrium point and nears zero. Therefore, small angle approximation can be made: $\cos\theta\approx 1$, $\sin\theta\approx\theta$, $(\frac{d\theta}{dt})^2\approx 0$. With u representing the input force F, the expressions () can be simplified as follows:

$$(M+m)\ddot{x}+m\ell\ddot{\theta}=u,$$
$$\left(I+m\ell^2\right)\ddot{\theta}+m\ell\ddot{x}=mg\ell\theta.$$

The frequency response of the linear motor is measured by using a dynamic signal analyzer Agilent 35670A. Agilent 35670A is a FFT type frequency spectrum/network analyzer with 4 channels. This standard apparatus can measure frequency spectrum, network, time domain and amplitude domain in the range of 0–100 KHz and can analyze frequency response, octave, harmonic distortion and order spectrum. Agilent 35670A requires that the input is an analog signal, but the displacement of the linear motor's mover given by a linear encoder which has a resolution of 5 μm is digital. So a TMS320F2812 DSP is used to decode and count the digital count value into analog voltage through DAC7731. In this way, Agilent 35670A can sweep sine to the linear motor.

The pedestal of the IP was mounted on the linear motor's mover without the pendulum while the sweeping process was in progress thus the mass M (including the mass of the angle encoder) of the pedestal is taken into account. The result of the measurement is the motor's frequency response within 1–100 Hz. Using fitting function of MATLAB, the transfer function of the linear motor was obtained; where the input is voltage u and the output was displacement x.

$$G(s)=\frac{X(s)}{U(s)}=\frac{1.869}{s^2+12.32s+0.4582}. \tag{5.107}$$

Applying inverse Laplace transformation to () and the result can be expressed as:

$$1.869u=\ddot{x}+12.32\dot{x}+0.4582x. \tag{5.108}$$

5.9.3 *Modeling of dDIP*

According to (5.108), we can get:

$$\ddot{x}=-12.32\dot{x}-0.4582x+1.869u. \tag{5.109}$$

Combining (5.106) and (5.109) we can obtain the following equation:

$$\ddot{\theta} = \frac{0.4582ml}{I+ml^2}x + \frac{12.32ml}{I+ml^2}\dot{x} + \frac{mgl}{I+ml^2}\theta - \frac{1.869ml}{I+ml^2}u. \tag{5.110}$$

Four state variables are chosen as follows:

$$x_1 = x, \qquad x_2 = \dot{x}, \qquad x_3 = \theta, \qquad x_4 = \dot{\theta}.$$

So the state vector is,

$$X = \begin{bmatrix} x \\ \dot{x} \\ \theta \\ \dot{\theta} \end{bmatrix} \tag{5.111}$$

and the state space description of the dDIP is

$$\begin{aligned} \dot{X} &= Ax + Bu \\ &= \begin{bmatrix} 0 & 1 & 0 & 0 \\ -0.4582 & -12.32 & 0 & 0 \\ 0 & 0 & 0 & 1 \\ a & b & c & 0 \end{bmatrix} \begin{bmatrix} x \\ \dot{x} \\ \theta \\ \dot{\theta} \end{bmatrix} + \begin{bmatrix} 0 \\ 1.869 \\ 0 \\ d \end{bmatrix} u, \end{aligned} \tag{5.112}$$

where

$$a = \frac{0.4582ml}{I+ml^2}, \qquad b = \frac{12.32ml}{I+ml^2},$$
$$c = \frac{mgl}{I+ml^2}, \qquad d = -\frac{1.869ml}{I+ml^2}.$$

The parameters are $m = 0.1$ kg, $l = 0.2415$ m and $g = 9.8$ m/s^2.

Choosing the outputs as follows:

$$y_1 = x, \qquad y_2 = \dot{x}, \qquad y_3 = \theta, \qquad y_4 = \dot{\theta}.$$

So the output vector is:

$$Y = CX = \begin{bmatrix} 1 & 0 & 0 & 0 \\ 0 & 1 & 0 & 0 \\ 0 & 0 & 1 & 0 \\ 0 & 0 & 0 & 1 \end{bmatrix} \begin{bmatrix} x \\ \dot{x} \\ \theta \\ \dot{\theta} \end{bmatrix}. \tag{5.113}$$

5.9.4 *Optimal Control of Turbo-Generator System*

A basic element in power generation is the turbo-generator, that is shown in Fig. 5.52, the dynamic model of which has six states, two inputs and two outputs [15, 17, 21, 28].

Using appropriate data, the system matrices are given by

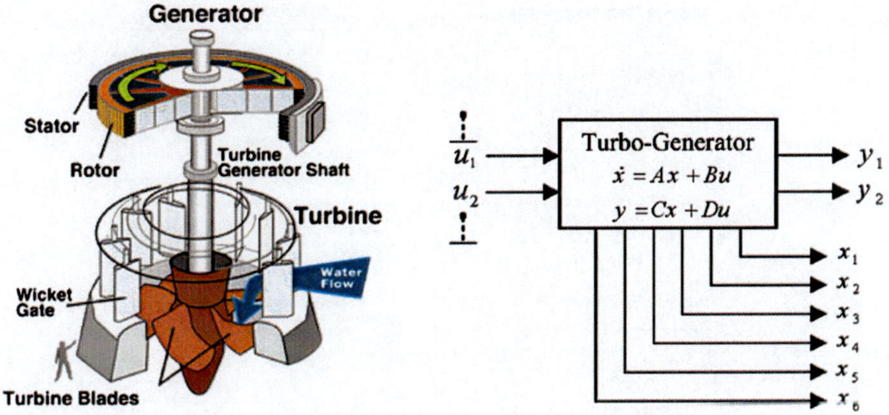

Fig. 5.52 Turbo-generator system: Physical and block-diagram

$$A = \begin{bmatrix} -18.4456 & 4.2263 & -2.2830 & 0.2260 & 0.4220 & -0.0951 \\ -4.0977 & -6.0706 & 5.6825 & -0.6966 & -1.2246 & 0.2873 \\ 1.4449 & 1.4336 & -2.6477 & 0.6092 & 0.8979 & -0.2300 \\ -0.0093 & 0.2302 & -0.5002 & -0.1764 & -6.3152 & 0.1350 \\ -0.0464 & -0.3489 & 0.7238 & 6.3117 & -0.6886 & 0.3645 \\ -0.0602 & -0.2361 & 0.2300 & 0.0915 & -0.3214 & -0.2087 \end{bmatrix},$$

$$B = \begin{bmatrix} -0.2748 & 3.1463 \\ -0.0501 & -9.3737 \\ -0.1550 & 7.4296 \\ 0.0716 & -4.9176 \\ -0.0814 & -10.2648 \\ 0.0244 & 13.7943 \end{bmatrix}, \qquad C^t = \begin{bmatrix} 0.5971 & 3.1013 \\ -0.7697 & 9.3422 \\ 4.8850 & -5.6000 \\ 4.8608 & -0.7490 \\ -9.8177 & 2.9974 \\ -8.8610 & 10.5719 \end{bmatrix}.$$

The eigenvalues of the system are computed using MATLAB command line **eig**(A) are given by -15.8730, -10.3872, $-0.3493 \pm j6.3444$, -1.0444, -0.2346. All system poles are to the left-hand side of imaginary axis of the complex plane, hence, the system is stable.

Using the MATLAB command line: $lqr(A, B, Q, R)$ with $Q = I_6$, $R = I_2$, the optimal state-feedback gain is:

$$K = \begin{bmatrix} -0.0077 & -0.0280 & -0.0916 & 0.0269 & -0.0344 & 0.0261 \\ -0.0037 & -0.0597 & 0.1593 & 0.6495 & -1.0641 & 1.0248 \end{bmatrix}$$

resulting in the closed-loop eigenvalues as

$$-19.5001 \pm j1.8608, \quad -7.1847, \quad -2.3323 \pm j5.6329, \quad -1.0094.$$

The optimal output trajectories are plotted in Fig. 5.53.

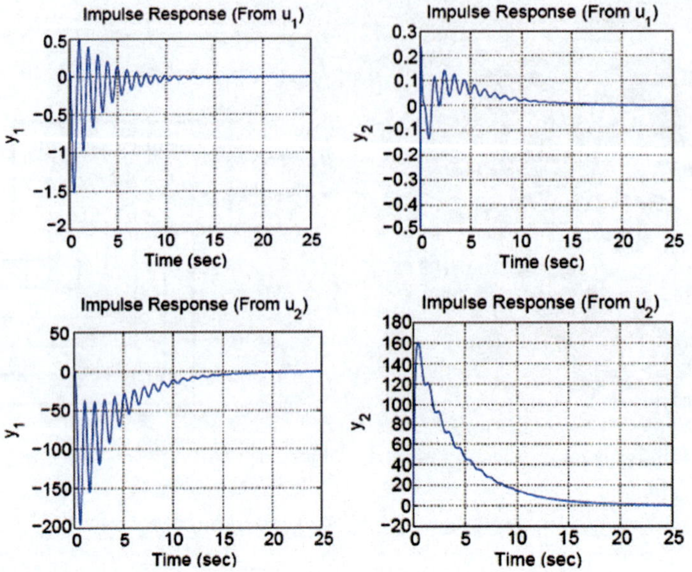

Fig. 5.53 Optimal output trajectories

5.10 Digital Control of Uninterruptible Power Supplies

The ultimate goal of uninterruptible power supplies (UPS) system is to supply constant amplitude sinusoidal voltage and constant frequency to load without any interruption in case of a main power failure [8, 22, 23, 26]. The quality of the UPS output voltage is defined by the total harmonic distortion (THD). The most common UPS configuration consists of a battery bank and a static rectifier-inverter-filter that produce a low total harmonic distortion sinusoidal output voltage that supplies the critical load. For such application, system performances are usually measured in terms of transient response and waveform distortions under sudden changes in load parameters [23, 33].

With the cost reduction of microcontrollers and digital signal processors (DSP), the use of digital control technique in power converter has increased. However, high power converters are usually operated at low switching frequencies in order to reduce switching losses. Therefore, advanced control strategies are required to overcome these complications [22, 32, 43].

To design the closed loop control, the model of the system has an important task in the conception of the controller. Some linear models for single phase PWM inverter system have been reported in literature [22, 33]. The output voltage and its derivative, that is proportional to the capacitor current, can be used as the state variables, as well as the output voltage and the inductor current. However, modelling errors and unmodeled dynamics are quite common. They may be a result of simplifications on the model, which can degrade the performance of the system [43].

Many discrete time controllers used to control a single phase inverters in UPS applications were reported in literature, such as predictive control [8, 13], repetitive control [25, 53], optimal state feedback [50] and selective harmonic compensation [26, 40]. Even if most of these schemes offered high performance feedback control results, they still relay on high switching frequencies and involve considerable computational over heads. In this paper, a single phase UPS with a low switching frequency is proposed in order to minimize switching losses and improve system efficiency. An adaptive linear quadratic regulator for single-phase UPS application is proposed. The regulator is a useful tool in modern optimal control design. For the proposed controller, a recursive least square estimator identifies the plant parameters which are used to compute the regulator gains periodically. The quadratic cost function parameter is chosen in order to reduce the energy of the control signal. Only the output voltage can be measured and the inductor current is not measurable. As a result, an observer is used to estimate the inductor current. Using a suitable filter, the effect of disturbances on the response of the system will be decreased. The simulations were carried out using MATLAB Simulink.

5.10.1 Plant Description

The single-phase PWM inverter is shown in Fig. 5.54, the LC filter and the resistive load R are considered to be the plant of the system.

The inverter is controlled by the unipolar PWM. The power switches are turned on and off at the carrier frequency. The plant can be modeled by the state space variable v_C and i_L:

$$\begin{bmatrix} \dot{v}_c \\ i_L \end{bmatrix} = \begin{bmatrix} \frac{-1}{RC} & \frac{1}{C} \\ \frac{-1}{L} & 0 \end{bmatrix} \begin{bmatrix} v_c \\ i_L \end{bmatrix} + \begin{bmatrix} 0 \\ \frac{1}{L} \end{bmatrix} u, \qquad y = \begin{bmatrix} 1 & 0 \end{bmatrix} \begin{bmatrix} v_c \\ i_L \end{bmatrix}, \tag{5.114}$$

or

$$\dot{x} = Ax + Bu, \qquad y = Cx. \tag{5.115}$$

Then, a discrete time model of the plant obtained by the forward method and sample time T_s is given by:

$$x(k+1) = A_d x(k) + B_d u(k), \qquad y(k) = C_d x(k), \tag{5.116}$$

where

$$\begin{aligned} x(k) &= \begin{bmatrix} v_c(k) & \hat{i}_L(k) \end{bmatrix}^T, \\ A_d &= I + T_s A, \qquad B_d = TsB. \end{aligned} \tag{5.117}$$

5.10.2 LQR Design

The adaptive linear quadratic regulator controller has the objective of tracking the discrete sinusoidal $r(k)$ reference in each sample instant.

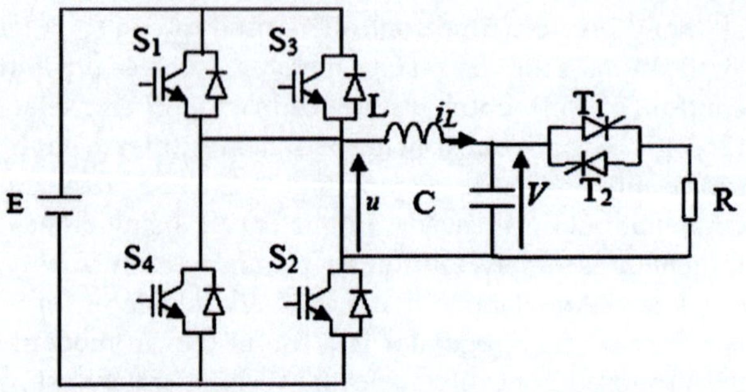

Fig. 5.54 Inverter, filter and load

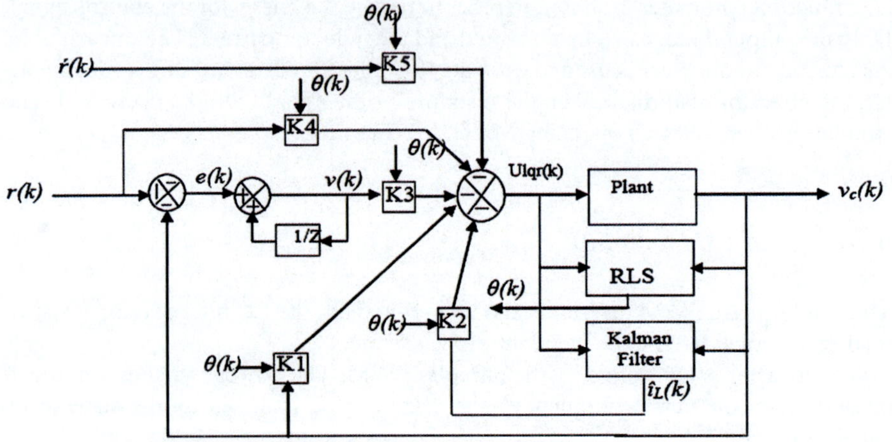

Fig. 5.55 Block diagram of the control system

The system output $y(k)$ is the capacitor voltage in the discrete form $v_c(k)$. The state variables used in the (LQR) are the measured output voltage $v_c(k)$, the estimated inductor current $\hat{i}_L(k)$, the integrated tracking error $v(k)$; all with a feedback action and the discrete reference $r(k)$ and its derivative $\dot{r}(k)$ with a feed forward action. Each state variable has weighting Ki tuned according to $\theta(k)$, which contains the plant parameters identified by the RLS estimator. The control system shown in Fig. 5.55 is therefore proposed. Then, in the proposed system, the state vector $z(k)$ is defined as:

$$z(k) = \begin{bmatrix} v_c(k) & \hat{i}_L(k) & v(k) & r(k) & \dot{r}(k) \end{bmatrix}^T, \tag{5.118}$$

and the LQR control signal is given by

$$u_{\text{LQR}}(k) = -Kz(k). \tag{5.119}$$

To design the optimal gains $K_1, K_2, \ldots, K_5$, the system must be represented in the form:

$$z(k+1) = Gz(k) + Hu_{\text{LQR}}(k), \tag{5.120}$$

where each state variable is calculated by a difference equation. The two first variables of vector $z(k)$ are obtained by (5.116). The signal $v(k)$ is:

$$v(k+1) = e(k+1) + v(k), \tag{5.121}$$

where the error is given by:

$$e(k) = r(k) - y(k). \tag{5.122}$$

From (5.116), (5.121) and (5.122) results the difference equation for

$$\begin{aligned} v(k+1) = v(k) + r(k) + T_s\dot{r}(k) - C_d A_d x(k) \\ - C_d B_d u_{\mathrm{LQR}}(k). \end{aligned} \tag{5.123}$$

The continuous time reference variables are:

$$\begin{bmatrix} \dot{r}(t) \\ \dot{r}(t) \end{bmatrix} = \begin{bmatrix} 0 & 1 \\ -\omega^2 & 0 \end{bmatrix} \begin{bmatrix} r(t) \\ \dot{r}(t) \end{bmatrix}, \qquad \dot{r} = Rr. \tag{5.124}$$

This system generates a sinusoidal reference when initiated with initial values:

$$\begin{aligned} r(0) &= 0, \\ \dot{r}(0) &= wV_p, \end{aligned}$$

where V_P is the sine wave amplitude and w is the angular frequency.

In the discrete form, using a sample period T_S, the subsystem (5.124) is given by:

$$n(k+1) = R_d n(k), \tag{5.125}$$

where

$$n(k) = \begin{bmatrix} r(k) & \dot{r}(k) \end{bmatrix}, \tag{5.126}$$

$$R_d = I + T_s R. \tag{5.127}$$

Then, using the state equations (5.116), (5.123) and (5.125), the closed-loop system representation becomes:

$$\begin{aligned} \begin{bmatrix} x(k+1) \\ v(k+1) \\ n(k+1) \end{bmatrix} &= \begin{bmatrix} A_d & 0 & 0 \\ -C_d A_d & 1 & C_d R_d \\ 0 & 0 & R_d \end{bmatrix} \begin{bmatrix} x(k) \\ v(k) \\ n(k) \end{bmatrix} \\ &\quad + \begin{bmatrix} B_d \\ -C_d B_d \\ 0 \end{bmatrix} u_{\mathrm{LQR}}(k), \\ y(k) &= \begin{bmatrix} C_d & 0 & 0 \end{bmatrix} \begin{bmatrix} x(k) & v(k) & n(k) \end{bmatrix}^T. \end{aligned} \tag{5.128}$$

The optimal gains of the control law (3.100) are those that minimize the following cost function:

$$J = \frac{1}{2} \sum_{k=0}^{\infty} \{ z^T(k) Q z(k) + u^T(k) R_u u(k) \}, \tag{5.129}$$

where Q and R_u are chosen as positive definite matrixes that set the weighting of states and the control signal respectively.

The K gains can be obtained through the evaluating the Riccati equations [44], as follows:

$$S(k) = G^T S(k+1)G + Q - \left[H^T S(k+1)G\right]^T \times \left[R_u + H^T S(k+1)H\right]^{-1}\left[H^T S(k+1)G\right], \tag{5.130}$$

$$K(k) = R_u^{-1} H^T \left(G^T\right)^{-1}\left(S(k) - Q\right). \tag{5.131}$$

A good flexibility in the design of the controller is provided by the selection of Q and R_u matrixes.

5.10.3 Recursive Least-Squares Estimator

To estimate the plant parameters when the load conditions are variable, a RLS algorithm is used [3]. The discrete plant model with a zero order hold is given by:

$$\frac{y(z)}{u(z)} = \frac{\theta}{z^2 + \theta_1 z + \theta_2}. \tag{5.132}$$

The difference equation of the estimated output is:

$$y(k) = -\theta_1 y(k-1) - \theta_2 y(k-2) + \theta_3 u(k-2), \tag{5.133}$$

or

$$\hat{y}(k) = \theta^T(k)\Psi(k-1), \tag{5.134}$$

where

$$\theta(k) = \begin{bmatrix}\theta_1 & \theta_2 & \theta_3\end{bmatrix}, \tag{5.135}$$

and

$$\Psi(k) = \begin{bmatrix}-y(k-1) & -y(k-2) & u(k-2)\end{bmatrix}. \tag{5.136}$$

The RLS gains are calculated using:

$$L(k) = \frac{p(k-1)\Psi(k)}{1 + \Psi^T k p(k-1)\Psi(k)}. \tag{5.137}$$

The RLS covariance matrix is given by:

$$p(k) = p(k-1) - \frac{p(k-1)\Psi(k)\Psi^T(k)p(k-1)}{1 + \Psi^T(k)p(k-1)\Psi(k)}, \tag{5.138}$$

and the plant parameters θ are recursively obtained by:

$$\hat{\theta}(k) = \hat{\theta}(k-1) + L(k)\left[y(k) - \Psi^T\hat{\theta}(k-1)\right], \tag{5.139}$$

where:

$$\hat{A}_d = \begin{bmatrix} 0 & -\hat{\theta}_2 \\ 1 & -\hat{\theta}_1 \end{bmatrix}, \qquad \hat{B}_d = \begin{bmatrix} \hat{\theta}_3 \\ 0 \end{bmatrix}, \qquad C_d = \begin{bmatrix} 0 & 1 \end{bmatrix}. \tag{5.140}$$

Then, it is possible to identify the plant parameters to a range of different loads through the substitution of matrixes (5.140) into system (5.128) and proceed there often with the LQR gains design in real time.

5.10.4 Kalman Filter

Since only the output voltage is measured, a Kalman filter [44, 55] is used to estimate the inductor current state.

$$\begin{aligned} x(k+1) &= A_d x(k) + B_d u(k) + w(k), \\ y(k) &= C_d x(k) + v(k). \end{aligned} \tag{5.141}$$

The random variables $w(k)$ and $v(k)$ represent the process and measurement noise respectively. They are assumed to be independent of each other and with normal probability distributions such that:

$$\begin{aligned} \mathbb{E}\big[w(k)^T \quad w(k)\big] &= R_w > 0, \\ \mathbb{E}\big[v(k)^T \quad v(k)\big] &= R_v > 0, \\ \mathbb{E}\big[w(k)^T \quad v(k)\big] &= 0. \end{aligned} \tag{5.142}$$

In practice, the process noise covariance and measurement noise covariance matrices might change with each time step or measurement. However, here, it is assumed that they are presented below [55].

The Kalman gains are given by:

$$K_G(k) = \big(M(k)C_d^T\big)\big(C_d M(k) C_d^T + R_v\big)^{-1}, \tag{5.143}$$

and the estimated variable, the inductor current, is

$$i_L = \hat{x}_2(k) = \begin{bmatrix} 0 & 1 \end{bmatrix} \hat{x}(k). \tag{5.144}$$

The following recursive equations are used:

$$P_K(k) = M(k) - K_G(k) C_d M(k), \tag{5.145}$$

and

$$M(k) = \big(A_d P_K(k) A_d^T\big) + \big(B_d R_w B_d^T\big). \tag{5.146}$$

After each time and measurement update pair, the process is repeated with the previous posterior estimates used to project or predict the new a priori estimates.

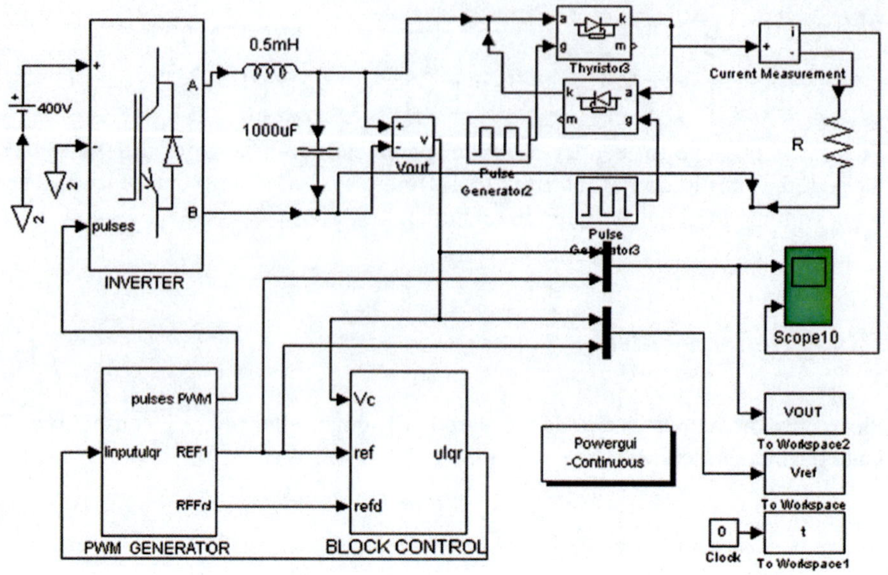

Fig. 5.56 System Simulink blockdiagram

5.10.5 Simulation Results

The simulation work is carried out according to the proposed block diagram presented in Fig. 5.56. The inverter system controlled by linear quadratic regulator algorithm is realized in order to study the output voltage (Vc) performance under linear and nonlinear loads. The plant controller parameters, algorithm constants and other system specifications are presented in Table 5.1.

For a linear load, the input and output voltage waveforms, estimated and measured inductor currents as well as estimated parameters are shown in Figs. 5.57, 5.58 and 5.59, respectively.

A linear load output voltage and current with values of R and K (gains) taken from Table 5.1 are illustrated in Fig. 5.60 and the output voltage frequency spectrum is presented in Fig. 5.61. From this spectrum, the THD is calculated and the obtained value is 1.12% showing a high quality output voltage.

For a nonlinear load, the output voltage, the output current and the output voltage frequency spectrum are shown in Figs. 5.62 and 5.63, respectively. The THD obtained from the voltage spectrum is equal to 1.61% proving a high quality output voltage. Figure 5.64 depicts the transient response of the output voltage compared to the reference. One notices that the dynamic time vanishes in brief time. Figure 5.65 shows the output voltage tracking the reference voltage efficiently in the case of linear load disturbance. From this figure, it is clear that the proposed LQR regulator is efficient.

Table 5.1 System parameters

DC input voltage	$E = 400$ V
Reference voltage	$V_{\text{ref}} = 320$ V (peak), 60 Hz
Sample time	$T_s = 1/18000$ s
States weightings	$Q = \text{diag}[50\ \ 100\ \ 150\ \ 1\ \ 1]$
Control weighting	$R_u = 100$
For linear load:	
Filter inductance	$L = 5.3$ mH
Filter capacitance	$C = 80$ μF
Linear load	$R = 6\ \Omega$
LQR gains $K = [8.0177\ \ 36.0875\ \ 1.0127\ \ -10.0251\ \ -0.0031]$	
For nonlinear load:	
Nonlinear load rated resistive load phase commutated at angle 45°	
Filter inductance	$L = 0.5$ mH
Filter capacitance	$C = 1000$ μF
LQR gains $K = [9.0097\ \ 3.4099\ \ -1.0096\ \ -10.8201\ \ -0.0036]$	
Switching frequency	$f = 1500$ Hz

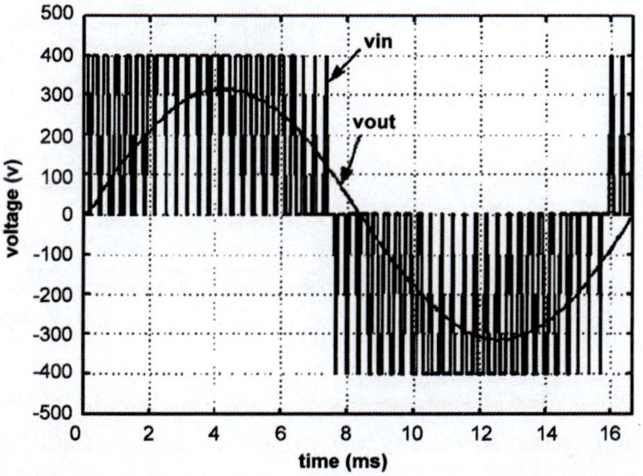

Fig. 5.57 Input and output voltage for a linear load

5.11 Model Predictive Control Method

The control techniques known collectively as MPC (Model Predictive Control) essentially consist of applying the first element of the control sequence obtained as the solution of an optimal control problem which is solved at each sampling time. Due to its ability to deal with multivariable systems and transport delays, and to handle constraints by explicit including them in the optimization problem [49], MPC strategies have become widely employed in industry. Stability requirements for predictive control laws have already been established when no uncertainties or disturbances are present [41]. However, predictive controllers may suffer from infeasibility problems

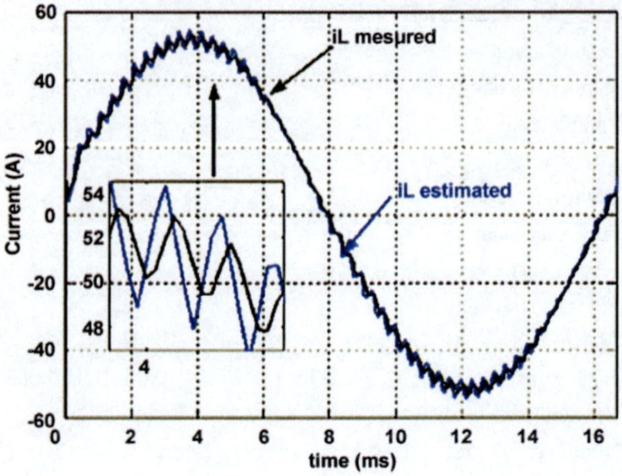

Fig. 5.58 Measured and estimated inductor current for a linear load

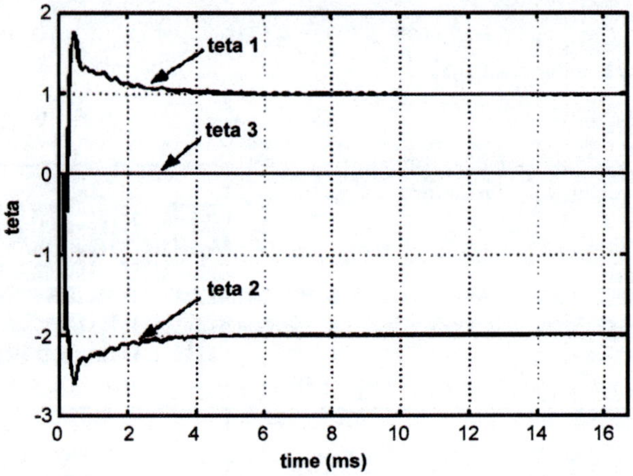

Fig. 5.59 The estimated parameters for a linear load

in the presence of disturbances, possibly leading to violation of system constraints and system instability, even if the controller stabilizes the system in the nominal case [12].

Among the possible approaches proposed to deal with this problem, one could cite min–max optimization [37, 51], and constraint restriction [12, 20]. Badgwell [4] points out that min–max MPC has an increased computational burden associated with the usual optimization problem solved by MPC at each sampling time. This does not occur with the restricted constraint formulation, as the nominal optimization problem is solved considering modified constraints, which can be obtained off-line. In this section, a robust predictive state regulator is designed for a nonlinear, sixth-order model of a three-degree-of-freedom (3DOF) helicopter subject to bounded disturbances and physical restrictions on its maneuvering space. Constraints are assumed to be convex polyhedral sets and the robustness is achieved by the use of the restricted constraints formulation presented in [12], which ensures fea-

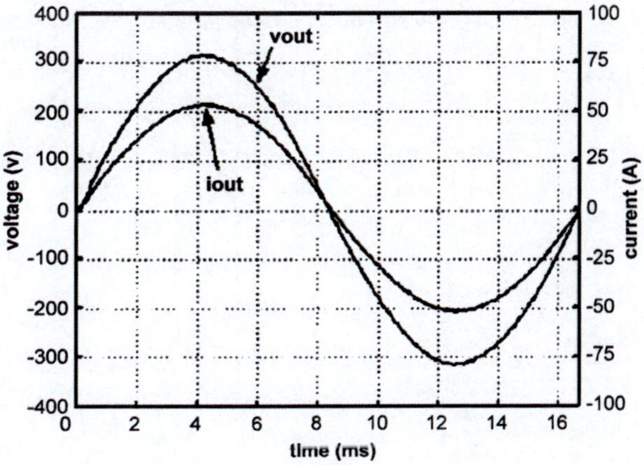

Fig. 5.60 Output voltage and current for a linear load

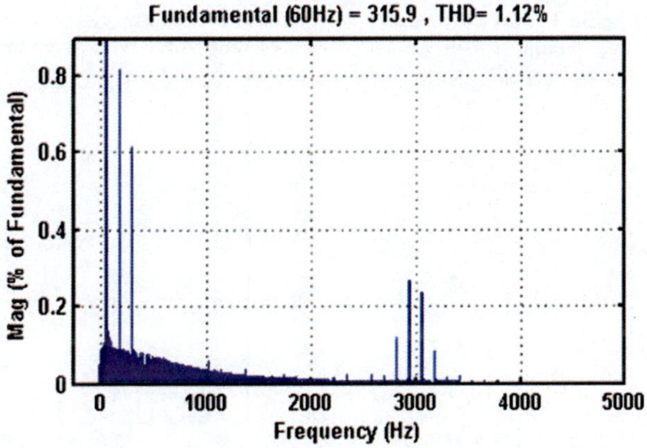

Fig. 5.61 Spectral analysis of the output voltage for a linear load

sibility and constraint fulfillment in spite of the existence of unknown but bounded disturbances. The computer routines used to calculate the modified restrictions were based on algorithms provided in [34] and employed some operations on polyhedra already implemented in the Multi-Parametric Toolbox (MPT) for MATLAB [36]. For comparison purposes, a nominal predictive control law is also considered. Simulations results are provided to illustrate that, in the presence of disturbances, while the robust predictive control law effectively guarantees that none of the system constraints is violated, the nominal predictive control law fails to do so.

5.11.1 Predictive Control Formulation

The robust predictive control formulation adopted herein was proposed in [12]. It concerns the regulation of time-invariant discrete-time linear systems subject to a disturbance input. The disturbance is assumed to be unknown but must belong to a compact set:

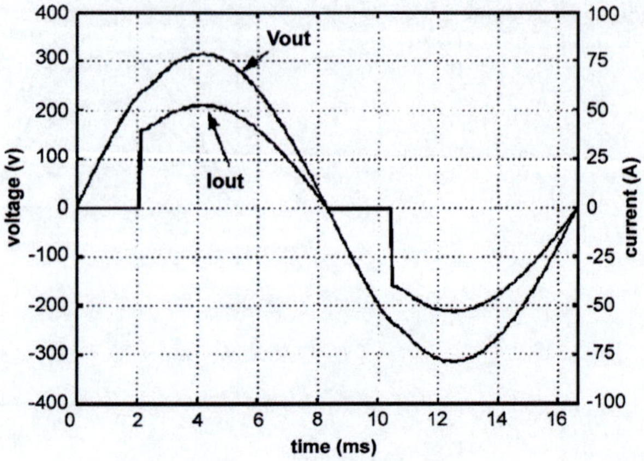

Fig. 5.62 Output voltage and current for a non-linear load

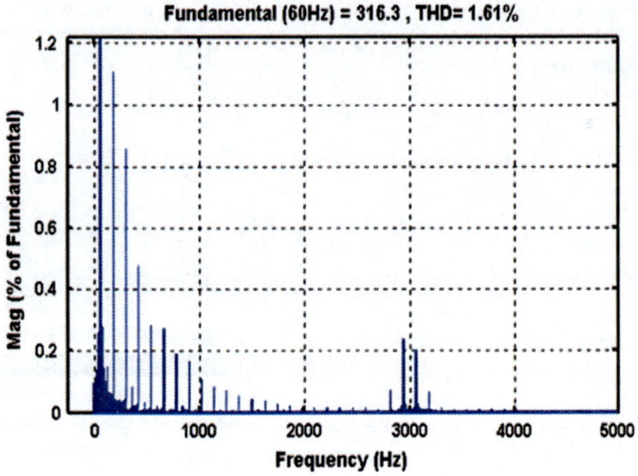

Fig. 5.63 Spectral analysis of the output voltage for a non-linear load

$$x(k+1) = Ax(k) + Bu(k) + Ew(k), \tag{5.147}$$

$$x(k) \in \mathcal{X} \subset \Re^n, \quad \forall k \geq 0, \tag{5.148}$$

$$u(k) \in \mathcal{U} \subset \Re^p, \quad \forall k \geq 0, \tag{5.149}$$

$$w(k) \in \mathcal{W} \subset \Re^m, \quad \forall k \geq 0. \tag{5.150}$$

It is also assumed that the pair (A, B) is stabilizable, $\mathcal{U}$ is compact and $\mathcal{X}$, $\mathcal{U}$, $\mathcal{W}$ contain the origin as an interior point. In this section, two predictive control laws were considered: NPC (nominal predictive control) and RPC (robust predictive control). It must be noted that the NPC algorithm does not take into account the effects of the disturbances affecting the system. On the other hand, the RPC algorithm performs a nominal optimization but modifies the original constraints to ensure their fulfillment in spite of the unknown disturbances [12]. These modifications involve the use of the following set operations. Let

$$\mathcal{A}, \mathcal{B} \subset \Re^n, \qquad \mathcal{F} \subset \Re^p, \qquad M \in \Re^{p \times n},$$

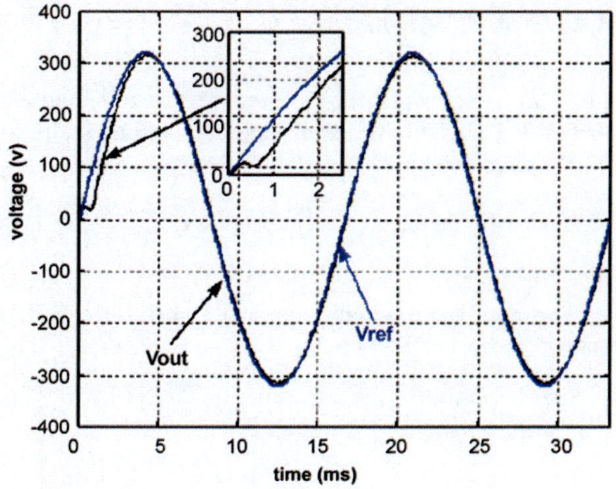

Fig. 5.64 Transient response of the output voltage for a linear load

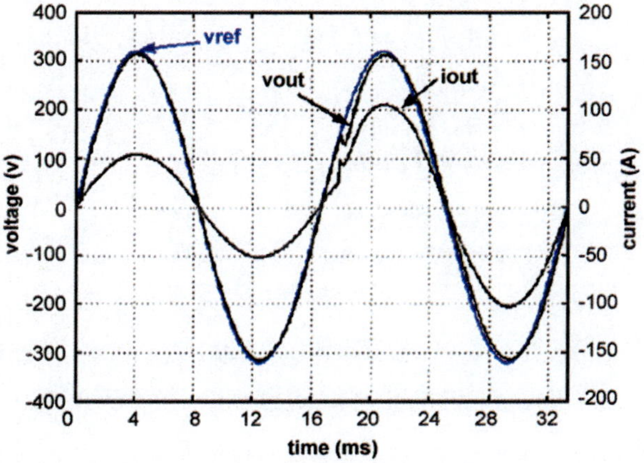

Fig. 5.65 Reference voltage, output voltage and current with linear load disturbance (from $R = 6\ \Omega$ to $R = 3\ \Omega$)

then

$$\mathcal{A}\,\mathcal{B} := \left\{a \in \Re^n \,|\, a + b \in \mathcal{A}, \forall b \in \mathcal{B}\right\},$$

$$\mathcal{A} \oplus \mathcal{B} := \left\{a + b \in \Re^n \,|\, a \in \mathcal{A}, b \in \mathcal{B}\right\},$$

$$\mathcal{LM}(M, \mathcal{A}) := \left\{Ma \in \Re^p \,|\, a \in \mathcal{A}\right\},$$

$$\mathcal{LM}^{-1}(M, \mathcal{F}) := \left\{a \in \Re^p \,|\, Ma \in \mathcal{F}\right\}.$$

Next, we summarize the NPC algorithm.

5.11.2 NPC Algorithm

Let $x(k+j|k)$ represent the predicted system state at instant $k+j$, computed at instant k, based on the actual state $x(k)$ and on the future control moves. Define the cost function

$$J\big(C[k]\big) = J\big(C(k)\big) = \sum_{j=0}^{N-1} c^t(k+j\,|\,k)\Psi c(k+j\,|\,k). \tag{5.151}$$

The nominal predictive control algorithm can be described by the following steps:

Step 1: Minimize the cost function (5.151), with $0 < \Psi = \Psi^t$ by considering the control sequence

$$C(k) = \big[c^t(k\,|\,k) \quad \cdots \quad c^t(k+N-1\,|\,k)\big]^t$$

subject to the constraints defined in Eqs. (5.147) to (5.150).

$$x(k+j+1\,|\,k) = Ax(k+j\,|\,k) + Bu(k+j\,|\,k), \qquad x(k\,|\,k) = x(k), \tag{5.152}$$

$$\begin{aligned} u(k+j\,|\,k) &= Kx(k+j\,|\,k) + c(k+j\,|\,k), \\ c(k+j\,|\,k) &= 0, \quad \forall j \geq N \end{aligned} \tag{5.153}$$

$$\begin{aligned} x(k+j\,|\,k) &\in \mathcal{X}_j, \quad 0 \leq j \leq N, \\ x(k+j\,|\,k) &\in \mathcal{U}_j, \quad 0 \leq j \leq N-1 \end{aligned} \tag{5.154}$$

$$\begin{aligned} \mathcal{X}_j &\in \mathcal{X}, \quad 0 \leq j \leq N-1, \qquad \mathcal{X}_N = \Im^h_\infty, \\ \mathcal{U}_j &\in \mathcal{U}, \quad 0 \leq j \leq N-1. \end{aligned} \tag{5.155}$$

Step 2: Let

$$C^*(k) = \big[c^{*t}(k|k) \quad \cdots \quad c^{*t}(k+N-1|k)\big]^t$$

be the optimal control sequence resulting from the optimization in Step 1. Apply

$$u(k) = Kx(k) + c^*(k|k)$$

to the plant.

Step 3: Set $k \to k+1$ and return to *Step 3*.

Remark 5.10 The set $\Im^h_\infty$ is the maximal positively invariant subset of

$$\mathcal{X}^h = \mathcal{X} \cap \mathcal{LM}^{-1}(K, \mathcal{U})$$

for the system under the nominal linear feedback, that is, the set of all states which satisfy state and control constraints (under nominal linear feedback) and for which the next state remains in such set. In the way, the set $\tilde{\Im}^h_\infty$ is the maximal robust positively invariant subset of

$$\mathcal{X}^h = \mathcal{X} \cap \mathcal{LM}^{-1}(K, \mathcal{U})$$

for the system under the nominal linear feedback, that is, the set of all states which satisfy state and control constraints (under nominal linear feedback) and for which the next state remains in such set, for all admissible disturbances. It can be arbitrarily select the gain matrix K defines a nominal linear state feedback $u(k) = Kx(k)$ as long as the resulting closed-loop system is stable. If K is taken as the unconstrained LQR gain minimizing the cost

$$\sum_{j=0}^{\infty} x^t(k)Qx(k) + u^t(k)Ru(k), \quad 0 \leq Q^t = Q,\ 0 < R^t = R \tag{5.156}$$

and selecting the weight matrix $\Psi = R + B^t\mathcal{P}B$ where $0 < \mathcal{P}^t = \mathcal{P}$ is the unique solution of the discrete ARE associated with the LQR problem, then it can be shown [12] that the minimization of (5.151) subject to (5.152)–(5.155) is equivalent to the minimization of (5.156) subject to the same constraints.

5.11.3 RPC Algorithm

The robust predictive control algorithm is identical to the NPC algorithm, except by the replacement of constraints (5.155) by

$$\begin{aligned} &\mathcal{X}_0 = \mathcal{X}, \qquad \mathcal{X}_N = \Im^h_\infty \approx \mathcal{R}_N, \qquad \mathcal{U}_0 = \mathcal{U}, \\ &\mathcal{X}_j = \mathcal{X} \approx \mathcal{R}_j, \quad 0 \leq j \leq N-1, \end{aligned} \tag{5.157}$$

$$\begin{aligned} &\mathcal{U}_j = \mathcal{U} \approx \mathcal{LM}(K, \mathcal{R}_j), \quad 0 \leq j \leq N-1, \\ &\mathcal{R}_j = \bigoplus_{m=0}^{j-1} \mathcal{LM}\big((A + BK)^m E, \mathcal{W}\big), \quad \forall j \geq 1. \end{aligned} \tag{5.158}$$

The main property of the RPC algorithm, see [12], is that, if the optimization problem has a solution for the initial state $x(0)$, then it will be feasible for all time, all state and control constraints will be fulfilled, the nonlinear predictive control law asymptotically approaches the nominal linear control law $u(k) = Kx(k)$, and the system state is asymptotically steered to a neighborhood of the origin $\mathcal{R}_\infty = \lim_{j\to\infty} \mathcal{R}_j$ for all admissible disturbances.

5.11.4 Implementation Details

If the constraints are defined by convex polyhedral sets, optimization problem (5.151)–(5.155) reduces to a quadratic programming format. To describe this format, we let the state and control constraints are defined by

$$\begin{aligned} &\mathcal{X}_j : S_{x[k+j]} x[k+j\,|\,k] \leq r_{x[k+j]}, \\ &\mathcal{U}_j : S_{u[k+j]} x[k+j\,|\,k] \leq r_{u[k+j]} \end{aligned}$$

and compute the following matrices

$$\mathcal{S} = \begin{bmatrix} \mathcal{S}_X \mathsf{H}_x(A+BK, B) \\ \mathcal{S}_U \mathsf{H}_u(A+BK, B) \end{bmatrix},$$

$$r = \begin{bmatrix} r_X \\ r_U \end{bmatrix} - \begin{bmatrix} \mathcal{S}_X \mathsf{F}_x(A+BK) \\ \mathcal{S}_U \mathsf{F}_u(K, A+BK) \end{bmatrix} x[k|k],$$

$$\mathcal{S}_X = \text{blockdiag}\left[\mathcal{S}_{x[k+1]} \quad \mathcal{S}_{x[k+2]} \quad \cdots \quad \mathcal{S}_{x[k+N]}\right],$$

$$\mathcal{S}_U = \text{blockdiag}\left[\mathcal{S}_{u[k]} \quad \mathcal{S}_{u[k+1]} \quad \cdots \quad \mathcal{S}_{u[k+N-1]}\right],$$

$$\mathsf{H}_x(\Phi, B) = \begin{bmatrix} B & 0 & \cdots & 0 \\ \Phi & B & \cdots & 0 \\ \vdots & \vdots & \ddots & \vdots \\ \Phi^{N-1}B & \Phi^{N-1}B & \cdots & B \end{bmatrix},$$

$$r_X = \begin{bmatrix} r_{x[k+1]} \\ r_{x[k+2]} \\ \vdots \\ r_{x[k+N]} \end{bmatrix}, \qquad r_U = \begin{bmatrix} r_{x[k]} \\ r_{x[k+1]} \\ \vdots \\ r_{x[k+N-1]} \end{bmatrix},$$

$$\mathsf{H}_u(K, \Phi, B) = \begin{bmatrix} I & \cdots & 0 & 0 \\ KB & \cdots & 0 & 0 \\ \vdots & \ddots & \vdots & \vdots \\ K\Phi^{N-3}B & \cdots & I & 0 \\ K\Phi^{N-2}B & \cdots & KB & I \end{bmatrix},$$

$$\mathsf{F}_x(\Phi) = \begin{bmatrix} \Phi \\ \Phi^2 \\ \vdots \\ \Phi^N \end{bmatrix}, \qquad \mathsf{F}_u(K, \Phi) = \begin{bmatrix} K \\ K\Phi \\ \vdots \\ K\Phi^{N-1} \end{bmatrix}.$$

Then the quadratic programming problem can be expressed as

$$\begin{aligned} C^*(k) &= \arg\min_{C[k]}\{C^t[k]\hat{\Psi}C[k]\} \\ \text{subject to} \quad SC[k] &\le r, \quad \hat{\Psi} = \text{diag}_N\{\Psi\}. \end{aligned} \tag{5.159}$$

5.12 LQGR Design

In this section, we review what is known as Linear Quadratic Gaussian theory or LQG theory for brevity. By including Gaussian white noise in the LQ paradigm linear optimal feedback systems based on output feedback rather than state feedback may be found.

Fig. 5.66 LQG feedback

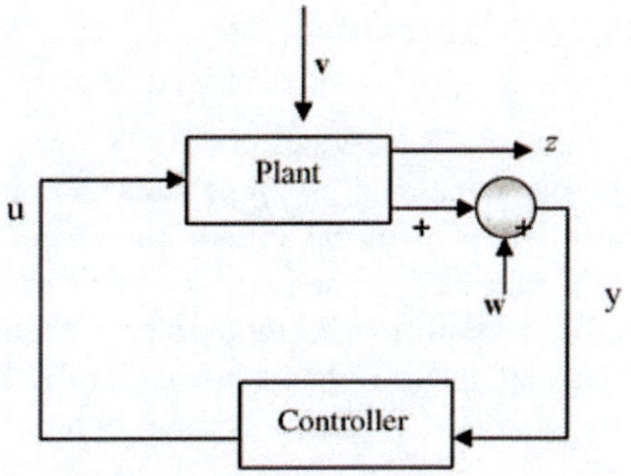

5.12.1 Introduction

In what follows, we consider the system

$$\begin{aligned} \dot{\mathbf{x}}(t) &= A\mathbf{x}(t) + B\mathbf{u}(t) + \Gamma\mathbf{v}(t), \\ \mathbf{z}(t) &= G\mathbf{x}(t) + D\mathbf{u}(t), \\ \mathbf{y}(t) &= C\mathbf{x}(t) + \mathbf{w}(t). \end{aligned} \tag{5.160}$$

The measured output $y(t)$ is available for feedback and $z(t)$ is the controlled output. The signals $\mathbf{v}$ and $\mathbf{w}$ are zero-mean Gaussian plant and measurement white noise processes with

$$\left.\begin{aligned} \mathbb{E}\mathbf{v}(t)\mathbf{v}^t(t) &= \mathsf{V}\delta(t-s) \\ \mathbb{E}\mathbf{v}(t)\mathbf{w}^t(t) &= 0 \\ \mathbb{E}\mathbf{w}(t)\mathbf{w}^t(t) &= \mathsf{W}\delta(t-s) \end{aligned}\right\} \quad \forall t, s \in \Re \tag{5.161}$$

where the power spectrum matrices $0 \le \mathsf{V}^t = \mathsf{V}$, $0 \le \mathsf{R}^t = \mathsf{W}$, are sometimes referred to as the intensity matrices of the two white noise processes, respectively.

We do not go into the theory of stochastic processes in general and that of white noise in particular, but refer to texts such as [5, 56]. The initial state $x(0)$ is assumed to be a random vector. The various assumptions define the state $\mathbf{x}(t)$, $t \in \Re$, and the controlled output $\mathbf{z}(t)$, $t \in \Re$, as random processes. As a result, also the quadratic error expression

$$\mathbf{z}^t(t)\mathsf{Q}\mathbf{z}(t) + \mathbf{u}^t(t)\mathsf{R}\mathbf{u}(t), \quad t \ge 0 \tag{5.162}$$

is a random process. The problem of controlling the system such that the integrated expected value

$$\int_0^T \mathbb{E}\big[\mathbf{z}^t(t)\mathsf{Q}\mathbf{z}(t) + \mathbf{u}^t(t)\mathsf{R}\mathbf{u}(t)\big]\,dt, \tag{5.163}$$

is minimal is the *stochastic linear regulator problem*. The time interval $[0; T]$ at this point is taken to be finite but eventually we consider the case that $T \to \infty$. At any time t the entire past measurement signal $y(s)$, $s \le t$, is assumed to be available for feedback. Figure 5.66 clarifies the situation.

5.12.2 Kalman Filter

A fundamental limitation of LQR is imposed by the need to measure the entire state. In many applications some states are not measurable, that is, there are no currently available sensors capable of measuring these states. In many other applications, the cost of including sensors for measuring the entire state is prohibitive or undesirable. Therefore, a methodology is needed for designing controllers when only partial state measurements are available. The Kalman filter is an optimal estimator of state, where optimal is defined in terms of minimizing the mean square estimation error. The Kalman filter estimates the state of a system given a set of known inputs and a set of measurements.

We learned before that the dynamical system

$$\dot{\hat{\mathbf{x}}}(t) = A\hat{\mathbf{x}}(t) + B\mathbf{u}(t) + L\big[\mathbf{y}(t) - C\hat{\mathbf{x}}(t)\big], \quad t \in \Re \tag{5.164}$$

can act as an observer for system (5.160) under the nominal conditions ($\mathbf{v}(t) \equiv 0$ and $\mathbf{w}(t) \equiv 0$) and hence reproduce the state $\mathbf{x}$ asymptotically, that is $\hat{\mathbf{x}} \xrightarrow{t\to\infty} \mathbf{x}$. The matrix L is the observer gain. We take advantage of this salient feature and connect the observer (5.164) to the noisy system (5.160). Differentiation of

$$e(t) = \hat{\mathbf{x}}(t) - \mathbf{x}(t)$$

leads to the error differential equation

$$\dot{e}(t) = (A - LC)Ae(t) - \Gamma\mathbf{v}(t) + L\mathbf{w}(t), \quad t \in \Re. \tag{5.165}$$

Owing to the two noise terms on the right-hand side the error now no longer converges to zero, even if the error system is stable. Suppose that the error system is stable. It is well known [58] that as $t \to \infty$, the *error covariance matrix*

$$\mathbb{E}\big[e^t(t)e(t)\big]$$

converges to a constant steady-state value Y that satisfies the linear Lyapunov matrix equation

$$(A - LC)\mathrm{Y} + \mathrm{Y}(A - LC)^t + \Gamma\mathrm{V}\Gamma^t + L\mathrm{Y}L^t = 0. \tag{5.166}$$

It is an easy task following arguments from Lyapunov theory that as a function of the gain matrix L the steady-state error covariance matrix Y is *minimal* if L is chosen as

$$L = \mathrm{Y}C^t\mathrm{W}^{-1}. \tag{5.167}$$

It should be noted that *minimal* means here that if $\hat{\mathrm{Y}}$ is the steady-state error covariance matrix corresponding to any other observer gain $\hat{L}$ then $\hat{\mathrm{Y}} \geq \mathrm{Y}$. This inequality is to be taken in the sense that $\hat{\mathrm{Y}} - \mathrm{Y} \geq 0$.

A consequence of this result is that the gain (5.167) minimizes the steady-state mean square state reconstruction error

$$\lim_{t\to\infty} \mathbb{E}\big[e^t(t)e(t)\big].$$

Actually, the gain minimizes the weighted mean square construction error

$$\lim_{t\to\infty} \mathbb{E}\big[e^t(t)\mathsf{W}_e e(t)\big]$$

for any nonnegative-definite weighting matrix W_e.

Substitution of the optimal gain matrix (5.167) into the Lyapunov equation (5.166) yields

$$A\mathsf{Y} + \mathsf{Y}A^t + \Gamma\mathsf{V}\Gamma^t - \mathsf{Y}C^t\mathsf{W}^{-1}C\mathsf{Y} = 0. \tag{5.168}$$

This is another algebraic matrix Riccati equation or the Dual Riccati. The observer (5.160) with the gain chosen as in (5.167) and the covariance matrix Y the nonnegative-definite solution of the Riccati equation (5.168) is the famous Kalman filter [31].

Remark 5.11 A significant result is system theory is that the optimal regulator and the Kalman filter are dual in the following sense. Given the regulator problem of Chap. 5, replace A with A^t, B with C^t, D with Γ^t, Q with V, and R with W. Then the regulator Riccati equation (5.16) becomes the observer Riccati equation (5.168), its solution X becomes Y, the state feedback gain K is the transpose of the observer gain L, and the closed-loop system matrix $A - BK$ is the transpose of the error system matrix $A - LC$. By matching substitutions the observer problem may be transposed to a regulator problem.

Next we review several properties of the Kalman filter.

1. Assume that the system

$$\begin{aligned} \dot{\mathbf{x}}(t) &= A\mathbf{x}(t) + \Gamma\mathbf{v}(t), \\ \mathbf{y}(t) &= C\mathbf{x}(t) \end{aligned} \tag{5.169}$$

   is stabilizable and detectable and the noise intensity matrices V and W are positive-definite. By duality to the regulator, the algebraic Riccati equation (5.168) has a unique nonnegative-definite symmetric solution Y. If the system (5.169) is controllable rather than just stabilizable then Y is positive-definite. It is important to note that if the system (5.169) is not detectable then no observer with a stable error system exists. If the system is not stabilizable (with d as input) then there exist observers that are not stable but are immune to the state noise d. Hence, stability of the error system is not guaranteed. Matrix W needs to be positive-definite to prevent the Kalman filter from having infinite gain. If V is not positive-definite, then there may be unstable modes that are not excited by the state noise and, hence, are not stabilized in the error system.
2. The minimal value of the steady-state weighted mean-square state reconstruction error

$$\lim_{t\to\infty} \mathbb{E}\big[e^t(t)\mathsf{W}_e e(t)\big] = \mathrm{Tr}[\mathsf{Y}\mathsf{W}_e].$$

3. The minimal value of the mean square reconstruction error is achieved by the observer gain matrix

$$L = \mathsf{Y}C^t\mathsf{W}^{-1}.$$

4. The error system

$$\dot{e}(t) = (A - LC)Ae(t), \quad t \in \Re \tag{5.170}$$

is stable, that is, all the eigenvalues of the matrix $A - LC$ have strictly negative real parts. As a consequence also the observer

$$\dot{\hat{\mathbf{x}}}(t) = A\hat{\mathbf{x}}(t) + B\mathbf{u}(t) + L[\mathbf{y}(t) - C\hat{\mathbf{x}}(t)], \quad t \in \Re \tag{5.171}$$

is stable.
5. Note that the implementation of Kalman filter requires a system noise spectral density matrix, a measurement noise spectral density matrix, an initial condition on the state estimate, and an initial estimation error covariance matrix.

5.12.3 Solution of the Stochastic Linear Regulator Problem

We consider the stochastic linear regulator problem that consists of minimizing

$$\int_0^T \mathbb{E}[\mathbf{z}^t(t)\mathsf{Q}\mathbf{z}(t) + \mathbf{u}^t(t)\mathsf{R}\mathbf{u}(t)]\,dt, \tag{5.172}$$

for the system

$$\begin{aligned} \dot{\mathbf{x}}(t) &= A\mathbf{x}(t) + B\mathbf{u}(t) + \Gamma\mathbf{v}(t), \\ \mathbf{z}(t) &= G\mathbf{x}(t), \\ \mathbf{y}(t) &= C\mathbf{x}(t) + \mathbf{w}(t) \end{aligned} \tag{5.173}$$

and discuss several versions:

1. *Noise-free state*: When the noise signal $\mathbf{v}(t)$ is absent and the state $\mathbf{x}(t)$ may be directly measurable, then for $T \to \infty$ the performance index is minimized by the control law

$$\mathbf{u}(t) = -K\mathbf{x}(t) = -\mathsf{R}^{-1}B^t\mathsf{X}\mathbf{x}(t) \tag{5.174}$$

where the symmetric $n \times n$ matrix X is the nonnegative-definite solution of the algebraic matrix Riccati equation (ARE)

$$\mathsf{X}A + A^t\mathsf{X} + G^t\mathsf{Q}G - \mathsf{X}B\mathsf{R}^{-1}B^t\mathsf{X} = 0. \tag{5.175}$$

2. *State feedback*: If white noise signal $\mathbf{v}(t)$ is present, then obviously the state and input cannot be driven to 0, and the integrated generalized square error (5.172) does not converge to a finite number as $T \to \infty$. It is proved [6, 7, 55] that the state feedback law (5.174) minimizes the rate at which (5.172) approaches ∞, that is, it minimizes

$$\lim_{T\to\infty} \int_0^T \mathbb{E}[\mathbf{z}^t(t)\mathsf{Q}\mathbf{z}(t) + \mathbf{u}^t(t)\mathsf{R}\mathbf{u}(t)]\,dt. \tag{5.176}$$

This limit equals the steady-state mean square error index steady-state mean square error

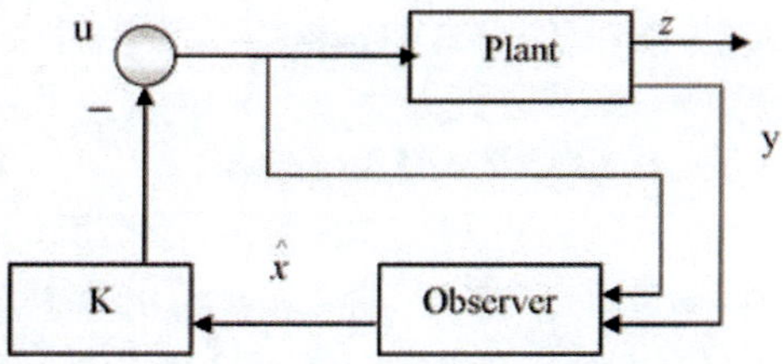

Fig. 5.67 Estimator-based feedback control

$$\lim_{t\to\infty} \mathbb{E}\big[\mathbf{z}^t(t)\mathbf{Q}\mathbf{z}(t) + \mathbf{u}^t(t)\mathbf{R}\mathbf{u}(t)\big]. \tag{5.177}$$

Hence, the state feedback law minimizes the steady-state mean square error.

3. *Output feedback*: The interesting situation is that the state cannot be accessed for measurement. The state may be optimally estimated, however, with the help of the Kalman filter. Then the solution of the stochastic linear regulator problem with output feedback (rather than state feedback) is to replace the state $\mathbf{x}(t)$ in the state feedback law (5.174) with the estimated state $\hat{\mathbf{x}}(t)$. Thus, the optimal controller is given by

$$\begin{aligned} \dot{\hat{\mathbf{x}}}(t) &= A\hat{\mathbf{x}}(t) + B\mathbf{u}(t) + L\big[\mathbf{y}(t) - C\hat{\mathbf{x}}(t)\big], \quad t \in \Re, \\ \mathbf{u}(t) &= -K\hat{\mathbf{x}}(t). \end{aligned} \tag{5.178}$$

The controller minimizes the steady-state mean square error (5.177) under output feedback. The feedback gain K and the observer gain L follow from the foregoing analysis, respectively. Figure 5.67 shows the arrangement of the closed-loop system.

Using the estimated state as if it were the actual state is known as *certainty equivalence*. It splits state estimation and control input selection thereby leading to the idea frequently referred to as the *separation principle*. It follows that the closed-loop system that results from interconnecting the plant (5.173) with the feedback controller (5.178) is stable—under the stabilizability-detectability assumptions. To clarify this, we substitute of $\mathbf{u}(t) = -K\hat{\mathbf{x}}(t)$ into (5.173) yields with the further substitution $\hat{\mathbf{x}}(t) = \mathbf{x}(t) - e(t)$

$$\dot{\mathbf{x}}(t) = (A - BK)A\mathbf{x}(t) - BKe(t) + \Gamma L\mathbf{v}(t) \tag{5.179}$$

which together with (5.165) yields

$$\begin{bmatrix} \dot{\mathbf{x}}(t) \\ \dot{e}(t) \end{bmatrix} = \begin{bmatrix} A - BK & -BK \\ 0 & A - LC \end{bmatrix} \begin{bmatrix} \mathbf{x}(t) \\ e(t) \end{bmatrix} + \begin{bmatrix} \Gamma\mathbf{v}(t) \\ -\Gamma\mathbf{v}(t) + L\mathbf{w}(t) \end{bmatrix}. \tag{5.180}$$

It is a straightforward task to show that the eigenvalues of this system are the eigenvalues of the closed-loop system. Simple inspection shows that these eigenvalues consist of the eigenvalues of $A - BK$ (the regulator poles) together with the eigenvalues of $A - LC$ (the observer poles). If the plant (5.173) has order n then the feedback controller also has order n). Hence, there are $2n$ closed-loop poles.

5.13 MATLAB Hints

5.13.1 LQR in MATLAB

The command **[K, P, E] = lqr(A, B, Q, R, N)** solves the ARE (5.62) and computes the optimal state-feedback gain matrix K given in (5.60) that minimizes the LQR criteria (5.58) for the continuous-time system (5.55). It also returns the poles E of the closed-loop system (5.64).

5.14 Questions and MATLAB Problems

5.14.1 Questions

Q5.1 Suppose that $\mathcal{P}_1$ and $\mathcal{P}_2$ are two symmetric positive-definite solutions to the ARE (5.62). Show that $\mathcal{P}_1$ and $\mathcal{P}_2$ satisfy $(A - B\mathcal{R}^{-1}B^t\mathcal{P}_2)(\mathcal{P}_1 - \mathcal{P}_2) + (\mathcal{P}_1 - \mathcal{P}_2)(A - B\mathcal{R}^{-1}B^t\mathcal{P}_2) = 0$ and argue that $\mathcal{P}_1 = \mathcal{P}_2$.

Q5.2 Derive a solution to the optimal control problem involving a performance index $J_\alpha = \int_0^\infty e^{2\alpha t}[x^t(t)\mathcal{Q}z(t) + u^t(t)\mathcal{R}u(t)]\,dt$, and show that the associated closed-loop eigenvalues have real parts less than $-\alpha$.

Q5.3 Let $(sI - A)^{-1}b = [P_{n-l}(s) \quad \cdots \quad P_0(s)]'/a(s)$. Show that the common roots of the $n + 1$ polynomials $\{P_{n-l}(s), \dots, P_0(s), a(s)\}$ specify exactly the uncontrollable natural frequencies of $\{A, b\}$.

Q5.4 A linear time-invariant system is described by

$$A = \begin{bmatrix} 1 & 0 & 0 \\ 0 & 2 & 0 \\ 0 & 0 & -1 \end{bmatrix}, \qquad B = \begin{bmatrix} 1 \\ 1 \\ 0 \end{bmatrix}, \qquad C^t = \begin{bmatrix} 1 \\ 0 \\ 1 \end{bmatrix}.$$

Evaluate the eigenvalues of the system and examine their structural properties. Compute the controllability and observability matrices. Apply a linear state-feedback with auxiliary input and discuss the effect of feedback on the controllability, observability and closed-loop eigenvalues. Repeat the foregoing effort for the case of constant output feedback with auxiliary input.

Q5.5 Consider a linear time-invariant system

$$\dot{x} = Ax + Bu, \qquad y = Cx.$$

The problem of interest is to choose $u = -Kx + v$ such that $y \to y_d$ as $t \to \infty$ where y_d is a constant set point. Give a detailed analysis of the problem and establish the required conditions.

Q5.6 Given the system

$$\dot{x} = Ax + Bu + Ew, \qquad y = Cx$$

where w is a constant disturbance. The objective is to regulate the system output (to the origin) in spite of the disturbance w using the integral error feedback

$$\dot{\eta} = y(t), \qquad u(t) = -K_1 x - K_2 \eta.$$

Establish the conditions to fulfill the objective.

5.14.2 MATLAB Problems

P5.1 For the linearized model of the Reverse osmosis (RO) plant discussed in Sect. 5.4, design and evaluate an observer-based feedback controller by selecting the observer eigenvalues distinctly different from the controller eigenvalues. Plot the state responses for different cases and comment on the results.

P5.2 For the linearized model of the Reverse osmosis (RO) plant discussed in Sect. 5.4, design and evaluate an optimal linear quadratic regulator with equal weighting for the state and input. Plot the output responses to unit step input and compare on the same graph the open-loop and the closed-loop responses.

P5.3 A linearized model of a vertical takeoff and landing (VTOL) aircraft [Dorf] has the matrices

$$A = \begin{bmatrix} -0.0389 & 0.0271 & 0.0188 & -0.4555 \\ 0.0482 & -1.0100 & 0.0019 & -4.0208 \\ 0.1024 & 0.3681 & -0.7070 & 1.4200 \\ 0 & 0 & 1 & 0 \end{bmatrix},$$

$$B = \begin{bmatrix} 0.4422 & 0.1291 \\ 3.5446 & -7.5922 \\ -6.0214 & 4.4900 \\ 0 & 0 \end{bmatrix}, \qquad C^t = \begin{bmatrix} 1 & 0 \\ 0 & 0 \\ 0 & 0 \\ 0 & 1 \end{bmatrix}.$$

Evaluate the structural properties of the system. Design stabilizing state-feedback, observer-based feedback and LQR controllers and compare among the three cases.

P5.4 Consider the turbo-generator system treated in Example 5.5. Design stabilizing observer-based feedback controller and plot the input–output trajectories. Compare the results of both design cases.

P5.5 A helicopter is a twin rotor aircraft that is lifted and propelled by one or more horizontal rotors, each rotor consisting of two or more rotor blades. Helicopters are classified as rotorcraft or rotary-wing aircraft to distinguish them from fixed-wing aircraft because the helicopter achieves lift with the rotor blades which rotate around a mast. Hover is the operating state in which the lifting rotor has no velocity relative to the air, either vertical or horizontal. Equations of motion of the helicopter during hovering conditions are obtained using the momentum theory which applies the basic theory of fluid mechanics, conservation of mass, momentum and energy. General vertical flight involves axial flow with respect to the rotor. Vertical flight implies axial symmetry of the rotor and hence that the velocities and loads on the rotor are independent of the azimuth position. Axial symmetry greatly simplifies the dynamics and aerodynamics of the helicopter rotor. The following eight-order linear system models the small-perturbation rigid body motion of a helicopter about the hover condition [14]:

$$\dot{x} = Ax + Bu,$$
$$y = Cx$$

where

$$x := \begin{matrix} \text{Pitch attitude (rad)} \\ \text{Roll attitude (rad)} \\ \text{Body roll rate (rad s}^{-1}\text{)} \\ \text{Body pitch rate (rad s}^{-1}\text{)} \\ \text{Body yaw rate (rad s}^{-1}\text{)} \\ \text{Forward velocity (ft s}^{-1}\text{)} \\ \text{Lateral velocity (ft s}^{-1}\text{)} \\ \text{Normal velocity (ft s}^{-1}\text{)} \end{matrix} = \begin{bmatrix} \theta \\ \phi \\ p \\ q \\ r \\ u \\ v \\ w \end{bmatrix},$$

$$A = \begin{bmatrix} A_1 & A_2 \end{bmatrix},$$

$$A_1 = \begin{bmatrix} 0 & 0 & 0 & 0.9986 \\ 0 & 0 & 1.0000 & -0.0032 \\ 0 & 0 & -11.5705 & -2.5446 \\ 0 & 0 & 0.4394 & -1.9982 \\ 0 & 0 & -2.0409 & -0.4590 \\ -32.1036 & 0 & -0.5034 & 2.2970 \\ 0.1022 & 32.0578 & -2.3470 & -0.5036 \\ -1.9110 & 1.7138 & -0.0040 & -0.0574 \end{bmatrix},$$

$$A_2 = \begin{bmatrix} 0.0534 & 0 & 0 & 0 \\ 0.0595 & 0 & 0 & 0 \\ -0.0636 & 0.1068 & -0.0949 & 0.0071 \\ 0 & 0.0167 & 0.0185 & -0.0012 \\ -0.7350 & 0.0193 & -0.0046 & 0.0021 \\ 0 & -0.0212 & -0.0212 & 0.0158 \\ 0.8349 & 0.0212 & -0.0379 & 0.0004 \\ 0 & 0.0140 & -0.0009 & -0.2905 \end{bmatrix},$$

$$B = \begin{bmatrix} 0 & 0 & 0 & 0 \\ 0000 & & & \\ 0.1243 & 0.0828 & -2.7525 & -0.0179 \\ -0.0364 & 0.4751 & 0.0143 & 0 \\ 0.3045 & 0.0150 & -0.4965 & -0.2067 \\ 0.2877 & -0.5445 & -0.0164 & 0 \\ -0.0191 & 0.0164 & -0.5445 & 0.2348 \\ -4.8206 & -0.0004 & 0 & 0 \end{bmatrix},$$

with inputs

$$u := \begin{bmatrix} \theta_{od} \\ \theta_{ls} \\ \theta_{lc} \\ \theta_{ot} \end{bmatrix} = \begin{matrix} \text{Main rotor collective (deg)} \\ \text{Longitudinal cyclic (deg)} \\ \text{Lateral cyclic (deg)} \\ \text{Tail rotor cyclic (deg)} \end{matrix} \tag{5.181}$$

$$C = \begin{bmatrix} C_1 & C_2 \end{bmatrix},$$

$$C_1 = \begin{bmatrix} 0 & 0 & 0 & 0 \\ 1 & 0 & 0 & 0 \\ 0 & 1 & 0 & 0 \\ 0 & 0 & 0 & 0.535 \\ 0 & 0 & 1 & 0 \\ 0 & 0 & 0 & 1 \end{bmatrix}, \qquad C_2 = \begin{bmatrix} 0 & 0.0595 & 0.0533 & -0.9968 \\ 0 & 0 & 0 & 0 \\ 0 & 0 & 0 & 0 \\ 1 & 0 & 0 & 0 \\ 0 & 0 & 0 & 0 \\ 0 & 0 & 0 & 0 \end{bmatrix}$$

and outputs

$$y = \begin{bmatrix} \dot{h} \\ \theta \\ \phi \\ \dot{\psi} \\ q \\ p \end{bmatrix} = \begin{matrix} \text{Heave velocity (ft s}^{-1}\text{)} \\ \text{Pitch attituded (rad)} \\ \text{Roll attitude (rad)} \\ \text{Heading rate (ft s}^{-1}\text{)} \\ \text{Body pitch rate (rad s}^{-1}\text{)} \\ \text{Body roll rate (rad s}^{-1}\text{)} \end{matrix} \tag{5.182}$$

Develop state-feedback controllers based pole assignment and optimal control and compare their closed-loop behavior. Comment on the result.

5.15 Notes and References

The analysis presented in this chapter made extensive use of the standard textbooks [2, 6, 7, 10, 16, 39, 47, 52, 54, 57].

References

1. Anderson, W.: Controlling Electrohydraulic Systems. Dekker, New York (1988)
2. Anderson, B.O.D., Moore, J.B.: Linear Optimal Control—Linear Quadratic Methods. Prentice-Hall, Englewood Cliffs (1990)
3. Astrom, K.J., Wittenmark, V.E.: Adaptive Control, 2nd edn. Prentice-Hall, New York (1995)
4. Badgwell, T.A.: Robust model predictive control. Int. J. Control **68**, 797–818 (1997)
5. Bagchi, A.: Optimal Control of Stochastic Systems. Prentice Hall International, Englewood Cliffs (1993)
6. Brockett, R.W.: Finite Dimensional Linear Systems. John Wiley and Sons, New York (1970)
7. Burl, J.B.: Linear Optimal Control, 3rd edn. Prentice-Hall, New York (1998)
8. Buso, S., Fasolo, S., Mattavelli, P.: Uninterruptible power supply multiloop control employing digital predictive voltage and current regulators. In: Proc. IEEE APEC, pp. 907–913 (2001)
9. Byeongil, K., Chang, Y., Lee, M.H.: System identification and 6-DOF controller design of unmanned model helicopter. JSME Int. J. Ser. C **49**(4), 1048–1057 (2006)

10. Callier, F.M., Desoer, C.A.: Multivariable Feedback Systems. Springer-Verlag, New York (1982)
11. Chen, C.-T.: Introduction to Linear Systems Theory. Holt, Rinehart and Winston, New York (1970)
12. Chisci, L., Rossiter, J.A., Zappa, C.: Systems with persistent disturbances: Predictive control with restricted constraints. Automatica **37**, 1019–1028 (2001)
13. Cho, J., Lee, S., Mok, H., Choe, G.: Modified deadbleat controller for UPS with 3-phase PWM inverter. In: Conf. Rec. IEEE-IAS Annu. Meeting, pp. 2208–2215 (1999)
14. Edwards, C., Spurgeon, S.K.: Sliding mode stabilization of uncertain systems using only output information. Int. J. Control **62**(5), 1129–1144 (1995)
15. Fredriksson, J., Egardt, B.: Backstepping control with local LQ performance applied to a turbocharged diesel engine. In: Proc. 40th IEEE Conference on Decision and Control, vol. 1, pp. 111–116 (2001)
16. Freudenberg, J.S., Looze, D.P.: Frequency Domain Properties of Scalar and Multivariable Feedback Systems. Springer-Verlag, Berlin (1988)
17. Friedrich, I., Liu, C.S., Oehlerking, D.: Coordinated EGR-rate model-based controls of turbocharged diesel engines via an intake throttle and an EGR valve. In: IEEE Conference on Vehicle Power and Propulsion, VPPC '09, 7–10 Sept. 2009, pp. 340–347
18. Fukushima, N., Arslan, M.S., Hagiwara, I.: An optimal control method based on the energy flow equation. IEEE Trans. Control Syst. Technol. **17**(4), 866–875 (2009)
19. Ghazy, M.A.: Variable Structure Control for Electrohydraulic Position Servo System. In: Proc. Industrial Electronics Conference, IECON, vol. 1, pp. 2194–2198 (2001)
20. Gossner, J.R., Kouvaritakis, B., Rossiter, J.A.: Stable generalized predictive control with constraints and bounded disturbances. Automatica **33**, 551–568 (1997)
21. Haiyan, W.: Control oriented dynamic modeling of a turbocharged diesel engine. In: Sixth Int. Conference on Intelligent Systems Design and Applications—ISDA '06, vol. 2, 16–18 Oct. 2006, pp. 142–145
22. Haneyoshi, T., Kawamura, A.: Waveform compensation of PWM inverter with cyclic fluctuating loads. IEEE Trans. Industrial Application **24**(4), 582–589 (1988)
23. IEEE recommended practice for emergency and standby power systems for industrial and commercial applications. IEEE-446-1995, 1995
24. Jaho Seo, J., Venugopala, R., Kenne, J.-P.: Feedback linearization based control of a rotational hydraulic drive. Control Eng. Pract. **15**, 2007 pp.
25. Jensen, U.B., Enjeti, P.N., Blaabjerg, F.: A new space vector based control method for UPS systems powering a nonlinear performance programmable AC power source with low harmonic distortion using DSP-based repetitive control technique. IEEE Trans. Power Electron. **12**, 715–725 (1997)
26. Jouanne, A.V., Enjeti, P.N., Lucas, D.J.: DSP control of high power UPS systems feeding nonlinear loads. IEEE Trans. Industrial Electronics **43**, 121–125 (1996)
27. Jovanovic, M.: Nonlinear control of an electrohydraulic velocity servosystem. Proc. Am. Control Conf. **1**, 588–593 (2002)
28. Jung, M., Glover, K., Christen, U.: Comparison of uncertainty parameterisations for robust control of turbocharged diesel engines. Control Eng. Pract. **13**(1), 15–25 (2005)
29. Kaddissi, C., Kenne, J.-P., Saad, M.: Identification and real-time control of an electrohydraulic servo system based on nonlinear backstepping. IEEE/ASME Trans. Mechatron. **12**, 12–22 (2007)
30. Kaliath, T.: Linear Systems. Prentice-Hall, New York (1980)
31. Kalman, R.E., Bucy, R.S.: New results in linear filtering and prediction theory. J. Basic Engineering, Trans. ASME Series D **83**, 95–108 (1961)
32. Karam, S., Mahdi, J.K.: Application of adaptive LQR with repetitive control for UPS systems. In: Proc. the 2003 IEEE Industry Applications Conference, pp. 1124–1129
33. Kawamura, A., Yokoyama, T.: Comparison of five control methods for digitally feedback controlled PWM inverters. In: Proc. the 1991 EPE Conference, pp. 035–040
34. Kerrigan, E.C.: Robust constraint satisfaction: Invariant sets and predictive control, PhD thesis, St. John's College, Cambridge, UK (2000)

35. Kim, N., Cha, S., Peng, H.: Optimal control of hybrid electric vehicles based on Pontryagin's minimum principle. IEEE Trans. Control Syst. Technol. **18**, 1–12 (2010)
36. Kvasnica, M., Grieder, P., Baotić, M.: Multi-parametric toolbox (MPT), http://control.ee.ethz.ch/~mpt/ (2004)
37. Lee, J.H., Yu, Z.: Worst case formulations of model predictive control for systems with bounded parameters. Automatica **33**, 763–781 (1997)
38. Liao, H.H., Roelle, M.J., Chen, J.S., Park, S., Gerdes, J.C.: Implementation and analysis of a repetitive controller for an electro-hydraulic engine valve system. IEEE Trans. Control Syst. Technol. **18**, 1–12 (2010)
39. Maciejowski, J.M.: Multivariable Feedback Design. Addison-Wesley, England (1989)
40. Mattavelli, P.: Synchronous-frame harmonic control for high performance power supplies. IEEE Trans. Industrial Application **37**, 864–872 (2001)
41. Mayne, D.Q., Rawlings, J.B., Rao, C.V., Scokaert, P.O.M.: Constrained model predictive control: Stability and optimality. Automatica **36**(6), 789–814 (2000)
42. Misgeld, B.J.E., Werner, J., Hexamer, M.: Robust and self-tuning blood flow control during extracorporeal circulation in the presence of system parameter uncertainties. Med. Biol. Eng. Comput. **43**(5), 589–598 (2005)
43. Montagner, V.F., Carati, E.G.: An adaptive linear quadratic regulator with repetitive controller applied to uninterruptible power supplies. In: Proc. the 2000 IEEE Industry Applications Conference, pp. 2231–2236
44. Ogata, K.: Discrete-Time Control Systems. Prentice-Hall, New York (1987)
45. Ogata, K.: MATLAB for Control Engineers. Prentice-Hall, New York (2008)
46. Pipeleers, G., Demeulenaere, B., Al-Bender, F., Schutter, J., Swevers, J.: Optimal performance tradeoffs in repetitive control: Experimental validation on an active air bearing setup. IEEE Trans. Control Syst. Technol. **17**(4), 970–979 (2009)
47. Postlethwaite, I., MacFarlane, A.G.J.: A Complex Variable Approach to the Analysis of Linear Multivariable Feedback Systems. Springer-Verlag, Berlin (1979)
48. Qi, Q., Deng, S.: Multivariable control-oriented modeling of a direct expansion (DX) air conditioning (A/C) system. Int. J. Refrig. **31**(5), 841–849 (2008)
49. Rossiter, J.A.: Model-Based Predictive Control: A Practical Approach. CRC Press, Boca Raton (2003)
50. Ryan, M.J., Brunsicle, W.E., Lorenz, R.D.: Control topology option for a single-phase UPS inverters. IEEE Trans. Industrial Application **33**(4), 493–501 (1997)
51. Scokaert, P.O.M., Mayne, D.Q.: Min-max feedback model predictive control for constrained linear systems. IEEE Trans. Autom. Control **43**, 1136–1142 (1998)
52. Skogestad, S., Postlethwaite, I.: Multivariable Feedback Control: Analysis and Design. John Wiley & Sons, Chichester, England (1996)
53. Tzou, Y.Y., Ou, R.S., Jung, S.L., Chang, M.Y.: High performance programmable AC power source with low harmonic distortion using DSP-based repetitive control technique. IEEE Trans. Power Electron. **12**, 715–725 (1997)
54. Vardulakis, A.I.G.: Linear Multivariable Control—Algebraic Analysis and Synthesis Methods. John Wiley & Sons, Chichester, England (1991)
55. Welch, G., Bishop, G.: An Introduction to the Kalman Filter. University of North Carolina at Chapel Hill, Chapel Hill, pp. 2759–3175 (2003)
56. Wong, E.: Introduction to Random Processes. Springer-Verlag, New York (1983)
57. Wonham, W.M.: Linear Multivariable Control—A Geometric Approach. Springer-Verlag, Berlin (1974)
58. Yano, K., Higashikawa, S., Terashima, K.: Motion control of liquid container considering an inclined transfer path. Control Eng. Pract. **10**, 465–472 (2002)
59. Yuan, Q.H., Li, P.Y.: Robust optimal design of unstable valves. IEEE Trans. Control Syst. Technol. **15**(6), 1065–1074 (2007)

Chapter 6
Applications II

6.1 Introduction

Feedback control has played a vital role in the advance of engineering and science. In addition to its extreme importance in space-vehicle systems, missile-guidance systems, robotic systems, and the like, automatic control has become an important and integral part of modern manufacturing and industrial processes. For example, automatic control is essential in the numerical control of machine tools in the manufacturing industries, in the design of autopilot systems in the aerospace industries, and in the design of cars and trucks in the automobile industries. It is also essential in such industrial operations as controlling pressure, temperature, humidity, viscosity, and flow in the process industries. Since advances in the theory and practice of automatic control provide the means for attaining optimal performance of dynamic systems, improving productivity, relieving the drudgery of many routine repetitive manual operations, and more, most engineers and scientists must now have a good understanding of this field, see [15, 16, 29] for different technical views.

6.2 Control of Shaping Process of Automobile Belt

To maintain stable tension and uniform distribution for winding string during the winding shaping process of automobile belt, a computer control system is proposed. In this system, the string tension can be measured by a tension sensor and regulated by a magnetic powder brake. Simultaneously, the velocity of the shaping model shaft that is driven by a *DC* motor can be measured by an opto-coder. For realizing optimal performance, the observer-based *LQR* controller is applied to this system. The feedback is determined by minimizing the cost function on the *LQR* rule, while a full order or a reduced-order observer is used to estimate system states besides be measurable states: shaft velocity and the string tension. Simulation results are given for these control strategies, and the disturbance rejection ability is examined.

The computer control system for the winding shaping process is shown in Fig. 6.1. When the system is started, the shaft of the shaping model turns. The

M.S. Mahmoud, Y. Xia, *Applied Control Systems Design*,
DOI 10.1007/978-1-4471-2879-3_6,

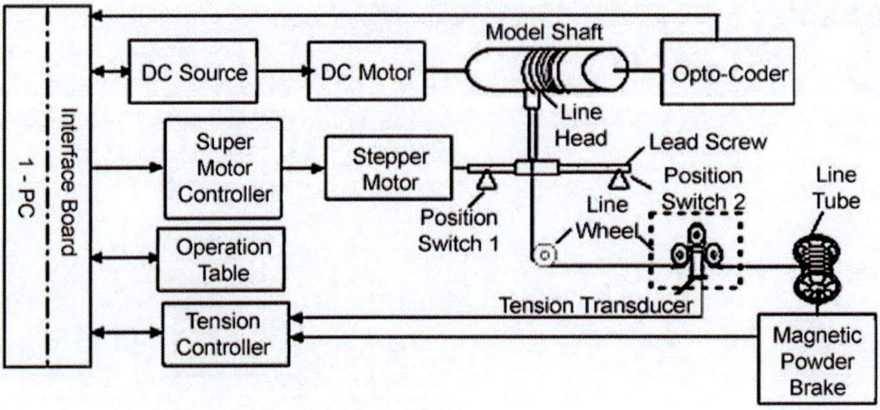

Fig. 6.1 Winding shaping system for automobile belt

string from a line tube is winded around the gum sheet which is mounted outside of the model shaft, and its moving step is controlled by a stepper motor [24]. The string tension is adjusted by the voltage of the magnetic powder brake with a feedback signal from the tension sensor. At the same time, the shaft is driven by a *DC* motor whose velocity is measured by an opto-coder. All the operations including string tension and shaft velocity, are controlled by an industrial *PC* computer. The design goal of the system is to maintain stable and uniform string tension and shaft velocity during whole winding process, and to quickly reach the demanded states after starting.

It is known that the tension regulation is not easy because it is sensitive to the velocity variation and the surrounding interferences, so proper control strategy should be found. There are some researches are related to tension control. In [48], some ideas of tension control are applied to a web machine. The torque and velocity control were used to rewinding roller to get desired results. A tension control system is proposed in [31] using an active dancer roller, which is suitable for the production of wire and sheet materials. Modeling and controller design with tension feedback, output feedback and state feedback with an observer were discussed. In [42], a modeling method is proposed for web tension profile in a paper machine, which could be built based on string model, 2D-connection model and finite element model. Fault-tolerant control is used in [39] for winding machine in processes such as sheet and film processes of steel industry. A modeling and control method of winding systems is presented in [30] for elastic webs. Robust $\mathcal{H}_\infty$ and linear parameter varying control were used to get the desired result.

It was shown in [24] that a PID controller can work smoothly, but took time to be stable. So the control algorithms should be further improved. Along similar lines, the $\mathcal{H}_2$-optimal digital control [11, 25] yielded good responses simulation studies despite it demands too large controls for the *DC* motor and magnetic powder brake.

In this section, the model of the winding system of the automobile belt is introduced, and a feedback control system is designed to minimize the cost function on the *LQR* rule. The full-order and reduced-order observer are used to estimate system states besides the measurable states: shaft velocity and the string tension. Simulation results are given for the optimal feedback control based on the full-order and reduced-order observer, and the disturbance rejection ability is examined.

6.2.1 System Model

In the sequel, we provide definition of the related variables.

Parameters

Constant: J_r—combined inertia of shaping model and the motor (3.2 kg m^2); β_1—viscous friction of main shaft (0.2 N m/s); K_1—motor torque constant (0.15 N m/A); L—motor armature inductance (3.6 mH); R—motor armature resistance (1 Ω); K_e—motor velocity constant (1.2 V s); K_g—gear ratio (20 : 1); J_2—inertia of string tube (0.4 kg m^2); β_2—viscous friction of string tube (0.02 N m/s); K_s—damping constant of magnetic brake (0.08 V s/(N–m)); K_F—torque constant of magnetic brake (1.2 V/(N–m)); K_L—spring constant of winding string (8×10^4 N/m); r_1—radius of shaping model (0.3 m); r_2—radius of string tube (0.2 m).

Differential Equations

For the main shaft velocity ω_t, it relates to, motor torque ($K_g K_I I$, where I is armature current) and string tension T:

$$J_1 \frac{d\omega_1}{dt} + \beta_1 \omega_1 = K_g K_1 I - T r_1. \tag{6.1}$$

And the armature current I conforms to:

$$L \frac{dl}{dt} + R.I + K_e K_g \omega_1 = U_M \tag{6.2}$$

where U_M is the motor control voltage.

The velocity ω_2 of the string tube is related to the string tension T and the magnetic brake friction torque F. We have

$$J_2 \frac{d\omega_2}{dt} + \beta_2 \omega_2 = T r_2 - F. \tag{6.3}$$

The magnetic brake friction torque is adjusted by the control voltage U_F:

$$K_s \frac{dF}{dt} + K_F F = U_F. \tag{6.4}$$

Let x_1 be the position on the winding model, and x_2 be position on the string tube. Then string tension T is related to the string deformation $x_1 - x_2$ and spring constant K_L of the winding string [6]:

$$T = K_L(x_1 - x_2), \tag{6.5}$$

$$\begin{aligned} &\frac{dx_1}{dt} = r_1\omega_1, \qquad \frac{dx_2}{dt} = r_2\omega_2, \\ &\frac{dT}{dt} = K_L\left(\frac{dx_1}{dt} - \frac{dx_2}{dt}\right) = K_L(r_1\omega_1 - r_2\omega_2). \end{aligned} \tag{6.6}$$

State Space Model

Let $x_1 = \omega_1$, $x_2 = I$, $x_3 = \omega_2$, $x_4 = F$, $x_5 = T$, and outputs are $y_1 = \omega_1$, $y_2 = T$. From (3.95)–(3.100), we get

$$\begin{aligned} &\begin{bmatrix} \dot{x}_1 \\ \dot{x}_2 \\ \dot{x}_3 \\ \dot{x}_4 \\ \dot{x}_5 \end{bmatrix} = \begin{bmatrix} -\frac{\beta_1}{J_1} & \frac{K_g K_1}{J_1} & 0 & 0 & -\frac{r_1}{J_1} \\ -\frac{K_e K_g}{L} & -\frac{R}{L} & 0 & 0 & 0 \\ 0 & 0 & -\frac{\beta_2}{J_2} & \frac{1}{J_2} & \frac{r_2}{J_2} \\ 0 & 0 & 0 & -\frac{K_F}{K_s} & 0 \\ K_L r_1 & 0 & -K_L r_2 & 0 & 0 \end{bmatrix} \begin{bmatrix} x_1 \\ x_2 \\ x_3 \\ x_4 \\ x_5 \end{bmatrix} \begin{bmatrix} 0 & 0 \\ \frac{1}{L} & 0 \\ 0 & 0 \\ 0 & \frac{1}{K_s} \\ 0 & 0 \end{bmatrix} \begin{bmatrix} U_M \\ U_F \end{bmatrix}, \\ &\begin{bmatrix} y_1 \\ y_2 \end{bmatrix} = \begin{bmatrix} x_1 \\ x_5 \end{bmatrix} = \begin{bmatrix} 1 & 0 & 0 & 0 & 0 \\ 0 & 0 & 0 & 0 & 1 \end{bmatrix} X. \end{aligned} \tag{6.7}$$

Substitute all parameters, A and B matrices become

$$A = \begin{bmatrix} -0.0625 & 0.9375 & 0 & 0 & -0.09375 \\ -3333.3 & -277.78 & 0 & 0 & 0 \\ 0 & 0 & -0.05 & -2.5 & 0.5 \\ 0 & 0 & 0 & -12 & 0 \\ 24000 & 0 & -16000 & 0 & 0 \end{bmatrix},$$

$$B = \begin{bmatrix} 0 & 0 \\ 277.78 & 0 \\ 0 & 0 \\ 0 & 10 \\ 0 & 0 \end{bmatrix}, \qquad C = \begin{bmatrix} 1 & 0 & 0 & 0 & 0 \\ 0 & 0 & 0 & 0 & 1 \end{bmatrix}.$$

By checking the rank of $[B \;\; AB \;\; A^2B \;\; A^3B \;\; A^4B]$ and $[C \;\; CA \;\; CA^2 \;\; CA^3 \;\; CA^4]$, the system is controllable and observable.

Open-Loop Step Response

The open-loop tension and velocity step responses are showed in Fig. 6.2. We can see that velocity reach stable stale very quickly, but tension has much oscillation and takes time to be stable. This response is consistent to that of real system.

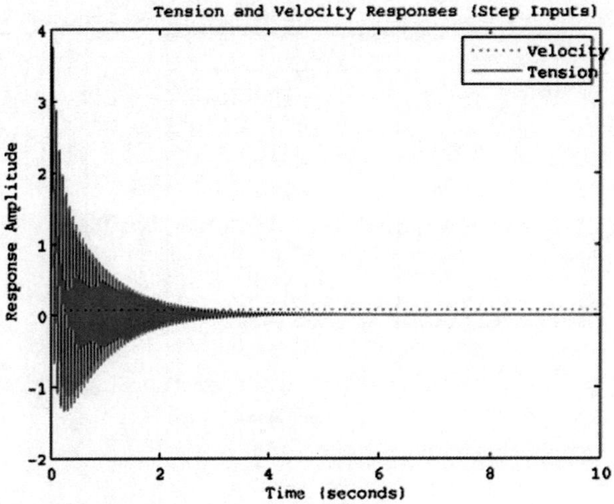

Fig. 6.2 Open-loop step responses

6.2.2 State-Feedback and LQR Control

State feedback is applied in control

$$u = -Kx + Pr, \tag{6.8}$$

where K is a feedback matrix and P is the feed forward for tracking. We want to design a state feedback to achieve good tracking and disturbance rejection ability.

6.2.3 Pole Placement

For $\dot{x} = Ax + Bu$ and $u = -Kx + Pr$, we have

$$\dot{x} = (A - BK)x + BPr. \tag{6.9}$$

Now we should properly choose the eigenvalues of $A - BK$ so that the system is stable and can quickly reach the stable values. Although we can place the eigenvalues at any places on the left hand of the polar plane because the system is controllable, the outputs of the resulted controller may be too large to be realized in real system. An optimal design method is to use *LQR* optimal control to get the feedback K.

6.2.4 LQR Optimal Control

The controller can be designed using MATLAB function "*lqr*", which calculates the optimal gain matrix K such that the state-feedback law $u = -Kx$ minimizes the cost function

$$J = \int_0^{\infty} \left(x^t q x + u^t r u\right) dt. \tag{6.10}$$

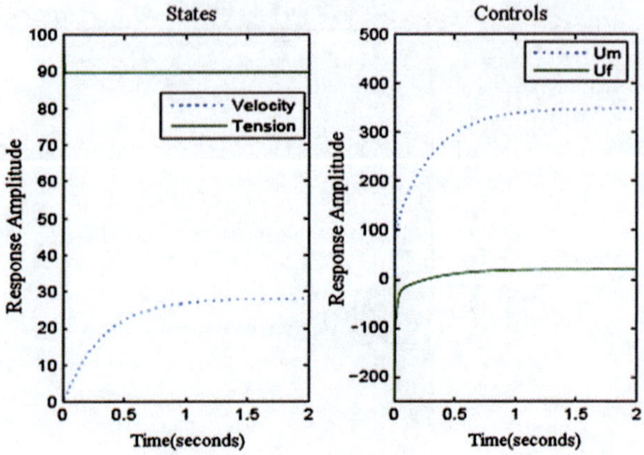

Fig. 6.3 Responses for both step inputs under LQR control

Choose

$$q = \begin{bmatrix} 1 & 0 & 0 & 0 & 0 \\ 0 & 10 & 0 & 0 & 0 \\ 0 & 0 & 10 & 0 & 0 \\ 0 & 0 & 0 & 10 & 0 \\ 0 & 0 & 0 & 0 & 1500 \end{bmatrix}, \qquad r = \begin{bmatrix} 1 & 0 \\ 0 & 1 \end{bmatrix}.$$

Using MATLAB, this leads to

$$K = \begin{bmatrix} 3263 & 4.737 & -2183 & 9.477 & 36.06 \\ 398.2 & 0.3412 & -274 & 6.456 & 2.495 \end{bmatrix}.$$

With the simulation model in MATLAB "simulink" environment, we get the system responses (shown in Fig. 6.3) for both step inputs of velocity and tension. The steady values of the responses are adjusted by the feedforward P (3500 and 0.5) so that they are between 60 N to 70 N for tension, and 16 to 22 radls (15 Q~2 IOrpm) for velocity. From the figure, the responses seem to be pretty fast and stable, but the controls have large initial negative values. For the control of magnetic powder brake, no negative voltage is allowed. For *DC* motor source, it is better to use positive voltage. So saturation elements were added in front of the controls to the system. But the system becomes unstable with this saturation, as shown in Fig. 6.4. An improved method is to use ramp velocity input with upper bounded. With trial-and-error method for choosing the *LQR* parameters, batter results are gotten with

$$q = \begin{bmatrix} 0.05 & 0 & 0 & 0 & 0 \\ 0 & 10 & 0 & 0 & 0 \\ 0 & 0 & 10 & 0 & 0 \\ 0 & 0 & 0 & 10 & 0 \\ 0 & 0 & 0 & 0 & 1000 \end{bmatrix}, \qquad r = \begin{bmatrix} 1 & 0 \\ 0 & 100 \end{bmatrix},$$

$$K = \begin{bmatrix} 5986 & 6.170 & -3999 & 12.52 & 96.74 \\ 5.810 & 0.0045 & -4.708 & 0.325 & 0.060 \end{bmatrix}$$

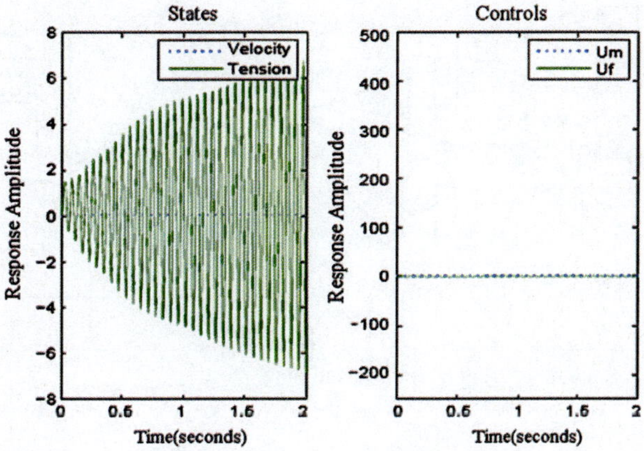

Fig. 6.4 Responses with saturation

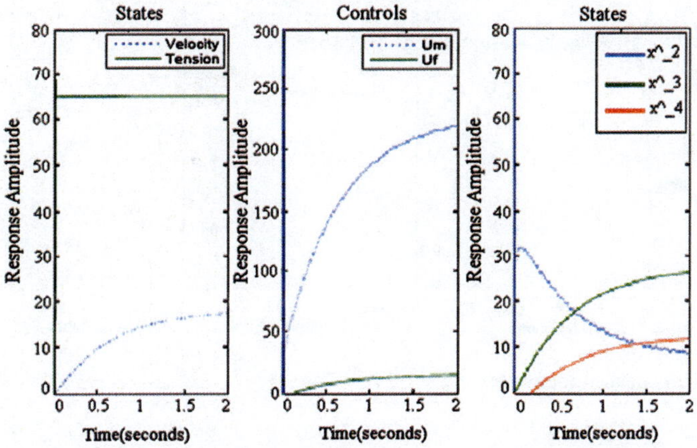

Fig. 6.5 System responses under ramp velocity input

and P are chosen with 6500 and 0.1. Figure 6.5 shows the results with these parameters. We can see that responses are pretty good and the controls have no negative values.

6.2.5 Disturbance Rejection

To examine the disturbance rejection ability, some pulses are added to the velocity input with amplitude 1 tenth of the upper value of the velocity input, as shown in Fig. 6.6. Also some white noises are added to the controls U_m and U_f to the system. The system still has good ability to remove them as shown in Fig. 6.7, even with large noise. We can see that the tension has also some jump noise, but returns to original values very quickly. So it has fast regulation speed, but the control U_m has much larger value.

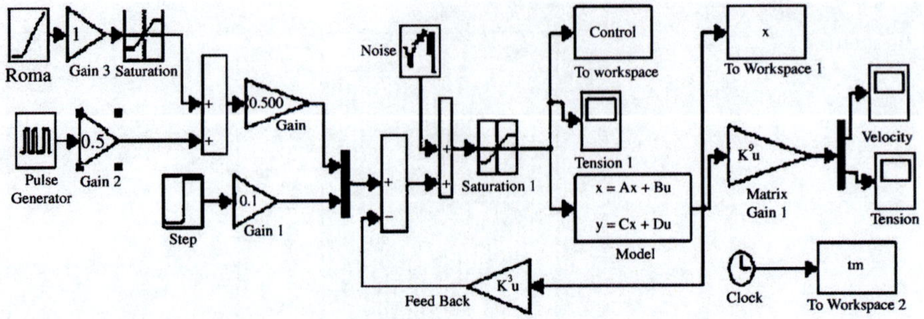

Fig. 6.6 Disturbance (pulses and noise) testing

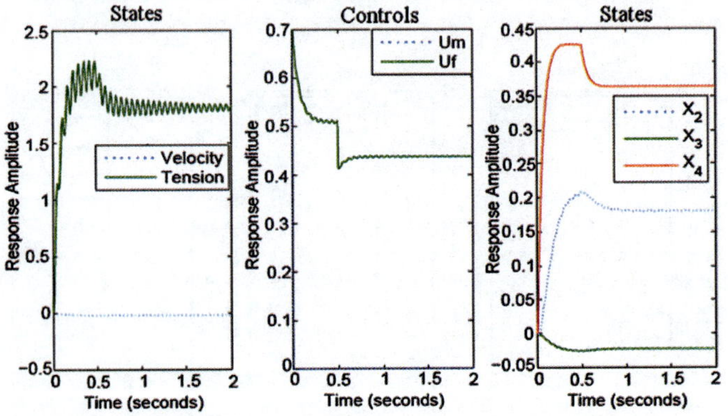

Fig. 6.7 Results of disturbance testing

6.2.6 Observer-Based Feedback

Figure 6.8 shows an observer-based state feedback control system. The full-order observer takes it form as

$$\dot{\hat{x}} = (A - GC)\hat{x}\begin{bmatrix} B & G \end{bmatrix}\begin{bmatrix} u & y \end{bmatrix}^t, \tag{6.11}$$

$$\hat{y} = C\dot{\hat{x}}. \tag{6.12}$$

The system is observable so that the eigenvalues of the error dynamic matrix $A - GC$ could be assigned negatively. Normally we choose these eigenvalues with real parts 3–5 times larger than the real parts of eigenvalues of $P = A - BK$. After getting the estimated states $\hat{x}$, the state feedback control is realized. Figure 6.9 shows the system step responses with real parts of eigenvalues of $(A - GC)$ equal to 2 times of real parts of eigenvalues of $(A - BK)$. Figure 6.10 shows the estimated errors of the observer. We found that negative real part of eigenvalues of $(A - GC)$ should be properly selected, since too high negative real part of these eigenvalues may cause the controls to oscillate. In this case, we found that 2 times of real parts of eigenvalues of $(A - BK)$ are appropriate.

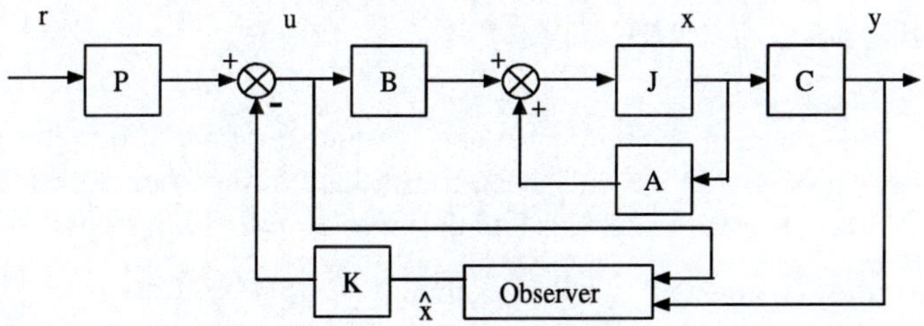

Fig. 6.8 Observer-based state feedback

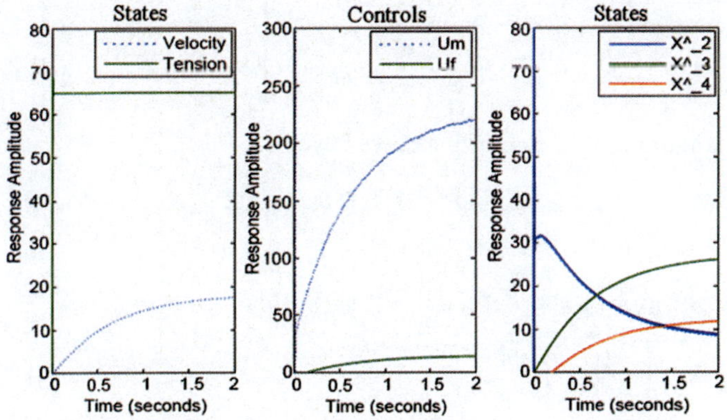

Fig. 6.9 Responses with full-order observer-based state feedback

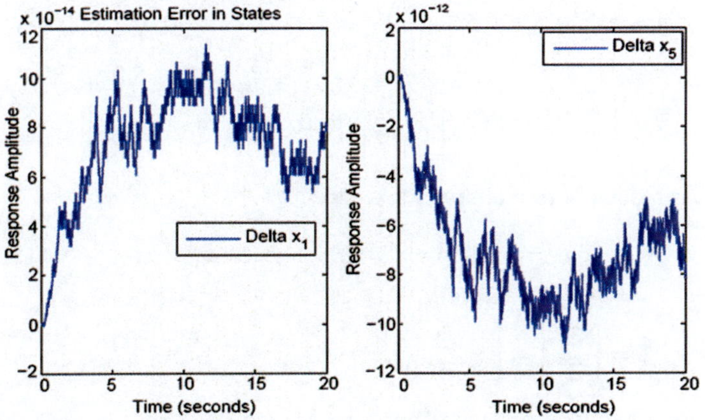

Fig. 6.10 Estimated errors of full-order observer

6.2.7 Reduced-Order Observer

Two states of the, system, x_1 (velocity) and x_5 (tension) can be directly measured through the sensors so that we only need to estimate three other states. Therefore, the reduced-order observer is applied and designed as following steps.

1) Consider the change of state coordinate: $Z = Px$. Let $p = \begin{bmatrix} C \\ R \end{bmatrix}$, $C = \begin{bmatrix} 1 & 0 & 0 & 0 & 0 \\ 0 & 0 & 0 & 0 & 1 \end{bmatrix}$ and R can be chosen arbitrarily so that P^1 exists. Here we choose

$$R = \begin{bmatrix} 0 & 1 & 0 & 0 & 0 \\ 0 & 0 & 1 & 0 & 0 \\ 0 & 0 & 0 & 1 & 0 \end{bmatrix}, \qquad P = \begin{bmatrix} 1 & 0 & 0 & 0 & 0 \\ 0 & 0 & 0 & 0 & 1 \\ 0 & 1 & 0 & 0 & 0 \\ 0 & 0 & 1 & 0 & 0 \\ 0 & 0 & 0 & 1 & 0 \end{bmatrix}.$$

2) After changing the state coordinate, we have

$$\dot{Z} = PAP^{-1}Z + PBu = \bar{A}Z + \bar{B}u, \tag{6.13}$$

$$y = CP^{-1}Z = \bar{C}Z. \tag{6.14}$$

3) Now we partition $\bar{A}$ and $\bar{B}$

$$\bar{A} = \begin{bmatrix} \bar{A}_{11} & \bar{A}_{12} \\ \bar{A}_{21} & \bar{A}_{22} \end{bmatrix},$$

$$\bar{A}_{11} = \begin{bmatrix} -0.0625 & -0.0938 \\ 24000 & 0 \end{bmatrix}, \quad \bar{A}_{12} = \begin{bmatrix} 0.938 & 0 & 0 \\ 0 & -16000 & 0 \end{bmatrix},$$

$$\bar{A}_{21} = \begin{bmatrix} -3333 & 0 \\ 0 & 0.5 \\ 0 & 0 \end{bmatrix}, \quad \bar{A}_{22} = \begin{bmatrix} -2.778 & 0 & 0 \\ 0 & -0.05 & -2.5 \\ 0 & 0 & -12 \end{bmatrix},$$

$$\bar{B} = \begin{bmatrix} \bar{B}_1 \\ \bar{B}_2 \end{bmatrix}, \quad \bar{B}_1 = \begin{bmatrix} 0 & 0 \\ 0 & 0 \end{bmatrix}, \quad \bar{B}_2 = \begin{bmatrix} 277.8 & 0 \\ 0 & 0 \\ 0 & 100 \end{bmatrix}.$$

4) Then the reduced-order observer takes form of

$$\dot{V} = (\bar{A}_{22} - G\bar{A}_{12})V + \big[(\bar{A}_{22} - G\bar{A}_{12})G + \bar{A}_{21} - G\bar{A}_{11}\big]y + (\bar{B}_2 - G\bar{B}_1)u, \tag{6.15}$$

$$\hat{Z} = \begin{bmatrix} \hat{Z}_1 \\ \hat{Z}_2 \end{bmatrix} = \begin{bmatrix} y \\ V + Gy \end{bmatrix}. \tag{6.16}$$

5) Finally, we get the estimated states

$$\hat{x} = P^{-1}Z = Z. \tag{6.17}$$

The same as that in full-order observer, state feedback, control can be realized by using $\hat{x}$. That is to place the eigenvalues of $\bar{A}_{22} - G\bar{A}_{l2}$ at desired positions.

Figure 6.12 shows the results with the real part of eigenvalues of $\bar{A}_{22} - G\bar{A}_{12}$ (3-element vector) equals to 5 times of real part of eigenvalues of $\{A - BK(2:4)\}$

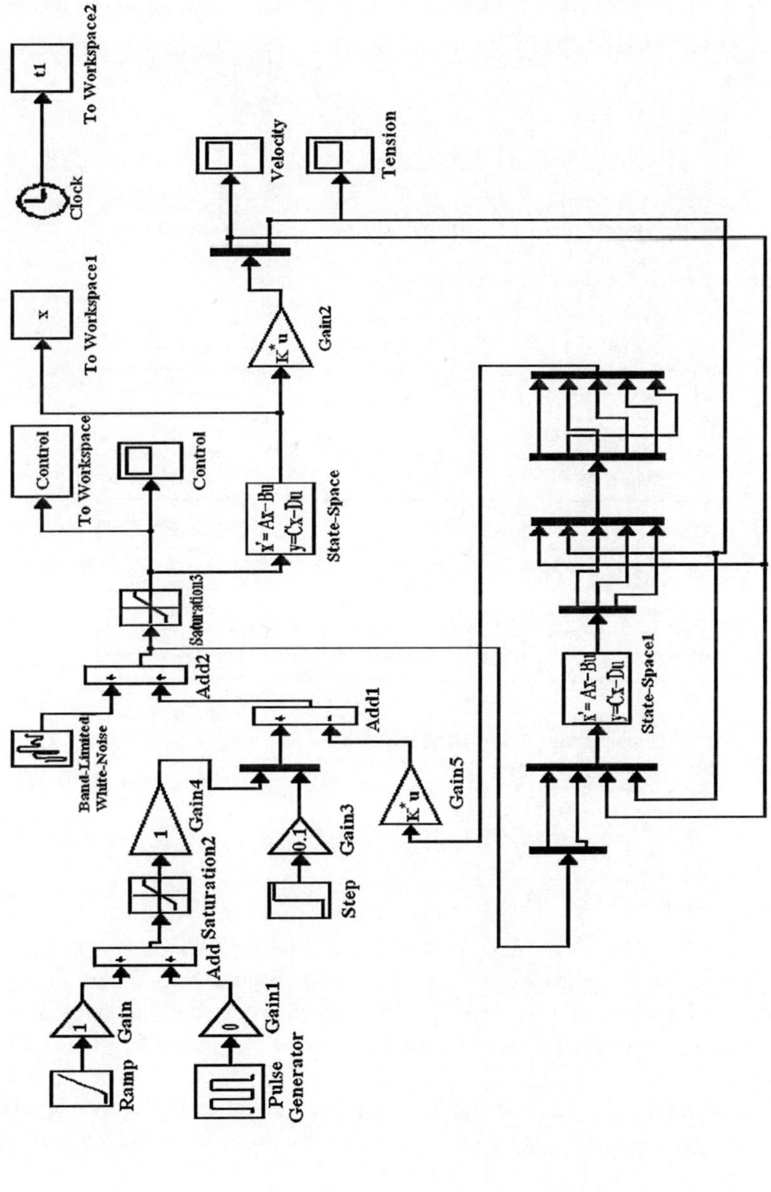

Fig. 6.11 Reduced-order observer-based state feedback

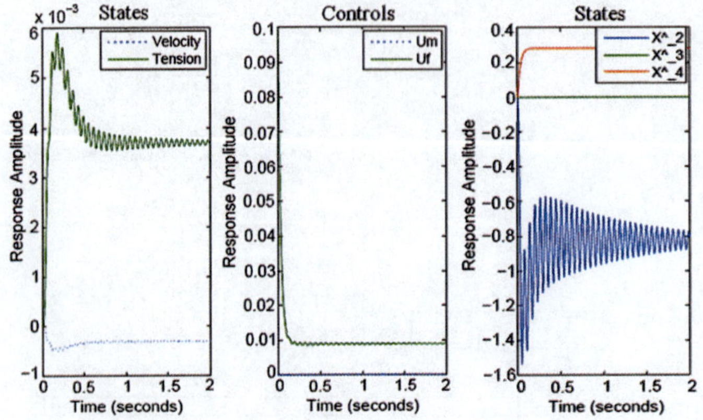

Fig. 6.12 Response of reduced-order observer-based state feedback

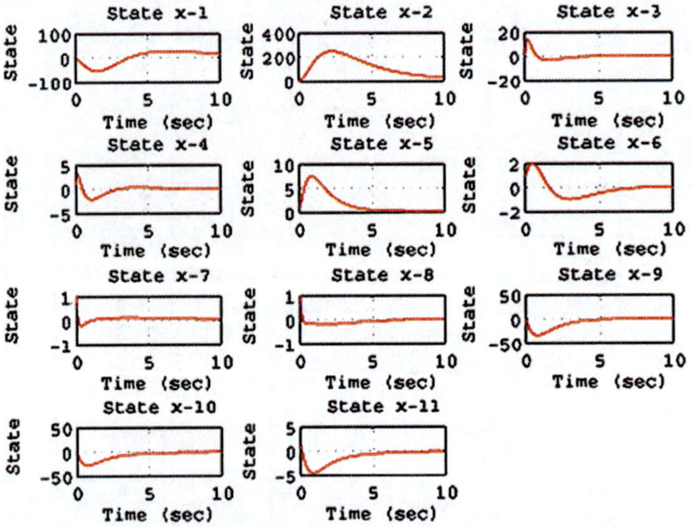

Fig. 6.13 Response of state-feedback system

(the 2nd, 3rd and 4th eigenvalues of the $A - BK$). Figure 6.13 shows the estimated errors. We can see that the system responses are good, but the controls have some noise, especially with the larger values of negative real parts of eigenvalues of $(\bar{A}_{22} - G\bar{A}_{12})$.

It was observed that there is less oscillation in the tension and velocity responses when the negative eigenvalues of $(\bar{A}_{22} - G\bar{A}_{12})$ are chosen farther away from the imaginary axis. However, when the eigenvalues are too far away from the imaginary axis, the system may cause another problem: small high-frequency oscillation. And it will even be unstable. So the real part of eigenvalues of $(\bar{A}_{22} - G\bar{A}_{12})$ should be properly chosen.

6.3 An Unmanned Helicopter

An unmanned aerial vehicle (UAV) is an aircraft that flies without a human crew on board the aircraft. These vehicles have wide applications in remote sensing and explorations. To distinguish UAVs from missiles, a UAV is defined as a reusable, uncrewed vehicle capable of controlled, sustained, level flight and powered by a jet or reciprocating engine. Therefore, cruise missiles are not considered UAVs, because, like many other guided missiles, the vehicle itself is a weapon that is not reused, even though it is also unmanned and in some cases remotely guided.

There are a wide variety of UAV shapes, sizes, configurations, and characteristics. Historically, UAVs were simple drones (remotely piloted aircraft), but autonomous control is increasingly being employed in UAVs. UAVs come in two varieties: some are controlled from a remote location, and others fly autonomously based on pre-programmed flight plans using more complex dynamic automation systems.

Currently, military UAVs perform reconnaissance as well as attack missions [50]. While many successful drone attacks on militants have been reported, they are also prone to collateral damage and/or erroneous targeting, as with many other weapon types [50]. UAVs are also used in a small but growing number of civil applications, such as firefighting or nonmilitary security work, such as surveillance of pipelines. UAVs are often preferred for missions that are too "dull, dirty, or dangerous" for manned aircraft.

6.3.1 Linearized Model

The numerical values of a linearized state-space model are given by [50]:

$$A = \begin{bmatrix} A_1 & A_2 \\ A_3 & A_4 \end{bmatrix}, \qquad A_1 = \begin{bmatrix} -0.126 & 0 & 0 & 0 & 0 & -32.2 \\ 0 & -0.425 & 0 & 0 & 32.2 & 0 \\ -0.168 & 0.087 & 0 & 0 & 0 & 0 \\ -0.082 & -0.052 & 0 & 0 & 0 & 0 \\ 0 & 0 & 1 & 0 & 0 & 0 \\ 0 & 0 & 0 & 1 & 0 & 0 \end{bmatrix},$$

$$A_2 = \begin{bmatrix} -32.2 & 0 & 0 & 0 & 0 \\ 0 & 32.2 & 0 & 0 & 0 \\ 36.71 & 161.11 & 0 & 0 & 0 \\ 63.58 & -19.49 & 0 & 0 & 0 \\ 0 & 0 & 0 & 0 & 0 \\ 0 & 0 & 0 & 0 & 0 \end{bmatrix}, \qquad A_3 = \begin{bmatrix} 0 & 0 & 0 & -1 & 0 & 0 \\ 0 & 0 & -1 & 0 & 0 & 0 \\ 0 & 0 & 0 & 0 & 0 \\ 0 & 0 & -1.33 & 0 & 0 \\ 0 & 0 & 0 & 0 & 0 \end{bmatrix},$$

$$A_4 = \begin{bmatrix} -3.444 & 0.829 & 0 & 0 & 0 \\ 0.361 & -3.444 & 0 & 0 & 0 \\ 0 & 9.64 & -0.76 & 8.42 & 0 \\ 0 & 0 & 0.057 & -5.51 & -44.873 \\ 0 & 0 & 0 & 1.816 & -11.02 \end{bmatrix},$$

$$B = \begin{bmatrix} 0 & 0 & 0 & 0 \\ 0 & 0 & 0 & 0 \\ 0 & 0 & 0 & 0 \\ 0 & 0 & 0 & 0 \\ 0 & 0 & 0 & 0 \\ 0 & 0 & 0 & 0 \\ -0.8417 & 2.8231 & 0 & 0 \\ -2.4090 & -0.3511 & 0 & 0 \\ 0 & 0 & 70.5041 & 0 \\ 0 & 0 & 23.6260 & 44.8734 \\ 0 & 0 & 0 & 0 \end{bmatrix}, \qquad C^t = \begin{bmatrix} 1 & 0 & 0 & 0 \\ 0 & 1 & 0 & 0 \\ 0 & 0 & 0 & 0 \\ 0 & 0 & 0 & 0 \\ 0 & 0 & 0 & 0 \\ 0 & 0 & 0 & 0 \\ 0 & 0 & 0 & 0 \\ 0 & 0 & 0 & 0 \\ 0 & 0 & 1 & 0 \\ 0 & 0 & 0 & 1 \\ 0 & 0 & 0 & 0 \end{bmatrix}.$$

The open-loop eigenvalues are

$$\begin{aligned} &-1.5725 \pm 12.2567j, \quad -8.2845 \pm 8.5844j, \quad -1.8659 \pm 8.2757j, \\ &0.2458 \pm 0.0277j, \quad -0.5262 \pm 0.0754j, \quad -0.7223 \end{aligned}$$

which means that the system is unstable.

6.3.2 *Stabilization Schemes*

To stabilize the system, one way is to employ a state-feedback controller, the gain of which can be determined by using the pole-placement technique through MATLAB software. With the desired eigenvalues being

$$-8.4, \quad -8.3, \quad -0.3, \quad -0.6, \quad -0.9, \quad -1, \quad -1.1, \quad -1.9, \quad -1.8, \quad -1.7, \quad -1.6$$

the gain matrix is given by

$$K = 10^{-3} \times \begin{bmatrix} 3.9 & -2.5 & 3.41 & 61.9 & -38.4 & -7.5 & -544.8 & -3045 & 0 & -0.1 & 0.8 \\ -5.6 & 12.8 & 127.5 & -192.2 & 11 & 126 & 2695.5 & -467.9 & 0 & -0.5 & 2.7 \\ -2.5 & 0.8 & 51.7 & 52.1 & 74.9 & 101.7 & 467.3 & 1061.4 & 12.1 & 92.1 & 144.9 \\ 1 & 0.8 & -33.1 & -64.5 & -25.3 & -91.9 & -567.9 & 978.7 & -5.2 & -350.4 & 36.6 \end{bmatrix}.$$

The corresponding trajectories of the closed-loop state-feedback control system are plotted in Fig. 6.13, which show that the stabilizing behavior.

6.4 Reverse Osmosis Desalination Plant

The production of fresh water for drinking, domestic, agricultural, landscape and industrial uses by desalination of sea and brackish waters has become a major issue in the regions suffering from the scarcity of natural fresh water supplies [18]. This has resulted in a demand for the desalination systems. In the last years, significant advances in the membrane technology have allowed an essential improvement in the filtering quality and simultaneously a general reduction of costs such that Reverse Osmosis plants have today lower energy consumption, investment cost, space requirements and maintenance than other desalination methods such as distillation, ion exchange and solar humidification.

Reverse Osmosis (RO) is a process used for demineralization of water to clean brackish water or to desalt seawater. When we try to separate pure water and a salt solution through a semi permeable membrane, the pure water diffuses through the membrane and dilutes the salt solution. The membrane rejects most of the dissolved salts, while allowing the water to permeate. This phenomenon is known as *natural osmosis*. As water passes through the membrane, the pressure on the dilute side drops, and the pressure of the concentrated solution rises. The osmotic flux continues until equilibrium is reached, where the net water flux through the membrane becomes zero at equilibrium; the liquid level in the saline water will be higher than that on the waterside. The amount of water passing in either direction will be equal. The hydrostatic pressure difference achieved is equal to the effective driving force causing the flow, called osmotic pressure. This pressure is a strong function of the solute concentration and the temperature, and depends on the type of ionic species present. Applying a pressure in excess of the osmotic pressure to the saline water section slows down the osmotic flow, and forces the water to flow from the salt solution into the waterside. Therefore, the direction of flow is reversed, and that is why this separation process is called *reverse osmosis*.

The process consists in recovering water from a saline solution pressurized by pumping it into a closed vessel to a point greater than the osmotic pressure of the solution. Thus, the solution is pressed against a membrane so that it is separated from the solutes (the dissolved material). The portion of water that passes through the membrane reducing strongly the solute concentration is called permeate. The remaining water (re tented) is discharged with a high salt concentration.

Reverse osmosis (RO) plant is described in detail in [1, 2, 18]. It is shown that the RO plant is modeled and a description about the modeling is given. In what follows, a linear state space model proposed in [18] is utilized to design a control system. Most of the RO plants includes a pre-treatment unit, where the salt concentration of permeate (or also permeate conductivity) is controlled by adjusting the pH value of the feed. However, plants for drink water purification do not include pH control and permeate conductivity is a non-controlled variable. In order to be able to adjust the permeate conductivity, a bypass valve, which allows mixing a small amount of feed water with permeate is included. This construction leads to a different system topology, which has not been studied much.

With the increasing energy awareness and scarcity, it is becoming more desirable to operate plants very close to target. That is over production and over purification is not economically justifiable if the plant can be operated closer to specification. For this purpose, controllers must be adapted in order to continuously and automatically adjust operating conditions to meet variable demand. The performance of RO plants is quite sensitive to the quality of the feed and plant operating conditions. This means that a RO plant requires a very efficient pretreatment process and an accurate control system to maintain its operation close to the optimum conditions, which results in increased productivity and prolongs the life of the membranes due to the reduction of membrane fouling.

6.4.1 Reverse Osmosis Modeling

A basic RO system normally consists of four main subsystems: pretreatment, high-pressure pump, membrane assembly (RO unit) and post-treatment (see Fig. 6.14). Salty feed water is first pretreated to avoid membrane fouling. It then passes through filter cartridges (a safety device) and is sent through the membrane modules (permeators) by a high-pressure pump. Because of the high pressure, pure water permeates through the membranes and the salty water becomes concentrated (retained or brine). The water product flows directly from the permeators into the post treatment unit, and the retentate (at high pressure) is discharged, usually, after passing through an energy recovery system. Pretreatment is important in RO plants because suspended particles must be removed in order to maintain the membrane surfaces continuously clean. Thus, pretreatment consists of fine filtration and the addition of chemicals to inhibit precipitation and the growth of microorganisms. The pH value of the feed water is also adjusted in this unit. The high-pressure pump supplies the pressure needed to enable the water to pass through the membrane and have the salts rejected. This pressure range is from 15 to 25 bars for drinking and brackish water and from 54 to 80 bars for seawater. The membrane assembly consists of a pressure vessel and several membrane units such that feed water is pressurized against the membrane. The membrane must be able to resist the entire pressure drop across it. The semi-permeable membranes vary in their ability to pass fresh water and reject the passage of salts. Finally, the post-treatment consists of stabilizing the water and preparing it for distribution. This post-treatment might consist of the removing gases such as hydrogen sulfide, adding minerals and adjusting the pH value.

In a typical RO desalination plant, there are basically four variables of interest: (1) flow rate of permeate, (2) salt content of permeate, (3) trans-membrane pressure, and (4) pH of feed water. The first two outputs are obviously important because they are production targets. Trans-membrane pressure must not be allowed to exceed an upper limit since that could cause membrane rupture and the pH of feed water should be within bounds to extend membrane life. Therefore, only the first two variables are selected as outputs as they are important. In the case of small plants, pretreatment

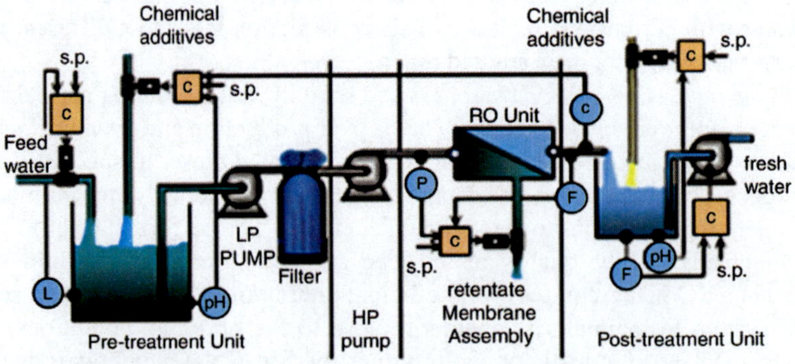

Fig. 6.14 Schematic of RO plant

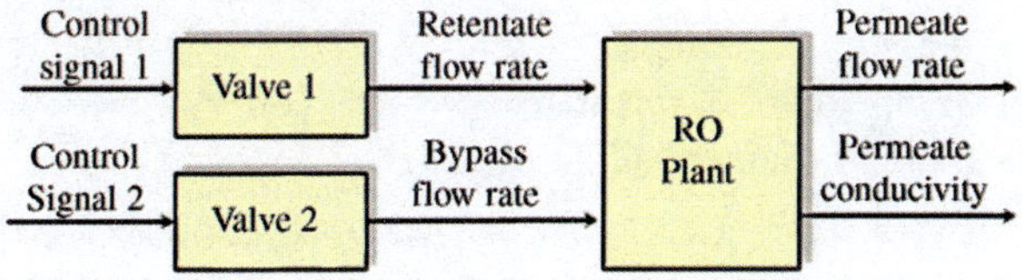

Fig. 6.15 Blockdiagram of RO system

units are very simple and normally pH control of feed water is not implemented. Permeate conductivity can be modified by using a bypass pipeline, which allows the mix of a small amount of feed water with the product, if the quality requirements for the product that allows for by recycling a small amount of retentate. The input/output representation with feed water bypass is illustrated in Fig. 6.15.

The plant under consideration has a capacity in nominal operation of about 900 l/h (that is, 0.25 l/s) for an inlet of 0.625 l/s. The flow rate of concentrate in 0.375 l/s, that is, 60% retentate and 40% permeate. The bypass flow rate is about 8% of the feed water, that is, 0.04 l/s for the nominal operation. The range for permeate flow rate is given by 0.021 l/s $< qp <$ 0.433 l/s for a valve opening varying between 100% < áp < 10%. Notice that this valve may not be close in order that the plant works at all. The maximum water purity is obtained by a closed bypass valve and a valve in the retentate stream closed up to 90% (10% valve opening). The normal operating point is 50% valve opening for both valves. Under these conditions, the permeate flow rate is 0.250 l/s, the retentate 0.375 l/s and the permeate conductivity 425 μS/cm (283 ppm). In order to put the set point, for example, at 0.20 l/s it is necessary to open the retentate valve up to 60%. Once the valve is fixed to this value it is not possible to modify this flow rate by using the bypass valve. On the contrary, modifying the bypass valve, the conductivity can be adjusted to other reference value.

6.4.2 Linear Discrete Model

In terms of the system variables
Inputs:

- Retentate valve opening u_1.
- Bypass valve opening as u_2.

Outputs:

- Permeate flow rate y_1.
- Permeate conductivity as y_2.

State variables:

- Temperature of feed water.
- Salinity of feed water.
- Pressure of retentate.
- Temperature of permeate.
- pH of feed water.

- Pressure of permeate.
- Temperature of retentate.
- Salinity of retentate.
- pH of retentate.

A discrete-time state space linear model was obtained from sampled-data for a sampling time of 0.15 s at the operating point mentioned above. The general equations are given by,

$$A = \begin{bmatrix} 0.201 & 0.01 & 0 & 0 & 0 & 0 & 0 & 2e\text{-}4 & -0.001 \\ -3.301 & -0.129 & 0 & 0 & 0 & 0 & 0 & 0.001 & 0.001 \\ 0 & 0 & 0.757 & 0 & 0 & 0 & 0 & 0 & 0.113 \\ 0 & 0 & 0 & 0.955 & 0.116 & 0 & 0 & 0.01 & -0.062 \\ 0 & 0 & 0 & -0.545 & 0.573 & 0 & 0 & 0.11 & -0.606 \\ 0 & 0 & 0 & 0 & 0 & 0.859 & 0.056 & 0 & 0.004 \\ 0 & 0 & 0 & 0 & 0 & -1.833 & 0.043 & 0 & 0.024 \\ 0 & 0 & 0 & 0 & 0 & 0 & 0 & 0.905 & 0 \\ 0 & 0 & 0 & 0 & 0 & 0 & 0 & 0 & 0.286 \end{bmatrix},$$

$$B^t = \begin{bmatrix} -8.02e\text{-}5 & -0.001 & 0 & -0.002 & -0.041 & 0 & 0 & -0.632 & 0 \\ -8.42e\text{-}5 & -0.001 & 0.009 & -0.002 & -0.041 & 1.7e4 & 0.002 & 0 & 0.057 \end{bmatrix},$$

$$C = \begin{bmatrix} 222.53 & 12.46 & 0.668 & 0 & 0 & 0 & 0 & 0 & 0 \\ 0 & 0 & 0 & -21.784 & -7.624 & 1209.53 & 3705.56 & 0 & 0 \end{bmatrix}.$$

The open-loop eigenvalues $\lambda(A)$ are given by

$$0.0360 \pm j0.0761,\ 0.7570,\ 0.7640 \pm 0.1639j,\ 0.7535,\ 0.1485,\ 0.9050,\ 0.2860$$

which indicate that the discrete-time model is stable since all eigenvalues lies within the unit disc in the complex plane. Moreover, the model is both controllable and observable.

For the state-feedback control design, we employ MATLAB file *place* to compute the gain matrix. By repeated application, we select the gain of least norm. This is expressed by

$$K = \begin{bmatrix} -10.089 & 0.330 & -15.436 & -0.237 & -2.154 & 4.625 & 0.357 & -1.398 & -0.065 \\ 2.172 & 14.076 & 19.32 & -1.511 & 0.261 & -5.773 & -0.436 & 0.169 & 3.133 \end{bmatrix},$$

and the associated closed-loop eigenvalues

$$\begin{bmatrix} 0.036 & 0.1481 & 0.357 & 0.264 & 0.764 & 0.7535 & 0.35 & 0.2 & 0.276 \end{bmatrix}.$$

The corresponding state trajectories are displayed in Figs. 6.16–6.20.

6.5 Turbocharged Diesel Engine

In recent years more stringent requirements on performance, fuel conservation and low emissions have paved way for increased complicated engine performance. Strategies like exhaust gas recirculation and turbo charging have been devised to

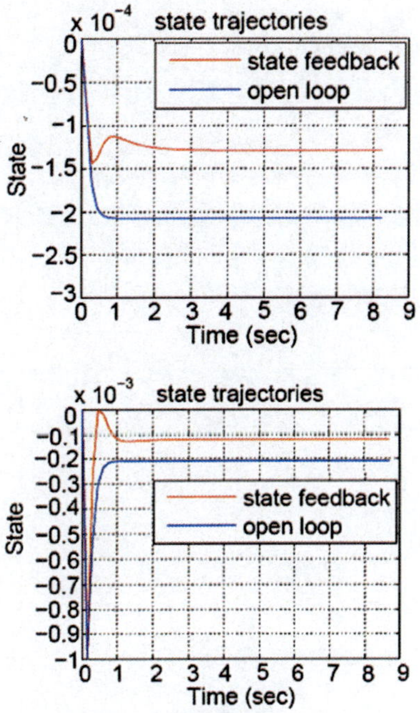

Fig. 6.16 Feedback trajectories of states x_1 and x_2

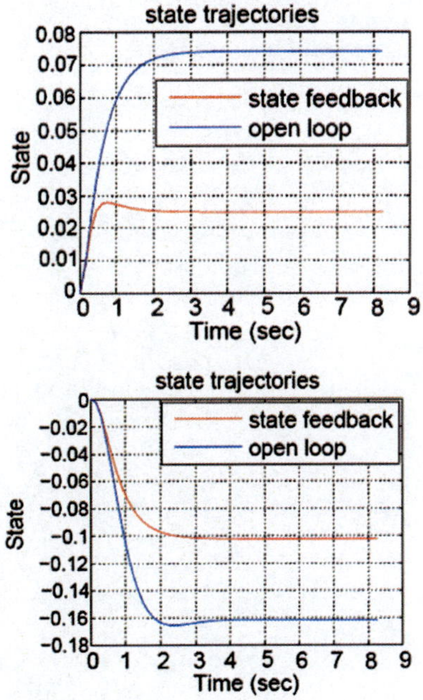

Fig. 6.17 Feedback trajectories of states x_3 and x_4

cope up with the requirements. These give us a great bit of freedom to control the behavior of the engine. Previous practices used these in a suboptimal way since the devices used to control these features affect many different parts of the engine

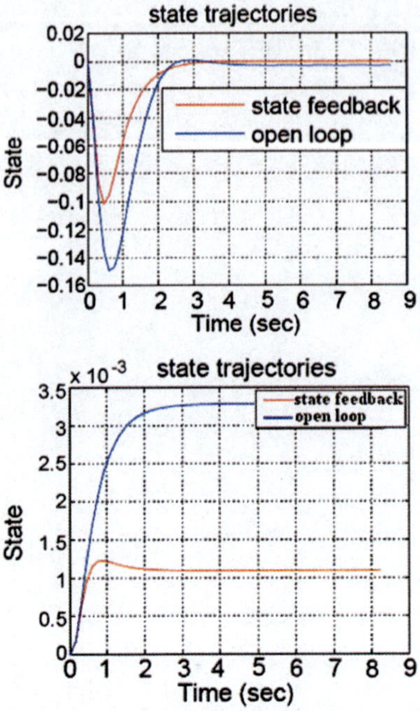

Fig. 6.18 Feedback trajectories of states x_5 and x_6

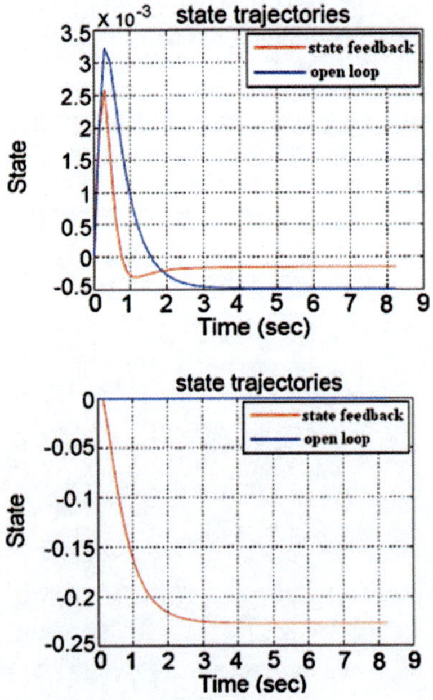

Fig. 6.19 Feedback trajectories of states x_7 and x_8

through the cross-couplings in the system. The development of an optimal coordinated strategy often takes more time than available in a production cycle. In order to fully extract the potential of these devices, we consider this as a multivariable

Fig. 6.20 Feedback trajectory of state x_9

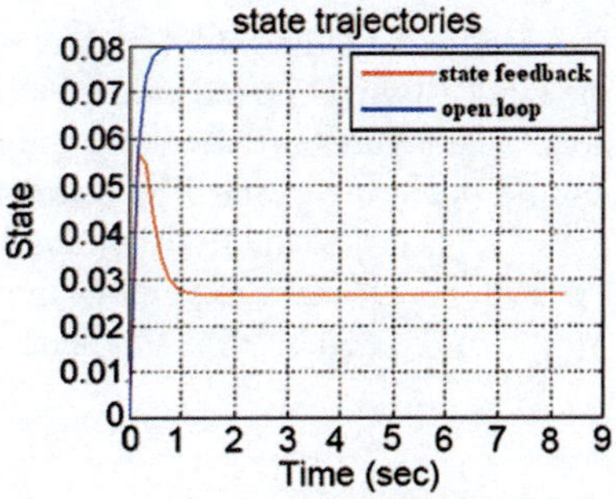

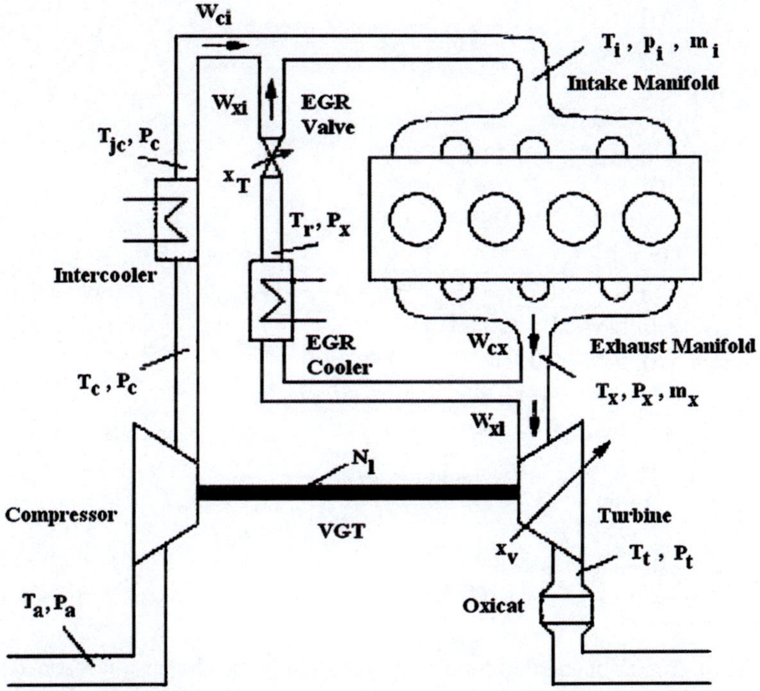

Fig. 6.21 Schematic diagram of the TDE model

control problem. A multivariable approach to this will yield a better performance. Turbochargers mainly find their applications in racing cars,automobiles, aircrafts and gas turbines. Diesel (compression ignition) engines hold a significant advantage over spark ignited (gasoline) engines in fuel economy. Moreover, diesel engines have lower feed-gas emissions of the regulated exhaust gases, but the after-treatment devices for diesel engines are far less efficient than the conventional three way catalysts for spark ignition engines.

In this section, the plant to be controlled is a turbocharged passenger car diesel engine equipped with exhaust gas recirculation and a variable geometry turbine as shown in Fig. 6.21. Turbocharger increases the power density of the engine by forcing air into the cylinders, which allows injection of additional fuel without reaching the smoke limit. The turbine, which is driven by the energy in the exhaust gas, has a variable geometry that allows the adaptation of the turbine efficiency based on the engine operating point. The second feedback path from the exhaust to the intake manifold is due to the EGR, which is controlled by the EGR valve. The recirculated exhaust gas replaces oxygen in the inlet charge, thereby reducing the temperature profile of the combustion and hence the emissions of oxides and nitrogen.

6.5.1 Dynamic Modeling

In terms of the data

$$\mathbf{A} = \begin{bmatrix} A_1 & A_2 \\ A_3 & A_4 \end{bmatrix},$$

$$\mathbf{A}_1 = \begin{bmatrix} -0.4125 & -0.0248 & 0.0741 \\ 101.5873 & -7.2651 & 2.7608 \\ 0.0704 & 0.0085 & -0.0741 \end{bmatrix},$$

$$\mathbf{A}_2 = \begin{bmatrix} 0.0089 & 0 & 0 \\ 2.8608 & 0 & 0 \\ -0.0089 & 0 & 0.0200 \end{bmatrix}, \quad \mathbf{A}_3 = \begin{bmatrix} 0.0878 & 0.2672 & 0 \\ -1.8414 & 0.0990 & 0 \\ 0 & 0 & 0 \end{bmatrix},$$

$$\mathbf{A}_4 = \begin{bmatrix} -0.3674 & 0.0044 & 0.3962 \\ 0 & -0.0343 & -0.0330 \\ -359 & 187.5364 & -87.0316 \end{bmatrix},$$

$$\mathbf{B} = \begin{bmatrix} -0.0042 & 0.0064 \\ -1.0360 & 1.5894 \\ 0.0042 & 0 \\ 0.1261 & 0 \\ 0 & -0.0168 \\ 0 & 0 \end{bmatrix}, \quad \mathbf{C} = \begin{bmatrix} 0 & 0 & 0 & 0 & 0 & 3.6 \\ 0 & 0 & 0 & 1 & 0 & 0 \end{bmatrix}.$$

Numerical simulation of the control designs using the linear-quadratic regulator (LQR) and linear-quadratic Gaussian regulator (LQGR) are summarized in terms of the feedback gains and the associated bounds:

$$L_{\ell qr} = \begin{bmatrix} -0.8195 & -0.1731 & -0.1973 & 1.1521 & -0.9907 & -0.0028 \\ 5.14277 & 0.3250 & 0.3654 & 0.7437 & 0.1943 & 0.0025 \end{bmatrix},$$

$$\|L_{\ell qr}\| = 5.2748, \quad \gamma_+ = 6.3472,$$

$$L_{\ell qgr} = \begin{bmatrix} -0.341 & -0.0628 & -0.0950 & 0.4114 & -0.3772 & -0.0009 \\ 1.9763 & 0.1176 & 0.1655 & 0.2696 & 0.0982 & 0.0009 \end{bmatrix},$$

$$\|L_{\ell qgr}\| = 2.0334, \quad \gamma_+ = 6.3472.$$

The numerical clearly suggests that the control design based on the mixed $\mathcal{H}_2/\mathcal{H}_\infty$ yields the best compromise. However, it requires, excessive computations compared with LQR, $\mathcal{H}_2$ and $\mathcal{H}_\infty$. The corresponding state trajectories are plotted in Figs. 6.22–6.27.

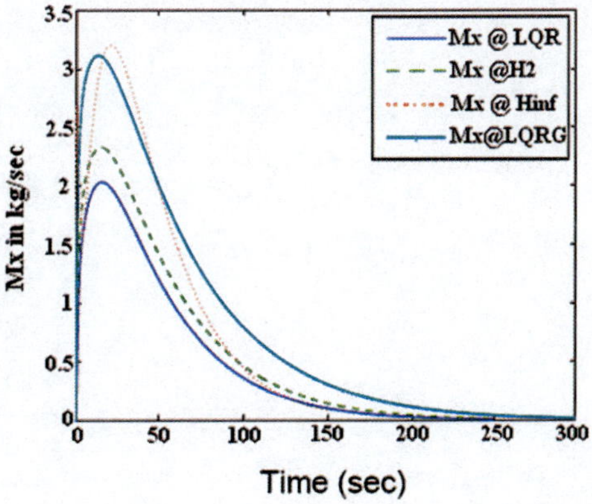

Fig. 6.22 Response of state 1

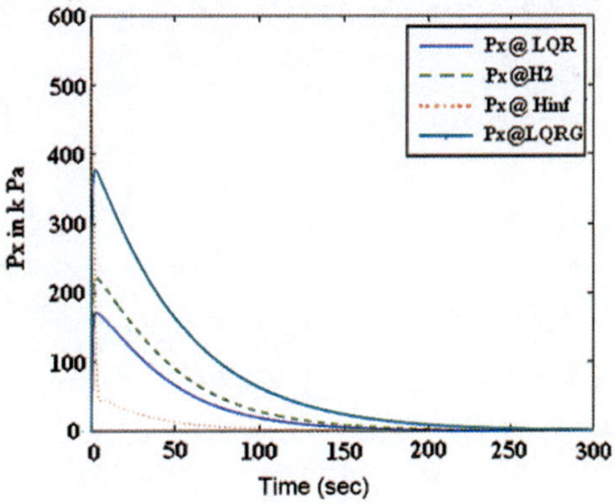

Fig. 6.23 Response of state 2

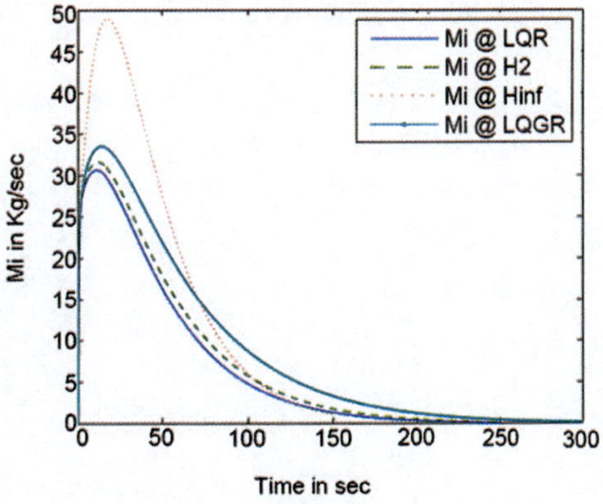

Fig. 6.24 Response of state 3

6.6 A Rotational Hydraulic Drive

Modern internal combustion engines equipped with variable valve actuation systems are proven to achieve better combustion characteristics. By appropriately varying the valve timing, one can increase fuel economy, boost power output and reduce

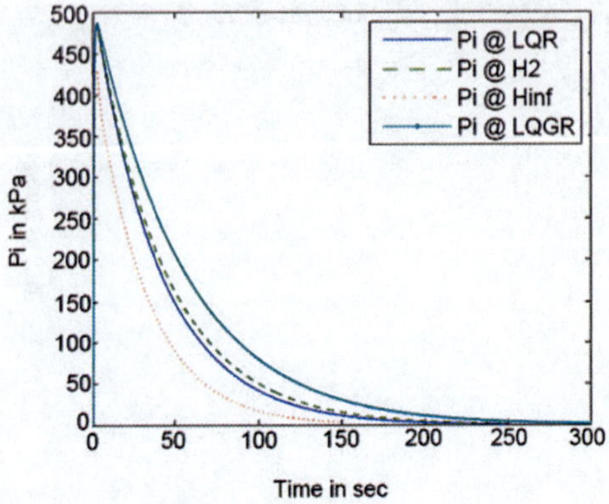

Fig. 6.25 Response of state 4

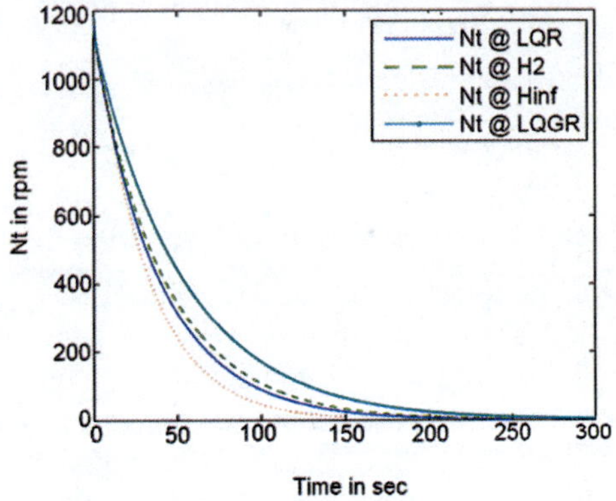

Fig. 6.26 Response of state 5

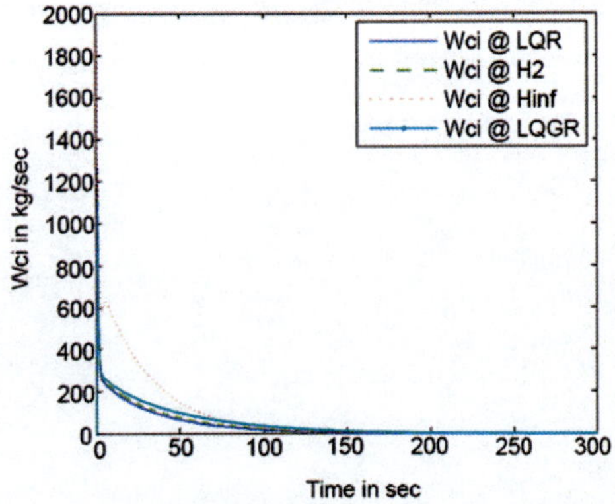

Fig. 6.27 Response of state 6

emissions [43]. Related studies are reported in [38, 51]. In particular, the problem of optimizing plug-in hybrid electric vehicle (PHEV) power management is studied in [38] by using stochastic dynamic over a distribution of drive cycles, rather than a single cycle and explicitly trades off fuel and electricity usage. Linear feedback controllers are developed in [33] for an electro-hydraulic valve system (EHVS) and a repetitive feed-forward controller is added to improve the tracking performance. The problem of power management of hybrid electric vehicles (HEVs) is treated in [28] via the Pontryagin's minimum principle as a viable real-time strategy. By employing performance index including fuel consumption, exhaust emission, or acceleration performance over the whole driving-cycle information, global optimal results are reported in [34]. Application of robust control design for the physical geometric design of electrohydraulic valves is reported in [53], where it is shown that viscosity effect is exclusively utilized in the nominal optimal design, whereas both the viscosity effect and the non orifice flux effect are needed in the robust optimal design. In [12], two-controller structure is proposed for a generic EPS system addressing motor torque and steering motion, by applying $\mathcal{H}_2$ and $\mathcal{H}_\infty$ design methods, respectively. An improved optimal control method based on the energy equation of the controlled system is presented in [16]. The work [44] experimentally demonstrates the implications of this trade off by applying a recently developed repetitive controller design approach to reduce the error motion of the spindle's axis of rotation on an active air bearing setup. In [51], the problem of inventory control is studied using high gain (sliding mode) adaptive control to handle the system uncertainties caused by modeling errors and unmeasured disturbances.

On another research direction pertinent to the present paper, electro-hydraulic servo-systems (EHSS) find extensive industrial applications ranging from hydraulic stamping, injection molding presses to aerospace flight-control actuators. EHSSs serve as highly efficient drive systems because they posses a high power/mass ratio, quick response, high stiffness and high load capabilities. To maximize the advantages of hydraulic systems and to meet increasingly precise performance with high accuracy and fast response, high performance servo controllers are required. However, traditional linear controllers have performance limitations due to the presence of nonlinear dynamics in EHSS, specifically, a square-root relationship between the differential pressure that drives the flow of the hydraulic fluid, and the flow rate. To achieve near-optimality, keep the methodology simple.

6.6.1 System Model

The system under consideration for this study is a rotational hydraulic drive and the set-up is generic and allows for ample extension of the results herewith to other electro-hydraulic systems including double-acting cylinders [8]. Referring to Fig. 6.28, a DC electric motor drives a pump, which delivers oil at a constant supply pressure from the oil tank to each component of the system. The oil is used for the operation of the hydraulic actuator and is returned through the servo-valve to the oil tank at atmospheric pressure. An accumulator and a relief valve are used to maintain a constant supply pressure from the output of the pump. The electro-hydraulic system includes two Moog Series 73 servo-valves which control the movement of the rotary actuator and the load torque of the system. These servo valves are operated by voltage signals generated by an Opal-RT real-time digital control system. The actuator and load are both hydraulic motors connected by a common shaft. One servo-valve regulates the flow of hydraulic fluid to the actuator and the other reg-

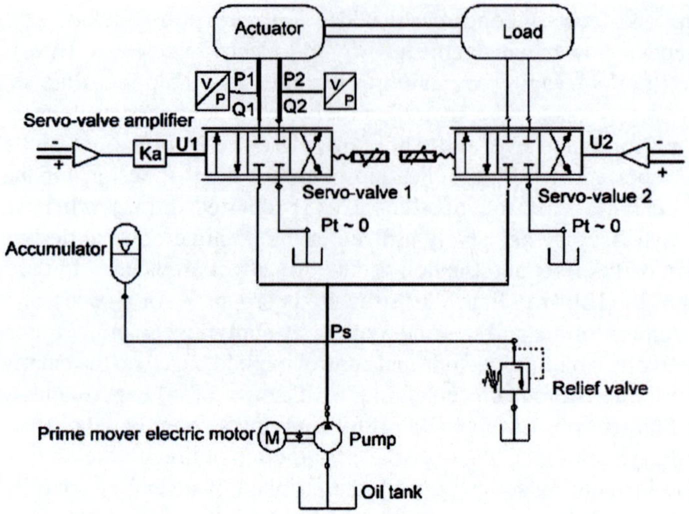

Fig. 6.28 Functional diagram

ulates the flow to the load. The actuator operates in a closed-loop while the load operates open-loop, with the load torque being proportional to the command voltage to the load servo-valve. While the actuator and load chosen for this study are rotary drives, the exact same set-up could be used with a linear actuator and load, and thus, they are represented as generic components in Fig. 6.28.

Using the *angular displacement, angular velocity, differential pressure* P_1 and *differential pressure* P_2 as the system states, the signals from *servo valves 1, 2* as the control inputs whereas *the angular velocity* and *angular displacement* as the outputs, a linearized model of an electro-hydraulic system about the origin ($\mathbf{x}_1 = 0$, $\mathbf{x}_2 = 0$, $\mathbf{x}_3 = 0$, $\mathbf{x}_4 = 0$) can be cast into the form

$$\begin{aligned} \dot{\mathbf{x}} &= \mathbf{A}\mathbf{x} + \mathbf{B}\mathbf{u}, \\ \mathbf{y} &= \mathbf{C}\mathbf{x}, \end{aligned} \tag{6.18}$$

$$\mathbf{A} = \begin{bmatrix} 0 & \omega_M & 0 & 0 \\ 0 & -\gamma\Omega_h/\alpha & \Omega_h/\alpha & 0 \\ 0 & -\alpha\Omega_h & -\Omega_h c_L & \alpha\Omega_h \\ 0 & 0 & 0 & -1/\tau_v \end{bmatrix},$$
$$\mathbf{B} = \begin{bmatrix} 0 & 0 \\ 0 & -\Omega_h/\alpha \\ 0 & 0 \\ 1/\tau_v & 0 \end{bmatrix}, \quad \mathbf{C}^t = \begin{bmatrix} 0 & 1 \\ \omega_M & 1 \\ 0 & 1 \\ 0 & 1 \end{bmatrix}. \tag{6.19}$$

Using typical data [26, 40], the different parameters are $\tau_v = 0.01$ s, $\omega_M = 173.45$ r/s, $\alpha = 4.7408$ s, $c_L = 0.077$, $\gamma = 0.5432$, $\omega_h = 138.68$ r/s. By evaluating the model matrices given by (4.70), it is readily seen that the linearized system is unstable as it has eigenvalue at the origin and has internal oscillations due to a pair of complex.

6.6.2 LQR: Continuous and Discrete Control

With $Q = 10 \times I_4$, $R = I_2$, the output response is depicted in Fig. 6.29. Typical simulation results are plotted in Figs. 6.30–6.33 for three sets of Q matrix with $R = I_2$. With respect to the norm of the gain matrix and the time taken by the states to settle to steady state, it is concluded that the case (ii) has yielded optimum results. In case (i), the number of oscillations in the states are more. Whereas, moving from case (ii) to case (iii), there is no significant reduction in the number of oscillations. There is a decrease in the settling time of the states, but the gain K is increasing significantly. In the step response of the system, we observe that in each case the second output of the system, that is, the angular velocity settles to zero after a finite interval of time despite the input being at unit step. This is because the signal from the two servo valves is treated as a positioning input. This means that if any constant input is applied at either of the inputs of the system, the actuator shaft moves to a distinct position and stops. Its angular position remains at that finite value while the angular velocity reduces to zero when the actuator shaft has stopped moving.

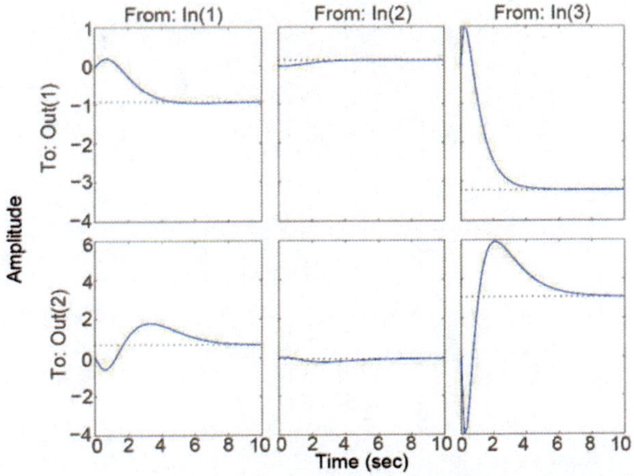

Fig. 6.29 Output response

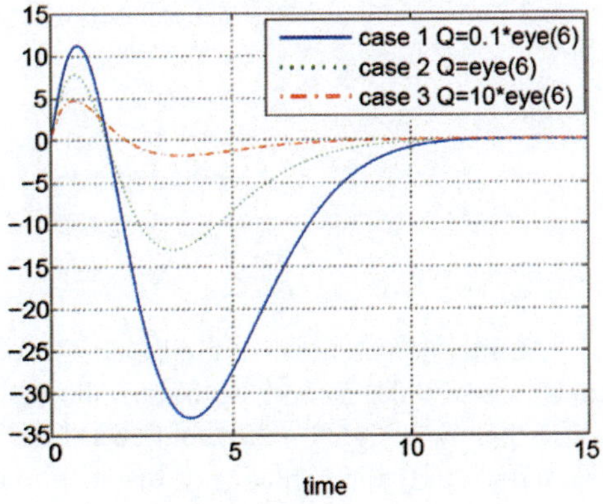

Fig. 6.30 Comparison of state x_1

Fig. 6.31 Comparison of state x_2

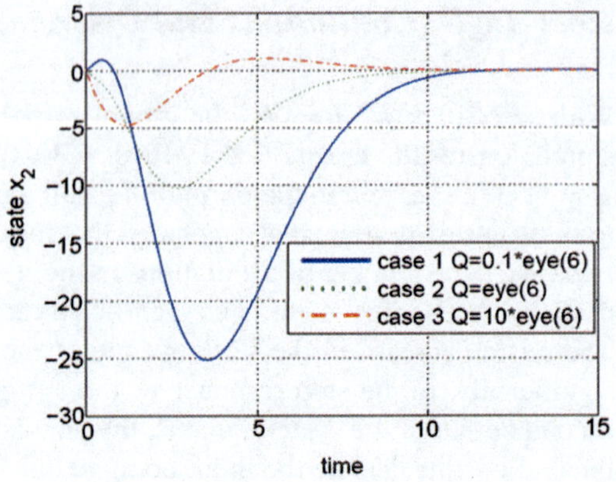

Fig. 6.32 Comparison of state x_3

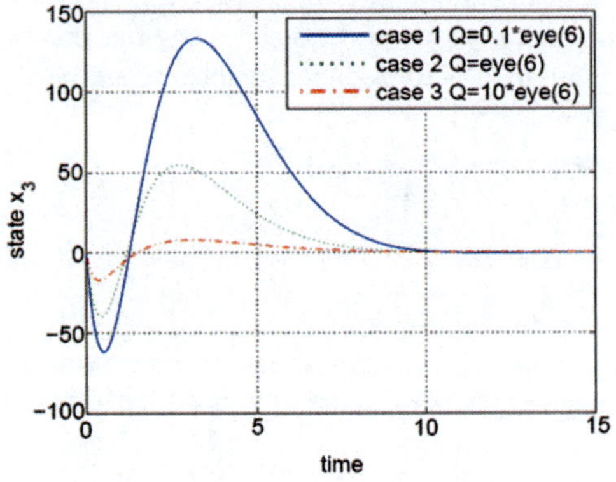

Fig. 6.33 Comparison of state x_4

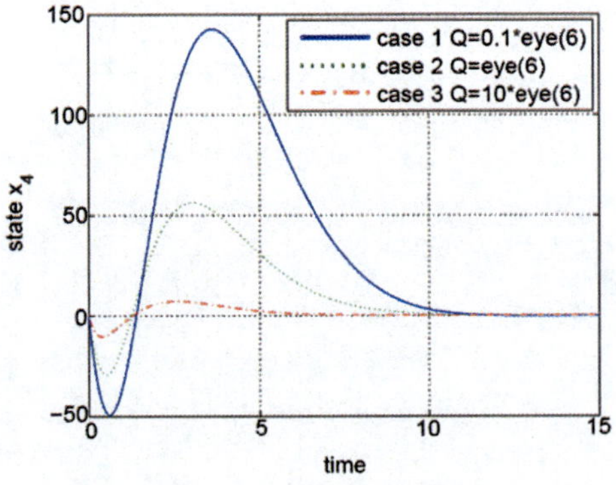

Turning now to the discrete LQR. The continuous system matrices were sampled at a rate of 0.01 seconds to obtain the discrete model. Simulation was carried out such that the weight on the inputs was kept constant and the weight on the states was varied to study the behavior of the system in three different cases. Optimum results

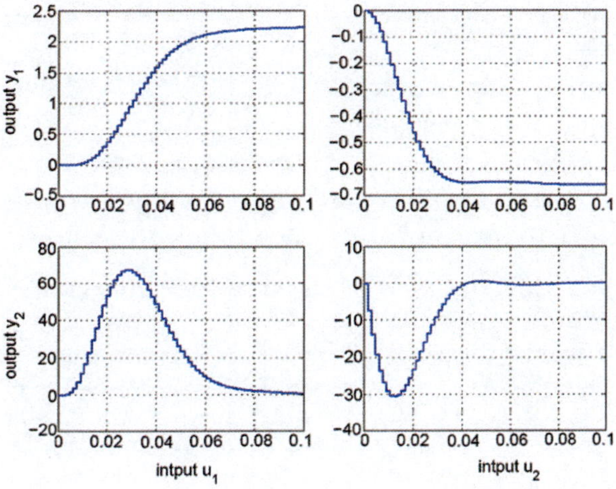

Fig. 6.34 Step response of system using *DLQR*

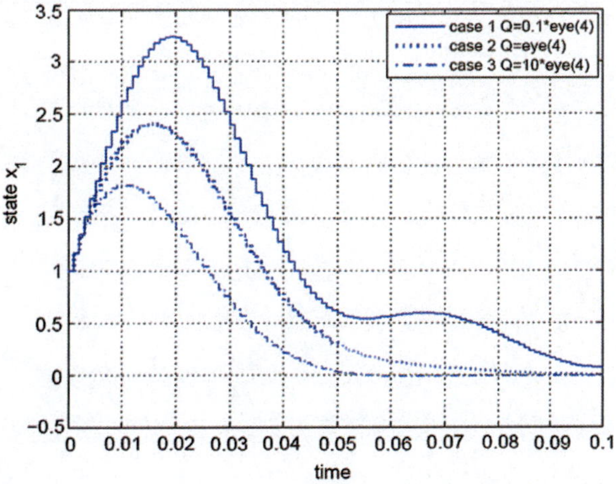

Fig. 6.35 Comparison of state x_1 trajectories

were found using the following weighting matrices: $Q = 10 * I_{4\times4}$, $R = I_{2\times2}$, the output response is depicted in Fig. 6.34.

Of all the 3 cases simulated above, it is noted that the controller gain K is the largest in the third case, while the settling time is the least in the third case. Hence, as we increase the controller gain, the settling time decreases. The response of the *DLQR* regulator is similar to the *LQR* regulator, the only difference being the control that is applied at discrete instants equal to the sampling time of the system model. Just as in the continuous *LQR* all the have been weighted equally in each case while implementing the discrete regulator. In the above simulation, the matrix R was used to weight the control input applied. The matrix Q was used to weight the states of the system. Typical simulation results are plotted in Figs. 6.35–6.38 for three sets of Q matrix with $R = I_2$.

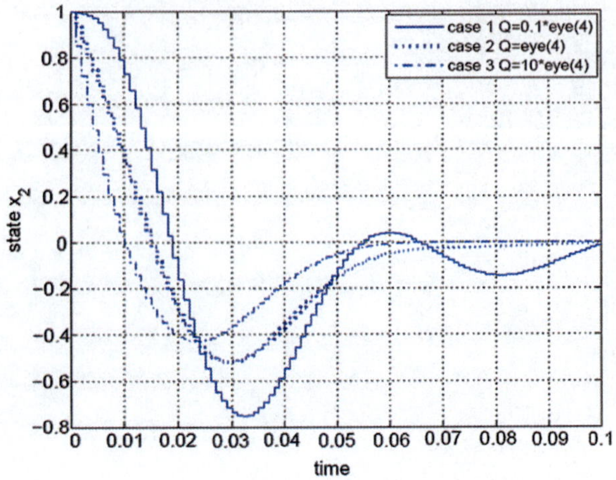

Fig. 6.36 Comparison of state x_2 trajectories

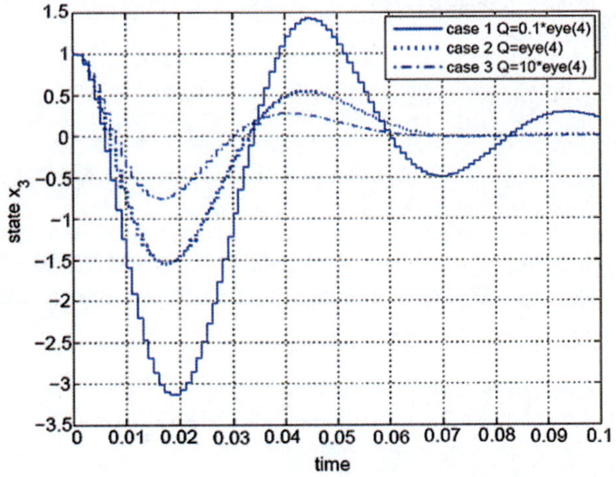

Fig. 6.37 Comparison of state x_3 trajectories

6.7 The Falling Film Evaporator

Based on the system description in the foregoing chapter, the numerical values of the system matrices are given by:

$$A = \begin{bmatrix} 0.9704 & 0.0043 & -0.0170 & 0.0003 & 0.0532 & -0.0197 \\ -0.0026 & 0.9415 & -0.0719 & 0.1236 & 0.0009 & 0.0520 \\ 0.0110 & 0.1548 & 0.9501 & -0.0793 & -0.0004 & 0.0078 \\ -0.0045 & -0.1885 & 0.1947 & 0.8364 & -0.0154 & -0.3347 \\ -0.1001 & -0.0280 & 0.0252 & -0.0889 & 0.8270 & 0.1096 \\ 0.0730 & -0.0335 & 0.0432 & -0.0503 & 0.2329 & 0.0905 \end{bmatrix}, \quad (6.20)$$

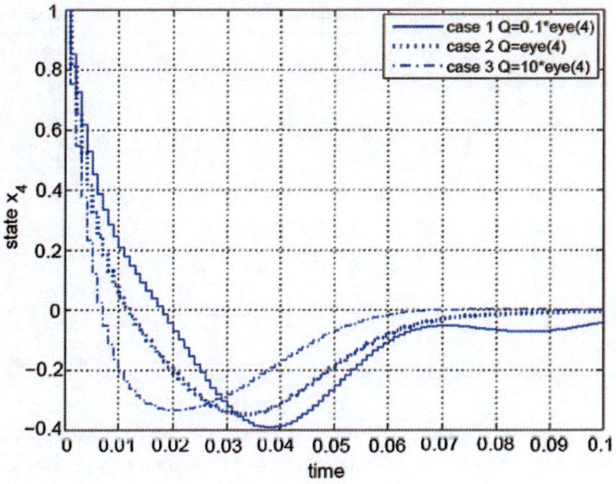

Fig. 6.38 Comparison of state x_4 trajectories

$$B = \begin{bmatrix} -0.0001 & 0.0001 & 0.0028 \\ -0.0050 & 0.0015 & 0.0001 \\ 0.0049 & -0.0005 & 0.0009 \\ -0.0062 & 0.0001 & 0.0005 \\ -0.0034 & -0.0015 & 0.0025 \\ -0.0030 & 0.0005 & -0.0084 \end{bmatrix}, \tag{6.21}$$

$$C = \begin{bmatrix} 18.4197 & 28.9777 & 34.3101 & 5.3853 & 2.0093 & 0.9821 \\ -6.2863 & -46.2754 & 7.2586 & 21.0611 & 0.0672 & 15.1997 \\ -52.2695 & 0.3603 & -0.1112 & 0.7214 & 0.6543 & -0.0106 \end{bmatrix}. \tag{6.22}$$

6.7.1 State Feedback Design

In what follows, we provide the simulation results of state-feedback design. We start with the continuous case:

A. *Continuous Case*: The pole placement method was used for the design of state feedback using MATLAB along with the desired eigenvalues as

$$\mathbf{v} = \begin{bmatrix} -1 & -2 & -3 & -2.5 & -1.5 & -5 \end{bmatrix}. \tag{6.23}$$

This yields the feedback gain matrix as

$$K = 10^3 \begin{bmatrix} -0.7351 & 0.0164 & 0.7176 & 0.5049 & -0.3379 & -0.2327 \\ -2.9280 & 4.7762 & 6.1816 & 0.6691 & 1.0067 & 0.2010 \\ 0.2584 & 0.5098 & 0.6744 & -0.0169 & 0.2578 & 0.0672 \end{bmatrix}. \tag{6.24}$$

The ensuing state trajectories of the closed-loop system are depicted in Figs. 6.39, 6.40, 6.41.

B. *Discrete Case*: Using a sampling period of 0.1 s, and performing similar MATLAB simulation, the obtained results are plotted in Figs. 6.42, 6.43, 6.44.

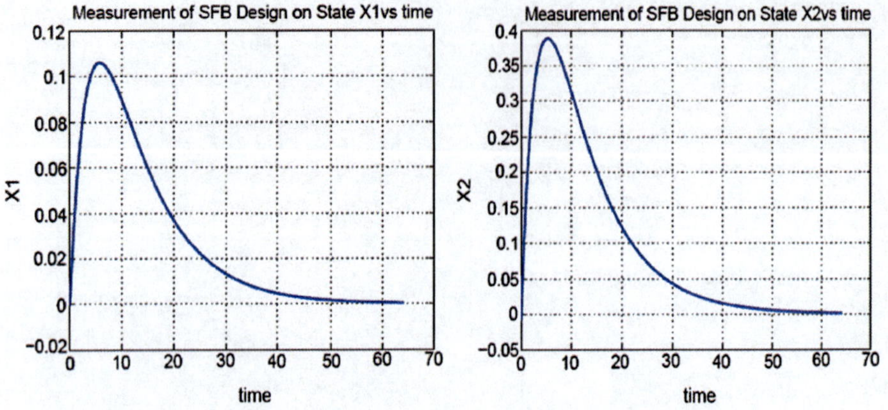

Fig. 6.39 Trajectory of state variables x_1 (*left*) and x_2 (*right*): Continuous-case

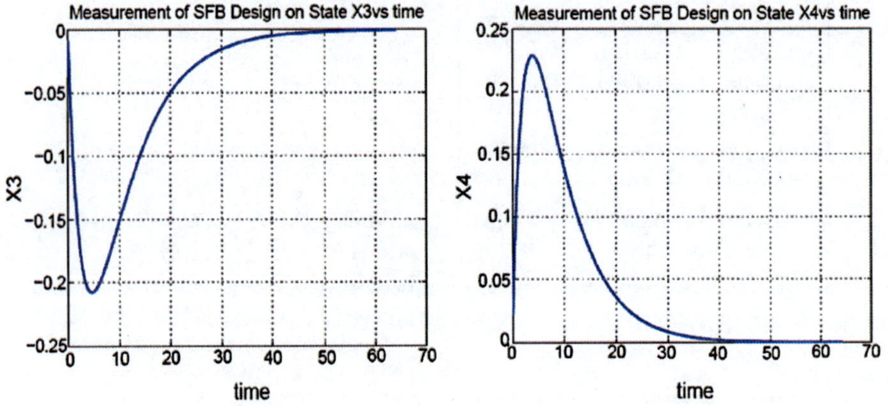

Fig. 6.40 Trajectory of state variables x_3 (*left*) and x_4 (*right*): Continuous-case

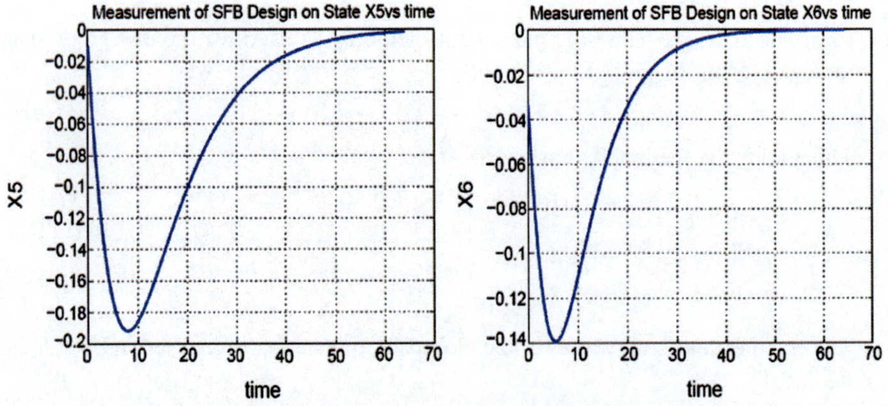

Fig. 6.41 Trajectory of state variables x_5 (*left*) and x_6 (*right*): Continuous-case

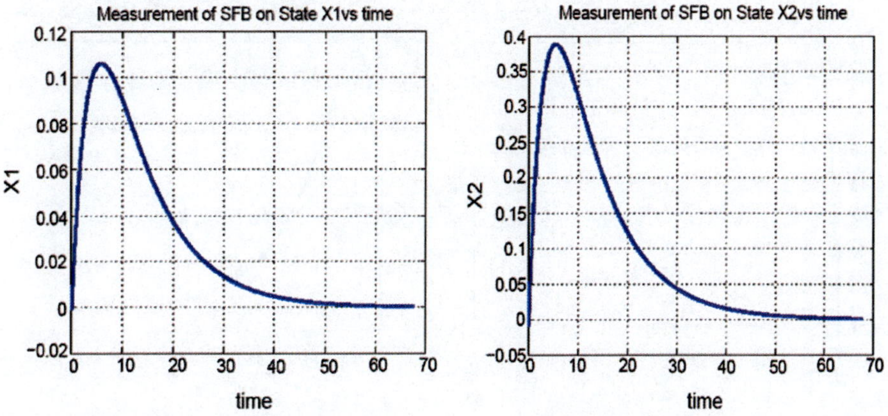

Fig. 6.42 Trajectory of state variables x_1 (*left*) and x_2 (*right*): Discrete-case

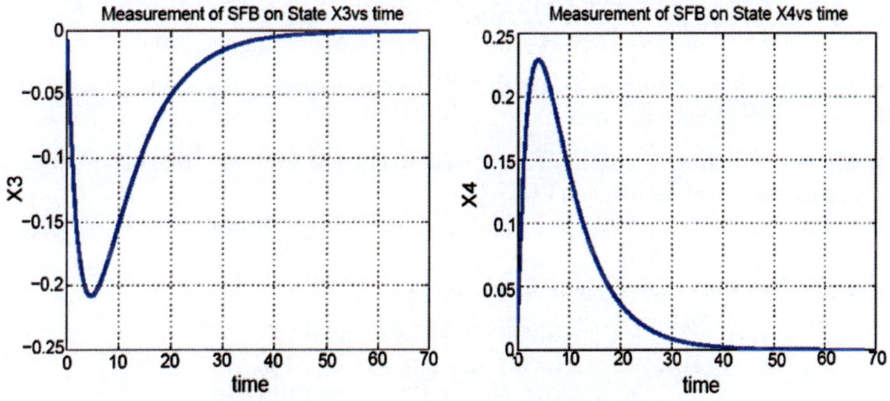

Fig. 6.43 Trajectory of state variables x_3 (*left*) and x_4 (*right*): Discrete-case

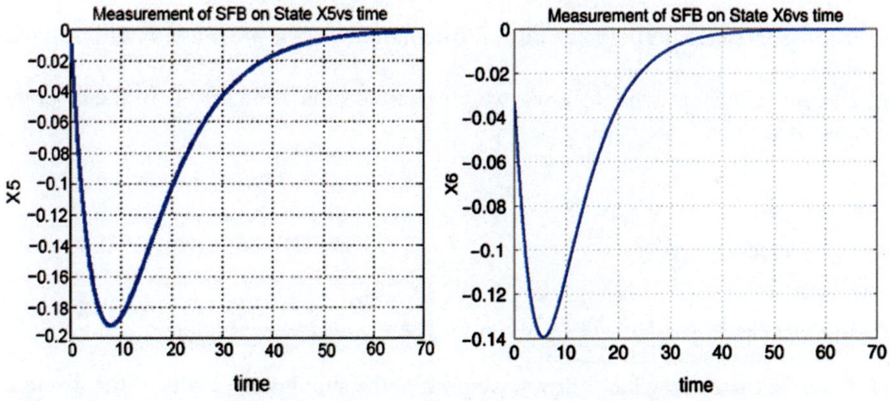

Fig. 6.44 Trajectory of state variables x_5 (*left*) and x_6 (*right*): Discrete-case

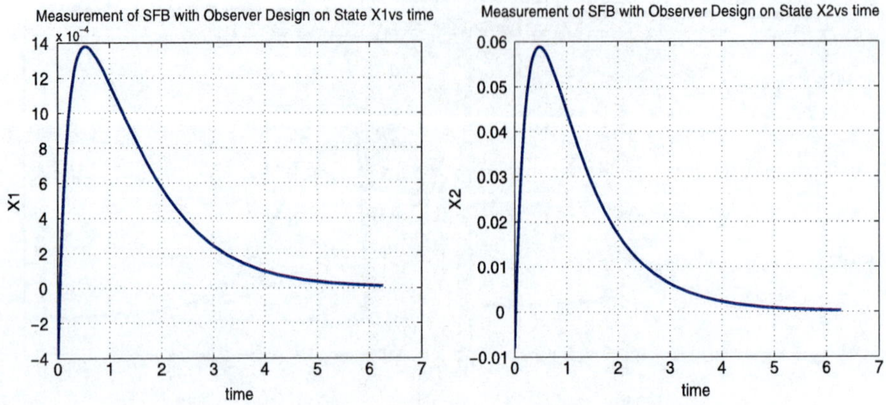

Fig. 6.45 Trajectory of state variables x_1 (*left*) and x_2 (*right*): Continuous observer-based

6.7.2 Observer Feedback Design

Now, we provide the simulation results of observer-based feedback design and start with the continuous case:

A. *Continuous Case*: In addition to the state-feedback design results, the eigenvalues for the observer is taken as

$$v2 = \begin{bmatrix} -1 & -2 & -3 & -2.5 & -1.5 & -5 \end{bmatrix}. \tag{6.25}$$

This yields the observer gain matrix as,

$$L = \begin{bmatrix} -0.0083 & -0.0412 & -0.1470 \\ 0.4337 & -0.8842 & 0.3158 \\ -0.3959 & 1.0391 & -0.1852 \\ 1.6329 & -2.3090 & 1.5799 \\ -1.4840 & 0.4975 & -3.8431 \\ -0.8085 & 0.6047 & -1.1962 \end{bmatrix}. \tag{6.26}$$

The associated simulation results of the continuous case are plotted in Figs. 6.45, 6.46, 6.47.

B. *Discrete Case*: In what follows, we present the simulation results of the discrete case, see Figs. 6.48, 6.49, 6.50.

6.7.3 LQR Designs

A. *Continuous Case*: In what follows, we present the simulation results of the continuous case, see Figs. 6.51–6.54.

B. *Discrete Case*: In what follows, we present the simulation results of the discrete case, see Figs. 6.55–6.58.

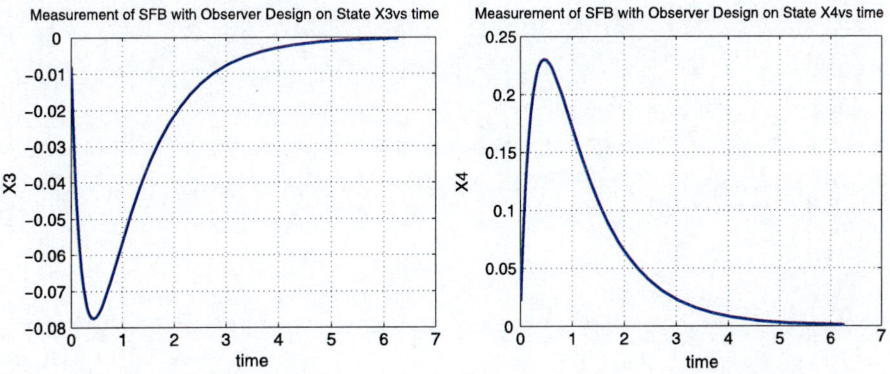

Fig. 6.46 Trajectory of state variables x_3 (*left*) and x_4 (*right*): Continuous observer-based

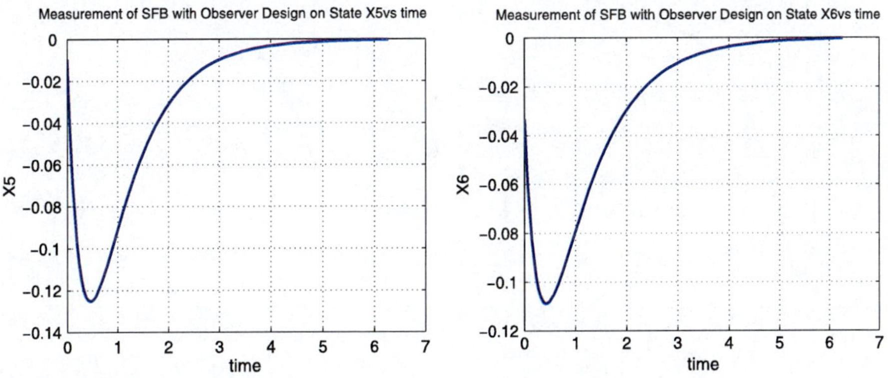

Fig. 6.47 Trajectory of state variables x_5 (*left*) and x_6 (*right*): Continuous observer-based

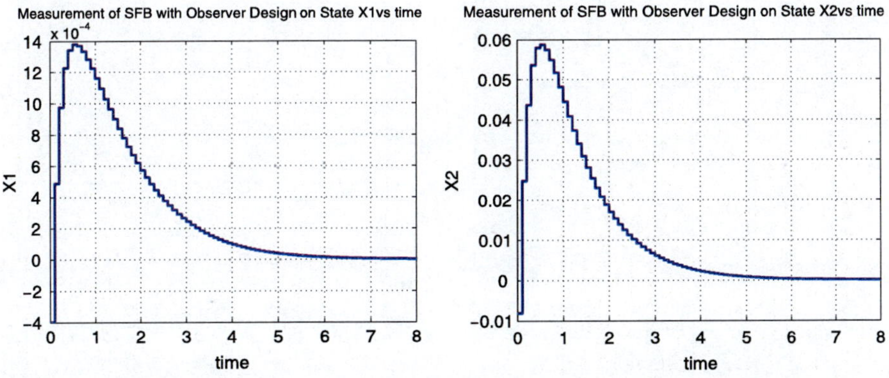

Fig. 6.48 Trajectory of state variables x_1 (*left*) and x_2 (*right*): Discrete observer-based

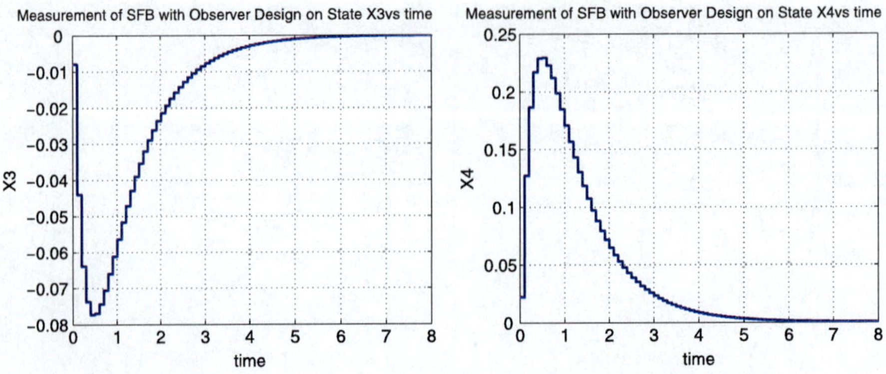

Fig. 6.49 Trajectory of state variables x_3 (*left*) and x_4 (*right*): Discrete observer-based

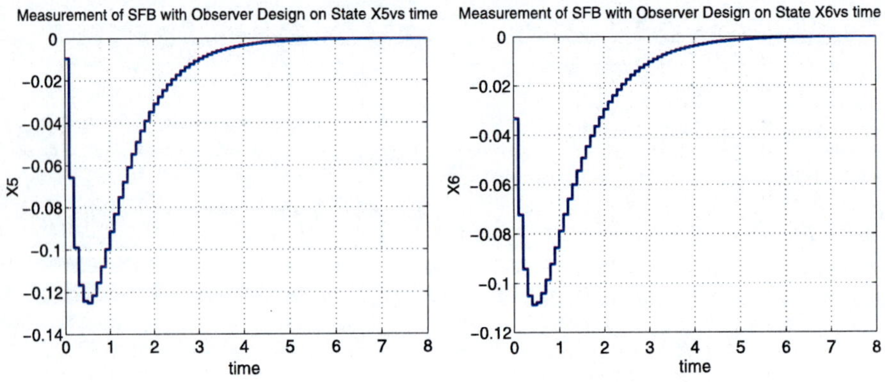

Fig. 6.50 Trajectory of state variables x_5 (*left*) and x_6 (*right*): Discrete observer-based

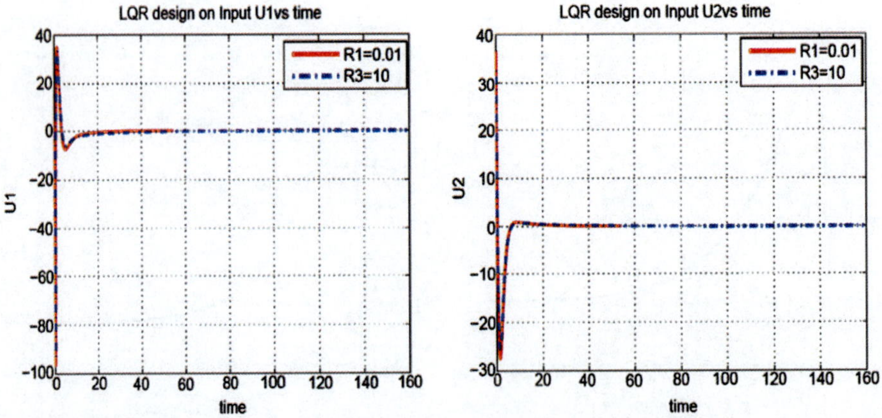

Fig. 6.51 Trajectories of input u_1 (*left*) and input u_2 (*right*): Continuous LQR

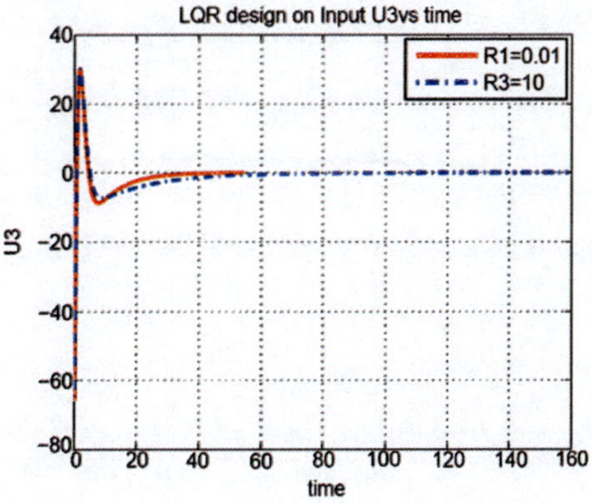

Fig. 6.52 Trajectory of input u_3: Continuous LQR

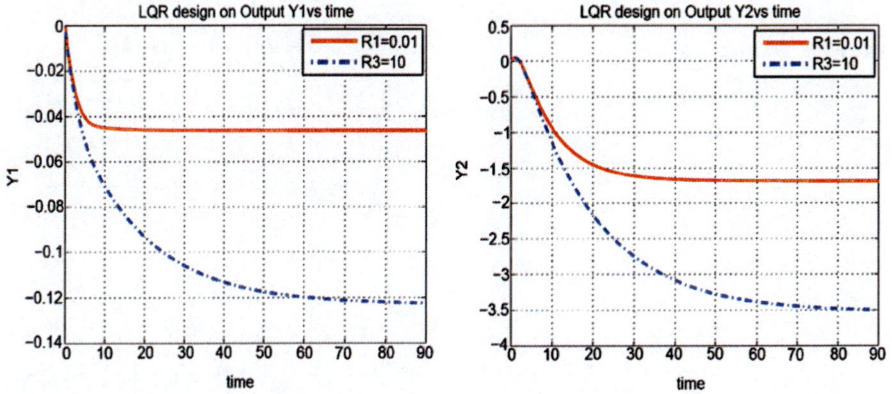

Fig. 6.53 Trajectory of output y_1 (*left*) and output y_2 (*right*): Continuous LQR

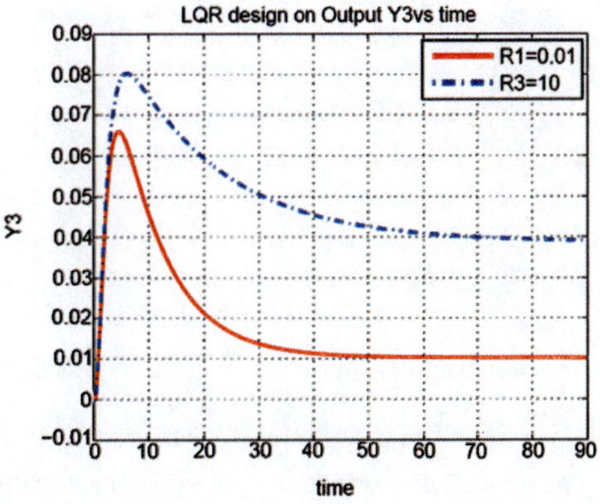

Fig. 6.54 Trajectory of output y_3: Continuous LQR

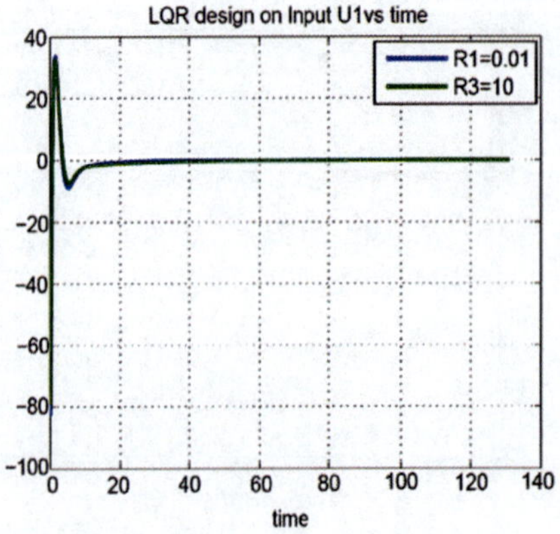

Fig. 6.55 Trajectories of input u_1 (*left*) and input u_2 (*right*): Discrete LQR

Fig. 6.56 Trajectory of input u_3: Discrete LQR

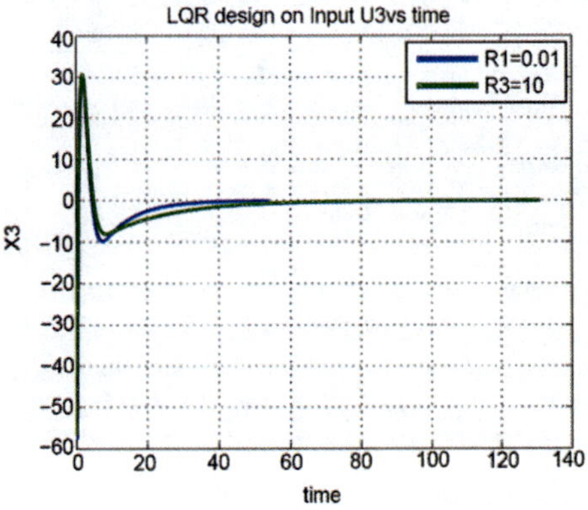

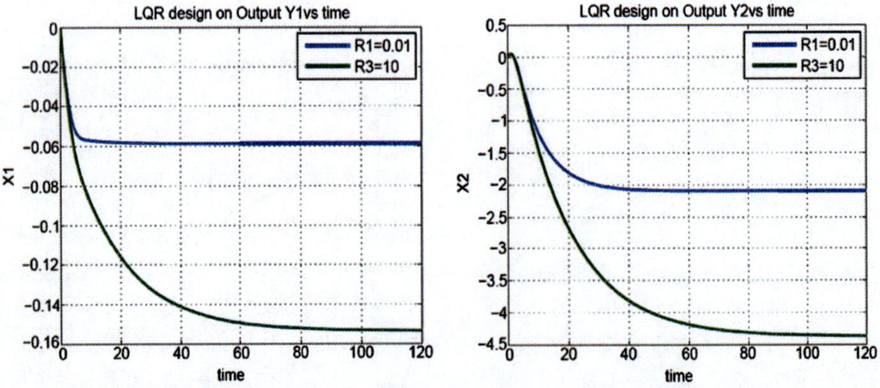

Fig. 6.57 Trajectories of output y_1 (*left*) and output y_2 (*right*): Discrete LQR

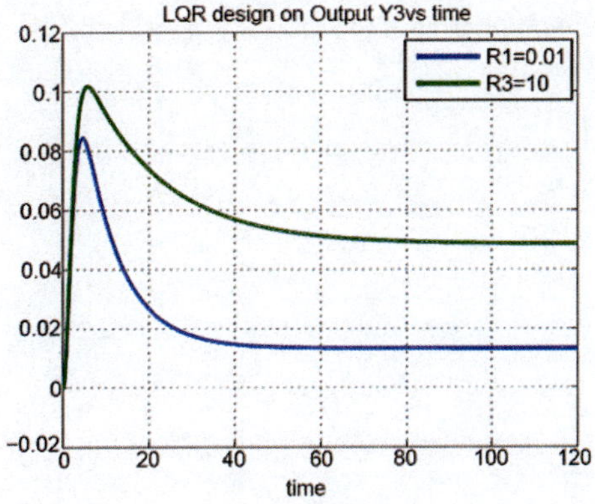

Fig. 6.58 Trajectory of output y_3: Discrete LQR

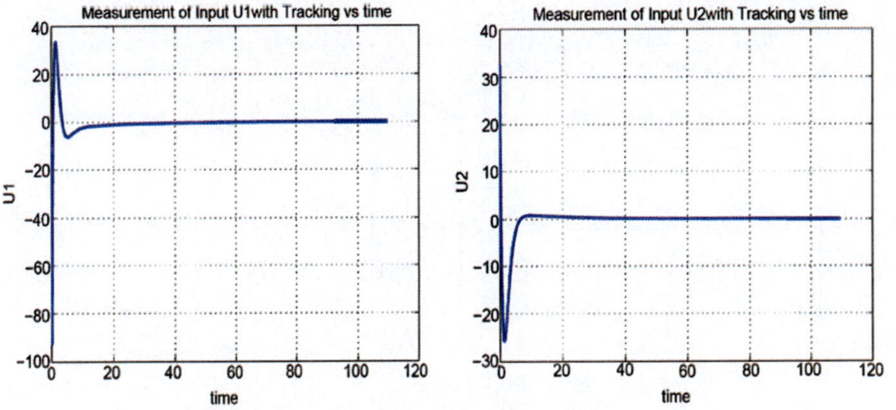

Fig. 6.59 Trajectories of input u_1 (*left*) and input u_2 (*right*): Tracking control

6.7.4 Tracking Control

The simulation results are depicted in Figs. 6.59–6.63.

6.8 Vapor Compression Cycle Systems

In what follows, the identified state-space models [**A**, **B**, **C**, **D**] based on two distinct cases: a) two output pressures and b) four output temperatures.

6.8.1 Model with Two Output Pressures

The state-space model based on two output pressures are given below. We note that the differential pressure, rather than actual pressures P_2 was used for identification.

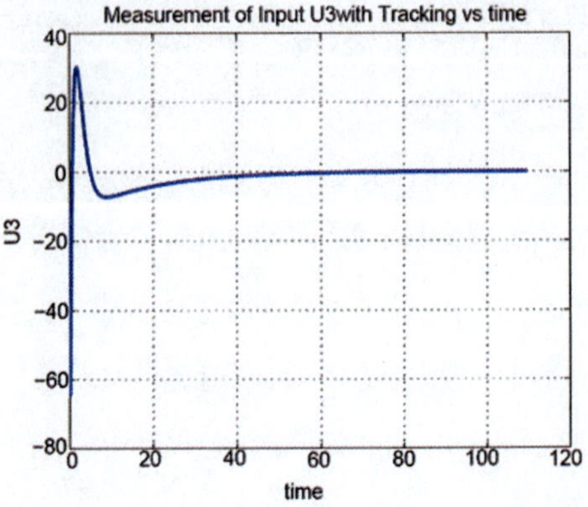

Fig. 6.60 Trajectory of input u_3: Tracking control

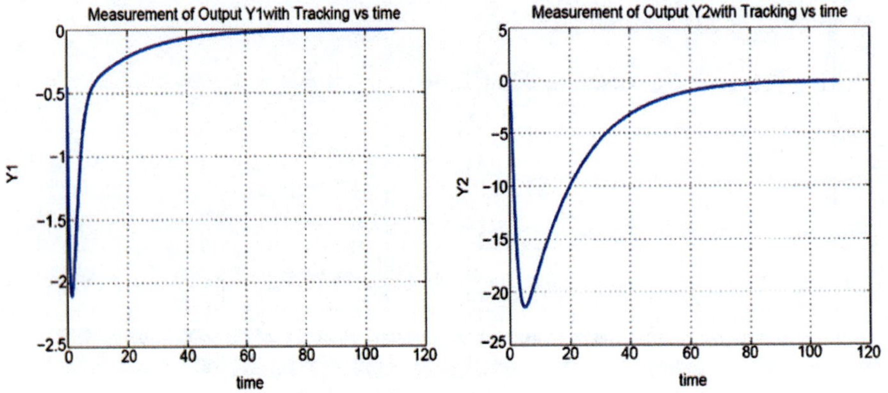

Fig. 6.61 Trajectories of output y_1 (*left*) and output y_2 (*right*): Tracking control

$$\mathbf{A} = \begin{bmatrix} A_1 & A_2 \\ A_3 & A_4 \end{bmatrix},$$

$$\mathbf{A}_1 = \begin{bmatrix} 0.9284 & 0.0352 & 0.2495 \\ -0.0029 & 0.5362 & -0.3899 \\ 0.2765 & -0.2607 & -0.1668 \end{bmatrix},$$

$$\mathbf{A}_2 = \begin{bmatrix} 0.0821 & 0.2620 & 0.0920 \\ 0.5040 & 0.0600 & -0.3541 \\ -0.1809 & -0.1152 & -0.5602 \end{bmatrix},$$

$$\mathbf{A}_3 = \begin{bmatrix} 0.1730 & 0.5656 & -0.1452 \\ -0.0759 & -0.2315 & 0.1099 \\ -0.0635 & 0.1477 & 0.3053 \end{bmatrix},$$

$$\mathbf{A}_4 = \begin{bmatrix} -0.0886 & -0.6043 & 0.0872 \\ 0.3524 & -0.8646 & -0.0532 \\ -0.0822 & -0.0022 & -0.7786 \end{bmatrix},$$

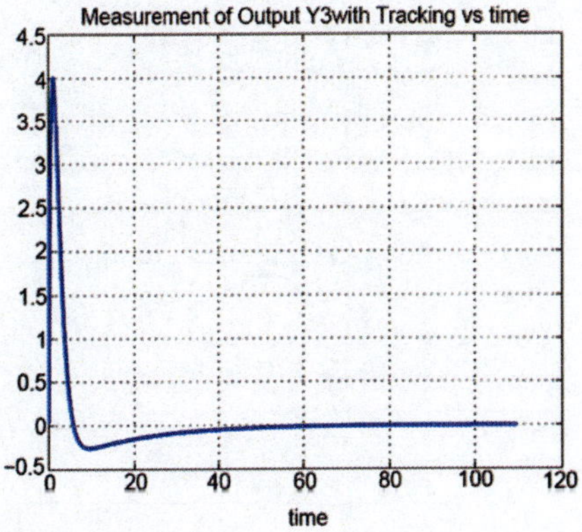

Fig. 6.62 Trajectory of output y_3: Tracking control

$$\mathbf{B} = \begin{bmatrix} -0.0101 & -0.0001 & 0.0377 \\ 0.0094 & -0.0000 & 0.0313 \\ 0.0500 & 0.0004 & -0.6057 \\ 0.0242 & 0.0003 & -0.3703 \\ -0.0143 & -0.0002 & 1.1068 \\ 0.0288 & 0.0004 & 0.1241 \end{bmatrix},$$

$$\mathbf{C} = \begin{bmatrix} C_1 & C_2 \end{bmatrix},$$

$$\mathbf{C}_1 = \begin{bmatrix} 267.7117 & 88.6283 & 19.3981 \\ 88.3195 & -236.8275 & 61.0602 \end{bmatrix},$$

$$\mathbf{C}_2 = \begin{bmatrix} 54.6803 & 26.2659 & 2.4993 \\ -73.9526 & -30.2428 & 37.3458 \end{bmatrix},$$

$$\mathbf{D} = \begin{bmatrix} 0 & 0 & 0 \\ 0 & 0 & 0 \end{bmatrix}.$$

6.8.2 Model with Four Output Temperatures

In this case, the identified state-space $[\mathbf{A}, \mathbf{B}, \mathbf{C}, \mathbf{D}]$ system model is based on four output temperatures and is given below.

$$\mathbf{A} = \begin{bmatrix} 0.9953 & 0.0002 & 0.0024 & 0.0004 \\ -0.0024 & 0.9955 & 0.0003 & -0.0096 \\ -0.0032 & -0.0087 & 0.9705 & 0.0131 \\ -0.0093 & -0.0262 & -0.0137 & 0.8923 \end{bmatrix},$$

$$\mathbf{B} = \begin{bmatrix} -0.0003 & -0.0000 & 0.0273 \\ -0.0002 & 0.0000 & 0.0260 \\ -0.0007 & 0.0000 & -0.1399 \\ -0.0018 & 0.0001 & 0.1591 \end{bmatrix},$$

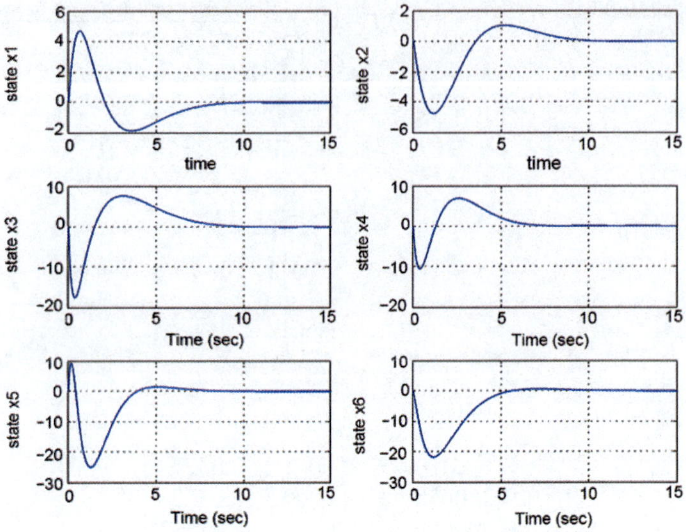

Fig. 6.63 State trajectories of system using *LQR*

$$\mathbf{C} = \begin{bmatrix} 71.4946 & 9.8191 & -6.6912 & 1.7531 \\ -14.0400 & 31.8341 & -4.8913 & 0.8771 \\ -12.8821 & -0.9395 & -3.9774 & 7.0485 \\ -22.2656 & -10.7576 & -11.4903 & -1.8273 \end{bmatrix}.$$

In the following, we present the results of simulating the closed-loop system with two output pressures using different controllers. The corresponding results for the case with four output temperatures are left as an exercise for the reader.

6.8.3 *LQR* Simulation Results: Continuous Case

For the simulation, we selected the weighting matrix R was kept constant and the matrix Q was varied in three different cases. The optimum results were obtained at $Q = 10 \times I_{4\times 4}$, $R = I_{2\times 2}$, which yields $\|K\| = 1.1889 \times 10^4$, Tr$(P)$ =7.1017 $\times 10^7$.

In our system, the six states have the same amount of significance. Hence, they have been weighted equally in each case. In the above simulation, the matrix R was used to weight the control input applied. The matrix Q was used to weight the states of the system. Simulation was carried out such that the weight on the inputs was kept constant and the weight on the states was varied to obtain the optimum results.

With respect to the norm of the gain matrix and the time taken by the states to settle to steady state, it is concluded that the case (iii) has yielded optimum results. In case (i), the number of oscillations in the states are more. Whereas, moving from case (ii) to case (iii) there is significant reduction in the number of oscillations. There is a decrease in the settling time of the states, but the gain K is increasing

Fig. 6.64 Comparison of state x_1

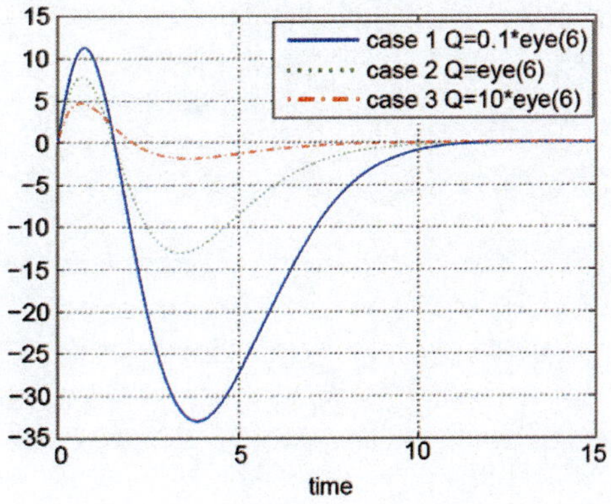

Fig. 6.65 Comparison of state x_2

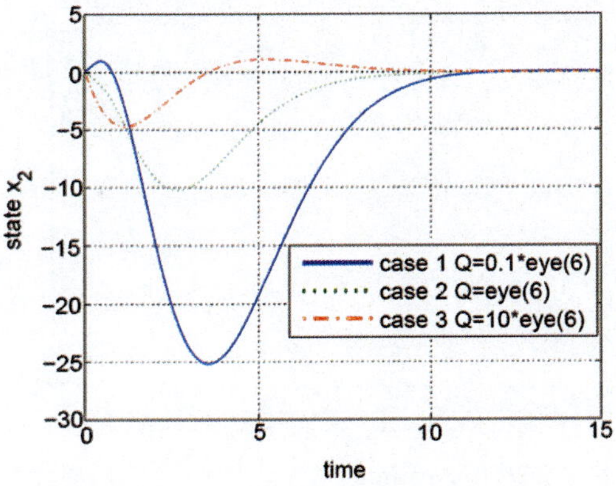

Fig. 6.66 Comparison of state x_3

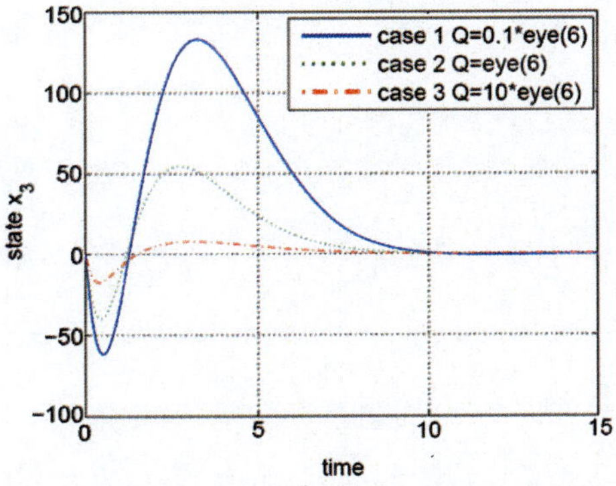

correspondingly. However, keeping in mind the stability of the system, case (iii) was found to be most suitable. The corresponding plots of state trajectories are presented in Figs. 6.64–6.69.

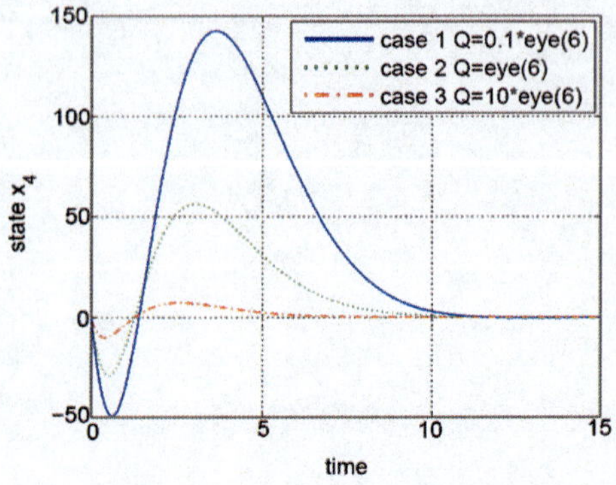

Fig. 6.67 Comparison of state x_4

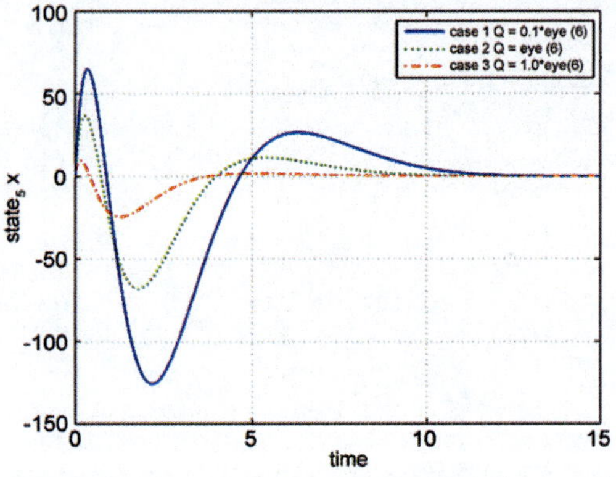

Fig. 6.68 Comparison of state x_5

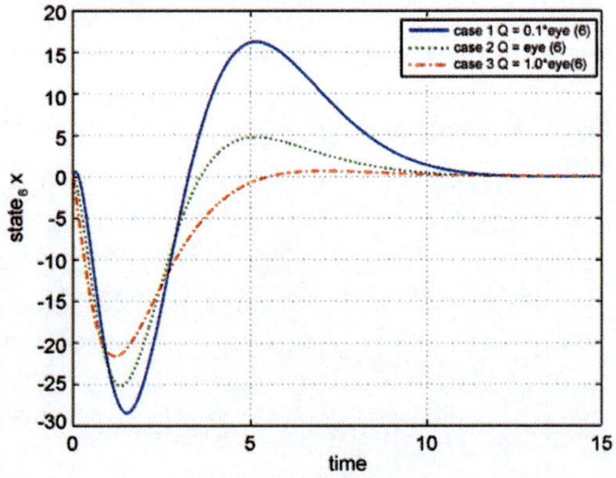

Fig. 6.69 Comparison of state x_6

6.8.4 LQR Simulation Results: Discrete Case

Simulation was carried out such that the weight on the inputs was kept constant and the weight on the states was varied to study the behavior of the system in three different cases. Optimum results were found using the weighting matrices $Q = I_{4\times 4}$, $R = I_{2\times 2}$. It is found that $\|K\| = 1.1294 \times 10^4$, $\mathrm{Tr}(P) = 7.1082 \times 10^7$.

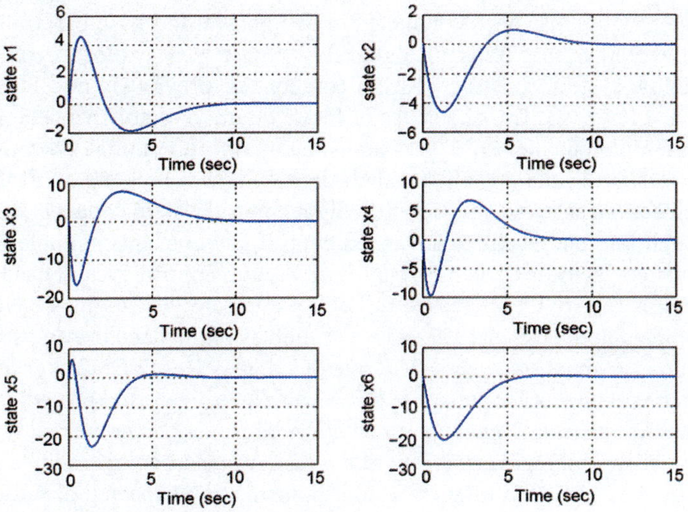

Fig. 6.70 State trajectories of system using *DLQR*

Of all the three cases simulated above, it is noted that the controller gain K is the largest in the third case, while the settling time is also the least in the third case. Hence, as we increase the controller gain, the settling time decreases. The response of the *DLQR* regulator is similar to the *LQR* regulator, the only difference being the control that is applied at discrete instants equal to the sampling time of the system model.

Just as in the continuous *LQR* all the have been weighted equally in each case while implementing the discrete regulator. In the above simulation, the matrix R was used to weight the control input applied. The matrix Q was used to weight the states of the system.

6.9 Stabilization of F-8 Fly-by-Wire Aircraft

NASA has been conducting research in digital fly-by-wire technology in a program called the NASA F-8 Digital Fly-By-Wire Program (DFBW). The broad objective of this program is to provide the technology required for implementation of advanced, reliable, DFBW flight control systems which will permit greater operational capability and increased performance of future aircraft. This program is being conducted jointly by the Dryden Flight Research Center, Edwards, CA, and the Langley Research Center, Hampton, VA. The program makes use of a test aircraft, an F-8 Crusade naval fighter aircraft, which has been modified by removal of the mechanical flight control system and its replacement with an electronic flight control system. In this modification, the pilot's mechanical linkages to primary actuator slide valves on the aircraft's control surfaces were replaced by electrical connections to secondary electro-hydraulic actuators which are then used to operate the primary actuator slide valves. The program has been conducted in two phases. In Phase I [I], pilot acceptability and technical feasibility of digital fly-by-wire were explored using a single channel digital system constructed from components previously developed for the Apollo Space Program. The objectives of Phase I1 are to establish a design base for practical multiple channel DFBW systems using a triplex digital system designed around three state-of-the-art, off-the-shelf digital flight computers, to flight test the system and certain selected space shuttle flight control system concepts, and to conduct research into and evaluate advanced control law concepts suitable for digital implementation. A triplex analog fly-by-wire backup control system has been used in both phases to provide increased reliability and safety of flight. Phase I flights were completed in the fall of 1973. Phase I1 flights commenced in August 1976 and will continue for about two years. The role of Langley Research Center in this program, which will be discussed herein, has been to investigate and promote advanced control laws for possible flight experimentation. This work is motivated by the much greater flexibility and logic capability of digital systems as compared to analog systems and by the increased complexity and sophistication expected of future aircraft flight control systems. Future control systems are expected to provide active control for modes of motion that are today either accomplished passively or not at all. For example, active controls for control configured vehicles (CCV) are being proposed to provide control over aircraft which are statically unstable aerodynamically, to modify span-wise wing lift distribution to reduce drag or provide structural load relief during high g maneuvering flight, to provide lower acceleration levels for pilot and passengers during wind turbulence, to provide flutter mode control, etc. Langley has promoted the integration of such concepts into an advanced control law package suitable for flight test. Flight tests of such a package, described in 121, will be conducted early in the Phase I1 program. Langley has also promoted advanced control concepts based on adaptation of the control system to the changing external environment of the airplane or to the failure of control system components internal to the aircraft. The purpose of this paper is to provide background material for the adaptive control law study papers that follow.

6.9.1 Linearized Model

We have relied on references [5, 19] for the following nonlinear model for the F8 aircraft longitudinal flight dynamics. The desired operating point corresponds to an altitude of 30,000 ft, again as in the references [5, 19]. The lift coefficients are complicated nonlinear functions of the angles of attack and elevator angle. For simplicity, we have again followed references [5, 19].

The F-8 is an "old-fashioned" aircraft that has been used by NASA as part of their digital "fly-by-wire" research program. We have modified the equations of motion by including a large "flaperon" on the wing so as to obtain two control variables in the longitudinal dynamics of the F-8. This flaperon does not exist in the F-8 aircraft. However, such surfaces exist in other recent aircraft, for example, the X-29, and provide some additional flexibility for precision maneuvers.

It has been assumed that the aircraft is flying at a constant altitude in equilibrium flight allows us to linearize the nonlinear equations of motion. In doing so, the longitudinal dynamics decouple from the lateral dynamics. The variables needed to characterize the longitudinal motion are as follows:

- Horizontal velocity $v(t)$,
- Pitch angle ϑ,
- Pitch rate, $q = \dot{\vartheta}$,
- Angle of attack α,
- Flight path angle $\beta = \vartheta - \alpha$,
- Elevators $\delta e(t)$, and
- Flaperons $\delta f(t)$.

The measurements are the pitch and flight path angles, $y(t) = [\vartheta \ \ \beta]$. The effect of wind gust disturbances, which primarily corrupt the angle of attack, is modeled as the output of a shaping filter driven with unit intensity white noise, $d(t)$.

The stabilization of the nonlinear airplane could be achieved in principle also by using linear feedback. The linearized, longitudinal equations of the F-8 aircraft are as follows:

$$\dot{x}(t) = \mathbf{A}x(t) + \mathbf{B}u(t) + \mathbf{L}d(t),$$
$$y(t) = \mathbf{C}x(t) + v(t),$$

where

$$\mathbf{A} = \begin{bmatrix} 0.0 & 0.0 & 1.0 & 0.0 & 0.0 \\ 1.50 & -1.50 & 0.0 & 0.0057 & 1.50 \\ -12.0 & 12.0 & -0.60 & -0.0344 & -12.0 \\ -0.852 & 0.290 & 0.0 & -0.0140 & -0.290 \\ 0.0 & 0.0 & 0.0 & 0.0 & -0.730 \end{bmatrix},$$

$$\mathbf{B} = \begin{bmatrix} 0.00 & 0.00 \\ 0.16 & 0.80 \\ -19.0 & -3.0 \\ -0.0115 & -0.0087 \\ 0.00 & 0.00 \end{bmatrix},$$

$$\mathbf{C}^t = \begin{bmatrix} 1 & 0 \\ 0 & 1 \\ 0 & 0 \\ 0 & 0 \\ 0 & 0 \end{bmatrix}, \quad \mathbf{L} = \begin{bmatrix} 0.00 \\ 0.00 \\ 0.00 \\ 0.00 \\ 1.1459 \end{bmatrix}, \quad x(t) = \begin{bmatrix} \vartheta(t) \\ \beta(t) \\ q(t) \\ v(t) \\ x_d(t) \end{bmatrix}$$

and $v(t)$ is white noise with an Intensity of $\mu = 0.01$ deg^2/s.

6.9.2 Simulation Results

The following are the MATLAB simulation results of the control techniques applied for the "f-8 fly-by-wire" aircraft stabilization. Each set of seven graphs in-

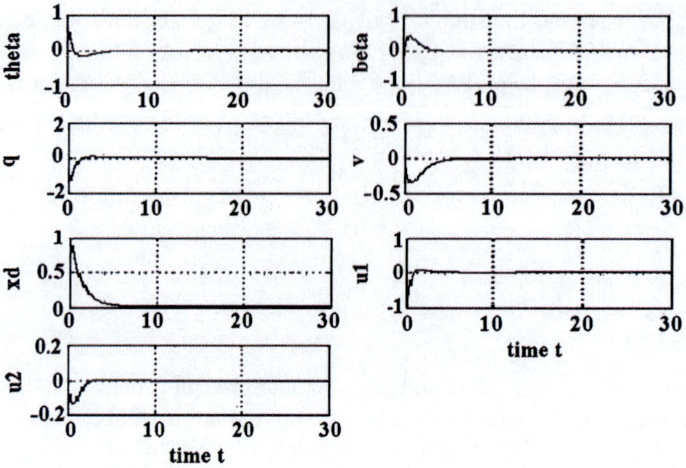

Fig. 6.71 LQR results

cludes graphs of the five states and two graphs of the controller inputs. Simulation results have been shown for the linear quadratic regulator control (LQR) and linear quadratic Gaussian control (LQGR).

It can be observe that the two control schemes are stabilizing the aircraft, however the LQG control has yielded the most suitable path with long period. On the other hand, depending upon the tolerance level of the state variables, the LQR Control seems to be the most unsuitable method of stabilizing the aircraft with a high overshoot factor which can lead the aircraft to a stall region. The corresponding state trajectories are plotted for LQR in Fig. 6.71 and LQGR in Fig. 6.72.

6.10 Air Conditioning System

A linearized dynamic model for a direct expansion (DX) A/C system was utilized. The physical system consists of six states, two inputs and two outputs. The model was developed to be able to capture the transient characteristics of the DX A/C system. This paper represents the work of designing different types of controllers such as State-feedback, Observer-based feedback, tracking control and integral control. The simplified schematic of the model is shown in Fig. 6.73.

The dynamic model, written in state-space representation which was suitable for designing multivariable control, was linearized at steady state operating points. The linearized model has been validated by comparing the model simulation results with the experimental data obtained from an experimental DX A/C system. The developed model was used in designing different multi-input multi-output (MIMO) controllers to simultaneously control indoor air temperature and humidity in a space served by a DX A/C system.

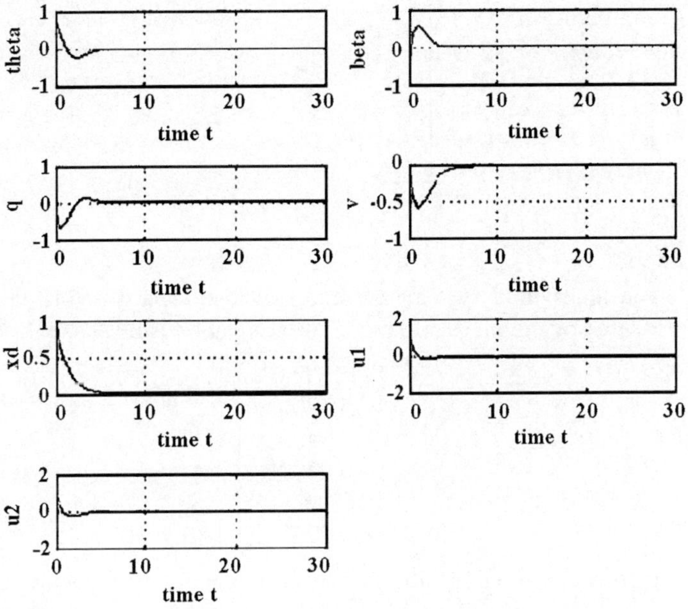

Fig. 6.72 LQGR results

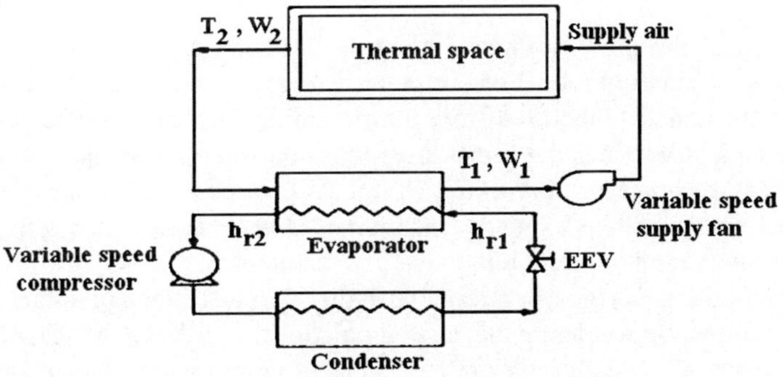

Fig. 6.73 The schematic diagram of the experimental DX A/C system

The system matrices A, B and C are as follows:

$$A = \begin{bmatrix} -5.731 & 0 & 0.0756 & 4.1883 & -5287 & 5287 \\ 0.0045 & -0.0045 & 0 & 0 & 0 & 0 \\ 0 & 4.6577 & -12.692 & 8.0346 & 0 & 0 \\ 0.0139 & 0.0067 & 0.0206 & -0.0412 & 0 & 0 \\ 0.00016 & 0 & 0 & 0 & 0 & 0 \\ 0 & 0 & 0 & 0 & 0.0045 & -0.0045 \end{bmatrix},$$

$$B = \begin{bmatrix} 55.035 & 0 \\ -0.098 & 0 \\ 172.5 & 0 \\ 0 & -5.931 \\ 0 & 0 \\ -0.00003 & 0 \end{bmatrix},$$

$$C = \begin{bmatrix} 0 & 1 & 0 & 0 & 0 & 0 \\ 0 & 0 & 0 & 0 & 0 & 1 \end{bmatrix}.$$

All the eigenvalues of the system have the negative real parts, so that the DX A/C system represented by the linearized model was asymptotically stable.

$$\lambda = \begin{bmatrix} -12.7050 \\ -5.5902 \\ -0.1436 \\ -0.0299 \\ -4.6144\text{e-}16 \\ -0.0045 \end{bmatrix}. \tag{6.27}$$

The system is fully controllable with rank 6. Also, it is observable with rank 6.

6.10.1 State-Feedback

With state space design, we remain in the time domain and thus work directly with the differential equation model of our plant. It is important to realize that whether we work with transfer functions or with differential equations in state space form, the mathematics describes the same thing and the forms can be interchanged. The major advantage however of working with a state space model of a system is that the internal system state is explicitly maintained over time, where as with a transfer function, only the input output relationship is maintained.

We would like to design a controller such that the closed loop poles are at certain desired locations. So we define the desired pole locations. Using MATLAB, we got the gain matrix K. The closed state feedback response to step change is shown in Fig. 6.74. The characteristic polynomial for this closed-loop system is the determinant of $(sI - (A - BK))$. Since the matrices A and $B * K$ are both 6 by 6 matrices, there will be 6 poles for the system. By using full-state feedback, we can place the poles anywhere we want. We could use the MATLAB function *place* to find the control matrix, K, which will give the desired poles.

K has as many elements (degrees of freedom) as there are poles. This means that we can place the closed loop poles anywhere as long as the system is controllable from the input. Calculating the feedback gain matrix K and then converting the gain back so that it is applicable to the original state vector. It uses the extra degrees of freedom provided by these inputs to not only place the eigenvalues of the closed loop system but to also 'shape' the eigenvectors such that the closed loop system is 'well-conditioned'.

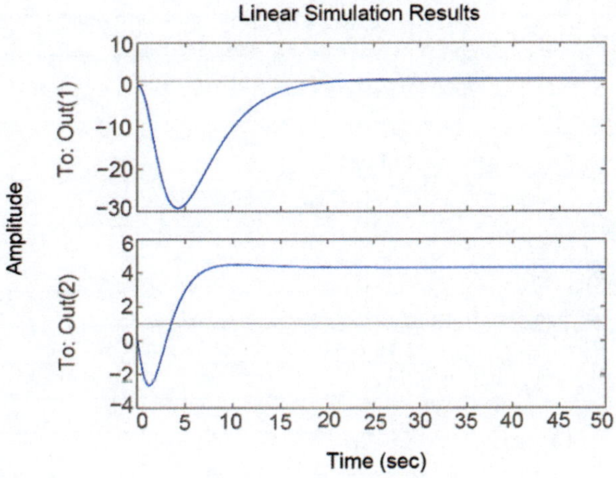

Fig. 6.74 Output trajectories by state-feedback controller

6.10.2 Observer-Based Feedback

Previously, we designed controllers using full state feedback. The state however is not usually directly available through measurements. The idea behind the estimator is to place a model of the plant in parallel with the actual plant and to drive them both with the same input. If the model's initial state vector is set equal to the plant's initial state vector then the state estimate (generated by the model) will track the actual state vector. However, there are always uncertainties in the plant model and in practice, without feedback, the state estimate would diverge from the true state. The solution is to use the measurement $y(t)$ and to compare it with the model's predicted measurement and use the difference between the two to modify the state estimate in such a way that it converges to the true state vector. We can build an observer to estimate them, while measuring only the output $y(t) = \mathbf{C}x(t)$.

The observer is basically a copy of the plant; it has the same input and almost the same differential equation. An extra term compares the actual measured output $y(t)$ to the estimated output; this will cause the estimated states$\hat{x}(t)$ to approach the values of the actual states $x(t)$. The error dynamics of the observer are given by the poles of $(A - L * C)$.

First, we need to choose the observer gain L. Since we want the dynamics of the observer to be much faster than the system itself, we need to place the poles at least five times farther to the left than the dominant poles of the system. If we want to use *place*, we need to put the three observer poles at different locations. The corresponding state trajectories are plotted in Fig. 6.75.

6.10.3 Tracking Control

Recall the state space feedback, we don't compare the output to the reference; instead we measure all the states, multiply by the gain vector K, and then subtract this result from the reference. There is no reason to expect that $K * x(t)$ will be equal

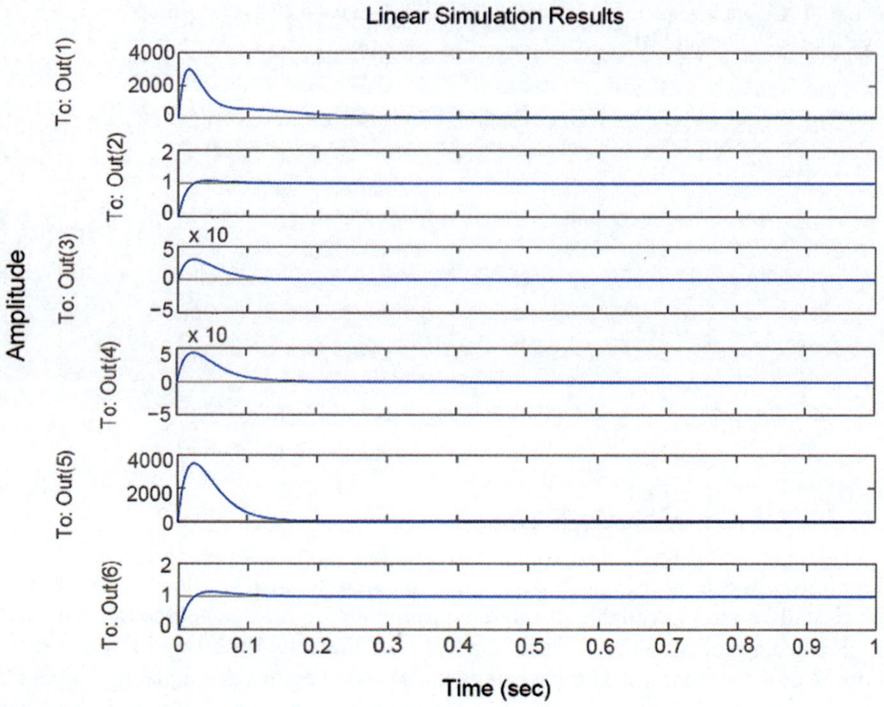

Fig. 6.75 State trajectories by observer-based controller

Fig. 6.76 Output trajectories by tracking controller

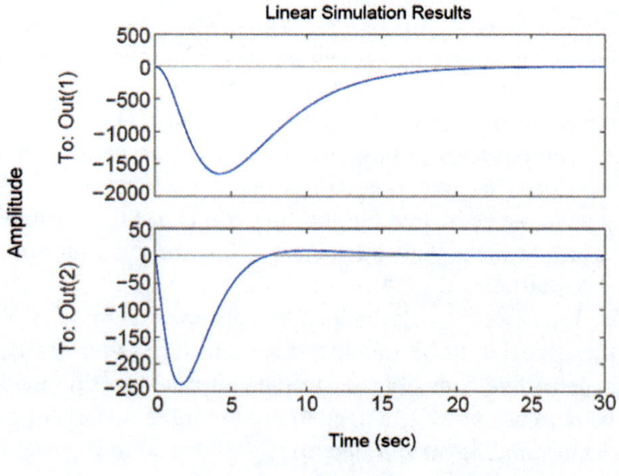

to the desired output. To eliminate this problem, we can scale the reference input to make it equal to $K * x(t)$ steady state. This scale factor is often called Nbar. If we want to find the response of the system under state feedback with this introduction of the reference, we simply note the fact that the input is multiplied by this new factor, Nbar. Now a step can be tracked reasonably well. The corresponding state trajectories are plotted in Fig. 6.76.

6.11 Three-Degree-of-Freedom Helicopter Model

Predictive control strategies have been widely used in industry for their ability to handle operational constraints. It is known that the presence of disturbances may cause predictive controllers to lose feasibility and to violate system constraints. This section addresses the implementation of a state-space predictive control law with restricted constraints to ensure feasibility and constraint fulfillment in spite of the existence of unknown but bounded disturbances.

6.11.1 Linearized Model

In the sequel, we consider a nonlinear, sixth order, three-degree-of-freedom (3DOF) helicopter model, see Fig. 6.77. Our objective is to achieve state regulation subject to bounded disturbances as well as state and control polyhedral constraints. The constraints on the maneuvering space are assumed to be convex polyhedral sets. As illustrated in [35, 36], the model is composed by the helicopter body, which is a small arm with one propeller at each end, and the helicopter arm, which connects the body to a fixed base. Although the system cannot exhibit translational motion, as it is fixed in a support, it can rotate freely about three axes. The helicopter position is characterized by the pitch, travel and elevation angles. The pitch movement corresponds to the rotation of the helicopter body about the helicopter arm, the travel movement corresponds to the rotation of the helicopter arm about the vertical axis and the elevation movement corresponds to the rotation of the helicopter arm about the horizontal axis. The control variables are the input voltages to the power amplifiers that drive each one of the two DC motors connected to the helicopter propellers. The maximum input voltage to the amplifiers is 5 V. Three digital encoders provide measurements of the helicopter angles. Encoder resolution is about 0.044° for travel

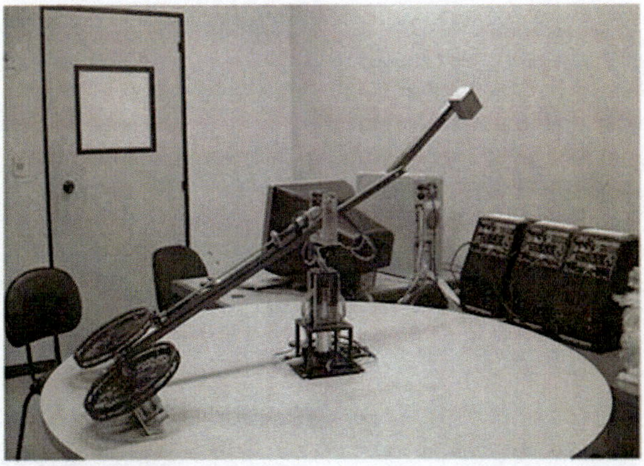

Fig. 6.77 Quanser 3DOF helicopter

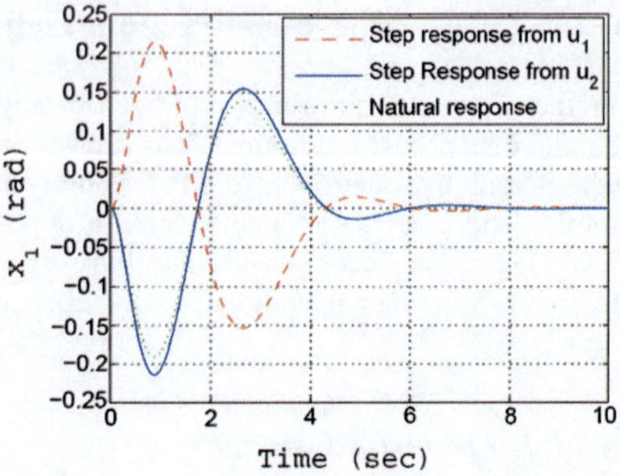

Fig. 6.78 Trajectories of pitch angle

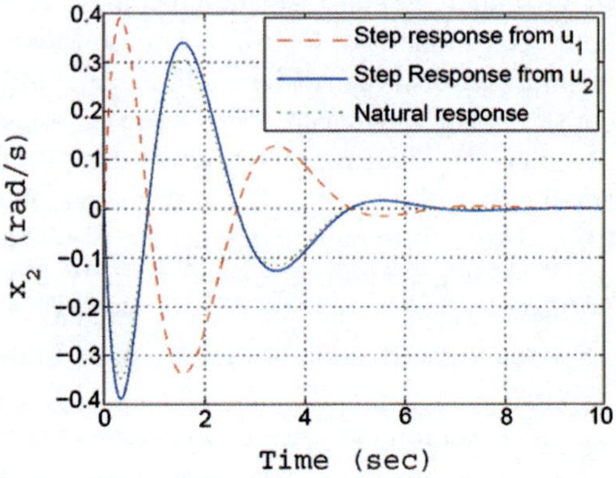

Fig. 6.79 Trajectories of pitch rate

angle and 0.088° for pitch and elevation angles. The original nonlinear model has x_1 is the pitch angle (in rad), x_2 is the pitch rate (in rad/s), x_3 is the elevation angle (in rad), x_4 is the elevation rate (in rad/s), x_5 is the travel angle (in rad), x_6 is the travel rate (in rad/s), u_1 is the front motor amplifier input voltage (in V), and u_2 is the back motor amplifier input voltage (in V).

An approximate linear model obtained by applying a first-order Taylor series expansion around a given equilibrium point $\bar{x} = [0 \;\; 0 \;\; -0:122 \;\; 0 \;\; 0 \;\; 0]^t$, $\bar{u} = [2.804, 2.804]^t$, which corresponds to helicopter hovering seven degrees below the horizontal, can be expressed as:

$$\dot{\hat{x}} = \begin{bmatrix} 0 & 1 & 0 & 0 & 0 & 0 \\ 0 & 0 & 0 & 0 & 0 & 0 \\ 0 & 0 & 0 & 1 & 0 & 0 \\ 0 & 0 & -1.192 & 0 & 0 & 0 \\ 0 & 1 & 0 & 0 & 0 & 1 \\ -1.257 & 0 & 0 & 0 & 0 & 0 \end{bmatrix} \hat{x} + \begin{bmatrix} 0 & 0 \\ 2.806 & -2.806 \\ 0 & 0 \\ 0.395 & 0.395 \\ 0 & 0 \\ 0 & 0 \end{bmatrix} \hat{u}.$$

The corresponding state trajectories are plotted in Figs. 6.78–6.83 for different cases.

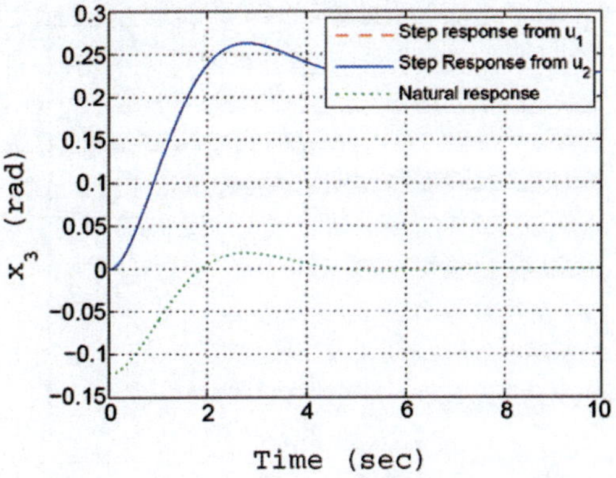

Fig. 6.80 Trajectories of elevation angle

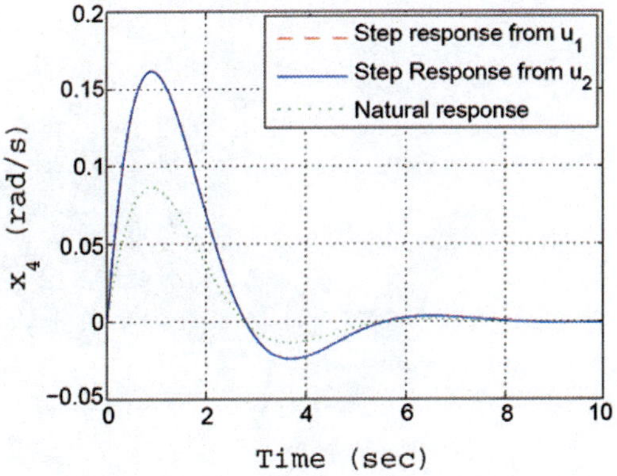

Fig. 6.81 Trajectories of elevation rate

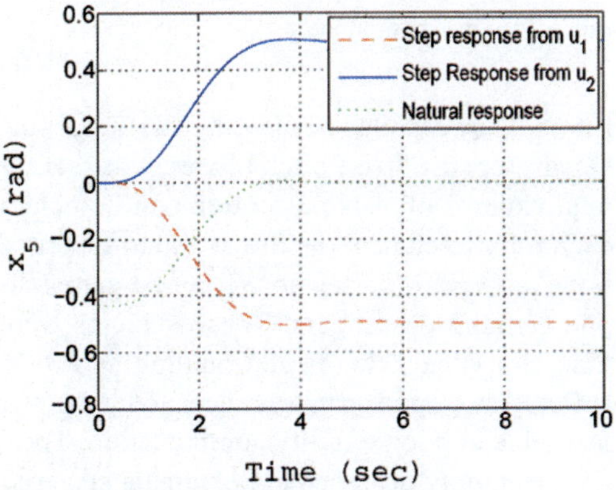

Fig. 6.82 Trajectories of travel angle

6.12 PID Control of a Quadrotor Unmanned Air Vehicle

A Quadrotor, also called a Quadrotor helicopter or Quadrocopter, is an aircraft that is lifted and propelled by four rotors, see Fig. 6.84. Quadrotors are classified as rotorcraft, as opposed to fixed-wing aircraft, because their lift is derived from four

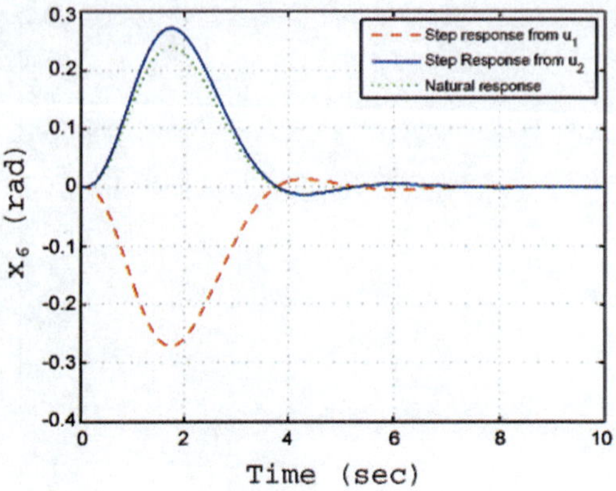

Fig. 6.83 Trajectories of travel rate

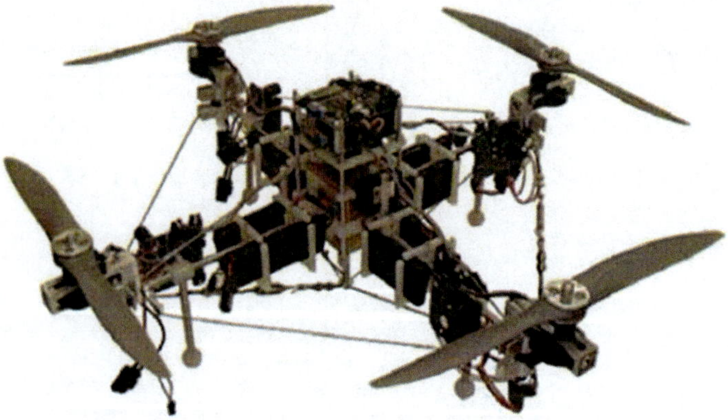

Fig. 6.84 Quadrotor model

rotors. They can also be classified as helicopters, though unlike standard helicopters, Quadrotors use fixed-pitch blades, whose rotor pitch does not vary as the blades rotate. Control of vehicle motion can be achieved by varying the relative speed of each rotor to change the thrust and torque produced by each. There are two generations of Quadrotor designs. The first generation Quadrotors were designed to carry one or more passengers. These vehicles were among the first successful heavier-than-airvertical takeoff and landing (VTOL) vehicles. However, early prototypes suffered from poor performance, and latter prototypes required too much pilot work load, due to poor stability augmentation. The more recent generation of Quadrotors are commonly designed to be unmanned aerial vehicles (UAVs). These vehicles use an electronic control system and electronic sensors to stabilize the aircraft. With their small size and agile maneuverability, these Quadrotors can be flown indoors as well as outdoors. There are a lot of advantages of the current generation of Quadrotors, versus comparably scale helicopters. For instance, Quadrotors do not require mechanical linkages to vary the rotor blade pitch angle as they spin. This simplifies the design of the vehicle, and reduces maintenance time and cost. Moreover, the use of four rotors allows each individual rotor to have a smaller diameter than the equivalent helicopter rotor, for a given vehicle size, allowing them to store less kinetic energy during flight. This reduces the damage caused should the rotors hit any objects. For small scale UAVs, this makes the vehicles safer to interact with in close proximity.

Unmanned Aerial Vehicles (UAVs) are defined as aircrafts without the onboard presence of pilots [50]. UAVs have been used to perform intelligence, surveillance, and reconnaissance missions. The technological promise of UAVs is to serve across the full range of missions. UAVs have several basic advantages over manned systems including increased maneuverability, reduced cost, reduced radar signatures, longer endurance, and less risk to crews. Vertical take-off and landing type UAVs exhibit even further maneuverability features. Such vehicles are to require little human intervention from take-off to landing. UAVs have potential for fulfilling many civil and military applications including surveillance, intervention in hostile environments, air pollution monitoring, and area mapping [10].

Unmanned aerial vehicles (UAV) have shown a growing interest thanks to recent technological projections, especially those related to instrumentation. They made possible the design of powerful systems (mini drones) endowed with real capacities of autonomous navigation at reasonable cost.

6.12.1 Introduction

In this section, we are studying the behavior of the quadrotor. This flying robot presents the main advantage of having quite simple dynamic features. Indeed, the quadrotor is a small vehicle with four propellers placed around a main body. The main body includes power source and control hardware. The four rotors are used to controlling the vehicle. The rotational speeds of the four rotors are independent. Thanks to this independence, it's possible to control the pitch, roll and yaw attitude of the vehicle. Then, its displacement is produced by the total thrust of the four rotors whose direction varies according to the attitude of the quadrotor. The vehicle motion can thus be controlled. There have been numerous projects involving quadrotors to date, with the first known hover reported in [32]. Recent interest in the quadrotor concept has been sparked by commercial remote control versions, such as the DraganFlyer IV [14]. Many groups [4, 9, 21, 45] have seen significant success in developing autonomous quadrotor vehicles. Nowadays, the mini-drones invade several application domains [20]: safety (monitoring of the airspace, urban and interurban traffic); natural risk management (monitoring of volcano activities); environmental protection (measurement of air pollution and forest monitoring); intervention in hostile sites (radioactive workspace and mine clearance), management of the large infrastructures (dams, high-tension lines and pipelines), agriculture and film production (aerial shooting).

In contrast to terrestrial mobile robots, for which it is often possible to limit the model to kinematics, the control of aerial robots (quadrotor) requires dynamics in order to account for gravity effects and aerodynamic forces [3].

In general, existing quadrotor dynamic models are developed on the hypothesis of a unique rigid body which is a restrictive hypothesis that does not account for the fact that the system is composed of five rigid bodies: four rotors and a crossing body frame. This makes the explanation of several aspects, like gyroscopic effects, very difficult. Additionally, simplification hypotheses are generally introduced early in the model development and leads in general to misleading interpretations.

6.12.2 Dynamic Modeling

A quadrotor is an under actuated aircraft with fixed pitch angle four rotors as shown in Fig. 6.85. Modeling a vehicle such as a quadrotor is not an easy task because of its complex structure. The aim is to develop a model of the vehicle as realistically as possible. In the quadrotor, there are four rotors with fixed angles which represent four input forces that are basically the thrust generated by each propeller as shown in Fig. 6.85. The collective input (u_1) is the sum of the thrusts of each motor. Pitch movement is obtained by increasing (reducing) the speed of the rear motor while reducing (increasing) the speed of the front motor. The roll movement is obtained similarly by increasing (reducing) the speed of the right motor while reducing (increasing) the speed of the left motor. The yaw movement is obtained by increasing (decreasing) the speed of the front and rear motors together while decreasing (increasing) the speed of the lateral motors together. This should be done while keeping the total thrust constant.

Each of the controller inputs affects certain side of the quadrotor model, u_2 here affects the rotation in the roll angle while u_3 affect the pitch angle and u_4 control the yaw angle during the flying process and u_1 affect the altitude (z-axis) for this model. Each rotor produces moments as well as vertical forces. These moments have been experimentally observed to be linearly dependent on the forces for low

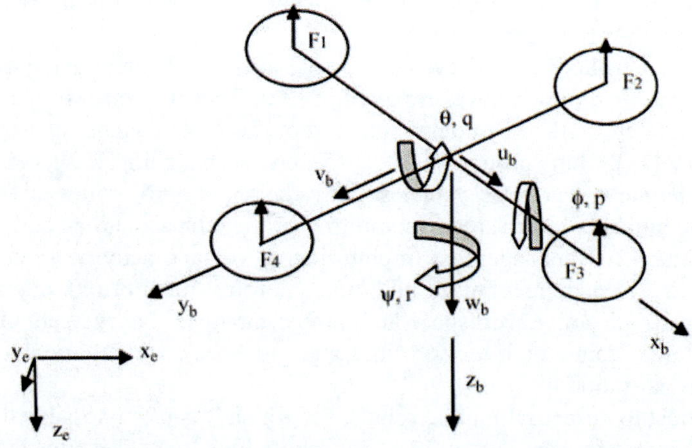

Fig. 6.85 The quadrotor schematic

speeds. There are four input forces and six output states $(x, y, z, \theta, \psi, \phi)$ therefore the quadrotor is an under-actuated system. The rotation direction of two of the rotors are clockwise while the other two are counterclockwise, in order to balance the moments and produce yaw motions as needed.

The compensation of this torque in the center of gravity is established thanks to the use of contra rotating rotors 1–3 and 2–4. Recall that rotors 2 and 4 turn counterclockwise while rotors 1 and 3 turn clockwise. In order to move the quadrotor model from the earth to a fixed point in the space, the mathematical design should depend on the direction cosine matrix as follows:

$$R_{zky} = \begin{bmatrix} C_\varphi C_\theta & C_\varphi S_\theta S_\psi - S_\varphi C_\psi & C_\varphi S_\theta C_\psi + S_\varphi S_\psi \\ C_\varphi S_\theta & S_\varphi S_\theta S_\psi + C_\varphi C_\psi & S_\varphi S_\theta C_\psi - C_\varphi S_\psi \\ -S_\theta & C_\theta S_\psi & C_\varphi C_\psi \end{bmatrix} \tag{6.28}$$

where

- $S_\phi = \sin\phi$, $C_\psi = \cos\psi$, etc.
- R is the matrix transformation.
- φ is the Roll angle.
- θ is the Pitch angle.
- ψ is the Yaw angle.

The dynamic model of the quadrotor helicopter can be obtained via a Lagrange approach and a simplified model is given as follow [5].

The equations of motion can be written using the force and moment balance.

$$\begin{aligned} \ddot{x} &= u_1(\cos\phi \sin\theta \cos\psi + \sin\phi \sin\psi) - K_1\dot{x}/m, \\ \ddot{y} &= u_1(\sin\phi \sin\theta \cos\psi + \cos\phi \sin\psi) - K_2\dot{y}/m, \\ \ddot{z} &= u_1(\cos\phi \cos\psi) - g - K_3\dot{z}/m, \end{aligned} \tag{6.29}$$

where

- x: Forward position in earth axes.
- y: Lateral position in earth axes.
- z: Vertical position in earth axes.
- K_i: The Drag Coefficients for the system.

The K_i's given above are the drag coefficients. In the following we assume the drag is zero, since drag is negligible at low speeds. The center of gravity is assumed to be at the middle of the connecting link. As the center of gravity moves up (or down) d units, then the angular acceleration becomes less sensitive to the forces, therefore stability is increased. Stability can also be increased by tilting the rotor forces towards the center. This will decrease the roll and pitch moments as well as the total vertical thrust.

For convenience, we will define the inputs to be:

$$\begin{aligned} U_1 &= (Th_1 + Th_2 + Th_3 + Th_4)/m, \\ U_2 &= l(-Th_1 - Th_2 + Th_3 + Th_4)/I_1, \\ U_3 &= l(-Th_1 + Th_2 + Th_3 - Th_4)/I_2, \\ U_4 &= C(Th_1 + Th_2 + Th_3 + Th_4)/I_3, \end{aligned} \tag{6.30}$$

where

- u_1: Vertical thrust generated by the four rotors.
- u_2: Pitching moment.
- u_3: Yawing moment.
- u_4: Rolling moment.
- Th_i: The thrusts generated by four rotors.
- I_i: The moments of inertia with respect to the axes,

where Th_i's are thrusts generated by four rotors and can be considered as the real control inputs to the system, and C the force to moment scaling factor. And I_i's are the moment of inertia with respect to the axes. Therefore, the equations of Euler angles become:

$$\begin{aligned} \ddot{\theta} &= u_2 - lK_4\dot{\theta}/I_1, \\ \ddot{\psi} &= u_3 - lK_5\dot{\psi}/I_2, \\ \ddot{\varphi} &= u_4 - K_6\dot{\varphi}/I_3, \end{aligned} \tag{6.31}$$

where (x, y, z) are three positions; (θ, φ, ψ) three Euler angles, representing pitch, roll and yaw, respectively; g the acceleration of gravity; l the half length of the helicopter; m the total mass of the helicopter; I_i's the moments of inertia with respect to the axes; K_i's the drag coefficients.

This quadrotor helicopter model has six outputs $(x, y, z, \theta, \psi, \varphi)$ while it only has four independent inputs, therefore the quadrotor is an under-actuated system. We are not able to control all of the states at the same time. A possible combination of controlled outputs can be x, y, z and φ in order to track the desired positions, move to an arbitrary heading and stabilize the other two angles, which introduces stable zero dynamics into the system [3, 45]. A good controller should be able to reach a desired position and a desired yaw angle while keeping the pitch and roll angles constant.

By applying Pythagoras theorem and implementing some assumptions and cancellations as follows:

1. The quadrotor structure is symmetrical and rigid.
2. The Inertia matrix (I) of the vehicle is very small and to be neglected.
3. The center of mass is placed at the origin o.
4. The propellers are rigid.
5. Thrust and drag are proportional to the square of the propellers speed.

These above equations have been established assuming that the structure is rigid and the gyroscopic effect resulting from the propellers rotation has been neglected.

The φ and ψ can be extracted in the following expressions:

$$\begin{aligned} \varphi_d &= \tan^{-1}\left(\frac{y_d - y}{x_d - x}\right), \\ \psi_d &= \tan^{-1}\left(\frac{z_d - z}{\sqrt{(x_d - x)^2 + (y_d - y)^2}}\right), \end{aligned} \tag{6.32}$$

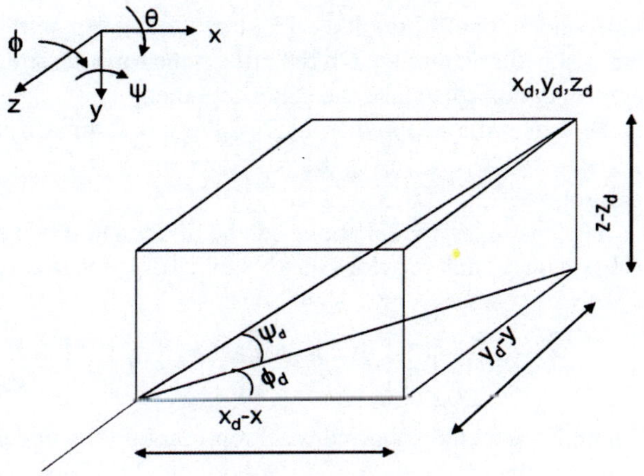

Fig. 6.86 The quadrotor angles movements

where

- φ is the desired yaw angle.
- ψ is the desired roll angle.

By supplying the four motors with the required voltage, the system will be on, the thrust here is directly proportional with these voltages, whenever increasing the voltage, the thrust for the motor increase and vice versa. The profile of quadrotor angle movements is depicted in Fig. 6.86.

6.12.3 PID Control Design

The PID design are pointed out in many references, such as [17], that PID controllers can be used only for plants with relatively small time delay for high performance devices like the quadrotor. This controller takes many structures but the most important one as in the following form:

$$u(t) = K_p\left[e(t) + \frac{1}{T_i}\int_0^t e(\tau)\,d\tau + T_d\frac{de(t)}{dt}\right] \tag{6.33}$$

where $u(t)$ is the input signal to the plant model, the error signal $e(t)$ is defined as

$$e(t) = r(t) - y(t) \tag{6.34}$$

and $r(t)$ is the reference input signal.

In this section, the PID controller for the quadrotor is developed based on the fast response. Using this approach as a recursive algorithm for the control-laws synthesis, all the calculation stages concerning the tracking errors are simplified.

One other aspect of the controller selection depends on the method of control of the UAV. It can be mode-based or non-mode based. For the mode based controller, independent controllers for each state are needed, and a higher level controller decides how these interact. On the other hand for a non-mode based controller, a single controller controls all of the states together.

However, the adopted control strategy is summarized in the control of two subsystems; the first relates to the position control while the second is that of the attitude control.

The quadrotor model above can be divided into two subsystems: a fully-actuated subsystem $S1$ that provides the dynamics of the vertical position z and the yaw angle (z and ψ).

$$\begin{bmatrix} \ddot{z} \\ \ddot{\phi} \end{bmatrix} = \begin{bmatrix} u_1 \cos(\phi)\cos(\psi) - g \\ u_4 \end{bmatrix} + \begin{bmatrix} -K_3\dot{z}/m \\ -K_6\dot{\phi}/I_3 \end{bmatrix}. \tag{6.35}$$

An under-actuated subsystem $S2$ representing the under-actuated subsystem which gives the dynamic relation of the horizontal positions (x, y) with the pitch and roll angles.

$$\begin{bmatrix} \ddot{x} \\ \ddot{y} \end{bmatrix} = \begin{bmatrix} u_1 \cos\phi & u_1 \sin\phi \\ u_1 \sin\phi & -u_1 \cos\phi \end{bmatrix} \begin{bmatrix} \sin\phi \cos\psi \\ \sin\psi \end{bmatrix} + \begin{bmatrix} -K_1\dot{x}/m \\ -K_2\dot{y}/m \end{bmatrix} \tag{6.36}$$

and

$$\begin{bmatrix} \ddot{\theta} \\ \ddot{\psi} \end{bmatrix} = \begin{bmatrix} u_2 \\ u_3 \end{bmatrix} + \begin{bmatrix} -lK_4\dot{\theta}/I_1 \\ -lK_5\dot{\psi}/I_2 \end{bmatrix}. \tag{6.37}$$

Since drag is very small at low speeds, the drag terms in the above equations can be considered as small disturbances to the system.

The PID control is applied to the equations above with inputs u_1, u_2, u_3, u_4 and outputs ϕ, θ, ψ and Z_d. Though these methods were rather successful in local analysis of nonlinear systems affine in control they usually fail to work for a global analysis and nonlinear systems that are nonaffine in control [41].

For the fully-actuated subsystem, we can construct a rate bounded PID controllers to move states z and ϕ, θ, ψ to their desired values.

6.12.4 Simulation Results

The nominal parameters and the initial conditions of the quadrotor for simulation are:

$$\begin{aligned}
&I_1 = I_2 = 1.25\ \text{Ns}^2/\text{rad},\\
&I_3 = 2.5\ \text{Ns},\\
&K_1 = K_2 = K_3 = 0.010\ \text{Ns/m},\\
&K_4 = K_5 = K_6 = 0.010\ \text{Ns/m},\\
&m = 2\ \text{kg}, \quad l = 0.2\ \text{m}, \quad g = 9.8\ \text{m/s}^2.
\end{aligned}$$

The proposed control algorithm shown in Fig. 6.87 which is composed of all controllers, inputs, speed reference and the inner relationships of the thrust, the quadro-

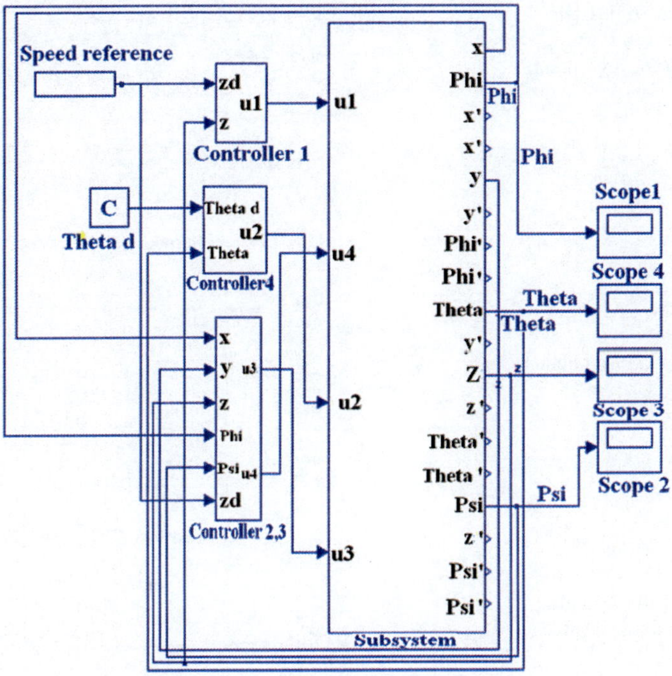

Fig. 6.87 Simulation model with PID controllers for the quadrotor

Fig. 6.88 The z-axis moving to the desired z-point

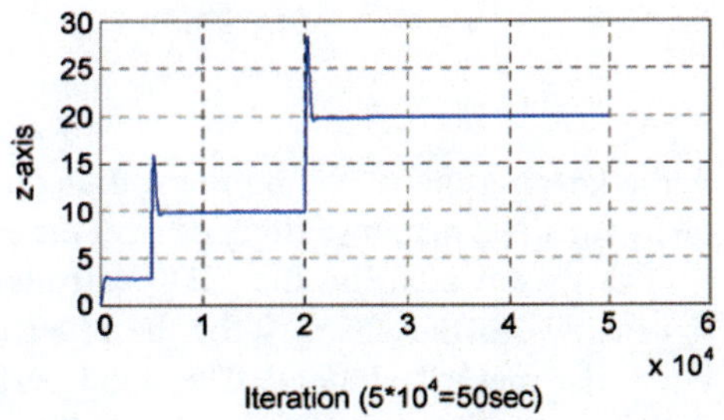

tor system is supplied by a step function for the altitude and (z-axis) which is subject to the three step inputs at (3, 10, 20) and the response yields as can be seen in Fig. 6.88 which is contains some transient overshot and another for the Yaw angle (ψ) which is subjected to step input after 5 second as shown in Fig. 6.90 and the roll angle (f) which is respond after 3 second as it can be seen in Fig. 6.89, the pitch angle response is shown in Fig. 6.91 which 5% overshot when subjected to step input these transient perturbation are due to many reasons such as a certain of some mechanical parameters in the design and the simplification of controller design.

The simulation results show that the PID controllers are able to robustly stabilize the quadrotor helicopter and move it to a desired position with a desired yaw angle

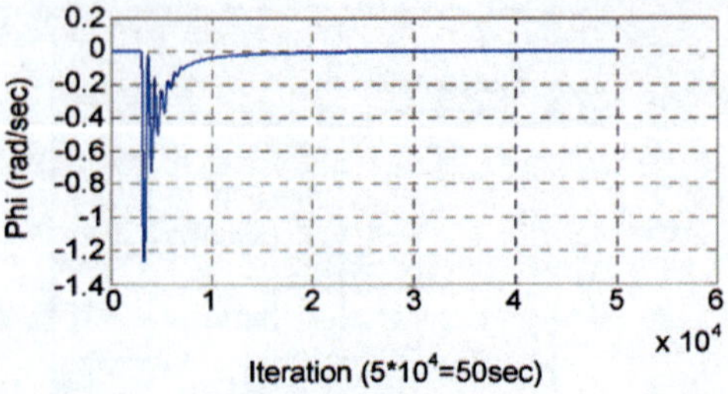

Fig. 6.89 Roll (ϕ) angle after 3 seconds to start moving to the desired point

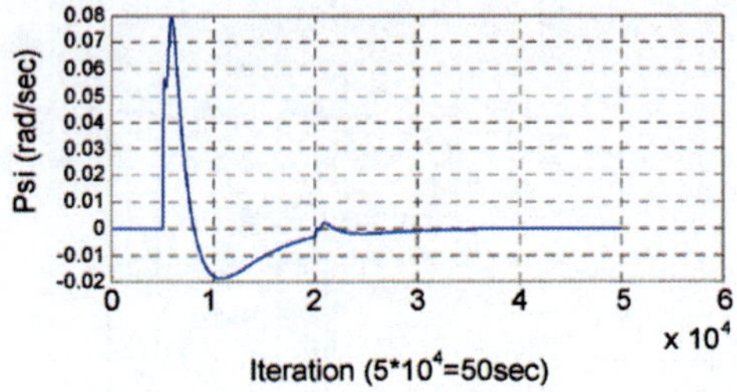

Fig. 6.90 Yaw (ψ) angle after 5 seconds to start moving to the desired point

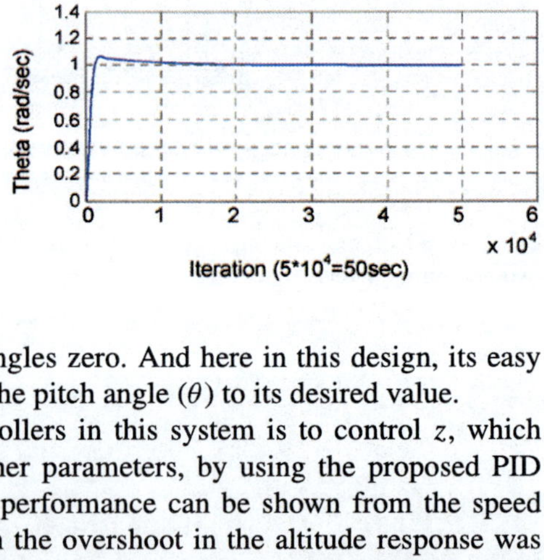

Fig. 6.91 Pitch (θ) angle start moving to the desired point

while keeping the pitch and the roll angles zero. And here in this design, its easy and with a fast response time, can get the pitch angle (θ) to its desired value.

The reason of using the PID controllers in this system is to control z, which is sensitive to the changes for the other parameters, by using the proposed PID controller method strategy. The good performance can be shown from the speed of response of the quadrotor; although the overshoot in the altitude response was removed, the transient response of the system became faster. The same speed of response can be also seen in the yaw, pitch and roll angles control of Figs. 6.88, 6.89, 6.90.

6.13 Design of an Aircraft Controller

The Lockheed L-1011 TriStar, commonly referred to as just L-1011 (pronounced "ell-ten-eleven") or TriStar, is a medium-to-long range, three-engine, wide body passenger jet airliner, see Fig. 6.92. It was the third wide-body airliner to enter commercial operations, following the Boeing 747 and the McDonnell Douglas DC-

Fig. 6.92 L-1011 tristar structure

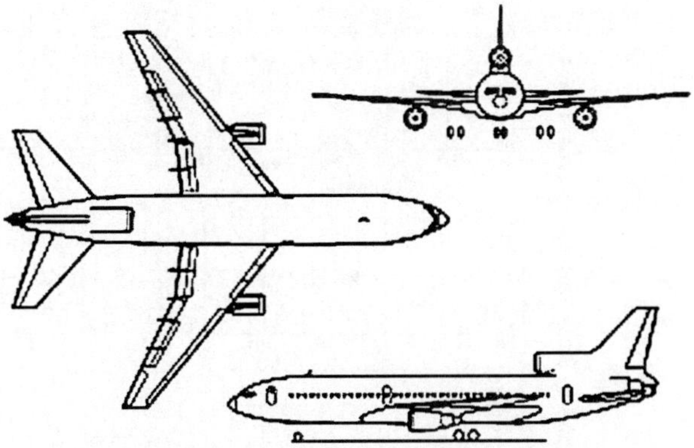

10. Between 1968 and 1984, Lockheed manufactured a total of 250 TriStars. The design featured a twin-aisle interior with a maximum of 400 passengers, a three engine layout, low noise emissions (in the early 1970s, Eastern Air Lines nicknamed the L-1011 "The WhisperLiner"), improved reliability, and efficient operation. The L-1011 featured a highly advanced autopilot system and was the first wide-body to receive FAA certification for Cat-IIIc auto-landing, which approved the TriStar for completely blind landings in zero-visibility weather performed by the aircraft's autopilot. The L-1011 used an Inertial Navigation System (INS) to operate its navigation needs. This included aligning the navigation system by entering current coordinates of longitude and latitude. It also had a unique Direct Lift Control (DLC) system, which allowed for smooth approaches when landing. DLC helps maintain the descending glide slope on final approach by automatically deploying spoiler panels on the wings. Thus, rather than maintaining the descent by adjusting pitch, DLC helps control the descent while maintaining a more consistent pitch angle, using four redundant hydraulic systems.

6.13.1 Linearized Model

A sixth order model of an aircraft is hereafter selected for the purpose of control design. A linearized model of the Lockheed L1011 TriStar aircraft at a cruise flight condition has the system matrices:

$$A = \begin{bmatrix} -20 & 0 & 0 & 0 & 0 & 0 \\ 0.337 & -1 & 0 & 0.249 & -1.12 & -5.2 \\ 0 & 1 & 0 & 0 & 1 & 0 \\ -0.744 & -0.032 & 0 & -0.154 & -0.0042 & 1.54 \\ 0 & 0 & 0 & 0 & -25 & 0 \\ 0.02 & 0 & 0.0386 & -0.996 & -0.00029 & -0.117 \end{bmatrix},$$

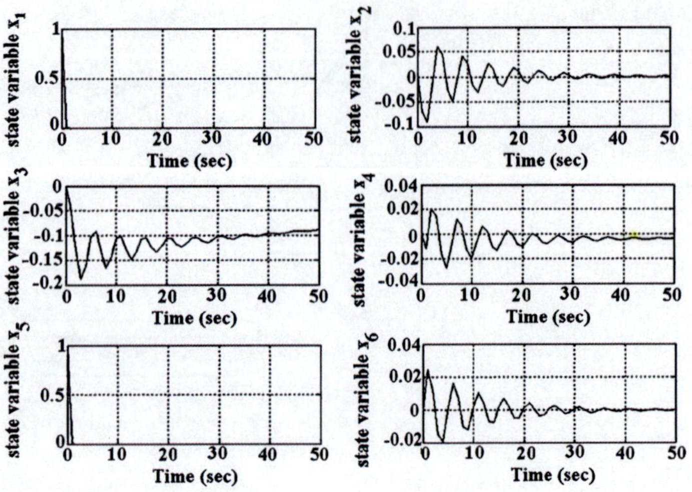

Fig. 6.93 State trajectories without input

$$B = \begin{bmatrix} 20 & 0 \\ 0 & 0 \\ 0 & 0 \\ 0 & 0 \\ 0 & 25 \\ 0 & 0 \end{bmatrix}, \qquad C = \begin{bmatrix} 0 & 0 & 1 & 0 & 0 & 0 \\ 0 & 0 & 0 & 0 & 0 & 1 \end{bmatrix}$$

where the state variables are $\partial_r =$ rudder deflection, $\partial_a =$ aileron deflection, $\varphi =$ bank angle (rad), $r =$ yaw rate (rad/s), $p =$ roll rate (rad/s) and $\beta =$ sideslip angle (rad). The input variables are $\partial_{rc} =$ rudder command (rad), $\partial_{ac} =$ rudder command (rad) and the output measured variables are $\varphi =$ bank angle (rad) and $\beta =$ sideslip angle (rad). The eigenvalues of the A matrix were found to be

$$-0.0882 \pm +1.2695j, \quad -1.0855, \quad -0.0092, \quad -20.00, \quad -25.000.$$

Two of the eigenvalues have complex parts which cause oscillations in the system response, see Fig. 6.93.

6.13.2 Simulation Results

In what follows, we present the simulation results of feedback control design. This is subsumed of state-feedback, observed-based feedback, tracking control and LQR design. The corresponding state trajectories are plotted in Fig. 6.94 under state-feedback.

The corresponding output trajectories are plotted in Fig. 6.95 under state-feedback. In Fig. 6.96, the corresponding state trajectories are plotted under observer-based feedback. The output trajectories under observer-based feedback are shown in Fig. 6.97 and a comparison of state trajectories is provided in Fig. 6.98.

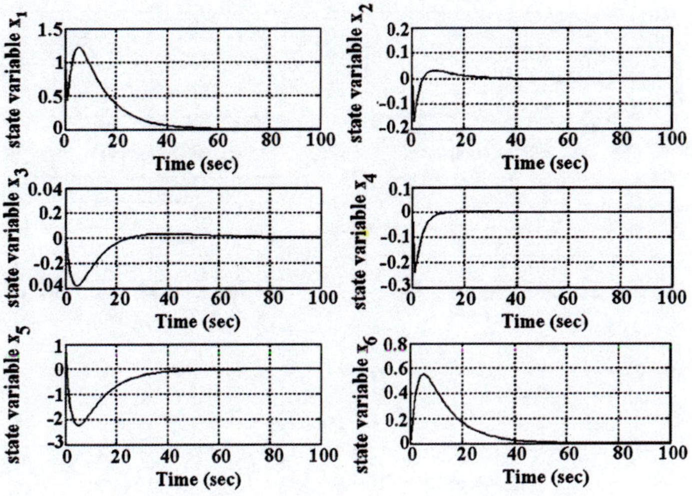

Fig. 6.94 State trajectories with state-feedback

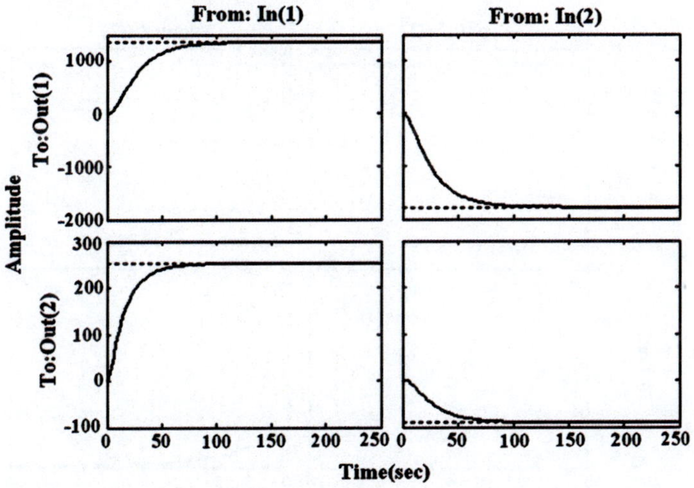

Fig. 6.95 Output trajectories with state-feedback

The corresponding state trajectories with tracking control, integral control and LQR are plotted in Figs. 6.99–6.101, respectively.

6.14 Motion Control Design of Liquid Container

In the casting and steel industry, containers with melted metal are transferred over long distances from the furnace. To achieve higher degrees of automatic operation, optimal motion control is always considered. It is important to shorten the trans-

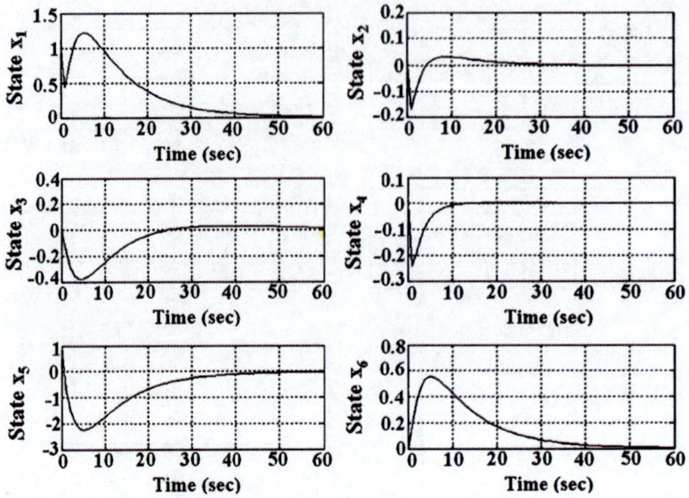

Fig. 6.96 State trajectories with observer-based feedback

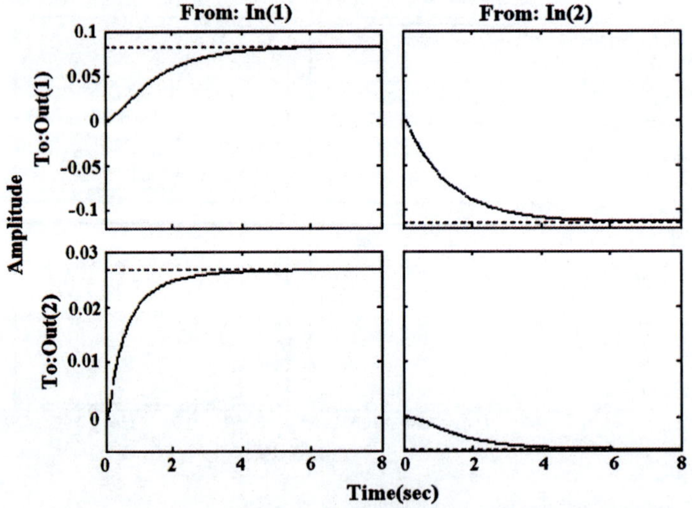

Fig. 6.97 Output trajectories with observer-based feedback

portation time in order to increase productivity. However, transfer at high speed causes molten metal to slosh in both the ladle and the molds. This sloshing phenomenon deteriorates the quality of the mold due to impurity and excessive cooling of the molten metal. Besides that, it can be dangerous as overflow can happen as well [52]. Many papers have been published about control of the sloshing in liquids. One of the studies considered it as a problem of suppressing liquid oscillations [49]. Another used the idea of jerk reduction to decrease the sloshing due to jerk move-

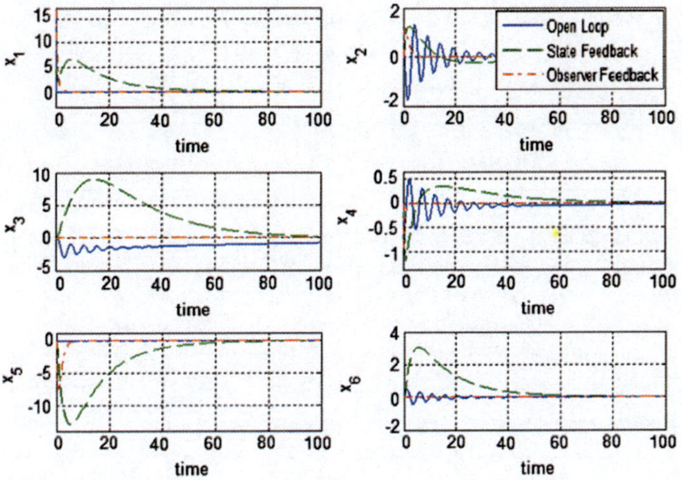

Fig. 6.98 A comparison of state trajectories

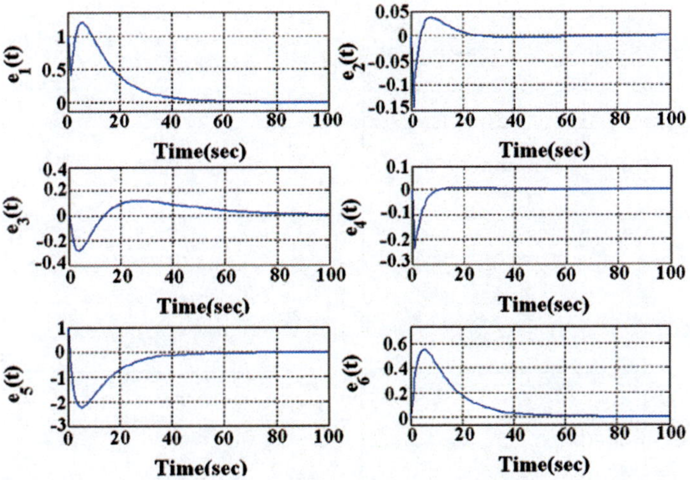

Fig. 6.99 State trajectories with tracking control

ments by optimal control [23]. Different control approaches are also applied. PID control and observer based control have been applied with great success in 2009 [7]. Not many studies have applied control to both the motion of the liquid container as well as the sloshing phenomenon at the same time [52]. In 2002 however, Yano et al. applied robust control to both the sloshing as well as the motion of the liquid container [52]. In this study, a similar model to Yano's will be used and several control techniques will be applied to study their effect on the system.

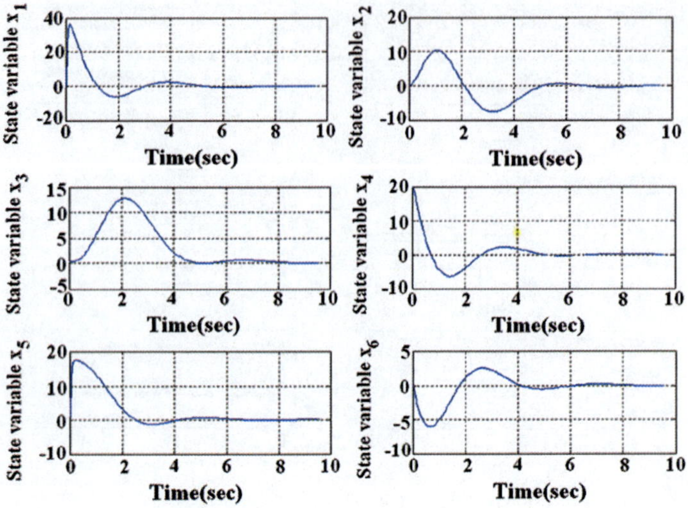

Fig. 6.100 State trajectories with integral control

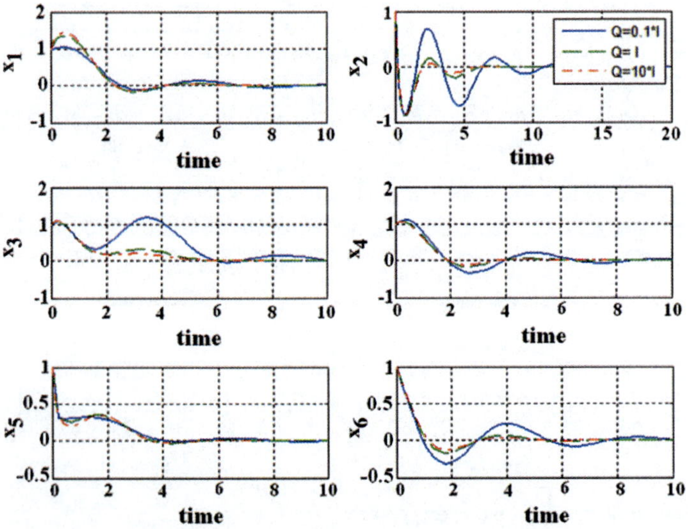

Fig. 6.101 State trajectories with LQR control

6.14.1 Dynamic Model

In our model, we assume that the transfer path is a straight line, see Fig. 6.102. Thus, as long as there are no sudden changes in the acceleration we can model the three-dimensional container as a two dimensional container. Therefore, the given sloshing model is described as a pendulum-type sloshing model [52]. Adding the rotational

Fig. 6.102 A sloshing model

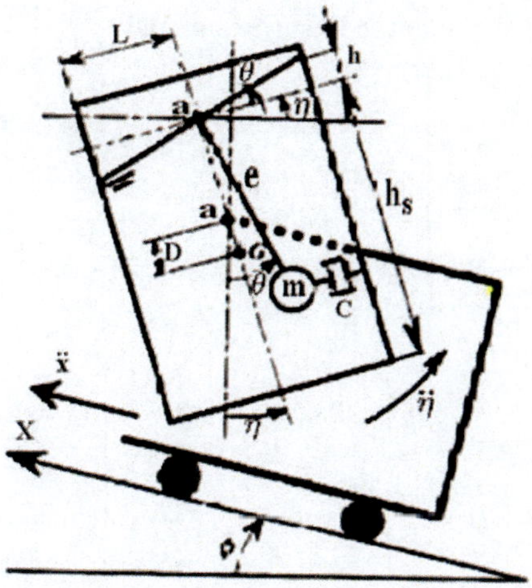

motion to the pendulum-type sloshing model gives a new model that describes both the transfer of the container as well as the rotation that causes the sloshing [52]. The following diagram shows the container with all the variables that affect the model.

The rotational motion of the system around point O is given by

$$J\frac{d^2\theta}{dt^2} = -c\frac{\ell\, d(\theta-\eta)}{dt}\ell\cos^2\theta - mg\ell\sin\theta + m\ddot{x}\ell\cos\phi\cos\theta$$
$$- m\ddot{x}\ell\sin\phi\sin\theta - m\ell D\frac{d^2\eta}{dt^2}\cos\theta$$

where $J = m\ell^2$ is the moment of inertia. A linearization about $\theta \approx 0$, with some simplification [52] yields

$$\ddot{\theta} = -\frac{c}{m}(\dot{\theta}-\dot{\eta}) - \frac{g}{\ell}\theta + \frac{1}{\ell}\ddot{x}\cos\phi. \tag{6.38}$$

The motor used is a DC servo-motor where the input is voltage and the output is velocity, which in turn applied to the contrainer transfer function. The model is described by

$$G_m(s) = \frac{K_m}{\tau_m + 1}$$

where K_m and τ_m are the motor gain and time-constant, respectively. The container rotation is described by the transfer function

$$G_r(s) = \frac{K_r\omega_n^2}{s^2 + 2\xi\omega_n s + \omega_n^2}.$$

Taking $\mathbf{x} = [\theta, \dot{\theta}, \eta, \dot{\eta}, \ddot{\eta}, x, \dot{x}]^t$, $\mathbf{Y} = [h, x]^t$ and $\mathbf{u} = [u_1, u_2]^t$ as the state, output and control input, respectively, along with the parameter values as

$$K_m = 0.0912, \quad \tau_m = 0.0227, \quad K_r = 0.5807, \quad \xi = 0.3778, \quad \omega_n = 41.446,$$
$$h_s = 0.14, \quad \phi = 5.0, \quad \ell = 0.0442, \quad c = 1.88, \quad m = 2.744, \quad D = 0.02$$

we obtain the linearized model

$$\dot{\mathbf{x}} = \begin{bmatrix} 0 & 1 & 0 & 0 & 0 & 0 & 0 \\ -221.72 & -0.6851 & 0 & 0.6851 & 0 & 0 & -992.878 \\ 0 & 0 & 0 & 1 & 0 & 0 & 0 \\ 0 & 0 & 0 & 0 & 1 & 0 & 0 \\ 0 & 0 & 0 & -1717.77 & -31.3166 & 0 & 0 \\ 0 & 0 & 0 & 0 & 0 & 0 & 1 \\ 0 & 0 & 0 & 0 & 0 & 07 & -44.053 \end{bmatrix} \mathbf{x}$$

$$+ \begin{bmatrix} 0 & 0 \\ 90.55 & 0 \\ 0 & 0 \\ 0 & 0 \\ 0 & -997.51 \\ 0 & 0 \\ 4.0176 & 0 \end{bmatrix} \hat{u},$$

$$\mathbf{y} = \begin{bmatrix} 0.25 & 0 & -0.25 & 0 & 0 & 0 & 0 \\ 0 & 0 & 0 & 0 & 0 & 1 & 0 \end{bmatrix} \mathbf{x}.$$

When a liquid container is transferred on an inclined transfer path, a transferring machine with one degree of freedom may cause overflow and contamination of the molten metal in terms of only acceleration control for linear container transfer. Evaluation of the structural properties of the linearized model shows that the model is both controllable and observable, however is unstable with internal oscillations since the eigenvalues are $0, 0, -0.3426 \pm 14.8863, -15.6583 \pm 28.3743, -44.0530$. In the following sections, we examine the feedback control design and simulation of different schemes.

6.14.2 State Feedback Design

The design objective is to stabilize the linearized sloshing model and remove the internal oscillations. With the aid of MATLAB, we place the closed-loop poles at $[-1, -2, -3, -5, -6, -8, -11]$ to obtain the state-feedback controller gain as

$$\mathbf{K} = \begin{bmatrix} -1.2297 & 0.1821 & 0.2100 & 0.1332 & 0.0149 & 0.2149 & -10.6701 \\ -0.0627 & -0.0035 & -0.1145 & 1.6334 & 0.0137 & -0.0323 & -0.0270 \end{bmatrix}.$$

The ensuing state-trajectories are plotted in Fig. 6.103. The nonzero values of x_3 and x_6 are justified by the application of the final value theorem (see the Appendix) which yields

$$\begin{aligned} x_{ss} &= -(A + BK)^{-1} B\mathbf{r} \\ &= \begin{bmatrix} 0 & 0 & 1.8120 & 0 & 0 & 6.4231 & 0 \\ 0 & 0 & 12.0565 & 0 & 0 & 11.7794 & 0 \end{bmatrix}^t. \end{aligned}$$

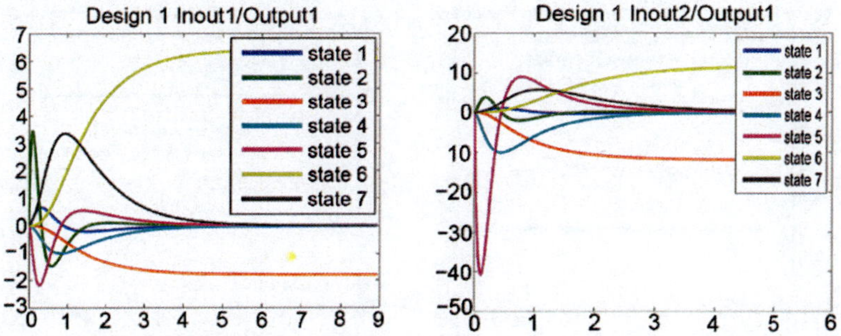

Fig. 6.103 State trajectories: Input 1 (*left*), Input 2 (*right*)

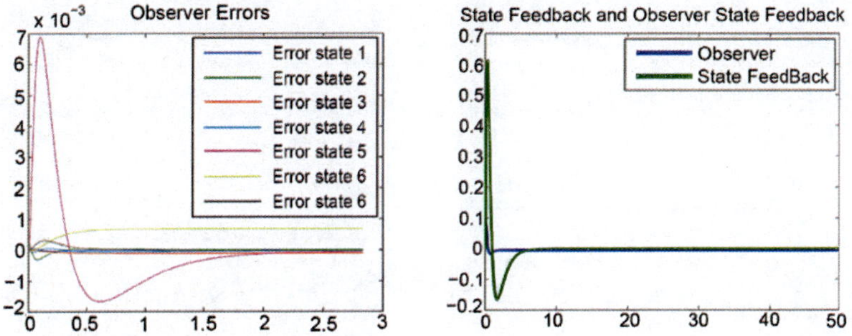

Fig. 6.104 Observeration errors (*left*), a comparison of state x_1 (*right*)

6.14.3 Observer-Based Feedback Design

Relying on the measured states and build an observer. Using MATLAB, placing the poles at [−5, −60, −15, −105, −3, −40, −5.5], which sufficiently high than the controller poles, gives the observer gain as

$$\mathbf{L} = 10^3 \begin{bmatrix} 0.4 & -3.2 & -0.5 & -15.7 & 1239.1 & -0.0002 & 0.0001 \\ 0.0003 & -0.0004 & 0.0001 & -0.0002 & -0.0008 & 0.0001 & -0.0005 \end{bmatrix}^t.$$

In Figs. 6.104, 6.105, the observation error as well as a comparison between state-feedback and observer-based feedback are plotted.

6.14.4 LQR Design

Seeking to optimize the performance, we considered the LQR design with different weighting matrices. The simulation results of the closed-loop step response for two distinct cases:

- Cheap control: $Q = 0.1 * I_7$, $R = I_2$,
- Expensive control: $Q = \text{diag}([10 \;\; 250 \;\; 10 \;\; 10 \;\; 10 \;\; 110 \;\; 10])$, $R = I_2$,

are presented in Fig. 6.106.

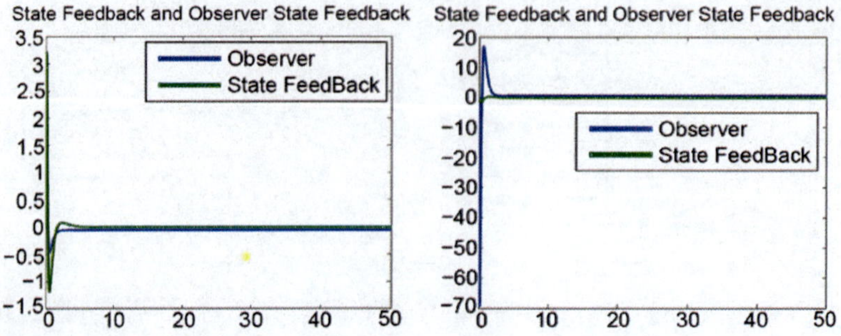

Fig. 6.105 A comparison of state x_3 (*left*), a comparison of state x_5 (*right*)

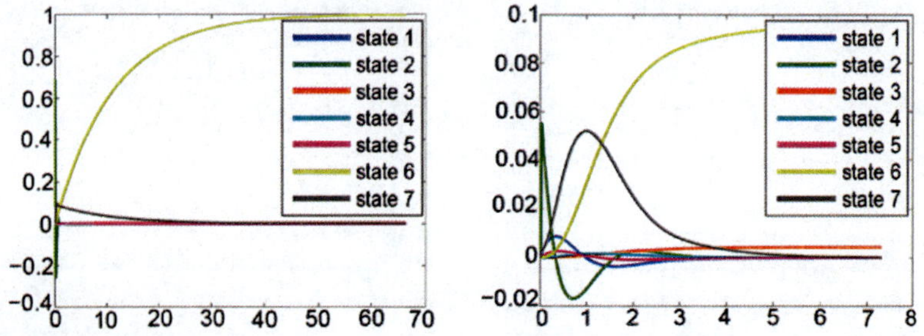

Fig. 6.106 Closed-loop response: expensive LQ control (*left*), cheap LQ control (*right*)

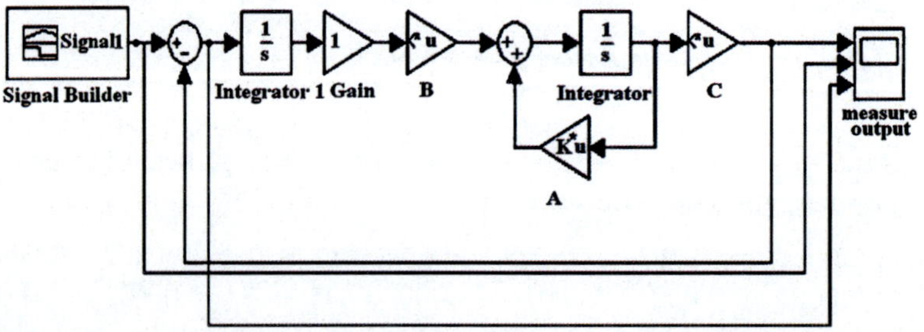

Fig. 6.107 A Simulink sloshing model

6.14.5 Tracking Control Design

There are two types of tracking control, one in which a signal is set to track zero (regulator) and the other case is when the signal is set to track a reference signal or trajectory. In both cases, the difference between the reference signal and the output is the error that is used to drive the system. The error is integrated before being used as a control signal in order to eliminate steady state errors. A *Simulink* model was created to track a reference signal, see Fig. 6.107.

The outputs shown below, Fig. 6.108 include the measured outputs of the system, the reference trajectory and the control inputs.

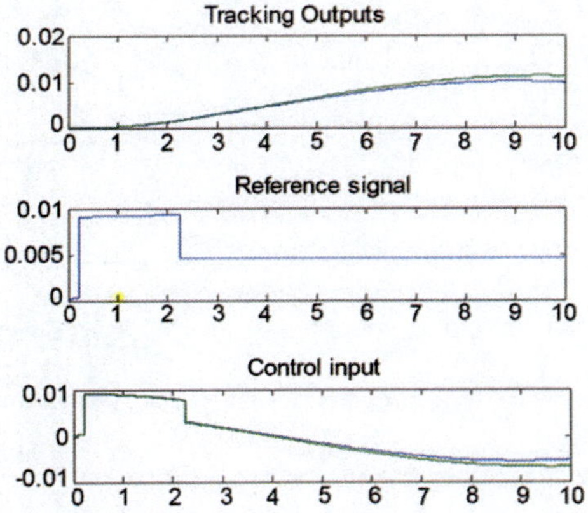

Fig. 6.108 Tracking trajectories of input and output

6.15 Vertical Motion Control of Marine Vehicles

Marine vehicles are designed to operate with adequate reliability and stability, and in order to accomplish this, it is essential to control the motion of the submarine. This control task consists in making the submarine to follow, as closely as possible, a desired trajectory, which can be defined in terms of submarine's depth, velocity, pitch angle and pitch rate. In most submarine's operational conditions, the desired trajectory is slowly varying due to the motion induced by the waves. The desired trajectory can be achieved in the face of uncertainty of the system to be controlled even in the presence of the uncontrollable external disturbances acting on the system in the form of waves, wind and current in the water. From the control system design perspective, the characterization of the disturbances acting on the submarine is essential to design good performance submarine motion controllers and to understand limitations that may prevent the design achieving the performance specifications. In this section, we apply optimal control techniques and simulations represent the achievement of stability.

The vertical dynamics of a submarine traveling a few meters below the surface of a sea are considered here. The purpose here is to control the depth and pitch of the submarine by the use of optimal control. Figure 6.109 illustrates the model of the submarine and its related parameters.

6.15.1 Dynamic Model

The vertical plane behavior of the submarine is modeled by the simplified linear time-invariant equations:

$$\dot{\mathbf{x}} = \begin{bmatrix} 6.5 \times 10^{-4} & -0.2502 & -0.008 & -0.0014 \\ 0 & 1 & 0 & 0 \\ 0 & 0 & -3.25 & 0 \end{bmatrix} \mathbf{x}$$

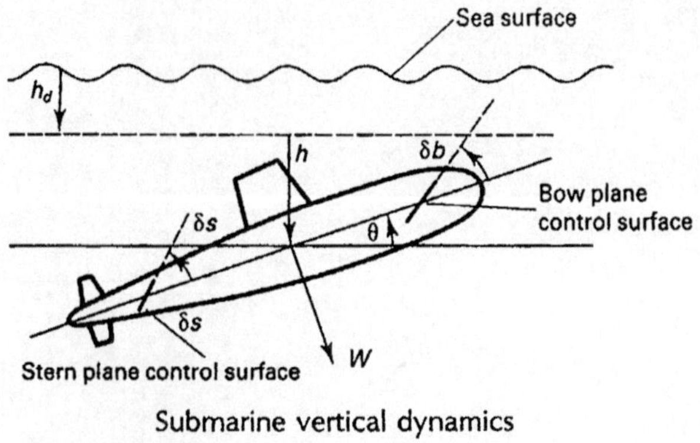

Fig. 6.109 A submarine model

$$+\begin{bmatrix} -0.0348 & -0.0686 \\ 0.0369 & -0.0109 \\ 0 & 0 \\ 0 & 0 \end{bmatrix}\hat{u} + \begin{bmatrix} 1 & 0 \\ 0 & 1 \\ 0 & 0 \\ 0 & 0 \end{bmatrix}\hat{w},$$

$$\mathbf{y} = \begin{bmatrix} 0 & 0 & 1 & 0 \\ 0 & 0 & 0 & 1 \end{bmatrix}\mathbf{x} + \mathbf{v}$$

where $\mathbf{x} = [W \;\; q \;\; \theta \;\; h]^t$ and $\mathbf{u} = [\delta b \;\; \delta s]^t$ are the state and control input respectively, and

- W = downwards velocity at right angles to the submarine main axis (m/s).
- q = pitch rate (rad/s).
- θ = pitch (rad).
- h = depth wrt the datum depth (h_d) below the main sea level (m).
- δb = bow control plane angle (rad).
- δs = stern control plane angle (rad).
- $\mathbf{y} = [\theta \;\; h]^t$ is the output.
- $\hat{w}$ = process noise of the sea.
- $\mathrm{v}\hat{v}$ = measurement noise of the sea.

The open loop model of submarine used here is unstable since $\lambda(A) = \{-0.2074 \pm 0.0927i, -0.0647, 0.1074\}$. In order to stabilize the system, optimal control methods are used in he sequel.

6.15.2 LQR Design

After several experimentations, it is found that using $Q = 0.0125 * I_4$, $R = 10^4 I_2$, the MATLAB command *lqr* gives the feedback gain matrix

$$\mathbf{K} = \begin{bmatrix} 0.0165 & 3.5451 & 1.4563 & -0.0462 \\ -0.0049 & -1.0482 & -0.4306 & 0.0137 \end{bmatrix}$$

and the associated closed-loop eigenvalues as

$$\lambda(A + BK) = \{-0.0647, -0.1076, -0.2073 \pm 0.0927i\}.$$

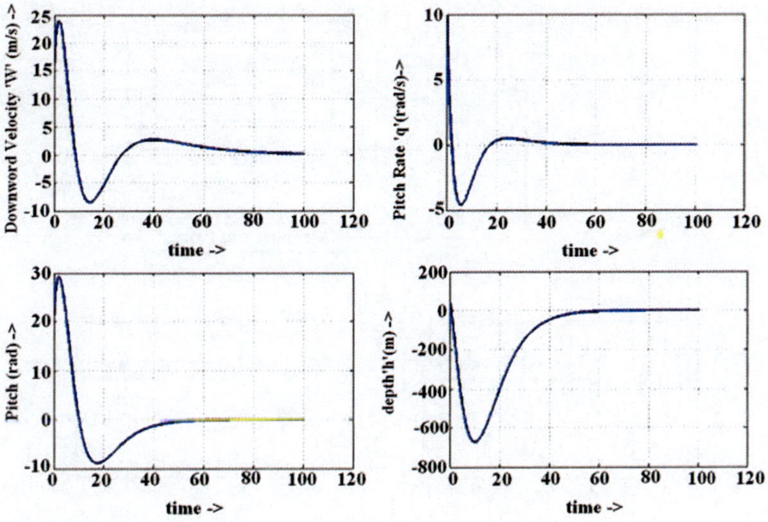

Fig. 6.110 State trajectories under LQR

Fig. 6.111 State trajectories under LQR

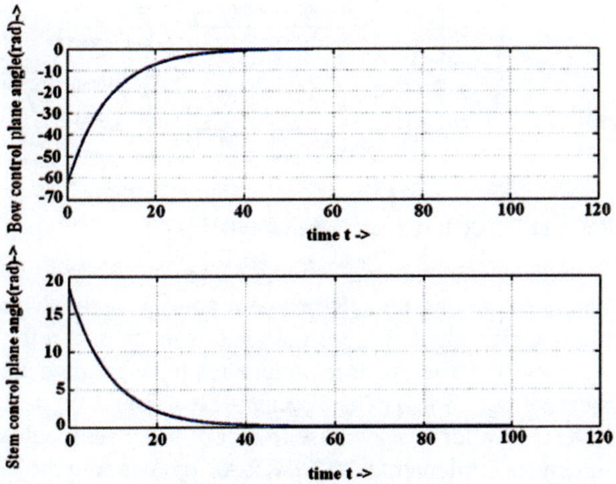

The resulting optimal state and the optimal control input trajectories are depicted in Figs. 6.110, 6.111.

6.15.3 LQGR Design

In most of the practical systems, all the states of the system are not available for feedback. LQ Gaussian control is a method of designing feedback control laws for linear systems with additive Gaussian processes that minimize a given quadratic cost function. This is achieved through the application of separation principle con-

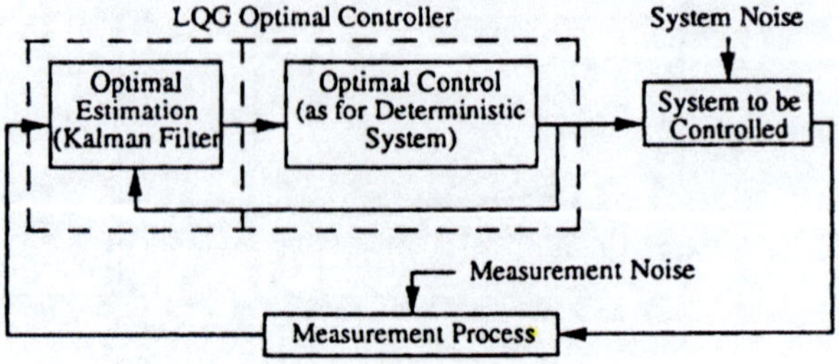

Fig. 6.112 LQGR block diagram

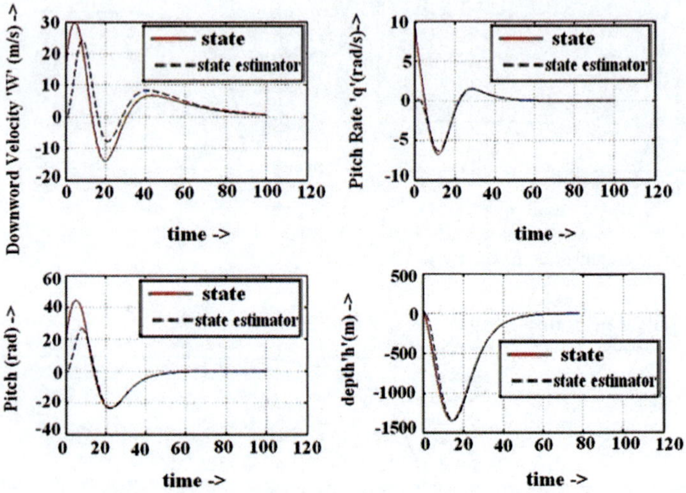

Fig. 6.113 LQGR state and state estimates

sisting of solving a LQR problem and the optimal linear state estimator (Kalman filter) which gives the estimate $\hat{x}$ of the state x, refer to Fig. 6.112 for details. In Kalman filter, the two free parameters to be selected are known as process noise and measurement noise. The Kalman filter should be designed such that the resulting LQG controller is very close to that of full state feedback control. The LQG control scheme is implemented in MATLAB by invoking the separation principle as it guarantees that the closed loop poles of the estimator will not appear in the closed loop poles of the system under control thereby finding the LQR solution using the MATLAB command *lqr* function and state estimates using the Kalman filter function MATLAB command *kalman*.

This yields the Kalman gain matrix as

$$\mathbf{L} = \begin{bmatrix} 0.0331 & -0.0247 \\ 0.0046 & -0.0024 \\ 0.0324 & -0.0286 \\ 0.2859 & 0.4218 \end{bmatrix}$$

and the corresponding trajectories of state and state-estimates are depicted in Fig. 6.113 which clearly indicate nice asymptotic behavior in reproducing the states.

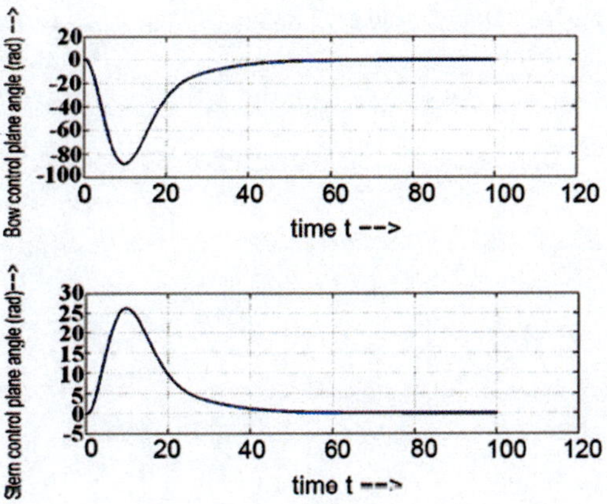

Fig. 6.114 LQGR control input

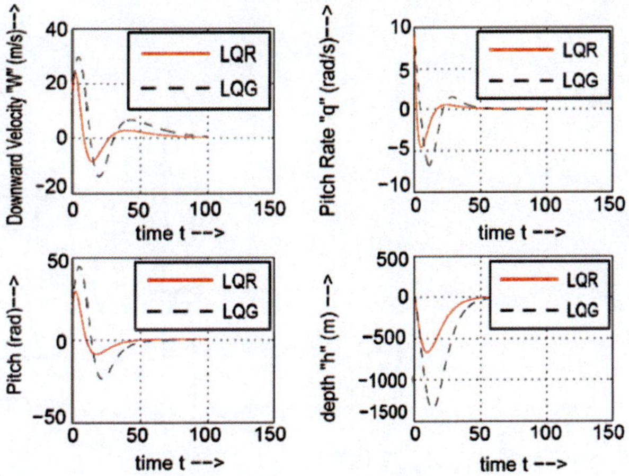

Fig. 6.115 LQR and LQGR state trajectories

The control input trajectories of the LQGR scheme are shown in Fig. 6.114.

For the purpose of comparison, we provide in Figs. 6.115 and 6.116 the optimal state and input trajectories based on LQR and LQGR.

6.16 Pitch Control of Wind Turbines

Since the 1990s, the wind energy industry has been growing rapidly. The wind power generation technology had developed from stall-controlled to variable speed pitch regulated. And wind turbine has demanded better performance of controller [22, 46, 47].

With the increasing of capacity of wind turbines, pitch-control technique of large wind turbine has become a key technique of wind energy. Pitch-control can not only output power steadily, but also make wind turbine have better starting and braking performance. Additionally, using optimized control algorithm can lower

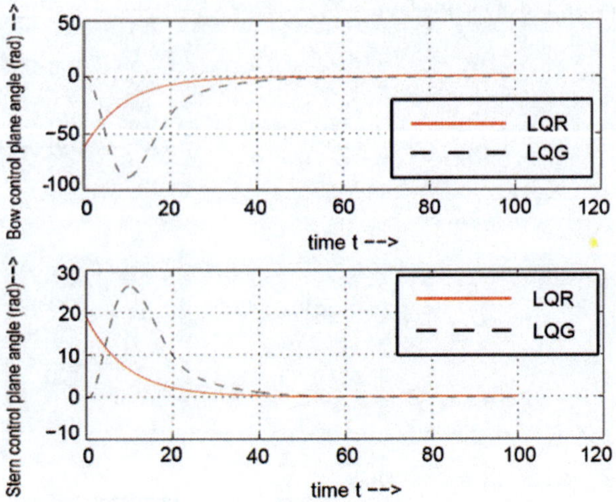

Fig. 6.116 LQR and LQGR input trajectories

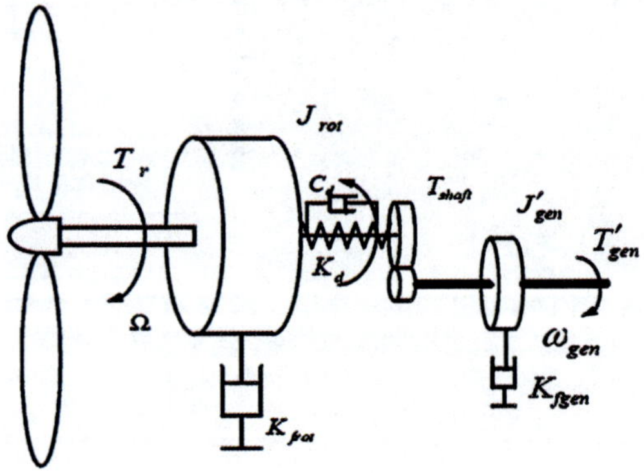

Fig. 6.117 Wind turbine model

load and torque ripple of wind turbine, extending the life of wind turbine. At present, in China most wind turbine is controlled by PID algorithm, which cannot have a satisfy effect. Abroad researchers have proposed many advanced control theory and strategy about pitch-control. Senjyu et al. had applied GPC control method to pitch-control [47]. This is wind speed predict model based on average wind speed and standard deviation, having pitch controlled according to predicted wind speed.

6.16.1 Simulation of Wind Turbine

The equivalent model of wind turbine is shown in Fig. 6.117. The aerodynamic torque gained by blade from wind energy [13]:

$$T_r = \frac{1}{2}\frac{\pi\rho R^2 C_p(\beta,\lambda)}{\lambda}V^2 \tag{6.39}$$

in which, ρ is the density of air (kg/m^3), R is the radius of rotor (m), V is the wind speed (m/s), β is the pitch angle (degree), λ is tip speed ratio, $\lambda = \Omega R/V$, Ω is the rotor speed, C_p is power conversion coefficient, which indicates wind turbine's efficiency of converting wind energy to usable mechanism power. C_p is function of tip speed ratio λ and blade pitch angle β. C_p can be written as [13, 27]:

$$C_P(\lambda, \beta) = 0.22\left(\frac{116}{\lambda_i} - 0.4\beta - 5\right) \cdot e - \frac{22.5}{\lambda_i} \tag{6.40}$$

in which λ_i satisfies:

$$\frac{1}{\lambda_i} = \frac{1}{\lambda + 0.08 \cdot \beta} - \frac{0.035}{\beta^2 + 1}.$$

Although wind turbine is a nonlinear model, at some point near by it can be treated as linear model. Linearizing torque T_r at point (V_0, Ω_0, β_0) nearby:

$$T_r = T_r(V_0, \Omega_0, \beta_0) + \alpha \Delta V + \gamma \Delta \Omega + \zeta \Delta \beta. \tag{6.41}$$

In which, $\Delta\Omega = \Omega - \Omega_0$, $\Delta V = V - V_0$, $\Delta\beta = \beta - \beta_0$

$$\alpha = \left.\frac{\partial T_r}{\partial V}\right|_{(V_0,\Omega_0,\beta_0)}, \qquad \gamma = \left.\frac{\partial T_r}{\partial \Omega}\right|_{(V_0,\Omega_0,\beta_0)}, \qquad \zeta = \left.\frac{\partial T_r}{\partial \beta}\right|_{(V_0,\Omega_0,\beta_0)}.$$

Let state variable q_1 and q_2 are blade angle and rotor angle respectively (calculated in low speed shaft. Tshaft is the reaction torque on the shaft. Then:

$$T_{\text{shaft}} = K_d(q_1 - q_2) + C_d(\dot{q}_1 - \dot{q}_2), \tag{6.42}$$

$$\Delta T_{\text{shaft}} = K_d(\Delta q_1 - \Delta q_2) + C_d(\Delta \dot{q}_1 - \Delta \dot{q}_2), \tag{6.43}$$

$$J_{\text{rot}} \ddot{q}_1 = T_r - T_{\text{shaft}} - K_{f\text{rot}} \Omega, \tag{6.44}$$

$$J_{\text{gen}} \Delta \dot{q}_2 = \Delta T_{\text{shaft}} - \Delta T_{\text{gen}} - K_{f\text{gen}} \Delta \omega_{\text{gen}}. \tag{6.45}$$

Above, K_d is elastic coefficient of propeller shaft, C_d is damping coefficient on propeller shaft, J_{rot} and J_{gen} are rotation inera of low speed side and generator (calculated in low speed side), $K_{f\text{rot}}$, $K_{f\text{rot}}$ are friction coefficient of low speed side and high speed side respectively. T_{shaft_0} is counter torque at working point (V_0, Ω_0, β_0). The speed acceleration is 0, so:

$$T_r(V_0, \Omega_0, \beta_0) = T_{\text{shaft}_0} + K_{f\text{rot}} \Omega_0. \tag{6.46}$$

Then:

$$J_{\text{rot}} \ddot{q}_1 = \Delta T_r - \Delta T_{\text{shaft}} - K_{f\text{rot}} \Delta \Omega. \tag{6.47}$$

Let

$$x_1 = \Delta \dot{q}_1,$$

$$x_2 = K_d(\Delta q_1 - \Delta q_2),$$

$$x_3 = \Delta \dot{q}_2.$$

Then:

$$J_{\text{rot}}\dot{x}_1 = (\gamma - C_d - K_{f\text{rot}})x_1 - x_2 + C_d x_3 + \zeta\Delta\beta + \alpha\Delta V, \tag{6.48}$$

$$\dot{x}_2 = K_d(x_1 - x_3). \tag{6.49}$$

According to the torque equation of generation:

$$J_{\text{gen}}\dot{x}_3 = C_d x_1 + x_2 - (C_d + K_{f\text{gen}})x_3 - \Delta T_{\text{gen}}. \tag{6.50}$$

In state equation form:

$$\begin{cases} \dot{x} = Ax + Bu + \Gamma u_D \\ y = Cx + Du \end{cases} \tag{6.51}$$

where,

$$A = \begin{bmatrix} \frac{(\gamma - C_d - K_{f\text{rot}})}{J_{\text{rot}}} & \frac{-1}{J_{\text{rot}}} & \frac{C_d}{J_{\text{rot}}} \\ K_d & 0 & -K_d \\ \frac{C_d}{J_{\text{gen}}} & \frac{1}{J_{\text{gen}}} & \frac{-C_d - K_{f\text{gen}}}{J_{\text{gen}}} \end{bmatrix},$$

$$B = \begin{bmatrix} \frac{\zeta}{J_{\text{rot}}} & 0 \\ 0 & 0 \\ 0 & \frac{-1}{J_{\text{gen}}} \end{bmatrix}, \quad \Gamma = \begin{bmatrix} \frac{\alpha}{J_{\text{rot}}} \\ 0 \\ 0 \end{bmatrix}, \quad C = \begin{bmatrix} 0 & 0 & 1 \end{bmatrix}, \quad D = 0$$

input $u = \lfloor \Delta\beta, \Delta T_{\text{gen}} \rfloor$, disturbance quantity $u_D = \Delta V$.

At present pitch actuator has hydraulic and electric two forms. For simplicity, pitch actuator can be simplified to a first-order inertia model, no matter it is hydraulic or electric actuator. The pitch actuator transmission function is:

$$Act(s) = \frac{1}{\tau_\beta s + 1}. \tag{6.52}$$

6.16.2 Pitch Control of Wind Turbine

After connected to the grid, wind turbine can work in two modes: one mode is when wind speed is slower that rated wind speed, another is when faster. When wind speed is slow, wind turbine output power is smaller than rated power. So the pitch angle is set to 0° and wind turbine runs in optimal tip speed by controlling generator speed, in order to absorb as much wind energy as possible. While wind speed is faster than rated speed, the output power will excess rated power. Because the electrical and mechanical limitation of wind turbine, the rotator speed and output power cannot excess rated value. So, when output power is larger than rated power, pitch angle should be increased to smaller wind energy utilization efficiency. When output power is smaller than rated power, pitch angle will be decreased to maintain the output power at about rated power nearby.

Nowadays variance speed pitch-control wind turbine always has its electromagnetic torque given value constant, maintaining output power by regulating generator speed. The most common method is adopting PI control to regulate generator speed. This method is simple and easily applied in engineering. However, PI control may have overshoot problems, which makes pitch actuator complicated and easily fatigued.

LQR is linear quadrics regular, whose control object is linear system given by state space form in modern control theory. And its object function is object states and quadrics function which controls input. LQR optimal control is designing state feedback controller G. In order to minimize the quadrics object function J, and also G is decided only by weight matrix Q and R, the selection of Q and R is very important. LQR theory is a relatively mature theory in modern control theory. It provided an efficient analysis method for multi-variable feedback system. Object function J included state variable and input variable, which requires state variable and input variable to be small. In the pitch-control system, input value is the error of pitch angle. Because of large inertia of blade, rapid pitch-control would damage pitch regulated mechanism and aggravate the friction of pitch-control shaft. So, having some limitation to input energy will be reasonable. Additionally, choosing torque variation as state variable can suppress torque ripple as much as possible in LQR optimal control. Then the life of wind turbine can be extended.

The linear state model is given by

$$\begin{cases} \dot{x} = Ax + Bu + \Gamma u_D \\ y = Cx \end{cases} \tag{6.53}$$

and the objective function is

$$J = \frac{1}{2}\int_{t0}^{tf} \left[x^t(t)Qx(t) + u^t(t)Ru(t)\right] dt \tag{6.54}$$

where Q is positive semidefinite matrix, R is positive definite matrix, Q and R are weighted matrix for state variable and input variable, respectively. $x(t)$ is n-dimension state variable, $u(t)$ is m-dimension input variable. According to control theory, in order to minimize object function, optimal control is:

$$u(t) = Gx(t), \quad G = -R^{-1}B^t P,$$

where P is Riccati function:

$$-PA - A^t P + PBR^{-1}B^t P - Q = 0. \tag{6.55}$$

Positive definite symmetric solution. The LQR control diagram is shown as Fig. 6.118. In engineering application, state variable cannot be measured usually. So it needs to design a state observer to estimate state variable value. Figure 6.118 is the diagram used in actual application. Because there is a disturbance variable u_d in wind turbine model, only using LQR control cannot regulate generator speed very well. And the disturbance from disturbance variable should be minimized as much as possible. Disturbance Accommodating Control (DAC) is a good method

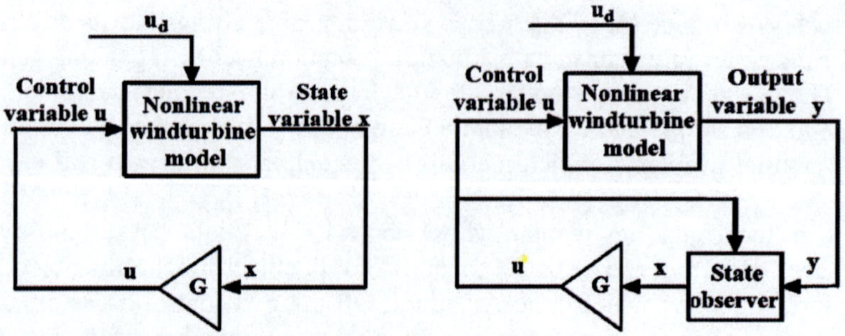

Fig. 6.118 LQR diagram

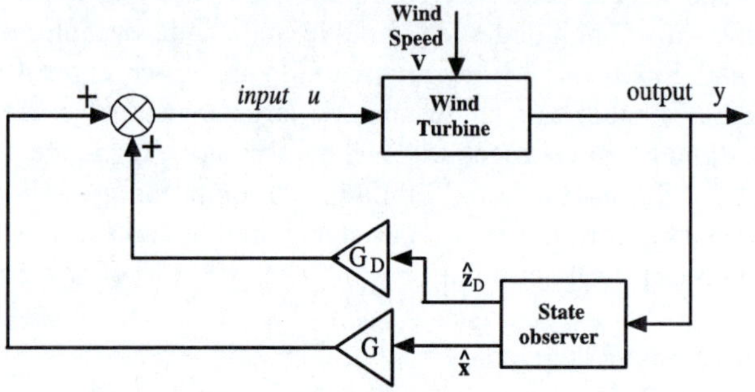

Fig. 6.119 Disturbance correction control diagram

to solve this problem. DAC control is a reconstructed disturbance model method based on state observer. The disturbance variable is reconstructed and is part of state feedback, can decrease or neutralize the disturbance effect. This paper adopted LQR method with DAC, which means that through LQR optimal control having a optimal feedback matrix G, then using DAC method to estimate disturbance variable and eliminating the disturbance from disturbance variable. DAC diagram is shown as Fig. 6.119. Using state observer to estimate state variable and disturbance variable, disturbance can be eliminated. Presume the disturbance variable has forms as below:

$$\begin{cases} u_d = \Theta z_D(t), \\ \dot{z}_D(t) = F z_D(t); \quad z_D(0) = z^0 D \end{cases} \tag{6.56}$$

$z^0 D$ is unknown, presume Θ and F is already known. According to DAC control theory, state feedback should contain the feedback of disturbance:

$$u(t) = Gx(t) + G_D z_D(t). \tag{6.57}$$

Replace $u(t)$ in the state function with the up function, we have:

$$\dot{x}(t) = (A + BG)x(t) + (BG_D + \Gamma\Theta)z_D(t). \tag{6.58}$$

To eliminate the disturbance, it requires $BG_D + \Gamma\Theta = 0$, then it can be considered as a system without disturbance. If the system cannot satisfy $BG_D + \Gamma\Theta = 0$, then choosing G_D to make $\|BG_D + \Gamma\Theta\|_{\text{minimum}}$.

Because state variable $x(t)$ and $z_D(t)$ cannot be measured directly, designing state observer is needed to predict state variable and disturbance variable. Wind turbine's state observer's math model:

$$\begin{cases} \dot{\hat{x}} = A\hat{x}(t) + Bu(t) + \Gamma\hat{u}_d(t) + K_x(y(t) - \hat{y}(t)), \\ \hat{y} = C\hat{x}(t); \quad \hat{x}(0) = 0. \end{cases} \tag{6.59}$$

Disturbance observer:

$$\begin{cases} \dot{\hat{x}}_D = F\hat{z}_D(t) + K_D(y(t) - \hat{y}(t)), \\ \hat{u}_D = \Theta\hat{z}_D(t). \end{cases} \tag{6.60}$$

Designing appropriate Kx and KD can let:

$$\begin{aligned} \lim_{t\to\infty} e_x(t) &= \lim_{t\to\infty}\big(x(t) - \hat{x}(t)\big) = 0, \\ \lim_{t\to\infty} e_D(t) &= \lim_{t\to\infty}\big(z_D(t) - \hat{z}_D(t)\big) = 0. \end{aligned} \tag{6.61}$$

Disturbance state function can be written as:

$$\dot{e}(t) = (\bar{A} - \bar{K}\bar{C})e(t). \tag{6.62}$$

Where, $e(t) = [\, e_x^t \;\; e_D^t \,]^t$, $\bar{A} = \left[\begin{smallmatrix} A & \Gamma\Theta \\ 0 & F \end{smallmatrix}\right]$, $\bar{C} = [\, c \;\; 0 \,]$, $\bar{K} = \left[\begin{smallmatrix} K_x \\ K_D \end{smallmatrix}\right]$.

According to the formula above, error's expression can be solved:

$$e(t) = e^{(\bar{A} - \bar{K}\bar{C})t}e(0). \tag{6.63}$$

If system $(\bar{A} \;\; \bar{C})$ is measurable, then $(\bar{A} - \bar{K}\bar{C})$ can have any poles configuration, letting $e(t)$ damping to 0 rapidly. Feedback control principal became:

$$u(t) = G\hat{x}(t) + G_D\hat{z}_D(t). \tag{6.64}$$

To verify the control performance of LQR algorithm based on disturbance correlation, a numeric simulation was performed on MATLAB Simulink. The wind turbine model parameter is: rated power 650 kW, rotor diameter 43 m, gear box transmission ratio 43.16, rotor rated speed 42 rpm. LQR algorithm based on disturbance correction and PI regulation method were simulated.

Choosing work point at $V_0 = 17$ m/s, $\Omega_0 = 42$ rpm, $\beta_0 = 13.35$ in LQR algorithm and linearizing at this point. Then wind turbine's state function is function (6.51), where:

$$A = \begin{bmatrix} -0.198 & -3.108 \times 10^{-6} & -3.108 \times 10^{-5} \\ 2.69 \times 10^{7} & 0 & -2.69 \times 10^{7} \\ 1.56 \times 10^{-4} & 1.56 \times 10^{-5} & -0.0624 \end{bmatrix},$$

$$B = \begin{bmatrix} -7.5 \times 10^{-3} \\ 0 \\ 0 \end{bmatrix},$$

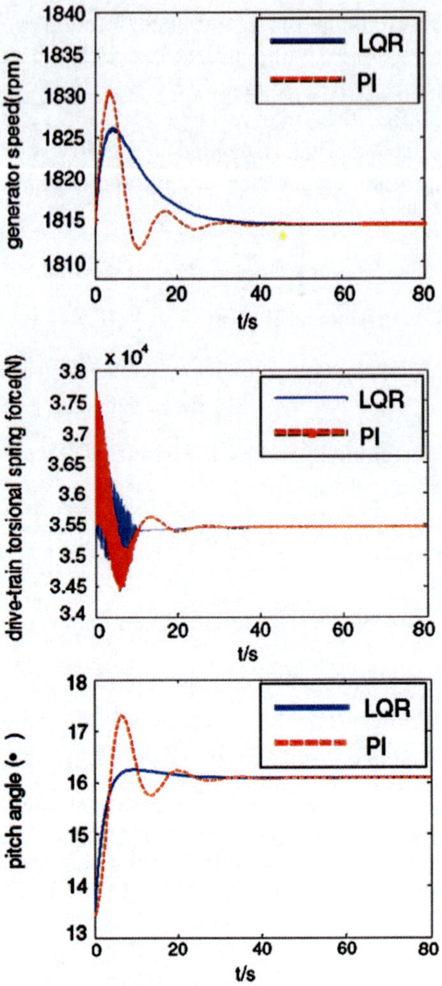

Fig. 6.120 Simulation waveform of LQR algorithm

choosing $R = 1$, $Q = \begin{bmatrix} 1 & 0 & 0 \\ 0 & 1\times10^{-12} & 0 \\ 0 & 0 & 50 \end{bmatrix}$. From matrix A, B, Q and R, state feedback matrix: $K = [2.2219 \;\; 1.6905 \times 10^{-8} \;\; -1.3289]$.

In the simulation, wind speed stepped from 17 m/s to 18 m/s at $t = 0$ moment. In PI regulation, $K_p = 8$, $K_I = 1.5$, simulation result is shown as Fig. 6.120. From the simulation we can tell, PI regulation method has a lager overshoot, while LQR algorithm has a much smaller one. In Fig. 6.120, LQR algorithm can decrease the elastic force on drive link. In Fig. 6.120, after adopting LQR algorithm, the overshoot can be very small, which can reduce the action of pitch actuator. While PI regulation has a larger overshoot, pitch angle fluctuated for a moment, which is harmful for pitch actuator.

6.17 LQR in MATLAB

The command **[K, P, E]** = **lqr(A, B, Q, R, N)** solves the ARE (5.62) and computes the optimal state-feedback gain matrix K given in (5.60) that minimizes the LQR criteria (5.58) for the continuous-time system (5.55). It also returns the poles E of the closed-loop system (5.64).

6.18 Questions

Q5.1 Suppose that $\mathcal{P}_1$ and $\mathcal{P}_2$ are two symmetric positive-definite solutions to the ARE (5.62). Show that $\mathcal{P}_1$ and $\mathcal{P}_2$ satisfy $(A - B\mathcal{R}^{-1}B^t\mathcal{P}_2)(\mathcal{P}_1 - \mathcal{P}_2) + (\mathcal{P}_1 - \mathcal{P}_2)(A - B\mathcal{R}^{-1}B^t\mathcal{P}_2) = 0$ and argue that $\mathcal{P}_1 = \mathcal{P}_2$.

Q5.2 Derive a solution to the optimal control problem involving a performance index $J_\alpha = \int_0^\infty e^{2\alpha t}[x^t(t)\mathcal{Q}z(t) + u^t(t)\mathcal{R}u(t)]\,dt$, and show that the associated closed-loop eigenvalues have real parts less than $-\alpha$.

6.18.1 MATLAB Problems

1. For the linearized model of the Reverse osmosis (RO) plant discussed in Sect. 5.4, design and evaluate an observer-based feedback controller by selecting the observer eigenvalues distinctly different from the controller eigenvalues. Plot the state responses for different cases and comment on the results.
2. For the linearized model of the Reverse osmosis (RO) plant discussed in Sect. 5.4, design and evaluate an optimal linear quadratic regulator with equal weighting for the state and input. Plot the output responses to unit step input and compare on the same graph the open-loop and the closed-loop responses.
3. Consider a quadruple-tank process, depicted in Fig. 6.121, which consists of four interconnected water tanks and two pumps. Its manipulated variables are voltages to the pumps and the controlled variables are the water levels in the two lower tanks. The quadruple tank system presents a typical multi-input-multi-output (MIMO) system that is widely used in control system labs. An appropriate state-space model is given by:

$$\frac{dh_1}{dt} = -\frac{a_1}{A_1}\sqrt{2gh_1} + \frac{a_3}{A_1}\sqrt{2gh_3} + \frac{\gamma_1 k_1}{A_1}v_1 + \frac{d}{A_1},$$
$$\frac{dh_2}{dt} = -\frac{a_2}{A_2}\sqrt{2gh_2} + \frac{a_4}{A_2}\sqrt{2gh_4} + \frac{\gamma_2 k_2}{A_2}v_2 - \frac{d}{A_2},$$
$$\frac{dh_3}{dt} = -\frac{a_3}{A_3}\sqrt{2gh_3} + \frac{(1-\gamma_2)k_2}{A_3}v_2,$$
$$\frac{dh_4}{dt} = -\frac{a_4}{A_4}\sqrt{2gh_4} + \frac{(1-\gamma_1)k_1}{A_4}v_1,$$

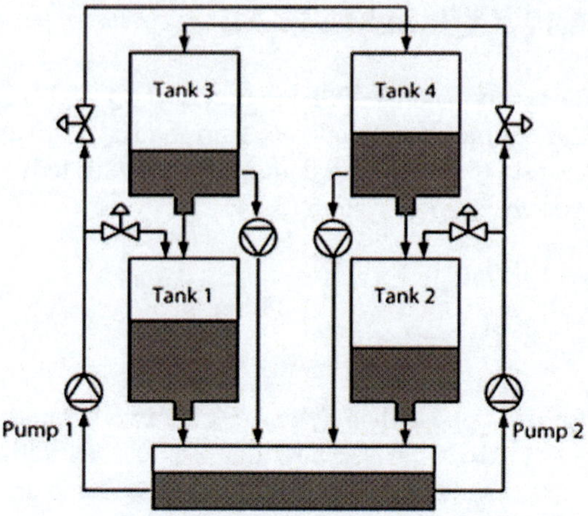

Fig. 6.121 Four tank model

Fig. 6.122 A twin-rotor helicopter when going upward along z-axis

$$\frac{dv_1}{dt} = -\frac{v_1}{\tau_1} + \frac{1}{\tau_1}u_1,$$
$$\frac{dv_2}{dt} = -\frac{v_2}{\tau_2} + \frac{2}{\tau_2}u_2.$$

By linearizing the model around the point $h^o = [11.4 \quad 11.6 \quad 5.3 \quad 4.0]^t$, $v^o = [0.5 \quad 0.5]^t$. The remaining data is $a_i = [2.10, 2.14, 2.2, 2.3]$, cm^2 $A_i = 730$, $\gamma_1 = 0.30$, $\gamma_2 = 0.35$, $k_j = [7.45, 7.30]$, $g = 9.81$, $\tau_i = [2.0, 2.1]$.

It is desired to undertake control studies using alternative control strategies. Provide simulations to compare among various controllers.

4. A twin-rotor helicopter when going upward along z-axis is depicted in Fig. 6.122. The objective is to control the azimuth and elevation angles and the height. The system is underactuated because it has two actuators and three degrees of freedom. The model has two inputs and three outputs. The outputs of the system include azimuth angle ϕ (position plane about the vertical axis), the elevation angle θ (position in the vertical plane about horizontal axis) and the height h (position along the vertical axis (z-axis)). The voltages u_1 and u_2 to the main and tail rotors served as inputs to the system.

The complete set of equations describing the helicopter process during its motion along z-axis is given by

$$\begin{aligned}
\frac{d\phi}{dt} &= \dot{\phi}, \\
\frac{d\dot{\phi}}{dt} &= [J_A + J_L]^{-1}[2J_L \cos\theta \sin\theta\dot{\theta}\dot{\phi} + \tau_\phi], \\
\frac{d\theta}{dt} &= \dot{\theta}, \\
\frac{d\dot{\theta}}{dt} &= \frac{1}{|d|}\big[-d_{22}\{\tau_\theta - J_L \cos\theta \sin\theta\dot{\phi}^2 - ml_c g\cos\theta\} \\
&\quad - d_{12}\{\tau_h + ml_c \sin\theta\dot{\theta}^2 - m - Km_s\}\big], \\
\frac{dh}{dt} &= \dot{h}, \\
\frac{d\dot{h}}{dt} &= \frac{1}{|d|}\big[-d_{12}\{\tau_\theta - J_L \cos\theta \sin\theta\dot{\phi}^2 - ml_c g\cos\theta\} \\
&\quad \times \{\tau_h + ml_c \sin\theta\dot{\theta}^2 - m - Km_s\}\big], \\
\frac{d\omega_1}{dt} &= -\frac{1}{T}\omega_1 + \frac{1}{k_1 T_1}u_1, \\
\frac{d\omega_2}{dt} &= -\frac{1}{T}\omega_2 + \frac{1}{k_2 T_2}u_2
\end{aligned}$$

here $K = \frac{2}{3}$, $k = \frac{3}{4}$ and $\phi =$ azimuth angle, $\dot{\phi} =$ rate of change of azimuth angle, $\theta =$ elevation angle, $\dot{\theta} =$ rate of change of elevation angle, $h =$ height, $\dot{h} =$ rate of change of height, $\omega_1 =$ angular velocity of the mail rotor, $\omega_2 =$ angular velocity of the tail rotor. By linearizing the above model around the equilibrium point $\phi_0 = 0.1$ rad, $\theta_0 = 0$ rad, $h_0 = 0.05$ m, $u_{1,0} = 0.7788$ V, $u_{2,0} = 1.2548$ V and taking

$$\begin{aligned}
x &= [\phi \quad \dot{\phi} \quad \theta \quad \dot{\theta} \quad h \quad \dot{h} \quad \omega_1 \quad \omega_2]^t, \\
u &= [u_1 \quad u_2]^t, \qquad y = [\phi \quad \theta \quad h]^t
\end{aligned}$$

as the state, input and output vectors, a linearized model has the following matrices

$$A = \begin{bmatrix}
0 & 1 & 0 & 0 & 0 & 0 & 0 & 0 \\
0 & 0 & 157.55 & 0 & 0 & 0 & 0 & 0 \\
0 & 0 & 0 & 0 & 1 & 0 & 0 & 0 \\
0 & 0 & 0 & 0 & 0 & 0 & 2.9741 & -0.218 \\
0 & 0 & 0 & 0 & 0 & 1 & 0 & 0 \\
0 & 0 & 0 & 0 & 0 & 0 & 0.0418 & 0.5115 \\
0 & 0 & 0 & 0 & 0 & 0 & -0.2 & 0 \\
0 & 0 & 0 & 0 & 0 & 0 & 0 & -0.4
\end{bmatrix},$$

$$B = \begin{bmatrix}
0 & 0 & 0 & 0 & 0 & 0 & 3.6364 & 0 \\
0 & 0 & 157.55 & 0 & 0 & 0 & 0 & 9.0909
\end{bmatrix}^t,$$

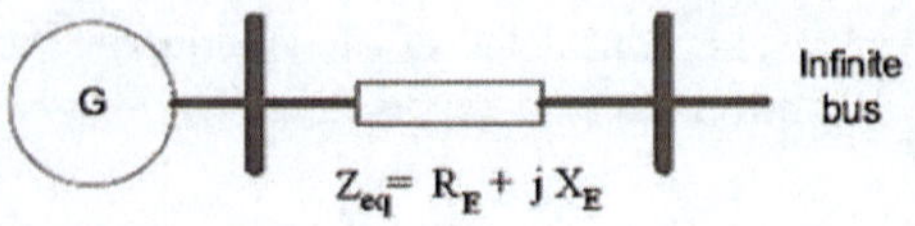

Fig. 6.123 Single machine and infinite bus system

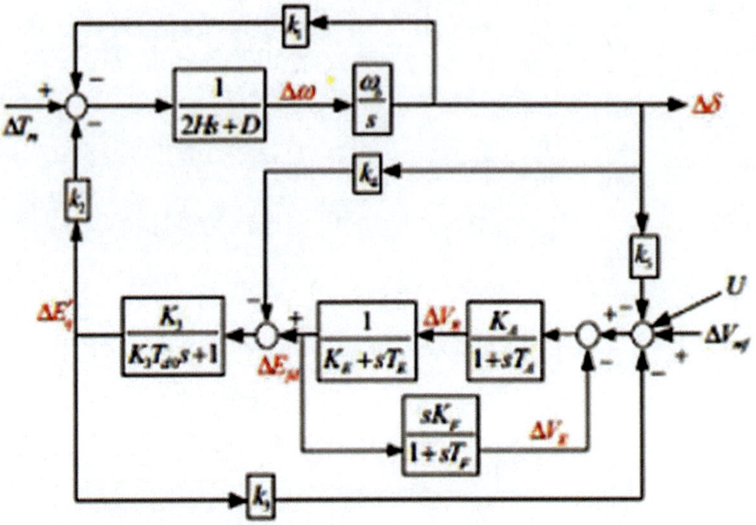

Fig. 6.124 Block diagram of SMIB with exciter and AVR

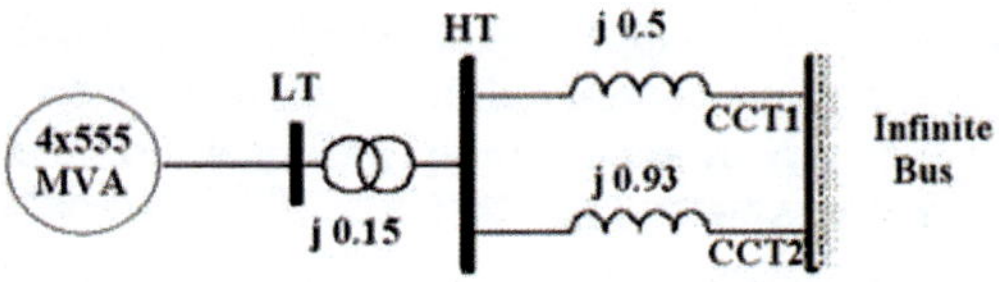

Fig. 6.125 Power system representation

$$C^t = \begin{bmatrix} 1 & 0 & 0 & 0 & 0 & 0 & 0 & 0 \\ 0 & 0 & 1 & 0 & 0 & 0 & 0 & 0 \\ 0 & 0 & 0 & 0 & 1 & 0 & 0 & 0 \end{bmatrix}.$$

It is desired to undertake control studies using alternative control strategies. Provide simulations to compare among various controllers.

5. Consider the problem of designing power system stabilizer (PSS) for a single machine and infinite bus (SMIB) system based on linear control techniques. A schematic representation of this system is shown in Fig. 6.123.

   A standard block diagram including the effect of excitation is given in Fig. 6.124. As a typical case, consider the following case in Fig. 6.125 along with data values

   (i) *Post fault system condition*

$$P = 0.9, \qquad Q = 0.3, \qquad E_t = 1.0\angle 36^\circ, \qquad E_B = 0.995\angle 0^\circ,$$
$$f = 60,$$

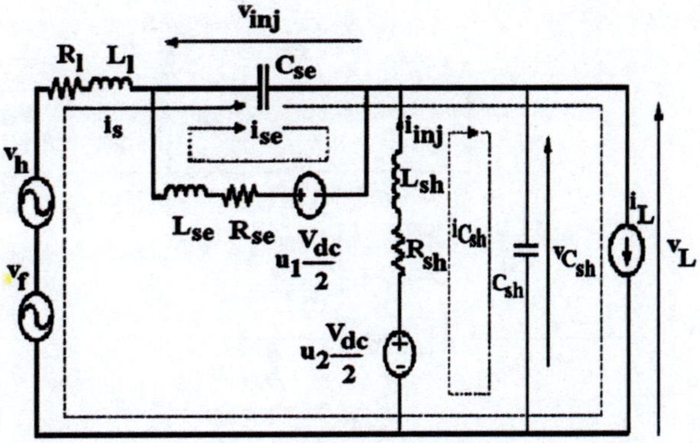

Fig. 6.126 Single phase representation of UPQC

(ii) *Generator parameters*

$$H = 3.5\ \text{MWs/MVA},\quad L_d = 1.81,\quad X_d = 1.81,\quad L_q = 1.76,$$
$$X_q = 1.76,\quad L_\ell = 0.15,\quad X'_d = 0.3,\quad L'_d = 0.3,$$
$$X_\ell = 0.16,\quad L'_q = 1.81,\quad X_d = 1.81,\quad L_q = 1.76,$$
$$T'_{do} = 1.76,\quad L_d = 0.65,\quad R_a = 0.003,\quad L''_d = 0.23,$$
$$T'_{do} = 8.0\ \text{s},\quad L''_q = 0.25,\quad K_D = 0,\quad T'_{qo} = 1.0\ \text{s},$$
$$T''_{do} = 0.03\ \text{s},\quad T''_{qo} = 0.07\ \text{s},$$

(iii) *IEEE type-1 excitation system*

$$K_A = 50,\quad T_A = 0.05,\quad K_E = -0.05.0,\quad T_E = 0.5,$$
$$K_F = 0.05,\quad T_F = 0.5,$$

(iv) *Constants*

$$K_1 = 0.84982,\quad K_2 = 1.0286,\quad K_3 = 0.38618,\quad K_4 = 1.55310,$$
$$K_5 = -0.1315,\quad K_6 = 0.49934.$$

6. In recent years, the increasing use of power electronic devices has led to the deterioration of power quality (PQ) due to harmonic generations. On the other hand, a stable supply voltage has always been desired for smooth operations of many industrial power plants. Recent research has shown that the unified power quality conditioners (UPQCs), an integration of series and shunt active filters, can be utilized to solve most PQ problems simultaneously. This motivates us to develop comprehensive and cost-effective controllers that cannot only be implemented easily but also fully utilize the UPQC to solve a wide range of PQ problems. Different control approaches for the UPQC have been proposed. The most common approach focuses on extracting and injecting distorted components, that is, harmonics (from sampled supply voltage and load current), into the network. This aims to make the load voltage and supply current undistorted. A single-phase representation of UPQC is shown in Fig. 6.126.

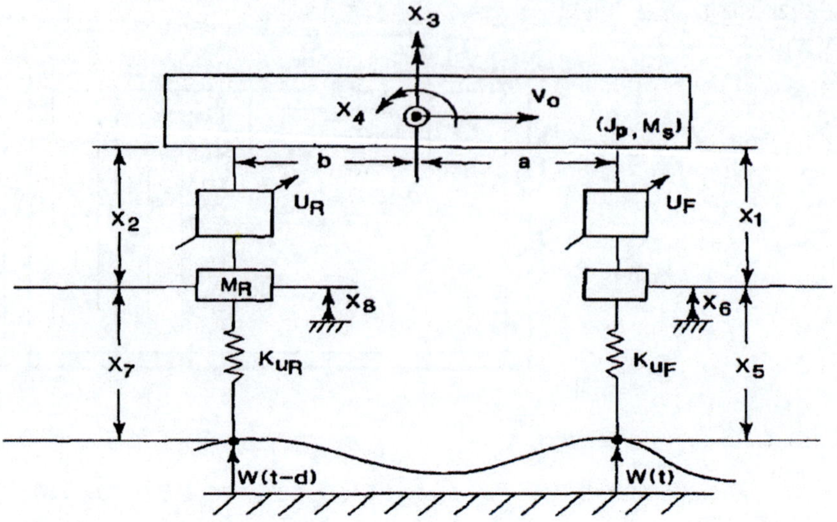

Fig. 6.127 Vehicle suspension system

Apply basic electrical circuit laws to develop an appropriate state-space model and explore the structural properties. Design suitable state and observer-based feedback controllers to improve the system behavior.

7. The study of ride quality evaluates the passenger's response to road/terrain irregularities with the objective of improving comfort and road isolation while maintaining wheel/ground contact. Ride problems mainly arise from vehicle vibrations, which may be induced by variety of sources including external factors, such as roadway roughness or aerodynamic forces, or they may be internally generated forces produced by vehicle subsystems, such as the engine, powertrain, or the suspension mechanisms. Usually the surface irregularity acts as a major source that excites the vibration of the vehicle. Passenger comfort in a road vehicle depends on a combination of vertical motion (heave) and angular motion (pitch). Suspension elements between the wheels and the vehicle body generate vertical forces which excite both heave and pitch motions. Suspension system designs are mostly based on ride analysis. Vehicle suspensions using various types of springs, dampers, and linkages with tailored flexibility in various directions have been developed over the last century since the beginning of the automobile age. Active suspensions, with proper control design, can give an improved overall suspension performance. A schematic diagram of the vehicle with an active suspension system is shown in Fig. 6.127.

   In the modeling process, rigid bodies of masses m_f and m_R represent respectively the front and rear equivalent mass of the wheel assembly, upper control arm, lower control arm and the associated linkages. The front and rear tire stiffness are denoted by K_{uF} and K_{uR}, respectively. The frame and body structure of the vehicle is characterized by the mass m_s and the pitch moment of inertia J_p (about a body-fixed coordinate system, centered at the vehicle's centre of gravity). The state variables of this model are: x_1 = front suspension deflection, x_2 =

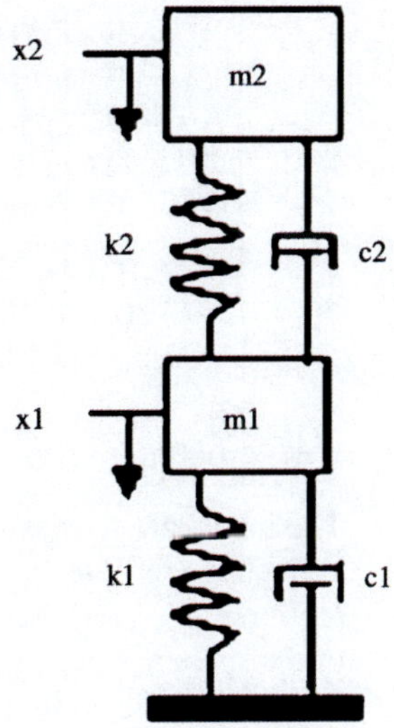

Fig. 6.128 One-quarter car model

rear suspension deflection, $x_3 =$ vertical velocity of the sprung mass, $x_4 =$ pitch rate of the sprung mass, $x_5 =$ front tire deflection, $x_6 =$ vertical velocity of the front unsprung mass, $x_7 =$ rear tire deflection and $x_8 =$ vertical velocity of the rear unsprung mass. Two actuators are placed in between the sprung mass m_s and the unsprung masses m_R and m_F. The active control forces of the front and rear actuators are denoted by u_F and u_R, respectively. Only the first and second states can be measured.

Develop an appropriate state-space model and examine the structural properties. Design suitable state, observer-based feedback controllers and linear-quadratic regulator to improve the system behavior.

8. Figure 6.128 shows a simplified linear quarter car model. Develop a state-space model in terms of $x_1, \dot{x}_1, x_2, \dot{x}_2$ as the state vector and u_1, u_2 as the force inputs to masses m_1, m_2.

   Examine the structural properties of the model and design linear feedback controllers using $m_1 = 60$ kg, $m_2 = 375$ kg, $k_1 = 200\ \text{kN m}^{-1}$, $k_2 = 15\ \text{kN m}^{-1}$, $c_1 = 7\ \text{Ns m}^{-1}$ and $c_2 = 1425\ \text{Ns m}^{-1}$. Moreover, a good-quality road with length $L = 100$ m is considered for a vehicle speed range $40 \to 180\ \text{km h}^{-1}$.
9. The longitudinal motion of a flexible bomber aircraft is modeled as a second order short-period mode, a second-order fuselage bending mode, and two first-order control-surface actuators. The sixth order system is described by the following linear, time-invariant, state-space representation

$$A = \begin{bmatrix} 0.4158 & 1.025 & -0.00267 & -0.0001106 & -0.08021 & 0 \\ -5.5 & -0.8302 & -0.06549 & -0.0039 & -5.115 & 0.809 \\ 0 & 0 & 0 & 1 & 0 & 0 \\ -1040 & -78.35 & -34.83 & -0.6214 & -865.6 & -631 \\ 0 & 0 & 0 & 0 & -75 & 0 \\ 0 & 0 & 0 & 0 & 0 & -100 \end{bmatrix},$$

$$B = \begin{bmatrix} 0 & 0 \\ 0 & 0 \\ 0 & 0 \\ 0 & 0 \\ 75 & 0 \\ 0 & 100 \end{bmatrix},$$

$$C = \begin{bmatrix} -1491 & -146.43 & -40.2 & -0.9412 & -1285 & 564.66 \\ 0 & 1.0 & 0 & 0 & 0 & 0 \end{bmatrix},$$

$$D = \begin{bmatrix} 0 & 0 \\ 0 & 0 \end{bmatrix}.$$

The inputs are the desired elevator deflection (rad), $u_1(t)$, and the desired canard deflection (rad), $u_2(t)$, while the outputs are the sensor location's normal acceleration m/s^2, $y_1(t)$, and the pitch-rate (rad/s), $y_2(t)$. Test the structural properties of the system. Then proceed to design feedback controllers and compare the corresponding closed-loop state trajectories.

References

1. Al-Bastaki, N.M., Abbas, A.: Modeling an industrial reverse osmosis unit. Desalination **126**, 33–39 (1999)
2. Alatiqi, I., Ettouney, H., El-Dessouky, H.: Process control in water desalination industry: An overview. Desalination **126**, 15–32 (1999)
3. Altug, E., Ostrowski, J.P., Mahony, R.: Control of a quadrotor helicopter using visual feedback. In: Proc. 2002 IEEE Int. Conference on Robotics and Automation, vol. 1, pp. 72–77 (2002)
4. Altug, E., Ostrowski, J.P., Taylor, C.J.: Quadrotor control using dual camera visual feedback, In: ICRA, Taipei, September 2003
5. Anderson, W.: Controlling Electrohydraulic Systems. Dekker, New York (1988)
6. Bagchi, A.: Optimal Control of Stochastic Systems. Prentice Hall International, Englewood Cliffs (1993)
7. Bandyopadhyay, B., Ghandi, P.S., Kurodo, S.: Sliding mode observer based sliding mode controller for slosh free motion through PID scheme. IEEE Trans. Ind. Electron. **56**(9), 3432–3442 (2009)
8. Burl, J.B.: Linear Optimal Control, 3rd edn. Prentice-Hall, New York (1998)
9. Castillo, P., Dzul, A., Lozano, R.: Real-time stabilization and tracking of a four-rotor mini rotorcraft. IEEE Trans. Control Syst. Technol. **12**(4), 510–516 (2004)
10. Castillo, P., Lozano, R., Dzul, A.: Stabilization of a mini rotorcraft with four rotors. IEEE Control Syst. Mag. **25**, 45–50 (2005)
11. Chen, T., Francis, B.A.: Optimal Sampled-Data Control Systems. Springer, London (1995)
12. Chen, X., Yang, T., Chen, X., Zhou, K.: A generic model-based advanced control of electric power-assisted steering systems. IEEE Trans. Control Syst. Technol. **16**(6), 1289–1300 (2008)
13. Datta, R., Ranganathan, V.T.: Variable-speed wind power generation using doubly fed wound rotor induction machine. A comparison with alternative schemes[J]. IEEE Trans. Energy Convers. **17**, 414–421 (2002)
14. Dragan Fly-Innovations, www.rctoys.com (2003)
15. Franklin, G.F., Powell, J.D., Naeini, A.E.: Feedback Control of Dynamic Systems, 4th edn. Prentice-Hall, New York (2002)
16. Fukushima, N., Arslan, M.S., Hagiwara, I.: An optimal control method based on the energy flow equation. IEEE Trans. Control Syst. Technol. **17**(4), 866–875 (2009)

17. Gaeid, K.S., Mohamed, H.A.F., Ping, H.W., Hassan, L.H.: PID controller for induction motors with faults, University of Malaya & University of Nottingham Malaysia Campus, In: The 2nd International Conference on Control, Instrumentation & Mechatronic, CIM-2009
18. Gambier, A., Krasnik, A., Badreddin, E.: Dynamic modelling of a small reverse osmosis desalination plants for advanced control purposes. In: Proc. of the 2007 American Control Conference. New York, July 11–13, pp. 4854–4859 (2007)
19. Ghazy, M.A.: Variable structure control for electrohydraulic position servo system. In: Proc. Industrial Electronics Conference, IECON, vol. 1, pp. 2194–2198 (2001)
20. Guenard, N., Hamel, T., Moreau, V.: Modélisation et élaboration de commande de stabilisation de vitesse et de correction d'assiette pour un drone. In: CIFA, 2004
21. Hamel, T., Mahoney, R., Lozano, R., Ostrowski, E.: Dynamic modelling and configuration stabilization for an X4-flyer. In: The 15éme IFAC World Congress, Barcelona, Spain, 2002
22. Hangzhi, Y.: Control Technology of Wind Turbine, Mechanics Industry Press, Beijing (2002)
23. Hoshijima, K., Ikeda, M.: Vibration suppression control for mechanical transfer systems by jerk reduction. Int. J. Control. Autom. Syst. **5**(6), 614–620 (2007)
24. Hu, C., Cheng, X.: Electric drive system for winding shaper of automobile belt, In: Proc. IEEE Conf. on Electric Machines and Systems, Shengyyang, China, 2001
25. Hu, C., Meng, M., Liu, P.X., Wang, X.: Optimal digital control system design for winding shaping process of automobile belt. In: IEEE Canadian Conference on Electrical and Computer Engineering, Montreal, pp. 1763–1766 (2003)
26. Jaho Seo, J., Venugopala, R., Kenne, J.-P.: Feedback linearization based control of a rotational hydraulic drive. Control Eng. Pract. **15**, 235–241 (2007)
27. Kanellos, F.D., Hatziargyriou, N.D.: A new control scheme for variable speed wind turbines using neural networks[C]. In: Power Engineering Society Winter Meeting, 2002, vol. 1. IEEE, pp. 260–365 (2002)
28. Kim, N., Cha, S., Peng, H.: Optimal control of hybrid electric vehicles based on Pontryagin's minimum principle. IEEE Trans. Control Syst. Technol. **21**, 73–82 (2010)
29. Ko, H.S., Lee, K.Y., Kim, H.C.: An intelligent based LQR controller design to power system stabilization. Electr. Power Syst. Res. **71**, 1–9 (2004)
30. Koc, H., Knittel, D., de Mathelin, M., Abba, G.: Modeling and robust control of winding systems for elastic webs. IEEE Trans. Control Syst. Technol. **10**(2), 197–208 (2002)
31. Kuribayashi, K., Nakajima, K.: Active dancer roller system for tension control of wine and sheet. In: Proc. IFAC Power Systems, vol. 4, pp. 1747–1752 (1985)
32. Lambermont, P.: Helicopters and Autogyros of the World (1958)
33. Liao, H.H., Roelle, M.J., Chen, J.S., Park, S., Gerdes, J.C.: Implementation and analysis of a repetitive controller for an electro-hydraulic engine valve system. IEEE Trans. Control Syst. Technol. **21**, 12–22 (2010)
34. Lin, C.C., Peng, H., Grizzle, J.W., Kang, J.: Power management strategy for a parallel hybrid electric truck. IEEE Trans. Control Syst. Technol. **11**(6), 839–849 (2003)
35. Lopes, R.V., Galvão, R.K.H., Milhan, A.P., Becerra, V.M., Yoneyama, T.: Modelling and constrained predictive control of a 3DOF helicopter. In: XVI Congresso Brasileiro de Automática, Salvador, Brazil, paper 123, vol. 1, pp. 429–434 (2006)
36. Maia, M.H., Galvão, R.K.H.: Robust constrained predictive control of a 3DOF helicopter model with external disturbances. In: ABCM Symposium Series in Mechatronics, vol. 3, pp. 19–26 (2008)
37. Mayne, D.Q., Rawlings, J.B., Rao, C.V., Scokaert, P.O.M.: Constrained model predictive control: Stability and optimality. Automatica **36**(6), 789–814 (2000)
38. Moura, S.J., Fathy, H.K., Callaway, D.S., Stein, J.L.: A stochastic optimal control approach for power management in plug-in hybrid electric vehicles. IEEE Trans. Control Syst. Technol. **21**, 1–11 (2010)
39. Noura, H., Sauter, D., Hamelin, F., Theilliol, D.: Fault-tolerant control in dynamic systems: Application to a winding machine. IEEE Control Syst. Mag. **7**, 33–49 (2000)
40. Ogata, K.: MATLAB for Control Engineers. Prentice-Hall, New York (2008)

41. Olfati-Saber, R.: Nonlinear control of underactuated mechanical systems with application to robotics and aerospace vehicles. Ph.D. thesis in Electrical Engineering and Computer Science, Massachusetts Institute of Technology (February 2001)
42. Parola, M., Vuorinen, S., Linna, H., Kaljunen, T., Beletski, N.: Modelling the web tension profile in a paper machine, http://www.vtt.fi/tte/informationcarriers/Oxford2001.pdf (2001)
43. Peterson, K.S., Stefanopoulou, A.G.: Rendering the electromechanical valve actuator globally asymptotically stable. In: Proc. IEEE Conf. Decision Control, pp. 1753–1758 (2003)
44. Pipeleers, G., Demeulenaere, B., Al-Bender, F., Schutter, J., Swevers, J.: Optimal performance tradeoffs in repetitive control: Experimental validation on an active air bearing setup. IEEE Trans. Control Syst. Technol. **17**(4), 970–979 (2009)
45. Pounds, P., Mahony, R., Hynes, P., Roberts, J.: Design of a fourrotor aerial robot. In: Australian Conference on Robotics and Automation, Auckland, November 2002
46. Sakamoto, R., Senjyu, T., Kinjo, T.: Output power leveling of wind turbine generator for all operating regions by pitch angle control. In: Power Engineering Society General Meeting, vol. 1, pp. 45–52 (2005)
47. Senjyu, T., Sakamoto, R., Urasaki, N., Funabashi, T.: Output power leveling of wind turbine generator for all operating regions by pitch angle control. IEEE Trans. Energy Convers. **21**, 467–475 (2006)
48. Shelton, J.J.: Dynamics of web tension control with velocity or torque control. In: Proc. American Control Conference, New York, pp. 1423–1427 (1986)
49. Terashima, K., Schmidt, G.: Motion control of a cart-based container considering suppression of liquid oscillations. In: Int. Symposium on Industrial Electronics, 25–27 May, pp. 275–280 (1994)
50. UAVs, New world vistas: Air and space for the 21st century. Human Systems and Biotechnology Systems **7**, 17–18 (1997)
51. Wang, J., Ydstie, E.B.: Robust inventory control system. IEEE Trans. Control Syst. Technol. **15**(4), 768–774 (2007)
52. Yano, K., Higashikawa, S., Terashima, K.: Motion control of liquid container considering an inclined transfer path. Control Eng. Pract. **10**, 465–472 (2002)
53. Yuan, Q.H., Li, P.Y.: Robust optimal design of unstable valves. IEEE Trans. Control Syst. Technol. **15**(6), 1065–1074 (2007)

Chapter 7
Robust Control Design

7.1 Introduction

It is commonly agreed practice that a successfully designed control system should be always able to maintain stability and performance level in spite of uncertainties in system dynamics and/or in the working environment to a reasonable degree [3, 25]. Design requirements such as gain margin and phase margin in using conventional frequency-domain techniques are solely for the purpose of robustness [4]. During the period of 1960s and 1970s when system models could be much more accurately described and design methods were mainly mathematical optimizations in time-domain, the robustness issue was not that prominently considered. However, due to its importance the research on robust design has been going on all the time. A breakthrough came in the late 1970s and early 1980s with the pioneering work of [23, 24] on the theory, now known as the $\mathcal{H}_\infty$ optimal control theory. The $\mathcal{H}_\infty$ optimization approach and μ-synthesis/analysis method are well developed and elegant. They provide systematic design procedures of robust controllers for linear systems, though the extension into nonlinear cases is being actively researched. Many books have since been published on $\mathcal{H}_\infty$ and related theories and methods [2, 6, 7, 10, 11, 21, 22, 25, 26].

On a parallel development, the application of optimal control theory to the practical design of multivariable control systems attracted much attention during the period 1960–1980. This theory considers linear finite-dimensional systems represented in state space form, with quadratic performance criteria. The system may be affected by disturbances and measurement noise represented as stochastic processes, in particular, by Gaussian white noise. The theoretical results obtained for this class of design methods are known under the general name of LQG theory [1, 15]. The deterministic part is called LQ theory which was discussed in Chap. 5. Since 1980 the theory has been further refined under the name of $\mathcal{H}_2$ theory [5], in the wake of the attention for the so-called $\mathcal{H}_\infty$ control theory.

In the present chapter, we present a short overview of a number of results of $\mathcal{H}_2$ and $\mathcal{H}_\infty$ theories with an eye to using them for control system design. Since robustness of a feedback control system is very important in control engineering practice.

M.S. Mahmoud, Y. Xia, *Applied Control Systems Design*,
DOI 10.1007/978-1-4471-2879-3_7, © Springer-Verlag London Limited 2012

In actual control problems, there are always disturbances due to the environment and uncertainties due to the imperfect model used in the controller design. Clearly, it is desirable for the controlled system to have certain robustness against these disturbances and uncertainties. To assess the robustness, first of all, a proper measure is needed. Norm measures to signals and systems are introduced, which can be regarded as the basis of robust control. In what follows, the symbols $\mathcal{L}$ and $\mathcal{H}$ are due to Lebesgue and Hardy, respectively.

7.1.1 Norm Measures of Signals

The size of a signal $f(t)$ is usually measured in its $\mathcal{L}_p$-norm defined as

$$\left\| f(t) \right\|_p = \left(\int_{-\infty}^{\infty} \left| f(t) \right|^p dt \right) \tag{7.1}$$

where p is a positive integer. Three key properties of a norm:

1. $\|f(t)\|_p \geq 0$ and $\|f(t)\|_p = 0$, iff $u = 0$.
2. $\alpha\|f(t)\|_p = |\alpha|\|f(t)\|_p$, $\forall$ scalars α.
3. $\|f(t) + g(t)\|_p \leq \|f(t)\|_p + \|g(t)\|_p$.

The following norms are commonly used:

1. The $\mathcal{L}_1$-norm: $\|f(t)\|_1 = \int_{-\infty}^{\infty} |f(t)|\, dt$.
2. $\mathcal{L}_2$-norm, the measure of signal power: $\|f(t)\|_2 = \sqrt{\int_{-\infty}^{\infty} f^2(t)\, dt}$.
3. $\mathcal{L}_\infty$-norm, the least upper bound of $|f(t)|$: $\|f(t)\|_\infty = \sup_t |f(t)|$.

Observe that in case of vector signal $x(t)$, which is a function of time $t \geq 0$, the $\mathcal{L}_2$-norm becomes

$$\left\| x(t) \right\|_2 = \sqrt{\int_0^{\infty} x^t(t) x(t)\, dt}.$$

7.1.2 Norm Measures of Systems

The size of a system in a transfer function form is usually measured by its $\mathcal{H}_2$- and $\mathcal{H}_\infty$-norms. Hereafter, we consider a transfer function matrix $G(s)$ with each element being *strictly proper and stable* transfer function.

1. $\mathcal{H}_2$-norm is defined by

$$\left\| G(s) \right\|_2 = \sqrt{\frac{1}{2\pi j} \int_{-j\infty}^{j\infty} \left| G(j\omega) \right|^2 d\omega}. \tag{7.2}$$

Standard algebraic manipulations lead to

$$\|G(s)\|_2 = \sqrt{\frac{1}{2\pi} \operatorname{Tr} \int_{-\infty}^{+\infty} G(j\omega)G^*(j\omega)\, d\omega}$$
$$= \sqrt{\operatorname{Tr} \int_0^{+\infty} G(t)G^t(t)\, dt} \qquad (7.3)$$

where $G^*(j\omega)$ is the complex conjugate transpose of $G(j\omega)$ and $G(t)$ is the inverse Laplace transform of $G(s)$. The $\mathcal{H}_2$-norm is in fact a measure of the square root of the integral squared value of the output when the input is an impulse signal. In stochastic system terminology, the $\mathcal{H}_2$-norm is the root mean square value of the output signal when the input is white noise.

2. $\mathcal{H}_\infty$-norm is generally defined by

$$\|G(s)\|_\infty = \sup_{u(t)\neq 0} \frac{\|y(t)\|_2}{\|u(t)\|_2} \qquad (7.4)$$

where $u(t)$ and $y(t)$ are the input and output of the system, respectively, and sup denotes supremum, the least upper bound. For stable systems, the $\mathcal{H}_\infty$-norm of the system can be computed from

$$\|G(s)\|_\infty = \sup_\omega |G(j\omega)|. \qquad (7.5)$$

It is readily seen that the $\mathcal{H}_\infty$-norm is in fact the peak value of the magnitude of the frequency response.

It is well-known that the condition

$$\|G(s)\|_{\mathcal{H}_\infty} \le \gamma$$

is satisfied if and only if there exists a matrix $P = P^t \ge 0$ that meets the following criteria:

1. It is a solution of

$$PA + A^t P + C^t C + \gamma^{-2} P B B^t P = 0. \qquad (7.6)$$

2. The matrix

$$A + \gamma^{-2} B B^t P \qquad (7.7)$$

is stable.

The above result is frequently known as the *Bounded Real Lemma* [9]. An interpretation of $\|G(s)\|_{\mathcal{H}_\infty}$ is that it is the *energy gain* from the input u to the output y, that is

$$\|G(s)\|_{\mathcal{H}_\infty} := \max_{u(t)\neq 0} \frac{\int_0^\infty y^t(t)y(t)\, dt}{\int_0^\infty u^t(t)u(t)\, dt}. \qquad (7.8)$$

This implies that $\|G(s)\|_{\mathcal{H}_\infty}$ achieves the maximum gain using a worst case input signal that is essentially a sinusoid at frequency ω^* with input direction that yields $\sup_\omega \sigma_M(G(j\omega^*))$ as the amplification.

The following properties of norms are given without proofs.

1. $\|y(t)\|_2 \le \|G(s)\|_\infty \|u(t)\|_2$.
2. $\|y(t)\|_\infty \le \|G(s)\|_2 \|u(t)\|_\infty$.
3. $\|G_1(s)G_2(s)\|_\infty \le \|G_1(s)\|_\infty \|G_2(s)\|_\infty$.

7.1.3 Significance of $\mathcal{H}_2$-Norm

Consider a MIMO system described by

$$Y(s) = G(s)U(s) \tag{7.9}$$

where $U(s) \in \Re^m$ and $Y(s) \in \Re^p$ are the input and output vectors in the s-domain. Let us apply a unit impulse, $\delta(t)$, in input channel i at a time. The output vector becomes

$$Y_i(s) = G(s)E_i, \quad E_i = \begin{bmatrix} 0 & \cdots & 0 & 1 & 0 & \cdots & 0 \end{bmatrix}^t. \tag{7.10}$$

Thus,

$$\|y_i(t)\|_2^2 = \int_0^{+\infty} \mathrm{Tr}\big[y_i(t)y_i^t(t)\big]\,dt. \tag{7.11}$$

By virtue of Parseval's theorem (see the Appendix), we have

$$\|y_i(t)\|_2^2 = \frac{1}{2\pi}\int_{-\infty}^{+\infty} \mathrm{Tr}\big[y_i(j\omega)y_i^*(j\omega)\big]\,d\omega. \tag{7.12}$$

On using (7.9), we obtain

$$\begin{aligned} \|y_i(t)\|_2^2 &= \frac{1}{2\pi}\int_{-\infty}^{+\infty} \mathrm{Tr}\big[G(j\omega)E_iE_i^tG^*(j\omega)\big]\,d\omega \\ &= \frac{1}{2\pi}\int_{-\infty}^{+\infty} \big[|G_{i1}(j\omega)|^2 + \cdots + |G_{ip}(j\omega)|^2\big]\,d\omega. \end{aligned} \tag{7.13}$$

Manipulating

$$\sum_{i=1}^{m}\|y_i(t)\|_2^2 = \frac{1}{2\pi}\int_{-\infty}^{+\infty}\left[G(j\omega)\left(\sum_{i=1}^{m}E_iE_i^t\right)G^*(j\omega)\right]d\omega. \tag{7.14}$$

Since it can be shown that $\left(\sum_{i=1}^{m} E_iE_i^t\right)I_m$, it follows from (7.14) that

$$\begin{aligned} \sum_{i=1}^{m}\|y_i(t)\|_2^2 &= \frac{1}{2\pi}\int_{-\infty}^{+\infty}\big[G(j\omega)G^*(j\omega)\big]\,d\omega \\ &= \|G(s)\|_2^2. \end{aligned} \tag{7.15}$$

In brief, minimizing the $\mathcal{H}_2$-norm of the transfer function matrix is equivalent o minimizing the sum of squares of $\mathcal{L}_2$-norm of outputs due to unit impulse in each input channel.

In what follows, we provide a connection to the root-mean squares (RMS) response. In the time-domain,

$$y(t) = \int_0^t G(t-\tau)u(\tau)\,d\tau. \tag{7.16}$$

Let each input be an independent zero-mean white noise with unit intensity, that is,

$$\mathbb{E}\big[u(t)u^t(\tau)\big] = I\delta(t-\tau). \tag{7.17}$$

The mean square response can be written as

$$\mathbb{E}\big[y^t(t)y(t)\big] = \mathbb{E}\big[\mathrm{Tr}\big(y(t)y^t(t)\big)\big]. \tag{7.18}$$

A little algebra yields

$$\begin{aligned}
\mathbb{E}\big[y^t(t)y(t)\big] &= \mathbb{E}\left(\mathrm{Tr}\left[\int_0^t G(t-\tau)u(\tau)\,d\tau \int_0^t u^t(\alpha)G^t(t-\alpha)\,d\alpha\right]\right) \\
&= \mathrm{Tr}\left[\int_0^t \int_0^t G(t-\tau)\mathbb{E}u(\tau)u^t(\alpha)G^t(t-\alpha)\,d\tau\,d\alpha\right] \\
&= \mathrm{Tr}\left[\int_0^t \int_0^t G(t-\tau)G^t(t-\alpha)\delta(\tau-\alpha)\,d\tau\,d\alpha\right] \\
&= \mathrm{Tr}\left[\int_0^t G(t-\alpha)G^t(t-\alpha)\,d\alpha\right].
\end{aligned} \tag{7.19}$$

Letting $\nu = t-\alpha$, we get

$$\mathbb{E}\big[y^t(t)y(t)\big] = \int_0^t \mathrm{Tr}\big[G(\nu)G^t(\nu)\big]\,d\nu \tag{7.20}$$

and therefore

$$\lim_{t\to\infty} \mathbb{E}\big[y^t(t)y(t)\big] = \int_0^\infty \mathrm{Tr}\big[G(\nu)G^t(\nu)\big]\,d\nu. \tag{7.21}$$

Using Parseval's theorem (see the Appendix)

$$\begin{aligned}
\lim_{t\to\infty} \mathbb{E}\big[y^t(t)y(t)\big] &= \frac{1}{2\pi}\int_{-\infty}^\infty \mathrm{Tr}\big[G(j\omega)G^*(\omega)\big]\,d\omega \\
&= \big\|G(s)\big\|_{\mathcal{H}_2}^2.
\end{aligned} \tag{7.22}$$

In brief, minimizing the $\mathcal{H}_2$-norm of the transfer function matrix is equivalent to minimizing the RMS of outputs in the statistical steady-state due to independent zero-mean white-noise inputs.

7.1.4 Significance of $\mathcal{H}_\infty$-Norm

Let us define the input as

$$u(t) = a_u e^{j\omega t}, \quad a_u = \begin{bmatrix} a_{u1} & a_{u2} & \cdots & a_{um} \end{bmatrix}^t. \tag{7.23}$$

In steady state, we have

$$y(t) = a_y e^{j\omega t}, \quad a_y = \begin{bmatrix} a_{y1} & a_{y2} & \cdots & a_{yp} \end{bmatrix}^t \tag{7.24}$$

where a_u, a_y are in general complex vectors. Thus,

$$a_y = G(j\omega)a_u. \tag{7.25}$$

Using the definition of the maximum singular value

$$\frac{\|a_y\|_2}{\|a_u\|_2} \le \|G(j\omega)\|_{\mathcal{H}_\infty} \quad \Longrightarrow \quad \sup_{a_u} \frac{\|a_y\|_2}{\|a_u\|_2} = \sigma_M\big(G(j\omega)\big) \le \|G(s)\|_{\mathcal{H}_\infty} \tag{7.26}$$

which provides an alternative statement of (7.4).

Following a different route to express the $\mathcal{H}_\infty$-norm starts by recalling

$$\begin{aligned} \|y(t)\|_{\mathcal{L}_2}^2 &= \frac{1}{2\pi} \int_{-\infty}^{\infty} y^*(j\omega) y(j\omega)\, d\omega \\ &= \frac{1}{2\pi} \int_{-\infty}^{\infty} \|y(j\omega)\|_2^2\, d\omega. \end{aligned} \tag{7.27}$$

Since $y(j\omega) = G(j\omega)u(j\omega)$, then for zero initial conditions of inputs and outputs

$$\begin{aligned} \|y(j\omega)\|_2 &\le \sigma_M\big(G(j\omega)\big) \|u(j\omega)\|_2 \\ &\le \|G(j\omega)\|_{\mathcal{H}_\infty} \|u(j\omega)\|_2. \end{aligned} \tag{7.28}$$

The substitution of (7.28) into (7.27) yields

$$\begin{aligned} \|y(t)\|_{\mathcal{L}_2}^2 &\le \frac{1}{2\pi} \int_{-\infty}^{\infty} \|G(j\omega)\|_{\mathcal{H}_\infty} \|u(j\omega)\|_2^2\, d\omega \\ &= \|G(j\omega)\|_{\mathcal{H}_\infty} \frac{1}{2\pi} \int_{-\infty}^{\infty} \|u(j\omega)\|_2^2\, d\omega. \end{aligned} \tag{7.29}$$

Using Parseval's theorem again,

$$\|y(t)\|_{\mathcal{L}_2} \le \|G(j\omega)\|_{\mathcal{H}_\infty} \|u(t)\|_{\mathcal{L}_2}. \tag{7.30}$$

This means that if

$$\|G(j\omega)\|_{\mathcal{H}_\infty} \le \gamma \tag{7.31}$$

then

$$\frac{\|y(t)\|_{\mathcal{L}_2}}{\|u(t)\|_{\mathcal{L}_2}} \le \gamma \tag{7.32}$$

which is often used in the sequel as

$$\|y(t)\|_{\mathcal{L}_2}^2 - \gamma^2 \|u(t)\|_{\mathcal{L}_2}^2 < 0 \quad \forall u(t) \in \mathcal{L}_2. \tag{7.33}$$

In the next sections, we proceed to solve the $\mathcal{H}_2$ and $\mathcal{H}_\infty$ control problems.

7.2 $\mathcal{H}_2$ Control

Recall that we introduced the linear quadratic Gaussian (LQG) problem in Chap. 5. In this section we cast the LQG problem as a special case of a larger class of problems, which lately has become known as $\mathcal{H}_2$ optimization. It must be emphasized that this approach allows to remove the stochastic ingredient of the LQG formulation. In many applications, it is difficult to establish the precise stochastic properties of disturbances and noise signals. Very often in the application of the LQG problem to control system design the noise intensities V and W play the role of design parameters rather than that they model reality.

Hereafter, the stochastic element is eliminated by recognizing that the performance index for the LQG problem may be represented as a system norm—the $\mathcal{H}_2$-norm. To introduce this point of view, we consider the stable system

$$\begin{aligned} \dot{x}(t) &= Ax(t) + \Gamma v(t), \\ y(t) &= Cx(t), \quad t \in \Re, \end{aligned} \tag{7.34}$$

and the associated transfer function $G(s) = C(sI - A)^{-1}\Gamma$. In (7.34), the disturbance signal $v(t)$ is white noise with covariance function $\mathbb{E}[v(t)v^t(s)] = V\delta(t-s)$. Thus, the output y of the system is a stationary stochastic process with spectral density matrix

$$S_y(f) = G(j2\pi f)VG^t(-j2\pi f), \quad f \in \Re.$$

It follows that the mean square output is given by

$$\begin{aligned} \mathbb{E}\left[y^t(t)y(t)\right] &= \mathrm{Tr}\left[\int_{-\infty}^{\infty} S_y(f)\,df\right] \\ &= \mathrm{Tr}\left[\int_{-\infty}^{\infty} G(j2\pi f)VG^t(-j2\pi f)\,df\right]. \end{aligned} \tag{7.35}$$

Recall that the quantity

$$\|G\|_2^2 = \mathrm{Tr}\left[\int_{-\infty}^{\infty} G(j2\pi f)VG^t(-j2\pi f)\,df\right] \tag{7.36}$$

is called earlier $\mathcal{H}_2$*-norm* of system (7.34). If the disturbance v has unit intensity ($\mathsf{V} = I$) then the mean square output $\mathbb{E}[y^t(t)y(t)]$ equals precisely the square of the $\mathcal{H}_2$-norm of system.

The impulse response matrix (inverse Laplace transform of $G(s)$) of system (7.34) is given by

$$g(t) = Ce^{At}\Gamma.$$

Obviously, if A is not a stability matrix them $g(t)$ is unbounded, hence the $\mathcal{H}_2$-norm is infinite. So consider in the sequel that A is a stability matrix. Therefore,

$$\|G\|_{\mathcal{H}_2}^2 = \mathrm{Tr}\left[\int_{-\infty}^{\infty} g^t(t)g(t)\,dt\right]$$

$$
\begin{aligned}
&= \mathrm{Tr}\left[\int_0^\infty \Gamma^t e^{A^t t} C^t C e^{At} \Gamma \, dt\right] \\
&= \mathrm{Tr}\left[\Gamma^t \int_0^\infty e^{A^t t} C^t C e^{At} \, dt \, \Gamma\right] \\
&= \mathrm{Tr}\left[\Gamma^t \mathbf{Y} \Gamma\right].
\end{aligned} \tag{7.37}
$$

The matrix $\mathbf{Y}$ obviously satisfies

$$
\begin{aligned}
A^t\mathbf{Y} + \mathbf{Y}A &= \int_0^\infty \left[A^t e^{A^t t} C^t C e^{At} + e^{A^t t} C^t C e^{At} A\right] dt \\
&= \left[e^{A^t t} C^t C e^{At}\right]_{t=0}^{t=\infty} \\
&= C^t C.
\end{aligned} \tag{7.38}
$$

That is, $\mathbf{Y}$ satisfies the Lyapunov equation

$$
A^t\mathbf{Y} + \mathbf{Y}A + C^t C = 0
$$

and as A is a stability matrix, the solution of (7.38) is well-known to be unique.

7.2.1 Control Example 7.1

Consider the following LTI system

$$
\begin{aligned}
\dot{x} &= \begin{bmatrix} 0 & 1 & 0 \\ 0 & 0 & 1 \\ -2 & -5 & -1 \end{bmatrix} x + \begin{bmatrix} 0 \\ 0 \\ 1 \end{bmatrix} u, \\
y &= x.
\end{aligned}
$$

Simple computations give the transfer function $G(s)$ as

$$
G(s) = \begin{bmatrix} (s^3 + s^2 + 5s + 2)^{-1} \\ (s^3 + s^2 + 5s + 2)^{-1} \\ (s^3 + s^2 + 5s + 2)^{-1} \end{bmatrix}.
$$

To compute $\|G\|_2$, we first solve

$$
A^t\mathbf{Y} + \mathbf{Y}A + C^t C = 0
$$

to yield

$$
\mathbf{Y} = \begin{bmatrix} 2.4167 & 2.4167 & 0.2500 \\ \bullet & 5.7500 & 0.5833 \\ \bullet & \bullet & 1.0833 \end{bmatrix}.
$$

Thus

$$
\|G\|_2 = \sqrt{\mathrm{Tr}\left[B^t \mathbf{Y} B\right]} = 1.0408.
$$

Incidently, invoking the m-file *h2norm* of MATLAB we find $\|G\|_2 = 1.041$.

Fig. 7.1 Feedback system with stochastic inputs and outputs

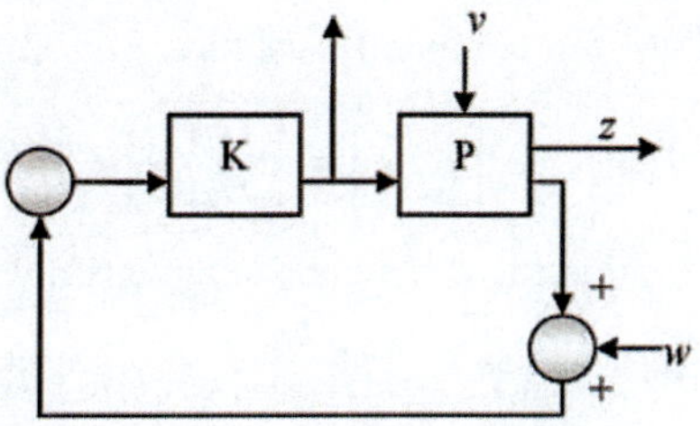

7.2.2 $\mathcal{H}_2$ Optimization

Proceeding further, we provide strong link between the LQG and $\mathcal{H}_2$ optimization problems. For this purpose, we rewrite the time-domain LQG problem into an equivalent frequency-domain $\mathcal{H}_2$ optimization problem. While the LQG problem requires state space realizations, the $\mathcal{H}_2$-optimization problem is in terms of transfer matrices. To simplify the expressions to come we assume that $\mathsf{Q} = I$ and $\mathsf{R} = I$, that is, the LQG performance index is

$$\lim_{t\to\infty} \mathbb{E}\big[z^t(t)z(t) + u^t(t)u(t)\big]. \tag{7.39}$$

This situation should not cause any loss of generality since by scaling and transformation of variables $z(t)$, $u(t)$, the performance index may always be brought into this form. For the open-loop system

$$\begin{aligned} \dot{x}(t) &= Ax(t) + Bu(t) + \Gamma v(t), \\ z(t) &= Gx(t), \\ y(t) &= Cx(t) + w(t) \end{aligned} \tag{7.40}$$

we solve for $z(t)$ and $y(t)$ in terms of $v(t)$, $w(t)$ and $u(t)$, to get

$$\begin{aligned} z(t) &= G(sI - A)^{-1}\Gamma v(t) + G(sI - A)^{-1}Bu(t) \\ &= P_{11}(s)v(t) + P_{12}(s)u(t), \end{aligned} \tag{7.41}$$

$$\begin{aligned} y(t) &= C(sI - A)^{-1}\Gamma v(t) + C(sI - A)^{-1}Bu(t) + w(t) \\ &= P_{21}(s)v(t) + P_{22}(s)u(t). \end{aligned} \tag{7.42}$$

In Fig. 7.1, the interconnection of feedback system with compensator K subject to stochastic inputs and outputs is shown, from which we have

$$\begin{aligned} u(t) &= -\mathsf{K}y(t) \quad \Longrightarrow \\ u(t) &= -\big(I + \mathsf{K}(P_{22})^{-1}\big)\mathsf{K}P_{21}v(t) - \big(I + \mathsf{K}(P_{22})^{-1}\big)\mathsf{K}w(t) \\ &= H_{21}(s)v(t) + H_{22}(s)w(t). \end{aligned} \tag{7.43}$$

From (7.41), we get

$$\begin{aligned} z(t) &= P_{11} - P_{12}\big(I + \mathsf{K}(P_{22})^{-1}\big)\mathsf{K}P_{21}v(t) - P_{12}\big(I + \mathsf{K}(P_{22})^{-1}\big)\mathsf{K}w(t) \\ &= H_{11}(s)v(t) + H_{12}(s)w(t). \end{aligned} \tag{7.44}$$

Written compactly, we have

$$\begin{bmatrix} z(t) \\ u(t) \end{bmatrix} = \begin{bmatrix} H_{11}(s) & H_{12}(s) \\ H_{21}(s) & H_{22}(s) \end{bmatrix} \begin{bmatrix} v(t) \\ w(t) \end{bmatrix} = H(s) \begin{bmatrix} v(t) \\ w(t) \end{bmatrix}. \tag{7.45}$$

Evaluating the mean square error, we get

$$\begin{aligned} \lim_{t\to\infty} \mathbb{E}\big[z^t(t)z(t) + u^t(t)u(t)\big] &= \lim_{t\to\infty} \mathbb{E}\left(\begin{bmatrix} z(t) \\ u(t) \end{bmatrix}^t \begin{bmatrix} z(t) \\ u(t) \end{bmatrix}\right) \\ &= \mathrm{Tr}\left[\int_{-\infty}^{\infty} G(j2\pi f)G^t(-j2\pi f)\,df\right] \\ &= \|H\|_2^2. \end{aligned} \tag{7.46}$$

This leads to the basic result that *solving the LQG problem amounts to minimizing the $\mathcal{H}_2$ norm of the closed-loop system with (v, w) input and (z, u) as output.*

It is interesting to recast the LQG problem using a generalized plant in state-space form as follows:

$$\begin{aligned} \dot{x}(t) &= Ax(t) + [\Gamma \quad 0] \begin{bmatrix} v(t) \\ w(t) \end{bmatrix} + Bu(t), \\ \begin{bmatrix} z(t) \\ u(t) \\ y(t) \end{bmatrix} &= \begin{bmatrix} G \\ 0 \\ C \end{bmatrix} x(t) + \begin{bmatrix} 0 & 0 \\ 0 & 0 \\ 0 & I \end{bmatrix} \begin{bmatrix} v(t) \\ w(t) \end{bmatrix} + \begin{bmatrix} 0 \\ I \\ 0 \end{bmatrix} u(t). \end{aligned} \tag{7.47}$$

7.2.3 The Standard $\mathcal{H}_2$ Problem

The performance of a feedback system can be quantified in terms of the closed-loop gain from the disturbance inputs to the reference outputs. The system 2-norm represents an average gain and can be used as a performance function for an optimal control problem.

In what follows, the standard $\mathcal{H}_2$ control is described by the block diagram in Fig. 7.2 where the objective is that of selecting the controller K such that it

1. stabilizes the resulting closed-loop system, and
2. minimizes the $\mathcal{H}_2$-norm of the closed-loop system (with w as input and z as output),

where $\mathbf{G(s)}$ is the transfer function of the form:

$$\mathbf{G(s)} = \begin{bmatrix} A & B_1 & B_2 \\ C_1 & 0 & D_{12} \\ C_2 & D_{21} & 0 \end{bmatrix}. \tag{7.48}$$

For convenience, we consider the partition of $\mathbf{G(s)}$ according to

$$\begin{bmatrix} z \\ y \end{bmatrix} = \begin{bmatrix} G_{11} & G_{12} \\ G_{21} & G_{22} \end{bmatrix} \begin{bmatrix} w \\ u \end{bmatrix}. \tag{7.49}$$

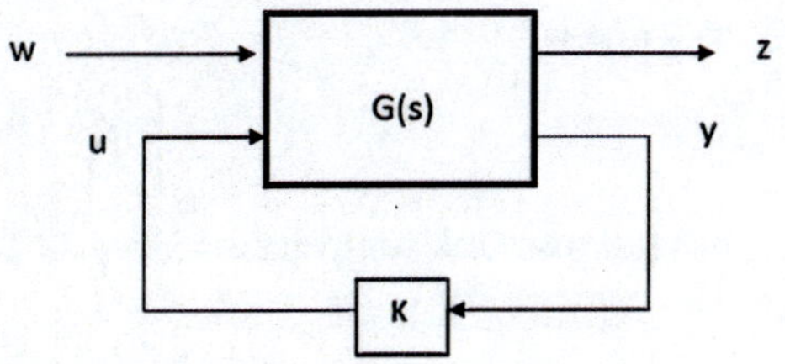

Fig. 7.2 The standard $\mathcal{H}_2$ problem

The closed-loop system

$$z = F(G, K)w$$

has the transfer function $F(G, K)$ given by

$$F(G, K) = G_{11} + G_{12}(I - KG_{22})^{-1}KG_{21}.$$

The $\mathcal{H}_2$-optimal control problem consists of finding a causal controller K which stabilizes the plant $\mathbf{G(s)}$ *while minimizing the cost function*

$$J_2(K) = \|F(G, K)\|_2^2$$

which is a standard convex optimization problem.

Adopting the state-space framework, the underlying $\mathcal{H}_2$-optimal problem is most conveniently solved in the time-domain. We will assume that $\mathbf{G(s)}$ can be cast into the generalized state-space representation [5]:

$$\begin{aligned}
\dot{x}(t) &= Ax(t) + B_1w(t) + B_2u(t), \\
z(t) &= C_1x(t) + D_{11}w(t) + D_{12}u(t), \\
y(t) &= C_2x(t) + D_{21}w(t) + D_{22}u(t).
\end{aligned}$$

The $\mathcal{H}_2$-optimal problem may be solved by reducing it to an LQG problem. The derivation necessitates the introduction of some assumptions, which are listed in the summary that follows. They are natural assumptions for LQG problems. To proceed for the solution, we consider the generalized plant has the following model

$$\begin{aligned}
\dot{x}(t) &= Ax(t) + B_1w(t) + B_2u(t), \\
z(t) &= C_1x(t) + D_{12}u(t), \\
y(t) &= C_2x(t) + D_{21}w(t) + D_{22}u(t)
\end{aligned} \tag{7.50}$$

for which we assume that the following conditions hold:

- The system

  $$\dot{x}(t) = Ax(t) + B_2u(t), \qquad z(t) = C_1x(t)$$

  is stabilizable and detectable.
- The system

  $$\dot{x}(t) = Ax(t) + B_1w(t), \qquad y(t) = C_2x(t)$$

  is stabilizable and detectable.

- The matrix

$$\begin{bmatrix} A - sI & B_1 \\ C_2 & D_{21} \end{bmatrix}$$

has full row rank for every $s = j\omega$ and D_{21} has full row rank.
- The matrix

$$\begin{bmatrix} A - sI & B_2 \\ C_1 & D_{12} \end{bmatrix}$$

has full column rank for every $s = j\omega$ and D_{12} has full column rank.

Under the foregoing assumptions, the optimal dynamic output (observer-based) feedback controller is

$$\begin{aligned} \dot{\hat{x}}(t) &= A\hat{x}(t) + B_2u(t) + L\big[y(t) - C_2\hat{x} - D_{22}u(t)\big], \\ u(t) &= -K\hat{x}(t). \end{aligned} \tag{7.51}$$

The observer and state-feedback gain matrices are given by

$$L = \big(\mathsf{Y}C_2^t + B_1D_{21}^t\big)\big(D_{21}D_{21}^t\big)^{-1}, \qquad K = \big(D_{12}D_{12}^t\big)^{-1}\big(B_2^t\mathsf{X} + D_{12}^tC_1\big) \tag{7.52}$$

where the matrices X and Y are the unique, symmetric positive-definite solutions of the algebraic Riccati equations

$$A^t\mathsf{X} + \mathsf{X}A + C_1^tC_1 - \big(\mathsf{X}B_2 + C_1^tD_{12}\big)\big(D_{12}D_{12}^t\big)^{-1}\big(B_2^t\mathsf{X} + D_{12}^tC_1\big) = 0, \tag{7.53}$$

$$A\mathsf{Y} + \mathsf{Y}A^t + B_1B_1^t - \big(\mathsf{Y}C_2^t + B_1D_{21}^t\big)\big(D_{21}D_{21}^t\big)^{-1}\big(C_2\mathsf{Y} + D_{21}B_1^t\big) = 0. \tag{7.54}$$

It should be noted that the condition that D_{12} has full column rank means that there is *direct feedthrough* from the input u to the error signal z. Likewise, the condition that D_{21} has full row rank means that the noise w is directly fed through to the observed output y.

The H_2 optimization problem and its solution are discussed at length in [20, 26].

7.2.4 Control Example 7.2

Consider the system of the form (7.50) with the following data

$$A = \begin{bmatrix} 1 & 1 \\ 0 & 1 \end{bmatrix}, \qquad B_1 = \begin{bmatrix} \sigma & 0 \\ \sigma & 0 \end{bmatrix}, \qquad B_2 = \begin{bmatrix} 0 \\ 1 \end{bmatrix}$$

$$C_1 = \begin{bmatrix} \beta & \beta \\ 0 & 0 \end{bmatrix}, \qquad D_{12} = \begin{bmatrix} 0 \\ 1 \end{bmatrix},$$

$$C_2 = \begin{bmatrix} 1 & 0 \end{bmatrix}, \qquad D_{21} = \begin{bmatrix} 0 & 1 \end{bmatrix}.$$

It is noted that $D_{11} = 0$, $D_{22} = 0$, $D_{12}^tC_1 = 0$, $(D_{12}D_{12}^t)^{-1} = I$ and therefore there are no cross-terms in the control Riccati equation (7.54), which is

$$A^t\mathsf{X} + \mathsf{X}A + C_1^tC_1 - \mathsf{X}B_2B_2^t\mathsf{X} = 0.$$

The stabilizing solution is easily verified to be

$$\mathsf{X} = \alpha \begin{bmatrix} 2 & 1 \\ 1 & 1 \end{bmatrix}, \quad \alpha = \left(2 + \sqrt{4 + \beta^2}\right)$$

leading to $K = \alpha[1 \ \ 1]$. Similarly, since $B_1 D_{21}^t = 0$, the measurement and process noise are uncorrelated and the Kalman filter Riccati equation is

$$A\mathsf{Y} + \mathsf{Y}A^t + B_1 B_1^t - \mathsf{Y}C_2^t C_2 \mathsf{Y} = 0.$$

It is easy to check that the stabilizing solution is

$$\mathsf{Y} = \nu \begin{bmatrix} 1 & 1 \\ 1 & 2 \end{bmatrix}, \quad \nu = \left(2 + \sqrt{4 + \sigma^2}\right)$$

leading to $L = \nu\left[\begin{smallmatrix} 1 \\ 1 \end{smallmatrix}\right]$. Thus, the optimal observer-based feedback controller (7.51) can be cast as

$$K(s) = \left[\begin{array}{c|c} A - B_2 K - L C_2 & L \\ \hline -K & 0 \end{array}\right].$$

Since

$$A - B_2 K - L C_2 = \begin{bmatrix} 1 - \nu & 1 \\ -(\alpha + \nu) & 1 - \alpha \end{bmatrix}$$

simple computations yield

$$K(s) = \frac{\alpha\nu(1 - 2s)}{s^2 + (\alpha + \nu - 2)s + \alpha\nu}.$$

The optimal cost is given by

$$\sqrt{\mathrm{Tr}\left(B_1^t \mathsf{X} B_1\right) + \mathrm{Tr}\left(K\mathsf{Y}K^t\right)} = \sqrt{5\alpha\left(\sigma^2 + \alpha\nu\right)}.$$

The optimal cost is monotonically increasing in both β and σ.

Seeking to a proper real rational transfer function $K(s)$, an alternative procedure can derived for the solution of the H_2 control problem of the plant

$$\begin{aligned} \dot{x}(t) &= Ax(t) + B_1 w(t) + B_2 u(t), \\ z(t) &= C_1 x(t) + D_{12} u(t), \\ y(t) &= C_2 x(t) + D_{21} w(t) \end{aligned} \tag{7.55}$$

under the following conditions:

- The system

  $$\dot{x}(t) = Ax(t) + B_2 u(t), \qquad y(t) = C_2 x(t)$$

  is stabilizable and detectable.
- $D_{12}^t D_{12} > 0$ and $D_{21} D_{21}^t > 0$.

- For all ω

$$\text{rank}\begin{bmatrix} A - j\omega I & B_2 \\ C_1 & D_{12} \end{bmatrix} = \text{number of columns}, \tag{7.56}$$

$$\text{rank}\begin{bmatrix} A - j\omega I & B_1 \\ C_2 & D_{21} \end{bmatrix} = \text{number of rows}. \tag{7.57}$$

The solution consists of two steps: First, we solve the two algebraic Riccati equations

$$\begin{aligned}
&(A - B_2(D_{12}D_{12}^t)^{-1}D_{12}^tC_1)^t\mathsf{S} + \mathsf{S}(A - B_2(D_{12}D_{12}^t)^{-1}D_{12}^tC_1) \\
&\quad - \mathsf{S}B_2(D_{12}D_{12}^t)^{-1}B_2^t\mathsf{S} + C_1^t(I - D_{12}(D_{12}D_{12}^t)^{-1}D_{12}^t)C_1 = 0, \\
&(A - B_1D_{21}^t(D_{21}^tD_{21})^{-1}C_2)\mathsf{Z} + \mathsf{Z}(A - B_1D_{21}^t(D_{21}^tD_{21})^{-1}C_2)^t \\
&\quad - \mathsf{Z}C_2^t(D_{21}^tD_{21})^{-1}C_2\mathsf{Z} + B_1(I - D_{21}^t(D_{21}^tD_{21})^{-1}D_{21})B_1^t = 0.
\end{aligned} \tag{7.58}$$

Second, the optimal transfer function $K(s)$ is given by

$$\begin{aligned}
K(s) &= -N_1(sI - A - B_2N_1 - N_2C_2)^{-1}N_2, \\
N_1 &= -(D_{12}^tD_{12})^{-1}(B_2^t\mathsf{S} + D_{12}^tC_1), \\
N_2 &= -(\mathsf{Z}C_2^t + B_1D_{21}^t)(D_{21}^tD_{21})^{-1}.
\end{aligned}$$

7.2.5 *Control Example 7.3*

Consider the system of the form (7.50) with the following data

$$A = \begin{bmatrix} -5 & 2 & -4 \\ 0 & -3 & 0 \\ 0 & 7 & -1 \end{bmatrix}, \quad B_1 = \begin{bmatrix} 7 \\ -3 \\ 1 \end{bmatrix}, \quad B_2 = \begin{bmatrix} 6 \\ 8 \\ -5 \end{bmatrix},$$

$$C_1 = \begin{bmatrix} -2 \\ 9 \\ 4 \end{bmatrix}, \quad D_{12} = 1, \quad D_{11} = 0,$$

$$C_2 = \begin{bmatrix} 6 \\ 3 \\ -1 \end{bmatrix}, \quad D_{21} = 2, \quad D_{22} = 0.$$

Expressing the transfer function $\mathbf{G(s)}$ as:

$$\mathbf{G(s)} = \begin{bmatrix} A & B_1 & B_2 \\ C_1 & 0 & D_{12} \\ C_2 & D_{21} & 0 \end{bmatrix} := \begin{bmatrix} \mathbf{G_{11}(s)} & \mathbf{G_{12}(s)} \\ \mathbf{G_{21}(s)} & \mathbf{G_{22}(s)} \end{bmatrix}$$

we compute the respective elements as

$$\mathbf{G}_{11}(\mathbf{s}) = \frac{-37s^2 - 2509s - 669}{s^3 + 9s^2 + 23s + 15}, \qquad \mathbf{G}_{12}(\mathbf{s}) = \frac{s^3 + 49s^2 + 339s + 1455}{s^3 + 9s^2 + 23s + 15},$$
$$\mathbf{G}_{21}(\mathbf{s}) = \frac{2s^3 + 50s^2 + 113s + 597}{s^3 + 9s^2 + 23s + 15}, \qquad \mathbf{G}_{22}(\mathbf{s}) = \frac{65s^2 + 488s - 865}{s^3 + 9s^2 + 23s + 15}. \tag{7.59}$$

Since $\lambda A = \{-1, -3, -5\}$ so A is stable, we can take $N_1 = 0$ and $N_2 = 0$. Therefore, the matrix

$$\mathbf{M}(\mathbf{s}) = \left[\begin{array}{c|cc} A & 0 & B_2 \\ \hline 0 & 0 & 1 \\ -C_2 & 1 & 0 \end{array}\right] := \begin{bmatrix} \mathbf{M}_{11}(\mathbf{s}) & \mathbf{M}_{12}(\mathbf{s}) \\ \mathbf{M}_{21}(\mathbf{s}) & \mathbf{M}_{22}(\mathbf{s}) \end{bmatrix}$$

from which we compute

$$\mathbf{M}_{11}(\mathbf{s}) = 0, \qquad \mathbf{M}_{12}(\mathbf{s}) = 0, \qquad \mathbf{M}_{21}(\mathbf{s}) = 0,$$
$$\mathbf{M}_{22}(\mathbf{s}) = \frac{-65s^2 - 488s + 865}{s^3 + 9s^2 + 23s + 15}.$$

It follows for a proper real rational transfer function $Q(s)$ that the controller $\mathbf{K}(\mathbf{s})$ can be written as

$$\begin{aligned} \mathbf{K}(\mathbf{s}) &= \mathbf{M}_{11}(\mathbf{s}) + \mathbf{M}_{12}(\mathbf{s})Q(s)\big[I - \mathbf{M}_{22}(\mathbf{s})Q(s)\big]^{-1}\mathbf{M}_{21}(\mathbf{s}) \\ &= Q(s)\big[I - \mathbf{M}_{22}(\mathbf{s})Q(s)\big]^{-1} \\ &= Q(s)\left[I - \frac{-65s^2 - 488s + 865}{s^3 + 9s^2 + 23s + 15}Q(s)\right]^{-1} \\ &= \frac{(s^3 + 9s^2 + 23s + 15)Q(s)}{s^3 + 9s^2 + 23s + 15 + (65s^2 + 488s - 865)Q(s)}. \end{aligned} \tag{7.60}$$

From (7.59) and (7.60), we compute the controlled system as

$$\begin{aligned} \mathbf{T}(\mathbf{s}) &= \mathbf{G}_{11}(\mathbf{s}) + \mathbf{G}_{12}(\mathbf{s})K(s)\big[I - \mathbf{G}_{22}(\mathbf{s})K(s)\big]^{-1}\mathbf{G}_{21}(\mathbf{s}), \\ &= \frac{-37s^2 - 2509s - 669}{s^3 + 9s^2 + 23s + 15} + \frac{s^3 + 49s^2 + 339s + 1455}{s^3 + 9s^2 + 23s + 15} \\ &\quad \times Q(s)\frac{2s^3 + 50s^2 + 113s + 597}{s^3 + 9s^2 + 23s + 15}. \end{aligned} \tag{7.61}$$

Control example illuminates the fact that the transfer function of the controlled systems can be written in as specific form. It turns out that using the controller

$$\mathbf{K}(\mathbf{s}) = \mathbf{M}_{11}(\mathbf{s}) + \mathbf{M}_{12}(\mathbf{s})Q(s)\big[I - \mathbf{M}_{22}(\mathbf{s})Q(s)\big]^{-1}\mathbf{M}_{21}(\mathbf{s})$$

the transfer function of the controlled system can be cast into the form

$$\mathbf{T}(\mathbf{s}) = \Phi_{11}(s) + \Phi_{12}(s)Q(s)\Phi_{21}(s) \tag{7.62}$$

where $\Phi_{ij}(s)$ is given by

$$\begin{bmatrix} \Phi_{11}(s) & \Phi_{12}(s) \\ \Phi_{21}(s) & \Phi_{22}(s) \end{bmatrix} = \left[\begin{array}{cc|cc} A + B_2N_1 & -B_2N_1 & B_1 & B_2 \\ 0 & A + N_2C_2 & B_1 + N_2D_{21} & 0 \\ \hline C_1 + D_{12}N_1 & -D_{12}N_1 & D_{11} & D_{12} \\ 0 & C_2 & D_{21} & 0 \end{array}\right].$$

7.2.6 Control Example 7.4

We reexamine the system previously treated in Example 7.3. We learned before that A is stable. It is not difficult to see that the pairs (A, B_2) and (A, C_2) are stabilizable and detectable, respectively. Also, by default, $D_{12}^t D_{12} = 1 > 0$ and $D_{21} D_{21}^t = 4 > 0$. Simple computation shows for all ω that

$$\text{rank} \begin{bmatrix} A - j\omega I & B_2 \\ C_1 & D_{12} \end{bmatrix} = \text{rank} \begin{bmatrix} -5 - j\omega & 2 & -4 & 6 \\ 0 & -3 - j\omega & 0 & 8 \\ 0 & 7 & -1 - j\omega & -5 \\ -2 & 9 & 4 & 1 \end{bmatrix} = 4,$$

$$\text{rank} \begin{bmatrix} A - j\omega I & B_1 \\ C_2 & D_{21} \end{bmatrix} = \text{rank} \begin{bmatrix} -5 - j\omega & 2 & -4 & 7 \\ 0 & -3 - j\omega & 0 & -3 \\ 0 & 7 & -1 - j\omega & 1 \\ 6 & 3 & -1 & 1 \end{bmatrix} = 4.$$

Therefore, the assumptions in (7.57) are satisfied. Proceeding to look at the Riccati equations (7.58), we first see that

$$C_1^t\big(I - D_{12}(D_{12}D_{12}^t)^{-1}D_{12}^t\big)C_1 = C_1^t(1-1)C_1 = 0,$$

$$B_1\big(I - D_{21}^t(D_{21}^t D_{21})^{-1}D_{21}\big)B_1^t = B_1(1-1)B_1^t = 0.$$

Hence, the solutions of (7.58) are $\mathsf{S} = 0$ and $\mathsf{Z} = 0$. This is turn leads to

$$\begin{aligned} N_1 &= -(D_{12}^t D_{12})^{-1}(B_2^t \mathsf{S} + D_{12}^t C_1) = -(D_{12}^t D_{12})^{-1} D_{12}^t C_1 \\ &= -C_1 = \begin{bmatrix} -2 & -9 & 4 \end{bmatrix}, \\ N_2 &= -(\mathsf{Z} C_2^t + B_1 D_{21}^t)(D_{21}^t D_{21})^{-1} = -B_1 D_{21}^t (D_{21}^t D_{21})^{-1} \\ &= -B_1/2 = \begin{bmatrix} -3.5 & 1.5 & -0.5 \end{bmatrix}^t. \end{aligned}$$

The optimal transfer function $K(s)$ is given by

$$\begin{aligned} K(s) &= -N_1(sI - A - B_2N_1 - N_2C_2)^{-1}N_2 \\ &= \frac{18.5s^2 + 12.5s + 334.5}{s^3 + 65s^2 + 2275s + 9665} \end{aligned}$$

which can be realized by the state-space model

$$\begin{aligned} \dot{\xi}(t) &= \begin{bmatrix} -14 & -62 & -24.5 \\ 25 & -70.5 & -33.5 \\ -13 & 50.5 & 19.5 \end{bmatrix} \xi(t) + \begin{bmatrix} -3.5 \\ 1.5 \\ -0.5 \end{bmatrix} y(t), \\ u(t) &= \begin{bmatrix} -2 & -9 & 4 \end{bmatrix} \xi(t). \end{aligned} \tag{7.63}$$

Next, we address the $\mathcal{H}_\infty$ control problem.

7.3 $\mathcal{H}_\infty$ Control

In this section, we introduce what is known as $\mathcal{H}_\infty$-optimization as a design tool for linear multivariable control systems. $\mathcal{H}_\infty$-optimization amounts to the minimization of the ∞-norm of a relevant frequency response function. The name derives from the fact that mathematically the problem may be set in the space $\mathcal{H}_\infty$ (the Hardy space), which consists of all bounded functions that are analytic in the right-half complex plane. We do not go to this length, however. $\mathcal{H}_\infty$-optimization resembles $\mathcal{H}_2$-optimization, where the criterion is the 2-norm. Because the 2- and ∞-norms have different properties the results naturally are not quite the same. An important aspect of $\mathcal{H}_\infty$ optimization is that it allows to include robustness constraints explicitly in the criterion.

In the sequel, we let the controller K and plant $\mathbf{G(s)}$ be real, rational and proper. We assume that the state space models of K and $\mathbf{G(s)}$ are available and that their realizations are assumed to be stabilizable and detectable. In this regard, the optimal $\mathcal{H}_\infty$ control problem [22, 25] is to find all the admissible controllers K such that T_{zv} is minimized. It should be noted that the optimal $\mathcal{H}_\infty$ controllers are generally not unique for MIMO systems. Furthermore, finding the optimal $\mathcal{H}_\infty$ controller is often both theoretically and numerically complicated. This is certainly in contrast with the standard $\mathcal{H}_2$ theory, in which the optimal controller is unique and can be obtained by solving two Riccati equations without iterations. Knowing the achievable optimal (minimum) $\mathcal{H}_\infty$ norm may be useful theoretically since it sets a limit on what we can achieve.

7.3.1 *Two Hamiltonians*

In practice however, it is often not necessary and sometimes even undesirable to design an optimal controller and it is usually much cheaper to obtain controllers that are very close in norm sense to the optimal ones, which will be called suboptimal controllers. A suboptimal controller may have some other properties which are better than the optimal controller (for example, lower bandwidth). The realization of the transfer matrix $\mathbf{G(s)}$ is taken to be of the form

$$\mathbf{G(s)} = \begin{bmatrix} A & B_1 & B_2 \\ C_1 & 0 & D_{12} \\ C_2 & D_{21} & 0 \end{bmatrix}. \tag{7.64}$$

The following standard assumptions are considered hold: The matrix block D_{22} is assumed to be zero so that G_{22} is strictly proper, also D_{11} is assumed to be zero in order to guarantee that $\mathcal{H}_2$ control problem is properly posed. We also assume that (A, B_1) is controllable and (C_1, A) is observable and that (A, B_2) is stabilizable and (C_2, A) is detectable.

It is known that the $\mathcal{H}_\infty$ solution involves the following two Hamiltonian matrices [20, 21, 25]:

$$\mathcal{H}_\infty = \begin{bmatrix} A & \gamma^{-2}B_1B_1^* - B_2B_2^* \\ -C_1C_1^* & -A^* \end{bmatrix}, \tag{7.65}$$

$$\mathbf{J}_\infty = \begin{bmatrix} A^* & \gamma^{-2}C_1C_1^* - C_2C_2^* \\ -B_1B_1^* & -A \end{bmatrix}. \tag{7.66}$$

Now it is clear that if the performance level γ approaches infinity, then these two Hamiltonian matrices become similar to the $\mathcal{H}_2$ Hamiltonian matrices. The transfer function from v to z can be written as

$$\mathbf{T_{zv}} = \left[\begin{array}{cc|c} A & B_2F_\infty & B_1 \\ -Z_\infty L_\infty C_2 & \hat{A}_\infty & -Z_\infty L_\infty D_{21} \\ \hline C_1 & D_{12}F_\infty & 0 \end{array}\right]. \tag{7.67}$$

7.3.2 LMI Framework

In this section, we provide an LMI-formulation of the H_2 and H_∞ control design. We direct attention to alternative techniques for computing the state-feedback controller $\mathbf{u} = L\mathbf{x}$. The closed-loop system is described by

$$\begin{aligned} \dot{\mathbf{x}}_s(t) &= A_s\mathbf{x}_s(t) + \Gamma\mathbf{w}(t), \\ \mathbf{z}(t) &= G_s\mathbf{x}_s(t) + \Phi\mathbf{w}(t), \end{aligned} \tag{7.68}$$

$$A_s = A + BL, \qquad G_s = G + DL. \tag{7.69}$$

Designing an $\mathcal{H}_2$ controller is approached via convex analysis. Suppose a Lyapunov function for the closed-loop system (7.68) is selected as

$$V(x_s) = x_s^t \mathcal{P} x_s(t), \quad 0 < \mathcal{P}^t = \mathcal{P} \in \Re^{n\times n}. \tag{7.70}$$

Along the solutions of the closed-loop system (7.68) with $w(t) \equiv 0$, we obtain

$$\dot{V}(\mathbf{x}_s) = \mathbf{x}_s^t(\mathcal{P}A_s + A_s^t\mathcal{P})\mathbf{x}_s. \tag{7.71}$$

From the Lyapunov theorem, the closed-loop system (7.68) is internally asymptotically stable if

$$A_s^t\mathcal{P} + \mathcal{P}A_s < 0 \tag{7.72}$$

is satisfied. The objective of this paper is to develop LMI-based characterization of the two optimization problems:

A) The $\mathcal{H}_2$-norm optimization in which it is required to find the state-feedback gain L that ensures the stability of closed-loop system (7.68) and keeps the $\mathcal{H}_2$-norm of the transfer function $T_{zw}(s)$ from $\mathbf{w}$ to $\mathbf{z}$ as small as possible.

B) The $\mathcal{H}_\infty$-norm optimization in which it is required to find the state-feedback gain L that ensures the stability of closed-loop system (7.68) and keeps the $\|\mathbf{z}\|_2 < \gamma\|\mathbf{w}\|_2$ for a prescribed attenuation level $\gamma > 0$.

7.3.3 $\mathcal{H}_2$ Design

Provided matrix A_s is Hurwitz for given L with $\Phi \equiv 0$, $\Psi \equiv 0$, the square of the $\mathcal{H}_2$-norm of the transfer function $H_{zw}(s)$ can be expressed in terms of the solution of a Lyapunov equation (controllability Grammian) such that the corresponding minimization problem with respect L is given by

$$\min\{\mathrm{Tr}[C_s^t \mathcal{P}_s C_s] : A_s \mathcal{P}_s + \mathcal{P}_s A_s^t + \Gamma\Gamma^t = 0\} \tag{7.73}$$

where Tr[.] denotes the trace operator. Since $\mathcal{P}_s < \mathcal{P}$ for any $\mathcal{P}$ satisfying

$$A_s \mathcal{P} + \mathcal{P} A_s^t + \Gamma\Gamma^t < 0 \tag{7.74}$$

it is readily verified that $\|H_{zw}(s)\|_2^2 = \mathrm{Tr}[C_s^t \mathcal{P} C_s] < \nu$ if and only if there exists $\mathcal{P} > 0$ satisfying (7.74) and $\mathrm{Tr}[C_s^t \mathcal{P} C_s] < \nu$. Introducing an auxiliary parameter $\mathcal{W}$, and in line of [16] the following analytical result is obtained.

Theorem 7.1 *Matrix A_s is stable and $\|H_{zw}(s)\|_2^2 < \nu$ for a prescribed ν if and only if there exist matrices $\hat{\mathcal{P}}$, $\mathcal{W}$ such that*

$$\mathrm{Tr}(\mathcal{W}) < \nu, \tag{7.75}$$

$$\begin{bmatrix} A_s^t \hat{\mathcal{P}} + \hat{\mathcal{P}} A_s & \hat{\mathcal{P}}\Gamma \\ \bullet & -I \end{bmatrix} < 0, \qquad \begin{bmatrix} \hat{\mathcal{P}} & C_s^t \\ \bullet & \mathcal{W} \end{bmatrix} > 0. \tag{7.76}$$

The main design result is summarized by the following theorem.

Theorem 7.2 *System (7.68)–(7.69) is stable with $\|H_{zw}(s)\|_2^2 < \nu$ for a prescribed ν if and only if there exist matrices $0 < \mathcal{X}$, $0 < \mathcal{Y}$ and $\mathcal{W}$ such that*

$$\mathrm{Tr}(\mathcal{W}) < \nu, \tag{7.77}$$

$$\begin{bmatrix} A\mathcal{X} + \mathcal{X}A^t + B\mathcal{Y} + \mathcal{Y}^t B^t & \Gamma \\ \bullet & -I \end{bmatrix} < 0, \tag{7.78}$$

$$\begin{bmatrix} \mathcal{X} & \mathcal{X}G^t + \mathcal{Y}^t D^t \\ \bullet & \mathcal{W} \end{bmatrix} > 0. \tag{7.79}$$

Moreover, the controller gain is $L = \mathcal{Y}\mathcal{X}^{-1}$.

Proof A congruent transformation [16–18] via diag$[\mathcal{X} \quad I]$, $\mathcal{X} = \hat{\mathcal{P}}^{-1}$ on (7.76) yields (7.79). □

7.3.4 $\mathcal{H}_\infty$ Design

In what follows, we consider the $\mathcal{H}_\infty$-norm optimization problem. It follows from robust control theory [25] that the solution of this problem corresponds to determining the controller parameters that guarantees the feasibility of

$$\dot{V}(\mathbf{x}_s) + \mathbf{z}^t\mathbf{z} - \gamma^2\mathbf{w}^t\mathbf{w} < 0. \tag{7.80}$$

The design result is summarized by the following theorem.

Theorem 7.3 *System (7.68) is asymptotically stable with γ-disturbance attenuation if there exist matrices $0 < \mathcal{X}$, $0 < \mathcal{Y}$, and scalar $\gamma > 0$ satisfying the following LMI*

$$\begin{bmatrix} \Pi_o & \Gamma & \Pi_c \\ \bullet & -\gamma^2 I & \Phi^t \\ \bullet & \bullet & -I \end{bmatrix} < 0, \tag{7.81}$$

$$\begin{aligned} \Pi_o &= A\mathcal{X} + \mathcal{X}A^t + B\mathcal{Y} + \mathcal{Y}^t B^t, \\ \Pi_c &= \mathcal{X}G^t + \mathcal{Y}^t D^t. \end{aligned} \tag{7.82}$$

Moreover, the controller gain is $L = \mathcal{Y}\mathcal{X}^{-1}$.

Proof With the aid of (7.71), we express inequality (7.80) in the form

$$\begin{aligned} &\mathbf{x}_s^t[\mathcal{P}A_s + A_s^t\mathcal{P}]\mathbf{x}_s + [C_s\mathbf{x}_s + \Phi\mathbf{w}]^t[C_s\mathbf{x}_s + \Phi\mathbf{w}] \\ &\quad + 2\mathbf{x}_s^t\mathcal{P}\Gamma - \gamma^2\mathbf{w}^t\mathbf{w} < 0. \end{aligned} \tag{7.83}$$

Inequality (7.83), by Schur complements, is equivalent to

$$\begin{bmatrix} \mathcal{P}A_s + A_s^t\mathcal{P} & \mathcal{P}B_s & C_s^t \\ \bullet & -\gamma^2 I & D_s^t \\ \bullet & \bullet & -I \end{bmatrix} < 0 \tag{7.84}$$

for any $[\mathbf{x}_s, \mathbf{w}] \neq 0$. Applying the congruent transformation diag$[\mathcal{X} I]$, $\mathcal{X} = \hat{\mathcal{P}}^{-1}$ to (7.84) and using $L\mathcal{X} = \mathcal{Y}$, we readily obtain LMI (7.81) subject to (7.82), which concludes the proof. □

7.3.5 Mixed $\mathcal{H}_2$–$\mathcal{H}_\infty$ Synthesis

Considering system (7.85), the mixed $\mathcal{H}_2$–$\mathcal{H}_\infty$ synthesis problem deals with the problem of finding the state-feedback controller which minimizes the $\mathcal{H}_2$ norm of the transfer function $T_{zw}(s)$ and subject to the $\mathcal{H}_\infty$-norm constrained by the bound γ.

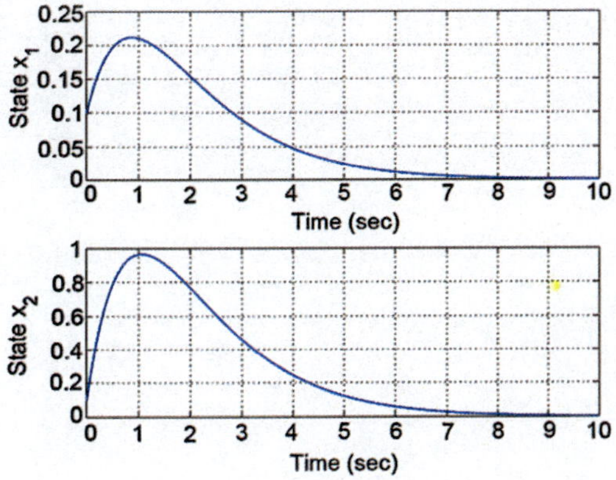

Fig. 7.3 State trajectories using LQR

7.4 Control Design of Hydraulic Pumping System

The first design method is the linear quadratic regulator discussed earlier in Chap. 6. For the simulation, we selected the weighting matrix R was kept constant and the matrix Q was varied in three different cases. The optimum results were obtained at.

$$Q = 0.1 \times I_{2\times2}, \qquad R = 1.$$

In our system, the two states have the same amount of significance. Hence, they have been weighted equally in each case. In the above simulation, the matrix R was used to weight the control input applied. The matrix Q was used to weight the states of the system. Simulation was carried out such that the weight on the inputs was kept constant and the weight on the states was varied to obtain the optimum results. With respect to the Norm of the Gain matrix and the time taken by the states to settle to steady state, it is concluded that the case shown in Fig. 7.3 has yielded optimum results.

Turning to the discrete-time LQR, the simulation was carried out such that the weight on the inputs was kept constant and the weight on the states was varied to study the behavior of the system in three different cases. Optimum results were found using the following weighting matrices:

$$Q = I_{2\times2}, \qquad R = 1.$$

Of all the cases simulated, it is noted that the controller gain K is the largest in the case shown in Fig. 7.4, while the settling time is also the least in the third case. Hence, as we increase the controller gain, the settling time decreases. The response of the *DLQR* regulator is similar to the *LQR* regulator, the only difference being the control that is applied at discrete instants equal to the sampling time of the system model.

Just as in the continuous *LQR* all the have been weighted equally in each case while implementing the discrete regulator. In the above simulation, the matrix R was used to weight the control input applied. The matrix Q was used to weight the states of the system.

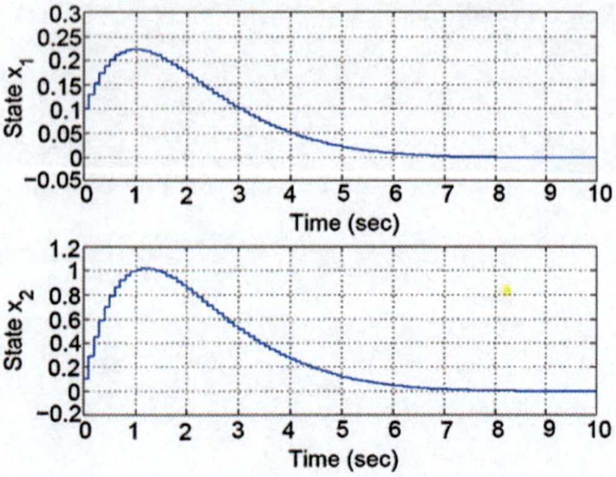

Fig. 7.4 States of the system using DLQR

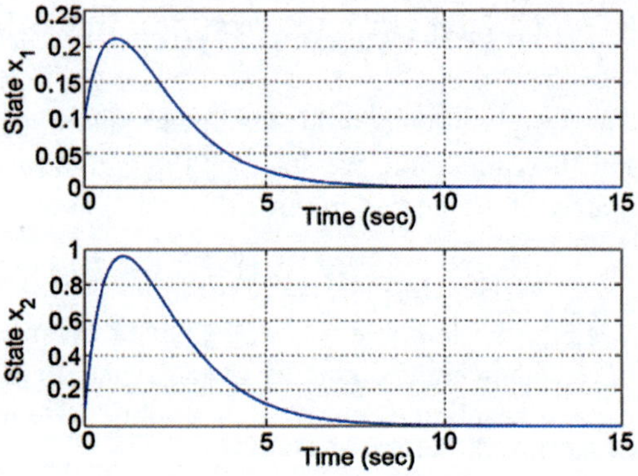

Fig. 7.5 State trajectories using LQGR

7.4.1 LQGR Control

The simulation of the (LQG) as carried out using a combination of the Linear Quadratic Regulator and a Kalman state estimator. The noises in consideration were the input noise and the plant noise. The LQG regulator was designed for three different cases by varying the LQR gains. The state trajectories are plotted in Fig. 7.5. The weighting matrix R was kept constant and the and the matrix Q was varied in three different cases. There were considerable oscillations observed in the states before they settle to steady state in the first case, they were reduced in the second case and found to be minimum in the third case. The controller gain K required to control the system was also found to be minimum in the third case.

7.4.2 $\mathcal{H}_2$ Optimal Control

The performance of a feedback system can be quantified in terms of the closed-loop gain from the disturbance inputs to the reference outputs. The system 2-norm

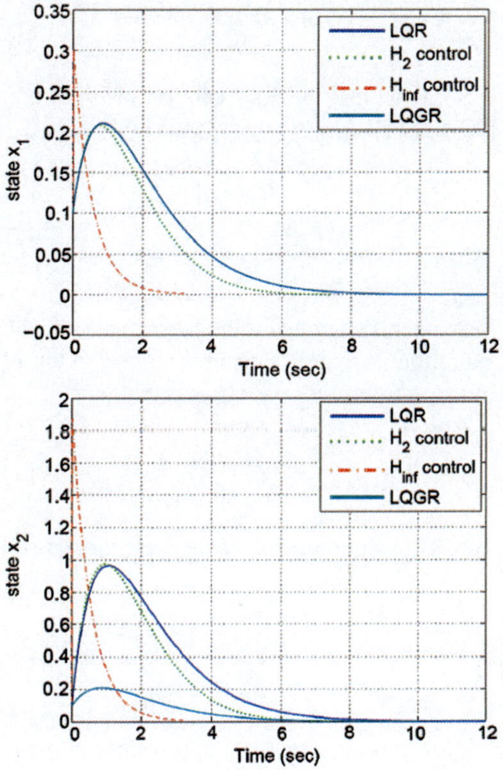

Fig. 7.6 Comparison of state trajectories using different controllers

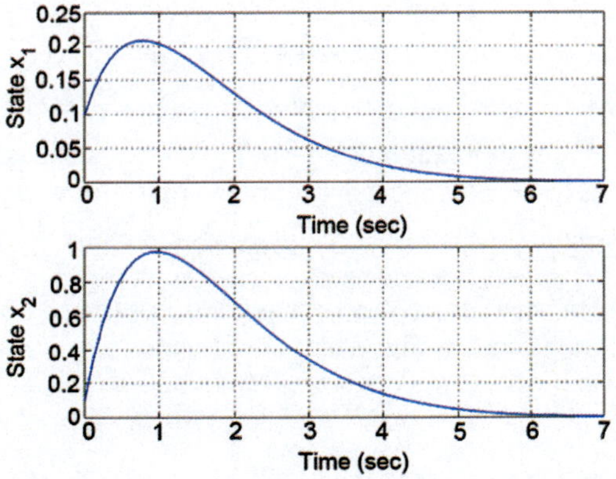

Fig. 7.7 State trajectories using $\mathcal{H}_2$ control

represents an average gain and can be used as a performance function for an optimal control problem. In what follows, the simulation for various values of D_{12} and D_{21} the value of gain matrix K for the most optimal controller was found to be at $D_{12} = 0$, $D_{21} = 20 \times I_{2\times 2}$. The state trajectories are plotted in Fig. 7.7.

7.4.3 $\mathcal{H}_\infty$ Control

It is known that the $\mathcal{H}_\infty$ solution involves the following two Hamiltonian matrices [20, 21, 25]:

$$\mathcal{H}_\infty = \begin{bmatrix} A & \gamma^{-2}B_1B_1^* - B_2B_2^* \\ -C_1C_1^* & -A^* \end{bmatrix},$$

$$\mathbf{J}_\infty = \begin{bmatrix} A^* & \gamma^{-2}C_1C_1^* - C_2C_2^* \\ -B_1B_1^* & -A \end{bmatrix}.$$

Now it is clear that if the performance level γ approaches infinity, then these two Hamiltonian matrices become similar to the $\mathcal{H}_2$ Hamiltonian matrices. The transfer function from v to z can be written as

$$\mathbf{T_{zv}} = \left[\begin{array}{c|c|c} A & B_2F_\infty & B_1 \\ -Z_\infty L_\infty C_2 & \hat{A}_\infty & -Z_\infty L_\infty D_{21} \\ C_1 & D_{12}F_\infty & 0 \end{array}\right].$$

The $\mathcal{H}_\infty$ solution involves the following two Hamiltonian matrices:

$$\mathcal{H}_\infty = \begin{bmatrix} A & \gamma^{-2}B_1B_1^* - B_2B_2^* \\ -C_1C_1^* & -A^* \end{bmatrix},$$

$$\mathbf{J}_\infty = \begin{bmatrix} A^* & \gamma^{-2}C_1C_1^* - C_2C_2^* \\ -B_1B_1^* & -A \end{bmatrix}.$$

Now it is clear that if γ approaches infinity, then these two Hamiltonian matrices become similar to the $\mathcal{H}_2$ Hamiltonian matrices. The transfer function from v to z can be written as

$$\mathbf{T_{zv}} = \left[\begin{array}{c|c|c} A & B_2F_\infty & B_1 \\ -Z_\infty L_\infty C_2 & \hat{A}_\infty & -Z_\infty L_\infty D_{21} \\ C_1 & D_{12}F_\infty & 0 \end{array}\right].$$

On simulation using MATLAB, for various values of D_{12}, the value of gain matrix K for the most appropriate controller was found to be at

$$D_{12} = 20, \qquad D_{21} = 0.02 \times I_{3\times 3}.$$

The state trajectories are plotted in Fig. 7.8. A comparison of the state trajectories are plotted in Fig. 7.6.

7.5 Vapor Compression Cycle Systems

In Chap. 6, we provided identified state-space models [**A**, **B**, **C**, **D**] based on two distinct output cases and presented typical control design and simulation results using LQR methods in the continuous-domain and discrete-domain. Here, we present the simulation results based on $\mathcal{H}_2$, $\mathcal{H}_\infty$ and LQGR designs. Then we compare between the controlled states arising from these designs.

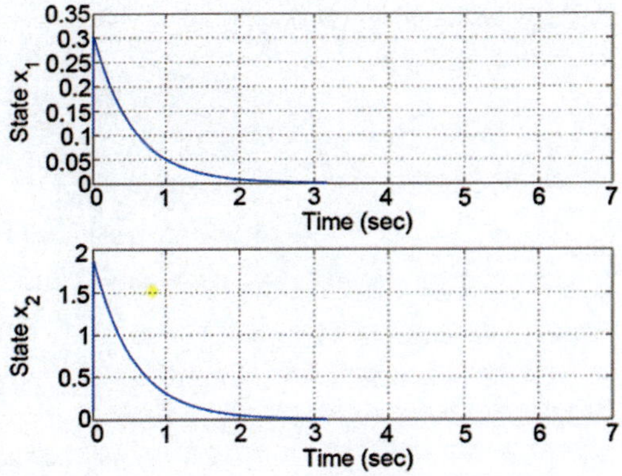

Fig. 7.8 State trajectories using $\mathcal{H}_\infty$ control

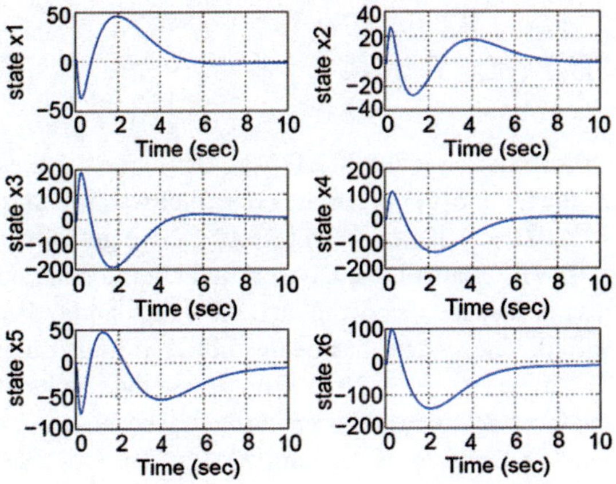

Fig. 7.9 State trajectories of system at $D_{21} = 20 \times I_{2\times2}$

7.5.1 $\mathcal{H}_2$ Results

On simulation, for various values of D_{12} and D_{21} the value of gain matrix K for the most optimal controller was found to be

$$D_{12} = 0, \qquad D_{21} = 20 \times I_{2\times2}$$

we get $\text{Norm}(K) = 5.1796 \times 10^5$. The corresponding state trajectories are plotted in Fig. 7.9.

In $\mathcal{H}_2$ control, we have carried out the simulation of the system considering the matrix D_{12} to be zero and the values of the matrix D_{21} have been varied. The system has been studied in five different cases assuming the values of D_{21} to be $0.002 \times I_{2\times2}$, $0.02 \times I_{2\times2}$, $0.2 \times I_{2\times2}$, $2 \times I_{2\times2}$ and $20 \times I_{2\times2}$. Comparing the results of the cases above, we observe that the feedback gain required in the first four cases is very high. With respect to the response of the system and the control input required, it is concluded that the we get the best response at $D_{12} = 0$, $D_{21} = 20 \times I_{2\times2}$. The corresponding state trajectories are plotted in Fig. 7.10.

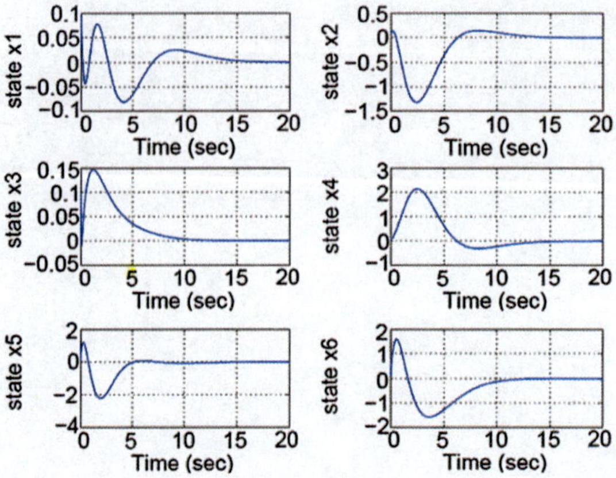

Fig. 7.10 States of system at $D_{12} = 20 \times I_{2\times 2}$

7.5.2 $\mathcal{H}_\infty$ Results

On simulation using MATLAB, for various values of D_{12}, the value of gain matrix K for the most appropriate controller was found to be:

At $D_{12} = 20$, $D_{21} = 0.02 \times I_{3\times 3}$, we get $\text{Norm}(K) = 1.0115 \times 10^5$.

In H_∞ control, we have carried out the simulation of the system considering the matrix D_{21} to be constant at $0.02 \times I_{3\times 3}$ and the values of the matrix D_{12} have been varied. The system has been studied in four different cases assuming the values of D_{12} to be 0.02, 0.2, 2, 20. Comparing the results of the cases above, we observe that the feedback gain required in the first case is very high. The last 3 cases do not have much difference in the gain K. But it is observed that the number of oscillations have been drastically reduced in the last case. Therefore, based on the response of the system and the control input required, it is concluded that the we get the best response at $D_{12} = 20$, $D_{21} = 0.02 \times I_{3\times 3}$.

7.5.3 LQGR Results

The simulation of the Linear Quadratic Gaussian regulator was carried out using a combination of the Linear Quadratic Regulator and a Kalman state estimator. The noises in consideration were the input noise and the plant noise. The LQG regulator was designed for three different cases by varying the LQR gains. The weighting matrix R was kept constant and the and the matrix Q was varied in three different cases. There were considerable oscillations observed in the states before they settle to steady state in the first case, they were reduced in the second case and found to be minimum in the third case. The controller gain K required to control the system was also found to be minimum in the third case. The corresponding state trajectories are plotted in Fig. 7.11.

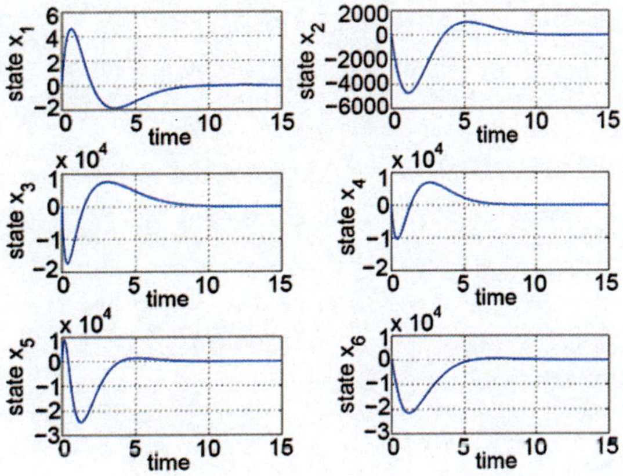

Fig. 7.11 States of system at Q_3

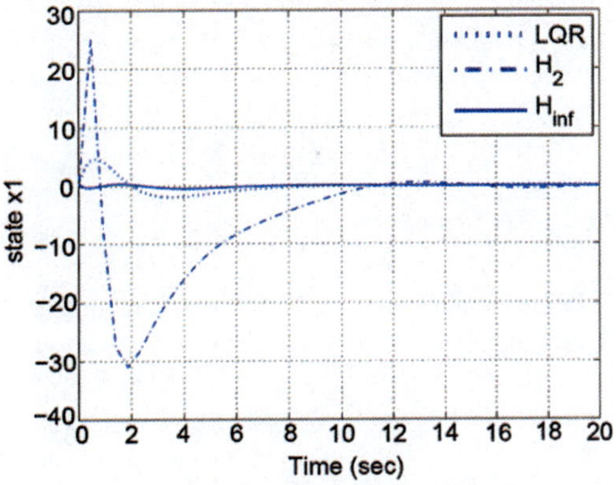

Fig. 7.12 Comparison of state x_1

7.5.4 A Comparative Study

The controller gain in the three cases is 1.1889×10^4 for *LQR*, 5.1796×10^5 for H_2 control and 1.0115×10^5 for $H_{\rm inf}$ control. It is noted that the gain requirements for H_2 and H_∞ control techniques do not vary much but are ten times the gain required by LQR controller. Therefore, it is concluded that the H_∞ controller is best suited for the control of the Vapor compression system though its gain requirement is higher than that of *LQR*. The corresponding state trajectories for different methods put together are plotted in Figs. 7.12–7.17.

The Kalman Filter was used on the system to filter out the noises acting on the system at the inputs, on the states and at the output. The Kalman Filer was implemented along with the Linear Quadratic Regulator to form a Linear Quadratic Gaussian Regulator. The *LQG* regulator is a more complex version of the *LQR* regulator that can be used for a system with Gaussian noises acting on it.

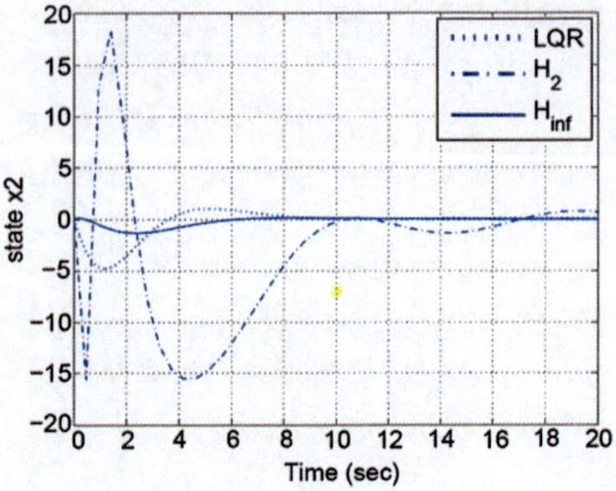

Fig. 7.13 Comparison of state x_2

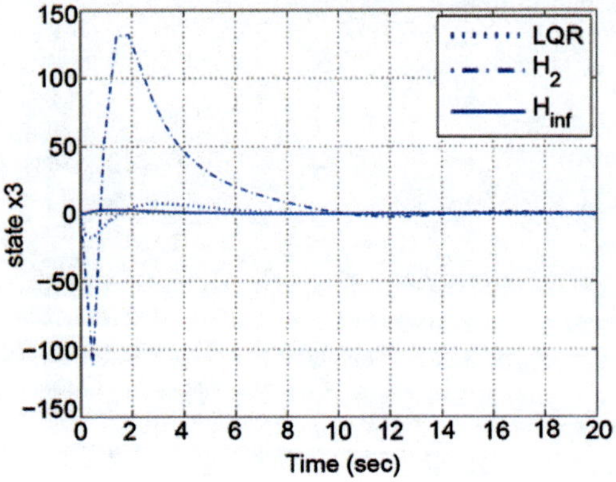

Fig. 7.14 Comparison of state x_3

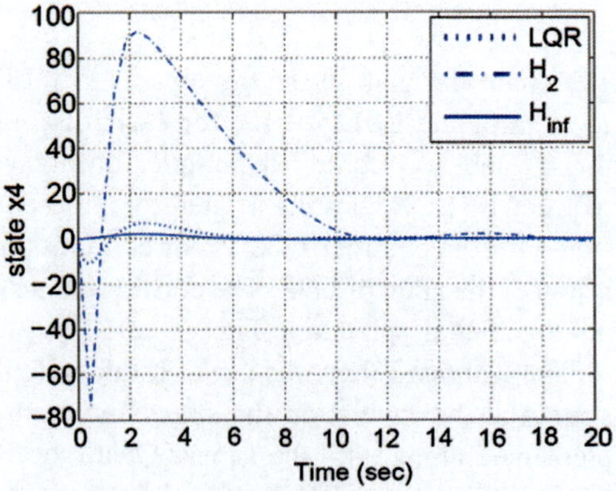

Fig. 7.15 Comparison of state x_4

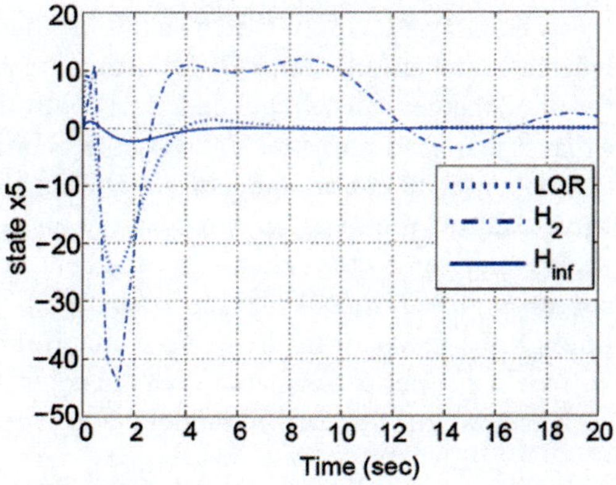

Fig. 7.16 Comparison of state x_5

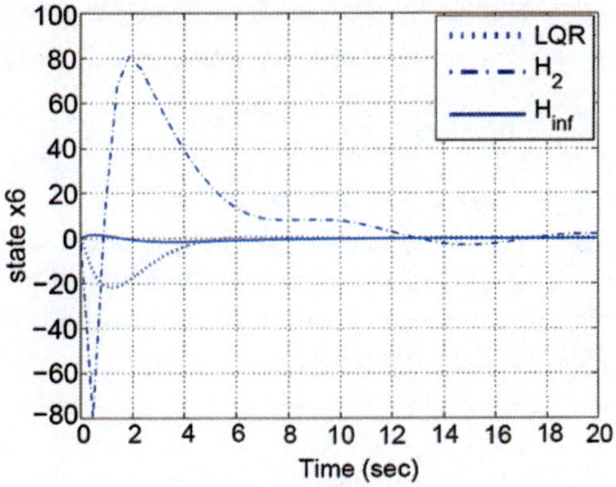

Fig. 7.17 Comparison of state x_6

7.6 Robust Control of Turbo Diesel Engine

Modern diesel engines are typically equipped with the VGT and EGR and both introduce feedback loops from exhaust to intake manifold. The recirculated exhaust gas is cooled down in the EGR cooler and its mass flow is controlled via the EGR valve. Both the EGR valve and the VGT are pneumatically actuated and fitted with position sensors. An intercooler reduces the temperature of the compressed air coming from the compressor. In addition to the standard production type sensors, for mass air flow (MAF) and manifold absolute pressure (MAP), the engine is equipped with various temperature and pressure sensors as well as with a turbocharger speed and inline shaft torque sensor. Exhaust gas recirculation (EGR) combined with the variable geometry turbocharging provides an important avenue for NOx emission reduction.

Reference is made Sect. 6.5 where we discussed a class of turbo diesel engines. In what follows, we consider a typical turbocharger consisting of an exhaust gas driven turbine that, by means of a mechanical shaft, is able to transfer its kinetic energy to the compressor impeller. The impeller imparts this energy to the air, which is turned into density increase in the compressor diffuser. The variable geometry turbocharging is accomplished by a turbine that has a system of movable guide vanes located on the turbine stator. By adjusting the guide vanes, the exhaust gas energy to the turbocharger can be regulated, thus controlling the compressor mass airflow and exhaust manifold pressure. The variable geometry turbocharger (VGT) actuator is typically used to control the intake manifold absolute pressure (MAP) and the EGR valve controls the mass air flow (MAF) into the engine. Both the EGR and VGT paths are driven by the exhaust gas and hence constitute an inherently multivariable control problem. Recall that the effect of the EGR and VGT actuators is coupled through the pressure in the exhaust manifold, therefore a co-ordinated approach will yield a better performance than the control strategies using SISO techniques.

An appropriate linearized model that can be conveniently cast into the format

$$\begin{aligned}\dot{\mathbf{x}}(t) &= A\mathbf{x}(t) + B\mathbf{u}(t) + \Gamma\mathbf{w}(t),\\ \mathbf{z}(t) &= G\mathbf{x}(t) + D\mathbf{u}(t) + \Phi\mathbf{w}(t),\\ \mathbf{y}(t) &= C\mathbf{x}(t) + \Psi\mathbf{w}(t),\end{aligned} \tag{7.85}$$

where $\mathbf{x}(t) \in \Re^n$, $\mathbf{u}(t) \in \Re^m$, $\mathbf{y}(t) \in \Re^p$, $\mathbf{z}(t) \in \Re^q$ and $\mathbf{w}(t) \in \Re^q$ are the state, the control input, the measured output, the controlled output and the external disturbance vectors. The matrices A, B, C, G, D, F, Φ, Ψ are real constants, the numerical values of which are given in the simulation section. In system (7.85), the states components are m_x = mass at the exhaust manifold, p_x = pressure at the exhaust manifold, m_i = mass at the intake manifold, p_i = pressure at the exhaust manifold, N_t = turbocharger shaft speed and W_{ci} = compressor mass flow. The system inputs are u_1 = exhaust gas recirculation (EGR) actuator position and u_2 = variable geometry turbocharger (VGT) actuator (vanes) position, whereas the system outputs which are y_1 = intake manifold absolute pressure (MAP) and y_2 = intake mass air flow (MAF).

7.6.1 Robust Simulation Results

Modern internal combustion engines equipped with variable valve actuation systems are proven to achieve better combustion characteristics. By appropriately varying the valve timing, one can increase fuel economy, boost power output and reduce emissions [19]. In particular, the problem of optimizing plug-in hybrid electric vehicle (PHEV) power management is studied in [12] by using stochastic dynamic over a distribution of drive cycles, rather than a single cycle and explicitly trades off fuel and electricity usage. Linear feedback controllers are developed in [13] for an electro-hydraulic valve system (EHVS) and a repetitive feed-forward controller is added to improve the tracking performance. The problem of power management of hybrid electric vehicles (HEVs) is treated in [14] via the Pontryagin's minimum principle as a viable real-time strategy. By employing performance index including fuel consumption, exhaust emission, or acceleration performance over the whole driving-cycle information, global optimal results are reported in [19].

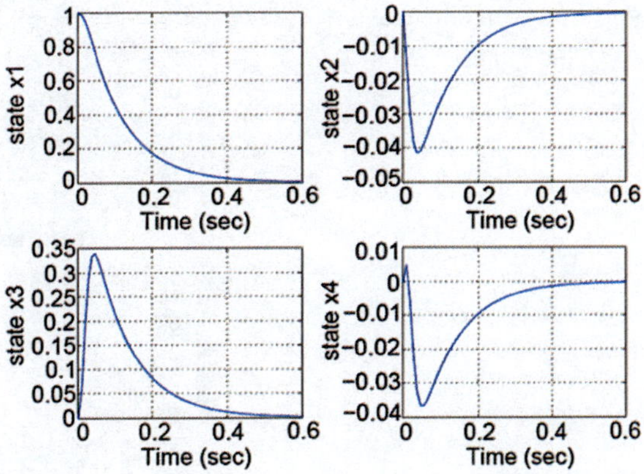

Fig. 7.18 States of system at $D_{21} = 20 \times I_{2\times2}$

Using typical data [19], the different parameters are $\tau_v = 0.01$ s, $\omega_M = 173.45$ r/s, $\alpha = 4.7408$ s, $c_L = 0.077$, $\gamma = 0.5432$, $\omega_h = 138.68$ r/s. By evaluating the model matrices given by (4.70), it is readily seen that the linearized system is unstable as it has eigenvalue at the origin and has internal oscillations due to a pair of complex.

$\mathcal{H}_2$ Control

On simulation, for various values of D_{12} and D_{21} the value of gain matrix K for the most optimal controller was found to be at $D_{12} = 0$, $D_{21} = 20 \times I_{2\times2}$, we get

$$\mathbf{K} = \begin{bmatrix} -0.0009 & -4.7608 & -0.9451 & -2.3892 \\ 0.0500 & -0.8939 & 0.9765 & 1.3927 \end{bmatrix}, \qquad \|K\| = 5.409. \qquad (7.86)$$

The corresponding state trajectories are plotted in Fig. 7.18.

In $\mathcal{H}_2$ control we have carried out the simulation of the system considering the matrix D_{12} to be zero and the values of the matrix D_{21} have been varied. The system has been studied in five different cases assuming the values of D_{21} to be $0.002 \times I_{2\times2}$, $0.02 \times I_{2\times2}$, $0.2 \times I_{2\times2}$, $2 \times I_{2\times2}$ and $20 \times I_{2\times2}$. Comparing the results of the cases above, we observe that the feedback gain required in the first four cases is very high. With respect to the response of the system and the control input required, it is concluded that the we get the best response at $D_{12} = 0$, $D_{21} = 20 \times I_{2\times2}$.

$\mathcal{H}_\infty$ Control

On simulation using MATLAB, for various values of D_{12}, the value of gain matrix K for the most appropriate controller was found to be:

At $D_{12} = 20$, $D_{21} = 0.02 \times I_{3\times3}$, we get

$$\mathbf{K} = \begin{bmatrix} -0.0000 & 0.0000 & -0.0499 & -0.0000 \\ 0.9199 & 1.9248 & -0.3750 & -0.9967 \end{bmatrix}, \qquad \|K\| = 2.384. \qquad (7.87)$$

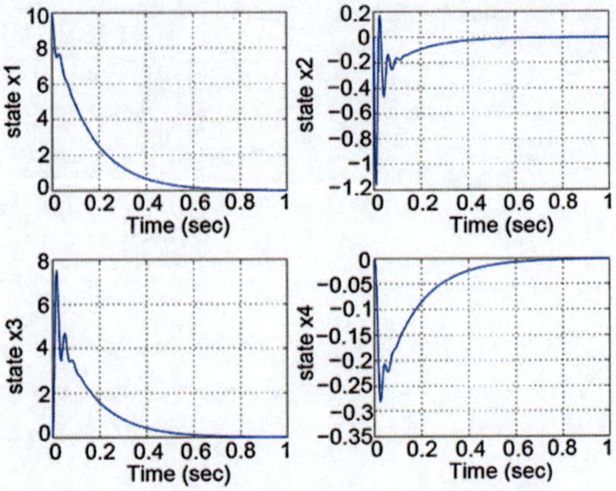

Fig. 7.19 States of system at $D_{12} = 20 \times I_{2\times 2}$

In H_∞ control we have carried out the simulation of the system considering the matrix D_{21} to be constant at $0.02 \times I_{3\times 3}$ and the values of the matrix D_{12} have been varied. The system has been studied in four different cases assuming the values of D_{12} to be 0.02, 0.2, 2, 20. Comparing the results of the cases above, we observe that the feedback gain required in the first case is very high. The last 3 cases do not have much difference in the gain K. But it is observed that the number of oscillations have been drastically reduced in the last case. Therefore, based on the response of the system and the control input required, it is concluded that the we get the best response at $D_{12} = 20$, $D_{21} = 0.02 \times I_{3\times 3}$, see the corresponding state trajectories as plotted in Fig. 7.19.

7.6.2 Kalman Filter

The Kalman gains and estimation error variances are generated for this system with various values for the plant noise spectral density matrix, the measurement noise spectral density matrix and the initial estimation error covariance matrix. In addition, the system is simulated and the simulated measurements are put into the Kalman filter to yield the estimated states, which are then compared with the actual states. The estimates approach the actual states as shown in the plots.

The plant states depart from a predictable pattern more rapidly when the plant noise spectral density is increased. As a consequence, the steady-state estimation errors increase, since less averaging can be performed on the measurements. The variation in measurement efficiency is more visible in the output y_2. The Kalman filter simulation was carried out taking into consideration the system noise and measurement noise. The initial estimation error covariance matrix B_v was defined as

$$\mathbf{B_v} = \begin{bmatrix} 1 & 1 \\ 0 & 0 \\ 1 & 1 \\ 10 & 0 \end{bmatrix}.$$

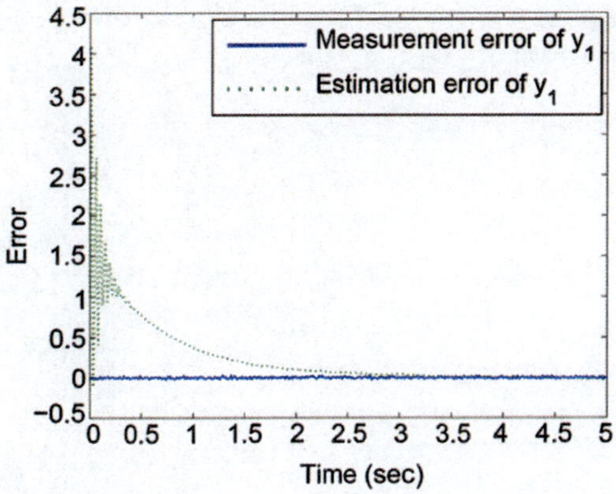

Fig. 7.20 Estimation error of y_1

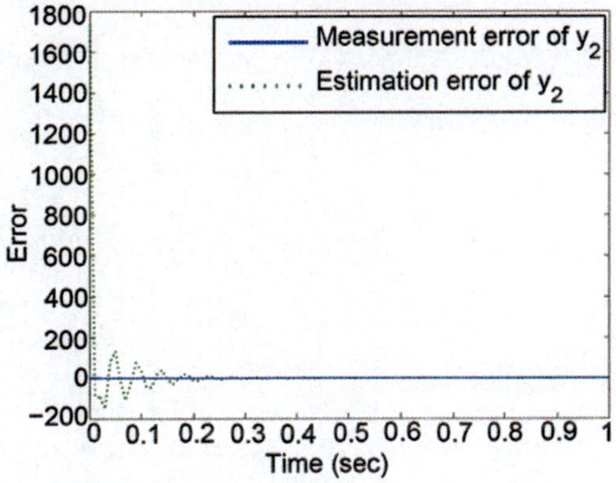

Fig. 7.21 Estimation error of y_2

The spectral density of the plant noise S_w and the measurement noise S_v was selected as 1 each. The random noise was generated and supplied to the system while the response of the system was studied. The real and estimated states of the system were plotted for comparison. The measurement error and estimation error at the two inputs were also plotted. The estimation output error trajectories are plotted in Figs. 7.20, 7.21.

7.6.3 LQGR Control

The simulation of the Linear Quadratic Gaussian regulator was carried out using a combination of the Linear Quadratic Regulator and a Kalman state estimator. The noises in consideration were the input noise and the plant noise. The LQG regulator was designed for three different cases by varying the LQR gains. The weighting matrix R was kept constant and the and the matrix Q was varied in three different cases. There were considerable oscillations observed in the states before they settle to steady state in the first case, they were reduced in the second case and found to

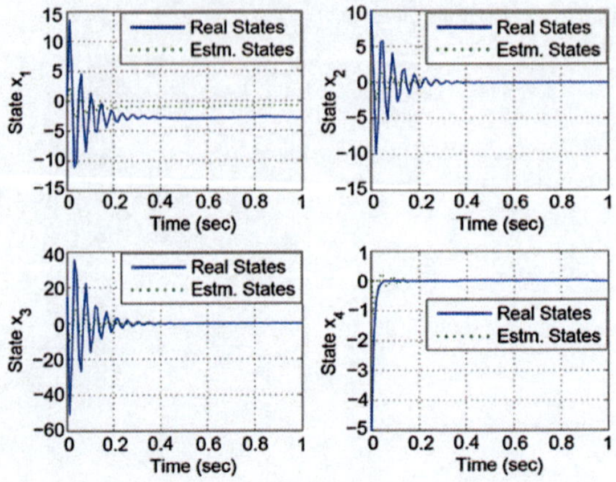

Fig. 7.22 Real versus estimated states

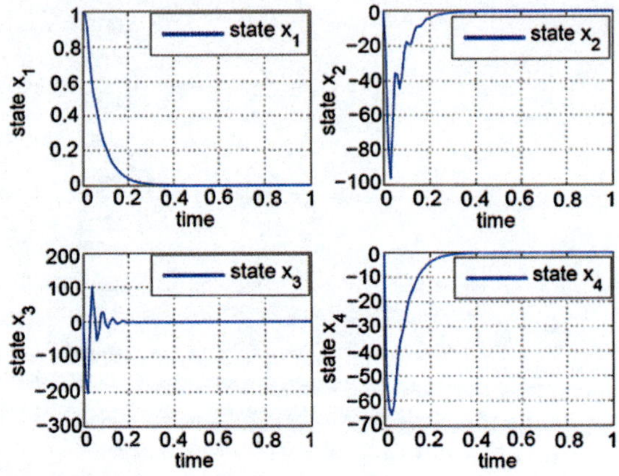

Fig. 7.23 States of system at $Q = 10 * I_{4\times4}$

be minimum in the third case. The controller gain K required to control the system was also found to be minimum in the third case. A comparison between the real and estimated state trajectories are plotted in Fig. 7.22.

The corresponding state, output and controlled input trajectories are plotted in Figs. 7.23–7.25.

Comparisons

The above system of a Electro-Hydraulic actuator was studied structurally and various optimal control techniques were applied to it. The system response in each case was studied with respect to its time requirement and the control energy required while implementing each strategy.

The comparative results of the three techniques namely, LQR, $\mathcal{H}_2$ and H_∞ have been plotted as shown in Fig. 7.26.

The dashed line represents the response of the LQR regulated system. The dot-dashed line represents the response of the $\mathcal{H}_2$ regulated system and the solid line

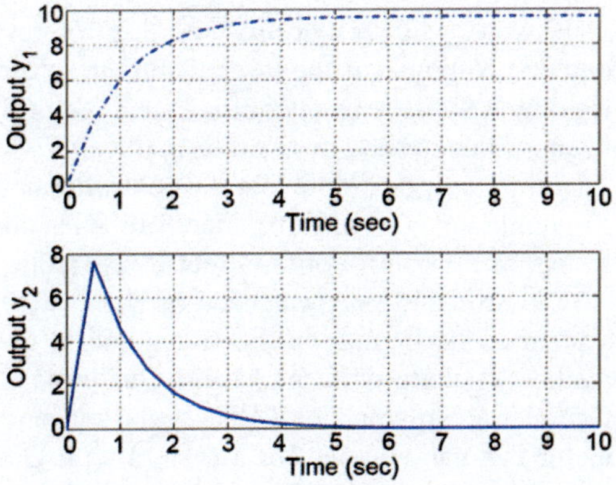

Fig. 7.24 Outputs of the system

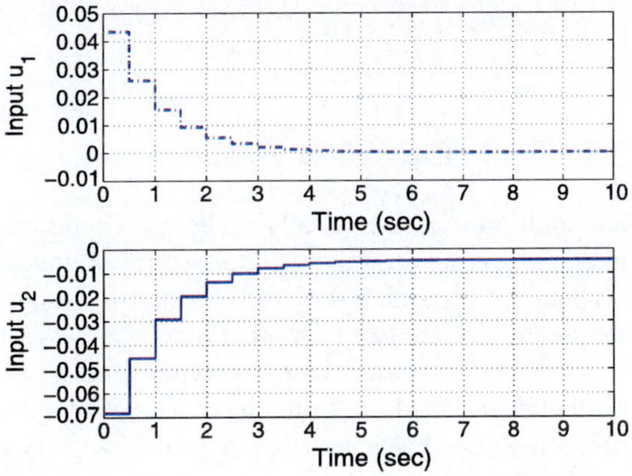

Fig. 7.25 Inputs of the system

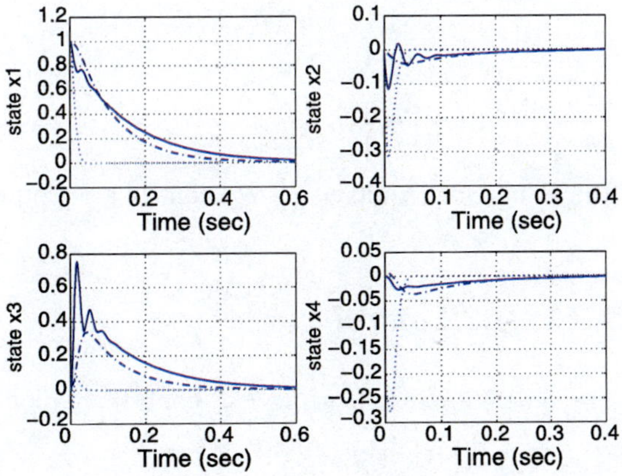

Fig. 7.26 Comparison of state responses

represents the response of the H_∞ regulated system. The controller gain in the three cases is 3.4045 for *LQR*, 5.409 for $\mathcal{H}_2$ control and 2.384 for H_{inf} control. It is noted that the gain requirements for the control techniques do not vary much.

However with respect to the speed of response it is noted that LQR has the quickest response of the three techniques compared above. The hydraulic actuator is commonly used in operations such as tilting of the ailerons or horizontal stabilizers of the aircraft or in other applications where the time factor is very critical. Therefore, it is desirable that the controller has a fast response with less oscillations and minimum control gain. Therefore, it is concluded that the LQR controller is best suited for the control of the Electro-Hydraulic actuator.

The Kalman Filter was used on the system to filter out the noises acting on the system at the inputs, on the states and at the output. The Kalman Filer was implemented along with the Linear Quadratic Regulator to form a Linear Quadratic Gaussian Regulator. The *LQG* regulator is a more complex version of the *LQR* regulator that can be used for a system with Gaussian noises acting on it. The *MPC* control was implemented on the system. It was found that *MPC* control is not feasible for the given system bearing in mind the system characteristics and the control requirements of the system.

7.7 The Falling Film Evaporator

Evaporation is used basically in the dairy industry for the concentration of products like milk, skimmed milk etc. Concentration involves the removal of water from the product. To minimize the cost evaporation is usually performed in multiple effect evaporators where two or more effects operate at progressively lower boiling points. In this type of arrangement, the vapor produced in the previous effect can be used as the heating medium in the next. The evaporator considered here is a four falling film effects and has a water evaporation capacity of 800 kg/h. The evaporators most commonly are used in the split effect mode, where only the third effect and the finishing effect are used.

7.7.1 $\mathcal{H}_2$ Control

The simulation results using $\mathcal{H}_2$ control are plotted in Figs. 7.27–7.32.

7.7.2 $\mathcal{H}_\infty$ Control

The simulation results using $\mathcal{H}_\infty$ control are plotted in Figs. 7.33–7.38.

7.8 Integral Control and Robust Tracking

In the state-space design methods discussed so far, no mention has been made of integral control, and no design examples have produced a compensation containing

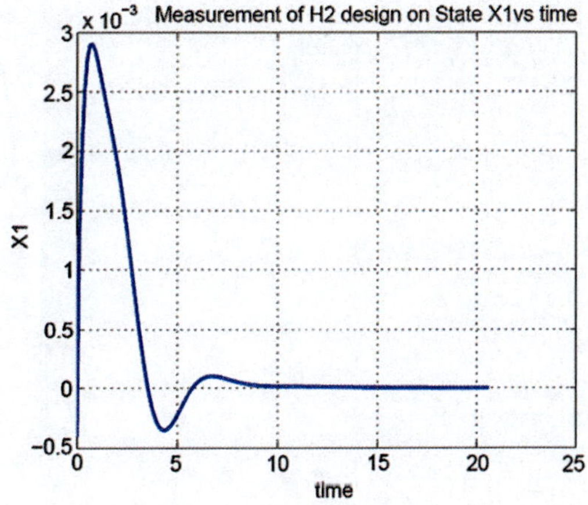

Fig. 7.27 The firth state trajectory using $\mathcal{H}_2$ control

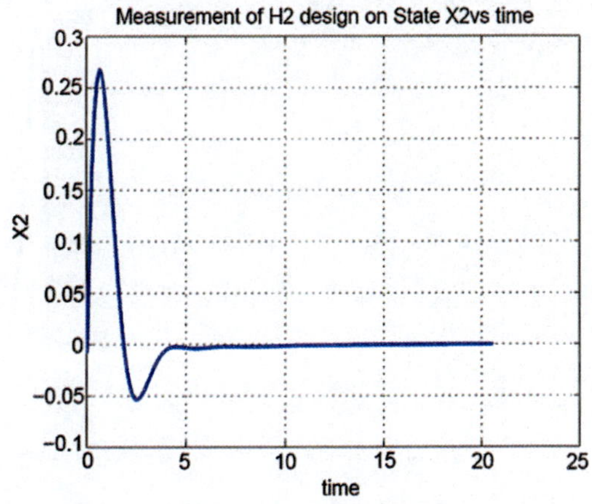

Fig. 7.28 The second state trajectory using $\mathcal{H}_2$ control

an integral term. In the sequel, we show how integral control can be introduced by a direct method of adding the integral of the system error to the equations of motion. Integral control is a special case of tracking a signal that does not go to zero in the steady-state. Then we introduce a general method for robust tracking that will present the internal model principle. This solves an entire class of tracking problems and disturbance-rejection controls.

7.8.1 Integral Control

Reference is made to Fig. 7.39. We start with a rudimentary solution to integral control by augmenting the state vector with the desired dynamics. For the system

$$\begin{aligned} \dot{\mathbf{x}} &= \mathbf{F}\mathbf{x} + \mathbf{G}u + \mathbf{G}_1 w, \\ y &= \mathbf{H}\mathbf{x}, \end{aligned} \tag{7.88}$$

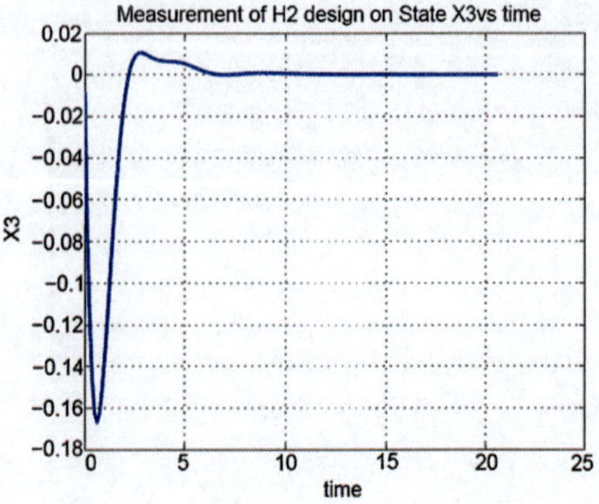

Fig. 7.29 The third state trajectory using $\mathcal{H}_2$ control

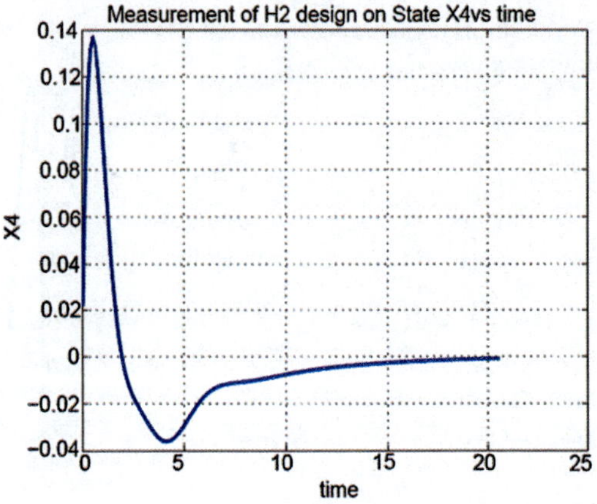

Fig. 7.30 The fourth state trajectory using $\mathcal{H}_2$ control

we can feed back the integral of the error, $e = y - r$, as well as the state of the plant, $\mathbf{x}$, by augmenting the plant state with the extra (integral) state x_I, which obeys the differential equation

$$\dot{x}_I = \mathbf{Hx} - r (= e),$$

which leads to

$$x_I = \int^t e\, dt.$$

The augmented state equations become

$$\begin{bmatrix} \dot{x}_I \\ \dot{\mathbf{x}} \end{bmatrix} = \begin{bmatrix} 0 & \mathbf{H} \\ \mathbf{0} & \mathbf{F} \end{bmatrix} \begin{bmatrix} x_I \\ \mathbf{x} \end{bmatrix} + \begin{bmatrix} 0 \\ \mathbf{G} \end{bmatrix} u - \begin{bmatrix} 1 \\ \mathbf{0} \end{bmatrix} r + \begin{bmatrix} 0 \\ \mathbf{G}_1 \end{bmatrix} w, \tag{7.89}$$

and the feedback law is

$$u = -\begin{bmatrix} K_1 & \mathbf{K}_0 \end{bmatrix} \begin{bmatrix} x_I \\ \mathbf{x} \end{bmatrix} = -\mathbf{K} \begin{bmatrix} x_I \\ \mathbf{x} \end{bmatrix}.$$

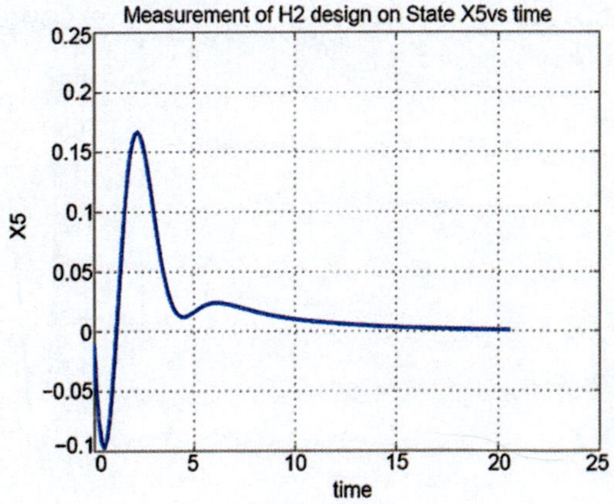

Fig. 7.31 The fifth state trajectory using $\mathcal{H}_2$ control

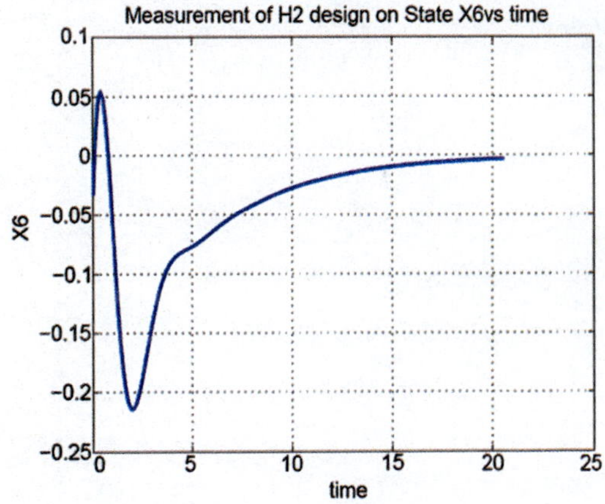

Fig. 7.32 The sixth state trajectory using $\mathcal{H}_2$ control

With this revised definition of the system, we can apply the design techniques from Chap. 5 in a similar fashion; they will result in the control structure shown in Fig. 7.39.

7.8.2 Control Example 7.4

Consider the motor speed system described by

$$u = -\mathbf{K}\begin{bmatrix} x_I \\ \mathbf{x} \end{bmatrix}$$

that is, $F = -3$, $G = 1$, and $H = 1$. Design the system to have integral control and two poles at $s = -5$. It is required to design an estimator with pole at $s = -10$. The disturbance enters at the same place as the control. Then we evaluate the tracking and disturbance rejection responses.

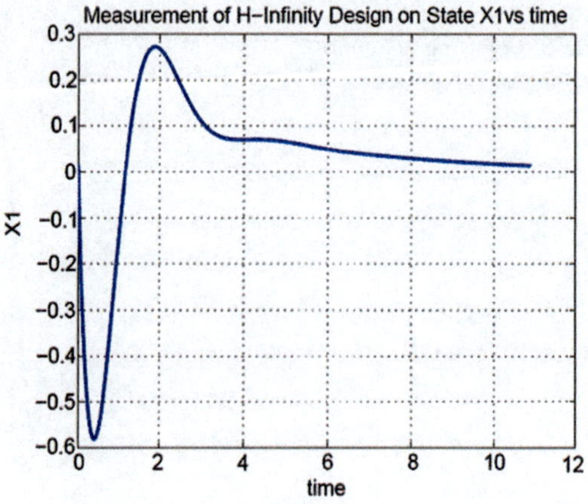

Fig. 7.33 The first state trajectory using $\mathcal{H}_\infty$ control

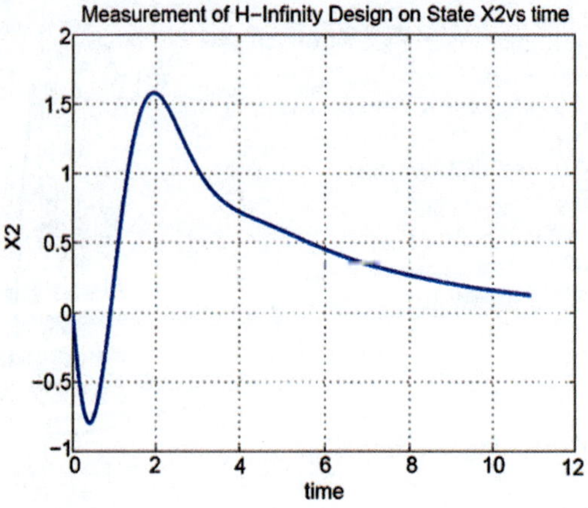

Fig. 7.34 The second state trajectory using $\mathcal{H}_\infty$ control

We proceed by noting that the pole-placement requirement is equivalent to

$$p_c = [-5; -5].$$

The augmented system description including the disturbance w is

$$\begin{bmatrix} \dot{x}_I \\ \dot{x} \end{bmatrix} = \begin{bmatrix} 0 & 1 \\ 0 & -3 \end{bmatrix} \begin{bmatrix} x_I \\ x \end{bmatrix} + \begin{bmatrix} 0 \\ 1 \end{bmatrix} (u + w) - \begin{bmatrix} 1 \\ 0 \end{bmatrix} r.$$

Therefore, we can find $\mathbf{K}$ from

$$\det\left(s\mathbf{I} - \begin{bmatrix} 0 & 1 \\ 0 & -3 \end{bmatrix} + \begin{bmatrix} 0 \\ 1 \end{bmatrix} \mathbf{K} \right) = s^2 + 10s + 25.$$

A little algebra gives

$$s^2 + (3 + K_0)s + K_1 = s^2 + 10s + 25$$

which yields

$$\mathbf{K} = \begin{bmatrix} K_1 & K_0 \end{bmatrix} = \begin{bmatrix} 25 & 7 \end{bmatrix}.$$

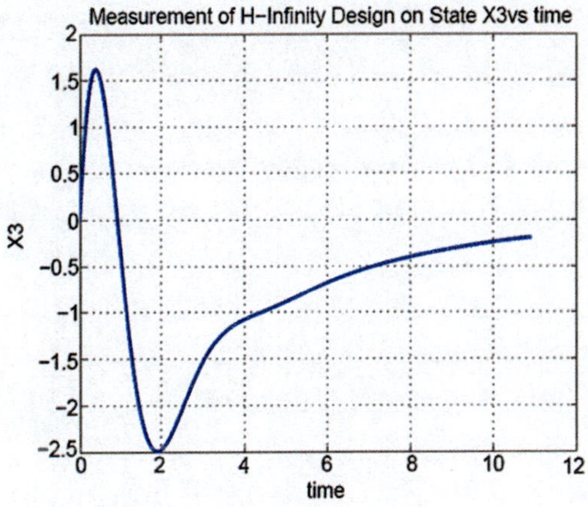

Fig. 7.35 The third state trajectory using $\mathcal{H}_\infty$ control

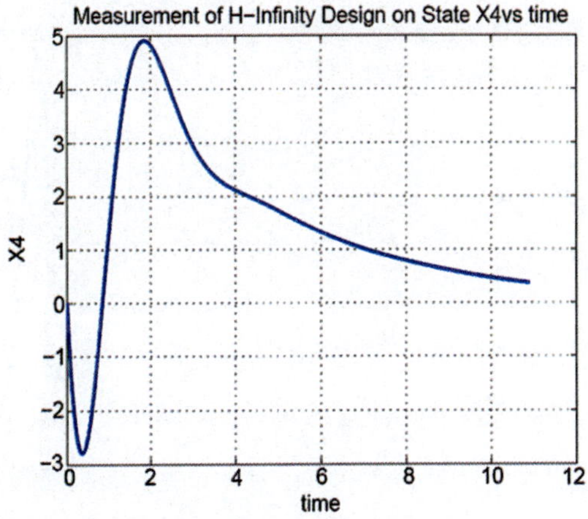

Fig. 7.36 The fourth state trajectory using $\mathcal{H}_\infty$ control

The system is shown with feedbacks in Fig. 7.40 along with a disturbance input w. On the other hand, the estimator gain $L = 7$ is obtained from

$$\alpha_e(s) = s + 10 = s + 3 + L.$$

The estimator equation is of the form

$$\begin{aligned}\dot{\hat{x}} &= (F - LH)\hat{x} + Gu + Ly \\ &= -10\hat{x} + u + 7y\end{aligned}$$

and

$$u = -K_0\hat{x} = -7\hat{x}.$$

The step response y_1 due to a step reference input r, and the output disturbance response y_2 due to a step disturbance input w are shown in Fig. 7.41(a) and the associated control efforts (u_1 and u_2) are shown in Fig. 7.41(b). As noted, the system tracks the step reference input and rejects the step disturbance asymptotically.

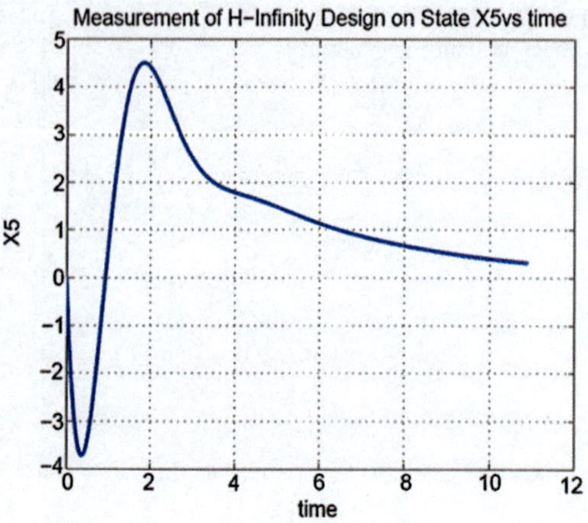

Fig. 7.37 The fifth state trajectory using $\mathcal{H}_\infty$ control

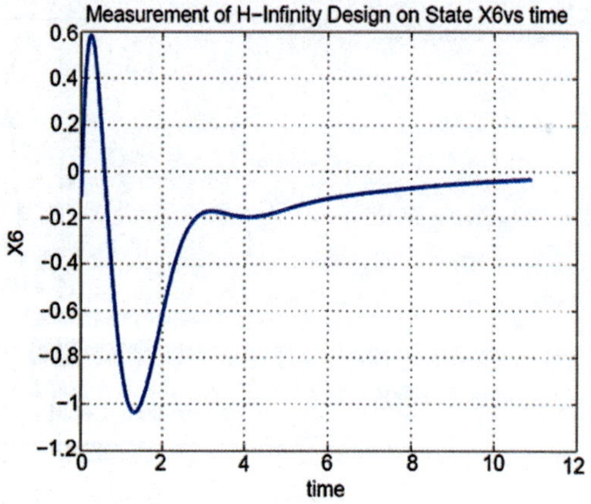

Fig. 7.38 The sixth state trajectory using $\mathcal{H}_\infty$ control

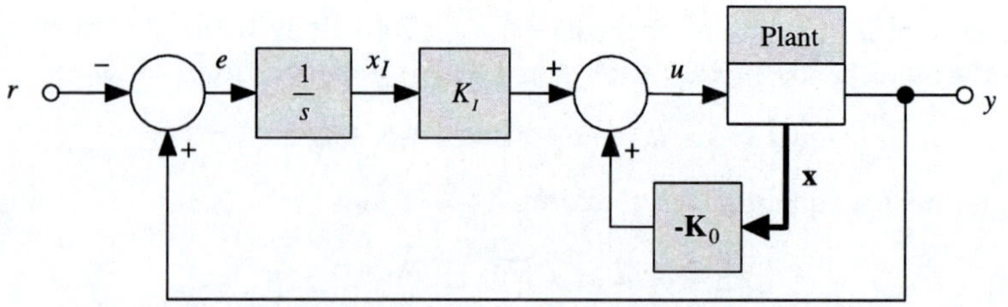

Fig. 7.39 Integral control structure

7.8.3 The Error-Space Approach

In what follows, we present a more analytical approach to giving a control system the ability to track (with zero steady-state error) a nondecaying input and to reject (with zero steady-state error) a nondecaying disturbance such as a step, ramp, or

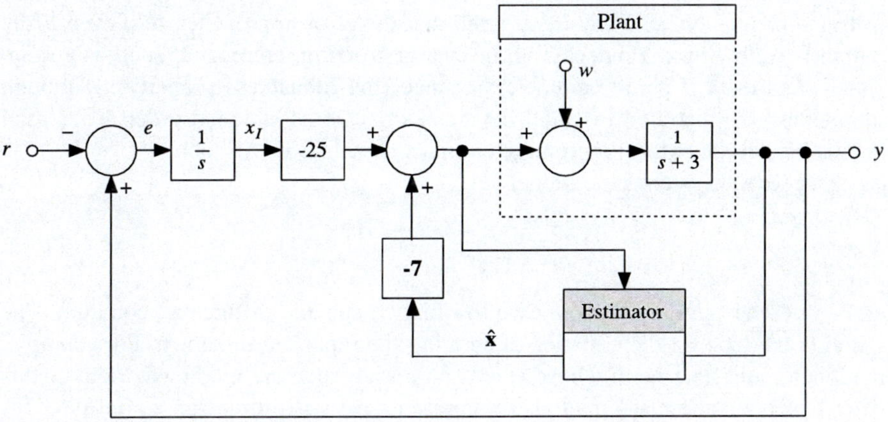

Fig. 7.40 Integral control example

Fig. 7.41 Transient response for motor speed system: (**a**) Step responses, (**b**) Control efforts

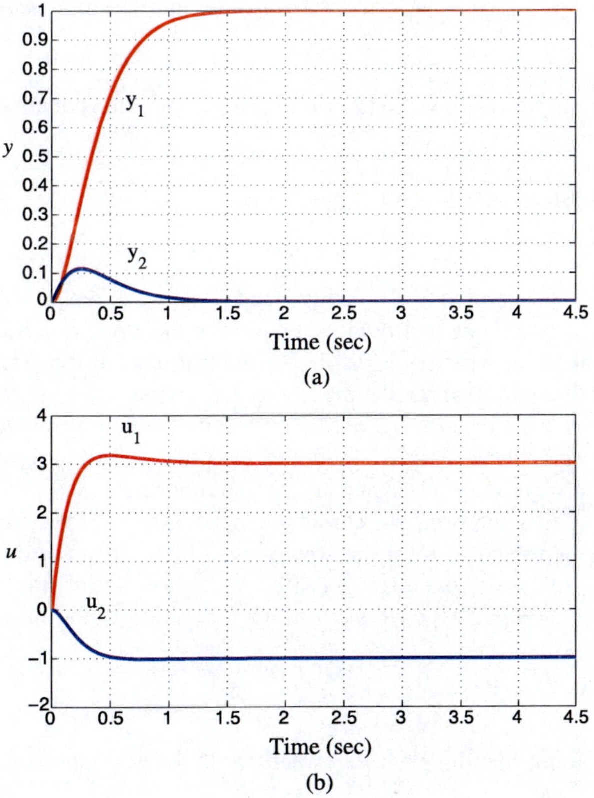

sinusoidal input. The method is based on including the equations satisfied by these external signals as part of the problem formulation and solving the problem of control in an *error space* so we are assured that the error approaches zero even if the output is following a nondecaying, or even a growing, command (such as a ramp signal) and even if some parameters change (the robustness property). Although the method is illustrated in detail for signals that satisfy second-order differential equations, but the extension to more complex signals is not difficult.

Consider the system state model

$$\begin{aligned} \hat{\mathbf{x}} &= \mathbf{F}\mathbf{x} + \mathbf{G}u + \mathbf{G}_1 w, \\ y &= \mathbf{H}\mathbf{x} \end{aligned} \tag{7.90}$$

and a reference signal that is known to satisfy a specific differential equation. The initial conditions on the equation generating the input are unknown. For example, the input could be a ramp whose slope and initial value are unknown. Plant disturbances of the same class may also be present. We wish to design a controller for this system so that the closed-loop system will have specified poles and will have the ability to track input command signals and to reject disturbances of the type described without steady-state error. We will develop the results only for second-order differential equations. Define the reference input to satisfy the relation

$$\ddot{r} + \alpha_1 \dot{r} + \alpha_2 r = 0 \tag{7.91}$$

and the disturbance to satisfy exactly the same equation:

$$\ddot{w} + \alpha_1 \dot{w} + \alpha_2 w = 0. \tag{7.92}$$

The (tracking) error is defined as

$$e = y - r. \tag{7.93}$$

The problem of tracking r and rejecting w can be seen as an exercise in designing a control law to provide regulation of the error, which is to say that the error e tends to zero as time gets large. The control must also be structurally stable or robust, in the sense that regulation of e to zero in the steady-state occurs even in the presence of "small" perturbations of the original system parameters. Note that in practice we never have a perfect model of the plant and the values of parameters are virtually always subject to some change, so robustness is always very important.

We know that the command input satisfies (7.90), and we would like to eliminate the reference from the equations in favor of the error. We begin by replacing r in (7.91) with the error in (7.93). When we do this, the reference cancels because of (7.90), and we have the formula for the error in terms of the state

$$\begin{aligned} \ddot{e} + \alpha_1 \dot{e} + \alpha_2 e &= \ddot{y} + \alpha_1 \dot{y} + \alpha_2 y \\ &= \mathbf{H}\ddot{\mathbf{x}} + \alpha_1 \mathbf{H}\dot{\mathbf{x}} + \alpha_2 \mathbf{H}\mathbf{x}. \end{aligned} \tag{7.94}$$

Replacing the plant state vector with the error-space state defined by

$$\xi := \ddot{\mathbf{x}} + \alpha_1 \dot{\mathbf{x}} + \alpha_2 \mathbf{x} \tag{7.95}$$

and similarly, we replace the control with the control in error space, defined as

$$\mu := \ddot{u} + \alpha_1 \dot{u} + \alpha_2 u. \tag{7.96}$$

With these definitions we can replace (7.94) with

$$\ddot{e} + \alpha_1 \dot{e} + \alpha_2 e = \mathbf{H}\xi. \tag{7.97}$$

The state equation for ξ is given by

$$\dot{\xi} = \frac{d^3}{dt^3}\mathbf{x} + \alpha_1 \ddot{\mathbf{x}} + \alpha_2 \dot{\mathbf{x}} = \mathbf{F}\xi + \mathbf{G}\mu. \tag{7.98}$$

Notice that the disturbance as well as the reference cancels from (7.98). Note that (7.97) and (7.98) now describe the overall system in an error space. In standard state-variable form, the equations are

$$\dot{\mathbf{z}} = \mathbf{A}\mathbf{z} + \mathbf{B}\mu, \tag{7.99}$$

where

$$\mathbf{z} = \begin{bmatrix} e & \dot{e} & \xi^t \end{bmatrix}^t \tag{7.100}$$

and

$$\mathbf{A} = \begin{bmatrix} 0 & 1 & \mathbf{0} \\ -\alpha_2 & -\alpha_1 & \mathbf{H} \\ \mathbf{0} & \mathbf{0} & \mathbf{F} \end{bmatrix}, \qquad \mathbf{B} = \begin{bmatrix} 0 \\ 0 \\ \mathbf{G} \end{bmatrix}. \tag{7.101}$$

The error system $(\mathbf{A}, \mathbf{B})$ can be given arbitrary dynamics by state feedback if it is controllable. If the plant $(\mathbf{F}, \mathbf{G})$ is controllable and does not have a zero at any of the roots of the reference-signal characteristic equation

$$\alpha_r(s) = s^2 + \alpha_1 s + \alpha_2,$$

then the error system $(\mathbf{A}, \mathbf{B})$ is controllable. Therefore, there exists a control law of the form

$$\mu = -\begin{bmatrix} K_2 & K_1 & \mathbf{K}_0 \end{bmatrix} \begin{bmatrix} e \\ \dot{e} \\ \xi \end{bmatrix} = -\mathbf{K}\mathbf{z}, \tag{7.102}$$

such that the error system has arbitrary dynamics by pole placement. We now need to express this control law in terms of the actual process state $\mathbf{x}$ and the actual control. We combine (7.95), (7.96), and (7.102) to get the control law in terms of u and $\mathbf{x}$ (we write $u^{(2)}$ to mean d^2u/dt^2):

$$(u + \mathbf{K}_0\mathbf{x})^{(2)} + \sum_{i=1}^{2} \alpha_i (u + \mathbf{K}_0\mathbf{x})^{(2-i)} = -\sum_{i=1}^{2} K_i e^{(2-i)}. \tag{7.103}$$

The structure for implementing (7.103) is very simple for tracking constant inputs. In that case the equation for the reference input is $\dot{r} = 0$. In terms of u and $\mathbf{x}$, the control law (7.103) reduces to

$$\dot{u} + \mathbf{K}_0\dot{\mathbf{x}} = -K_1 e. \tag{7.104}$$

Here, we only need to integrate to reveal the control law and the action of integral control:

$$u = -K_1 \int^t e\, d\tau - \mathbf{K}_0\mathbf{x}. \tag{7.105}$$

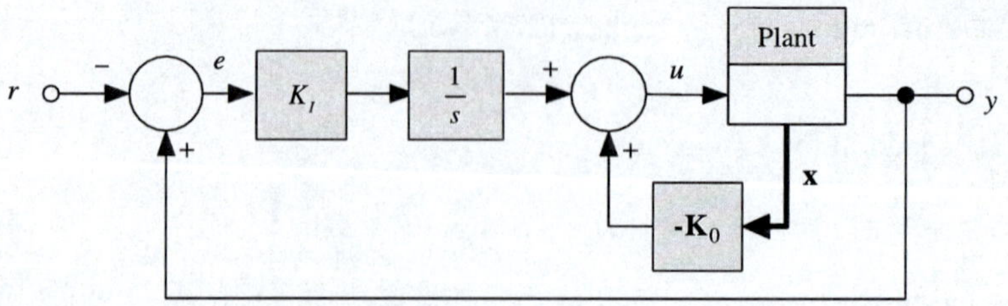

Fig. 7.42 Integral control using the internal model approach

A block diagram of the system, shown in Fig. 7.42, clearly shows the presence of a pure integrator in the controller. In this case, the only difference between the internal model method of Fig. 7.42 and the ad hoc method of Fig. 7.39 is the relative location of the integrator and the gain.

A more complex problem that clearly shows the power of the error-space approach to robust tracking is posed by requiring that a sinusoid be tracked with zero steady-state error. The problem arises, for instance, in the control of a mass-storage disk-head assembly.

7.8.4 Control Example 7.5

A simple normalized model of a computer disk-drive servomechanism [8] is given by the equations

$$\mathbf{F} = \begin{bmatrix} 0 & 1 \\ 0 & -1 \end{bmatrix}; \qquad \mathbf{G} = \begin{bmatrix} 0 \\ 1 \end{bmatrix};$$

$$\mathbf{G}_1 = \begin{bmatrix} 0 \\ 1 \end{bmatrix}; \qquad \mathbf{H} = \begin{bmatrix} 1 & 0 \end{bmatrix}; \qquad J = 0.$$

Because the data on the disk is not exactly on a centered circle, the servo must follow a sinusoid of radian frequency ω_0. determined by the spindle speed. It is required to give the structure of a controller for this system that will follow the given reference input with zero steady-state error. Then, setting $\omega_0 = 1$, consider that the desired closed-loop poles are at $-1 \pm \sqrt{3}$ and $-\sqrt{3} \pm j1$. Finally, demonstrate the tracking and disturbance rejection properties of the system using MATLAB or Simulink.

We attend to the problem in steps. Initially, the reference input satisfies the differential equation $\ddot{r} = -\omega_0^2 r$ so that $\alpha_1 = 0$ and $\alpha_2 = \omega_0^2$. With these values the error-state matrices according to (7.101) are

$$\mathbf{A} = \begin{bmatrix} 0 & 1 & 0 & 0 \\ -\omega_0^2 & 0 & 1 & 0 \\ 0 & 0 & 0 & 1 \\ 0 & 0 & 0 & -1 \end{bmatrix}, \qquad \mathbf{B} = \begin{bmatrix} 0 \\ 0 \\ 0 \\ 1 \end{bmatrix}.$$

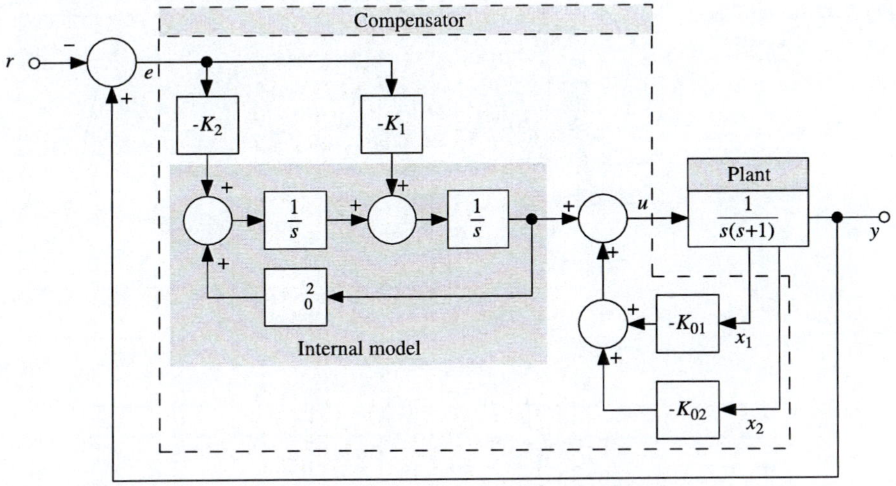

Fig. 7.43 Structure of the tracking compensator for the servomechanism

Simple calculations show that the characteristic equation of $\mathbf{A} - \mathbf{BK}$ is

$$s^4 + (1 + K_{02})s^3 + (\omega_0^2 + K_{01})s^2 + [K_1 + \omega_0^2(1 + K_{02})]s + K_{01}\omega_0^2 K_2 = 0$$

from which the gain may be selected by pole assignment. The compensator implementation from (7.103) has the structure shown in Fig. 7.43, which clearly shows the presence of the oscillator with frequency ω_0 (known as the *internal model of the input generator*) in the controller. This is a particular case of the *internal model principle*, which requires that a model of the external or exogenous signal be in the controller for robust tracking and disturbance rejection [8].

Next, we assume that $\omega_0 = 1$ rad/s and the desired closed-loop poles are as given above. If

$$pc = [-1 + j\sqrt{3}; -1 - j\sqrt{r}; -\sqrt{3} + j; -\sqrt{3} - j]$$

then the feedback gain is

$$\mathbf{K} = [K_2 \quad K_1 \; : \; \mathbf{K}_0] = [2.0718 \quad 16.3923 \; : \; 13.9282 \quad 4.4641]$$

which results in the controller

$$\dot{\mathbf{x}}_c = \mathbf{A}_c \mathbf{x}_c + \mathbf{B}_c e,$$
$$u = \mathbf{C}_c \mathbf{x}_c$$

with

$$\mathbf{A}_c = \begin{bmatrix} 0 & 1 \\ -1 & 0 \end{bmatrix}, \qquad \mathbf{B}_c = \begin{bmatrix} -16.323 \\ -2.0718 \end{bmatrix}, \qquad \mathbf{C}_c = [1 \quad 0].$$

The controller frequency response is shown in Fig. 7.44 and shows a gain of infinity at the rotation frequency of $\omega_0 = 1$ rad/s. The frequency response from r to e, that is, the sensitivity function $\mathcal{S}(s)$, is shown in Fig. 7.45 and shows a sharp

Fig. 7.44 Controller frequency response

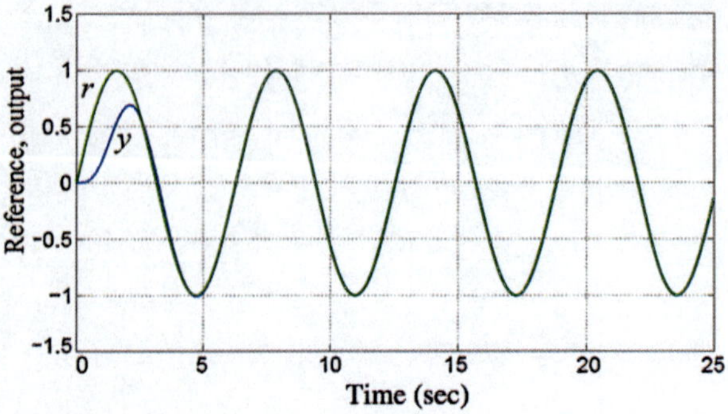

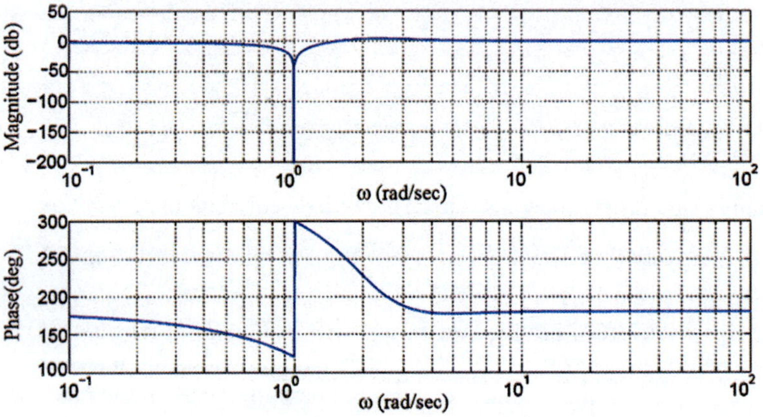

Fig. 7.45 Sensitivity function frequency response

notch at the rotation frequency $\omega_0 = 1$ rad/s. The same notch is also present in the frequency response of the transfer function from w to y. In Fig. 7.46, a Simulink simulation diagram for the system is shown. Although the simulations can also he clone in MATLAB, it is more instructive to use the interactive graphical environment of Simulink. Simulink also provides the capability to add nonlinearities and carry out robustness studies efficiently. The tracking properties of the system are shown in Fig. 7.47 showing asymptotic tracking property of the system. The disturbance rejection properties of the system are illustrated in Fig. 7.48 depicting asymptotic disturbance rejection of sinusoidal disturbance input. The closed-loop frequency response, that is, the complementary transfer function $T(s)$, for the robust servomechanism is shown in Fig. 7.49. As seen from the figure, the frequency response from r to y is unity at $\omega_0 = 1$ rad/s as expected.

The zeros of the system from r to e are located at $\pm j$, $-2.7321 \pm j2.5425$. The robust tracking properties are due to the presence of the blocking zeros at $\pm j$. The zeros from w to y, both blocking zeros, are located at $\pm j$. The robust disturbance rejection properties are due to the presence of these blocking zeros.

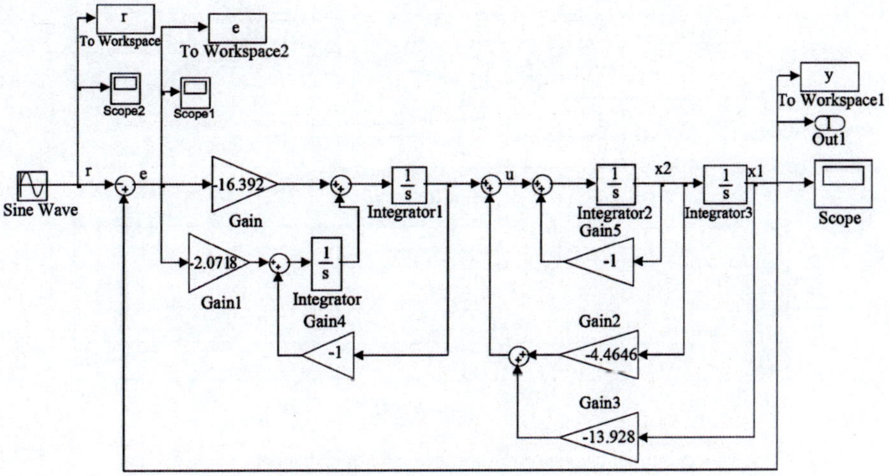

Fig. 7.46 Simulink block diagram for robust servomechanism

Fig. 7.47 Tracking properties for robust servomechanism

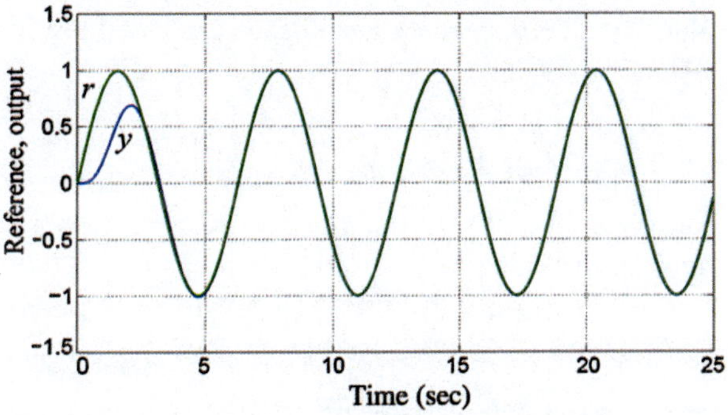

Fig. 7.48 Disturbance rejection properties for robust servomechanism

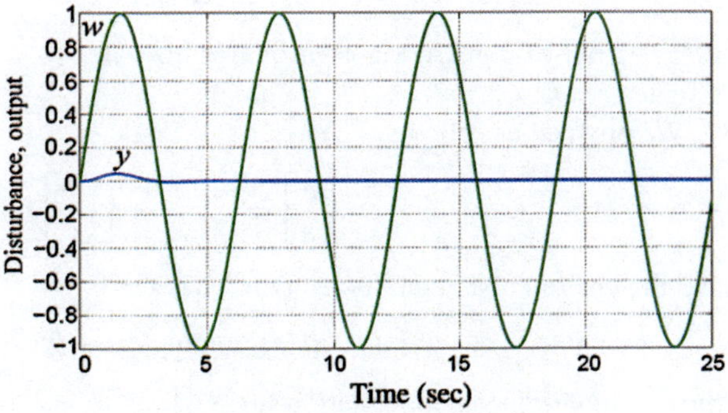

From the nature of the pole-placement problem, the state $\mathbf{z}$ in (7.100). will tend toward zero for all perturbations in the system parameters as long as $\mathbf{A}-\mathbf{BK}$ remains stable. Notice that the signals that are rejected are those that satisfy the equations

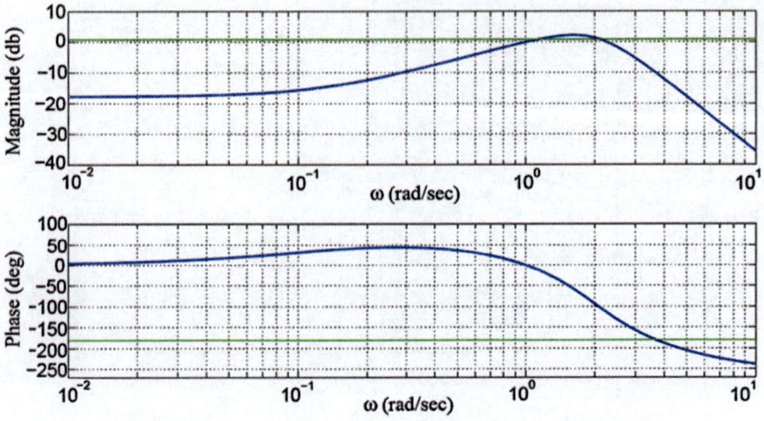

Fig. 7.49 Closed-loop frequency response for robust servomechanism

with the values of α_1 actually implemented in the model of the external signals. The method assumes that these are known and implemented exactly. If the implemented values are in error, then a steady-state error will result.

7.8.5 Control Example 7.6

Consider the system

$$H(s) = \frac{1}{s+3}$$

which has the state-variable description

$$F = -3, \qquad G = 1, \qquad H = 1.$$

It is desired to construct a controller with poles at $s = -5$ to track an input that satisfies $\dot{r} = 0$.

We proceed as follows. The error system is

$$\begin{bmatrix} \dot{e} \\ \dot{z} \end{bmatrix} = \begin{bmatrix} 0 & 1 \\ 0 & -3 \end{bmatrix} \begin{bmatrix} 0 \\ 1 \end{bmatrix} \mu$$

and the desired characteristic equation has the form

$$\alpha_c(s) = s^2 + 10s + 25, \tag{7.106}$$

then the pole-placement equation for K is

$$\det[s\mathbf{I} - \mathbf{A} + \mathbf{BK}] = \alpha_c(s). \tag{7.107}$$

By (7.106) and (7.107), we have

$$s^2 + (3 + K_0)s + K_1 = s^2 + 10s + 25,$$

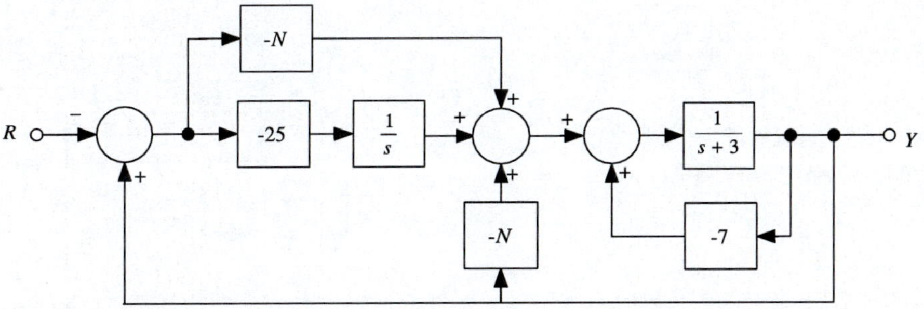

Fig. 7.50 Example of internal model with feedforward

which gives

$$\mathbf{K} = \begin{bmatrix} 25 & 7 \end{bmatrix} = \begin{bmatrix} K_1 & K_0 \end{bmatrix},$$

and the system is implemented as shown in Fig. 7.50. The transfer function from r to e for this system, the sensitivity function,

$$\frac{E(s)}{R(s)} = \mathcal{S}(s) = -\frac{s(s+10)}{s^2+10s+25}$$

shows a blocking zero at $s = 0$, which prevents the constant input from affecting the error. The closed-loop transfer function, that is, the complementary sensitivity function is

$$\frac{Y(s)}{R(s)} = \mathcal{T} = \frac{5}{s^2+10s+25}.$$

The structure of Fig. 7.50 permits us to add a feedforward of the reference input, which provides one extra degree of freedom in zero assignment. If we add a term proportional to r in (7.104), then

$$u = -K_1 \int^t e(\tau)\,d\tau - \mathbf{K}_0\mathbf{x} + Nr. \tag{7.108}$$

This relationship has the effect of creating a zero at $-K_1/N$. The location of this zero can he chosen to improve the transient response of the system. For actual implementation, we can rewrite (7.108) in terms of e to get

$$u = -K_1 \int^t e(\tau)\,d\tau - \mathbf{K}_0\mathbf{x} + N(y - e). \tag{7.109}$$

The block diagram for the system is shown in (7.108). It is easy to see that the overall transfer function becomes

$$\frac{Y(s)}{R(s)} = \frac{Ns+25}{s^2+10s+25}.$$

Notice that the DC gain is unity for any value of N and that through our choice of N we can place the zero at any real value to improve the dynamic response. A natural strategy for locating the zero is to have it cancel one of the system poles, in this case

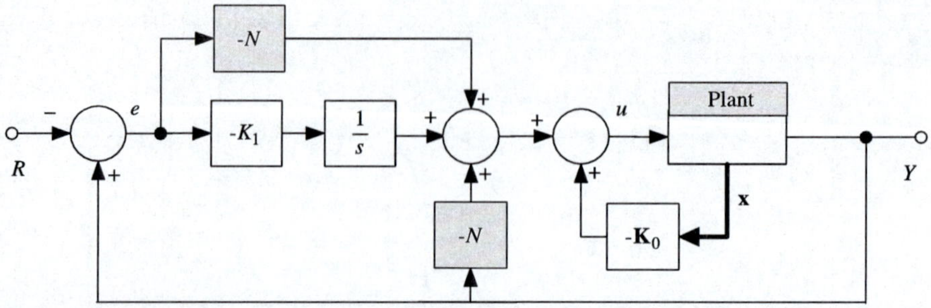

Fig. 7.51 Internal model as integral control with feedforward

Fig. 7.52 Step responses with integral control and feedforward

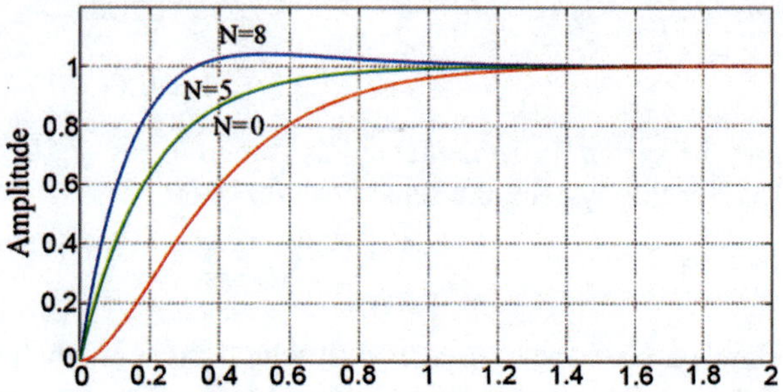

at $s = -5$. The step response of the system is shown in Fig. 7.52 for $N = 5$, as well as for $N = 0$ and 8. With the understanding that one pole can be canceled in integral control designs, we make sure to choose one of the desired control poles such that it is both real and able to be canceled through the proper choice of N.

7.8.6 The Extended Estimator

Recall that the discussion of robust control so far has used a control based on full-state feedback. If the state is not available, then as in the regular case, the full-state feedback, $\mathbf{Kx}$, can be replaced by the estimates, $\mathbf{K\hat{x}}$, where the estimator is built as before. As a final look at ways to design control with external inputs, in this section we develop a method for tracking a reference input and rejecting disturbances. The method is based on augmenting the estimator to include estimates from external signals in a way that permits us to cancel out their effects on the system error.

Suppose the plant is described by the equations

$$\begin{aligned}\dot{\mathbf{x}} &= \mathbf{Fx} + \mathbf{G}u + \mathbf{G}w,\\ y &= \mathbf{Hx},\\ e &= \mathbf{Hx} - r.\end{aligned} \tag{7.110}$$

Fig. 7.53 Block diagram of a system for tracking and disturbance rejection with extended estimator: Equivalent disturbance

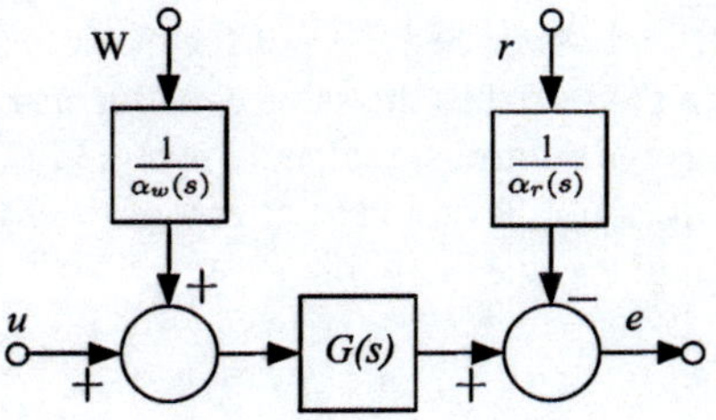

Fig. 7.54 Block diagram of a system for tracking and disturbance rejection with extended estimator: Block diagram for *design*

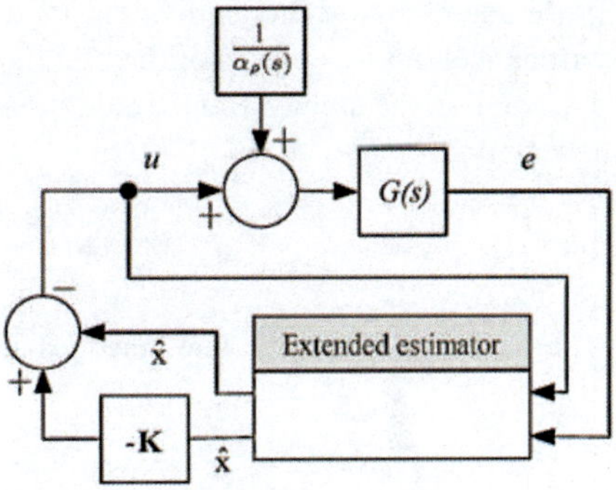

Fig. 7.55 Block diagram of a system for tracking and disturbance rejection with extended estimator: Block diagram for *implementation*

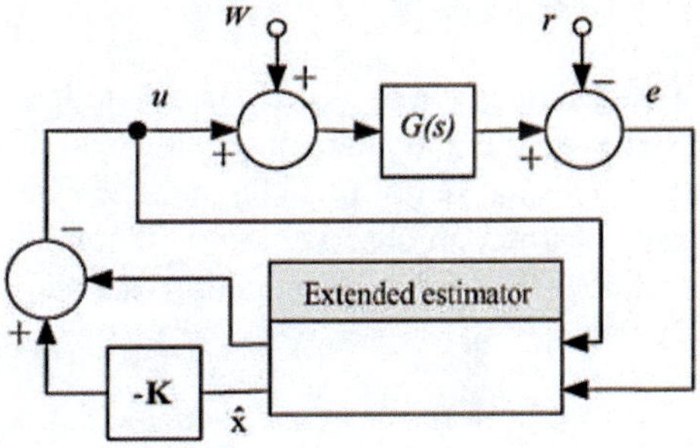

Furthermore, assume that both the reference r and the disturbance w are known to satisfy the equations[1]

$$\alpha_w(s)w = \alpha_\rho(s)w = 0, \tag{7.111}$$

$$\alpha_r(s)r = \alpha_\rho(s)r = 0, \tag{7.112}$$

where

$$\alpha_\rho(s) = s^2 + \alpha_1 s + \alpha_2$$

corresponding to polynomials $\alpha_w(s)$ and $\alpha_r(s)$ in Fig. 7.53. In general, we would select the equivalent disturbance polynomial $\alpha_\rho(s)$ in Fig. 7.54 to be the *least common multiple* of $\alpha_w(s)$ and $\alpha_r(s)$. The first step is to recognize that, as far as the

[1] Again we develop the results for a second-order equation in the external signals; the discussion can be extended to higher-order equations.

steady-state response of the output is concerned, there is an input-equivalent signal ρ that satisfies the same equation as r and w and enters the system at the same place as the control signal as shown in Fig. 7.54. As before, we must assume that the plant does not have a zero at any of the roots of (7.111). For our purposes here, we can replace (7.111) with

$$\begin{aligned}\dot{\mathbf{x}} &= \mathbf{F}\mathbf{x} + \mathbf{G}(u+\rho),\\ e &= \mathbf{H}\mathbf{x}.\end{aligned} \tag{7.113}$$

If we can estimate this equivalent input, we can add to the control a term $-\hat{\rho}$ that will cancel out the effects of the real disturbance and reference and cause the output to track r in the steady-state. To do this, we combine (7.111) and (7.113) into a state description to get

$$\begin{aligned}\dot{\mathbf{z}} &= \mathbf{A}\mathbf{z} + \mathbf{B}u,\\ e &= \mathbf{C}\mathbf{z},\end{aligned} \tag{7.114}$$

where $\mathbf{z} = [\,\rho \;\; \dot{\rho} \;\; \mathbf{x}^T\,]^T$. The matrices are

$$\begin{aligned}\mathbf{A} &= \begin{bmatrix} 0 & 1 & \mathbf{0} \\ -\alpha_2 & -\alpha_1 & \mathbf{0} \\ \mathbf{G} & \mathbf{0} & \mathbf{F} \end{bmatrix}, \qquad \mathbf{B} = \begin{bmatrix} 0 \\ 0 \\ \mathbf{G} \end{bmatrix},\\ \mathbf{C} &= \begin{bmatrix} 0 & 0 & \mathbf{H} \end{bmatrix}.\end{aligned} \tag{7.115}$$

The system given by (7.115) is not controllable since we cannot influence ρ from u. However, if $\mathbf{F}$ and $\mathbf{H}$ are observable and if the system $(\mathbf{F}, \mathbf{G}, \mathbf{H})$ does not have a zero that is also a root of (7.111), then the system of (7.115) will be observable, and we can construct an observer that will compute estimates of both the state of the plant and of ρ. The estimator equations are standard, but the control is not:

$$\begin{aligned}\dot{\hat{\mathbf{z}}} &= \mathbf{A}\hat{\mathbf{z}} + \mathbf{B}u + \mathbf{L}(e - \mathbf{C}\hat{z}),\\ u &= -\mathbf{K}\hat{\mathbf{x}} - \hat{\rho}.\end{aligned} \tag{7.116}$$

In terms of the original variables, the estimator equations are

$$\begin{bmatrix} \dot{\hat{\rho}} \\ \ddot{\hat{\rho}} \\ \dot{\hat{\mathbf{x}}} \end{bmatrix} = \begin{bmatrix} 0 & 1 & \mathbf{0} \\ -\alpha_2 & -\alpha_1 & \mathbf{0} \\ \mathbf{G} & \mathbf{0} & \mathbf{F} \end{bmatrix} \begin{bmatrix} \hat{\rho} \\ \dot{\hat{\rho}} \\ \hat{\mathbf{x}} \end{bmatrix} + \begin{bmatrix} 0 \\ 0 \\ \mathbf{G} \end{bmatrix} u + \begin{bmatrix} l_1 \\ l_2 \\ \mathbf{L}_3 \end{bmatrix} [e - \mathbf{H}\hat{\mathbf{x}}]. \tag{7.117}$$

The overall block diagram of the system for *design* is shown in Fig. 7.54. If we write out the last equation for $\hat{\mathbf{x}}$ in (7.117) and substitute (7.116), a simplification of sorts results because a term in $\hat{\rho}$ cancels out:

$$\begin{aligned}\dot{\hat{\mathbf{x}}} &= \mathbf{G}\hat{\rho} + \mathbf{F}\hat{\mathbf{x}} + \mathbf{G}(-\mathbf{K}\hat{\mathbf{x}} - \hat{\rho}) + \mathbf{L}_3(e - \mathbf{H}\hat{\mathbf{x}})\\ &= \mathbf{F}\hat{\mathbf{x}} + \mathbf{G}(-\mathbf{K}\hat{\mathbf{x}}) + \mathbf{L}_3(e - \mathbf{H}\hat{\mathbf{x}})\\ &= \mathbf{F}\hat{\mathbf{x}} + \mathbf{G}\bar{u} + \mathbf{L}_3(e - \mathbf{H}\hat{\mathbf{x}}).\end{aligned}$$

With the estimator of (7.117) and the control of (7.116), the state model is

$$\dot{\mathbf{x}} = \mathbf{F}\mathbf{x} + \mathbf{G}(-\mathbf{K}\hat{\mathbf{x}} - \hat{\rho}) + \mathbf{G}\rho. \tag{7.118}$$

In terms of the estimate errors, (7.230) can be rewritten as

$$\dot{\mathbf{x}} = (\mathbf{F} - \mathbf{GK})\mathbf{x} + \mathbf{GK}\tilde{\mathbf{x}} + \mathbf{G}\tilde{\rho}. \tag{7.119}$$

Because we designed the estimator to be stable, the values of $\tilde{\rho}$ and $\tilde{\mathbf{x}}$ go to zero in the steady-state, and the final value of the state is not affected by the external input. The block diagram of the system for implementation is drawn in Fig. 7.54. The steps involved in this process will now be illustrated.

7.8.7 Control Example 7.7

Construct an estimator to controlling the state [8] and canceling a constant bias at the output and track a constant reference in the motor speed system described by

$$\begin{aligned} \dot{x} &= -3x + u, \\ y &= x + w, \\ \dot{w} &= 0, \\ \dot{r} &= 0. \end{aligned} \tag{7.120}$$

Place the control pole at $s = -5$ and the two extended estimator poles at $s = -15$.

To begin, we design the control law by ignoring the equivalent disturbance. Rather, we notice by inspection that a gain of -2 will move the single pole from -3 to the desired -5. Therefore, $K = 2$. The system augmented with equivalent external input ρ, which replaces the actual disturbance w and the reference r, is given by

$$\begin{aligned} \dot{\rho} &= 0; \\ \dot{x} &= -3x + u + \rho, \\ e &= x. \end{aligned}$$

The extended estimator equations are

$$\begin{aligned} \dot{\hat{\rho}} &= l_i(e - \hat{x}), \\ \dot{\hat{x}} &= -3\hat{x} + u + \hat{\rho} + l_2(e - \hat{x}). \end{aligned}$$

The estimator error gain is found to be $\mathbf{L} = [225 \;\; 27]^t$ from the characteristic equation

$$\det \begin{bmatrix} s & l_1 \\ 1 & s + 3 + l_2 \end{bmatrix} = s^2 + 30s + 225.$$

A block diagram of the system is given in Fig. 7.56(top), and the step responses to input at the command r (applied at $t = 0$ s) and at the disturbance w (applied at $t = 0.5$ s) are shown in Fig. 7.56.

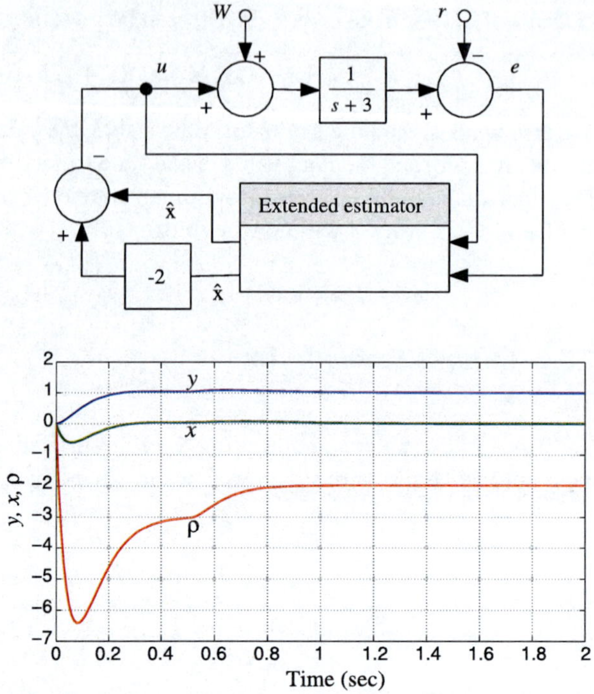

Fig. 7.56 Motor speed system with extended estimator: (*top*) Block diagram; (*bottom*) Command step response and disturbance step response

7.9 Questions

Q7.1 For the following linear time-invariant system

$$\dot{x} = Ax + Bd, \qquad e = Cx$$

consider the optimization problem

$$\sup_d J(d) = \sup_d \int_0^\infty \left[e^t e - \gamma^2 d^t d\right] dt < \infty.$$

Show that the solution of this optimization problem leads to relationships identical to those for the bounded real lemma.

7.10 Notes and References

The $\mathcal{H}_2$ optimal control problem is the modern version of what is commonly known as the linear quadratic Gaussian (LQG) problem. As indicated in the foregoing sections, minimization of the $\mathcal{H}_2$-norm of the closed loop transfer matrix can be given the stochastic interpretation of minimizing the expected value of the squared norm of the output, in case that the disturbance input is a standard white noise process. It is exactly the minimization of this expected value that the classical formulation of the LQG-problem deals with.

References

1. Anderson, B.O.D., Moore, J.B.: Linear Optimal Control—Linear Quadratic Methods. Prentice-Hall, Englewood Cliffs (1990)
2. Bosgra, O.H., Kwakernaak, H.: Design Methods for Control Systems. Dutch Institute of Systems and Control (2001)
3. Boyd, S.P., Barratt, C.H.: Linear Controller Design: Limits of Performance. Prentice-Hall, Englewood Cliffs (1991)
4. Doyle, J.C., Stein, G.: Multivariable feedback design: Concepts for a classical/modern synthesis. IEEE Trans. Autom. Control **26**, 4–16 (1981)
5. Doyle, J.C., Glover, K., Khargonekar, P.P., Francis, B.A.: State-space solutions to standard $\mathcal{H}_2$ and $\mathcal{H}_\infty$ control problems. IEEE Trans. Autom. Control **34**, 831–847 (1989)
6. Doyle, J.C., Francis, B.A., Tannenbaum, A.R.: Feedback Control Theory. Macmillan, New York (1992)
7. Francis, B.A.: A Course in $\mathcal{H}_\infty$ Control Theory. Springer-Verlag, Berlin (1987)
8. Franklin, G.F., Powell, J.D., Naeini, A.E.: Feedback Control of Dynamic Systems, 4th edn. Prentice-Hall, Englewood Cliffs (2002)
9. Green, M., Limebeer, D.J.N.: Linear Robust Control. Wiley, New York (1999)
10. Gu, Da-Wei, Petkov, P.H., Konstantinov, M.M.: Robust Control Design with MATLAB. Springer, London (2003)
11. Helton, J.W., Merino, O.: Classical Control Using $\mathcal{H}_\infty$ Methods. Society for Industrial and Applied Mathematics, Philadelphia (1998)
12. Hu, c., Meng, M.Q., Liu, P.X.: Observer based LQR control of shaping process of automobile belt. In: Proc. the 5th World Congress on Intelligent Control, China, June, pp. 3310–3314 (2004)
13. Jovanovic, M.: Nonlinear control of an electrohydraulic velocity servosystem. Proc. Am. Control Conf. **1**, 588–593 (2002)
14. Kaddissi, C., Kenne, J.-P., Saad, M.: Identification and real-time control of an electrohydraulic servo system based on nonlinear backstepping. IEEE/ASME Trans. Mechatron. **12**, 12–22 (2007)
15. Kwakernaak, H., Sivan, R.: Linear Optimal Control Systems. Wiley, New York (1972)
16. Mahmoud, M.S.: Resilient $\mathcal{L}_2$–$\mathcal{L}_\infty$ filtering of polytopic systems with state-delays. IET Control Theory Appl. **1**(1), 141–154 (2007)
17. Mahmoud, M.S., Al-Rayyah, A.Y.: Efficient parameterization to stability and feedback synthesis of linear time-delay systems. IET Control Theory Appl. **3**(8), 1107–1118 (2009)
18. Mahmoud, M.S., Xia, Y.: Robust filter design for piecewise discrete-time systems with time-varying delays. Int. J. Robust Nonlinear Control **20**, 540–560 (2010)
19. Moura, S.J., Fathy, H.K., Callaway, D.S., Stein, J.L.: A stochastic optimal control approach for power management in plug-in hybrid electric vehicles. IEEE Trans. Control Syst. Technol. **21**, 1–11 (2010)
20. Saberi, A., Sannuti, P., Chen, B.M.: $\mathcal{H}_\infty$ Optimal Control. Prentice-Hall, Englewood Cliffs (1995)
21. Sanchez-Pena, R.S., Sznaier, M.: Robust Systems. Theory and Applications. John Wiley and Sons, New York (1998)
22. Stoorvogel, A.A.: The $\mathcal{H}_\infty$ Control Problem: A State Space Approach. Prentice-Hall, Englewood Cliffs (1992)
23. Zames, G.: Feedback and optimal sensitivity: Model reference transformations, multiplicative seminorms and approximate inverses. IEEE Trans. Autom. Control **26**, 301–320 (1981)
24. Zames, G., Francis, B.A.: Feedback, minimax sensitivity, and optimal robustness. IEEE Trans. Autom. Control **28**, 585–600 (1983)
25. Zhou, K., Doyle, J.C.: Essentials of Robust Control. Prentice-Hall, Englewood Cliffs (1998)
26. Zhou, K., Doyle, J.C., Glover, K.: Robust and Optimal Control. Prentice-Hall, Upper Saddle River (1995)

Chapter 8
Adaptive Control

8.1 Introduction

In this chapter, we adopt the definition of adaptive control as the combination of a parameter estimator, which generates parameter estimates online, with a control law in order to control classes of plants whose parameters are completely unknown and/or could change with time in an unpredictable manner. The choice of the parameter estimator, the choice of the control law, and the way they are combined leads to different classes of adaptive control schemes. Adaptive control as defined above has also been referred to as *identifier-based* adaptive control in order to distinguish it from other approaches referred to as *non-identifier-based*, where similar control problems are solved without the use of an online parameter estimator.

The choice of adaptive control as a solution to a particular control problem involves understanding of the plant properties as well as of the performance requirements. Research in adaptive control has a long history of intense activities that involved debates about the precise definition of adaptive control, examples of instabilities, stability and robustness proofs, and applications. The material contained in this chapter relies on the basic references [8, 9, 19]. Adaptive control involves learning, and learning requires data which carry sufficient information about the unknown parameters. For such information to be available in the measured data, the plant has to be excited, and this may lead to transients which, depending on the problem under consideration, may not be desirable. Furthermore, in many applications there is sufficient information about the parameters, and online learning is not required. In such cases, linear robust control techniques may be more appropriate.

8.2 Preliminary Examples

In what follows, we present some typical examples to illustrate the notions of adaptive control schemes. The following two simple examples illustrate situations where adaptive control is superior to the traditional linear control.

M.S. Mahmoud, Y. Xia, *Applied Control Systems Design*,
DOI 10.1007/978-1-4471-2879-3_8, © Springer-Verlag London Limited 2012

8.2.1 Example 8.1

Consider the scalar plant

$$\dot{x} = ax + u,$$

where u is the control input and x the scalar state of the plant. The parameter a is unknown. We want to choose the input u so that the state x is bounded and driven to zero with time. If a is a known parameter, then the linear control law

$$u = -kx, \quad k > |a|,$$

can meet the control objective. In fact if an upper bound $\bar{a} \geq |a|$ is known, the above linear control law with $k > \bar{a}$ can also meet the control objective. On the other hand, if a changes so that $a > k > 0$, then the closed-loop plant will be unstable.

The conclusion is that in the absence of an upper bound for the plant parameter no linear controller could stabilize the plant and drive the state to zero. The adaptive control law

$$u = -kx, \quad \dot{k} = x^2,$$

guarantees that all signals are bounded and x converges to zero no matter what the value of the parameter a is. This simple example demonstrates that adaptive control is a potential approach to use in situations where linear controllers cannot handle the parametric uncertainty.

8.2.2 Example 8.2

As another example, consider the same example, Example 8.1, but with an external bounded disturbance d:

$$\dot{x} = ax + u + d.$$

The disturbance is unknown but can be approximated as

$$d = \sum_{i=1}^{N} \theta_i^* \phi_i(t, x),$$

where $\phi_i(t, x)$ are known functions and θ_i^* are unknown constant parameters. In this case if we use the linear control law

$$u = -kx$$

with $k > \bar{a} \geq |a|$, we can establish that $|x|$ is bounded and at steady state

$$|x| \leq \frac{d_0}{k - a},$$

where d_o is an upper bound for $|d|$. It is clear that by increasing the value of the controller gain k, we can make the steady-state value of x as small as we like. This will lead to a high gain controller, however, which is undesirable especially in the presence of high-frequency unmodeled dynamics. In principle, however, we cannot guarantee that x will be driven to zero for any finite control gain in the presence of nonzero disturbance d. The adaptive control approach is to estimate online the disturbance d and cancel its effect via feedback. The following adaptive control law can be shown to guarantee signal boundedness and convergence of the state x to zero with time:

$$u = -kx - \hat{d}, \quad \hat{d} = \sum_{i=1}^{N} \theta_i \phi_i(t, x), \quad \dot{\theta}_i = x\phi_i(t, x),$$

where $k > \bar{a} \geq |a|$, assuming of course that $\bar{a}$ is known; otherwise k has to be estimated, too.

It is readily seen that, in addition to stability, adaptive control techniques could be used to improve performance in a wide variety of situations where linear techniques would fail to meet the performance characteristics. This by no means implies that adaptive control is the most appropriate approach to use in every control problem.

8.2.3 Example 8.3

The design of autopilots for high-performance aircraft was one of the primary motivations for active research in adaptive control. Aircrafts operate over a wide range of speeds and altitudes, and their dynamics are nonlinear and conceptually time-varying. For a given operating point, the complex aircraft dynamics can be approximated by a linear model. For example, for an operating point i, the longitudinal dynamics of an aircraft model may be described by a linear system of the form [14–16]:

$$\begin{aligned} \dot{x} &= A_i x + B_i u, \qquad x(t_0) = x_0, \\ y &= C_i^t x + D_i U, \end{aligned} \tag{8.1}$$

where the matrices A_i, B_i, C_i, D_i are functions of the operating point i; x is the state; u is the input; and y is the measured outputs. As the aircraft goes through different flight conditions, the operating point changes, leading to different values for A_i, B_i, C_i, D_i. Because the measured outputs carry information about the state x and parameters, one may argue that, in principle, a sophisticated feedback controller could learn the parameter changes, by processing the outputs $y(t)$, and use the appropriate adjustments to accommodate them. This argument led to a feedback control structure on which adaptive control is based. The controller structure consists of a feedback loop and a controller with adjustable gains, as shown in Fig. 8.1. The way of adjusting the controller characteristics in response to changes in the plant and disturbance dynamics distinguishes one scheme from another.

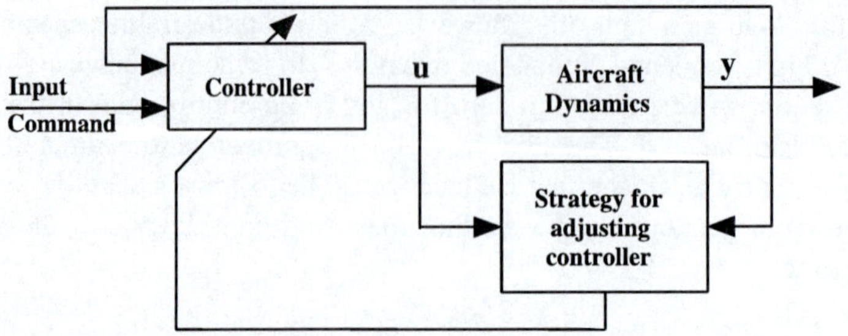

Fig. 8.1 General adaptive control structure for aircraft control

8.3 Adaptive Control Approaches

The class of adaptive control schemes studied in this chapter is characterized by the combination of an online parameter estimator, which provides estimates of the unknown parameters at each instant of time, with a control law that is motivated from the known parameter case. The way the parameter estimator, also referred to as adaptive law in the chapter, is combined with the control law gives rise to two different approaches. In the first approach, referred to as indirect adaptive control, the plant parameters are estimated online and used to calculate the controller parameters. Typically, at each time t, the estimated plant is formed and treated as if it is the true plant in calculating the controller parameters. This approach has also been referred to as explicit adaptive control, because the controller design is based on an explicit plant model. In the second approach, referred to as direct adaptive control, the plant model is parameterized in terms of the desired controller parameters, which are then estimated directly without intermediate calculations involving plant parameter estimates. This approach has also been referred to as implicit adaptive control because the design is based on the estimation of an implicit plant model. In [1–3, 5–7, 10–18, 20–23], pertinent issues have been thoroughly discussed.

8.3.1 Indirect Adaptive Control Approach

The basic structure of indirect adaptive control is shown in Fig. 8.2. The plant model $G(\theta^*)$ is parameterized with respect to some unknown parameter vector θ^*.

For example, for a linear time-invariant (LTI) single-input single-output (SISO) plant model, θ^* is a vector with the unknown coefficients of the numerator and denominator of the plant model transfer function. An online parameter estimator generates the estimate $\theta(t)$ of θ^* at each time t by processing the plant input u and output y. The parameter estimate $\theta(t)$ specifies an estimated plant model characterized by $G(\theta(t))$, which for control design purposes is treated as the "true"

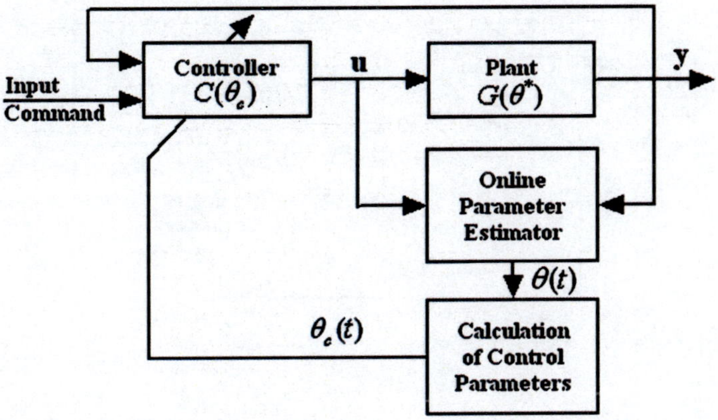

Fig. 8.2 Indirect adaptive control structure

plant model and is used to calculate the controller parameter or gain vector θ_c by solving a certain algebraic equation, $\theta_c(t) = F(\theta(t))$, that relates the plant parameters with the controller parameters at each time t. The form of the control law $C(\theta_c, (t))$ and algebraic equation $\theta_c(t) = F(\theta(t))$ is chosen to be the same as that of the control law $C(\theta_c^*)$ and equation $\theta_c^* = F(\theta^*)$, which could be used to meet the performance requirements for the plant model $G(\theta^*)$ if θ^* was known. It is, therefore, clear that with this approach, $C(\theta_c(t))$ is designed at each time t to satisfy the performance requirements for the estimated plant model $G(\theta(t))$ rather than for the actual plant $G(\theta^*)$. Therefore, the main problem in indirect adaptive control is to choose the class of control laws $C(\theta^c)$ and the class of parameter estimators that generate $\theta(t)$, as well as the algebraic equation $\theta_c = F(\theta)$, so that $C(\theta_c)$ meets the performance requirements for the plant model $G(\theta^*)$ with unknown θ^*.

8.3.2 *Direct Adaptive Control Approach*

Figure 8.3 shows the structure of direct adaptive control. In this case, the plant model $G(\theta^*)$ is parameterized in terms of the unknown controller parameter vector θ_c^*, for which $C(\theta_c^*)$ meets the performance requirements, to obtain the plant model $G_c(\theta_c^*)$ with exactly the same input/output (I/O) characteristics as $G(\theta^*)$.

The online parameter estimator is designed based on $G_c(\theta_c^*)$ instead of $G(\theta^*)$ to provide the direct online estimate $\theta_c(t)$ of θ_c^* at each time t, by processing the plant input u and output y. The estimate $\theta_c(t)$ is then used in the control law without intermediate calculations. The choice of the class of control laws $C(\theta_c)$ and parameter estimators that generate $\theta_c(t)$ so that the closed-loop plant meets the performance requirements is the fundamental problem in direct adaptive control. The properties

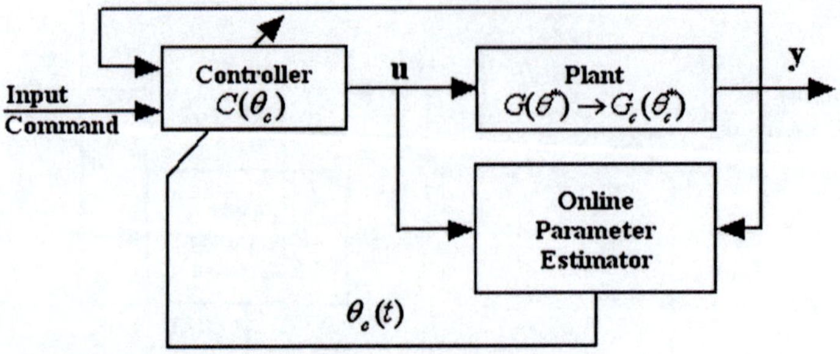

Fig. 8.3 Direct adaptive control structure

of the plant model $G(\theta^*)$ are crucial in obtaining the parameterized plant model $G_c(\theta_c^*)$ that is convenient for online estimation. As a result, direct adaptive control is restricted to certain classes of plant models. In general, not every plant can be expressed in a parameterized form involving only the controller parameters, which is also a suitable form for online estimation. As we show in Chap. 5, a class of plant models that is suitable for direct adaptive control for a particular control objective consists of all SISO LTI plant models that are minimum phase; that is, their zeros are located in $\mathrm{Re}[s] < 0$.

In general, the ability to parameterize the plant model with respect to the desired controller parameters is what gives us the choice to use the direct adaptive control approach. Note that Figs. 8.2 and 8.3 can be considered as having the exact same structure if in Fig. 8.3 we add the calculation block $\theta_c(t) = F(\theta_c(t)) = \theta_c(t)$. This identical-in-structure interpretation is often used in the literature of adaptive control to argue that the separation of adaptive control into direct and indirect is artificial and is used simply for historical reasons. In general, direct adaptive control is applicable to SISO linear plants which are minimum phase, since for this class of plants the parameterization of the plant with respect to the controller parameters for some controller structures is possible. Indirect adaptive control can be applied to a wider class of plants with different controller structures, but it suffers from a problem known as the stabilizability problem explained as follows: As shown in Fig. 8.2, the controller parameters are calculated at each time t based on the estimated plant. Such calculations are possible, provided that the estimated plant is controllable and observable or at least stabilizable and detectable. Since these properties cannot be guaranteed by the online estimator in general, the calculation of the controller parameters may not be possible at some points in time, or it may lead to unacceptable large controller gains. As we explain in Chap. 6, solutions to this stabilizability problem are possible at the expense of additional complexity. Efforts to relax the minimum-phase assumption in direct adaptive control and resolve the stabilizability problem in indirect adaptive control led to adaptive control schemes where both the controller and plant parameters are estimated online, leading to combined direct/indirect schemes that are usually more complex [6].

8.3.3 Comparisons

The principle behind the design of direct and indirect adaptive control shown in Figs. 8.2 and 8.3 is conceptually simple. The form of the control law is the same as the one used in the case of known plant parameters. In the case of indirect adaptive control the unknown controller parameters are calculated at each time t using the estimated plant parameters generated by the online estimator, whereas in the direct adaptive control case the controller parameters are generated directly by the online estimator. In both cases the estimated parameters are treated as the true parameters for control design purposes. This design approach is called certainty equivalence (CE) and can be used to generate a wide class of adaptive control schemes by combining different online parameter estimators with different control laws. The idea behind the CE approach is that as the parameter estimates $\theta_c(t)$ converge to the true ones θ_c^*, the performance of the adaptive controller $C(\theta_c)$ tends to that of $C(\theta_c^*)$ used in the case of known parameters. In some approaches, the control law is modified to include nonlinear terms, and this approach deviates somewhat from the CE approach. The principal philosophy, however, that as the estimated parameters converge to the unknown constant parameters the control law converges to that used in the known parameter case, remains the same.

8.4 Non-identifier-Based Adaptive Schemes

Another class of schemes that fit the generic structure given in Fig. 8.1 but do not involve online parameter estimators is referred to as non-identifier-based adaptive control schemes. In this class of schemes, the online parameter estimator is replaced with search methods for finding the controller parameters in the space of possible parameters, or it involves switching between different fixed controllers, assuming that at least one is stabilizing or uses multiple fixed models for the plant covering all possible parametric uncertainties or consists of a combination of these methods. We briefly describe the main features, advantages, and limitations of these non-identifier-based adaptive control schemes in the following subsections. Since some of these approaches are relatively recent and research is still going on, we will not discuss them further in the rest of the chapter.

8.4.1 Gain Scheduling

Let us consider the aircraft model (8.1), where for each operating point $i = 1, 2, \dots, N$ the parameters A_i, B_i, C_i, D_i are known. For each operating point i, a feedback controller with constant gains, say K_i, can be designed to meet the perfor-

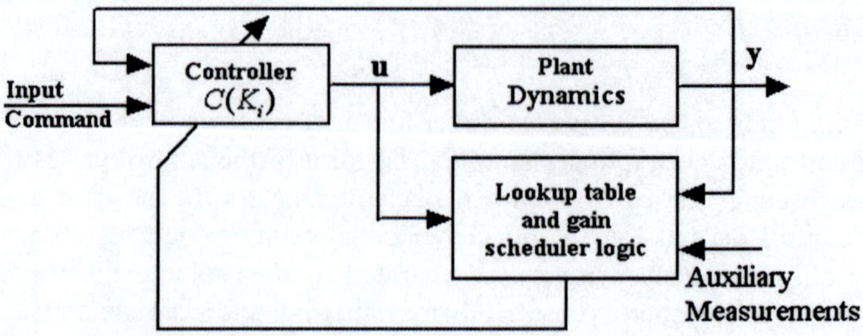

Fig. 8.4 Gain scheduling structure

mance requirements for the corresponding linear model. This leads to a controller, say $C(K_i)$, with a set of gains $K_1, K_2, \dots, K_N$ covering N operating points. Once the operating point, say i, is detected the controller gains can be changed to the appropriate value of K_i obtained from the precomputed gain set. Transitions between different operating points that lead to significant parameter changes may be handled by interpolation or by increasing the number of operating points. The two elements that are essential in implementing this approach are a lookup table to store the values of K_i and the plant measurements that correlate well with the changes in the operating points. The approach is called gain scheduling and is illustrated in Fig. 8.4.

The gain scheduler consists of a lookup table and the appropriate logic for detecting the operating point and choosing the corresponding value of K_i from the lookup table. With this approach, plant parameter variations can be compensated by changing the controller gains as functions of the input, output, and auxiliary measurements. The advantage of gain scheduling is that the controller gains can be changed as quickly as the auxiliary measurements respond to parameter changes. Frequent and rapid changes of the controller gains, however, may lead to instability [2]; therefore, there is a limit to how often and how fast the controller gains can be changed. One of the disadvantages of gain scheduling is that the adjustment mechanism of the controller gains is precomputed offline and, therefore, provides no feedback to compensate for incorrect schedules. A careful design of the controllers at each operating point to meet certain robustness and performance measures can accommodate some uncertainties in the values of the plant parameters A_i, B_i, C_i, D_i. Large unpredictable changes in the plant parameters, however, due to failures or other effects may lead to deterioration of performance or even to complete failure. Despite its limitations, gain scheduling is a popular method for handling parameter variations in flight control [9, 12] and other systems [17, 20, 23]. While gain scheduling falls into the generic definition of adaptive control, we do not classify it as adaptive control in this chapter due to the lack of online parameter estimation which could track unpredictable changes in the plant parameters.

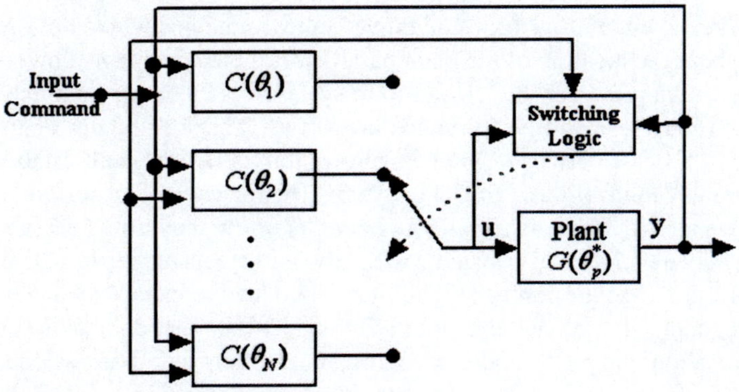

Fig. 8.5 Multiple models adaptive control with switching

8.4.2 Multiple Models and Search Methods

A class of non-identifier-based adaptive control schemes emerged over the years which do not explicitly rely on online parameter estimation [1, 3–5, 7, 10, 11, 13–16, 18]. These schemes are based on search methods in the controller parameter space [20] until the stabilizing controller is found or the search method is restricted to a finite set of controllers, one of which is assumed to be stabilizing [4, 7]. In some approaches, after a satisfactory controller is found it can be tuned locally using online parameter estimation for better performance [20, 22, 23]. Since the plant parameters are unknown, the parameter space is parameterized with respect to a set of plant models which is used to design a finite set of controllers so that each plant model from the set can be stabilized by at least one controller from the controller set. A switching approach is then developed so that the stabilizing controller is selected online based on the I/O data measurements. Without going into specific details, the general structure of this multiple model adaptive control with switching, as it is often called, is shown in Fig. 8.5.

In Fig. 8.5, N controllers are used to control a plant whose parameters θ_p^* are unknown or could change with time. In some approaches an a priori knowledge of where the elements of θ_p^* are located, such as lower and upper bounds, is used to parameterize the plant and generate a finite set of controllers so that for each possible plant there exists at least one stabilizing controller from the set of the N controllers. This by itself could be a difficult task in some practical situations where the plant parameters are unknown or change in an unpredictable manner. Furthermore, since there is an infinite number of plants within any given bound of parametric uncertainty, finding controllers to cover all possible parametric uncertainties may also be challenging. In other approaches [4, 7], it is assumed that the set of controllers with the property that at least one of them is stabilizing is available. Once the set of controllers with the stabilizing property is available the problem of finding the stabilizing one using I/O data has to be resolved. This is achieved by the use of a switching logic that differs in detail from one approach to another. While these methods provide another set of tools for dealing with plants with unknown parameters, they cannot replace the identifier-based adaptive control schemes where no assumptions are made about the location of the plant parameters. One advantage, however, is that once the switching is over, the closed-loop system is LTI, and it is much easier to analyze its robustness and performance properties. This LTI nature of the closed-loop system, at least between switches, allows the use of the well-established and powerful robust control tools for LTI systems [6] for controller design. These approaches are still at their infancy and it is not clear how they affect performance, as switching may generate bad transients with adverse effects on performance. Switching may also increase the controller bandwidth and lead to instability in the presence of high-frequency unmodeled dynamics. Guided by data that do not carry sufficient information about the plant model, the wrong controllers could be switched on over periods of time, leading to internal excitation and bad transients before the switching process settles to the right controller. Some of these issues may also exist in classes of identifier-based adaptive control, as such phenomena are independent of the approach used.

8.5 A Class of Parametric Models

Let us consider the first-order system

$$\dot{x} = -x + ax + bu,$$

where x, u are the scalar state and input, respectively, and a, b are the unknown constants we want to identify online using the measurements of x, u.

8.5.1 Static Parametric Model

The first step in the design of online parameter identification (PI) algorithms is to lump the unknown parameters in a vector and separate them from known signals, transfer functions, and other known parameters in an equation that is convenient for parameter estimation. For the above example, one such suitable parametric representation is obtained by expressing the above system as

$$x = \frac{1}{s+1}(ax + bu) = a\frac{1}{s+1}x + b\frac{1}{s+1}u,$$

and in the compact algebraic form

$$x = \theta^{*t}\phi,$$

where

$$\theta^* = [a, b]^t, \qquad \phi = \left[\frac{1}{s+1}x, \frac{1}{s+1}u\right]^t.$$

In the general case, this class of parameterizations is of the form

$$z = \theta^{*t}\phi, \tag{8.2}$$

where $\theta^* \in \Re^n$ is the vector with all the unknown parameters and $z \in R, \phi \in \Re^n$ are signals available for measurement. We refer to (8.2) as the linear "static" parametric model (SPM). The SPM may represent a dynamic, static, linear, or nonlinear system. Any linear or nonlinear dynamics in the original system are hidden in the signals z, ϕ that usually consist of the I/O measurements of the system and their filtered values.

8.5.2 Dynamic Parametric Model

Another parameterization of the above scalar plant is

$$x = \frac{1}{s+1}[a, b]\begin{bmatrix} x \\ u \end{bmatrix} = \frac{1}{s+1}\theta^{*t}\phi,$$
$$\theta^* = [a, b]^t, \qquad \phi = [x, u]^t.$$

In the general case, the above parametric model is of the form

$$z = W(q)\big(\theta^{*t}\big), \tag{8.3}$$

where $z \in R$, $\phi \in \Re^n$ are signals available for measurement and $W(q)$ is a known stable proper transfer function, where q is either the shift operator in discrete time (i.e., $q = z$) or the differential operator ($q = s$) in continuous time. We refer to (8.3) as the linear "dynamic" parametric model (DPM). The importance of the SPM and DPM as compared to other possible parameterizations is that the unknown parameter vector θ^* appears linearly. For this reason, we refer to (8.2) and (8.3) as linear in the parameters parameterizations.

As we will show later, this property is significant in designing online PI algorithms whose global convergence properties can be established analytically.

We can derive (8.2) from (8.3) if we use the fact that θ^* is a constant vector and redefine ϕ to obtain

$$z = \theta^{*t}\varphi, \qquad \varphi = W(q)\phi.$$

In a similar manner, we can filter each side of (8.2) or (8.3) using a stable proper filter and still maintain the linear in the parameters property and the form of SPM, DPM. This shows that there exist an infinite number of different parametric models in the form of SPM, DPM for the same parameter vector θ^*.

8.5.3 Bilinear Parametric Models

In some cases, the unknown parameters cannot be expressed in the form of the linear in the parameters models. In such cases, the PI algorithms based on such models cannot be shown to converge globally. A special case of nonlinear in the parameters models for which convergence results exist is when the unknown parameters appear in the special bilinear form

$$z = \rho^*\big(\theta^*\phi + z_1\big) \tag{8.4}$$

or

$$z = W(q)\rho^*(\theta^*\phi + z_1), \tag{8.5}$$

where $z \in R$, $\phi \in \mathfrak{R}^n$, $z_1 \in R$ are signals available for measurement at each time t, and $\rho^* \in \mathfrak{R}^n$, $\theta^* \in \mathfrak{R}^n$ are the unknown parameters. The transfer function $W(q)$ is a known stable transfer function. We refer to (8.4) and (8.5) as the bilinear static parametric model (B-SPM) and bilinear dynamic parametric model (B-DPM), respectively.

In some applications of parameter identification or adaptive control of plants of the form

$$\dot{x} = Ax + Bu,$$

whose state x is available for measurement, the following parametric model may be used:

$$\dot{x} = A_m x + (A - A_m)x + Bu,$$

where A_m is a stable design matrix; A, B are the unknown matrices; and x, u are signal vectors available for measurement. The model may be also expressed in the form

$$\dot{x} = A_m x + \Theta^{*t}\Phi,$$

where $\Theta^{*t} = [A - A_m, B]$, $\Phi = [x^t, u^t]^t$. We refer to this class of parametric models as state-space parametric models (SSPM). It is clear that the SSPM can be expressed in the form of the DPM and SPM. Another class of state-space models that appear in adaptive control is of the form

$$\dot{x} = A_m x + B\Theta^{*t}\Phi,$$

where B is also unknown but is positive definite, is negative definite, or the sign of each of its elements is known. We refer to this class of parametric models as bilinear state-space parametric models (B-SSPM). The B-SSPM model can be easily expressed as a set of scalar B-SPM or B-DPM.

The *PI* problem can now be stated as follows:

- For the SPM and DPM: Given the measurements $z(t)$, $\phi(t)$, generate $\theta(t)$, the estimate of the unknown vector θ^*, at each time t. The *PI* algorithm updates $\theta(t)$ with time so that as time evolves, $\theta(t)$ approaches or converges to θ^*. Since we are dealing with online *PI*, we would also expect that if θ^* changes, then the PI algorithm will react to such changes and update the estimate $\theta(t)$ to match the new value of θ^*.
- For the B-SPM and B-DPM: Given the measurements of z, z_1, and ϕ, generate the estimates $\theta(t)$, $\rho(t)$ of θ^*, ρ^*, respectively, at each time t the same way as in the case of SPM and DPM.
- For the SSPM: Given the measurements of x, u, that is, Φ, generate the estimate Θ of θ^* (and hence the estimates $\hat{A}(t)$, $\hat{B}(t)$ of A, B, respectively) at each time t the same way as in the case of SPM and DPM.

The online *PI* algorithms generate estimates at each time t, by using the past and current measurements of signals. Convergence is achieved asymptotically as time evolves. For this reason they are referred to as recursive *PI* algorithms to be distinguished from the nonrecursive ones, in which all the measurements are collected a priori over large intervals of time and are processed offline to generate the estimates of the unknown parameters.

Generating the parametric models (8.2)–(8.5) is a significant step in the design of the appropriate *PI* algorithms. Below, we present several examples that demonstrate how to express the unknown parameters in the form of the parametric models presented above.

8.5.4 Example 8.4

Consider the mass-spring-dashpot system shown in Fig. 8.6, where k is the spring constant, f is the viscous-friction or damping coefficient, M is the mass of the system, u is the forcing input, and x is the displacement of the mass M.

If we assume that the spring is "linear," that is, the force acting on the spring is proportional to the displacement, and the friction force is proportional to the velocity $\dot{x}$, using Newton's law we obtain the differential equation that describes the dynamics of the system as

$$M\ddot{x} = u - kx - f\dot{x}. \tag{8.6}$$

Let us assume that M, f, k are the constant unknown parameters that we want to estimate online. We can easily express (8.6) in the form of SPM by defining

$$\theta^* = [M, f, k]^t, \qquad z = u, \qquad \phi = [\ddot{x}, \dot{x}, x]^t.$$

However, in this formulation we are making the assumption that the vector $\phi = [\ddot{x}, \dot{x}, x]^t$ is available for measurement, which is true, provided that x and its first two derivatives are available for measurement. If not, the parametric model associated with $\phi = [\ddot{x}, \dot{x}, x]^t$ cannot be used for developing *PI* algorithms because ϕ is not available for measurement. Let us assume that only x, the displacement of the mass, is available for measurement. In this case, in order to express (8.6) in the form of the SPM, we filter both sides of (8.6) with the stable filter $\frac{1}{\Lambda(s)}$ where $\Lambda(s) = (s + \lambda)^2$ and $\lambda > 0$ is a constant design parameter we can choose arbitrarily, to obtain

$$\frac{Ms^2 + fs + k}{\Lambda(s)} = \frac{1}{\Lambda(s)}u. \tag{8.7}$$

Using (8.7), we can express the unknown parameters in the form of (8.2) as follows:

$$z = \theta^{*t}\phi,$$

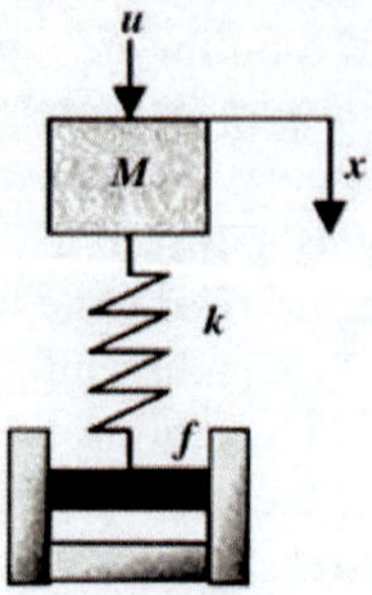

Fig. 8.6
Mass-spring-dashpot system

where

$$z = \frac{1}{\Lambda(s)} u,$$

$$\phi = \left[\frac{s^2}{\Lambda(s)} x, \frac{s}{\Lambda(s)} x, \frac{1}{\Lambda(s)} x \right]^t,$$

$$\theta^* = [M, f, k]^t.$$

In this case z, ϕ are available for measurement since they can be generated by filtering the measurements u and x, respectively. Another possible parametric model is

$$z = \theta^{*t} \phi,$$

where

$$z = \frac{s^2}{\Lambda(s)} x,$$

$$\phi = \left[\frac{1}{\Lambda(s)} u, \frac{s}{\Lambda(s)} x, -\frac{1}{\Lambda(s)} x \right]^t,$$

$$\theta^* = \left[\frac{1}{M}, \frac{f}{M}, \frac{k}{M} \right]^t.$$

In this model, the unknown parameters are rearranged to a different vector θ^*.

8.5.5 Example 8.5

Consider the cart with two inverted pendulums shown in Fig. 8.7, where M is the mass of the cart, m_1 and m_2 are the masses of the bobs, and l_1 and l_2 are the lengths of the pendulums, respectively. Using Newton's law and assuming small angular deviations of $|\alpha_1|$, $|\alpha_2|$, the equations of motion are given by

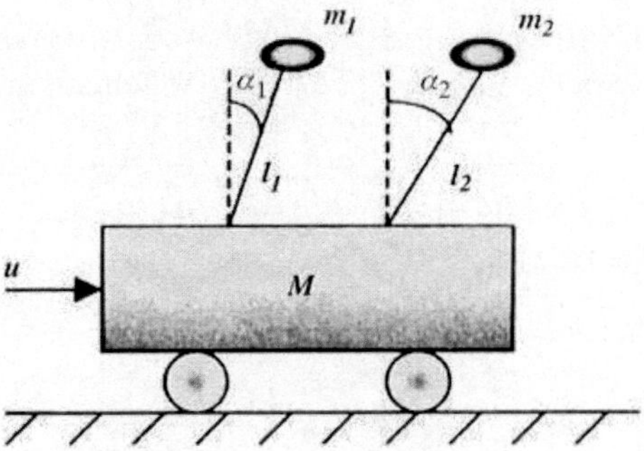

Fig. 8.7 Cart with two inverted pendulums

$$M\dot{v} = -m_1 g\alpha_1 - m_2 g\alpha_2 + u,$$
$$m_1(\dot{v} + l_1\ddot{\alpha}_1) = m_1 g\alpha_1,$$
$$m_2(\dot{v} + l_2\ddot{\alpha}_2) = m_2 g\alpha_2,$$

where v is the velocity of the cart, u is an external force, and g is the acceleration due to gravity. Letting α_1 be the output, i.e., $y = \alpha_1$, the system can be described by the differential equation

$$y^{(4)} + a_2 y^{(2)} + a_0 y = b_2 u^{(2)} + b_0 u,$$

where

$$a_2 = -\frac{g}{M}\left(\frac{M+m_1}{l_1} + \frac{M+m_2}{l_2}\right), \qquad a_0 = \frac{(M+m_1+m_2)g^2}{Ml_1 l_2},$$
$$b_2 = \frac{1}{Ml_1}, \qquad b_0 = \frac{g}{Ml_1 l_2}.$$

The above equation can be rewritten as

$$y^{(4)} = \theta^{*t} Y_0,$$

where

$$Y_0 = \left[u^{(2)}, u, -y^{(2)}, -t\right]^t,$$
$$\theta^* = [b_2, b_0, a_2, a_0]^t.$$

In order to avoid the use of differentiators, we filter each side with fourth-order stable filter $\frac{1}{\Lambda(s)}$, e.g., $\Lambda(a) = (s+\lambda)^4$, $\lambda > 0$, to obtain the SPM model

$$z = \theta^{*t}\phi,$$

where

$$z = \frac{s^4}{(s+\lambda)^4} y,$$
$$\phi = \left[\frac{s^2}{(s+\lambda)^4}u, \frac{1}{(s+\lambda)^4}u, -\frac{s^2}{(s+\lambda)^4}y, -\frac{1}{(s+\lambda)^4}y\right]^t,$$
$$\theta^* = [b_2, b_0, a_2, a_0]^t.$$

If in the above model we know that a_0 is nonzero, redefining the constant parameters as $\bar{b}_2 = \frac{b_2}{a_0}$, $\bar{b}_0 = \frac{b_0}{a_0}$, $\bar{a}_1 = \frac{a_1}{a_0}$, we obtain the following B-SPM:

$$z = \rho^*(\theta^{*t}\phi + z_1),$$

where

$$z = \frac{s^4}{(s+\lambda)^4}y, \qquad z_1 = \frac{1}{(s+\lambda)^4}y,$$
$$\phi = \left[\frac{s^2}{(s+\lambda)^4}u, \frac{1}{(s+\lambda)^4}u, -\frac{s^2}{(s+\lambda)^4}y\right]^t,$$
$$\theta^* = [\bar{b}_2, \bar{b}_0, \bar{a}_2]^t, \qquad \rho^* = a_0.$$

8.5.6 Example 8.6

Consider the second-order plant

$$\dot{x} = Ax + Bu,$$

where $x = [x_1, x_2]^t$, $u = [u_1, u_2]^t$, and

$$A = \begin{bmatrix} a_{11} & a_{12} \\ a_{21} & a_{22} \end{bmatrix}, \qquad B = \begin{bmatrix} b_{11} & b_{12} \\ b_{21} & b_{22} \end{bmatrix}$$

are matrices with unknown elements. The SSPM is generated as

$$\dot{x} = \begin{bmatrix} -a_m & 0 \\ 0 & -a_m \end{bmatrix} x + \begin{bmatrix} a_{11}+a_m & a_{12} \\ a_{21} & a_{22}+a_m \end{bmatrix} x + \begin{bmatrix} b_{11} & b_{12} \\ b_{21} & b_{22} \end{bmatrix} u,$$

where $a_m > 0$ is a design constant. The model may be also expressed as

$$\dot{x} = \begin{bmatrix} -a_m & 0 \\ 0 & -a_m \end{bmatrix} x + \Theta^{*t}\Phi,$$

where

$$\Theta^{*t} = \begin{bmatrix} a_{11}+a_m, & a_{12}, & b_{11}, & b_{12} \\ a_{21}, & a_{22}+a_m, & b_{21}, & b_{22} \end{bmatrix}, \qquad \Phi = [x_1, x_2, u_1, u_2]^t.$$

8.6 Parameter Identification

The purpose of this section is to present the design, analysis, and simulation of a wide class of algorithms that can be used for online parameter identification of continuous-time plants. The online identification procedure involves the following three steps.

1. Lump the unknown parameters in a vector $\theta*$ and express them in the form of the parametric model SPM, DPM, B-SPM, or B-DPM.

2. Use the estimate θ of θ^* to set up the estimation model that has the same form as the parametric model. The difference between the outputs of the estimation and parametric models, referred to as the estimation error, reflects the distance of the estimated parameters $\theta(t)$ from the unknown parameters θ^* weighted by some signal vector. The estimation error is used to drive the adaptive law that generates $\theta(t)$ online. The adaptive law is a differential equation of the form

$$\dot{\theta} = H(t)\varepsilon,$$

where ε is the estimation error that reflects the difference between $\theta(t)$ and θ^* and $H(t)$ is a time-varying gain vector that depends on measured signals. A wide class of adaptive laws with different $H(t)$ and ε may be developed using optimization techniques and Lyapunov-type stability arguments.
3. Establish conditions that guarantee that $\theta(t)$ converges to θ^* with time. This step involves the design of the plant input so that the signal vector $\phi(t)$ in the parametric model is *persistently exciting* (a notion to be defined later on), i.e., it has certain properties that guarantee that the measured signals that drive the adaptive law carry sufficient information about the unknown parameters. For example, for $\phi(t) = 0$, we have $z = \theta^{*t}\phi = 0$, and the measured signals ϕ, z carry no information about θ^*. Similar arguments could be made for ϕ that is orthogonal to θ^* leading to $z = 0$ even though $\theta \neq \theta^*$, etc.

   We demonstrate the three design steps using the following example of a scalar plant.

8.6.1 One-Parameter Case

Consider the first-order plant model

$$y = \frac{a}{s+2}u, \tag{8.8}$$

where a is the only unknown parameter and y and u are the measured output and input of the system, respectively.

1. *Parametric Model*: We write (8.8) as

$$y = a\frac{1}{s+2}u = au_f, \tag{8.9}$$

   where $u_f = \frac{1}{s+2}u$. Since u is available for measurement, u_f is also available for measurement. Therefore, (8.9) is in the form of the SPM

$$z = \theta^*\phi, \tag{8.10}$$

   where $\theta^* = a$ and $z = y$, $\phi = u_f$ are available for measurement.
2. *Parameter Identification Algorithm*: This step involves the development of an estimation model and an estimation error used to drive the adaptive law that generates the parameter estimates.

Estimation Model and Estimation Error: The estimation model has the same form as the SPM with the exception that the unknown parameter θ^* is replaced with its estimate at time t, denoted by $\theta(t)$, i.e.,

$$\hat{z} = \theta(t)\phi, \tag{8.11}$$

where $\hat{z}$ is the estimate of z based on the parameter estimate $\theta(t)$ at time t. It is obvious that the difference between z and $\hat{z}$ is due to the difference between $\theta(t)$ and θ^*. As $\theta(t)$ approaches θ^* with time we would expect that $\hat{z}$ would approach z at the same time. Note that the reverse is not true, i.e., $\hat{z}(t) = z(t)$ does not imply that $\theta(t) = \theta^*$. Since θ^* is unknown, the difference $\tilde{\theta} = \theta(t) - \theta^*$ is not available for measurement. Therefore, the only signal that we can generate, using available measurements, that reflects the difference between $\theta(t)$ and θ^* is the error signal

$$\varepsilon = \frac{z - \hat{z}}{m_s^2}, \tag{8.12}$$

which we refer to as the estimation error. $m_s^2 \geq 1$ is a normalization signal[1] designed to guarantee that $\frac{\phi}{m_s}$ is bounded. This property of m_s is used to establish the boundedness of m_s the estimated parameters even when ϕ is not guaranteed to be bounded. A straightforward choice for m_s in this example is $m_s^2 = 1 + \alpha\phi^2$, $\alpha > 0$. If ϕ is bounded, we can take $\alpha = 0$, that is, $m_s^2 = 1$. Using (8.11) in (8.12), we can express the estimation error as a function of the parameter error $\tilde{\theta} = \theta(t) - \theta^*$, i.e.,

$$\varepsilon = \frac{\tilde{\theta}\phi}{m_s^2}. \tag{8.13}$$

Equation (8.13) shows the relationship between the estimation error ε and the parameter error $\tilde{\theta}$. It should be noted that ε cannot be generated using (8.13) because the parameter error $\tilde{\theta}$ is not available for measurement. Consequently, (8.13) can be used only for analysis.

Adaptive Law: A wide class of adaptive laws or parameter estimators for generating $\theta(t)$, the estimate of θ^*, can be developed using (8.11)–(8.13). The simplest one is obtained by using the SPM (8.10) and the fact that ϕ is scalar to write

$$\theta(t) = \frac{z(t)}{\phi(t)}, \tag{8.14}$$

provided $\phi(t) \neq 0$. In practice, however, the effect of noise on the measurements of $\phi(t)$, especially when $\phi(t)$ is close to zero, may lead to erroneous parameter estimates. Another approach is to update $\theta(t)$ in a direction that minimizes a certain cost of the estimation error ε. With this approach, $\theta(t)$ is adjusted in a

---

[1] Note that any $m_s^2 \geq$ nonzero constant is adequate. The use of a lower bound 1 is without loss of generality.

direction that makes $|\varepsilon|$ smaller and smaller until a minimum is reached at which $|\varepsilon| = 0$ and updating is terminated.

Now, consider the cost criterion

$$J(\theta) = \frac{\varepsilon^2 m_s^2}{2} = \frac{(z - \theta\phi)^2}{2m_s^2}, \tag{8.15}$$

which we minimize with respect to θ using the gradient method to obtain

$$\dot{\theta} = -\gamma \frac{dJ(\theta)}{d\theta}, \tag{8.16}$$

where $\gamma > 0$ is a scaling constant or step size which we refer to as the adaptive gain and where $\frac{dJ(\theta)}{d\theta}$ is the gradient of J with respect to θ. In this scalar case,

$$\frac{dJ(\theta)}{d\theta} = \frac{dJ}{d\theta} = -\frac{(z - \theta\phi)}{m_s^2}\phi = -\varepsilon\phi,$$

which leads to the adaptive law

$$\dot{\theta} = \gamma\varepsilon\phi, \qquad \theta(0) = \theta_0. \tag{8.17}$$

3. *Stability and Parameter Convergence*: The adaptive law should guarantee that the parameter estimate $\theta(t)$ and the speed of adaptation $\dot{\theta}$ are bounded and that the estimation error ε gets smaller and smaller with time. These conditions still do not imply that $\theta(t)$ will get closer and closer to θ^* with time unless some conditions are imposed on the vector $\phi(t)$, referred to as the regressor vector.

   Let us start by using (8.13) and the fact that $\dot{\tilde{\theta}} = \dot{\theta} - \dot{\theta}^* = \dot{\theta}$ (due to θ^* being constant) to express (8.17) as

$$\dot{\tilde{\theta}} = -\gamma \frac{\phi^2}{m_s^2}\tilde{\theta}, \qquad \tilde{\theta}(0) = \tilde{\theta}_0. \tag{8.18}$$

   This is a scalar linear time-varying differential equation whose solution is

$$\tilde{\theta}(t) = e^{-\gamma \int_0^t \frac{\phi^2(\tau)}{m_s^2(\tau)} d\tau}\tilde{\theta}_0, \tag{8.19}$$

   which implies that for

$$\int_0^t \frac{\phi^2(\tau)}{m_s^2(\tau)} d\tau \geq \alpha_0 t \tag{8.20}$$

   and some $\alpha_0 > 0$, $\tilde{\theta}(t)$ converges to zero exponentially fast, which in turn implies that $\theta(t) \to \theta^*$ exponentially fast. It follows from (8.19) that $\theta(t)$ is always bounded for any $\phi(t)$ and from (8.18) that $\dot{\theta}(t) = \dot{\tilde{\theta}}(t)$ is bounded due to $\frac{\phi(t)}{m_s(t)}$ being bounded.

   Another way to analyze (8.18) is to use a Lyapunov-like approach as follows: We consider the function

$$V = \frac{\tilde{\theta}^2}{2\gamma}.$$

Then

$$\dot{V} = \frac{\tilde{\theta}}{\gamma}\frac{d\tilde{\theta}}{dt} = -\frac{\phi^2}{m_s^2}\tilde{\theta}^2 \leq 0$$

or, using (8.13),

$$\dot{V} = -\frac{\phi^2}{m_s^2}\tilde{\theta}^2 = -\varepsilon^2 m_s^2 \leq 0. \tag{8.21}$$

We should note that $\dot{V} = -\varepsilon^2 m_s^2 \leq 0$ implies that $\dot{V}$ is a negative semidefinite function in the space of $\tilde{\theta}$. $\dot{V}$ in this case is not negative definite in the space of $\tilde{\theta}$ because it can be equal to zero when $\tilde{\theta}$ is not zero. Consequently, if we apply the stability results of the Appendix, we can conclude that the equilibrium $\tilde{\theta}_e = 0$ of (8.18) is uniformly stable (u.s.) and that the solution of (8.18) is uniformly bounded (u.b.). These results are not as useful, as our objective is asymptotic stability, which implies that the parameter error converges to zero. We can use the properties of $V, \dot{V}$, however, to obtain additional properties for the solution of (8.18) as follows.

Since $V > 0$ and $\dot{V} \leq 0$, it follows that (see the Appendix) V is bounded, which implies that $\tilde{\theta}$ is bounded and V converges to a constant, i.e., $\lim_{t\to\infty} V(t) = V_\infty$. Let us now integrate both sides of (8.21). We have

$$\int_0^t \dot{V}(\tau)\,d\tau = -\int_0^t \varepsilon^2(\tau) m_s^2(\tau)\,d\tau$$

or

$$V(t) - V(0) = -\int_0^t \varepsilon^2(\tau) m_s^2(\tau)\,d\tau. \tag{8.22}$$

Since $V(t)$ converges to the limit V_∞ as $t \to \infty$, it follows from (8.22) that

$$\int_0^\infty \varepsilon^2(\tau) m_s^2(\tau)\,d\tau = V(0) - V_\infty < \infty,$$

i.e., εm_s is square integrable or $\varepsilon m_s \in L_2$. Since $m_s^2 \geq 1$, we have $\varepsilon^2 \leq \varepsilon^2 m_s^2$, which implies $\varepsilon \in L_2$. From (8.13), we conclude that $\frac{\tilde{\theta}\phi}{m_s} \in L_2$ due to $\varepsilon m_s \in L_2$. Using (8.17), we write

$$\dot{\theta} = \gamma \varepsilon m_s \frac{\phi}{m_s}.$$

Since $\frac{\phi}{m_s}$ is bounded and $\varepsilon m_s \in L_2 \cap L_\infty$, it follows (see Problem 2) that $\dot{\theta} \in L_2 \cap L_\infty$. In summary, we have established that the adaptive law (8.17) guarantees that (i) $\theta \in L_\infty$ and (ii) $\varepsilon, \varepsilon m_s, \dot{\theta} \in L_2 \cap L_\infty$ independent of the boundedness of ϕ. The L_2 property of ε, εm_s, and $\dot{\theta}$ indicates that the estimation error and the speed of adaptation $\dot{\theta}$ are bounded in the L_2 sense, which in turn implies that their average value tends to zero with time.

It is desirable to also establish that ε, εm_s, and $\dot{\theta}$ go to zero as $t \to \infty$, as such a property will indicate the end of adaptation and the completion of learning. Such a property can be easily established when the input u is bounded (see Problem 3).

The above properties still do not guarantee that $\theta(t) \to \theta^*$ as $t \to \infty$. In order to establish that $\theta(t) \to \theta^*$ as $t \to \infty$ exponentially fast, we need to restrict $\frac{\phi}{m_s}$ to be persistently exciting (PE), i.e., to satisfy

$$\frac{1}{T}\int_t^{t+T} \frac{\phi^2(\tau)}{m_s^2}\, d\tau \geq \alpha_0 > 0 \tag{8.23}$$

$\forall t \geq 0$ and some constants T, $\alpha_0 > 0$. The PE property of $\frac{\phi}{m_s}$ is guaranteed by choosing the input u appropriately. Appropriate choices of u for this particular example include (i) $u = c > 0$, (ii) $u = \sin\omega t$ for any $\omega \neq 0$ and any bounded input u that is not vanishing with time. The condition (8.23) is necessary and sufficient for exponential convergence of $\theta(t) \to \theta^*$.

4. The *PI* algorithm for estimating the constant a in the plant (8.8) can now be summarized as

$$\dot{\theta} = \gamma\varepsilon\phi, \qquad \theta(0) = \theta_0,$$
$$\varepsilon = \frac{(z - \hat{z})}{m_s^2}, \qquad \hat{z} = \theta\phi,$$
$$z = y, \qquad \phi = \frac{1}{s+2}u, \qquad m_s^2 = 1 + \phi^2,$$

where $\theta(t)$ is the estimate of the constant a in (8.8).

The above analysis for the scalar example carries over to the vector case without any significant modifications, as demonstrated in the next section. One important difference, however, is that in the case of a single parameter, convergence of the Lyapunov-like function V to a constant implies that the estimated parameter converges to a constant. Such a result cannot be established in the case of more than one parameter for the gradient algorithm.

8.6.2 Two-Parameters Case

Consider the plant model

$$y = \frac{b}{s+a}u, \tag{8.24}$$

where a, b are unknown constants. Let us assume that y, $\dot{y}$, u are available for measurement. We would like to generate online estimates for the parameters a, b.

1. *Parametric Model*: Since y, $\dot{y}$ are available for measurement, we can express (8.24) in the SPM form

$$z = \theta^{*t}\phi,$$

where $z = \dot{y}$, $\theta^* = [b, a]^t$, $\phi = [u, -y]^t$, and z, ϕ are available for measurement.

2. *Parameter Identification Algorithm*: It consists of:
   Estimation Model:

$$\hat{z} = \theta^t \phi,$$

where $\theta(t)$ is the estimate of θ^* at time t.
   Estimation Error:

$$\varepsilon = \frac{z - \hat{z}}{m_s^2} = \frac{z - \theta^t \phi}{m_s^2}, \tag{8.25}$$

where m_s is the normalizing signal such that $\frac{\phi}{m_s} \in L_\infty$. A straightforward choice for m_s is $m_s^2 = 1 + \alpha\phi^t\phi$ for any $\alpha > 0$.
   Adaptive Law: We use the gradient method to minimize the cost,

$$J(\theta) = \frac{\varepsilon^2 m_s^2}{2} = \frac{(z - \theta^t \phi)^2}{2m_s^2} = \frac{(z - \theta_1\phi_1 - \theta_2\phi_2)^2}{2m_s^2},$$

where $\phi_1 = u$, $\phi_2 = -y$, and set

$$\dot{\theta} = -\Gamma \frac{dJ(\theta)}{d\theta},$$

where

$$\frac{dJ(\theta)}{d\theta} = \left[\frac{\partial J}{\partial \theta_1}, \frac{\partial J}{\partial \theta_2}\right]^t,$$

$\Gamma = \Gamma^t > 0$ is the adaptive gain, and θ_1, θ_2 are the elements of $\theta = [\theta_1, \theta_2]^t$. Since

$$\frac{\partial J}{\partial \theta_1} = -\frac{(z - \theta^t \phi)}{m_s^2}\phi_1 = -\varepsilon\phi_1, \qquad \frac{\partial J}{\partial \theta_2} = -\frac{(z - \theta^t \phi)}{m_s^2}\phi_2 = -\varepsilon\phi_2,$$

we have

$$\dot{\theta} = \Gamma\varepsilon\phi, \qquad \theta(0) = \theta_0, \tag{8.26}$$

which is the adaptive law for updating $\theta(t)$ starting from some initial condition $\theta(0) = \theta_0$.
3. *Stability and Parameter Convergence*: As in the previous example, the equation for the parameter error $\tilde{\theta} = \theta - \theta^*$ is obtained from (8.25), (8.26) by noting that

$$\varepsilon = \frac{z - \theta^t \phi}{m_s^2} = \frac{\theta^{*t}\phi - \theta^t \phi}{m_s^2} = -\frac{\tilde{\theta}^t \phi}{m_s^2} = -\frac{\phi^t \tilde{\theta}}{m_s^2} \tag{8.27}$$

and $\dot{\tilde{\theta}} = \dot{\theta}$, i.e.,

$$\dot{\tilde{\theta}} = \Gamma\phi\varepsilon = -\Gamma\frac{\phi\phi^t}{m_s^2}\tilde{\theta}. \tag{8.28}$$

It is clear from (8.28) that the stability of the equilibrium $\tilde{\theta}_e = 0$ will very much depend on the properties of the time-varying matrix $-\frac{\Gamma\phi\phi^t}{m_s^2}$, which in turn depends on the properties of ϕ. For simplicity, let us assume that the plant is stable, i.e., $a > 0$. If we choose $m_s^2 = 1$, $\Gamma = \gamma I$ for some $\gamma > 0$ and a constant input $u = c_o > 0$, then at steady state $y = c_1 := \frac{c_0 b}{a} \neq 0$ and $\phi = [c_o, -c_1]^t$, giving

$$-\frac{\Gamma \phi \phi^t}{m_s^2} = -\gamma \begin{bmatrix} c_0^2 & -c_0 c_1 \\ -c_0 c_1 & c_1^2 \end{bmatrix} := A,$$

i.e.,

$$\dot{\tilde{\theta}} = A\tilde{\theta},$$

where A is a constant matrix with eigenvalues $0, -\gamma(c_0^2 + c_1^2)$, which implies that the equilibrium $\theta_e = 0$ is only marginally stable; i.e., $\tilde{\theta}$ is bounded but does not necessarily converge to 0 as $t \to \infty$. The question that arises in this case is what properties of ϕ guarantee that the equilibrium $\tilde{\theta}_e = 0$ is exponentially stable. Given that

$$\phi = H(s)u,$$

where for this example $H(s) = [1, -\frac{b}{s+a}]^t$, the next question that comes up is how to choose u to guarantee that ϕ has the appropriate properties that imply exponential stability for the equilibrium $\tilde{\theta}_e = 0$ of (8.28). Exponential stability for the equilibrium point $\tilde{\theta}_e = 0$ of (8.28) in turn implies that $\theta(t)$ converges to θ^* exponentially fast. As demonstrated above for the two-parameter example, a constant input $u = c_o > 0$ does not guarantee exponential stability. We answer the above questions in the following section.

8.7 Gradient Algorithms

The gradient algorithm is developed by using the gradient method to minimize some appropriate functional $J(\theta)$. Different choices for $J(\theta)$ lead to different algorithms. As in the scalar case, we start by defining the estimation model and estimation error for the SPM.

The estimate $\hat{z}$ of z is generated by the estimation model

$$\hat{z} = \theta^t \phi, \tag{8.29}$$

where $\theta(t)$ is the estimate of θ^* at time t. The estimation error is constructed as

$$\varepsilon = \frac{z - \hat{z}}{m_s^2} = \frac{z - \theta^t \phi}{m_s^2}, \tag{8.30}$$

where $m_s^2 \geq 1$ is the normalizing signal designed to bound ϕ from above. The normalizing signal often has the form $m_s^2 = 1 + n_s^2$, where $n_s \geq 0$ is referred to as the static normalizing signal designed to guarantee that $\frac{\phi}{m_s}$ is bounded from above. Some straightforward choices m, for n_s include

$$n_s^2 = \alpha \phi^t \phi, \quad \alpha > 0,$$

or

$$n_s^2 = \phi^t P\phi, \quad P = P^t > 0,$$

where α is a scalar and P is a matrix selected by the designer.

The estimation error (8.30) and the estimation model (8.29) are common to several algorithms that are generated in the following sections.

8.7.1 Gradient Algorithm with Instantaneous Cost Function

The cost function $J(\theta)$ is chosen as

$$J(\theta) = \frac{\varepsilon^2 m_s^2}{2} = \frac{(z - \theta^t\phi)^2}{2m_s^2}, \tag{8.31}$$

where m_s is the normalizing signal given by (8.30). At each time t, $J(\theta)$ is a convex function of θ and therefore has a global minimum. The gradient algorithm takes the form

$$\dot{\theta} = -\Gamma\frac{dJ}{d\theta}, \tag{8.32}$$

where $\Gamma = \Gamma^t > 0$ is a design matrix referred to as the adaptive gain. Since $\frac{dJ}{d\theta} = -\frac{(z-\theta^t\phi)\phi}{m_s^2} = -\varepsilon\phi$, we have

$$\dot{\theta} = \Gamma\varepsilon\phi. \tag{8.33}$$

The adaptive law (8.33) together with the estimation model (8.29), the estimation error (8.30), and filtered signals z, ϕ constitute the gradient parameter identification algorithm based on the instantaneous cost function.

Theorem 8.1 *The gradient algorithm* (8.33) *guarantees the following*:

1. $\varepsilon, \varepsilon m_s, \dot{\theta} \in L_2 \cap L_\infty$ *and* $\theta \in L_\infty$.
2. *If* $\frac{\phi}{m_s}$ *is PE, i.e.,* $\int_t^{t+T_0} \frac{\phi\phi^t}{m_s^2} d\tau > \alpha_0 T_0 I$ $\forall t \geq 0$ *and for some* $T_0, \alpha_0 > 0$, *then* $\theta(t) \to \theta^*$ *exponentially fast. In addition,*
$$\big(\theta(t) - \theta^*\big)^t \Gamma^{-1}\big(\theta(t) - \theta^*\big) \leq (1 - \gamma_1)^n \big(\theta(0) - \theta^*\big)^t \Gamma^{-1}\big(\theta(0) - \theta^*\big),$$
where $0 \leq t \leq nT_0$, $n = 0, 1, 2, \ldots$, *and*
$$\gamma_1 = \frac{2\alpha_0 T_0 \lambda_{\min}(\Gamma)}{2 + \beta^4 \lambda_{\max}^2(\Gamma) T_0^2}, \quad \beta = \sup_t \left|\frac{\phi}{m_s}\right|.$$
3. *If the plant model has stable poles and no zero-pole cancellations and the input* u *is sufficiently rich of order* $n + m + 1$, *i.e., it consists of at least* $\frac{n+m+1}{2}$ *distinct frequencies, then* ϕ, $\frac{\phi}{m_s}$ *are PE. Furthermore,* $|\theta(t) - \theta^*|$, ε, εm_s, $\dot{\theta}$ *converge to zero exponentially fast.*

8.7.2 Example 8.7

Consider the dynamics of a hard-disk drive servo system [8] given by

$$y = \frac{k_p}{s^2}(u+d),$$

where y is the position error of the head relative to the center of the track, k_p is a known constant, and

$$d = A_1 \sin(\omega_1 t + \varphi_1) + A_2 \sin(\omega_2 t + \varphi_2)$$

is a disturbance that is due to higher-order harmonics that arise during rotation of the disk drive. In this case, ω_1, ω_2 are the known harmonics that have a dominant effect and A_i, φ_i, $i = 1, 2$, are the unknown amplitudes and phases. We want to estimate d in an effort to nullify its effect using the control input u.

Using $\sin(a+b) = \sin a \cos b + \cos a \sin b$, we can express d as

$$d = \theta_1^* \sin \omega_1 t + \theta_2^* \cos \omega_1 t + \theta_3^* \sin \omega_2 t + \theta_4^* \cos \omega_2 t,$$

where

$$\begin{aligned} \theta_1^* &= A_1 \cos \varphi_1, & \theta_2^* &= A_1 \sin \varphi_1, \\ \theta_3^* &= A_2 \cos \varphi_2, & \theta_4^* &= A_2 \sin \varphi_2 \end{aligned} \tag{8.34}$$

are the unknown parameters. We first obtain a parametric model for

$$\theta^* = \left[\theta_1^*, \theta_2^*, \theta_3^*, \theta_4^*\right]^t.$$

We have

$$s^2 y = k_p u + k_p \theta^{*t} \psi,$$

where

$$\psi(t) = [\sin \omega_1 t, \cos \omega_1 t, \sin \omega_2 t, \cos \omega_2 t]^t.$$

Filtering each side with $\frac{1}{\Lambda(s)}$ where $\Lambda(s) = (s+\lambda_1)(s+\lambda_2)$ and $\lambda_1, \lambda_2 > 0$ are design constants, we obtain the SPM

$$z = \theta^{*t} \phi,$$

where

$$\begin{aligned} z &= \frac{s^2}{\Lambda(s)} y - k_p \frac{1}{\Lambda(s)} u, \\ \phi &= k_p \frac{1}{\Lambda(s)} \psi(t) = k_p \frac{1}{\Lambda(s)} [\sin \omega_1 t, \cos \omega_1 t, \sin \omega_2 t, \cos \omega_2 t]^t. \end{aligned}$$

Therefore, the adaptive law

$$\begin{aligned} \dot{\theta} &= \Gamma \varepsilon \phi, \\ \varepsilon &= \frac{z - \theta^t \phi}{m_s^2}, \quad m_s^2 = 1 + n_s^2, \quad n_s^2 = \alpha \phi^t \phi, \end{aligned}$$

where $\Gamma = \Gamma^t > 0$ is a 4×4 constant matrix, may be used to generate $\theta(t)$, the online estimate of θ^*. In this case, $\phi \in L_\infty$ and therefore we can take $\alpha = 0$, i.e., $m_s^2 = 1$. For $\omega_1 \neq \omega_2$, we can establish that ϕ is PE and therefore $\theta(t) \to \theta^*$ exponentially fast. The online estimate of the amplitude and phase can be computed using (8.34) as follows:

$$\tan\hat{\varphi}_1(t) = \frac{\theta_2(t)}{\theta_1(t)}, \qquad \tan\hat{\varphi}_2(t) = \frac{\theta_4(t)}{\theta_3(t)},$$
$$\hat{A}_1(t) = \sqrt{\theta_1^2(t) + \theta_2^2(t)}, \qquad \hat{A}_2(t) = \sqrt{\theta_3^2(t) + \theta_4^2(t)},$$

provided of course that $\theta_1(t) \neq 0$, $\theta_3(t) \neq 0$. The estimated disturbance

$$\hat{d}(t) = \hat{A}_1(t)\sin\big(\omega_1(t) + \hat{\varphi}_1(t)\big) + \hat{A}_2(t)\sin\big(\omega_2(t) + \hat{\varphi}_2(t)\big)$$

can then be generated and used by the controller to cancel the effect of the actual disturbance d.

8.7.3 Gradient Algorithm with Integral Cost Function

The cost function $J(\theta)$ is chosen as

$$J(\theta) = \frac{1}{2}\int_0^t e^{-\beta(t-\tau)}\varepsilon^2(t,\tau)m_s^2(\tau)\,d\tau,$$

where $\beta > 0$ is a design constant acting as a forgetting factor and

$$\varepsilon(t,\tau) = \frac{z(\tau) - \theta^t(t)\phi(\tau)}{m_s^2(\tau)}, \qquad \varepsilon(t,t) = \varepsilon, \qquad \tau \leq t,$$

is the estimation error that depends on the estimate of θ at time t and on the values of the signals at $\tau \leq t$. The cost penalizes all past errors between $z(\tau)$ and $\hat{z}(\tau) = \theta^t(t)\phi(\tau)$, $\tau \leq t$, obtained by using the current estimate of θ at time t with past measurements of $z(\tau)$ and $\phi(\tau)$. The forgetting factor $e^{-\beta(t-\tau)}$ is used to put more weight on recent data by discounting the earlier ones. It is clear that $J(\theta)$ is a convex function of θ at each time t and therefore has a global minimum. Since $\theta(t)$ does not depend on τ, the gradient of J with respect to θ is easy to calculate despite the presence of the integral. Applying the gradient method, we have

$$\dot{\theta} = -\Gamma\frac{dJ}{d\theta},$$

where

$$\frac{dJ}{d\theta} = -\int_0^t e^{-\beta(t-\tau)}\frac{z(\tau) - \theta^t(t)\phi(\tau)}{m_s^2(\tau)}\phi(\tau)\,d\tau.$$

This can be implemented as

$$\dot{\theta} = -\Gamma\big(R(t)\theta + Q(t)\big), \qquad \theta(0) = \theta_0,$$
$$\dot{R} = -\beta R + \frac{\phi\phi^t}{m_s^2}, \qquad R(0) = 0,$$
$$\dot{Q} = -\beta Q - \frac{z\phi}{m_s^2}, \qquad Q(0) = 0,$$

where $R \in \Re^{n\times n}$, $Q \in \Re^{n\times 1}$; $\Gamma = \Gamma^t > 0$ is the adaptive gain; n is the dimension of the vector θ^*; and m_s is the normalizing signal defined in (8.30).

Theorem 8.2 *The gradient algorithm with integral cost function guarantees that*

1. $\varepsilon, \varepsilon m_s, \dot{\theta} \in L_2 \cap L_\infty$ *and* $\theta \in L_\infty$.
2. $\lim_{t\to\infty} |\dot{\theta}(t)| = 0$.
3. *If* $\frac{\phi}{m_s}$ *is PE, then* $\theta(t) \to \theta^*$ *exponentially fast. Furthermore, for* $\Gamma = \gamma I$, *the rate of convergence increases with* γ.
4. *If* u *is sufficiently rich of order* $n + m + 1$, *i.e., it consists of at least* $\frac{n+m+1}{2}$ *distinct frequencies, and the plant is stable and has no zero-pole cancellations, then* ϕ, $\frac{\phi}{m_s}$ *are PE and* $\theta(t) \to \theta^*$ *exponentially fast.*

Theorem 8.2 indicates that the rate of parameter convergence increases with increasing adaptive gain. Simulations demonstrate that the gradient algorithm based on the integral cost gives better convergence properties than the gradient algorithm based on the instantaneous cost. The gradient algorithm based on the integral cost has similarities with the least-squares (LS) algorithms to be developed in the next section.

8.8 Least-Squares Algorithms

The LS method dates back to the eighteenth century, when Gauss used it to determine the orbits of planets. The basic idea behind LS is fitting a mathematical model to a sequence of observed data by minimizing the sum of the squares of the difference between the observed and computed data. In doing so, any noise or inaccuracies in the observed data are expected to have less effect on the accuracy of the mathematical model. The method is simple to apply and analyze in the case where the unknown parameters appear in a linear form, such as in the linear SPM

$$z = \theta^{*t}\phi. \tag{8.35}$$

We illustrate the use and properties of LS by considering the simple scalar example

$$z = \theta^*\phi + d_n,$$

where $z, \theta^*, \phi \in R$, $\phi \in L_\infty$, and d_n is a noise disturbance whose average value goes to zero as $t \to \infty$, i.e.,

$$\lim_{t\to\infty} \frac{1}{t}\int_0^t d_n(\tau)\,d\tau = 0.$$

In practice, d_n may be due to sensor noise or external sources, etc. We examine the following estimation problem: Given the measurements of $z(\tau)$, $\phi(\tau)$ for $0 \le \tau < t$, find a "good" estimate $\theta(t)$ of θ^* at time t. One possible solution is to calculate $\theta(t)$ as

$$\theta(t) = \frac{z(\tau)}{\phi(\tau)} = \theta^* + \frac{d_n(\tau)}{\phi(\tau)} \tag{8.36}$$

by using the measurements of $z(\tau)$, $\phi(\tau)$ at some $\tau < t$ for which $\phi(r) \neq 0$. Because of the noise disturbance, however, such an estimate may be far off from θ^*. For example, at the particular time τ at which we measured z and ϕ, the effect of $d_n(\tau)$ may be significant, leading to an erroneous estimate for $\theta(t)$ generated by (8.36).

A more appropriate approach is to choose the estimate $\theta(t)$ at time t to be the one that minimizes the square of all the errors that result from the mismatch of $z(\tau) - \theta(t)\phi(\tau)$ for $0 \le \tau \le t$. Hence, the estimation problem above becomes the following LS problem: Minimize the cost

$$J(\theta) = \frac{1}{2}\int_0^t \left|z(\tau) - \theta(t)\phi(\tau)\right|^2 d\tau \tag{8.37}$$

w.r.t. $\theta(t)$ at any given time t. The cost $J(\theta)$ penalizes all the past errors from $\tau = 0$ to t that are due to $\theta(t) \neq \theta^*$. Since $J(\theta)$ is a convex function over R at each time t, its minimum satisfies

$$\frac{dJ}{d\theta}(\theta) = -\int_0^t z(\tau)\phi(\tau)\,d\tau + \theta(t)\int_0^t \phi^2(\tau)\,d\tau = 0,$$

which gives the LS estimate

$$\theta(t) = \left(\int_0^t \phi^2(\tau)\,d\tau\right)^{-1} \int_0^t z(\tau)\phi(\tau)\,d\tau,$$

provided of course that the inverse exists. The LS method considers all past data in an effort to provide a good estimate for θ^* in the presence of noise d_n. For example, when $\phi(t) = 1$, $\forall t \ge 0$, we have

$$\lim_{t\to\infty} \theta(t) = \lim_{t\to\infty} \frac{1}{t}\int_0^t z(\tau)\,d\tau = \theta^* + \lim_{t\to\infty} \frac{1}{t}\int_0^t d_n(\tau)\,d\tau = \theta^*;$$

i.e., $\theta(t)$ converges to the exact parameter value despite the presence of the noise disturbance d_n.

Let us now extend this problem to the linear model (8.35). As in Sect. 8.7, the estimate $\hat{z}$ of z and the normalized estimation are generated as

$$\hat{z} = \theta^t\phi, \qquad e = \frac{z - \hat{z}}{m_s^2} = \frac{z - \theta^t\phi}{m_s^2},$$

where $\theta(t)$ is the estimate of θ^* at time t, and $m_s^2 = 1 + n_s^2$ is designed to guarantee $\frac{\phi}{m_s} \in L_\infty$. Below we present different versions of the LS algorithm, which correspond to different choices of the LS cost $J(\theta)$.

8.8.1 Recursive LS Algorithm with Forgetting Factor

Consider the function

$$J(\theta) = \frac{1}{2}\int_0^t e^{-\beta(t-\tau)}\frac{[z(\tau) - \theta^t(t)\phi(\tau)]^2}{m_s^2(\tau)}\,d\tau + \frac{1}{2}e^{-\beta t}(\theta - \theta_0)^t Q_0(\theta - \theta_0), \tag{8.38}$$

where $Q_0 = Q_0^t > 0$, $\beta \geq 0$ are design constants and $\theta_0 = \theta(0)$ is the initial parameter estimate. This cost function is a generalization of (8.37) to include possible discounting of past data and a penalty on the initial error between the estimate θ_0 and θ^*. Since $\frac{z}{m_s}, \frac{\phi}{m_s} \in L_\infty$, $J(\theta)$ is a convex function of θ over $\Re^n$ at each time t. Hence, any local minimum is also global and satisfies

$$\frac{dJ}{d\theta}(\theta(t)) = 0 \quad \forall t \geq 0.$$

The LS algorithm for generating $\theta(t)$, the estimate of θ^*, in (8.35) is therefore obtained by solving

$$\frac{dJ}{d\theta}(\theta) = e^{-\beta t}Q_0\big(\theta(t) - \theta_0\big) - \int_0^t e^{-\beta(t-\tau)}\frac{z(\tau) - \theta^t(t)\phi(\tau)}{m_s^2(\tau)}\phi(\tau)\,d\tau = 0 \tag{8.39}$$

for $\theta(t)$, which yields the nonrecursive LS algorithm

$$\theta(t) = P(t)\left[e^{-\beta t}Q_0\theta_0 + \int_0^t e^{-\beta(t-\tau)}\frac{z(\tau)\phi(\tau)}{m_s^2(\tau)}\,d\tau\right], \tag{8.40}$$

where

$$P(t) = \left[e^{-\beta t}Q_0 + \int_0^t e^{-\beta(t-\tau)}\frac{z(\tau)\phi^t(\tau)}{m_s^2(\tau)}\,d\tau\right]^{-1}, \tag{8.41}$$

is the so-called covariance matrix. Because $Q_0 = Q_0^t > 0$ and $\phi\phi^t$ is positive semidefinite, $P(t)$ exists at each time t. Using the identity

$$\frac{d}{dt}PP^{-1} = \dot{P}P^{-1} + P\frac{d}{dt}P^{-1} = 0$$

and $\varepsilon m_s^2 = z - \theta^t\phi$, and differentiating $\theta(t)$ w.r.t. t, we obtain the recursive LS algorithm with forgetting factor

$$\begin{aligned} \dot{\theta} &= P\varepsilon\phi, \qquad \theta(0) = \theta_0, \\ \dot{P} &= \beta P - P\frac{\phi\phi^t}{m_s^2}P, \qquad P(0) = P_0 = Q_0^{-1}. \end{aligned} \tag{8.42}$$

The stability properties of (8.42) depend on the value of the forgetting factor β, as discussed in the following sections. If $\beta = 0$, the algorithm becomes the pure LS algorithm discussed and analyzed in Sect. 8.8.2. When $\beta > 0$, stability cannot be established unless $\frac{\phi}{m_s}$ is PE. In this case (8.42) is modified, leading to a different algorithm.

The following theorem establishes the stability and convergence of θ to θ^* of the algorithm (8.42) in the case where $\frac{\phi}{m_s}$ is persistently excited (PE), see the Appendix.

Theorem 8.3 *If $\frac{\phi}{m_s}$ is PE, then the recursive LS algorithm with forgetting factor (8.42) guarantees that $P, P^{-1} \in L_\infty$ and that $\theta(t) \to \theta^*$ as $t \to \infty$. The convergence of $\theta(t) \to \theta^*$ is exponential when $\beta > 0$.*

Since the adaptive law (8.42) could be used in adaptive control where the PE property of $\frac{\phi}{m_s}$ cannot be guaranteed, it is of interest to examine the properties of (8.42) in the absence of PE. In this case, (8.42) is modified in order to avoid certain undesirable phenomena, as discussed in the following sections.

8.8.2 Pure LS Algorithm

When $\beta = 0$ in (8.38), the algorithm (8.42) reduces to

$$\begin{aligned} \dot{\theta} &= P\varepsilon\phi, \qquad \theta(0) = \theta_0, \\ \dot{P} &= -P\frac{\phi\phi^t}{m_s^2}P, \qquad P(0) = P_0, \end{aligned} \tag{8.43}$$

which is referred to as the pure LS algorithm.

Theorem 8.4 *The pure LS algorithm (8.43) guarantees that*

1. *$\varepsilon, \varepsilon m_s, \dot{\theta} \in L_2 \cap L_\infty$ and $\theta, P \in L_\infty$.*
2. *$\lim_{t\to\infty} \theta(t) = \bar{\theta}$, where $\bar{\theta}$ is a constant vector.*
3. *If $\frac{\phi}{m_s}$ is PE, then $\theta(t) \to \theta^*$ as $t \to \infty$.*
4. *If (8.35) is the SPM for the plant*

$$\begin{aligned} &y^{(n)} + a_{n-1}y^{(n-1)} + \cdots + a_1\dot{y} + a_0 y \\ &\quad = b_m u^{(m)} + \cdots + b_1\dot{u} + b_0 u \end{aligned} \tag{8.44}$$

with stable poles and no zero-pole cancellations, and u is sufficiently rich of order $n+m+1$, i.e., consists of at least $\frac{n+m+1}{2}$ distinct frequencies, then ϕ, $\frac{\phi}{m_s}$ are PE and therefore $\theta(t) \to \theta^*$ as $t \to \infty$.

8.8.3 Example 8.8

In order to get some understanding of the properties of the pure LS algorithm, let us consider the scalar SPM

$$z = \theta^*\phi,$$

where $z, \theta^*, \phi \in R$. Let us assume that $\phi \in L_\infty$. Then the pure LS algorithm is given by

$$\dot{\theta} = p\varepsilon\phi, \qquad \theta(0) = \theta_0,$$
$$\dot{P} = -p^2\phi^2, \qquad p(0) = p_0 > 0,$$
$$\varepsilon = z - \theta\phi = -\tilde{\theta}\phi.$$

Let us also take $\phi = 1$, which is PE, for this example. Then we can show by solving the differential equation via integration that

$$p(t) = \frac{p_0}{1 + p_0 t}$$

and $\tilde{\theta}(t) = \frac{\tilde{\theta}(0)}{1+p_0 t}$, i.e.,

$$\theta(t) = \theta^* + \frac{\theta(0) - \theta^*}{1 + p_0 t}.$$

It is clear that as $t \to \infty$, $p(t) \to 0$, leading to the so-called covariance wind-up problem. Since $\phi = 1$ is PE, however, $\theta(t) \to \theta^*$ as $t \to \infty$, with a rate of $\frac{1}{t}$ (not exponential) as predicted by Theorem 8.4. Even though $\theta(t) \to \theta^*$, the covariance windup problem may still pose a problem in the case where θ^* changes to some other value after some time. If at that instance $p(t) \cong 0$, leading to $\dot{\theta} \cong 0$, no adaptation will take place and $\theta(t)$ may not reach the new θ^*.

For the same example, consider $\phi(t) = \frac{1}{1+t}$ which is not PE since

$$\int_t^{t+T} \phi^2(\tau)\, d\tau = \int_t^{t+T} \frac{1}{(1+\tau)^2}\, d\tau = \frac{1}{1+t} - \frac{1}{1+t+T}$$

goes to zero as $t \to \infty$, i.e., it has zero level of excitation. In this case, we can show that

$$p(t) = \frac{p_0(1+t)}{1 + (1+p_0)t},$$
$$\theta(t) = \theta^* + \bigl(\theta(0) - \theta^*\bigr)\frac{1+t}{1 + (1+p_0)t}$$

by solving the differential equations above. It is clear that $p(t) \to \frac{p_0}{1+p_0}$ and $\theta(t) \to \frac{p_0\theta^* + \theta(0)}{1+p_0}$ as $t \to \infty$; i.e., $\theta(t)$ converges to a constant but not to θ^* due to lack of PE. In this case, p_o converges to a constant and no covariance wind-up problem arises.

8.8.4 Modified LS Algorithms

One way to avoid the covariance wind-up problem is to modify the pure LS algorithm using a covariance resetting modification to obtain

$$\dot{\theta} = P\varepsilon\phi, \qquad \theta(0) = \theta_0,$$
$$\dot{P} = -P\frac{\phi\phi^t}{m_s^2}P, \qquad P(t_r^+) = P_0 = \rho_0 I, \tag{8.45}$$
$$m_s^2 = 1 + n_s^2, \quad n_s^2 = \alpha\phi^t\phi, \quad \alpha > 0,$$

where t_r^+ is the time at which $\lambda_{\min}(P(t) \le \rho_1$ and $\rho_0 > \rho_1 > 0$ are some design scalars. Due to covariance resetting, $P(t) \ge \rho_1 I \; \forall t \ge 0$. Therefore, P is guaranteed to be positive definite for all $t \ge 0$. In fact, the pure LS algorithm with covariance resetting can be viewed as a gradient algorithm with time-varying adaptive gain P, and its properties are very similar to those of a gradient algorithm analyzed in the previous section. They are summarized by Theorem 8.5 in this section.

When $\beta > 0$, the covariance wind-up problem, i.e., $P(t)$ becoming arbitrarily small, does not exist. In this case, $P(t)$ may grow without bound. In order to avoid this phenomenon, the following modified LS algorithm with forgetting factor is used:

$$\dot{\theta} = P\varepsilon\phi,$$
$$\dot{P} = \begin{cases} \beta P - \frac{P\phi\phi^t P}{m_s^2} & \text{if } \|P(t)\| \le R_0, \\ 0 & \text{otherwise,} \end{cases} \tag{8.46}$$

where $P(0) = P_0 = P_0^t > 0$, $\|P_0\| \le R_0$, R_0 is a constant that serves as an upper bound for $\|P\|$, and $m_s^2 = 1 + n_s^2$ is the normalizing signal which satisfies $\frac{\phi}{m_s} \in L_\infty$.

The following theorem summarizes the stability properties of the two modified LS algorithms.

Theorem 8.5 *The pure LS algorithm with covariance resetting* (8.45) *and the modified LS algorithm with forgetting factor* (8.46) *guarantee that*

1. $\varepsilon, \varepsilon m_s, \dot{\theta} \in L_2 \cap L_\infty$ *and* $\dot{\theta} \in L_\infty$.
2. *If* $\frac{\phi}{m_s}$ *is PE, then* $\theta(t) \to \theta^*$ *as* $t \to \infty$ *exponentially fast.*
3. *If* (8.35) *is the SPM for the plant* (8.44) *with stable poles and no zero-pole cancellations, and* u *is sufficiently rich of order* $n+m+1$, *then* ϕ, $\frac{\phi}{m_s}$ *are PE, which guarantees that* $\theta(t) \to \theta^*$ *as* $t \to \infty$ *exponentially fast.*

8.8.5 Parameter Identification Based on DPM

Let us consider the DPM

$$z = W(s)\left[\theta^{*t}\psi\right].$$

This model may be obtained from $z = \theta^{*t}\phi$ by filtering each side with $W(s)$ and redefining the signals z, ϕ. Since θ^* is a constant vector, the DPM may be written as

$$z = W(s)L(s)\left[\theta^{*t}\phi\right], \tag{8.47}$$

where $\phi = L^{-1}(s)\psi$, $L(s)$ is chosen so that $L^{-1}(s)$ is a proper stable transfer function, and $W(s)L(s)$ is a proper strictly positive real (SPR) transfer function.

$$\hat{z} = W(s)L(s)\left[\theta^t\phi\right].$$

We form the normalized estimation error

$$\varepsilon = z - \hat{z} - W(s)L(s)\left[\varepsilon n_s^2\right], \tag{8.48}$$

where the static normalizing signal n_s is designed so that $\frac{\phi}{m_s} \in L_\infty$ for $m_s^2 = 1 + n_s^2$. If $W(s)L(s) = 1$, then (8.48) has the same expression as in the case of the gradient algorithm. Substituting for z in (8.48), we express ε in terms of the parameter error $\tilde{\theta} = \theta - \theta^*$:

$$\varepsilon = W(s)L(s)\left[-\tilde{\theta}^t \phi - \varepsilon n_s^2\right]. \tag{8.49}$$

For simplicity, let us assume that $W(s)L(s)$ is strictly proper and rewrite (8.49) in the minimum state-space representation form

$$\begin{aligned} \dot{e} &= A_c e + b_c\left(-\tilde{\theta}^t \phi - \varepsilon n_s^2\right), \\ \varepsilon &= c_c^t e, \end{aligned} \tag{8.50}$$

where $W(s)L(s) = c_c^t(sI - A_c)^{-1}b_c$. Since $W(s)L(s)$ is SPR, it follows that (see the Appendix) there exist matrices $P_c = \Re_c^t > 0$, $L_c = L_c^t > 0$, a vector q, and a scalar $\nu > 0$ such that

$$\begin{aligned} P_c A_c + A_c^t P_c &= -qq^t - \nu L_c, \\ P_c b_c &= c_c. \end{aligned} \tag{8.51}$$

The adaptive law for θ is generated using the Lyapunov-like function

$$V = \frac{e^t P_c e}{2} + \frac{\tilde{\theta}^t \Gamma^{-1} \tilde{\theta}}{2},$$

where $\Gamma = \Gamma^t > 0$. The time derivative $\dot{V}$ of V along the solution of (8.50) is given by

$$\dot{V} = -\frac{1}{2} e^t q q^t e - \frac{\nu}{2} e^t L_c e + e^t P_c b_c\left(-\tilde{\theta}^t \phi - \varepsilon n_s^2\right) + \tilde{\theta}^t \Gamma^{-1} \dot{\tilde{\theta}}.$$

Since $e^t P_c b_c = e^t c_c = \varepsilon$, it follows that by choosing $\dot{\tilde{\theta}} = \dot{\theta}$ as

$$\dot{\theta} = \Gamma \varepsilon \phi, \tag{8.52}$$

we get

$$\dot{V} = -\frac{1}{2} e^t q q^t e - \frac{\nu}{2} e^t L_c e - \varepsilon n_s^2 \le 0.$$

As before, from the properties of $V, \dot{V}$ we conclude that e, ε, $\theta \in L_\infty$ and $e, \varepsilon, \varepsilon n_s \in L_2$. These properties in turn imply that $\dot{\theta} \in L_2$. Note that without the use of the second equation in (8.51), we are not able to choose $\dot{\tilde{\theta}} = \dot{\theta}$ using signals available for measurement to make $\dot{V} \le 0$. This is because the state e in (8.50) cannot be generated since it depends on the unknown input $\tilde{\theta}^t \phi$. Equation (8.50) is used only for analysis.

The stability properties of the adaptive law (8.52) are summarized by the following theorem.

Theorem 8.6 *The adaptive law (8.52) guarantees that*

1. $\varepsilon, \theta \in L_\infty$ *and* $\varepsilon, \varepsilon n_s, \dot{\theta} \in L_2$.
2. *If* $n_s, \phi, \dot{\phi} \in L_\infty$ *and* ϕ *is PE, then* $\theta(t) \to \theta^*$ *exponentially fast.*

The adaptive law (8.52) is referred to as the adaptive law based on the SPR–Lyapunov synthesis approach.

Remark 8.7 The adaptive law (8.52) has the same form as the gradient algorithm even though it is developed using a Lyapunov approach and the SPR property. In fact, for $W(s)L(s) = 1$, (8.52) is identical to the gradient algorithm.

8.8.6 Parameter Identification Based on B-SPM

Consider the bilinear SPM described by

$$z = \rho^*\big(\theta^{*t}\phi + Z_0\big), \tag{8.53}$$

where z, z_0 are known scalar signals at each time t and ρ^*, θ^* are the scalar and vector unknown parameters, respectively. The estimation error is generated as

$$\hat{z} = \rho\big(\theta^t\phi + z_0\big),$$

$$\varepsilon = \frac{z - \hat{z}}{m_s^2},$$

where $\rho(t)$, $\theta(t)$ are the estimates of ρ^*, θ^*, respectively, at time t and where m_s is designed to bound ϕ, z_0 from above. An example of m_s with this property is $m_s^2 = 1 + \phi^t\phi + z_0^2$.

Let us consider the cost

$$J(\rho, \theta) = \frac{\varepsilon^2 m_s^2}{2} = \frac{(z - \rho^*\theta^t\phi - \rho\xi + \rho^*\xi - \rho^* z_0)}{2m_s^2},$$

where

$$\xi = \theta^t\phi + z_0$$

is available for measurement. Applying the gradient method we obtain

$$\dot{\theta} = -\Gamma_1 \frac{dJ}{d\theta}_{\theta} = \Gamma_1 \varepsilon\rho^*\phi,$$

$$\dot{\rho} = -\gamma \frac{dJ}{d\theta}_{\rho} = \gamma_1 \varepsilon\xi,$$

where $\Gamma_1 = \Gamma_1^t > 0$, $\gamma > 0$ are the adaptive gains. Since ρ^* is unknown, the adaptive law for θ cannot be implemented. We bypass this problem by employing the equality

$$\Gamma_1\rho^* = \Gamma_1|\rho^*|\operatorname{sgn}(\rho^*) = \Gamma \operatorname{sgn}(\rho^*),$$

where $\Gamma_1 = \Gamma_1|\rho^*|$. Since Γ_1 is arbitrary any $\Gamma = \Gamma^t > 0$ can be selected without having to know $|\rho^*|$. Therefore, the adaptive laws for θ, ρ, may be written as

$$\begin{aligned}
\dot{\theta} &= \Gamma \varepsilon \phi \operatorname{sgn}(\rho^*), \\
\dot{\rho} &= \gamma \varepsilon \xi, \\
\varepsilon &= \frac{z - \rho \xi}{m_s^2}, \qquad \xi = \theta^t \phi + z_0.
\end{aligned} \tag{8.54}$$

Theorem 8.8 *The adaptive law (8.54) guarantees that*

1. $\varepsilon, \varepsilon m_s, \dot{\theta}, \dot{\rho} \in L_2 \cap L_\infty$ *and* $\theta, \rho \in L_\infty$.
2. *If* $\frac{\varepsilon}{m_s}$, *then* $\rho(t) \to \bar{\rho}$ *as* $t \to \infty$, *where* $\bar{\rho}$ *is a constant.*
3. *If* $\frac{\varepsilon}{m_s} \in L_2$ *and* $\frac{\phi}{m_s}$ *is PE, then* $\theta(t)$ *converges to* θ^* *as* $t \to \infty$.
4. *If the plant (8.44) has stable poles with no zero-pole cancellations and u is sufficiently rich of order* $n + m + 1$, *then* ϕ, $\frac{\phi}{m_s}$ *are PE and* $\theta(t)$ *converges to* θ^* *as* $t \to \infty$.

Proof Consider the Lyapunov-like function

$$V = \frac{\tilde{\theta}^t \Gamma^{-1} \tilde{\theta}}{2} |\rho^*| + \frac{\tilde{\rho}^2}{2\gamma}.$$

Then

$$\dot{V} = \tilde{\theta}^t \phi \varepsilon |\rho^*| \operatorname{sgn}(\rho^*) + \tilde{\rho} \varepsilon \xi.$$

Using $|\rho^*| \operatorname{sgn}(\rho^*) = \rho^*$ and the expression

$$\begin{aligned}
\varepsilon m_s^2 &= \rho^* \theta^{*t} \phi + \rho^* z_0 - \rho \theta^t \phi - \rho z_0 \\
&= -\tilde{\rho} z_0 + \rho^* \theta^{*t} \phi - \rho \theta^t \phi + \rho^* \theta^t \phi - \rho^* \theta^t \phi \\
&= -\tilde{\rho}(z_0 + \theta^t \phi) - \rho^* \tilde{\theta}^t \phi = -\tilde{\rho} \xi - \rho^* \tilde{\theta}^t \phi,
\end{aligned}$$

we have

$$\dot{V} = \varepsilon(\rho^* \tilde{\theta}^t \phi + \tilde{\rho} \xi) = -\varepsilon^2 m_s^2 \leq 0,$$

which implies that $V \in L_\infty$ and therefore $\rho, \theta \in L_\infty$. Using similar analysis as in the case of the gradient algorithms for the SPM, we can establish (i) from the properties of V, $\dot{V}$ and the form of the adaptive laws.

(ii) We have

$$\begin{aligned}
\rho(t) - \rho(0) &= \int_0^t \dot{\rho}\, d\tau \leq \int_0^t |\dot{\rho}|\, d\tau \leq \gamma \int_0^t |\varepsilon m_s| \frac{|\xi|}{m_s}\, d\tau \\
&\leq \gamma \left(\int_0^t \varepsilon^2 m_s^2\, d\tau \right)^{1/2} \left(\int_0^t \frac{|\xi|^2}{m_s^2}\, d\tau \right)^{1/2} < \infty,
\end{aligned}$$

where the last inequality is obtained using the Schwarz inequality (see [6]). Since $\varepsilon m_s, \frac{\xi}{m_s} \in L_2$, the limit as $t \to \infty$ exists, which implies that $\dot{\rho} \in L_1$ and

$\lim_{t\to\infty} \rho(t) = \bar{\rho}$ for some constant $\bar{\rho}$. The proof of (iii) is long and is presented in the web resource [9]. The proof of (iv) is included in the proof of Theorem 8.1. □

The assumption that the sign of ρ^* is known can be relaxed, leading to an adaptive law for θ, ρ with additional nonlinear terms.

8.9 Parameter Projection

In many practical problems, we may have some a priori knowledge of where θ^* is located in $\Re^n$. This knowledge usually comes in terms of upper and/or lower bounds for the elements of θ^* or in terms of location in a convex subset of $\Re^n$. If such a priori information is available, we want to constrain the online estimation to be within the set where the unknown parameters are located. For this purpose, we modify the gradient algorithms based on the unconstrained minimization of certain costs using the gradient projection method presented in Sect. 3.5.18 as follows.

The gradient algorithm with projection is computed by applying the gradient method to the following minimization problem with constraints:

$$\min J(\theta) \quad \text{subject to } \theta \in S,$$

where S is a convex subset of $\Re^n$ with smooth boundary almost everywhere. Assume that S is given by

$$S = \{\theta \in \Re^n | g(\theta) \le 0\},$$

where $g : \Re^n \to R$ is a smooth function.

The adaptive laws based on the gradient method can be modified to guarantee that $\theta \in S$ by solving the constrained optimization problem given above to obtain

$$\begin{aligned}\dot{\theta} &= \Pr\left(-\Gamma \frac{dJ}{d\theta}\right) \\ &= \begin{cases} -\Gamma \frac{dJ}{d\theta} & \text{if } \theta \in S^0, \text{ or } \theta \in \delta(S) \text{ and } -(\Gamma \frac{dJ}{d\theta})^t := g \le 0, \\ -\Gamma \frac{dJ}{d\theta} + \Gamma \frac{\frac{dg}{d\theta}\frac{dg}{d\theta}^t}{\frac{dg}{d\theta}^t \Gamma \frac{dg}{d\theta}} \Gamma \frac{dJ}{d\theta} & \text{otherwise,} \end{cases}\end{aligned} \tag{8.55}$$

where $\delta(S) = \{\theta \in \Re^n |\, g(\theta) = 0\}$ and $S^0 = \{\theta \in \Re^n |\, g(\theta) < 0\}$ denote the boundary and the interior, respectively, of S and $\Pr(\cdot)$ is the projection operator.

The gradient algorithm based on the instantaneous cost function with projection follows from (8.55) by substituting for $\frac{dJ}{d\theta} = -\varepsilon\phi$ to obtain

$$\begin{aligned}\dot{\theta} &= \Pr(\Gamma\varepsilon\phi) \\ &= \begin{cases} \Gamma\varepsilon\phi & \text{if } \theta \in S^0, \text{ or } \theta \in \delta(S) \text{ and } (\Gamma\varepsilon\phi)^t \frac{dg}{d\theta} \le 0, \\ \Gamma\varepsilon\phi - \Gamma \frac{\frac{dg}{d\theta}\frac{dg}{d\theta}^t}{\frac{dg}{d\theta}^t \Gamma \frac{dg}{d\theta}} \Gamma\varepsilon\phi & \text{otherwise,} \end{cases}\end{aligned} \tag{8.56}$$

where $\theta(0) \in S$.

The pure LS algorithm with projection becomes

$$\dot{\theta} = \Pr(P\varepsilon\phi) = \begin{cases} P\varepsilon\phi & \text{if } \theta \in S^0 \text{ or } \theta \in \delta(S) \text{ and } (P\varepsilon\phi)^t \frac{dg}{d\theta} \leq 0, \\ P\varepsilon\phi - P\frac{\frac{dg}{d\theta}\frac{dg}{d\theta}^t}{\frac{dg}{d\theta}^t P \frac{dg}{d\theta}} P\varepsilon\phi & \text{otherwise,} \end{cases} \tag{8.57}$$

where $\theta(0) \in S$,

$$\dot{P} = \begin{cases} \beta P - P\frac{\phi\phi^t}{m_s^2}P & \text{if } \theta \in S^0 \text{ or } \theta \in \delta(S) \text{ and } (P\varepsilon\phi)^t \frac{dg}{d\theta} \leq 0, \\ 0 & \text{otherwise,} \end{cases}$$

and $P(0) = P_0 = P_0^t > 0$.

Theorem 8.9 *The gradient adaptive laws of Sect. 8.7 and the LS adaptive laws of Sect. 8.8 with the projection modifications given by (8.55) and (8.57), respectively, retain all the properties that are established in the absence of projection and in addition guarantee that $\theta(t) \in S$ $\forall t \geq 0$, provided $\theta(0) \in S$ and $\theta^* \in S$.*

8.9.1 Example 8.9

Let us consider the plant model

$$y = \frac{b}{s+a}u$$

where a, b are unknown constants that satisfy some known bounds, e.g., $b \geq 1$ and $20 \geq a \geq -2$. For simplicity, let us assume that $y, \dot{y}, u$ are available for measurement so that the SPM is of the form

$$z = \theta^{*t}\phi,$$

where $z = \dot{y}$, $\theta^* = [b, a]^t$, $\phi = [u, -y]^t$. In the unconstrained case the gradient adaptive law is given as

$$\dot{\theta} = \Gamma\varepsilon\phi, \quad \varepsilon = \frac{z - \theta^t\phi}{m_s^2},$$

where $m_s^2 = 1 + \phi^t\phi$; $\theta = [\hat{b}, \hat{a}]$; $\hat{b}$, $\hat{a}$ are the estimates of b, a, respectively. Since we know that $b \geq 1$ and $20 \geq a \geq -2$, we can constrain the estimates $\hat{b}, \hat{a}$ to be within the known bounds by using projection. Defining the sets for projection as

$$S_b = \{\hat{b} \in R \,|\, 1 - \hat{b} \leq 0\},$$
$$S_a^l = \{\hat{a} \in R \,|\, -2 - \hat{a} \leq 0\},$$
$$S_a^u = \{\hat{a} \in R \,|\, \hat{a} - 20 \leq 0\}$$

and applying the projection algorithm (8.55) for each set, we obtain the adaptive laws

$$\dot{\hat{b}} = \begin{cases} \gamma_1 \varepsilon u & \text{if } \hat{b} > 1 \text{ or } (\hat{b} = 1 \text{ and } \varepsilon u \geq 0), \\ 0 & \text{if } \hat{b} = 1 \text{ and } \varepsilon u < 0, \end{cases}$$

with $\hat{b}(0) \geq 1$, and

$$\dot{\hat{a}} = \begin{cases} -\gamma_2 \varepsilon y & \text{if } 20 > \hat{a} > -2 \text{ or } (\hat{a} = -2 \text{ and } \varepsilon y \leq 0) \\ & \quad \text{or } (\hat{a} = 20 \text{ and } \varepsilon y \geq 0), \\ 0 & \text{if } (\hat{a} = -2 \text{ and } \varepsilon y > 0) \text{ or } (\hat{a} = 20 \text{ and } \varepsilon y < 0), \end{cases}$$

with $\hat{a}(0)$ satisfying $20 \geq \hat{a}(0) \geq -2$.

8.9.2 Example 8.10

Let us consider the gradient adaptive law

$$\dot{\theta} = \Gamma \varepsilon \phi$$

with the a priori knowledge that $|\theta^*| \leq M_0$ for some known bound $M_0 > 0$. In most applications, we may have such a priori information. We define

$$S = \left\{ \theta \in \Re^n \middle| g(\theta) = \frac{\theta^t \theta}{2} - \frac{M_0^2}{2} \leq 0 \right\}$$

and use (8.55) together with $\frac{dg}{d\theta} = \theta$ to obtain the adaptive law with projection

$$\dot{\theta} = \begin{cases} \Gamma \varepsilon \phi & \text{if } |\theta| < M_0 \text{ or } (|\theta| = M_0 \text{ and } \phi^t \Gamma \theta \varepsilon \leq 0), \\ \Gamma \varepsilon \phi - \Gamma \frac{\theta \theta^t}{\theta^t \Gamma \theta} \Gamma \varepsilon \phi & \text{if } |\theta| = M_0 \text{ and } \phi^t \Gamma \theta \varepsilon > 0 \end{cases}$$

with $|\theta(0)| \leq M_0$.

8.10 Robust Parameter Identification

In the previous sections, we designed and analyzed a wide class of *PI* algorithms based on the parametric models

$$z = \theta^{*t} \phi \quad \text{or} \quad W(s) \theta^{*t} \phi.$$

These parametric models are developed using a plant model that is assumed to be free of disturbances, noise, unmodeled dynamics, time delays, and other frequently encountered uncertainties. In the presence of plant uncertainties, we are no longer able to express the unknown parameter vector θ^* in the form of the SPM or DPM where all signals are measured and θ^* is the only unknown term. In this case, the SPM or DPM takes the form

$$z = \theta^{*t} \phi + \eta \quad \text{or} \quad z = W(s) \theta^{*t} \phi + \eta, \tag{8.58}$$

where η is an unknown function that represents the modeling error terms. The following examples are used to show how (8.58) arises for different plant uncertainties.

8.10.1 Example 8.11

Let us consider the plant

$$y = \theta^*\big(1 + \mu\Delta_m(s)\big)u, \tag{8.59}$$

where μ is a small constant and $\Delta_m(s)$ is a proper transfer function with poles in the open left half s-plane. Since μ is small and $\Delta_m(s)$ is proper with stable poles, the term $\mu\Delta_m(s)$ can be treated as the modeling error term which can be approximated with zero. We can express (8.59) in the form of (8.58) as

$$y = \theta^* u + \eta,$$

where

$$\eta = \mu\theta^*\Delta_m(s)u$$

is the modeling error term.

For LTI plants, the parametric model with modeling errors is usually of the form

$$\begin{aligned} z &= \theta^{*t} u + \eta, \\ \eta &= \Delta_1(s)u + \Delta_2(s)y + d, \end{aligned} \tag{8.60}$$

where $\Delta_1(s)$, Δ_2 are proper transfer functions with stable poles and d is a bounded disturbance. The principal question that arises is how the stability properties of the adaptive laws that are developed for parametric models with no modeling errors are affected when applied to the actual parametric models with uncertainties.

8.10.2 Example 8.12

The following example demonstrates that the adaptive laws of the previous sections that are developed using parametric models that are free of modeling errors cannot guarantee the same properties in the presence of modeling errors. Furthermore, it often takes only a small disturbance to drive the estimated parameters unbounded.

Consider the scalar constant gain system

$$y = \theta^* u + d,$$

where d is a bounded unknown disturbance and $u \in L_\infty$. The adaptive law for estimating θ^* derived for $d = 0$ is given by

$$\dot{\theta} = \gamma\varepsilon u, \quad \varepsilon = y - \theta u, \tag{8.61}$$

where $\gamma > 0$ and the normalizing signal is taken to be 1. If $d = 0$ and $u, \dot{u} \in L_\infty$ then we can establish that (i) $\dot{\theta}, \theta, \varepsilon \in L_\infty$, (ii) $\varepsilon(t) \to 0$ as $t \to \infty$ by analyzing the parameter error equation

$$\dot{\tilde{\theta}} = -\gamma u^2 \tilde{\theta},$$

which is a linear time-varying differential equation. When $d \neq 0$, we have

$$\dot{\tilde{\theta}} = -\gamma u^2 \tilde{\theta} + \gamma d u. \tag{8.62}$$

In this case, we cannot guarantee that the parameter estimate $\theta(t)$ is bounded for any bounded input u and disturbance d. In fact, for $\theta^* = 2$, $\gamma = 1$,

$$u = (1+t)^{-1/2},$$
$$d(t) = (1+t)^{-1/4}\left(\frac{5}{4} - 2(1+t)^{-1/4}\right) \to 0 \quad \text{as } t \to \infty,$$

we have

$$y(t) = \frac{5}{4}(1+t)^{-1/4} \to 0 \quad \text{as } t \to \infty,$$
$$\varepsilon(t) = \frac{1}{4}(1+t)^{-1/4} \to 0 \quad \text{as } t \to \infty,$$
$$\theta(t) = (1+t)^{-1/4} \to \infty \quad \text{as } t \to \infty;$$

that is, the estimated parameter drifts to infinity with time even though the disturbance $d(t)$ disappears with time. This instability phenomenon is known as parameter drift. It is mainly due to the pure integral action of the adaptive law, which, in addition to integrating the "good" signals, integrates the disturbance term as well, leading to the parameter drift phenomenon.

Another interpretation of the above instability is that, for $u = (1+t)^{-1/2}$, the homogeneous part of (8.62), that is, $\dot{\tilde{\theta}} = -\gamma u^2 \tilde{\theta}$, is only uniformly stable, which is not sufficient to guarantee that the bounded input $\gamma d u$ will produce a bounded state $\tilde{\theta}$. If u is persistently exciting, i.e., $\int_t^{t+T_0} u^2(\tau)\,d\tau \geq \alpha_0 T_0$ for some $\alpha_0, T_0 > 0$ and $\forall t \geq 0$, then the homogeneous part of (8.62) is e.s. and the bounded input $\gamma d u$ produces a bounded state $\tilde{\theta}$.

If the objective is parameter convergence, then parameter drift can be prevented by making sure the regressor vector is PE with a level of excitation higher than the level of the modeling error. In this case, the plant input in addition to being sufficiently rich is also required to guarantee a level of excitation for the regressor that is higher than the level of the modeling error. This class of inputs is referred to as dominantly rich and is discussed in the following section.

8.10.3 Dominantly Rich Excitation

Let us revisit the example in Sect. 8.10.2 and analyze (8.62), that is,

$$\dot{\tilde{\theta}} = -\gamma u^2 \tilde{\theta} + \gamma d u, \tag{8.63}$$

when u is PE with level $\alpha_0 > 0$. The PE property of u implies that the homogeneous part of (8.63) is e.s., which in turn implies that

$$\left|\tilde{\theta}(t)\right| \leq e^{-\gamma\alpha_1 t}\left|\tilde{\theta}(0)\right| + \frac{1}{\alpha_1}\left(1 - e^{-\gamma\alpha_1 t}\right) \sup_{\tau \leq t}\left|u(\tau)d(\tau)\right|$$

for some $\alpha_1 > 0$ which depends on α_0. Therefore, we have

$$\lim_{t\to\infty} \sup_{\tau\geq t} \left|\tilde{\theta}(\tau)\right| \leq \frac{1}{\alpha_1} \lim_{t\to\infty} \sup_{\tau\geq t} \left|u(\tau)d(\tau)\right| = \frac{1}{\alpha_1} \sup_{\tau} \left|u(\tau)d(\tau)\right|. \tag{8.64}$$

The bound (8.64) indicates that the PI error at steady state is of the order of the disturbance; i.e., as $d \to 0$ the parameter error also reduces to zero. For this simple example, it is clear that if we choose $u = u_0$, where u_0 is a constant different from zero, then $\alpha_1 = \alpha_0 = u_0^2$; therefore, the bound for $|\tilde{\theta}|$ is $\sup_1 \frac{|d(t)|}{u_0}$. Thus, the larger the value of u_0 is, the smaller the parameter error. Large u_0 relative to $|d|$ implies large signal-to-noise ratio and therefore better accuracy of identification.

Definition 8.10 A sufficiently rich input u of order $n + m + 1$ for the dominant part of the plant

$$y = G_0(s)u + \Delta(\mu s)u + d \tag{8.65}$$

is called dominantly rich of order $n + m + 1$ if it achieves its richness with frequencies ω_i, $i = 1, 2, \dots, N$, where $N \geq \frac{n+m+1}{2}$, $|\omega_i| < O(\frac{1}{\mu})$,[2] $|\omega_i - \omega_j| > O(\mu)$, $i \neq j$, and $|u| > O(\mu) + O(d)$.

Lemma 8.11 *Let $H_0(s)$, $H_1(\mu s, s)$ satisfy the following assumptions:*

1. *The vectors $H_0(j\omega_1), H_0(j\omega_2), \dots, H_0(j\omega_{\bar{n}})$ are linearly independent on C^n for all possible $\omega_1, \omega_2, \dots, \omega_{\bar{n}} \in R$, where $\bar{n} := n + m + 1$ and $\omega_1 \neq \omega_k$ for $i \neq k$.*
2. *For any set $\{\omega_1, \omega_2, \dots, \omega_{\bar{n}}\}$ satisfying $|\omega_i - \omega_k| > O(\mu)$ for $i \neq k$ and $|\omega_i| < O(\frac{1}{\mu})$, we have $|\det(\bar{H})| > O(\mu)$, where $\bar{H} := [H_0(j\omega_1), H_0(j\omega_2), \dots, H_0(j\omega_{\bar{n}})]$.*
3. *$|H_1(j\mu\omega, j\omega)| \leq c$ for some constant c independent of μ and $\forall\omega \in R$.*

Then there exists a $\mu^* > 0$ such that for $\mu \in [0, \mu^*)$, ϕ is PE of order $n + m + 1$ with level of excitation $\alpha_1 > O(\mu)$, provided that the input signal u is dominantly rich of order $n + m + 1$ for the plant (8.65).

Consider the plant

$$y = \frac{b}{s+a}\left(1 + \frac{2\mu(s-1)}{(\mu s + 1)^2}\right)u,$$

where a, b are the unknown parameters and $\mu = 0.001$. The plant may be modeled as

$$y = \frac{b}{s+a}u$$

by approximating $\mu = 0.001 \cong 0$. The input $u = \sin\omega_0 t$ with $1 \ll \omega_0 \ll 1000$ would be a dominantly rich input of order 2. Frequencies such as $\omega_0 = 0.006$ rad/s or $\omega_0 = 900$ rad/s would imply that u is not dominantly rich even though u is sufficiently rich of order 2.

---

[2] A function $f(x)$ is of $O(\mu)$ $\forall x \in \Omega$ if there exists a constant $c \geq 0$ such that $\|f(x)\| \leq c|\mu|$ $\forall x \in \Omega$.

8.11 State-Space Identifiers

Let us consider the state-space plant model

$$\dot{x} = A_p x + B_p u, \tag{8.66}$$

where $x \in \Re^n$ is the state, $u \in \Re^m$ is the input vector, and $A_p \in \Re^{n\times n}$, $B_p \in \Re^{n\times m}$ are unknown constant matrices. We assume that x, u are available for measurement. One way to estimate the elements of A_p, B_p online is to express (8.66) as a set of n scalar differential equations and then generate n parametric models with scalar outputs and apply the parameter estimation techniques covered in the previous sections. Another more compact way of estimating A_p, B_p is to develop online estimators based on an SSPM model for (8.66) as follows.

We express (8.66) in the form of the SSPM:

$$\dot{x} = A_m x + (A_p - A_m)x + B_p u,$$

where A_m is an arbitrary stable matrix. The estimation model is then formed as

$$\dot{\hat{x}} = A_m \hat{x} + (\hat{A}_p - A_m)x + \hat{B}_p u = A_m(\hat{x} - x) + \hat{A}_p x + \hat{B}_p u$$

where $\hat{A}_p(t)$, $\hat{B}_p(t)$ are the estimates of A_p, B_p at time t, respectively. The above estimation model has been referred to as the series-parallel model in the literature [8, 9, 12]. The estimation error vector is defined as

$$\varepsilon = x - \hat{x} - (sI - A_m)^{-1}\varepsilon n_s^2$$

or

$$\varepsilon = x - \hat{x} - w,$$

$$\dot{w} = A_m w + \varepsilon n_s^2, \quad w(0) = 0,$$

where n_s^2 is the static normalizing signal designed to guarantee $\frac{x}{\sqrt{1+n_s^2}}, \frac{u}{\sqrt{1+n_s^2}} \in L_\infty$. A straightforward choice for n_s is $n_s^2 = x^t x + u^t u$. It is clear that if the plant model (8.66) is stable and the input u is bounded, then n_s can be taken to be equal to zero.

It follows that the estimation error satisfies

$$\dot{\varepsilon} = A_m \varepsilon - \tilde{A}_p x - \tilde{B}_p u - \varepsilon n_s^2,$$

where $\tilde{A}_p := \hat{A}_p - A_p$, $\tilde{B}_p := \hat{B}_p - B_p$ are the parameter errors.

The adaptive law for generating $\hat{A}_p$, $\hat{B}_p$ is developed by considering the Lyapunov function

$$V(\varepsilon, \tilde{A}_p, \tilde{B}_p) = \varepsilon^t P \varepsilon + \mathrm{tr}\left(\frac{\tilde{A}_p^t P \tilde{A}_p}{\gamma_1}\right) + \mathrm{tr}\left(\frac{\tilde{B}_p^t P \tilde{B}_p}{\gamma_2}\right), \tag{8.67}$$

where $\mathrm{tr}(A)$ denotes the trace of matrix A; $\gamma_1, \gamma_2 > 0$ are constant scalars; and $P = p^t > 0$ is chosen as the solution of the Lyapunov equation

$$PA_m + A_m P^t = -Q \tag{8.68}$$

for some $Q = Q^t > 0$, whose solution is guaranteed by the stability of A_m (see the Appendix). The time derivative $\dot{V}$ is given by

$$\dot{V} = \dot{\varepsilon}^t P\varepsilon + \varepsilon^t P\dot{\varepsilon} + \mathrm{Tr}\left(\frac{\dot{\tilde{A}}_p^t P\tilde{A}_p}{\gamma_1} + \frac{\tilde{A}_p^t P\dot{\tilde{A}}_p}{\gamma_1}\right) + \mathrm{Tr}\left(\frac{\dot{\tilde{B}}_p^t P\tilde{B}_p}{\gamma_2} + \frac{\tilde{B}_p^t P\dot{\tilde{B}}_p}{\gamma_2}\right).$$

Substituting for $\dot{\varepsilon}$, using (8.68), and employing the equalities $\mathrm{Tr}(A+B) = \mathrm{Tr}(A) + \mathrm{Tr}(B)$ and $\mathrm{Tr}(A^t) = \mathrm{Tr}(A)$ for square matrices A, B of the same dimension, we obtain

$$\dot{V} = -\varepsilon^t Q\varepsilon - 2\varepsilon^t P\tilde{A}_p x - 2\varepsilon^t P\tilde{B}_p u + 2\,\mathrm{tr}\left(\frac{\tilde{A}_p^t P\dot{\tilde{A}}_p}{\gamma_1} + \frac{\tilde{B}_p^t P\dot{\tilde{B}}_p}{\gamma_2}\right) - 2\varepsilon^t P\varepsilon n_s^2. \tag{8.69}$$

Using the equality $v^t y = \mathrm{tr}(vy^t)$for vectors v, y of the same dimension, we rewrite (8.69) as

$$\dot{V} = -\varepsilon^t Q\varepsilon + 2\,\mathrm{Tr}\left(\frac{\tilde{A}_p^t P\dot{\tilde{A}}_p}{\gamma_1} - \tilde{A}_p^t P\varepsilon x^t + \frac{\tilde{B}_p^t P\dot{\tilde{B}}_p}{\gamma_2} - \tilde{B}_p^t P\varepsilon u^t\right) - 2\varepsilon^t P\varepsilon n_s^2. \tag{8.70}$$

The obvious choice for $\dot{\hat{A}}_p = \dot{\tilde{A}}_p$, $\dot{\hat{B}}_p = \dot{\tilde{B}}_p$ to make $\dot{V}$ negative is

$$\dot{\hat{A}}_p = \gamma_1 \varepsilon x^t, \qquad \dot{\hat{B}}_p = \gamma_2 \varepsilon u^t, \tag{8.71}$$

which gives us

$$\dot{V} = -\varepsilon^t Q\varepsilon - 2\varepsilon^t P\varepsilon n_s^2 \le 0.$$

This implies that V, $\hat{A}_p$, $\hat{B}_p$, ε are bounded. We can also write

$$\dot{V} \le -|\varepsilon|^2 \lambda_{\min}(Q) - 2|\varepsilon n_s|^2 \lambda_{\min}(P) \le 0$$

and use similar arguments as in the previous sections to establish that $\varepsilon, \varepsilon n_s \in L_2$. From (8.71), we have that

$$\|\dot{\hat{A}}_p\| \le \gamma_1 |\varepsilon m_s| \frac{|x^t|}{m_s}, \qquad \|\dot{\hat{B}}_p\| \le \gamma_2 |\varepsilon m_s| \frac{|u^t|}{m_s},$$

where $m_s^2 = 1 + n_s^2$. Since $\frac{|x|}{m_s}, \frac{|u|}{m_s} \in L_\infty$ and $|\varepsilon m_s| \in L_2$, we can also conclude that $\|\dot{\hat{A}}_p\|, \|\dot{\hat{B}}_p\| \in L_2$.

We have, therefore, established that independent of the stability of the plant and boundedness of the input u, the adaptive law (8.71) guarantees that

- $\|\hat{A}_p\|, \|\hat{B}_p\|, \varepsilon \in L_\infty$,
- $\|\dot{\hat{A}}_p\|, \|\dot{\hat{B}}_p\|, \varepsilon, \varepsilon n_s \in L_2$.

These properties are important for adaptive control where the adaptive law is used as part of the controller and no a priori assumptions are made about the stability of the plant and boundedness of the input. If the objective, however, is parameter estimation, then we have to assume that the plant is stable and the input u is designed to be bounded and sufficiently rich for the plant model (8.66). In this case, we can take $n_s^2 = 0$.

Theorem 8.12 *Consider the plant model* (8.66) *and assume that* (A_p, B_p) *is controllable and* A_p *is a stable matrix. If each element* u_i, $i = 1, 2, \ldots, m$, *of vector* u *is bounded, sufficiently rich of order* $n+1$, *and uncorrelated, i.e., each* u_i *contains different frequencies, then* $\hat{A}_p(t)$, $\hat{B}_p(t)$ *generated by* (8.71) (*where* n_s *can be taken to be zero*) *converge to* A_p, B_p, *respectively, exponentially fast.*

8.11.1 Example 8.13

Consider the second-order plant

$$\dot{x} = Ax + Bu,$$

where $x = [x_1, x_2]$, $u = [u_1, u_2]^t$ is a bounded input vector, the matrices A, B are unknown, and A is a stable matrix. The estimation model is generated as

$$\dot{\hat{x}} = \begin{bmatrix} -a_m & 0 \\ 0 & -a_m \end{bmatrix} (\hat{x} \vdash x) + \begin{bmatrix} \hat{a}_{11} & \hat{a}_{12} \\ \hat{a}_{21} & \hat{a}_{22} \end{bmatrix} x + \begin{bmatrix} \hat{b}_{11} & \hat{b}_{12} \\ \hat{b}_{21} & \hat{b}_{22} \end{bmatrix} u,$$

where $\hat{x} = [\hat{x}_1, \hat{x}_2]^t$ and $a_m > 0$. The estimation error is given by

$$\varepsilon = x - \hat{x},$$

where $n_s = 0$ due to the stability of A and boundedness of u. The adaptive law (8.71) can be written as

$$\dot{\hat{a}}_{ij} = \gamma_1 \varepsilon_i x_j, \qquad \dot{\hat{b}}_{ij} = \gamma_2 \varepsilon_i u_j$$

for $i = 1, 2$, $j = 1, 2$, and adaptive gains γ_1, γ_2. An example of a sufficiently rich input for this plant is

$$u_1 = c_1 \sin 2.5t + c_2 \sin 4.6t,$$
$$u_2 = c_3 \sin 7.2t + c_4 \sin 11.7t$$

for some nonzero constants c_i, $i = 1, 2, 3, 4$.

The class of plants described by (8.66) can be expanded to include more realistic plants with modeling errors. The adaptive laws in this case can be made robust by using exactly the same techniques as in the case of SISO plants described in previous sections, and this is left as an exercise for the reader.

8.12 Adaptive Observers

Consider the LTI SISO plant

$$\begin{aligned} \dot{x} &= Ax + Bu, \qquad x(0) = x_0, \\ y &= C^t x, \end{aligned} \tag{8.72}$$

where $x \in \Re^n$. We assume that u is a piecewise continuous bounded function of time and that A is a stable matrix. In addition, we assume that the plant is completely controllable and completely observable. The problem is to construct a scheme that estimates both the plant parameters, i.e., A, B, C, as well as the state vector x using only I/O measurements. We refer to such a scheme as the adaptive observer.

A good starting point for designing an adaptive observer is the Luenberger observer used in the case where A, B, C are known. The Luenberger observer is of the form

$$\begin{aligned} \dot{\hat{x}} &= A\hat{x} + Bu + K(y - \hat{y}), \qquad \hat{x}(0) = \hat{x}_0, \\ \hat{y} &= C^t \hat{x}, \end{aligned} \tag{8.73}$$

where K is chosen so that $A - KC^t$ is a stable matrix, and guarantees that $\hat{x} \to x$ exponentially fast for any initial condition x_0 and any input u. For $A - KC^t$ to be stable, the existence of K is guaranteed by the observability of (A, C).

A straightforward procedure for choosing the structure of the adaptive observer is to use the same equation as the Luenberger observer (8.73), but replace the unknown parameters A, B, C with their estimates A, B, C, respectively, generated by some adaptive law. The problem we face with this procedure is the inability to estimate uniquely the $n^2 + 2n$ parameters of A, B, C from the I/O data. The best we can do in this case is to estimate the parameters of the plant transfer function and use them to calculate $\hat{A}$, $\hat{B}$, $\hat{C}$. These calculations, however, are not always possible because the mapping of the $2n$ estimated parameters of the transfer function to the $n^2 + 2n$ parameters of $\hat{A}$, $\hat{B}$, $\hat{C}$ is not unique unless (A, B, C) satisfies certain structural constraints. One such constraint is that (A, B, C) is in the observer form, i.e., the plant is represented as

$$\begin{aligned} \dot{x}_\alpha &= \left[-a_p \;\middle|\; \frac{I_{n-1}}{0} \right] x_\alpha + b_p u, \\ y &= [1, 0, \ldots, 0] x_\alpha, \end{aligned} \tag{8.74}$$

where $a_p = [a_{n-1}, a_{n-2}, \ldots, a_0]^t$ and $b_p = [b_{n-1}, b_{n-2}, \ldots, b_0]^t$ are vectors of dimension n, and $I_{n-1} \in \Re^{(n-1)\times(n-1)}$ is the identity matrix. The elements of a_p and b_p are the coefficients of the denominator and numerator, respectively, of the transfer function

$$\frac{y(s)}{u(s)} = \frac{b_{n-1}s^{n-1} + b_{n-2}s^{n-2} + \cdots + b_0 s}{s^n + a_{n-1}s^{n-1} + a_{n-2}s^{n-2} + \cdots + a_0 s} \tag{8.75}$$

and can be estimated online from I/O data using the techniques presented in the previous sections.

Since both (8.72) and (8.74) represent the same plant, we can focus on the plant representation (8.74) and estimate x_α instead of x. The disadvantage is that in a practical situation x may represent some physical variables of interest, whereas x_α may be an artificial state vector.

The adaptive observer for estimating the state x_α of (8.74) is motivated from the Luenberger observer structure (8.73) and is given by

$$\begin{aligned} \dot{\hat{x}} &= \hat{A}(t)\hat{x} + \hat{b}_p(t)u + K(t)(y - \hat{y}), \qquad \hat{x}(0) = \hat{x}_0, \\ \hat{y} &= [1, 0, \ldots, 0]\hat{x}, \end{aligned} \tag{8.76}$$

where $\hat{x}$ is the estimate of x_α,

$$\hat{A}(t) = \left[-\hat{a}_p(t) \,\middle|\, \frac{I_{n-1}}{0} \right], \qquad K(t) = a^* - \hat{a}_p(t),$$

$\hat{a}_p(t)$ and $\hat{b}_p(t)$ are the estimates of the vectors a_p and b_p, respectively, at time t, and $a^* \in \Re^n$ is chosen so that

$$A^* = \left[-a^* \,\middle|\, \frac{I_{n-1}}{0} \right]. \tag{8.77}$$

is a stable matrix that contains the eigenvalues of the observer.

A wide class of adaptive laws may be used to generate $\hat{a}_p(t)$ and $\hat{b}_p(t)$ online. As in Chap. 2, the parametric model

$$z = \theta^{*t}\phi \tag{8.78}$$

may be developed using (8.75), where

$$\theta^* = \left[b_p^t, a_p^t\right]^t$$

and z, ϕ are available signals, and used to design a wide class of adaptive laws to generate $\theta(t) = [\hat{b}_p^t(t), a_p^t(t)]^t$, the estimate of θ^*. As an example, consider the gradient algorithm

$$\dot{\theta} = \Gamma\varepsilon\phi, \qquad \varepsilon = \frac{z - \theta^t\phi}{m_s^2}, \tag{8.79}$$

where $m_s^2 = I + \alpha\phi^t\phi$ and $\alpha \geq 0$.

Theorem 8.13 *The adaptive observer described by* (8.76)–(8.79) *guarantees the following properties*:

1. *All signals are bounded.*
2. *If u is sufficiently rich of order $2n$, then the state observation error $|\hat{x} - x_a|$ and the parameter error $|\theta - \theta^*|$ converge to zero exponentially fast.*

Proof (i) Since A is stable and u is bounded, we have $x_a, y, \phi \in L_\infty$ and hence $m_s^2 = 1 + \alpha\phi^t\phi \in L_\infty$. The adaptive law (8.79) guarantees that $\theta \in L_\infty$ and $\varepsilon, \varepsilon m_s, \dot{\theta} \in L_2 \cap L_\infty$.

The observer equation may be written as

$$\dot{\hat{x}} = A^*\hat{x} + \hat{b}_p(t)u + \left(\hat{A}(t) - A^*\right)x_\alpha.$$

Since A^* is a stable matrix and $\hat{b}_p$, $\hat{A}$, u, x_α are bounded, it follows that $\hat{x} \in L_\infty$, which in turn implies that all signals are bounded.

(ii) The state observation error $\tilde{x} = \hat{x} - x_\alpha$ satisfies

$$\dot{\tilde{x}} = A^*\tilde{x} + \tilde{b}_p u - \tilde{a}_p y, \tag{8.80}$$

where $\tilde{b}_p = \hat{b}_p - b_p$, $\tilde{a}_p = \hat{a}_p - a_p$ are the parameter errors. Since for u sufficiently rich we have that $\theta(t) \to \theta^*$ as $t \to \infty$ exponentially fast, it follows that $\tilde{b}_p, \tilde{a}_p \to 0$ exponentially fast. Since $u, y \in L_\infty$, the error equation consists of a homogeneous part that is exponentially stable and an input that is decaying to zero. This implies that $\tilde{x} = \hat{x} - x_a \to 0$ as $t \to \infty$ exponentially fast. □

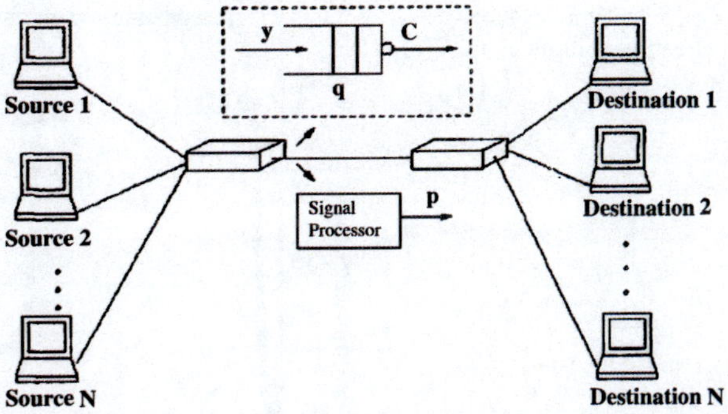

Fig. 8.8 Network topology

8.13 A Single Bottleneck Link Computer Network

The congestion control problem in computer networks has been identified as a feedback control problem. The network users adjust their sending data rates, in response to congestion signals they receive from the network, in an effort to avoid congestion and converge to a stable equilibrium that satisfies certain requirements: high network utilization, small queue sizes, small delays, fairness among users, etc. Many of the proposed congestion control schemes require that at each link the number of flows, N say, utilizing the link is known. Since the number of users varies with time, N is an unknown time-varying parameter, which needs to be estimated online. Estimation algorithms, which have been proposed in the literature, are based on point wise time division, which is known to lack robustness and may lead to erroneous estimates. In this study, we consider a simple estimation algorithm, which is based on online parameter identification.

We consider the single bottleneck link network shown in Fig. 8.8. It consists of N users which share a common bottleneck link through high bandwidth access links. At the bottleneck link, we assume that there exists a buffer, which accommodates the incoming packets. The rate of data entering the buffer is denoted by y, the queue size is denoted by q, and the output capacity is denoted by C. At the bottleneck link, we implement a signal processor, which calculates the desired sending rate p. This information is communicated to the network users, which set their sending rate equal to p. The desired sending rate p is updated according to the following control law:

$$\dot{p} = \begin{cases} \frac{1}{\hat{N}}[k_i(C-y) - k_q q] & \text{if } 1 < p < C, \\ \frac{1}{\hat{N}}[k_i(C-y) - k_q q] & \text{if } p = 1, \frac{1}{\hat{N}}[k_i(C-y) - k_q q] > 0, \\ \frac{1}{\hat{N}}[k_i(C-y) - k_q q] & \text{if } p = C, \frac{1}{\hat{N}}[k_i(C-y) - k_q q] < 0, \\ 0 & \text{otherwise,} \end{cases} \tag{8.81}$$

where $\hat{N}$ is an estimate of N which is calculated online and k_i, k_q are design parameters. Since N is changing with time, its estimate $\hat{N}$ has to be updated accordingly. In this study, we use online parameter estimation to generate $\hat{N}$. Since the sending rate of all users is equal to p, it follows that

$$y = Np. \tag{8.82}$$

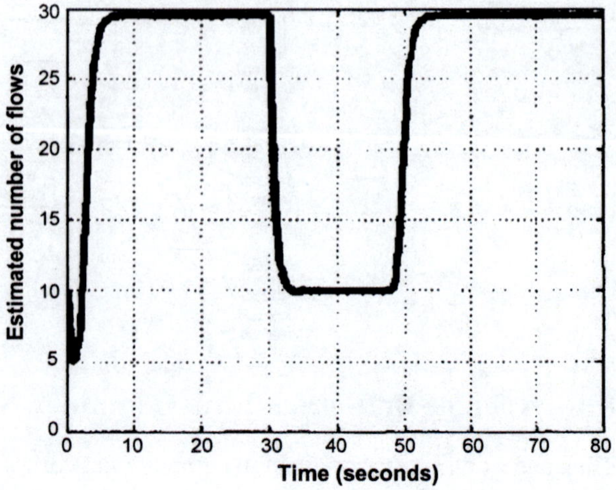

Fig. 8.9 Time response of the estimate of the number of flows

Since y and p are measured at the bottleneck link, (8.82) is in the form of an SPM with N as the unknown parameter. We also know that N cannot be less than 1. Using the results of the chapter, we propose the following online parameter estimator:

$$\dot{\hat{N}} = \begin{cases} \gamma\varepsilon p & \text{if } \hat{N} > 1 \text{ or } \hat{N} = 1 \text{ and } \varepsilon p \geq 0, \\ 0 & \text{otherwise,} \end{cases}$$
$$\varepsilon = \frac{y - \hat{N}p}{1 + p^2}, \tag{8.83}$$

where $\hat{N}(0) \geq 1$. We demonstrate the effectiveness of the proposed algorithm using simulations, which we conduct on the packet-level simulator $ns - 2$. We consider the network topology of Fig. 8.8 in our simulations. The bandwidth of the bottleneck link is set to 155 Mb/s, and the propagation delay of each link is set to 20 ms. The design parameters are chosen as follows: $\gamma = 0.1$, $k_i = 0.16$, $k_q = 0.32$. Initially 30 users utilize the network. The estimator starting with an initial estimate of 10 converges to 30. After $t = 30$ seconds 20 of these users stop sending data, while an additional 20 users enter the network at $t = 45$ s. The output of the estimator at the bottleneck link is shown in Fig. 8.9. We observe that the estimator accurately tracks the number of flows utilizing the network. In addition, we observe good transient behavior as the responses are characterized by fast convergence and no overshoots. The estimator results are obtained in the presence of noise and delays which were not included in the simple model (8.82). Since the number of parameters to be estimated is 1, the PE property is satisfied for $p \neq 0$, which is always the case in this example.

8.14 MATLAB Hints

A fundamental software toolkit is the Adaptive Control Toolbox for MATLAB and Simulink. It aims at designing, implementing, and analyzing parameter identifica-

Table 8.1 A short account of some of the MATLAB functions

ucgrad, ucgradbk	On-line continuous-time parameter estimation using generalized gradient method (including SPR-Lyapunov approach)
udgrad	On-line discrete-time parameter estimation using generalized gradient method (including SPR–Lyapunov approach)
ucrls	On-line continuous-time parameter estimation using least-squares method
udproj	On-line parameter estimation using projection algorithm
mrcpoly	MRC/MRAC design based on polynomial approach
udmracdr	Discrete-time direct MRAC
udmracidr	Discrete-time indirect MRAC
dmpc	Minimum prediction error control (MPE) design
umpcdrl	Direct MPEC for linear parametric models
umpcidr	Indirect MPEC

tion and adaptive control schemes. It is of interest to stress that the design and implementation can be performed for both continuous-time and discrete-time plants. This extensive toolbox includes most of the widely-accepted algorithms in the adaptive control literature. These include various gradient and least squares based parameter identification routines, model reference and pole placement adaptive control laws, parameter projection and robust modification algorithms. In Table 8.1 we provide a short account of some of the MATLAB functions [4].

8.15 Questions

1. Consider the third-order plant

$$y = G(s)u,$$

where

$$G(s) = \frac{b_2 s^2 + b_1 s + b_0}{s^3 + a_2 s^2 + a_1 s + a_0}.$$

a) Obtain parametric models for the plant in the form of SPM and DPM when $\theta^* = [b_2, b_1, b_0, a_2, a_1, a_0]^t$.
b) If a_0, a_1, and a_2 are known, i.e., $a_0 = 2$, $a_1 = 1$, and $a_2 = 3$, obtain a parametric model for the plant in terms of $\theta^* = [b_2, b_1, b_0]^t$.
c) If b_0, b_1, and b_2 are known, i.e., $b_0 = 2$, $b_1 = b_2 = 0$, obtain a parametric model in terms of $\theta^* = [a_2, a_1, a_0]^t$.

2. Consider the mass-spring-dashpot system of Fig. 8.6 described by (8.6) with x, u as the only signals available for measurement. Let us assume that $M = 100$ kg and f, k are the unknown constant parameters that we want to estimate online. Develop a parametric model for estimating the unknown parameters f, k. Specify any arbitrary parameters or filters used.

3. Consider the second-order ARMA model
$$y(k) = -1.3y(k-1) - a_2 y(k-2) + b_1 u(k-1) + u(k-2),$$
where the parameters a_2, b_1 are unknown constants. Express the unknown parameters in the form of a linear parametric model. Assume that $u(k)$, $y(k)$, and their past values are available for measurement.
4. Consider the fourth-order ARMA model
$$y(k+4) = a_1 y(k+3) - a_2 y(k) + b_1 u(k) + u(k+2),$$
where a_1, a_2, b_1 are unknown constants. Express the unknown parameters in the form of a linear parametric model. Assume that only the current and the past four values of the signals u and y, i.e., $u(k), \dots, u(k-4)$, $y(k), \dots, y(k-4)$, are available for measurement.
5. Consider the nonlinear system
$$\ddot{x} + 2\dot{x} + x = a_1 f_1(x) + a_2 f_2(x) + b_1 g_1(x)u + b_2 g_2(x)u,$$
where a_1, a_2, b_1, b_2 are unknown constants and x, $f_1(x)$, $f_2(x)$, $g_1(x)$, $g_2(x)$, u are available for measurement. Express the unknown parameters in the form of
   a) the linear SPM,
   b) the linear DPM.
6. Consider the following system described in the I/O form
$$y = K_p \frac{s+b}{s^2 + as + c} u,$$
where b, a, c, K_p are unknown constants. In addition, we know that $K_p > 0$ and only u and y are available for measurement. Express the unknown parameters in the form of the
   a) B-SPM,
   b) B-DPM,
   c) linear SPM,
   d) linear DPM.
7. Consider the nonlinear system
$$\dot{x} = f(x) + g(x)u,$$
where the state x and the input u are available for measurement and $f(x)$, $g(x)$ are smooth but unknown functions of x. In addition, it is known that $g(x) > 0$ $\forall x$. We want to estimate the unknown functions f, g online using neural network approximation techniques. It is known that there exist constant parameters W_f^*, W_g^*, referred to as weights, such that
$$f(x) \approx \sum_{i=1}^{m} W_{fi}^* \varphi_{fi}(x),$$
$$g(x) \approx \sum_{i=1}^{n} W_{gi}^* \varphi_{gi}(x),$$

where $\varphi_{fi}(\cdot)$, $\varphi_{gi}(\cdot)$ are some basis functions that are known and n, m are known integers representing the number of nodes of the neural network. Obtain a parameterization of the system in the form of SPM that can be used to identify the weights W_{fi}^*, W_{gi}^* online.

8. Consider the mass-spring-dashpot system of Fig. 8.6 described by (8.6) and the SPM with $\theta^* = [M, j, k]^t$ presented in Example 8.4.
   a) Generate the signals z, ϕ of the parametric model using the Adaptive Control Toolbox for $M = 100$ kg, $f = 0.15$ kg/s, $k = 7$ kg/s^2, $u(t) = 1 + \cos(\frac{\pi}{3}t)$, and $0 \leq t \leq 25$ s.
   b) The SPM in (a) is based on the assumption that M, f, k are unknown. Assume that M is known. Use the Adaptive Control Toolbox to generate the signals of the reduced SPM for the same values of M, f, k, $u(t) = 1 + \cos(\frac{\pi}{3}t)$, and $0 \leq t \leq 25$ s.
   c) Consider the SPM

   $$z = \theta^{*t}\phi$$

   and the estimation model

   $$\hat{z} = \theta^t(t)\phi.$$

   Find values for $\theta(t)$, $\phi(t)$ such that $z = \hat{z}$ but $\theta(t) \neq \theta^*$.
   d) Consider the adaptive law

   $$\dot{\theta} = \gamma\varepsilon\phi,$$

   where $\theta, \phi \in \Re^n$, $\frac{\phi}{m_s} \in L_\infty$, $\varepsilon m_s \in L_2 \cap L_\infty$, and $m_s \geq 1$. Show that $\dot{\theta} \in L_2 \cap L_\infty$.
   e) Show that if $u \in L_\infty$ in (8.8), then the adaptive law (8.17) with $m_s^2 = 1 + \alpha\phi^2, \alpha \geq 0$, guarantees that $\varepsilon, \varepsilon m_s, \dot{\theta} \in L_2 \cap L_\infty$ and that $\varepsilon(t)$, $\varepsilon(t)m_s(t)$, $\dot{\theta}(t) \to 0$ as $t \to \infty$.
   f) (a) Show that (8.23) is a necessary and sufficient condition for $\theta(t)$ in the adaptive law (8.17) to converge to θ^* exponentially fast.
      (b) Establish which of the following choices for u guarantee that ϕ in (8.17) is PE:
         (i) $u = c_0 \neq 0$, c_0 is a constant.
         (ii) $u = \sin t$.
         (iii) $u = \sin t + \cos 2t$.
         (iv) $u = \frac{1}{1+t}$.
         (v) $u = e^{-t}$.
         (vi) $u = \frac{1}{(1+t)^{1/2}}$.
      (c) In (b), is there a choice for u that guarantees that $\theta(t)$ converges to θ^* but does not guarantee that ϕ is PE?
9. Use the plant model (8.53) to develop the bilinear parametric model. Show all the steps.
10. In Theorem 8.1, assume that $\phi, \dot{\phi} \in L_\infty$. Show that the adaptive law (8.33) with $m_s^2 = 1 + n_s^2$, $n_s^2 = \alpha\phi^t\phi$, and $\alpha \geq 0$ guarantees that $\varepsilon(t)$, $\varepsilon(t)m_s(t)$, $\dot{\theta}(t) \to 0$ as $t \to \infty$.

11. Consider the SPM $z=\theta^{*t}\phi$ and the cost function
$$J(\theta)=\frac{1}{2}\int_0^t e^{-\beta(t-\tau)}\frac{(z(\tau)-\theta^t(t)\phi(\tau))^2}{m_s^2(\tau)}d\tau,$$
where $m_s^2=1+\phi^t\phi$ and $\theta(t)$ is the estimate of θ^* at time t.
   i. Show that the minimization of $J(\theta)$ w.r.t. θ using the gradient method leads to the adaptive law
$$\dot{\theta}(t)=\Gamma\int_0^t e^{-\beta(t-\tau)}\frac{z(\tau)-\theta^t(t)\phi(\tau)}{m_s^2(\tau)}\phi(\tau)d\tau,\quad \theta(0)=\theta_0.$$
   ii. Show that the adaptive law in part (a) can be implemented as
$$\dot{\theta}(t)=-\Gamma\big(R(T)\theta(t)+Q(t)\big),\qquad \theta(0)=\theta_0,$$
$$\dot{R}(t)=-\beta R(t)+\frac{\phi(t)\phi^t(t)}{m_s^2(t)},\qquad R(0)=0,$$
$$\dot{Q}(t)=-\beta Q(t)-\frac{z(t)\phi(t)}{m_s^2(t)},\qquad Q(0)=0,$$
   which is referred to as the integral adaptive law.
12. Consider the second-order stable system
$$\dot{x}=\begin{bmatrix}a_{11} & a_{12}\\ a_{21} & 0\end{bmatrix}x+\begin{bmatrix}b_1\\ b_2\end{bmatrix}u,$$
where x, u are available for measurement, $u\in L_\infty$ and a_{11}, a_{12}, a_{21}, b_1, b_2 are unknown parameters. Design an online estimator to estimate the unknown parameters. Simulate your scheme using $a_{11}=-0.25$, $a_{12}=3$, $a_{21}=-5$, $b_1=1$, $b_2=2.2$, and $u=10\sin 2t$. Repeat the simulation when $u=10\sin 2t+7\cos 3.6t$. Comment on your results.
13. Consider the nonlinear system
$$\dot{x}=a_1 f_1(x)+a_2 f_2(x)+b_1 g_1(x)u+b_2 g_2(x)u,$$
where $u,x\in R$; f_i, g_i are known nonlinear functions of x; and a_i, b_i are unknown constant parameters and $i=1,2$. The system is such that $u\in L_\infty$ implies $x\in L_\infty$. Assuming that x,u can be measured at each time t, design an estimation scheme for estimating the unknown parameters online.
14. Design and analyze an online estimation scheme for estimating θ^* in (8.47) when $L(s)$ is chosen so that $W(s)L(s)$ is biproper and SPR.
15. Design an online estimation scheme to estimate the coefficients of the numerator polynomial
$$Z(s)=b_{n-1}s^{n-1}+b_{n-2}s^{n-2}+\cdots+b_1 s+b_0$$
of the plant
$$y=\frac{Z(s)}{R(s)}u$$
when the coefficients of $R(s)=s^n+a_{n-1}s^{n-1}+\cdots+a_1 s+a_0$ are known. Repeat the same problem when $Z(s)$ is known and $R(s)$ is unknown.

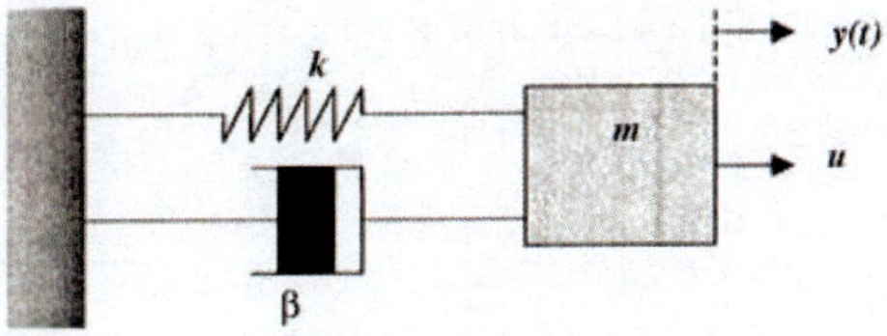

Fig. 8.10 The mass-spring-damper system for Problem 17

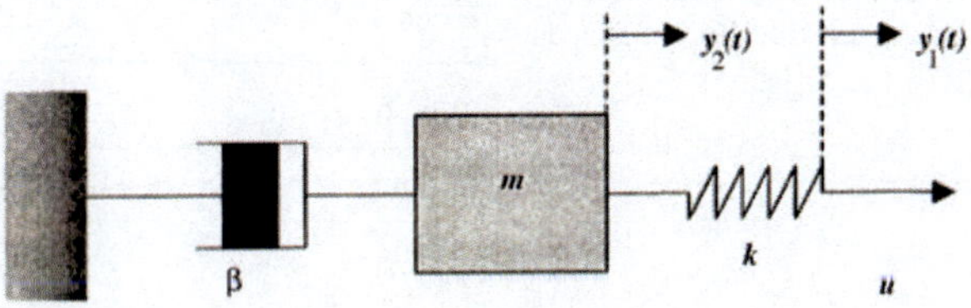

Fig. 8.11 The mass-spring-damper system for Problem 16

16. Consider the mass-spring-damper system shown in Fig. 8.10, where β is the damping coefficient, k is the spring constant, u is the external force, and $y(t)$ is the displacement of the mass m resulting from the force u.
    i. Verify that the equations of motion that describe the dynamic behavior of the system under small displacements are
    $$m\ddot{y} + \beta\dot{y} + ky = u.$$
    ii. Design a gradient algorithm to estimate the constants m, β, k when y, u can be measured at each time t.
    iii. Repeat (b) for an LS algorithm.
    iv. Simulate your algorithms in (b) and (c) assuming $m = 20$ kg, $\beta = 0.1$ kg/s, $k = 5$ kg/s^2, and inputs u of your choice.
    v. Repeat (d) when $m = 20$ kg for $0 \le t \le 20$ s and $m = 20(2 - e^{-0.01(r-20)})$ kg for $t \ge 20$ s.
17. Consider the mass-spring-damper system shown in Fig. 8.11.
    i. Verify that the equations of motion are given by
    $$k(y_1 - y_2) = u,$$
    $$k(y_1 - y_2) = m\ddot{y}_2 + \beta\dot{y}_2.$$
    ii. If y_1, y_2, u can be measured at each time t, design an online parameter estimator to estimate the constants k, m, and β.
    iii. We have the a priori knowledge that $0 \le \beta \le 1$, $k \ge 0.1$, and $m \ge 10$. Modify your estimator in (b) to take advantage of this a priori knowledge.
    iv. Simulate your algorithm in (b) and (c) when $\beta = 0.2$ kg/s, $m = 15$ kg, $k = 2$ kg/s^2, and $u = 5\sin 2t + 10.5$ kg m/s^2.
18. Consider the block diagram of a steer-by-wire system of an automobile shown in Fig. 8.12, where r is the steering command in degrees, θ_p is the pinion angle in degrees, and $\dot{\theta}$ is the yaw rate in degree/s. The transfer functions $G_0(s)$, $G_1(s)$ are of the form
$$G_0(s) = \frac{k_0\omega_0^2}{s^2 + 2\xi_0\omega_0 s + \omega_0^2(1 - k_0)},$$

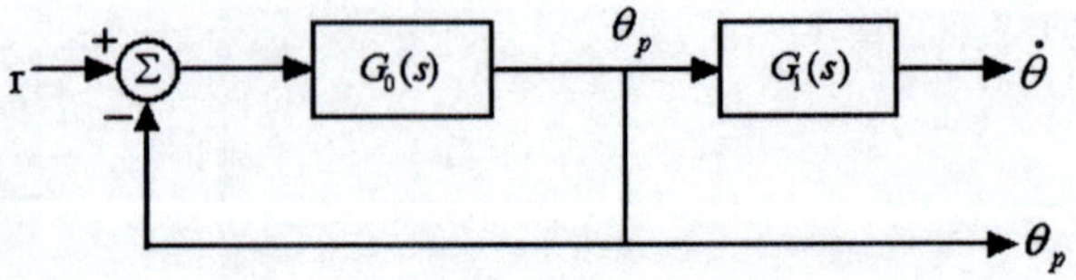

Fig. 8.12 Block diagram of a steer-by-wire system for Problem 18

Table 8.2 The values of the parameters at different speeds

Speed V	k_0	ω_0	ξ_0	k_1	ω_1	ξ_1
30 mph	0.81	19.75	0.31	0.064	14.0	0.365
60 mph	0.77	19.0	0.27	0.09	13.5	0.505

$$G_1(s) = \frac{k_1\omega_1^2}{s^2 + 2\xi_1\omega_1 s + \omega_1^2},$$

where $k_0, \omega_0, \xi_0, k_1, \omega_1, \xi_1$ are functions of the speed of the vehicle. Assuming that $r, \theta_p, \dot{\theta}$ can be measured at each time t, do the following:

i. Design an online parameter estimator to estimate $k_i, \omega_i, \xi_i, i = 0, 1$, using the measurements of $\theta_p, \dot{\theta}, r$.

ii. Consider the values of the parameters shown in Table 8.2 at different speeds: Assume that between speeds the parameters vary linearly. Use these values to simulate and test your algorithm in (a) when

A. $r = 10\sin 0.2t + 8$ degrees and $V = 20$ mph.

B. $r = 5$ degrees and the vehicle speeds up from $V = 30$ mph to $V = 60$ mph in 40 seconds with constant acceleration and remains at 60 mph for 10 seconds.

19. Consider the equation of the motion of the mass-spring-damper system given in Problem 16, i.e.,

$$m\ddot{y} + \beta\dot{y} + ky = u.$$

This system may be written in the form

$$y = \rho^*(u - m\ddot{y} - \beta\dot{y}),$$

where $\rho^* = \frac{1}{k}$ appears in a bilinear form with the other unknown parameters m, β. Use the adaptive law based on the bilinear parametric model to estimate ρ^*, m, β when u, y are the only signals available for measurement. Since $k > 0$, the sign of ρ^* may be assumed to be known. Simulate your adaptive law using the numerical values given in (d) and (e) of Problem 8.

20. The effect of initial conditions on the SPM can be modeled as

$$z = \theta^{*t}\phi + \eta_0,$$

$$\dot{\omega} = \Lambda\omega_0,$$

$$\eta_0 = C^t\omega_0,$$

where Λ is a transfer matrix with all poles in $R[s] < 0$ and $\omega_0 \in \Re^n$, $\eta_0 \in R$. Show that the properties of an adaptive law (gradient, LS, etc.) with $\eta_0 = 0$ are the same as those for the SPM with $\eta_0 \neq 0$.

21. Consider the system

$$y = e^{-\tau s}\frac{b}{(s-a)(\mu s+1)}u,$$

where $0 < \tau \ll 1$, $0 < \mu \ll 1$, and a, b, τ, μ are unknown constants. We want to estimate a, b online.
   i. Obtain a parametric model that can be used to design an adaptive law to estimate a, b.
   ii. Design a robust adaptive law of your choice to estimate a, b online.
   iii. Simulate your scheme for $a = -5$, $b = 100$, $\tau = 0.0$, $\mu = 0.001$, and different choices of the input signal u. Comment on your results.

22. The dynamics of a hard-disk drive servo system are given by

$$y = \frac{k_p}{s^2}\sum_{i=1}^{6}\frac{b_{1i}s + b_{0i}}{s^2 + 2\zeta_i\omega_i s + \omega_i^2}u,$$

where ω_i, $i = 1, \ldots, 6$, are the resonant frequencies which are large, i.e., $\omega_1 = 11.2\pi \times 10^3$ rad/s, $\omega_2 = 15.5\pi \times 10^3$ rad/s, $\omega_3 = 16.6\pi \times 10^3$ rad/s, $\omega_4 = 18\pi \times 10^3$ rad/s, $\omega_5 = 20\pi \times 10^3$ rad/s, $\omega_6 = 23.8\pi \times 10^3$ rad/s. The unknown constants b_{1i} are of the order of 10^4, b_{0i} are of the order of 10^8, and k_p is of the order of 10^7. The damping coefficients ζ_i are of the order of 10^{-2}.
   i. Derive a low-order model for the servo system. (Hint: Take $\frac{\alpha}{\omega_i} \cong 0$ and hence $\frac{\alpha^2}{\omega_i^2} \cong 0$ for α of the order of less than 10^3.)
   ii. Assume that the full-order system parameters are given as follows:

$$\begin{aligned} &b_{16} = -5.2 \times 10^4, && b_{01} = 1.2 \times 10^9, && b_{02} = 5.4 \times 10^8,\\ &b_{03} = -7.7 \times 10^8, && b_{04} = -1.6 \times 10^8, && b_{05} = -1.9 \times 10^8,\\ &b_{06} = 1.2 \times 10^9, && k_p = 3.4 \times 10^7, && \zeta_1 = 2.6 \times 10^{-2},\\ &\zeta_2 = 4.4 \times 10^{-3}, && \zeta_3 = 1.2 \times 10^{-2}, && \zeta_4 = 2.4 \times 10^{-3},\\ &\zeta_5 = 6.8 \times 10^{-3}, && \zeta_6 = 1.5 \times 10^{-2}. \end{aligned}$$

   Obtain a Bode plot for the full-order and reduced-order models.
   iii. Use the reduced-order model in (a) to obtain a parametric model for the unknown parameters. Design a robust adaptive law to estimate the unknown parameters online.

23. Consider the time-varying plant

$$\dot{x} = -a(t)x + b(t)u,$$

where $a(t)$, $b(t)$ are slowly varying unknown parameters; i.e., $|\dot{a}|$, $|\dot{b}|$ are very small.
   i. Obtain a parametric model for estimating a, b.
   ii. Design and analyze a robust adaptive law that generates the estimates $\hat{a}(t)$, $\hat{b}(t)$ of $a(t)$, $b(t)$, respectively.
   iii. Simulate your scheme for a plant with $a(t) = 5 + \sin\mu t$, $b(t) = 8 + \cos 2\mu t$ for $\mu = 0, 0.01, 0.1, 1, 5$. Comment on your results.

24. Consider the parameter error differential equation (8.62), i.e.,

$$\dot{\tilde{\theta}} = -\gamma u^2 \tilde{\theta} + \gamma d u.$$

Show that if the equilibrium $\tilde{\theta}_e = 0$ of the homogeneous equation

$$\dot{\tilde{\theta}} = -\gamma u^2 \tilde{\theta}$$

is exponentially stable, then the bounded input $\gamma d u$ will lead to a bounded solution $\tilde{\theta}(t)$. Obtain an upper bound for $|\tilde{\theta}(t)|$ as a function of the upper bound of the disturbance term $\gamma d u$.

25. Consider the system

$$y = \theta^* u + \eta$$
$$\eta = \Delta(s) u,$$

where y, u are available for measurement, θ^* is the unknown constant to be estimated, and η is a modeling error signal with $\Delta(s)$ being proper and analytic in $R[s] \geq -0.5$.

The input u is piecewise continuous.

i. Design an adaptive law with a switching σ-modification to estimate θ^*.
ii. Repeat (a) using projection.
iii. Simulate the adaptive laws in (a), (b) using the following values:

$$\theta^* = 5 + \sin 0.1t,$$
$$\Delta(s) = 10\mu \frac{\mu s - 1}{(\mu s + 1)^2}$$

for $\mu = 0, 0.1, 0.01$ and $u =$ constant, $u = \sin \omega_0 t$, where $\omega_0 = 1, 10, 100$. Comment on your results.

26. The linearized dynamics of a throttle angle θ to vehicle speed V subsystem are given by the third-order system

$$V = \frac{b p_1 p_2}{(s+a)(s+p_1)(s+p_2)} \theta + d,$$

where $p_1, p_2 > 20$, $1 \geq a > 0$, and d is a load disturbance.

(a) Obtain a parametric model for the parameters of the dominant part of the system.
(b) Design a robust adaptive law for estimating these parameters online.
(c) Simulate your estimation scheme when $a = 0.1$, $b = 1$, $p_1 = 50$, $p_2 = 100$, and $d = 0.02 \sin 5t$, for different constant and time-varying throttle angle settings θ of your choice.

27. Consider the parametric model

$$z = \theta^{*t} \phi + \eta,$$

where

$$\eta = \Delta_u(s) u + \Delta_y(s) y$$

and Δ_u, Δ_y are proper transfer functions analytic in $R[s] \geq -\frac{\delta_0}{2}$ for some known $\delta_0 > 0$.

a) Design a normalizing signal m_s that guarantees $\frac{\eta}{m_s} \in L_\infty$ when (i) Δ_u, Δ_y are biproper, (ii) Δ_u, Δ_y are strictly proper. In each case specify the upper bound for $\frac{|\eta|}{m_s}$.

b) Calculate the bound for $\frac{\eta}{m_s}$ when
   i. $\Delta_u(s) = \frac{e^{\tau s}-1}{(s+1)^2}$, $\Delta_y(s) = \mu \frac{s^2}{(s+1)^2}$,
   ii. $\Delta_u(s) = \frac{\mu s}{\mu s+2}$, $\Delta_y(s) = \frac{\mu s}{\mu s+1}$, where $0 < \mu \ll 1$ and $0 < \tau \ll 1$.

c) Design and simulate a robust adaptive law to estimate θ^* for the following example:

$$\theta^* = [1, 0.1]^t,$$
$$\phi = [u, y]^t,$$
$$z = \frac{s}{s+5} y,$$
$$\Delta_u(s) = \frac{e^{-\tau s} - 1}{s+2}, \qquad \Delta_y(s) = \frac{\mu(s-1)}{(\mu s+1)^2},$$

where $\tau = 0.01$, $\mu = 0.01$.

References

1. Anderson, B.O.D., Brinsmead, T., Liberson, D., Morse, A.S.: Multiple model adaptive control with safe switching. Int. J. Adapt. Control Signal Process. **15**, 445–470 (2001)
2. Astrom, K.J.: Theory and applications of adaptive control: A survey. Automatica **19**, 471–486 (1983)
3. Astrom, K.J., Wittenmark, B.: Adaptive Control. Addison-Wesley, Reading (1989)
4. Fidan, B., Ioannou, P.A.: Adaptive Control Toolbox: User's Guide. The Mathworks, Boston (2005)
5. Fu, M., Barmish, B.R.: Adaptive stabilization of linear systems via switching control. IEEE Trans. Autom. Control **31**, 1097–1103 (1986)
6. Goodwin, G.C., Sin, K.S.: Adaptive Filtering Prediction and Control. Prentice-Hall, Englewood Cliffs (1984)
7. Gupta, M.M. (ed.): Adaptive Methods for Control System Design. IEEE Press, Piscataway (1986)
8. Ioannou, P.A., Fidan, B.: Adaptive Control Tutorial, SIAM's Advances in Design and Control (2006)
9. Ioannou, P.A., Sun, J.: Robust Adaptive Control. Prentice-Hall, Englewood Cliffs (1996); also available online from http://www-rcf.usc.edu/~ioannou/Robust_Adaptive_Control.htm
10. Kanellakopoulos, I., Kokotovis, P.V., Morse, A.S.: Systematic design of adaptive controllers for feedback linearizable systems. IEEE Trans. Autom. Control **36**, 1241–1253 (1991)
11. Kreisselmier, G.: An indirect adaptive controller with a self-excitation capability. IEEE Trans. Autom. Control **34**, 524–528 (1989)
12. Landau, I.D.: Adaptive Control: The Model Reference Approach. Marcel Dekker, New York (1979)
13. Miller, D.E., Davison, E.J.: An adaptive controller which provides an arbitrarily good transient and steady-state response. IEEE Trans. Autom. Control **36**, 68–81 (1991)
14. Narendra, K.S., Balakrishnan, J.: Improving transient response of adaptive control systems using multiple models and switching. IEEE Trans. Autom. Control **39**, 1861–1866 (1994)

15. Narendra, K.S., Balakrishnan, J.: Adaptation and learning using multiple models, switching, and tuning. IEEE Control Syst. Mag. **15**, 37–51 (1995)
16. Narendra, K.S., Balakrishnan, J.: Adaptive control using multiple models. IEEE Trans. Autom. Control **42**, 171–187 (1997)
17. Narendra, K.S., Monopoli, R.V. (eds.): Applications of Adaptive Control. Academic Press, New York (1980)
18. Paul, A., Safonov, M.G.: Model reference adaptive control using multiple controllers and switching. In: Proc. the 42nd IEEE Conference on Decision and Control, vol. 4, pp. 3256–3261. IEEE Press, New York (2003)
19. Poularikas, A.D., Ramadan, Z.M.: Adaptive Filtering Primer with MATLAB. Taylor and Francis, New York (2006)
20. Stein, G.: Adaptive flight control: A pragmatic view. In: Narendra, K.S., Monopoli, R.V. (eds.) Applications of Adaptive Control. Academic Press, New York (1980)
21. Tsakalis, K.S., Ioannou, P.A.: A new indirect adaptive control scheme for time-varying plants. IEEE Trans. Autom. Control **35**, 697–705 (1990)
22. Tsakalis, K.S., Ioannou, P.A.: Linear Time Varying Systems: Control and Adaptation. Prentice-Hall, Englewood Cliffs (1993)
23. Wang, R., Safonov, M.G.: Stability of unfalsified adaptive control using multiple controllers. In: Proc. the American Control Conference, pp. 3162–3167 (2005)

Chapter 9
Appendix

9.1 Important Facts in Linear Algebra

We assume that the reader already has basic training in linear algebra and for a more complete introduction, the reader should resort to a book such as Strang [6]. Some familiarity with the numerical software MATLAB is also encouraged. In this book, we deal mostly with finite-dimensional linear spaces, which are also often called linear vector spaces. For generality, we consider the linear space to be n-dimensional. A linear space is typically denoted by the letter V (for "vector space"). Although most of the time we will deal with vectors of real numbers $\Re$, occasionally, we will encounter vectors of complex numbers $\mathbb{C}$. As a reminder, for instance, the eigenvalues or eigenvectors of a real matrix could be complex. For simplicity, our review will be conducted for linear spaces over the field of real numbers $\Re$ with the understanding that most definitions and results generalize to the complex case with little change.

9.1.1 Basic Notions

Definition 9.1 (A linear space or a vector space) A set (of vectors) V is considered a linear space over the field $\Re$ if its elements, called vectors, are closed under two basic operations: scalar multiplication and vector summation "+". That is, given any two vectors $v_1, v_2 \in V$ and any two scalars $\alpha, \beta \in \Re$, the linear combination $v = \alpha v_1 + \beta v_2$ is also a vector in V. Furthermore, the addition is commutative and associative, it has an identity 0, and each element has an inverse, "$-v$", such that $v + (-v) = 0$. The scalar multiplication respects the structure of $\Re$, that is, $\alpha(\beta)v = (\alpha\beta)v$, $1v = v$ and $0v = 0$. The addition and scalar multiplication are related by the distributive laws: $(\alpha + \beta)v = \alpha v + \beta v$ and $\alpha(v + u) = \alpha v + \alpha u$.

For example, $\Re^n$ is a linear space over the field of real numbers $\Re$. To be consistent, we always use a column to represent a vector:

M.S. Mahmoud, Y. Xia, *Applied Control Systems Design*,
DOI 10.1007/978-1-4471-2879-3_9,

$$\begin{bmatrix} x_1 & x_2 & \cdots & x_n \end{bmatrix}^t = \begin{bmatrix} x_1 \\ x_2 \\ \vdots \\ x_n \end{bmatrix} \in \Re^n \tag{9.1}$$

where $[x_1\ x_2\ \cdots\ x_n]^t$ means "the (row) vector $[x_1\ x_2\ \cdots\ x_n]$ transposed." Given two scalars $\alpha, \beta \in \Re$ and two vectors $x = [x_1\ x_2\ \cdots\ x_n]^t \in \Re^n$ and $y = [y_1\ y_2\ \cdots\ y_n]^t \in \Re^n$, their linear combination is a componentwise summation weighted by α and β:

$$\begin{aligned} \alpha x + \beta y &= \alpha \begin{bmatrix} x_1 & x_2 & \cdots & x_n \end{bmatrix}^t + \beta \begin{bmatrix} y_1 & y_2 & \cdots & y_n \end{bmatrix}^t \\ &= \begin{bmatrix} \alpha x_1 + \beta y_1 & \alpha x_2 + \beta y_2 & \cdots & \alpha x_n + \beta y_n \end{bmatrix}^t. \end{aligned} \tag{9.2}$$

We will now provide a brief review of basic notions and frequently used notation associated with a linear vector space V (that is, $\Re^n$).

Definition 9.2 (Subspace) A subset W of a linear space is called a subspace if the zero vector $\mathbf{0}$ is in W and $w = \alpha w_1 + \beta w_2 \in W$ for all $\alpha, \beta \in \Re$ and $w_1, w_2 \in W$.

Definition 9.3 (Spanned subspace) Given a set of vectors $S = \{v_i\}_{i=1}^m$, the subspace spanned by S is the set of all finite linear combinations $\sum_{i=1}^m \alpha_i v_i$ for all $[\alpha_1\ \alpha_2\ \cdots\ \alpha_n]^t$. This subspace is usually denoted by Span(S).

For example, the two vectors $v_1 = [1\ 0\ 0]^t$ and $v_2 = [1\ 1\ 0]^t$ span a subspace of $\Re^3$ whose vectors are of the general form $v = [x\ y\ 0]^t$.

Definition 9.4 (Linear independence) A set of vectors $S = \{v_i\}_{i=1}^m$ is linearly independent if

$$[\alpha_1 v_1 + \alpha_2 v_2 + \cdots + \alpha_m v_m] = 0$$

implies

$$\alpha_1 = \alpha_2 = \cdots = \alpha_m = 0.$$

On the other hand, a set of vectors $\{v_i\}_{i=1}^m$ is said to be linearly dependent if there exist $[\alpha_1\ \alpha_2\ \cdots\ \alpha_n] \in \Re$ not all zero such that

$$[\alpha_1 v_1 + \alpha_2 v_2 + \cdots + \alpha_m v_m] = 0.$$

Definition 9.5 (Basis) A set of vectors $B = \{b_i\}_{i=1}^n$ of a linear space V is said to be a basis if B is a linearly independent set and B spans the entire space V; that is, $V = \text{span}(B)$.

Properties of a basis: Suppose B and B' are two bases for a linear space V. Then:

1. B and B' contain exactly the same number of linearly independent vectors. This number, say n, is the dimension of the linear space V.

2. Let $B = \{b_i\}_{i=1}^n$ and $B' = \{b'_i\}_{i=1}^n$. Then each basis vector of B can be expressed as a linear combination of those in B'; that is,

$$b_j = a_{1j}b'_1 + a_{2j}b'_2 + \cdots + a_{nj}b'_n = \sum_{i=1}^{n} a_{ij}b'_i, \tag{9.3}$$

for some $a_{ij} \in R, i, j = 1, 2, \ldots, n$.
3. Any vector $v \in V$ can be written as a linear combination of vectors in either of the bases:

$$v = x_1b_1 + x_2b_2 + \cdots + x_nb_n = x_nb_n = x'_1b'_1 + x'_2b'_2 + \cdots + x'_nb'_n, \tag{9.4}$$

where the coefficients $\{x_i \in R\}_{i=1}^n$ and $\{x'_i \in R\}_{i=1}^n$ are uniquely determined and are called the coordinates of v with respect to each basis.

In particular, if B and B' are two bases for the linear space R^n, we may put the basis vectors as columns of two $n \times n$ matrices and also call them B and B', respectively:

$$B \doteq [b_1, b_2, \ldots, b_n], \qquad B' \doteq \left[b'_1, b'_2, \ldots, b'_n\right] \in R^{n\times n}. \tag{9.5}$$

Then we can express the relationship between them in the matrix form $B = B'A$ as

$$[b_1, b_2, \ldots, b_n] = \left[b'_1, b'_2, \ldots, b'_n\right] \begin{bmatrix} a_{11} & a_{12} & \cdots & a_{1n} \\ a_{21} & a_{22} & \cdots & a_{2n} \\ \vdots & \vdots & \ddots & \vdots \\ a_{n1} & a_{n2} & \cdots & a_{nn} \end{bmatrix}. \tag{9.6}$$

The role of the $n \times n$ matrix is to transform one basis (B') to the other (B). Since such a transformation can go the opposite way, the matrix A must be invertible. So we can also write $B' = BA^{-1}$.

If v is a vector in V, it can be expressed in terms of linear combinations of either basis as

$$v = x_1b_1 + x_2b_2 + \cdots + x_nb_n = x_1^tb_1^t + x_2^tb_2^t + \cdots + x_n^tb_n^t. \tag{9.7}$$

Thus, we have

$$v = [b_1, b_2, \ldots, b_n] \begin{bmatrix} x_1 \\ x_2 \\ \vdots \\ x_n \end{bmatrix} = \left[b_1^t, b_2^t, \ldots, b_n^t\right] \begin{bmatrix} a_{11} & a_{12} & \cdots & a_{1n} \\ a_{21} & a_{22} & \cdots & a_{2n} \\ \vdots & \vdots & \ddots & \vdots \\ a_{n1} & a_{n2} & \cdots & a_{nn} \end{bmatrix} \begin{bmatrix} x_1 \\ x_2 \\ \vdots \\ x_n \end{bmatrix}.$$

Since the coordinates of v with respect to B^t are unique, we obtain the following transformation of coordinates of a vector from one basis to the other:

$$\begin{bmatrix} x_1^t \\ x_2^t \\ \vdots \\ x_n^t \end{bmatrix} = \begin{bmatrix} a_{11} & a_{12} & \cdots & a_{1n} \\ a_{21} & a_{22} & \cdots & a_{2n} \\ \vdots & \vdots & \ddots & \vdots \\ a_{n1} & a_{n2} & \cdots & a_{nn} \end{bmatrix} \begin{bmatrix} x_1 \\ x_2 \\ \vdots \\ x_n \end{bmatrix}. \tag{9.8}$$

Let $x = [x_1, x_2, \dots, x_n]^t \in R^n$ and $x^t = [x_1^t, x_2^t, \dots, x_n^t]^t \in R^n$ denote the two coordinate vectors. We may summarize in matrix form the relationships between two bases and coordinates with respect to the bases as

$$B^t = BA^{-1}, \qquad x^t = Ax. \tag{9.9}$$

Be aware of the difference in transforming bases from transforming coordinates!

9.1.2 Inner Product and Orthogonality

Definition 9.6 (Inner product) A function is an inner product[1] if

1. $\langle u, \alpha v + \beta w\rangle = \alpha\langle u, v\rangle + \beta\langle u, w\rangle$, $R^n \times R^n \to R$,
2. $\langle u, v\rangle = \langle v, u\rangle$,
3. $\langle v, v\rangle \geq 0$, and $\langle v, v\rangle = 0 \Leftrightarrow v = 0$.

For each vector v, $\sqrt{\langle v, v\rangle}$ is called its norm.

The inner product is also called a metric, since it can be used to measure length and angles.

For simplicity, a standard basis is often chosen for the vector space R^n as the set of vectors

$$e_1 = [1, 0, 0, \dots, 0]^t, \qquad e_2 = [0, 1, 0, \dots, 0]^t, \qquad e_n = [0, 0, 0, \dots, 0, 1]^t. \tag{9.10}$$

The matrix $I = [e_1, e_2, \dots, e_n]$ with these vectors as columns is exactly the $n \times n$ identity matrix.

Definition 9.7 (Canonical inner product on R^n) Given any two vectors $x = [x_1, x_2, \dots, x_n]^t$ and $y = [y_1, y_2, \dots, y_n]^t$ in R^n, we define the canonical inner product to be

$$\langle x, y\rangle \doteq x^t y = x_1 y_1 + x_2 y_2 + \cdots + x_n y_n. \tag{9.11}$$

This inner product induces the standard 2-norm, or Euclidean norm, $\|\cdot\|_2$, which measures the length of each vector as

$$\|x\|_2 \doteq \sqrt{x^t x} = \sqrt{x_1^2 + x_2^2 + \cdots + x_n^2}. \tag{9.12}$$

Notice that if we choose another basis B^t related to the above standard basis I as $I = B^t A$, then the coordinates of the vectors x, y related to the new basis are x^t and y^t, respectively, and they relate to x, y by $x^t = Ax$ and $y^t = Ay$. The inner product in terms of the new coordinates becomes

$$\langle x, y\rangle = x^t y = \left(A^{-1}x^t\right)^t\left(A^{-1}y^t\right) = \left(x^t\right)A^{-t}A^{-1}\left(y^t\right). \tag{9.13}$$

[1] In some literature, an inner product is also called a dot product, denoted by $u \cdot v$. However, in this book, we will not use that name.

We denote this expression of the inner product with respect to the new basis by

$$\langle x^t, y^t \rangle_{A^{-t}A^{-1}} \doteq (x^t)^t A^{-t} A^{-1} (y^t). \tag{9.14}$$

This is called an induced inner product from the matrix A. Knowing the matrix $A^{-t}A^{-1}$, we can compute the canonical inner product directly using coordinates with respect to the nonstandard basis B^t.

Definition 9.8 (Orthogonality) Two vectors x, y are said to be orthogonal if their inner product is zero: $\langle x, y \rangle = 0$. This is often indicated as $x \perp y$.

9.1.3 Kronecker Product and Stack of Matrices

Definition 9.9 (Kronecker product of two matrices) Given two matrices $A \in \mathbb{R}^{m \times n}$ and $B \in \mathbb{R}^{k \times l}$, their Kronecker product, denoted by $A \otimes B$, is a new matrix

$$A \otimes B \doteq \begin{bmatrix} a_{11}B & a_{12}B & \cdots & a_{1n}B \\ a_{21}B & a_{22}B & \cdots & a_{2n}B \\ \vdots & \vdots & \ddots & \vdots \\ a_{n1}B & a_{n2}B & \cdots & a_{mn}B \end{bmatrix} \in \mathbb{R}^{mk \times nl}. \tag{9.15}$$

If A and B are two vectors, that is, $n = l = 1$, the product $A \otimes B$ is also a vector but of dimension mk.

In MATLAB, one can easily compute the Kronecker product by using the command $C = \mathrm{kron}(A, B)$.

Definition 9.10 (Stack of a matrix) Given an $m \times n$ matrix $A \in \mathbb{R}^{m \times n}$ the stack of the matrix A is a vector, denoted by A^s, in $\mathbb{R}^{mn}$ obtained by stacking its n column vectors, say $a_1, a_2, \dots, a_n \in \mathbb{R}^m$, in order

$$A^s \doteq \begin{bmatrix} a_1 \\ a_2 \\ \vdots \\ a_n \end{bmatrix} \in \mathbb{R}^{mn}. \tag{9.16}$$

As mutually inverse operations, A^s is called A "stacked," and A is called A^s "unstacked."

The Kronecker product and stack of matrices together allow us to rewrite algebraic equations that involve multiple vectors and matrices in many different but equivalent ways. For instance, the equation

$$u^t A v = 0 \tag{9.17}$$

for two vectors u, v and a matrix A of proper dimensions can be rewritten as

$$(v \otimes u)^t A^s = 0. \tag{9.18}$$

The second equation is particularly useful when A is the only unknown in the equation.

9.2 Linear Transformations and Matrix Groups

Linear algebra studies the properties of the linear transformations, or linear maps, between different linear spaces. Since such transformations can be represented as matrices, linear algebra to a large extent studies the properties of matrices.

Definition 9.11 (Linear transformation) A linear transformation from a linear (vector) space R^n to R^m is defined as a map $L : R^n \to R^m$ such that

- $L(x+y) = L(x) + L(y)$, $\forall x, y \in R^n$;
- $L(\alpha x) = \alpha L(x)$, $\forall x \in R^n$, $\alpha \in R$.

With respect to the standard bases of R^n and R^m, the map L can be represented by a matrix $A \in R^{m \times n}$ such that

$$L(x) = Ax, \quad \forall x \in \Re^n. \tag{9.19}$$

The ith column of the matrix A is then nothing but the image of the standard basis vector $e_i \in \Re^n$ under the map L; that is,

$$A = \big[L(e_1), L(e_2), \ldots, L(e_n)\big] \in \Re^{m \times n}.$$

The set of all (real) $m \times n$ matrices is denoted by $M(m,n)$. When viewed as a linear space, $M(m,n)$ can be identified as the space $\Re^{mn}$. When there is little ambiguity, we refer to a linear map L by its matrix representation A. If $n = m$, the set $M(n,n) \doteq M(n)$ forms an algebraic structure called a ring (over the field R). That is, matrices in $M(n)$ are closed under both matrix multiplication and summation: If A, B are two $n \times n$ matrices, so are $C = AB$ and $D = A + B$.

Linear maps or matrices that we encounter in computer vision often have a special algebraic structure called a group.

Definition 9.12 A group is a set G with an operation "$\circ$" on the elements of G that:

- is closed: if $g_1, g_2 \in G$, then also $g_1 \circ g_2 \in G$;
- is associative: $(g_1 \circ g_2) \circ g_3 = g_1 \circ (g_2 \circ g_3)$, for all $g_1, g_2, g_3 \in G$;
- has a unit element $e : e \circ g = g \circ e = g$, for all $g \in G$;
- is invertible: For every element $g \in G$, there exists an element $g^{-1} \in G$ such that $g \circ g^{-1} = g^{-1} = g^{-1} \circ g = e$.

Definition 9.13 (The general linear group $GL(n)$) The set of all $n \times n$ nonsingular (real) matrices with matrix multiplication forms a group. Such a group of matrices is usually called the general linear group and denoted by $GL(n)$.

Definition 9.14 (Matrix representation of a group) A group G has a matrix representation or can be realized as a matrix group if there exists an injective map

$$R : G \to GL(n); \qquad g \to R(g).$$

Note that a map $f(\cdot)$ is called injective if $f(x) \neq f(y)$ as long as $x \neq y$, which preserves the group structure of G. That is, the inverse and composition of elements in G are preserved by the map in the following way: Such a map is called a group homomorphism in algebra.

$$R(e) = I_{n \times n}, \qquad R(g \circ h) = R(g)R(h), \quad \forall g, h \in G. \tag{9.20}$$

Below, we identify a few important subsets of $M(n)$ that have special algebraic structures (as examples of matrix groups) and nice properties.

The group $GL(n)$ itself can be identified as the set of all invertible linear transformations from $\Re^n$ to $\Re^n$ in the sense that for every $A \in GL(n)$, we obtain a linear map

$$L : \Re^n \to \Re^n; \qquad x \to Ax. \tag{9.21}$$

Notice that if $A \in GL(n)$, then so is its inverse: $A^{-1} \in GL(n)$. We know that an matrix is invertible if and only if its determinant is nonzero. Therefore, we have

$$\det(A) \neq 0, \quad \forall A \in GL(n). \tag{9.22}$$

The general linear group, when matrices are considered to be known only up to a scalar factor, $GL(n)/R$, is referred to as the projective transformation group, whose elements are called projective matrices or homographies.

Matrices in $GL(n)$ of determinant $+1$ forming subgroup called the special linear group, denoted by $SL(n)$. That is, $\det(A) = +1$ for all $A \in SL(n)$. It is easy to verify that if $A \in SL(n)$, then so is A^{-1}, since $\det A^{-1} = \det A^{-1}$.

Definition 9.15 (The affine group $A(n)$) An affine transformation L from $\Re^n$ to $\Re^n$ is defined jointly by a matrix $A \in GL(n)$ and a vector $b \in \Re^n$ such that

$$L : \Re^n \to \Re^n; \qquad x \to Ax + b. \tag{9.23}$$

The set of all such affine transformations is called the affine group of dimension and is denoted by $A(n)$.

Notice that the map L so-defined is not a linear map from $\Re^n$ to $\Re^n$ unless $b = 0$. Nevertheless, we may "embed" this map into a space one dimension higher so that we can still represent it by a single matrix. If we identify an element $a \in \Re^n$ with $\left[\begin{smallmatrix} x \\ 1 \end{smallmatrix}\right] \in \Re^{n+1}$, then L becomes a map from $\Re^{n+1}$ to $\Re^{n+1}$ in the following sense:

Observe that this is the so-called homogeneous representation of x. Notice that this identification does not preserve the vector structure of $\Re^n$.

$$L : \Re^{n+1} \to \Re^{n+1}; \qquad \begin{bmatrix} x \\ 1 \end{bmatrix} \mapsto \begin{bmatrix} A & b \\ 0 & 1 \end{bmatrix} \begin{bmatrix} x \\ 1 \end{bmatrix}. \tag{9.24}$$

Thus, a matrix of the form

$$\begin{bmatrix} A & b \\ 0 & 1 \end{bmatrix} \in \Re^{(n+1 \times n+1)}, \qquad A \in GL(n),\ b \in \Re^n, \tag{9.25}$$

fully describes an affine map, and we call it an affine matrix. This matrix is an element in the general linear group $GL(n+1)$. In this way, $A(n)$ is identified as a subset (and in fact a subgroup) of $GL(n+1)$. The multiplication of two affine matrices in the set $A(n)$ is

$$\begin{bmatrix} A_1 & b_1 \\ 0 & 1 \end{bmatrix} \begin{bmatrix} A_2 & b_2 \\ 0 & 1 \end{bmatrix} = \begin{bmatrix} A_1 A_2 & A_1 b_2 + b_1 \\ 0 & 1 \end{bmatrix} \in \Re^{(n+1 \times n+1)}, \tag{9.26}$$

which is also an affine matrix in $A(n)$ and represents the composition of two affine transformations.

Given $\Re^n$ and its standard inner product structure, $\langle x, y\rangle = x^t y$, $\forall x, y \in \Re^n$, let us consider the set of linear transformations (or matrices) that preserve the inner product.

Definition 9.16 (The orthogonal group $O(n)$) An $n \times n$ matrix A (representing a linear map from $\Re^n$ to itself) is called orthogonal if it preserves the inner product, that is,

$$\langle Ax, Ay\rangle = \langle x, y\rangle, \quad \forall x, y \in \Re^n. \tag{9.27}$$

The set of all $n \times n$ orthogonal matrices forms the orthogonal group of dimension, and it is denoted by $O(n)$.

Obviously, $O(n)$ is a subset (and in fact a subgroup) of $GL(n)$. If R is an orthogonal matrix, we must have $\Re^t R = R\Re^t = I$. Therefore, the orthogonal group $O(n)$ can be characterized as

$$O(n) = \{R \in GL(n) | \Re^t R = I\}. \tag{9.28}$$

The determinant $\det(R)$ of an orthogonal matrix R can be either $+1$ or -1. The subgroup of $O(n)$ with determinant $+1$ is called the special orthogonal group and is denoted by $SO(n)$. That is, for any $R \in SO(n)$, we have $\det(R) = +1$. Equivalently, one may define $SO(n)$ as the intersection $SO(n) = O(n) \cap SL(n)$. In the case $n = 3$, the special orthogonal matrices are exactly the 3×3 rotation matrices.

The affine version of the orthogonal group gives the Euclidean (transformation) group.

Definition 9.17 (The Euclidean group $E(n)$) A Euclidean transformation L from $\Re^n$ to $\Re^n$ is defined jointly by a matrix $R \in O(n)$ and a vector $T \in \Re^n$ such that

$$L : \Re^n \to \Re^n; \qquad x \mapsto Rx + T. \tag{9.29}$$

The set of all such transformations is called the Euclidean group of dimension and is denoted by $E(n)$.

Obviously, $E(n)$ is a subgroup of $A(n)$. Therefore, it can also be embedded into a space one-dimension higher and has a matrix representation

$$\begin{bmatrix} R & T \\ 0 & 1 \end{bmatrix} \in \Re^{(n+1 \times n+1)}, \quad R \in O(n),\ T \in \Re^n. \tag{9.30}$$

If R further belongs to $SO(n)$, such transformations form the special Euclidean group, which is traditionally denoted by $SE(n)$. When $n = 3$, $SE(3)$, represents the conventional rigid-body motion in $\Re^3$, where R is the rotation of a rigid body and T is the translation (with respect to a chosen reference coordinate frame).

Since all the transformation groups introduced so far have natural matrix representations, they are matrix groups. To summarize their relationships, we have

$$SO(n) \subset O(n) \subset GL(n), \qquad SE(n) \subset A(n) \subset GL(n+1). \tag{9.31}$$

Since these groups themselves admit a differential structure, they belong to the Lie groups.

9.3 Matrix Algebra

In what follows, we let $A \in \mathbb{C}^{n\times n}$ and denote its conjugate transpose $\bar{A}^t := A^*$. Suppose A has p distinct eigenvalues, $\lambda_1, \dots, \lambda_p$, with the ith one being of multiplicity m_i. The sep of all eigenvalues of A is called *spectrum*, denoted by $\sigma(A)$. Let the characteristic polynomial of A be

$$\Delta_A(\lambda) := \det(\lambda I - A) = \lambda^n + c_{n-1}\lambda^{n-1} + \cdots + c_0. \tag{9.32}$$

Then we have the following:

1. $\det(A) = (-1)^n c_0 = \prod_{i=1}^p \lambda_i^{m_i}$. Furthermore,
$$\det(A) = \det\bigl(A^t\bigr), \quad \det(AB) = \det(A)\det(B), \quad \forall A, B \in \mathbb{C}^{n\times n}.$$
2. $\text{Trace}(A) = \sum_{i=1}^p a_{ii} = \sum_{i=1}^p m_i\lambda_i = (-1)^{n-1}c_{n-1}$. Moreover,
$$\text{Trace}(A+B) = \text{Trace}(A) + \text{Trace}(B).$$
3. If λ is an eigenvalue of A, its complex conjugate $\bar{\lambda}$ is an eigenvalue of A^*.
4. If A is real, so are c_i's. Therefore, if λ is an eigenvalue, so is $\bar{\lambda}$.
5. If A is both real and symmetric, $\sigma(A)$ is real.
6. In general, A is called Hermitian if $A = A^*$. For real matrix, this is synonymous to being symmetric. For a Hermitian matrix A, $\sigma(A)$ is real, and it admits a complete set of n orthogonal eigenvectors (even if the eigenvalues are not distinct). Denote the normalized versions of these eigenvectors by $x_1, \dots, x_n$ where x_i corresponds to the eigenvalue λ_i where we allow the possibility for $\lambda_i = \lambda_j$ for $i \neq j$. Then we can rewrite A as:
$$A = \sum_{i=1}^n \lambda_i x_i x_i^*,$$
which is known as the *spectral representation* of A.
7. If A is not Hermitian, but semisimple (that is, has a set of n linearly independent eigenvectors x_i's), it still admits a spectral representation, this time of the form
$$A = \sum_{i=1}^n \lambda_i x_i y_i^*,$$
where y_i^* is the ith row of M^{-1}, with $M = [x_1, \dots, x_n]$. This is known as the *eigenvector dyadic expansion* of A.

8. If a Hermitian matrix A has only positive (respectively, nonnegative) eigenvalues, it is called a *positive definite* (respectively, *nonnegative definite*) matrix, and this property is symbolically displayed as $A > 0$ (respectively, $A \geq 0$). A is said to be *negative definite* (respectively, *nonpositive definite*) if $-A > 0$ (respectively, $-A \geq 0$). A positive definite matrix A has the property that $x^*Ax > 0$ for all $x \in \mathbb{C}^n$ which is not zero.

9.3.1 Inverse of Block Matrices

Let A be a square matrix of appropriate dimension and partitioned in the form

$$A = \begin{bmatrix} A_1 & A_2 \\ A_3 & A_4 \end{bmatrix} \tag{9.33}$$

where both A_1 and A_4 are square matrices. If A_1 is invertible, then

$$\Delta_1 = A_4 - A_3 A_1^{-1} A_2$$

is called the Schur complement of A_1. Alternatively, if A_4 is invertible, then

$$\Delta_4 = A_1 - A_2 A_4^{-1} A_3$$

is called the Schur complement of A_4.

It is well-known that matrix A is invertible if and only if either

$$A_1 \quad \text{and} \quad \Delta_1 \quad \text{are invertible,}$$

or

$$A_4 \quad \text{and} \quad \Delta_4 \quad \text{are invertible.}$$

Specifically, we have the following equivalent expressions

$$\begin{bmatrix} A_1 & A_2 \\ A_3 & A_4 \end{bmatrix}^{-1} = \begin{bmatrix} \Upsilon_1 & -A_1^{-1} A_2 \Delta_1^{-1} \\ -\Delta_1^{-1} A_3 A_1^{-1} & \Delta_1^{-1} \end{bmatrix} \tag{9.34}$$

or

$$\begin{bmatrix} A_1 & A_2 \\ A_3 & A_4 \end{bmatrix}^{-1} = \begin{bmatrix} \Delta_4^{-1} & -\Delta_4^{-1} A_2 A_4^{-1} \\ -A_4^{-1} A_3 \Delta_4^{-1} & \Upsilon_4 \end{bmatrix} \tag{9.35}$$

where

$$\begin{aligned} \Upsilon_1 &= A_1^{-1} + A_1^{-1} A_2 \Delta_1^{-1} A_3 A_1^{-1}, \\ \Upsilon_4 &= A_4^{-1} + A_4^{-1} A_3 \Delta_4^{-1} A_2 A_4^{-1}. \end{aligned} \tag{9.36}$$

Important special cases are

$$\begin{bmatrix} A_1 & 0 \\ A_3 & A_4 \end{bmatrix}^{-1} = \begin{bmatrix} A_1^{-1} & 0 \\ -A_4^{-1} A_3 A_1^{-1} & A_4^{-1} \end{bmatrix} \tag{9.37}$$

and

$$\begin{bmatrix} A_1 & A_2 \\ 0 & A_4 \end{bmatrix}^{-1} = \begin{bmatrix} A_1^{-1} & -A_1^{-1} A_2 A_4^{-1} \\ 0 & A_4^{-1} \end{bmatrix}. \tag{9.38}$$

9.3.2 Matrix Inversion Lemma

Let $A \in \mathbb{R}^{n\times n}$ and $C \in \mathbb{R}^{m\times m}$ be nonsingular matrices. By using the definition of matrix inverse, it can be easily verified that

$$[A + BCD]^{-1} = A^{-1} - A^{-1}B\bigl[DA^{-1}B + C^{-1}\bigr]^{-1}DA^{-1}. \tag{9.39}$$

9.4 Range, Kernel, Rank and Eigenvectors of a Matrix

Let A be a general $m \times n$ matrix that also conveniently represents a linear map from the vector space $\Re^n$ to $\Re^n$.

Definition 9.18 (Range, span, null space, and kernel) Define the range or span of A, denoted by range(A) or span(A), to be the subspace of $\Re^m$ such that $y \in \text{range}(A)$ if and only if $y = Ax$ for some $x \in \Re^n$. Define the null space of A, denoted by null(A), to be the subspace of $\Re^n$ such that $x \in \text{null}(A)$ if and only if $Ax = 0$. When A is viewed as an abstract linear map, null(A) is also referred to as the kernel of the map, denoted by ker(A).

Notice that the range of a matrix A is exactly the span of all its column vectors; the null space of a matrix A is exactly the set of vectors which are orthogonal to all its row vectors (for a definition of orthogonal vectors see Definition 9.8). The notion of range or null space is useful whenever the solution to a linear equation of the form $Ax = b$ is considered. In terms of range and null space, this equation will have a solution if $b \in \text{range}(A)$ and will have a unique solution only if $\text{null}(A) = \emptyset$ (the empty set).

In MATLAB, the null space of a matrix can be computed using the command

$$>> \quad Z = \textbf{null}(A).$$

Definition 9.19 (Rank of a matrix) The rank of a matrix is the dimension of its range:

$$\text{rank}(A) \doteq \dim\bigl(\text{range}(A)\bigr). \tag{9.40}$$

Properties of matrix rank: For an arbitrary $m \times n$ matrix A, its rank has the following properties:

- $\text{rank}(A) = n - \dim(\text{null}(A))$.
- $0 \le \text{rank}(A) \le \min\{m, n\}$.
- rank(A) is equal to the maximum number of linearly independent column (or row) vectors of A.
- rank(A) is the highest order of a nonzero minor of A.
- *Sylvester's inequality*: Let B be an $n \times k$ matrix. Then AB is an $m \times k$ matrix and

$$\text{rank}(A) + \text{rank}(B) - n \le \text{rank}(AB) \le \min\bigl\{\text{rank}(A), \text{rank}(B)\bigr\}. \tag{9.41}$$

- For any nonsingular matrices $C \in \Re^{m\times m}$ and $D \in \Re^{n\times n}$, we have

$$\operatorname{rank}(A) = \operatorname{rank}(CAD). \tag{9.42}$$

In MATLAB, the rank of a matrix A is just

$$>> \quad \mathbf{rank}(A).$$

It should be noted that a minor of order k is the determinant of a $k \times k$ submatrix of A.

Definition 9.20 (Orthogonal complement to a subspace) Given a subspace S of $\Re^n$, we define its orthogonal complement to be the subspace $S^\perp \subseteq \Re^n$ such that $x \in S^\perp$ if and only if $x^t y = 0$ for all $y \in S$. We write $\Re^n = S \oplus S^\perp$.

The notion of orthogonal complement is used in this book to define the "coimage" of an image of a point or a line. Also, with respect to any linear map A from $\Re^n$ to $\Re^m$, the space $\Re^n$ can be decomposed as a direct sum of two subspaces,

$$\Re^n = \operatorname{null}(A) \oplus \operatorname{null}(A)^\perp,$$

and $\Re^m$ can be decomposed similarly as

$$\Re^m = \operatorname{range}(A) \oplus \operatorname{range}(A)^\perp.$$

We also have the following not so obvious relationships.

Theorem 9.21 *Let be a linear map from $\Re^n$ to $\Re^m$. Then:*

(a) $\operatorname{null}(A)^\perp = \operatorname{range}(A^t)$,
(b) $\operatorname{range}(A)^\perp = \operatorname{null}(A^t)$,
(c) $\operatorname{null}(A^t) = \operatorname{null}(AA^t)$,
(d) $\operatorname{range}(A) = \operatorname{range}(AA^t)$,

Proof To prove part c: $\operatorname{null}(AA^t) = \operatorname{null}(A^t)$, we have

- $AA^t x = 0 \Rightarrow \langle x, AA^t x\rangle = \|A^t x\|^2 = 0 \Rightarrow A^t x = 0$, hence $\operatorname{null}(AA^t) \subseteq \operatorname{null}(A^t)$.
- $A^t x = 0 \Rightarrow AA^t x = 0$; hence $\operatorname{null}(AA^t) \supseteq \operatorname{null}(A^t)$.

To prove part (d), $\operatorname{range}(AA^t) = \operatorname{range}(A)$, we first need to prove that $\Re^n$ is a direct sum of $\operatorname{range}(A^t)$ and $\operatorname{null}(A)$, that is, part (a) of the theorem. Part (b) can then be proved similarly. We prove this by showing that a vector x is in $\operatorname{null}(A)$ if and only if it is orthogonal to $\operatorname{range}(A^t)$: $x \in \operatorname{null}(A) \Leftrightarrow \langle Ax, y\rangle = 0, \forall y \Leftrightarrow \langle x, A^t y\rangle = 0$, $\forall y$. Hence, $\operatorname{null}(A)$ is exactly the subspace that is the orthogonal complement to $\operatorname{range}(A^t)$ (denoted by $\operatorname{range}(A^t)^\perp$). Therefore, $\Re^n$ is a direct sum of $\operatorname{range}(A^t)$ and $\operatorname{null}(A)$. Now to complete our proof of part (d), let $\operatorname{Img}_A(S)$ denote the image of a subspace S under the map A. Then we have $\operatorname{range}(A) = \operatorname{Img}_A(\Re^n) = \operatorname{Img}_A(\operatorname{range}(A^t)) = \operatorname{range}(AA^t)$ (in the second equality we used the fact that $\Re^n$ is a direct sum of $\operatorname{range}(A^t)$ and $\operatorname{null}(A)$). These relations are depicted by Fig. 9.1.

In fact, the same result holds even if the domain of the linear map is replaced by an infinite-dimensional linear space with an inner product (i.e., $\Re^n$ is replaced by

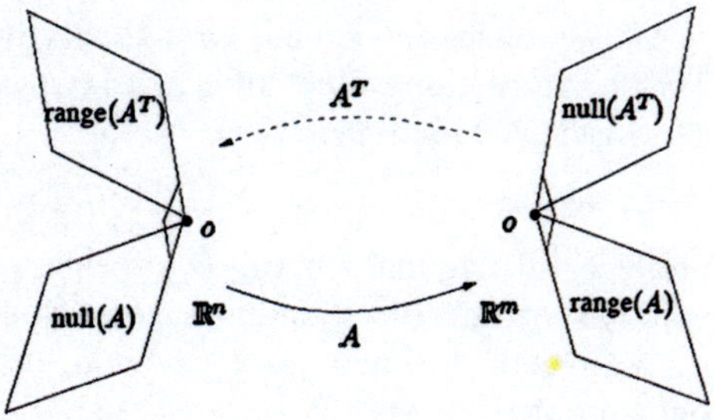

Fig. 9.1 The orthogonal decomposition of the domain and codomain of a linear map A

a Hilbert space). In that case, this theorem is also known as the *finite-rank operator fundamental lemma* [5]. We will later use this result to prove the singular value decomposition. But already it implies a result that is extremely useful in the study of multiple-view geometry. □

Lemma 9.22 (Rank reduction lemma) *Let $A \in \Re^{n\times n}$ be a matrix and let W be a matrix of the form*

$$W = \begin{bmatrix} M & 0 \\ AB & AA^t \end{bmatrix} \in \Re^{(m+n)\times(k+n)} \tag{9.43}$$

for some matrices $M \in \Re^{m\times k}$ and $B \in \Re^{n\times k}$. Then, regardless of what B is, we always have

$$\text{rank}(M) = \text{rank}(W) - \text{rank}(A). \tag{9.44}$$

The proof is easy using the fact $\text{range}(AB) \subseteq \text{range}(A) = \text{range}(AA^t)$ with the second identity from the previous theorem, and we leave the rest of the proof to the reader as an exercise.

A linear map from $\Re^n$ to itself is represented by a square $n \times n$ matrix A. For such a map, we sometimes are interested in subspaces of $\Re^n$ that are "invariant" under the map.This notion turns out to be closely related to the eigenvectors of the matrix A. More rigorously speaking, a subspace $S \subset \Re^n$ is invariant if $A(S) \subseteq S$.

Definition 9.23 (Eigenvalues and eigenvectors of a matrix) Let A be an $n \times n$ complex matrix in $C^{n\times n}$. A nonzero vector $v \in C^n$ is said to be its (right) eigenvector if

$$Av = \lambda v \tag{9.45}$$

for some scalar $\lambda \in C$; λ is called an eigenvalue of A. Similarly, a nonzero row vector $\eta^t \in C^n$ is called a left eigenvector of A if $\eta^t A = \lambda \eta^t$ for some $\lambda \in C$.

Although A will mostly be a real matrix in this book, to talk about its eigenvectors, it is more convenient to think of it as a complex matrix (with all entries that happen to be real).

Unless otherwise stated, an eigenvector by default means a right eigenvector. The set of all eigenvalues of a matrix A is called its spectrum, denoted by $\sigma(A)$. The MATLAB command

$$[V, D] = \mathbf{eig}(A)$$

produces a diagonal matrix D of eigenvalues and a full-rank matrix V whose columns are the corresponding eigenvectors, so that $AV = VD$.

We give the following facts about eigenvalues and eigenvectors of a matrix without a proof.

Properties of eigenvalues and eigenvectors: Given a matrix $A \in \Re^{n\times n}$, we have:

1. If $Av = \lambda v$, then for the same eigenvalue λ, there also exists a left eigenvector η^t such that $\eta^t A = \lambda \eta^t$ and vice versa. Hence, $\sigma(A) = \sigma(A^t)$.
2. The eigenvectors of A associated with different eigenvalues are linearly independent.
3. All its eigenvalues $\sigma(A)$ are the roots of the (characteristic) polynomial equation $\det(\lambda I - A) = 0$. Hence, $\det(A)$ is equal to the product of all eigenvalues of A.
4. If $B = PAP^{-1}$ for some nonsingular matrix P, then $\sigma(B) = \sigma(A)$.
5. If A is a real matrix, then $\lambda \in C$ is an eigenvalue implies that its conjugate λ is also an eigenvalue. Simply put, $\sigma(A) = \bar{\sigma}(A)$ for real matrices.

9.5 Symmetric and Skew-Symmetric Matrices

Definition 9.24 (Symmetric matrix) A matrix $S \in \Re^{n\times n}$ is called symmetric if $S^t = S$. A symmetric matrix S is called positive (semi-)definite, if $x^t Sx > 0$ (or $x^t Sx \geq 0$) for all $x \in \Re^n$, denoted by $S > 0$ (or $S \geq 0$).

Properties of symmetric matrices: If S is a real symmetric matrix, then:

1. All eigenvalues of S must be real, that is, $\sigma(S) \subset R$.
2. Let (λ, v) be an eigenvalue-eigenvector pair. If $\lambda_i \neq \lambda_j$, then $v_i \perp v_j$; that is, eigenvectors corresponding to distinct eigenvalues are orthogonal.
3. There always exist n orthonormal eigenvectors of S, which form a basis for $\Re^n$.
4. $S > 0$ $(S \geq 0)$ if $\lambda_i > 0$ $(\lambda_i \geq 0)$ $\forall i = 1, 2, \ldots, n$; i.e. S is positive (semi-)definite if all eigenvalues are positive (nonnegative).
5. $S \geq 0$ and $\lambda_1 \geq \lambda_2 \geq \cdots \geq \lambda_n$; then $\max_{\|x\|_2=1}\langle x, Sx\rangle = \lambda_1$ and $\min_{\|x\|_2=1}\langle x, Sx\rangle = \lambda_n$.

From point 3, we see that if $V = [v_1, v_2, \ldots, v_n] \in \Re^{n\times n}$ is the matrix of all the eigenvectors, and $\lambda = \text{diag}\{\lambda_1, \lambda_2, \ldots, \lambda_n\}$ is the diagonal matrix of the corresponding eigenvalues, then we can write

$$S = V\Lambda V^t,$$

where V is an orthogonal matrix. In fact, V can be further chosen to be in $SO(n)$ (that is, of determinant $+1$) if n is odd, since $V\Lambda V^t = (-V)\Lambda(-V)^t$ and $\det(-V) = (-1)^n \det(V)$.

Definition 9.25 (Induced 2-norm of a matrix) Let $A \in \Re^{m\times n}$. We define the induced 2-norm of A (as a linear map from $\Re^n$ to $\Re^m$) as

$$\|A\|_2 \doteq \max_{\|x\|_2=1} \|A\|_2 = \max_{\|x\|_2=1} \sqrt{\langle x, A^t A x\rangle}.$$

Similarly, other induced operator norms on A can be defined starting from different norms on the domain and codomain spaces on which A operates.

Let A be as above. Then $A^t A \in \Re^{n\times n}$ is clearly symmetric and positive semidefinite, so it can be diagonalized by a orthogonal matrix V. The eigenvalues, being nonnegative, can be written as σ_i^2. By ordering the columns of V so that the eigenvalue matrix Λ has decreasing eigenvalues on the diagonal, we see, from point 5 of the preceding fact, that $A^t A = V \operatorname{diag}\{\sigma_1^2, \sigma_2^2, \dots, \sigma_n^2\} V^t$ and

$$\|A\|_2 = \sigma_1.$$

The induced 2-norm of a matrix $A \in \Re^{m\times n}$ is different from the "2-norm" of A viewed as a vector in $\Re^{mn}$. To distinguish them, the latter one is conventionally called the Frobenius norm of A, precisely defined as $\|A\|_f = \sqrt{\sum_{i,j} a_{ij}^2}$. Notice that $\sum_{i,j} a_{ij}^2$ is nothing but the trace of $A^t A$ (or AA^t). Thus, we have

$$\|A\|_f = \sqrt{\operatorname{trace}(A^t A)} = \sqrt{\sigma_1^2, \sigma_2^2, \dots, \sigma_n^2}.$$

The inverse problem of retrieving from the symmetric matrix $S = A^t A$ is usually solved by Cholesky factorization. For the given S, its eigenvalues must be nonnegative. Thus, we have $S = V\Lambda V^t = A^t A$ for $A = \Lambda^{(\frac{1}{2})} V^t$, where $\Lambda^{(\frac{1}{2})} = \operatorname{diag}\{\sigma_1, \sigma_2, \dots, \sigma_n\}$ is the "square root" of the diagonal matrix λ. Since $\Re^t R = I$ for any orthogonal matrix, the solution for A is not unique: RA is also a solution. Cholesky factorization then restricts the solution to be an upper triangular matrix (exactly what we need for camera calibration in Chap. 6). In MATLAB, the Cholesky factorization is given by the command $A = \operatorname{chol}(S)$.

Definition 9.26 (Skew-symmetric matrix) A matrix $A \in \Re^{n\times n}$ is called skew-symmetric (or antisymmetric) if $A^t = -A$.

Properties of a skew-symmetric matrix: If A is a real skew-symmetric matrix, then:

1. All eigenvalues of A are either zero or purely imaginary, that is, of the form $i\omega$ for $i = \sqrt{-1}$ and some $\omega \in R$.
2. There exists an orthogonal matrix V such that

$$A = V\Lambda V^t, \tag{9.46}$$

where Λ is a block-diagonal matrix $\Lambda = \operatorname{diag}\{A_i, \dots, A_m, 0, \dots, 0\}$, where each A_i is a 2×2 real skew-symmetric matrix of the form

$$A_i = \begin{bmatrix} 0 & a_i \\ -a_i & 0 \end{bmatrix} \in \Re^{2\times 2}, \quad i = 1, 2, \dots, m. \tag{9.47}$$

From point 2, we conclude that the rank of any skew-symmetric matrix must be even. A commonly used skew-symmetric matrix in computer vision is associated with a vector $u \in \Re^3$, denoted by

$$\hat{u} = \begin{bmatrix} 0 & -u_3 & u_2 \\ u_3 & 0 & -u_1 \\ -u_2 & u_1 & 0 \end{bmatrix} \in \Re^{3\times 3}. \tag{9.48}$$

The reason for such a definition is that $\hat{u}v$ is equal to the conventional cross product $u \times v$ of two vectors in $\Re^3$. Then we have $\text{rank}(\hat{u}) = 2$ if $u \neq 0$ and the (left and right) null space of $\hat{u}$ is exactly spanned by the vector u itself. That is, $\hat{u}u = 0$ and $u^t\hat{u} = 0$. In other words, columns and rows of the matrix $\hat{u}$ are always orthogonal to u.

Obviously, $A^t\hat{u}A$ is also a skew-symmetric matrix. Then $A^t\hat{u}A = \hat{v}$ for some $v \in \Re^3$. We want to know what the relationship between v and A, u is.

Hat operator: If A is a 3×3 matrix of determinant 1, then we have

$$A^t\hat{u}A = \widehat{A^{-1}u}. \tag{9.49}$$

This is an extremely useful fact, which will be extensively used in our book. For example, this property allows us to "push" a matrix through a skew-symmetric matrix in the following way: $\hat{u}A = A^{-t}\hat{u}A = A^{-t}\widehat{A^{-1}u}$. We leave to the reader as an exercise to think about how this result needs to be modified when the determinant of A is not !, or when A is not even invertible.

9.6 Singular Value Decomposition

The singular value decomposition (SVD) is a useful tool to capture essential features of a matrix (that represents a linear map), such as its rank, range space, null space, and induced norm, as well as to "generalize" the concept of "eigenvalue– eigenvector" pair to non-square matrices. The computation of the SVD is numerically well conditioned, making it extremely useful for solving many linear-algebraic problems such as matrix inversion, calculation of the rank, linear least-squares estimate, projections, and fixed-rank approximations.

We give hereafter a complete description and proof for the singular value decomposition (SVD) theorem. Recall first that the inner product defined on $\mathbb{C}^n$ is $\langle x, y\rangle = x^*y$, $\forall x, y \in \mathbb{C}^n$. We now introduce the following important lemma.

Lemma 9.27 *Let $A \in \mathbb{C}^{m\times n}$ and A^* be its conjugate transpose. We then always have*:

$$\mathcal{N}(AA^*) = \mathcal{N}(A^*), \qquad \mathcal{R}(AA^*) = \mathcal{R}(A).$$

Proof To prove $\mathcal{N}(AA^*) = \mathcal{N}(A^*)$, we have:

1. $AA^*x = \theta \Longrightarrow \langle x, AA^*x\rangle = \|A^*x\|^2 = 0 \Longrightarrow A^*x = \theta$. Hence $\mathcal{N}(AA^*) \subseteq \mathcal{N}(A^*)$.
2. $A^*x = \theta \Longrightarrow AA^*x = \theta$. Hence $\mathcal{N}(AA^*) \supseteq \mathcal{N}(A^*)$.

To prove $\mathcal{R}(AA^*) = \mathcal{R}(A)$, we need to prove that $\mathbb{C}^n$ is a direct sum of $\mathcal{R}(A^*)$ and $\mathcal{N}(A)$. We prove this by showing that a vector x is in $\mathcal{N}(A)$ if and only if it is orthogonal to $\mathcal{R}(A^*)$:

$$x \in \mathcal{N}(A) \quad \Longrightarrow \quad \langle A^*x, y\rangle = 0, \quad \forall y \in \mathcal{R}(A^*) \quad \Longrightarrow \quad \langle x, Ay\rangle = 0, \quad \forall y.$$

Hence $\mathcal{N}(A)$ is exactly the subspace which is orthogonal supplementary to $\mathcal{R}(A^*)$ (sometimes denoted as $\mathcal{R}(A^*)^{\perp}$). Therefore, $\mathbb{C}^n$ is a direct sum of $\mathcal{R}(A^*)$ and $\mathcal{N}(A)$. Let $\Im_A(S)$ denote the image of a subspace S under the map A. Then we have:

$$\mathcal{R}(A) = \Im_A(\mathbb{C}^n) = \Im_A \mathcal{R}(A^*) = \mathcal{R}(AA^*)$$

where in the second equality we used the fact that $\mathbb{C}^n$ is a direct sum of $\mathcal{R}(A^*)$ and $\mathcal{N}(A)$. □

We are now ready to give a complete proof for the singular value decomposition theorem which is the following.

Theorem 9.28 (Singular value decomposition) *Let* $\mathsf{F} = \Re$ *or* $\mathbb{C}$. *Let* $A \in \mathsf{F}^{n\times n}$ *be a matrix of rank* r. *Then there exist matrices* $U \in \mathsf{F}^{m\times m}$ *and* $V \in \mathsf{F}^{n\times n}$, *and* $\Sigma_1 \in \Re^{n\times n}$ *such that*:

1. $V = [V_1 : V_2]$, $V_1 \in \mathsf{F}^{n\times r}$, *satisfies*:

   V is unitary, that is, $V^*V = I_{n\times n}$,
   $\mathcal{R}(V_1) = \mathcal{R}(A^*)$, *the columns of V_1 form an orthonormal basis of* $\mathcal{R}(A^*)$,
   $\mathcal{R}(V_2) = \mathcal{N}(A)$, *the columns of V_2 form an orthonormal basis of* $\mathcal{N}(A)$,
   The columns of V form a complete orthonormal basis of eigenvectors of A^*A.

2. $U = [U_1 : VU_2]$, $U_1 \in \mathsf{F}^{m\times r}$, *satisfies*:

   U is unitary, that is, $U^*U = I_{m\times m}$,
   $\mathcal{R}(U_1) = \mathcal{R}(A^*)$, *the columns of U_1 form an orthonormal basis of* $\mathcal{R}(A)$,
   $\mathcal{R}(U_2) = \mathcal{N}(A)$, *the columns of U_2 form an orthonormal basis of* $\mathcal{N}(A^*)$,
   The columns of U form a complete orthonormal basis of eigenvectors of AA^*.

3. $\Sigma_1 = \operatorname{diag}(\sigma_1, \ldots, \sigma_r) \in \Re^{r\times r}$ *such that* $\sigma_1 \geq \sigma_2 \cdots \sigma_r > 0$. $A \in \mathsf{F}^{m\times n}$ *has dyadic expansion*

$$A = U_1\Sigma_1 V_1^* \quad \Longrightarrow \quad A = \sum_{i=1}^{r} \sigma_i u_i v_i^*$$

   where u_i, v_i *are the columns of* U_1 *and* V_1, *respectively.*

4. $A \in \mathsf{F}^{m\times n}$ *has a singular value decomposition (SVD)*

$$A = U\Sigma V^*, \quad \Sigma = \begin{bmatrix} \Sigma_1 & 0 \\ 0 & 0 \end{bmatrix}.$$

Proof

1. $A \in \mathsf{F}^{m\times n}$ has rank r, hence the nonnegative (or, equivalently, positive semidefinite) Hermitian matrix AA^* has rank r according to Lemma 9.27. It has n nonnegative eigenvalues σ_i^2 ordered as
$$\sigma_1^2 \geq \sigma_2^2 \geq \cdots \geq \sigma_r^2 > 0 = \sigma_{r+1}^2 = \cdots = \sigma_n^2$$
to which corresponds a complete orthonormal eigenvector basis $(\nu)_{i=1}^2$ of AA^*. This family of vectors ($\in \mathsf{F}^m$) form the columns of a unitary $n \times n$ matrix, say, V. From Lemma 9.27, $\mathcal{R}(AA^*) = \mathcal{R}(A^*)$ and $\mathcal{N}(AA^*) = \mathcal{N}(A)$, the properties listed in 1 follow.
2. Define a diagonal matrix $\Sigma_1 = \mathrm{diag}(\sigma_1, \sigma_2, \ldots, \sigma_r) \in \Re^{r\times r}$. When then have $AA^*V_1 = V_1\Sigma_1^2$, hence
$$(AV_1\Sigma_1^{-1})^*(AV_1\Sigma_1^{-1}) = I_{r\times r}.$$
This defines an $m \times r$ matrix:
$$U_1 = AV_1\Sigma_1^{-1}.$$
It follows that $U_1^*U_1 = I_{r\times r}$. Since A^*A and AA^* both have exactly r nonzero eigenvalues, it follows that the columns of U_1 form an orthonormal basis for $\mathcal{R}(AA^*)$ and $\mathcal{R}(A)$. Thus, the properties of U_1 listed in 2 hold. Now define an $m \times (m-r)$ matrix U_2 with orthonormal columns which are orthogonal to columns of U_1. Then $U = [U_1 : U_2]$ is clearly an unitary matrix. From the proof of Lemma 9.27, columns of U_2 form an orthonormal basis of $\mathcal{N}(A^*)$ or $\mathcal{N}(AA^*)$. Therefore, columns of U_2 are all the eigenvectors corresponding to the zero eigenvalue. Hence, columns of U form a complete orthonormal basis of eigenvectors of AA^*. List 2 is then fully proven.
3. Since $U_1 = AV_1\Sigma_1^{-1}$, we have
$$A = U_1\Sigma_1V_1^*.$$
The dyadic expansion directly follows.
4. The singular value decomposition follows because
$$A[V_1 : V_2] = [U_1\Sigma_1 : 0] = [U_1 : U_2]\Sigma \quad \Longrightarrow \quad A = U\Sigma V^*.$$ □

9.6.1 Geometric Interpretation

Notice that in the *SVD* of a square matrix $A = U\Sigma V^t \in \Re^{n\times n}$, columns of $U = [u_1, u_2, \ldots, u_n]$ and columns of $V = [v_1, v_2, \ldots, v_n]$ form orthonormal bases for $\Re^n$. The *SVD* essentially states that if A (as a linear map) maps a point x to y, then coordinates of with respect to the basis U are related to coordinates of x with respect to the basis V by the diagonal matrix Σ that scales each coordinate by the corresponding singular value.

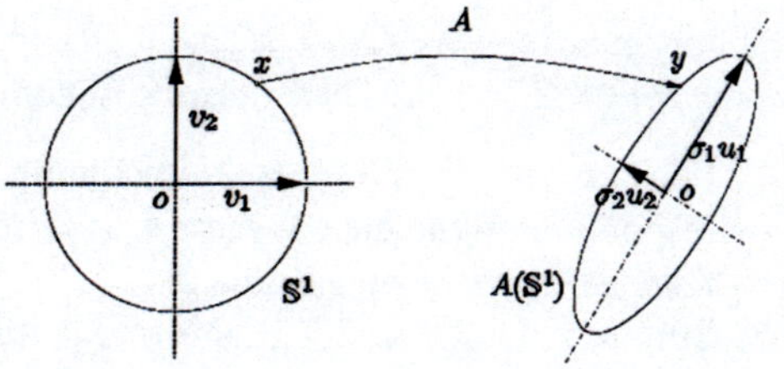

Fig. 9.2 The image of a unit sphere on the left under a nonsingular map $A \in \Re^{2\times 2}$ is an ellipsoid on the right

Theorem 9.29 *Let $A \in \Re^{n\times n} = U\Sigma V^t$ be a square matrix. Then A maps the unit sphere $S^{n-1} \doteq \{x \in \Re^n : \|x\|_2 = 1\}$ to an ellipsoid with semi-axes $\sigma_i u_i$, where u_i is the ith column of U.*

Proof Let x, y be such that $Ax = y$. The set $\{u_i\}_{i=1}^n$ is an orthonormal basis for $\Re^n$. With respect to such a basis x has coordinates

$$[\alpha_1, \alpha_2, \dots, \alpha_n]^t = \left[\langle v_1, x\rangle, \langle v_2, x\rangle, \dots, \langle v_n, x\rangle\right]^t.$$

That is, $x = \sum_{i=1}^n \alpha_i v_i$. With respect to the basis $\{u_i\}_{i=1}^n$, has coordinates

$$[\beta_1, \beta_2, \dots, \beta_n]^t = \left[\langle u_1, y\rangle, \langle u_2, y\rangle, \dots, \langle u_n, y\rangle\right]^t.$$

We also have $y = \sum_{i=1}^n \beta_i u_i = Ax = \sum_{i=1}^n \sigma_i u_i v_i^t x = \sum_{i=1}^n \sigma_i \langle v_i, x\rangle u_i$. Hence $\sigma_i \alpha_i = \beta_i$. Now $\|x\|_2^2 = \sum_{i=1} \alpha_i^2 = 1, \forall x \in S^{n-1}$, and so we have $\sum_{i=1}^n \beta_i^2/\sigma_i^2 = 1$, which implies that the point y satisfies the equation of an ellipsoid with semi-axes of length σ_i. This is illustrated in Fig. 9.2 for the case $n = 2$. □

9.6.2 Example A.1

The singular value decomposition of matrix

$$A = \begin{bmatrix} 0 \\ 3 \\ 4 \end{bmatrix}$$

is $A = U\Sigma V^*$ where

$$U = \begin{bmatrix} 0 & -0.6 & -0.8 \\ 0.6 & 0.64 & -0.84 \\ 0.8 & -0.48 & 0.36 \end{bmatrix}, \qquad \Sigma = \begin{bmatrix} 5 \\ 0 \\ 0 \end{bmatrix}, \qquad V = 1.$$

Since matrix A has rank $= 1$, it has a single nonzero singular value. Therefore, the spectral norm of A is $\sigma_M(A) = 5$.

9.6.3 Some Properties of the SVD

Let $A = U\Sigma V^*$ be the singular value decomposition of the $n \times m$ matrix A, with singular values $\sigma_1, \sigma_2, \dots, \sigma_r$, $r = \min(n, m)$. Denote the columns of the $n \times n$

unitary matrix U as u_i, $i = 1, 2, \ldots, n$, and those of the $m \times m$ unitary matrix V as v_i, $i = 1, 2, \ldots, m$. The following statements hold true:

1. For $i = 1, 2, \ldots, \min(n, m)$, the column vector u_i is an eigenvector of AA^* corresponding to the eigenvalue σ_i^2. Any remaining columns are eigenvectors corresponding to the eigenvalue 0.
2. Similarly, for $i = 1, 2, \ldots, \min(n, m)$, the column vector v_i is an eigenvector of A^*A corresponding to the eigenvalue σ_i^2. Any remaining columns are eigenvectors corresponding to the eigenvalue 0.
3. For $i = 1, 2, \ldots, \min(n, m)$, the vectors u_i and v_i satisfy

$$Av_i = \sigma_i u_i, \qquad A^* u_i = \sigma_i v_i.$$

4. Given a square $n \times n$ matrix A, the following properties hold [5]:

$$\begin{aligned}
&\sigma_M(A) = \max_{x \in \mathbb{C}^n} \frac{\|Ax\|_2}{\|x\|_2}, \\
&\sigma_m(A) = \min_{x \in \mathbb{C}^n} \frac{\|Ax\|_2}{\|x\|_2}, \\
&\sigma_m(A) \le \left|\lambda_i(A)\right| \le \sigma_M(A), \\
&\qquad \text{where } \lambda_i(A) \text{ is the } i\text{th eigenvalue of } A, \\
&\sigma_m(A) = \frac{1}{\sigma_M(A^{-1})}, \qquad \sigma_M(A) = \frac{1}{\sigma_m(A^{-1})}, \qquad \text{if } A^{-1} \text{ exists}, \\
&\sigma_M(\alpha A) = |\alpha| \sigma_M(A), \quad \text{with } \alpha \text{ any complex number}, \\
&\sigma_M(A+B) \le \sigma_M(A) + \sigma_M(B), \\
&\sigma_M(AB) \le \sigma_M(A)\sigma_M(B), \\
&\sigma_m(A) - \sigma_M(B) \le \sigma_m(A+B) \le \sigma_m(A) + \sigma_M(B), \\
&\max\bigl(\sigma_M(A), \sigma_M(B)\bigr) \le \sigma_M\bigl([AB]\bigr) \le \sqrt{2} \max\bigl(\sigma_M(A), \sigma_M(B)\bigr), \\
&\max_{i,j} |A_{ij}| \le \sigma_M(A) \le n \max_{i,j} |A_{ij}|, \\
&\qquad \text{with } A_{ij} \text{ the } (i, j) \text{ element of } A, \\
&\sum_{i=1}^{n} \sigma_i^2(A) = \mathrm{Tr}\bigl(A^* A\bigr).
\end{aligned}$$

Problems involving orthogonal projections onto invariant subspaces of $\mathbb{C}$, such as the linear least-squares (LLS) problem, can be easily solved using the *SVD*.

Definition 9.30 (Generalized (Moore–Penrose) inverse) Given a matrix $A \in \Re^{m \times n}$ of rank r with its SVD $A = U \Sigma V^t$, we then define the generalized inverse of A to be

$$A^{\perp} = V \Sigma^{\perp} U^t, \quad \Sigma^{\perp} = \begin{bmatrix} \Sigma_1^{-1} & 0 \\ 0 & 0 \end{bmatrix}_{n \times m}.$$

The generalized inverse is sometimes also called the pseudo-inverse.

In MATLAB, the pseudo-inverse of a matrix is computed by the command $X = \text{pinv}(A)$.

Properties of generalized inverse:

- $AA^{\perp}A = A$, $A^{\perp}AA^{\perp} = A^{\perp}$.

The generalized inverse can then be used to solve linear equations in general.

Proposition 9.31 (Least-squares solution of a linear systems) *Consider the problem $Ax = b$ with $A \in \Re^{m\times n}$ of rank $r \leq \min(m,n)$. The solution x^* that minimizes $\|Ax - b\|_2$ is given by $x^* = A^{\perp}b$.*

The following two results have something to do with the sensitivity of solving linear equations of the form $Ax = b$.

Proposition 9.32 (Perturbations) *Consider a nonsingular matrix $A \in \Re^{n\times n}$. Let δA be a full-rank perturbation. Then*

- $|\sigma_k(A + \delta A) - \sigma_k(A)| \leq \sigma_1(\delta A)$, $\forall k = 1, 2, \dots, n$.
- $\sigma_n(A\delta A) \geq \sigma_n(A)\sigma_n(\delta A)$,
- $\sigma_1(A^{-1}) = \frac{1}{\sigma_n(A)}$,

where σ_i denotes the ith singular value.

Proposition 9.33 (Condition number) *Consider the problem $Ax = b$, and consider a "perturbed" full-rank problem $(A + \delta A)(x + \delta x) = b$. Since $Ax = b$, then to first-order approximation, $\delta x = -A^{\dagger}\delta Ax$. Hence, $\|\delta x\|_2 \leq \|A^{\dagger}\|_2\|\delta A\|_2\|x\|_2$, from which*

$$\frac{\|\delta x\|_2}{\|x\|_2} \leq \left\|A^{\dagger}\right\|_2\|A\|_2\frac{\|\delta A\|_2}{\|A\|_2} \doteq k(A)\frac{\|\delta A\|_2}{\|A\|_2},$$

where $k(A) = \|A^{\dagger}\|_2\|A\|_2$ is called the condition number of A. It easy to see that $k(A) = \sigma_1/\sigma_n$ if A is invertible.

Last but not the least, one of the most important properties of the SVD is related to a fixed-rank approximation of a given matrix. Given a matrix A of rank r, we want to find a matrix B such that it has fixed rank $p < r$ and the Frobenius norm of the difference $\|A - B\|_f$ is minimal. The solution to this problem is given simply by setting all but the first p singular values to zero

$$B \doteq U\Sigma_{(p)}V^t,$$

where $\Sigma_{(p)}$ denotes the matrix obtained from Σ by setting to zero its elements on the diagonal after the pth *entry. The matrix B has exactly the same induced 2-norm of, that is, $\sigma_1(A) = \sigma_1(B)$, and satisfies the requirement on the rank.*

Proposition 9.34 (Fixed rank approximation) *Let A, B be defined as above. Then $\|A - B\|_f^2 = \sigma_{p+1}^2 + \cdots + \sigma_r^2$. Furthermore, such a norm is the minimum achievable.*

The proof is an easy exercise that follows directly from the properties of orthogonal projection and the properties of the SVD given above.

After we have gone through all the trouble proving this theorem, you must know that SVD has become a numerical routine available in many computational softwares such as MATLAB. Within MATLAB, to compute the SVD of a given $m \times n$ matrix A, simply use the command

$$>> \quad [U, S, V] = \mathbf{SVD(A)}$$

which returns matrices U, S, V satisfying $A = USV^*$ (where S represents Σ as defined above).

9.7 Gram–Schmidt and the QR Decomposition

A matrix in $GL(n)$ has n independent rows (or columns). A matrix in $O(n)$ has orthonormal rows (or columns). The Gram–Schmidt procedure can be viewed as a map from $GL(n)$ to $O(n)$, for it transforms a nonsingular matrix into an orthogonal one. Call $L_+(n)$ the subset of $GL(n)$ consisting of lower triangular matrices with positive elements along the diagonal. Such matrices form a subgroup of $GL(n)$.

Theorem 9.35 (Gram–Schmidt procedure) *For every $A \in GL(n)$, there exists a lower triangular matrix $L \in \Re^{n\times n}$ and an orthogonal matrix $E \in O(n)$ such that*

$$A = LE. \tag{9.50}$$

Proof Contrary to the convention of the book, for simplicity in this proof all vectors indicate row vectors. That is, if v is an n-dimensional row vector, it is of the form: $v = [v_1, v_2, \dots, v_n] \in \Re^n$. Denote the ith row vector of the given matrix A by a_i for $i = 1, 2, \dots, n$. The proof consists in constructing L and E iteratively from the row vectors a_i:

$$\begin{aligned}
l_1 &\doteq a_1 & &\to e_1 \doteq l_1/\|l_1\|_2,\\
l_2 &\doteq a_2 - (a_2, \epsilon_1)\epsilon_1 & &\to e_2 \doteq l_2/\|l_2\|_2,\\
&\vdots\ \ \vdots\ \vdots & &\ \ \vdots \qquad \vdots\\
l_n &\doteq a_n - \sum_{i=1}^{n-1} \langle a_{i+1}, e_i\rangle e_i & &\to e_n \doteq l_n/\|l_n\|_2.
\end{aligned}$$

Then $E = [e_1^t, \dots, e_n^t]^t$, and the matrix L is obtained as

$$L = \begin{bmatrix} \|l_1\|_2 & 0 & \cdots & 0\\ \langle a_2, e_1\rangle & \|l_2\|_2 & \cdots & 0\\ \vdots & \vdots & \ddots & \vdots\\ \langle a_2, e_1\rangle & \cdots & \langle a_n, e_{n-1}\rangle & \|l_n\|_2 \end{bmatrix}.$$

By construction is orthogonal; that is, $EE^t = E^tE = I$. □

Remark 9.36 The Gram–Schmidt's procedure has the peculiarity of being causal, in the sense that the ith row of the transformed matrix E depends only upon rows with index $J \leq i$ of the original matrix A. The choice of the name E for the orthogonal matrix above is not accidental.

There are a few useful variations to Gram–Schmidt procedure. By transposing $A = LE$, we get $A^t = E^t L^t \doteq QR$. Notice that $R = L^t$ is an upper triangular matrix. Thus, by applying Gram–Schmidt procedure to the transpose of a matrix, we can also decompose it into the form QR where Q is an orthogonal matrix and R an upper triangular matrix. Such a decomposition is called the QR decomposition. In MATLAB, this can be done by the command $[Q, R] = \mathrm{qr}(A)$. Furthermore, by inverting $A^t = E^t L^t$, we get $A^{-t} = L^{-t} E \doteq KE$. Notice that $K = L^{-t}$ is still an upper triangular matrix. Thus, we can also decompose any matrix into the form of an upper triangular matrix followed by an orthogonal There are a few useful variations to Gram–Schmidt procedure. By transposing, we get. Notice that is an upper triangular matrix. Thus, by applying Gram–Schmidt procedure to the transpose of a matrix, we can also decompose it into the form where is an orthogonal one. The latter one is the kind of "QR decomposition" we use in Chap. 6 for camera calibration.

9.8 Useful Formulae

In what follows, some standard formulae that are of common use in the book are presented.

9.8.1 Ackermann's Formula for Eigenvalue Assignment

In [1], a formula is provided to compute the feedback gain matrix K such that a set of eigenvalues of the linear system

$$\dot{x}(t) = Ax(t) + Bu(t)$$

at desired locations. The formula emerges by organizing a three-step procedure of converting the system matrices into controller canonical form, solving for the gain and converting the gain back. It has the form:

$$\begin{aligned} K &= [0, \ldots, 0, 1] P_c \alpha_c(A), \\ P_c &= \left[B, AB, A^2 B, \ldots, A^{n-1} B\right], \\ \alpha_c(A) &= A^n + \alpha_1 A^{n-1} + \alpha_2 A^{n-2} + \cdots + \alpha_n I \end{aligned}$$

where α_j are the coefficients of the desired characteristic polynomial

$$\begin{aligned} \alpha_c(s) &= (s - s_1)(s - s_2) \cdots (s - s_n) \\ &= s^n + \alpha_1 s^{n-1} + \alpha_2 s^{n-2} + \cdots + \alpha_n \end{aligned}$$

and $s_1, \ldots, s_n$ are the desired eigenvalues.

9.8.2 Parseval Formula

Let $f(t)$ be a continuous-time signal over the time-interval $[t_1, t_2]$. Then the *continuous-time signal energy* over the length $L = t_2 - t_1]$ is defined by

$$E_L = \int_{t_1}^{t_2} |f(t)|^2 \, dt$$

whereas the *total continuous-time signal energy* is given by

$$E_\infty = \int_{-\infty}^{\infty} |f(t)|^2 \, dt. \tag{9.51}$$

From Laplace transform methods [Gajic], we recall the frequency domain convolution which states that *the Frequency transform of a product of two signals in time is proportional to the convolution of their convolution of their Fourier transforms in the frequency domain*, that is for two continuous-time signals $x_1(t)$ and $x_2(t)$, we have

$$\begin{aligned}\mathcal{F}\big[x_1(t)x_2(t)\big] &= \frac{1}{2\pi} X_1(j\omega) * X_2(j\omega) \\ &= \frac{1}{2\pi} \int_{-\infty}^{\infty} X_1\big(j(\omega - \lambda)\big) * X_2(j\lambda)\, d\lambda \end{aligned} \tag{9.52}$$

where $\mathcal{F}[x]$ is the Fourier transform of x and $*$ stands for the convolution operation. Based thereon, we can now establish the relationship between signal energy in the time and frequency domains. Rewriting (9.52) as

$$\mathcal{F}\big[x_1(t)x_2(t)\big] = \int_{-\infty}^{\infty} x_1(t)x_2(t)e^{-j\omega t}\, dt. \tag{9.53}$$

Since (9.53) is valid for any ω, it must be valid for $\omega = 0$. In view of (9.51)–(9.53), we have

$$\int_{-\infty}^{\infty} x_1(t)x_2(t)e^{-j\omega t}\, dt = \frac{1}{2\pi} \int_{-\infty}^{\infty} X_1\big(j(\omega - \lambda)\big) * X_2(j\lambda)\, d\lambda. \tag{9.54}$$

Letting $x_1(t) = x_2(t) = x(t)$ with $x(t)$ being a real function such that $|x(t)|^2 = x^2(t)$, it follows that

$$\begin{aligned} E_\infty &= \int_{-\infty}^{\infty} x^2(t)\, dt = \frac{1}{2\pi} \int_{-\infty}^{\infty} X(j\lambda)X(j\lambda)\, d\lambda \\ &= \frac{1}{2\pi} \int_{-\infty}^{\infty} |X(j\lambda)|^2\, d\lambda = \int_{-\infty}^{\infty} |X(jf)|^2\, df. \end{aligned} \tag{9.55}$$

Note that $\lambda = 2\pi f$, is a dummy variable of integration, plays the role of the angular frequency. The quantity $|X(jf)|^2$ is known as the *energy spectrum*. The result established in (9.55) is known as *Parseval theorem* which has great importance in signal processing and communications.

9.8.3 Frobenius Formula

Let $\lambda_1, \lambda_2, \dots, \lambda_n$ are the eigenvalues of the $n \times n$ matrix A and let $f(x)$ be a function which is analytic inside a circle in the complex plane that contains all the λ_i. Then $f(\lambda_1), f(\lambda_2), \dots, f(\lambda_n)$ are the eigenvalues of the matrix function $f(A)$. For example, the state transition matrix $\Phi(t, t_o) = e^{A(t-t_o)}$ has eigenvalues $\gamma_i = e^{\lambda_i(t-t_o)}$. It can be verified that the eigenvectors of A and $\Phi(t, t_o)$ are the same.

9.9 Inequalities

All mathematical inequalities are proved for completeness. They are termed facts due to their high frequency of usage in the analytical developments.

9.9.1 Inequality 1

For any real matrices Σ_1, Σ_2 and Σ_3 with appropriate dimensions and $\Sigma_3^t \Sigma_3 \leq I$, it follows that

$$\Sigma_1 \Sigma_3 \Sigma_2 + \Sigma_2^t \Sigma_3^t \Sigma_1^t \leq \alpha \Sigma_1 \Sigma_1^t + \alpha^{-1} \Sigma_2^t \Sigma_2, \quad \forall \alpha > 0.$$

Proof This inequality can be proved as follows. Since $\Phi^t \Phi \geq 0$ holds for any matrix Φ, then take Φ as

$$\Phi = \left[\alpha^{1/2} \Sigma_1 - \alpha^{-1/2} \Sigma_2\right].$$

Expansion of $\Phi^t \Phi \geq 0$ gives $\forall \alpha > 0$

$$\alpha \Sigma_1 \Sigma_1^t + \alpha^{-1} \Sigma_2^t \Sigma_2 - \Sigma_1^t \Sigma_2 - \Sigma_2^t \Sigma_1 \geq 0$$

which by simple arrangement yields the desired result. □

9.9.2 Inequality 2

Let Σ_1, Σ_2, Σ_3 and $0 < R = R^t$ be real constant matrices of compatible dimensions and $H(t)$ be a real matrix function satisfying $H^t(t)H(t) \leq I$. Then for any $\rho > 0$ satisfying $\rho \Sigma_2^t \Sigma_2 < R$, the following matrix inequality holds:

$$\left(\Sigma_3 + \Sigma_1 H(t) \Sigma_2\right) R^{-1} \left(\Sigma_3^t + \Sigma_2^t H^t(t) \Sigma_1^t\right) \leq \rho^{-1} \Sigma_1 \Sigma_1^t + \Sigma_3 \left(R - \rho \Sigma_2^t \Sigma_2\right)^{-1} \Sigma_3^t.$$

Proof The proof of this inequality proceeds like the previous one by considering that

$$\Phi = \left[\left(\rho^{-1} \Sigma_2 \Sigma_2^t\right)^{-1/2} \Sigma_2 R^{-1} \Sigma_3^t - \left(\rho^{-1} \Sigma_2 \Sigma_2^t\right)^{-1/2} H^t(t) \Sigma_1^t\right].$$

Recall the following results

$$\rho\Sigma_2^t\Sigma_2 < R,$$
$$\left[R-\rho\Sigma_2^t\Sigma_2\right]^{-1} = \left[R^{-1}+R^{-1}\Sigma_2^t\left[\rho^{-1}I-\Sigma_2R^{-1}\Sigma_2^t\right]^{-1}\Sigma_2R^{-1}\right]\Sigma_2$$

and

$$H^t(t)H(t) \le I \quad \Longrightarrow \quad H(t)H^t(t) \le I.$$

Expansion of $\Phi^t\Phi \ge 0$ under the condition $\rho\Sigma_2^t\Sigma_2 < R$ with standard matrix manipulations gives

$$\begin{aligned}
&\Sigma_3R^{-1}\Sigma_2^tH^t(t)\Sigma_1^t + \Sigma_1H(t)\Sigma_2R^{-1}\Sigma_3^t + \Sigma_1H(t)\Sigma_2\Sigma_2^tH^t(t)\Sigma_1^t \\
&\quad \le \rho^{-1}\Sigma_1H(t)H^t(t)\Sigma_1^t + \Sigma_3^tR^{-1}\Sigma_2\left[\rho^{-1}I\Sigma_2\Sigma_2^t\right]^{-1}\Sigma_2R^{-1}\Sigma_3^t \quad \Longrightarrow \\
&\left(\Sigma_3+\Sigma_1H(t)\Sigma_2\right)R^{-1}\left(\Sigma_3^t+\Sigma_2^tH^t(t)\Sigma_1^t\right) - \Sigma_3R^{-1}\Sigma_3^t \\
&\quad \le \rho^{-1}\Sigma_1H(t)H^t(t)\Sigma_1^t + \Sigma_3^tR^{-1}\Sigma_2\left[\rho^{-1}I-\Sigma_2\Sigma_2^t\right]^{-1}\Sigma_2R^{-1}\Sigma_3^t \quad \Longrightarrow \\
&\left(\Sigma_3+\Sigma_1H(t)\Sigma_2\right)R^{-1}\left(\Sigma_3^t+\Sigma_2^tH^t(t)\Sigma_1^t\right) \\
&\quad \le \Sigma_3\left[R^{-1}+\Sigma_2\left[\rho^{-1}I-\Sigma_2\Sigma_2^t\right]^{-1}\Sigma_2R^{-1}\right]\Sigma_3^t + \rho^{-1}\Sigma_1H(t)H^t(t)\Sigma_1^t \\
&\quad = \rho^{-1}\Sigma_1H(t)H^t(t)\Sigma_1^t + \Sigma_3\left(R-\rho\Sigma_2^t\Sigma_2\right)^{-1}\Sigma_3^t
\end{aligned}$$

which completes the proof. □

9.9.3 *Inequality 3*

For any real vectors β, ρ and any matrix $Q^t = Q > 0$ with appropriate dimensions, it follows that

$$-2\rho^t\beta \le \rho^tQ\rho + \beta^tQ^{-1}\beta.$$

Proof Starting from the fact that

$$\left[\rho+Q^{-1}\beta\right]^tQ\left[\rho+Q^{-1}\beta\right] \ge 0, \quad Q > 0$$

which when expanded and arranged yields the desired result. □

9.9.4 *Inequality 4 (Schur Complements)*

Given a matrix Ω composed of constant matrices Ω_1, Ω_2, Ω_3 where $\Omega_1 = \Omega_1^t$ and $0 < \Omega_2 = \Omega_2^t$ as follows

$$\Omega = \begin{bmatrix} \Omega_1 & \Omega_3 \\ \Omega_3^t & \Omega_2 \end{bmatrix}.$$

We have the following results

(A) $\Omega \geq 0$ if and only if either

$$\begin{cases} \Omega_2 \geq 0, \\ \Pi = \Upsilon \Omega_2, \\ \Omega_1 - \Upsilon \Omega_2 \Upsilon^t \geq 0 \end{cases} \tag{9.56}$$

or

$$\begin{cases} \Omega_1 \geq 0, \\ \Pi = \Omega_1 \Lambda, \\ \Omega_2 - \Lambda^t \Omega_1 \Lambda \geq 0 \end{cases} \tag{9.57}$$

hold where Λ, Υ are some matrices of compatible dimensions.

(B) $\Omega > 0$ if and only if either

$$\begin{cases} \Omega_2 > 0, \\ \Omega_1 - \Omega_3 \Omega_2^{-1} \Omega_3^t > 0 \end{cases}$$

or

$$\begin{cases} \Omega_1 \geq 0, \\ \Omega_2 - \Omega_3^t \Omega_1^{-1} \Omega_3 > 0 \end{cases}$$

hold where Λ, Υ are some matrices of compatible dimensions.

In this regard, matrix $\Omega_3 \Omega_2^{-1} \Omega_3^t$ is often called the Schur complement $\Omega_1(\Omega_2)$ in Ω.

Proof (A) To prove (9.56), we first note that $\Omega_2 \geq 0$ is necessary. Let $z^t = [z_1^t \;\; z_2^t]$ be a vector partitioned in accordance with Ω. Thus, we have

$$z^t \Omega z = z_1^t \Omega_1 z_1 + 2 z_1^t \Omega_3 z_2 + z_2^t \Omega_2 z_2. \tag{9.58}$$

Select z_2 such that $\Omega_2 z_2 = 0$. If $\Omega_3 z_2 \neq 0$, let $z_1 = -\pi \Omega_3 z_2$, $\pi > 0$. Then it follows that

$$z^t \Omega z = \pi^2 z_2^t \Omega_3^t \Omega_1 \Omega_3 z_2 - 2\pi z_2^t \Omega_3^t \Omega_3 z_2$$

which is negative for a sufficiently small $\pi > 0$. We thus conclude $\Omega_1 z_2 = 0$ which then leads to $\Omega_3 z_2 = 0$, $\forall z_2$ and consequently

$$\Omega_3 = \Upsilon \Omega_2 \tag{9.59}$$

for some Υ.

Since $\Omega \geq 0$, the quadratic term $z^t \Omega z$ possesses a minimum over z_2 for any z_1. By differentiating $z^t \Omega z$ from (9.58) wrt z_2^t, we get

$$\frac{\partial (z^t \Omega z)}{\partial z_2^t} = 2\Omega_3^t z_1 + 2\Omega_2 z_2 = 2\Omega_2 \Upsilon^t z_1 + 2\Omega_2 z_2.$$

Setting the derivative to zero yields

$$\Omega_2 \Upsilon z_1 = -\Omega_2 z_2. \tag{9.60}$$

Using (9.59) and (9.60) in (9.58), it follows that the minimum of $z^t \Omega z$ over z_2 for any z_1 is given by

$$\min_{z_2} z^t \Omega z = z_1^t \left[\Omega_1 - \Upsilon \Omega_2 \Upsilon^t\right] z_1$$

which prove the necessity of $\Omega_1 - \Upsilon \Omega_2 \Upsilon^t \geq 0$.

On the other hand, we note that the conditions (9.56) are necessary for $\Omega \geq 0$ and since together they imply that the minimum of $z^t \Omega z$ over z_2 for any z_1 is non-negative, they are also sufficient.

Using similar argument, conditions (9.57) can be derived as those of (9.56) by starting with Ω_1.

The proof of (B) follows as direct corollary of (A). □

9.9.5 *Inequality 5*

For any quantities u and v of equal dimensions and for all $\eta_t = i \in \mathcal{S}$, it follows that the following inequality holds

$$\|u + v\|^2 \leq \left[1 + \beta^{-1}\right] \|u\|^2 + [1 + \beta] \|v\|^2 \tag{9.61}$$

for any scalar $\beta > 0$, $i \in \mathcal{S}$.

Proof Since

$$[u + v]^t [u + v] = u^t u + v^t v + 2u^t v. \tag{9.62}$$

It follows by taking norm of both sides of (9.62) for all $i \in \mathcal{S}$ that

$$\|u + v\|^2 \leq \|u\|^2 + \|v\|^2 + 2\left\|u^t v\right\|. \tag{9.63}$$

We know from the triangle inequality that

$$2\left\|u^t v\right\| \leq \beta^{-1} \|u\|^2 + \beta \|v\|^2. \tag{9.64}$$

On substituting (9.64) into (9.63), it yields (9.61). □

9.10 Lemmas

The basic tools and standard results that are utilized in robustness analysis and resilience design in the different chapters are collected hereafter.

Lemma 9.37 *The matrix inequality*

$$-\Lambda + S \Omega^{-1} S^t < 0 \tag{9.65}$$

holds for some $0 < \Omega = \Omega^t \in \Re^{n \times n}$, *if and only if*

$$\begin{bmatrix} -\Lambda & S\mathcal{X} \\ \bullet & -\mathcal{X}-\mathcal{X}^t+\mathcal{Z} \end{bmatrix} < 0 \tag{9.66}$$

holds for some matrices $\mathcal{X} \in \Re^{n\times n}$ *and* $\mathcal{Z} \in \Re^{n\times n}$.

Proof ($\Longrightarrow$) By Schur complements, inequality (9.65) is equivalent to

$$\begin{bmatrix} -\Lambda & S\Omega^{-1} \\ \bullet & -\Omega^{-1} \end{bmatrix} < 0. \tag{9.67}$$

Setting $\mathcal{X} = \mathcal{X}^t = \mathcal{Z} = \Omega^{-1}$, we readily obtain inequality (9.66).

($\Longleftarrow$) Since the matrix $[I \quad S]$ is of full rank, we obtain

$$\begin{aligned} &\begin{bmatrix} I \\ S^t \end{bmatrix}^t \begin{bmatrix} -\Lambda & S\mathcal{X} \\ \bullet & -\mathcal{X}-\mathcal{X}^t+\mathcal{Z} \end{bmatrix} \begin{bmatrix} I \\ S^t \end{bmatrix} < 0 \quad \Longleftrightarrow \\ &-\Lambda + S\mathcal{Z}S^t < 0 \quad \Longleftrightarrow \quad -\Lambda + S\Omega^{-1}S^t < 0, \quad \mathcal{Z} = \Omega^{-1} \end{aligned} \tag{9.68}$$

which completes the proof. □

Lemma 9.38 *The matrix inequality*

$$A\mathbb{P} + \mathbb{P}A^t + D^t\mathbb{R}^{-1}D + \mathbb{M} < 0 \tag{9.69}$$

holds for some $0 < \mathbb{P} = \mathbb{P}^t \in \Re^{n\times n}$, *if and only if*

$$\begin{bmatrix} A\mathcal{V} + \mathcal{V}^tA^t + \mathbb{M} & \mathbb{P} + A\mathcal{W} - \mathcal{V} & D^t\mathbb{R} \\ \bullet & -\mathcal{W} - \mathcal{W}^t & 0 \\ \bullet & \bullet & -\mathbb{R} \end{bmatrix} < 0 \tag{9.70}$$

holds for some $\mathcal{V} \in \Re^{n\times n}$ *and* $\mathcal{W} \in \Re^{n\times n}$.

Proof ($\Longrightarrow$) By Schur complements, inequality (9.69) is equivalent to

$$\begin{bmatrix} A\mathbb{P} + \mathbb{P}A^t + \mathbb{M} & D^t\mathbb{R} \\ \bullet & -\mathbb{R} \end{bmatrix} < 0. \tag{9.71}$$

Setting $\mathcal{V} = \mathcal{V}^t = \mathbb{P}$, $\mathcal{W} = \mathcal{W}^t = \mathbb{R}$, it follows from Lemma 9.37 with Schur complements that there exists $\mathbb{P} > 0$, $\mathcal{V}$, $\mathcal{W}$ such that inequality (9.70) holds.

($\Longleftarrow$) In a similar way, Schur complements to inequality (9.70) imply that:

$$\begin{aligned} &\begin{bmatrix} A\mathcal{V} + \mathcal{V}^tA^t + \mathbb{M} & \mathbb{P} + A\mathcal{W} - \mathcal{V} & D^t\mathbb{R} \\ \bullet & -\mathcal{W} - \mathcal{W}^t & 0 \\ \bullet & \bullet & -\mathbb{R} \end{bmatrix} < 0 \\ \Longleftrightarrow \quad &\begin{bmatrix} I \\ A \end{bmatrix} \begin{bmatrix} A\mathcal{V} + \mathcal{V}^tA^t + \mathbb{M} + D^t\mathbb{R}^{-1}D & \mathbb{P} + A\mathcal{W} - \mathcal{V} \\ \bullet & -\mathcal{W} - \mathcal{W}^t \end{bmatrix} \begin{bmatrix} I \\ A \end{bmatrix}^t < 0 \\ \Longleftrightarrow \quad &A\mathbb{P} + \mathbb{P}A^t + D^t\mathbb{R}^{-1}D + \mathbb{M} < 0, \quad \mathcal{V} = \mathcal{V}^t \end{aligned} \tag{9.72}$$

which completes the proof. □

Lemma 9.39 *Given any $x \in \Re^n$:*

$$\max\{[x^t RH\Delta Gx]^2 : \Delta \in \Re\} = x^t RHH^t Rxx^t G^t Gx.$$

Lemma 9.40 *Given matrices $0 \leq X = X^t \in \Re^{p\times p}$, $Y = Y^t < 0 \in \Re^{p\times p}$, $0 \leq Z = Z^t \in \Re^{p\times p}$, such that*

$$[\xi^t Y\xi]^2 - 4[\xi^t X\xi \quad \xi^t Z\xi]^2 > 0$$

for all $0 \neq \xi \in \Re^p$ is satisfied. Then there exists a constant $\alpha > 0$ such that

$$\alpha^2 X + \alpha Y + Z < 0.$$

Lemma 9.41 *For a given two vectors $\alpha \in \mathbb{R}^n$, $\beta \in \mathbb{R}^m$ and matrix $\mathbb{N} \in \mathbb{R}^{n\times m}$ defined over a prescribed interval Ω, it follows for any matrices $X \in \mathbb{R}^{n\times n}$, $Y \in \mathbb{R}^{n\times m}$, and $Z \in \mathbb{R}^{m\times m}$, the following inequality holds*

$$-2\int_\Omega \alpha^t(s)\mathbb{N}\beta(s)\,ds \leq \int_\Omega \begin{bmatrix} \alpha(s) \\ \beta(s) \end{bmatrix}^t \begin{bmatrix} X & Y - \mathbb{N} \\ Y^t - \mathbb{N}^t & Z \end{bmatrix} \begin{bmatrix} \alpha(s) \\ \beta(s) \end{bmatrix} ds$$

where

$$\begin{bmatrix} X & Y \\ Y^t & Z \end{bmatrix} \geq 0.$$

An algebraic version of Lemma 9.41 is stated below

Lemma 9.42 *For a given two vectors $\alpha \in \mathbb{R}^n$, $\beta \in \mathbb{R}^m$ and matrix $\mathbb{N} \in \mathbb{R}^{n\times m}$ defined over a prescribed interval Ω, it follows for any matrices $X \in \mathbb{R}^{n\times n}$, $Y \in \mathbb{R}^{n\times m}$, and $Z \in \mathbb{R}^{m\times m}$, the following inequality holds*

$$\begin{aligned} -2\alpha^t\mathbb{N}\beta &\leq \begin{bmatrix} \alpha \\ \beta \end{bmatrix}^t \begin{bmatrix} X & Y - \mathbb{N} \\ Y^t - \mathbb{N}^t & Z \end{bmatrix} \begin{bmatrix} \alpha \\ \beta \end{bmatrix} \\ &= \alpha^t X\alpha + \beta^t(Y^t - \mathbb{N}^t)\alpha + \alpha^t(Y - \mathbb{N})\beta + \beta^t Z\beta \end{aligned}$$

subject to

$$\begin{bmatrix} X & Y \\ Y^t & Z \end{bmatrix} \geq 0.$$

Lemma 9.43 *Let $0 < Y = Y^t$ and M, N be given matrices with appropriate dimensions. Then it follows that*

$$Y + M\Delta N + N^t\Delta^t M^t < 0, \quad \forall \Delta^t\Delta \leq I$$

holds if and only if there exists a scalar $\varepsilon > 0$ such that

$$Y + \varepsilon MM^t + \varepsilon^{-1}N^t N < 0.$$

In the following lemma, we let $X(z) \in \mathbb{R}^{n\times p}$ be a matrix function of the variable z. A matrix $X_*(z)$ is called the orthogonal complement of $X(z)$ if $X^t(z)X_*(z) = 0$ and $X(z)X_*(z)$ is nonsingular (of maximum rank).

Lemma 9.44 *Let $0 < L = L^t$ and X, Y be given matrices with appropriate dimensions. Then it follows that the inequality*

$$L(z) + X(z)PY(z) + Y^t(z)P^tX^t(z) > 0 \tag{9.73}$$

holds for some P and $z = z_o$ if and only if the following inequalities

$$X_*^t(z)L(z)X_*(z) > 0, \qquad Y_*^t(z)L(z)Y_*(z) > 0 \tag{9.74}$$

hold with $z = z_o$.

It is significant to observe that feasibility of matrix inequality (9.73) with variables P and z is equivalent to the feasibility of (9.74) with variable z and thus the matrix variable P has been eliminated from (9.73) to form (9.74). Using Finsler's lemma, we can express (9.74) in the form

$$L(z) - \beta X(z)X^t(z) > 0, \qquad L(z) - \beta Y(z)Y^t(z) > 0 \tag{9.75}$$

for some $\beta \in \mathbb{R}$.

The following is a statement of the reciprocal projection lemma.

Lemma 9.45 *Let $P > 0$ be a given matrix. The following statements are equivalent*:

(i) $\mathbb{M} + Z + Z^t < 0$,
(ii) *the LMI problem*

$$\begin{bmatrix} \mathbb{M} + P - (V + V^t) & V^t + Z^t \\ V + Z & -P \end{bmatrix} < 0$$

is feasible with respect to the general matrix V.

9.11 Linear Matrix Inequalities

It has been shown that a wide variety of problem arising in system and control theory can conveniently reduced to a few standard convex or quasiconvex optimization problems involving linear matrix inequalities (LMIs). The resulting optimization problems can then be solved numerically very efficiently using commercially available interior-point methods.

9.11.1 Basics

One of the earliest LMIs arises in Lyapunov theory. It is well-known that the differential equation

$$\dot{x}(t) = Ax(t) \tag{9.76}$$

has all of its trajectories converge to zero (stable) id and only if there exists a matrix $P > 0$ such that

$$A^t P + AP < 0. \tag{9.77}$$

This leads to the LMI formulation of stability, that is, *a linear time-invariant system is asymptotically stable if and only if there exists a matrix* $0 < P = P^t$ *satisfying the LMIs*

$$A^t P + AP < 0, \quad P > 0.$$

Given a vector variable $x \in \mathbb{R}^n$ and a set of matrices $0 < G_j = G_j^t \in \mathbb{R}^{n\times n}$, $j = 0, \dots, p$, then a basic compact formulation of a linear matrix inequality is

$$G(x) \triangleq G_0 + \sum_{j=1}^{p} x_j G_j > 0. \tag{9.78}$$

Notice that (9.78) implies that $v^t G(x) v > 0 \ \forall 0 \neq v \in \mathbb{R}^n$. More importantly, the set $\{x | G(x) > 0\}$ is convex. Nonlinear (convex) inequalities are converted to LMI form using Schur complements in the sense that

$$\begin{bmatrix} Q(x) & S(x) \\ \bullet & R(x) \end{bmatrix} > 0 \tag{9.79}$$

where $Q(x) = Q^t(x)$, $R(x) = R^t(x)$, $S(x)$ depend affinely on x, is equivalent to

$$R(x) > 0, \qquad Q(x) - S(x) R^{-1}(x) S^t(x) > 0. \tag{9.80}$$

More generally, the constraint

$$\mathrm{Tr}\big[S^t(x) P^{-1}(x) S(x)\big] < 1, \quad P(x) > 0$$

where $P(x) = P^t(x) \in \mathbb{R}^{n\times n}$, $S(x) \in \mathbb{R}^{n\times p}$ depend affinely on x, is handled by introducing a new (slack) matrix variable $Y(x) = Y^t(x) \in \mathbb{R}^{p\times p}$ and the LMI (in x and Y):

$$\mathrm{Tr}\, Y < 1, \qquad \begin{bmatrix} Y & S(x) \\ \bullet & P(x) \end{bmatrix} > 0. \tag{9.81}$$

Most of the time, our LMI variables are matrices. It should clear from the foregoing discussions that a quadratic matrix inequality (QMI) in the variable P can be readily expressed as linear matrix inequality (LMI) in the same variable.

9.11.2 Some Standard Problems

Here we provide some common convex problems that we encountered throughout the monograph. Given an LMI $G(x) > 0$, the corresponding LMI problem (LMIP) is to

find a feasible $x \equiv x^f$ *such that* $G(x^f) > 0$, or *determine that the LMI is infeasible*.

It is obvious that this is a convex feasibility problem. The generalized eigenvalue problem (GEVP) is to minimize the maximum generalized eigenvalue of a pair of matrices that depend affinely on a variable, subject to an LMI constraint. GEVP has the general form

$$\begin{aligned}&\text{minimize} \quad \lambda,\\&\text{subject to} \quad \lambda B(x) - A(x) > 0, \quad B(x) > 0, \quad C(x) > 0\end{aligned} \tag{9.82}$$

where A, B, C are symmetric matrices that are affine functions of x. Equivalently stated

$$\begin{aligned}&\text{minimize} \quad \lambda_M\big[A(x), B(x)\big],\\&\text{subject to} \quad B(x) > 0, \quad C(x) > 0\end{aligned} \tag{9.83}$$

where $\lambda_M[X, Y]$ denotes the largest generalized eigenvalue of the pencil $\lambda Y - X$ with $Y > 0$. This is problem is quasiconvex optimization problem since the constraint is convex and the objective , $\lambda_M[A(x), B(x)]$, is quasiconvex. The eigenvalue problem (EVP) is to minimize the maximum eigenvalue of a matrix that depend affinely on a variable, subject to an LMI constraint. EVP has the general form

$$\begin{aligned}&\text{minimize} \quad \lambda,\\&\text{subject to} \quad \lambda I - A(x) > 0, \quad B(x) > 0\end{aligned} \tag{9.84}$$

where A, B are symmetric matrices that are affine functions of the optimization variable x. This is problem is convex optimization problem.

EVPs can appear in the equivalent form of minimizing a linear function subject to an LMI, that is

$$\begin{aligned}&\text{minimize} \quad c^t x,\\&\text{subject to} \quad G(x) > 0\end{aligned} \tag{9.85}$$

where $G(x)$ is an affine function of x. Examples of $G(x)$ include

$$PA + A^t P + C^t C + \gamma^{-1} P B B^t P < 0, \quad P > 0.$$

It should be stressed that the standard problems (LMIPs, GEVPs, EVPs) are tractable, from both theoretical and practical viewpoints:

They can be solved in polynomial-time.
They can solved in practice very efficiently using commercial software.

9.11.3 The S-Procedure

In some design applications, we faced the constraint that some quadratic function be negative whenever some other quadratic function is negative. In such cases, this constraint can be expressed as an LMI in the data variables defining the quadratic functions. Let $G_0, \ldots, G_p$ be quadratic functions of the variable $\xi \in \mathbb{R}^n$:

$$G_j(\xi) \stackrel{\Delta}{=} \xi^t R_j \xi + 2u_j^t \xi + v_j, \quad j = 0, \ldots, p, \ R_j = R_j^t.$$

We consider the following condition on $G_0, \dots, G_p$:

$$G_0(\xi) \le 0 \quad \forall \xi \quad \text{such that } G_j(\xi) \ge 0, \ j = 0, \dots, p. \tag{9.86}$$

It is readily evident that if there exist scalars $\omega_1 \ge 0, \dots, \omega_p \ge 0$ such that

$$\forall \xi, \quad G_0(\xi) - \sum_{j=1}^{p} \omega_j G_j(\xi) \ge 0 \tag{9.87}$$

then inequality (9.86) holds. Observe that if the functions $G_0, \dots, G_p$ are affine, then Farkas lemma state that (9.86) and (9.87) are equivalent. Interestingly enough, inequality (9.87) can written as

$$\begin{bmatrix} R_0 & u_0 \\ \bullet & v_0 \end{bmatrix} - \sum_{j=1}^{p} \omega_j \begin{bmatrix} R_j & u_j \\ \bullet & v_j \end{bmatrix} \ge 0. \tag{9.88}$$

The foregoing discussions were stated for non strict inequalities. In case of strict inequality, we let $R_0, \dots, R_p \in \mathbb{R}^{n \times n}$ be symmetric matrices with the following qualifications

$$\xi^t R_0 \xi > 0 \quad \forall \xi \quad \text{such that } \xi^t G_j \xi \ge 0, \ j = 0, \dots, p. \tag{9.89}$$

Once again, it is obvious that there exist scalars $\omega_1 \ge 0, \dots, \omega_p \ge 0$ such that

$$\forall \xi, \quad G_0(\xi) - \sum_{j=1}^{p} \omega_j G_j(\xi) > 0 \tag{9.90}$$

then inequality (9.89) holds. Observe that (9.90) is an LMI in the variables $R_0, \omega_1, \dots, \omega_p$. It should be remarked that the **S**-procedure deals with non strict inequalities allows the inclusion of constant and linear terms. In the strict version, only quadratic functions can be used.

9.12 Lyapunov Map and Lyapunov Equation

An important type of linear equation that we will encounter in our book is of Lyapunov type:[2] find a matrix $X \in C^{n \times n}$ that satisfies the equation

$$AX + XB = 0 \tag{9.91}$$

for a given pair of matrices $A, B \in C^{n \times n}$. Although solutions to this type of equation can be difficult in general, simple solutions exist when both A and B have n independent eigenvectors. Suppose $\{u_i \in C^n\}_{i=1}^n$ are the n right eigenvectors of A, and $\{v_j \in C^n\}_{j=1}^n$ are the n left eigenvectors of B; that is,

$$Au_i = \lambda_i u_i; \qquad v_j^* B = \eta_j v_j^* \tag{9.92}$$

for eigenvalues λ_i, η_j for each i, j. Here v^* means the complex-conjugate and transpose of v, since v can be complex.

[2]It is also called Sylvester equation in some literature.

Lyapunov map: For the above matrix A and B, the n^2 eigenvectors of the Lyapunov map

$$L : X \mapsto AX + XB \tag{9.93}$$

are exactly $X_{ij} = u_i v_j^* \in C^{n \times n}$, and the corresponding eigenvalues are $\lambda_i + \eta_j \in C$, $i, j = 1, 2, \ldots, n$.

Proof The n^2 matrices $\{X_{ij}\}_{i,j=1}^n$ are linearly independent, and they must be all the eigenvectors of L. □

Due to this fact, any matrix X that satisfies the Lyapunov equation $AX + XB = 0$ must be in the subspace spanned by eigenvectors X_{ij} that have zero eigenvalues: $\lambda_i + \eta_j = 0$. In MATLAB, the command $X = \text{lyap}(A, B, C)$ solves the more general Lyapunov equation $AX + XB = -C$.

In this book, we often look for solutions X with extra requirements on its structure. For instance, X needs to be real and symmetric (Chap. 6), or X has to be a rotation matrix (Chap. 8). If so, we have only to take the intersection of the space of solutions to the Lyapunov equation with the space of symmetric matrices or rotation matrices.

9.13 Persistence of Excitation and Sufficiently Rich Inputs

We start with the following definition.

Definition 9.46 The vector $\phi \in R^n$ is persistently excited (PE) with level α_0 if it satisfies

$$\int_t^{t+T_0} \phi(\tau)\phi^t(\tau)\, d\tau \geq \alpha_0 T_0 I \tag{9.94}$$

for some $\alpha_0 > 0$, $T_0 > 0$ and $\forall t \geq 0$.

Since $\phi\phi^t$ is always positive semidefinite, the PE condition requires that its integral over any interval of time of length T_0 is a positive definite matrix.

Definition 9.47 The signal $u \in R$ is called sufficiently rich of order n if it contains at least $\frac{n}{2}$ distinct nonzero frequencies.

For example, $u = \sum_{i=1}^{10} \sin \omega_i t$, where $\omega_i \neq \omega_j$ for $i \neq j$ is sufficiently rich of order 20. A more general definition of sufficiently rich signals and associated properties may be found in [4].

Let us consider the signal vector $\phi \in R^n$ generated as

$$\phi = H(s)u, \tag{9.95}$$

where $u \in R$ and $H(s)$ is a vector whose elements are transfer functions that are strictly proper with stable poles.

Theorem 9.48 *Consider (9.95) and assume that the complex vectors*

$$H(j\omega_1), \ldots, H(j\omega_n)$$

are linearly independent on the complex space

$$C^n \quad \forall \omega_1, \omega_2, \ldots, \omega_n \in R, \quad \omega_i \neq \omega_j,\ i \neq j.$$

Then ϕ is PE if and only if u is sufficiently rich of order n.

Proof The proof of Theorem 9.48 can be found in [3, 4]. □

We demonstrate the use of Theorem 9.48 for the Example 8.3, where

$$\phi = H(s)u$$

and

$$H(s) = \begin{bmatrix} 1 \\ -\frac{b}{s+u} \end{bmatrix}.$$

In this case $n = 2$ and

$$H(j\omega_1) = \begin{bmatrix} 1 \\ -\frac{b}{j\omega_1 + a} \end{bmatrix}, \qquad H(j\omega_2) = \begin{bmatrix} 1 \\ -\frac{b}{j\omega_2 + a} \end{bmatrix}.$$

We can show that the matrix $[H(j\omega_1), H(j\omega_2)]$ is nonsingular, which implies that $H(j\omega_1)$, $H(j\omega_2)$ are linearly independent for any ω_1, ω_2 different than zero and $\omega_1 \neq \omega_2$.

Let us choose

$$u = \sin \omega_0 t$$

for some $\omega_0 \neq 0$ which is sufficiently rich of order 2. According to Theorem 9.48, this input should guarantee that ϕ is PE for the Example 8.3. Ignoring the transient terms that converge to zero exponentially fast, we can show that at steady state

$$\phi = \begin{bmatrix} \sin \omega_0 t \\ c_0 \sin(\omega_0 t + \varphi_0) \end{bmatrix},$$

where

$$c_0 = \frac{|b|}{\sqrt{\omega_0^2 + a^2}}, \qquad \varphi_0 = \arg\left(\frac{-b}{j\omega_0 + a}\right).$$

Now

$$\phi\phi^t = \begin{bmatrix} \sin^2 \omega_0 t & c_0 \sin \omega_0 t \sin(\omega_0 t + \varphi_0) \\ c_0 \sin \omega_0 t \sin(\omega_0 t + \varphi_0) & c_0^2 \sin^2(\omega_0 t + \varphi) \end{bmatrix}$$

and

$$\int_t^{t+T_0} \phi(\tau)\phi^t(\tau)\, d\tau = \begin{bmatrix} a_{11} & a_{12} \\ a_{12} & a_{22} \end{bmatrix},$$

where

$$a_{11} = \frac{T_0}{2} - \frac{\sin 2\omega_0(t+T_0) - \sin 2\omega_0 t}{4\omega_0},$$

$$a_{12} = c_0\frac{T_0}{2}\cos\varphi_0 + c_0\frac{\sin\varphi_0}{4\omega_0}\big(\cos 2\omega_0 t - \cos 2\omega_0(t+T_0)\big),$$

$$a_{22} = c_0^2\frac{T_0}{2} - c_0^2\frac{\sin 2(\omega_0(t+T_0)+\varphi_0) - \sin 2(\omega_0 t+\varphi_0)}{4\omega_0}.$$

Choosing $T_0 = \frac{\pi}{\omega_0}$ it follows that

$$a_{11} = \frac{T_0}{2}, \qquad a_{12} = \frac{T_0 c_0}{2}\cos\varphi_0, \qquad a_{22} = \frac{T_0}{2}c_0^2$$

and

$$\int_t^{t+T_0} \phi(\tau)\phi^t(\tau)\,d\tau = \frac{T_0}{2}\begin{bmatrix} 1 & c_0\cos\varphi_0 \\ c_0\cos\varphi_0 & c_0^2 \end{bmatrix},$$

which is a positive definite matrix. We can verify that for $\alpha_0 = \frac{1}{2}\frac{(1-\cos^2\varphi_0)c_0^2}{1+c_0^2} > 0$,

$$\int_t^{t+T_0} \phi(\tau)\phi^t(\tau)\,d\tau \geq T_0\alpha_0 I,$$

which implies that ϕ is PE.

Let us consider the plant model

$$y = \frac{b(s^2+4)}{(s+5)^3}u,$$

where b is the only unknown parameter. A suitable parametric model for estimating b is

$$z = \theta^*\phi,$$

where

$$z = y, \qquad \theta^* = b, \qquad \phi = \frac{s^2+4}{(s+5)^3}u.$$

In this case $\phi \in R$ and $H(s) = \frac{s^2+4}{(s+5)^3}$; i.e., $n = 1$ in Theorem 9.48. Let us use Theorem 9.48 to choose a sufficiently rich signal u that guarantees ϕ to be PE. In this case, according to the linear independence condition of Theorem 9.48 for the case of $n = 1$, we should have

$$\big|H(j\omega_0)\big| = \frac{4-\omega_0^2}{(25+\omega_0^2)^{3/2}} \neq 0$$

for any $\omega_0 \neq 0$. This condition is clearly violated for $\omega_0 = 2$, and therefore a sufficiently rich input of order 1 may not guarantee ϕ to be PE. Indeed, the input $u = \sin 2t$ leads to $y = 0$, $\phi = 0$ at steady state, which imply that the output y and regressor ϕ carry no information about the unknown parameter b. For this example $u = \sin\omega_0 t$ will guarantee ϕ to be PE, provided $\omega_0 \neq 2$. Also, $u =$ constant $\neq 0$ and $u = \sum_{i=1}^{m} \sin\omega_i t$, $m \geq 2$, will guarantee that ϕ is PE. In general, for each two unknown parameters we need at least a single nonzero frequency to guarantee PE, provided of course that $H(s)$ does not lose its linear independence as demonstrated by the above example.

The two-parameter case leads to the differential equation (8.28), which has exactly the same form as in the case of an arbitrary number of parameters. In the following section, we consider the case where θ^*, ϕ are of arbitrary dimension and analyze the convergence properties of equations of the form (8.28).

9.14 Notes and References

The analysis presented in this chapter made extensive use of the standard texts including [5].

References

1. Ackermann, J.: Der Entwurf Linearer Regelungssysteme im Zustandsraum, Regelungstech, Prozess-Datenverarb. **7**, 297–300 (1972)
2. Chiang, R.Y., Safonov, M.G.: User's Guide, Robust Control Toolbox. The Math Works, Natick, USA (1992)
3. Goodwin, G.C., Sin, K.S.: Adaptive Filtering Prediction and Control. Prentice-Hall, Englewood Cliffs (1984)
4. Ioannou, P.A., Fidan, B.: Adaptive Control Tutorial, SIAM's Advances in Design and Control (2006)
5. Noble, B., Daniel, J.: Applied Linear Algebra, 3rd edn. Prentice-Hall, New York (1988)
6. Strange, G., Introduction to Applied Mathematics. Wellesley, Cambridge (1986)

Index

M.S. Mahmoud, Y. Xia, *Applied Control Systems Design*,
DOI 10.1007/978-1-4471-2879-3,

M

N

O

P

Q

R

S

T

U

W